Name	Symbol	Atomic number	Atomic weight
Mercury	Hg	80	200.59
Molybdenum	Mo	42	95.94
Neodymium	Nd	60	144.24
Neon	Ne	10	20.183
Neptunium	Np	93	(237)
Nickel	Ni	28	58.71
Niobium	Nb	41	92.906
Nitrogen	N	7	14.0067
Nobelium	No	102	(254)
Osmium	Os	76	190.2
Oxygen	O	8	15.9994
Palladium	Pd	46	106.4
Phosphorus	P	15	30.9738
Platinum	Pt	78	195.09
Plutonium	Pu	94	(242)
Polonium	Po	84	(210)
Potassium	K	19	39.102
Praseodymium	Pr	59	140.907
Promethium	Pm	61	(147)
Protactinium	Pa	91	(231)
Radium	Ra	88	(226)
Radon	Rn	86	(222)
Rhenium	Re	75	186.2
Rhodium	Rh	45	102.905
Rubidium	Rb	37	85.47
Ruthenium	Ru	44	101.07
Samarium	Sm	62	150.35
Scandium	Sc	21	44.956
Selenium	Se	34	78.96
Silicon	Si	14	28.086
Silver	Ag	47	107.870
Sodium	Na	11	22.9898
Strontium	Sr	38	87.62
Sulfur	S	16	32.064
Tantalum	Ta	73	180.948
Technetium	Tc	43	(99)
Tellurium	Te	52	127.60
Terbium	Tb	65	158.924
Thallium	Tl	81	204.37
Thorium	Th	90	232.038
Thulium	Tm	69	168.934
Tin	Sn	50	118.69
Titanium	Ti	22	47.90
Tungsten	W	74	183.85
Uranium	U	92	238.04
Vanadium	V	23	50.942
Xenon	Xe	54	131.30
Ytterbium	Yb	70	173.04
Yttrium	Y	39	88.905
Zinc	Zn	30	65.37
Zirconium	Zr	40	91.22

Chemistry

HARPER'S CHEMISTRY SERIES

Under the Editorship of Stuart Alan Rice

Eugene H. Cordes
Riley Schaeffer

Indiana University

CHEMISTRY

HARPER & ROW, PUBLISHERS
New York, Evanston, San Francisco, London

Photographs on pp. 12–13, 314–315, 424–425, and 594–595 by Beckwith Studios. Quote on p. 17 from "Hard, Ain't It Hard." Words and music by Woody Guthrie. Copyright 1952, Ludlow Music, Inc., New York. Used by permission. Quote on p. 630 from *The Complete Poetry and Selected Prose of John Donne and the Complete Poetry of William Blake,* edited by John Hayward and Geoffrey Keynes, Random House, New York, 1941, p. 41. Copyright 1941, Random House, Inc., New York.

Standard Book Number: 06-041359-X

Library of Congress Catalog Card Number: 72-8261

Contents

17

History of the atmosphere and the origin of life 517

18

Chemistry of the atmosphere 539

19

Utilization of solar light energy 567

Part four Electron transfer reactions 594

20

Oxidation-reduction reactions 598

21

The life processes: Thermodynamic aspects 629

22

Life processes: Some kinetic considerations 653

Epilogue: A look back and a look forward 679

Appendix 683

Index 689

Preface

Chemistry is the outgrowth of an experimental effort in the teaching of general chemistry at Indiana University, Bloomington, a project in which we have been involved for several years. Our earliest experiences suggested that neither the classical descriptive approach to general chemistry nor the alternative to it, an elementary course in physical chemistry, adequately met the legitimate needs of our students. The group of students for which this book was designed had usually completed one year of general chemistry in high school and intended to continue in more advanced chemistry courses (usually organic chemistry); a substantial majority had career goals in either the field of medicine or biology. The group included some students who would continue as full-fledged chemistry majors, although they constituted a small fraction of the total class. We believe that this profile describes general chemistry classes at many colleges and universities. The considerable educational challenge of such a profile is summarized in the following set of questions:

1. How can the needs of a full-fledged chemistry major be met while simultaneously satisfying the requirements of medically or biologically oriented students, who generally constitute the bulk of a class?
2. How can the general chemistry course be structured to optimize the student's subsequent experiences in organic chemistry and other chemistry courses?
3. How can the variable quality of a student's previous introduction to chemistry be compensated for, thus providing well prepared students with something new and stimulating without, at the same time, discriminating against those with less solid backgrounds?
4. What evidence can be offered students that chemistry is something worth knowing about?

Let us briefly examine how our goals were realized in the organization of *Chemistry*. We have divided the book into four parts. Part One provides the bulk of the general chemistry foundation. By "general chemistry" we mean those aspects of chemistry that are of interest to all breeds of chemist: molecular structure, chemical bonding, energetics, and reaction mechanisms. Furthermore, we hold that general chemistry ought to be presented as such, and not distinguished as inorganic chemistry, organic chemistry, biochemistry, and so forth. The student will have ample opportunity, in more advanced courses, to learn about the increasingly artificial subdivisions of chemistry. We have conscientiously avoided the use of any terminology that would tend to fragment our concept of general chemistry.

We have begun our consideration of general chemistry with molecular structure. Much about this topic can be grasped on an intuitive basis without recourse to either concepts of quantum mechanics, which are difficult for beginning students and quite unessential for most students, or to concepts of mathematical methods, which the average student may not yet have mastered. From among the various possible starting points for a discussion of molecular structure, we have selected tetrahedral structures and, more specifically, the structural chemistry of carbon. We believe that compelling arguments support this choice: The structural chemistry of carbon is regular and systematic; it provides a maximal opportunity to give students a firm grasp of what molecules look like in three dimensions through discussions of the various classes of isomerism; it provides the basis for a discussion of the structural chemistry of many other elements on the right-hand side of the periodic table; tetrahedral geometry is excellent for introducing the concepts of symmetry; and the structural chemistry of carbon is fundamental to later considerations of molecules of biological interest and provides an introduction into organic chemistry, which the student may encounter in later courses. It is important to note that we are not simply beginning the course with "some organic chemistry"; the emphasis is entirely on molecular structure—no chemistry of functional groups is introduced, and the nomenclature of organic chemistry is reduced to a minimum.

The theme of molecular structure continues with a consideration of molecules based on elements other than carbon but whose geometries can be simply related to those of carbon compounds; these include nitrogen, oxygen, silicon, phosphorus, sulfur, and several others. Concomitant with the structural chemistry, some descriptive chemistry of these elements is introduced, consistent with our belief that descriptive chemistry is sufficiently important to be integrated into the discussion, rather than reserved for the later chapters, which are often ignored in the course of study. At the same time, some practice in simple quantitative calculations is provided.

Part One progresses next to a discussion of electronic structure and chemical bonding, which is treated in terms of both valence bond and molecular orbital theories. Whenever possible, we refer back to the topic of molecular geometries: How do theories of chemical bonding account for what we have learned about structural chemistry? In the subsequent chapter the argument is turned around: Nontetrahedral geometries are rationalized in terms of our development of chemical bonding. Consideration of molecular structures concludes with a treatment of polymers, both natural and synthetic. The first directly biologically oriented material is introduced here.

Our consideration of the fundamentals of general chemistry concludes with a qualitative examination of thermodynamics and a development of basic features of reaction mechanisms through consideration of hydrolysis of a broad spectrum of halides. Throughout, emphasis is placed on development of a sound understanding of molecular behavior. We intend that completion of Part One—material that ordinarily requires a complete semester to master—will leave the student thinking comfortably about molecules in three dimensions: how they look and what elements of symmetry are present; what stresses and strains they react to; and how these reactions may lead to chemical transformations. To achieve this goal, mathematical methodology has been intentionally sacrificed.

The remaining three parts of the book build on the first, always adding new information but consistently returning to the most basic concepts, both to emphasize their importance and to reinforce them in the mind of the student. Part Two deals with the chemistry of water and includes a detailed examination of this complicated liquid as well as a treatment of equilibria in aqueous solution, both solubilities and acid-base reactions. This part concludes with a consideration of the desalination of sea water, a topic that permits us to employ material presented earlier to a real-life situation, to introduce new information about membranes and membrane phenomena, to reinforce some aspects of thermodynamics, and to inform the student about a matter of current importance. Ordinarily, such material would be introduced at the beginning of the second semester, and it might easily be postponed to a later date at the instructor's discretion. Our experience suggests, however, that this material is difficult for the average student, perhaps because of the algebra involved, and that treatment of it early in the course gives him maximum opportunity to master it.

Part Three deals with the general topic of energy and light and ranges from physics to molecular biology. We begin by considering the nature of light and matter and how these are related, and continue with a discussion of how matter and light interact, that is, with a treatment of photochemistry, including some information on the nature of the visual process. Subsequently, we progress to nuclear chemistry and element synthesis, leading to a treatment of solar chemistry and the generation of light. We then return to photochemistry, this time in terms of the interaction of radiant energy from the sun with the earth's primitive atmosphere, and the resulting ramifications for earth history and the development of living forms. In the subsequent chapter, the present-day analog of early atmospheric photochemistry is developed, consisting of upper-atmosphere photochemistry and the generation of photochemical smog. Part Three concludes with additional material devoted to environmental and biological matters—the utilization of solar energy. In this regard, we discuss solar cells, solar stills, and photosynthesis. Throughout the entire unit, every opportunity is taken to reinforce our earlier development of electronic structure, chemical bonding, and thermodynamics.

Part Four, the final unit, begins with a rather straightforward consideration of oxidation–reduction reactions. Subsequently, biological redox reactions are developed in terms of both the generation and utilization of metabolic energy. We then return to the topic of biological macromolecules, relying on the previous description of their structures to develop their modes of biosynthesis. The student shall have achieved, by this point, a substantial familiarity with biomacromolecules, and the last chapter of the book focuses on some phenomena related to them—enzymatic catalysis and sensory reception. Again, we return to the first topic, molecular structure, in order to illustrate how the subtleties of molecular structure can dramatically affect biological responses through interaction with macromolecules.

Our intent is to inform the student, not entertain him. To the extent that matching student interests and career goals aids him in a mastery of the fundamentals of chemistry, we have attempted to introduce appropriate material. Response at Indiana University to the material in this book has been heartening; student interest in general chemistry has been markedly improved, and student performance in more advanced courses provides substantial evidence of increased mastery of the subject material as well.

We have noted that *Chemistry* is intended primarily for students with some background in chemistry who intend to pursue the subject at a more advanced level. Nevertheless, introduction of a modest quantity of elementary material permits use of the text with beginning students. Moreover, we consider the format of the book to be sufficiently flexible to permit a marked degree of innovation by individual instructors in order to meet more exactly the talents and interests of individual classes. Finally, we have endeavored to produce a book of modest size that will encourage the student rather than overwhelm him.

The merits of *Chemistry* are due in large part to the insight of our lovely and delightful editor, Caroline Lanford Eastman, to the diligent efforts of Jane Woodbridge in translating our written efforts into comprehensible English, and to the helpful suggestions and criticisms of many of our colleagues at Indiana University. Finally, we are indebted to several generations of general chemistry students for their forebearance with our early experimental efforts; to them we can only apologize for subjecting them to reams of handouts in various stages of completion and organization and thank them for pointing out the several instances in which our experiment was headed in the wrong direction. We trust we have gotten it right at last.

E. H. Cordes
R. Schaeffer

1

Something about chemistry and chemists

A dog jumped over the fence.
Ducks and chickens flew back over the fence.
Was the dog changed into the ducks and chickens?

Attributed to Prof. Arthur C. Lunn
Martin Kamen, *A Tracer Experiment*, 1964

Have you ever closely watched a chicken pecking her way about the barnyard? If not, you might want to the next time the opportunity arises: You can learn something about chemists from them. Any casual observer can tell you that chickens spend the great majority of their time actively searching about them, pecking away as they go, apparently in search of food. But expert chicken watchers (there are experts in almost everything these days) assure us that these chickens are searching for food only about 20 percent of the time. The remainder of their hunting-around activity is thought to be exploration pure and simple: The chicken wants to know more about its environment and goes about studying it in the only way it knows how. So do chemists. So in one way or another do other scientists, people generally, and animals generally. Inquisitive behavior concerning the world about us seems to be a trait that transcends the boundaries of species and intelligence. Chemistry is a body of knowledge that has resulted from probing nature in certain ways, some of which we describe in the pages that follow.

What is chemistry? About the only acceptable answer to that question is that chemistry is what chemists do. But it is getting harder every day to say what chemists do, because they are doing more and more different sorts of things. An increasing fraction of these things have more or less to do with other sciences: physics, biology, geology, and so forth. All the classical lines that once could be used to place people into various scientific disciplines have been gradually eroded away. There really are no plain ordinary chemists any more: Now we have organic geochemists, biophysical chemists, and other rare and exotic types. Perhaps one day each chemist will have his own title; such people may well know more about less than any other scientists in history. But it is a fact that the "information explosion" in chemistry and other areas of human inquiry has led to increasing specialization and difficulties in communication. More about these matters follows below. For the moment, let us examine some of the things that chemists do and why they do them. Perhaps some generalization about what chemistry is will emerge from this search.

WHAT DO CHEMISTS DO?

One thing that lots of chemists do all the time and many more do part of the time is *chemical synthesis*. The object of this kind of work is to use existing methods or to develop new ones for the purpose of creating new unknown molecules, to devise new procedures for making known ones, or to improve known procedures for making known compounds. All of these kinds of activity have their own special importance. Let us consider a few examples. To begin with: Why make new compounds? Don't we already have enough? In fact, about 300 new compounds are made each day. About 1.5 million are known so far, and we clearly will have a great many more. Some of these new compounds largely serve the immediate needs of chemists. That is, they are made so that their properties may be examined and additional insight into, for example, the nature of chemical bonding may be obtained. Studies of this type provide a source of new understanding of molecules at a very fundamental level. In so doing, they broaden the foundation on which more directly useful studies are built. Such studies might include, for example, the synthesis of molecules that might have useful kinds of biological activity. The sulfa drugs provide one example. As early as 1904, the great German chemist Paul Ehrlich realized that infectious diseases might be fought with chemical compounds that are more toxic to the infectious agent than to the host. He developed the use of certain dyes against African sleeping sickness as well as some compounds of arsenic against syphilis, but success during these early studies was limited. Great impetus for the discovery of useful antibiotics was provided by the deaths of many wounded in World War I due to secondary bacterial infections and by the worldwide flu epidemic in 1917–1918. Finally, in 1935 it was demonstrated that sulfanilamide was effective in combating bacterial infections. Now it was time for the synthetic chemist to use his talents. In the succeeding years, many thousands of molecules that were structurally related to sulfanilamide were synthesized and tested as antibacterials. Many proved to be useful; some were much more effective than sulfanilamide itself. Altogether, they saved countless lives and endless suffering. Studies related to these are still being carried on at great rate; the search for useful cancer drugs perhaps occupies more of this effort than any other area.

Every once in a while, chemists get lucky: A molecule synthesized for the purposes of chemists may also serve those of the public generally. For example, a molecular oddity termed 1-aminoadamantane,

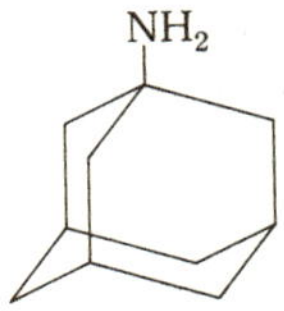

turned out to have antiviral activity and is marketed by Du Pont for this and other purposes. No really satisfactory antiviral agents, including 1-aminoadamantane, are currently available, and it would be exceedingly useful to have some. Perhaps further studies involving synthesis of compounds of this type will yield our first really good antiviral agent.

The synthesis of compounds of known structure and the improvement of existing synthetic procedures are also frequently worthwhile undertakings. For example, a new group of compounds called the *prostaglandins* has been discovered in man and other organisms within the last few years. These compounds appear to act as hormones and have a large number of *potential* medical uses. But they occur in living organisms in exceedingly small amounts; isolation from biological sources, as is done in the case of penicillin, for example, appears to be a hopeless way of obtaining sufficiently large quantities to be useful for medicinal purposes. Our only recourse is chemical synthesis. A related example is provided by the steroids. The development of methods for chemical synthesis and chemical modification of these crucial molecules of living systems has paved the way for numerous medical advances, including the development of antifertility pills.

The reasons for chemical synthesis and the successes that have been achieved in this area are numerous. But let us conclude with one exotic example and then move on to two other activities in which chemists spend their time. It is frequently of medical interest to find out how a given compound is metabolized in the human body: The more that we find out about such things, the easier it will be to detect abnormalities in pathological states and to work toward correcting them. What one needs, then, is some way to follow the fate of a particular chemical in the human body after it is administered in some way: injected, eaten, and so forth. One of the ways that we can do this is to use compounds that are labeled with a radioactive isotope: The fate of the radioactivity can be monitored directly with a suitable Geiger counter or related device. But care must be taken in such studies, because radioactivity is itself potentially harmful to humans. One isotope that appears to be almost ideally suited to such uses is carbon 11, an isotope that is similar to ordinary carbon but contains one less neutron in the nucleus. The synthesis of compounds containing carbon 11 is complicated by the fact that this isotope has a half-life of only 20 min; that is, after each 20 min, half of a given sample will have been lost through conversion to an isotope of boron by the process of radioactive decay. In fact, one of the reasons that this isotope is so useful is that it disappears so rapidly from the body; within a few hours, only a small fraction of the original sample remains. But a half-life so short poses a real synthetic problem. The carbon 11 must be collected from a cyclotron as carbon dioxide or another small molecule, converted to a synthetically useful form, all the steps of the synthesis must be carried out, and it must be administered to a patient—and, all the while, half of it is being lost each 20 min. Dramatically new synthetic methods must be devised to make carbon 11-labeled compounds medically useful, and their development will be expensive. Supposing that having these compounds would save 1000 lives a year, how much money should be spent in the research necessary to provide them? Supposing that one of these 1000 lives were your own, how much money should be spent?

Another principal activity of chemists is *analysis*. Like synthesis, analysis is carried out for a great many reasons. Suppose that we think a little about the reasons for analysis of samples that a geologist might provide. One reason is very direct and straightforward: Analysis provides the information necessary to know if important quantities of a particular material are present in a given sample. Does a particular sample contain sufficient iron to be worth mining? Sufficient gold to be worth extracting? Sufficient uranium to provide nuclear fuel at a reasonable cost? Those compounds characteristic of oil-containing regions? Answers to these questions are clearly important.

But analysis of geological samples serves other purposes as well. One of these is the study of the history of the earth. Dating—carbon 14 dating is the most familiar example—serves to establish the age of various types of structures and samples of geological origin. Other analytical procedures are also of use; examples are provided later in this book. Analytical procedures are also useful for tracing the history of life on earth. Extensive searches for those molecules characteristic of living organisms have been made in many geological samples, and useful information has been obtained. These investigations were extended to lunar samples obtained in the Apollo 11, 12, 14, 15, and 16 missions. A great deal of information concerning lunar history was obtained, although no evidence for present or past living forms was among it.

Chemical analysis of the great variety of life forms found on earth is a principal activity of chemists. Many important observations, both for chemistry itself and man in general, have resulted from such studies. Work in this area is frequently called *natural product chemistry*, since it focuses attention on those chemicals provided for us by nature. Just as the efforts of synthetic chemists sometimes lead to unexpected benefits, so do those of chemists involved in analysis of natural products. Let us consider just one recent example, which is important in that it relates to the treatment of leukemia, a major cause of death in children. Several of the drugs commonly used to halt this form of cancer were developed on the basis of the knowledge of how nucleic acid molecules are synthesized in higher organisms coupled with the efforts of chemists engaged in synthetic efforts. Limited successes have been obtained. A few years ago, a chemist involved in searching out those chemicals that occur in a certain class of sponges discovered a new class of natural compounds that, he speculated, might prove medicinally useful. One member of this class, arabinocytosine, has proved to be surprisingly effective in treatment of a type of childhood leukemia that had been utterly refractory to treatment with earlier drugs.

Perhaps this breakthrough will result in even better antileukemia drugs in the future. But would you have guessed that a useful anticancer drug would be found by someone investigating the chemistry of marine sponges? If the choice had been yours, would you have agreed to fund research in this area?

Thus far, we have confined our attention to analytical procedures in which the sample to be analyzed is actually in hand, even if it meant bringing it from the moon. In addition, we have the technology to study the compositions and chemistry of the sun and distant stars. Using sufficiently sophisticated instruments, it is possible to draw surprisingly detailed conclusions. The element technetium, for example, has been detected in distant stars by these methods; technetium is not found on earth, but its presence in stars provides useful information about one of the central questions in all chemistry—the origin of the elements. Specifically, it provides us with the information that element synthesis is an ongoing process in the universe, not the product of a single cataclysmic event at the "beginning."

One of the central features of chemical activity is the study of *chemical reactions;* this constitutes the dynamic aspect of chemistry. What kinds of chemical processes occur? How do chemical reactions proceed? What are the energy changes associated with chemical reactions? All these questions and more are under intense investigation. The answers will tell us something about how to make better rocket fuels, how to generate useful energy in a way that involves minimal environmental pollution, how the enzymes, the catalysts of life, function, and much, much more.

In the chapters of this book that follow, we shall have the opportunity to learn much about the results of chemical studies involving synthesis, analysis, investigation of the details of chemical reactions, and other matters in which chemists are frequently engaged. But let us pause just long enough to recognize that there are chemists who do not do experimental or theoretical chemistry at all, but who are nevertheless actively involved in the total chemical enterprise. These include the *teachers* of chemistry, who pass on information from previous generations of chemists to a new one; *communicators,* who are the interface between the chemists themselves and the general public (it is, after all, the public that is the final arbiter of the course taken by scientific research); *administrators,* who oversee the uses to which available resources for chemical research and development are put; *historians,* who remind us of the contributions of others and how they were achieved; and others. In one way or another, the activities of each of these groups of people lead to new information and insight of general importance. One last example will suffice.

During the first half of this century, mental illness was recognized as the most rapidly growing disease in the United States (or at least the most rapidly recognized one). By 1950, one-half of all hospital beds in this country were occupied by the mentally ill, and it was projected that this figure would rapidly rise to 80 percent, a catastrophic situation in terms of the delivery of adequate medical care. In fact, at present only about 20 percent of hospital beds are occupied by mental cases. What happened to reverse the trend toward increasing hospitalization of the mentally ill?

Historically oriented chemists discovered that extracts of the rauwolfia plants had been used by Hindus centuries ago for the treatment of a number of maladies, including epilepsy, chronic headaches, and snakebite, among others. Useless folk medicine or an important lead to useful products? In searching for an answer to this question, chemists discovered a number of complex molecules (alkaloids), several of which are medically useful and one of which, *reserpine*, is the most important tranquilizing drug for the mentally disturbed that we know of and certainly one of the most important drugs of any type, in terms of its impact on the health situation as a whole. Because of it, patients who would otherwise have had to have been hospitalized can lead relatively normal lives outside of medical facilities. It is important to remind ourselves that the most important of new information can result from what appear to be the least likely sources.

WHAT IS A CHEMIST?

From all that we have just said about what chemists do, we should be able to arrive at some feeling for what a chemist is. Like other scientists, chemists are concerned with nature. What, then, is the manner in which a chemist views nature? Perhaps the most distinguishing characteristic of a chemist's thinking is the emphasis he gives to the molecular makeup of matter. A long line of theoretical and experimental discoveries, dating well back into the nineteenth century, convinces the chemist that if he seeks to understand the nature and behavior of a particular substance he must concentrate on the nature of the molecules of which it is composed. He is interested in both the chemical changes that occur and in physical changes such as those between states of matter.

The kinetic molecular theory leads to certain general statements about the behavior of matter. It describes each of the three states of matter as collections of moving molecules. Gases are described as collections of molecules, widely separated and moving relatively long distances between collisions with one another. We recognize that weak forces exist between molecules (and shall have interest later in the nature of these forces), but the gaseous state is possible only when the kinetic energy of the molecules is sufficient to overcome these forces. If the temperature is lowered, the kinetic energy of the molecules decreases, and at some temperature the forces between molecules become important enough to overcome the kinetic energy: The isolated molecules condense to the liquid state. In liquids, molecules still retain sufficient kinetic energy to permit considerable freedom of motion, and the characteristic fluidity of liquids is a direct result of this freedom. When the temperature is reduced still further, the kinetic energy of the molecules eventually becomes too small to allow such freedom, and the molecules become locked into relatively restricted positions in the solid state; the characteristic rigidity of solids is the result. It is important to note that the forces between molecules remain unchanged as they move from the gaseous to the solid state. These forces are related to the composition and structure of the molecules. It is the change in kinetic energy that results in a changing balance between the competing factors.

The molecules of interest to the chemist range in complexity from simple diatomic molecules in the air around him to the highly complex molecules contained in his own body. He is concerned both with the individual molecules themselves and with the large collections of molecules normally studied in the laboratory, and also with the manner in which one molecule can be converted into another. He seeks the relationship between the observed properties in the laboratory and the fundamental composition of the molecules. Substantial progress has already been made in relating properties of matter to its molecular makeup, and the years ahead will no doubt see even more striking advances.

There are, then, two fundamental aspects to chemistry: the structure of molecules and their interconversions. These are, of course, intimately related topics, since the structure of a molecule determines its essential physical, chemical, and biological properties, and these include its chemical reactions. An understanding of the structure of molecules is essential to understanding chemical change or reaction. Nevertheless, understanding of chemical reactions is not based on understanding of structure alone; the dynamic aspect is an independent sector of chemistry.

Much more fundamental work remains to be done in the fields of both chemistry and physics before we can be satisfied with our ability to relate the properties of matter to the composition of the atoms themselves (as constructs of more fundamental particles). However, by concentrating only on the gross nature of atoms within molecules and on the ways in which they are held together, considerable understanding of chemistry can be achieved. The differences in construction of two molecules are as great to the chemist as the differences in construction of a Ferrari and a Ford are to the auto enthusiast, and the differences of behavior of both cars and molecules are equally dependent on their construction. The factors that hold atoms together in molecules are not conceptually greatly different from the nuts and bolts in a car, and at a useful level of description are no more difficult to understand. To achieve an understanding of chemical bonding adequate for many purposes, it is unnecessary to go much beyond the concept that atoms are held together by electron pair bonds without undue concern for the detailed nature of these bonds.

If we accept for the moment the fact that forces do exist that hold the components of a molecule together (and worry little about the exact nature of these forces), a great deal of chemistry can be immediately described. We shall first examine the numbers and types of atoms held within a particular molecule, and the geometric relationships that exist between the various parts. In brief, we shall speak of the structure of a molecule and of how the observable properties of molecules can often be simply related to the structures they have. Indeed, we can go still further and describe the changes that occur as one molecule is converted into another in terms of the changes in structure that take place as reaction occurs.

The field of chemistry merges with that of physics in the fact that both areas are concerned with the electronic structure of atoms and of molecules. Nevertheless, the emphasis that chemists and physicists give to this fact differs substantially. A primary goal of the physicist is to understand the atom itself in very sophisticated terms: the nature of the elementary particles of which it is composed, reactions and interactions of these elementary particles, the structure of the nucleus, the nature of forces responsible for maintenance of nuclear and atomic structure, a mathematical description of these very small systems, and similar problems. The chemist, on the other hand, is more interested in explaining the physical and chemical properties of the atoms and molecules composed of atoms by utilizing a knowledge of electronic configurations and energies. Thus, the atom and its constituents are of themselves a prime interest of the physicist, whereas the chemist views understanding of the atom as a basis on which to rationalize the behavior of more complicated species.

In most cases, the chemist is interested in the properties of molecules containing two or more atoms, and he must be concerned with the structural description of three-dimensional arrangements of atoms in space. The chemical, physical, and biological properties of molecules frequently depend in a very sensitive fashion on the exact three-dimensional arrangement of the constituent atoms, and the nature of chemical reactions is best understood in terms of the detailed structures of reactants and products. In a simple analogy, we may point out that all gloves bear a striking resemblance to one another, but both the size and shape become extremely important when one tries to put them on. For those molecules that contain only a few atoms, an adequate specification of their structure may require only a statement of which partners are involved in the formation of strong chemical (covalent or ionic) bonds. Thus, the pattern of chemical bonding may suffice to establish the complete structure. With more complex molecules, there may exist several possible arrangements in space (configurations or conformations), all of which correspond to the same basic pattern of formation of strong chemical bonds. In these instances, the three-dimensional structure may well be determined in part by weak interactions between various parts of the molecule or between the molecule and others in its immediate environment. For example, some chemists are concerned with the exceedingly complex macromolecules such as the proteins and nucleic acids that occur in all living materials. In comparison to ordinary molecules, these species may be exceptionally large; large enough in some cases that their maximum dimension might easily be detected with the unaided eye. For such molecules there exist a very large number of possible three-dimensional arrangements of the atoms, all of which correspond to a single arrangement of covalent bonds.

Furthermore, the biological activity of such molecules is almost always crucially dependent on an exact three-dimensional disposition for the atoms. Here indeed is a challenging area of investigation for the chemist interested in the structures of molecules. Here too is the area in which chemistry merges with biology. The biologist, concerned with all aspects of the life process, of necessity is concerned with the nature and function of these *biomacromolecules*. Again, the emphasis of the biologist and chemist differ just as did that of the chemist and physicist. The chemist's interest in biomacromolecules centers principally on the molecules themselves: the intimate details of the structure and synthesis, the nature of the interactions that account for and maintain the exact three-dimensional structure, means and mechanisms for affecting conversions of certain molecules into others. The biologist, on the other hand, is interested primarily in what the information deduced by the chemist reveals to him about life processes: the transmission of genetic information from parent to progeny, the forces responsible for the orderly development of an organism from embryo to adult, the basis of thought, learning, and deduction.

From all of the above, we should by now be coming to grips with what a chemist is: how he views nature, how his view is different from that of the physicist or biologist, and some of the kinds of activity in which he engages. We must ask one last question.

HOW DO CHEMISTS DO CHEMISTRY?

We began this introductory chapter with a question in quotation form. It really consists of two parts. First, an observation was made: A dog jumping over the fence resulted in some ducks and chickens jumping back over the fence. Second, one possible explanation to account for this observation was suggested: Did the dog change into the chickens and ducks? Most of us would probably think not and suggest a more sensible explanation: that the dog frightened the ducks and chickens across the fence. But suppose that we really wanted to *know* the answer to the question? The thing to do would be to look over the fence and actually watch the event in progress. This simple experiment would suffice to clarify the matter once and for all. All of this has something to do with chemistry, which is basically an *experimental* science, although theoretical methods have been contributing increasingly in recent years. But Leonardo da Vinci recognized the point several centuries ago: ". . . those sciences are vain and full of errors which are not born from experiment, the mother of all certainty. . . ."

From the point of view of chemistry, the question is how we can "look over the fence" to see what is happening when the systems that chemists study are so very small. This is, in fact, no small task. As it turns out, progress in chemistry is governed in part by progress in other sciences and in engineering (just as chemistry contributes its part to advances in related areas). The current rapid rate of increase of knowledge in chemistry reflects partially the power of the tools the modern chemist has at his disposal: high-speed computers, sophisticated spectrometers of many types, microscopes capable of viewing single atoms and molecules, highly automated data collection and treatment devices, radio telescopes, cyclotrons, and an almost endless list of related items, each suited to its own purposes and each, in some way, serving the needs of chemists. Much chemistry is still done, of course, with the standard apparatus usually associated with this science. Increasingly, however, the tools of physics and astronomy, as well as other sciences, are being brought to bear on problems of chemistry. In the pages that follow we shall have the opportunity to make casual acquaintance with some of these techniques.

WHAT IS WORTH KNOWING ABOUT CHEMISTRY?

In one way or another, chemistry lies near the heart of many matters of public concern: improvement of health care, conservation of natural resources, protection of the environment, provision of our everyday needs for food, clothing, and shelter, production of energy for constructive purposes, and much, much more. We cannot begin to describe all of the ways that chemistry is involved in the life of each of us. But one of our goals in this book is to establish the fact that it is so involved and in ways that we ought to know about and understand. So we are really faced with twin tasks: to learn some chemistry and to find out why it was worth the effort. Let us get on with it.

Suggestions for further reading

I. Asimov *A Short History of Chemistry,* Doubleday, New York, 1965.
J. F. Henahan *Men and Molecules,* Crown, New York, 1965.
P. A. Horigan *The Challenge of Chemistry,* McGraw-Hill, New York, 1970.
B. Jaffe *Crucibles: The Story of Chemistry,* Fawcett, New York, 1957.
W. F. Kieffer *Chemistry, A Cultural Approach,* Harper & Row, New York, 1971.
J. R. Partington *A Short History of Chemistry,* Harper & Row, New York, 1937.
L. Schubert and L. A. Veguilla-Berdecia *Chemistry and Society,* Allyn & Bacon, Boston, 1972.

Part one

FUNDAMENTALS OF CHEMISTRY

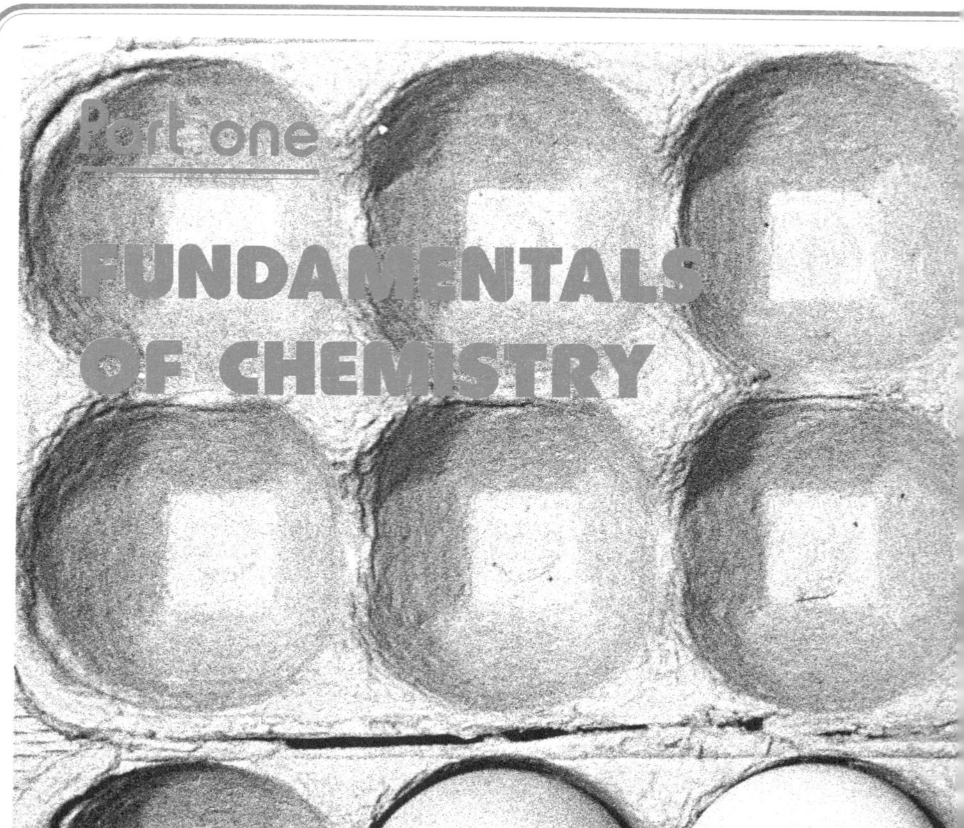

One of the most remarkable aspects of chemistry is the great subtlety in the link between the structure and behavior of molecules. It frequently turns out that very small changes, even in large molecules (perhaps the deletion or replacement of a single atom), radically affect the properties of these molecules, particularly in terms of their interactions with living matter. For example, below are the structural formulas for testosterone and estradiol; the former is a male and the latter a female sex hormone.

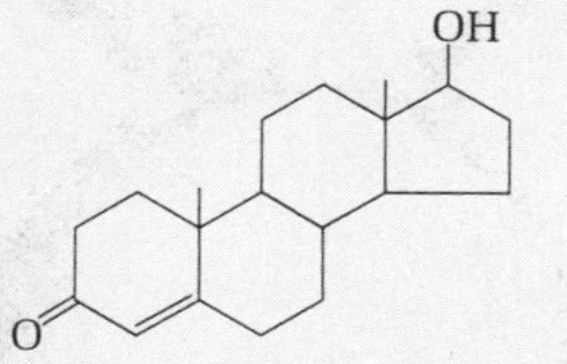

testosterone

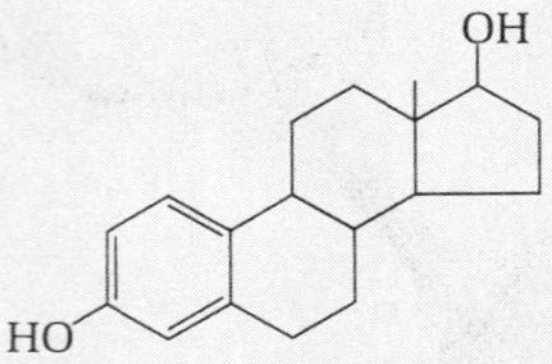

estradiol

These two compounds have dramatically different effects in living systems, yet they are obviously close relatives structurally. If we make additional small changes in structure, we can arrive at a new molecule:

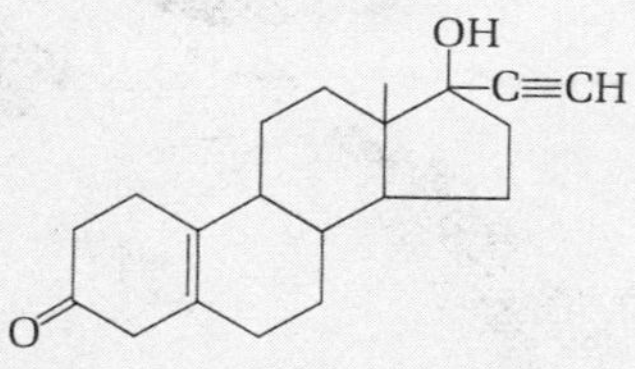

norethynodrel

This compound is neither a female nor a male sex hormone, but an antifertility drug. How can we account for such marked differences in activities of such closely related molecules? A great many examples of the subtle relationship between structure and function of molecules might be cited in addition to that just developed; in fact, several such examples will be found as we work our way through this book. Advances in whole areas of chemistry depend on understanding as much about the structure–function relationships of molecules as possible: the design of antibiotics, the search for better anticancer drugs, research in flavor chemistry, the acquisition of novel pesticides, and so forth. One of our long-range goals in this book is to develop some feeling for why molecules behave as they do. This involves understanding how molecules interact with each other.

We begin our search for this understanding in a very straightforward way—by simply forgetting all about electrons and how they act to hold atoms together in molecules and just asking questions about what molecules look like in three dimensions. Among the most important qualities of molecules that determine how they interact with others are size and shape. Our examination of this matter will include both simple molecules and some of great complexity. We shall see later that interaction of small molecules with large ones underlies some of the most interesting phenomena in chemistry and biology.

Important as size and shape of molecules are, they are not the whole story of chemistry, and we shall get around to worrying about electrons and chemical bonding. This new information will not only form the basis for additional understanding of intermolecular interactions, but it will let us account for some aspects of molecular size and shape. Essentially, theories of chemical bonding will provide explanations for why molecules look as they do. These theories also lead us into the last matter considered in this part of the book—why molecules react in the ways that they do.

In considering chemical reactions, we need to ask two kinds of questions. The first kind deals with the properties of the reactions themselves, the amount of energy released or consumed, for example. Such properties of reactions provide information concerning how a given chemical reaction might be employed usefully: to power a rocket, drive an automobile engine, or provide energy for a cardiac pacemaker. The second kind of question deals with how the reaction actually occurs. This information is also useful: It lets us understand something about how smog is formed and, therefore, how its formation may be prevented, how fertilizers are used by plants, how antibiotics act, and the like.

Overall, the material provided in the first part of this book describes the central areas of modern chemistry and lays the foundation for what is to come later. This material is, consequently, particularly important. Hence, the words of Prof. Henry Bent are probably particularly appropriate at the beginning of our study: "Nothing poorly learned is long relevant."

2

Molecular structures based on carbon

On chemistry:
It's hard and it's hard, ain't it hard, Good Lord

Woody Guthrie, "Hard, Ain't It Hard," 1952

The world around us is made up of molecules. Molecules are composed of atoms. For the moment, we accept as a postulate that molecules consist of *atoms held together by chemical bonds,* and further, that *these bonds are oriented in space in specific directions.* At a basic level, the understanding of many aspects of a particular molecule's chemistry is based on answering questions that follow directly from this postulate: Between what atoms are chemical bonds formed? What are the sizes of the atoms involved? What are the distances between atoms linked by chemical bonds? How are these chemical bonds oriented in space? From our answers, we can deduce two very important properties about a molecule—its *size* and its *shape.* These quantities partly determine the physical and chemical properties of the molecule, how it will taste and smell, whether it will prove beneficial, innocuous, or toxic toward living things, and how it will react with other molecules to yield new substances. Consequently, as chemists we must become deeply concerned with the determination of the size and shape of molecules. Let us discover the varieties of molecular structure that are allowed by our initial postulate.

The structures of compounds derived from *carbon* atoms, the element on which life on earth is based, is a good place to begin our discussion. The chemistry of carbon is the richest and most varied of all the elements. In part, this reflects the fact that carbon usually forms several chemical bonds simultaneously. To help us understand the spatial relationships of atoms in molecules, based on carbon or other elements, it is helpful to consider the *symmetry properties* of the molecule. Let us begin by discussing some fundamental aspects of symmetry.

SYMMETRY

Symmetry is important in most of the aspects of our lives; in day-to-day living, the word *symmetry* has two kinds of meaning. On the qualitative side, symmetry implies the quality of being well balanced or well proportioned. On the quantitative side, symmetry can also be thought of as deriving from certain definite and well-defined spatial properties of an object; these can be treated rigorously in purely mathematical terms. Certain kinds of symmetry are found in most of the objects in nature. For example, most animals are symmetrical in the sense that their left side looks much like their right side. Such objects are said to possess bilateral symmetry (see Figure 2-1).

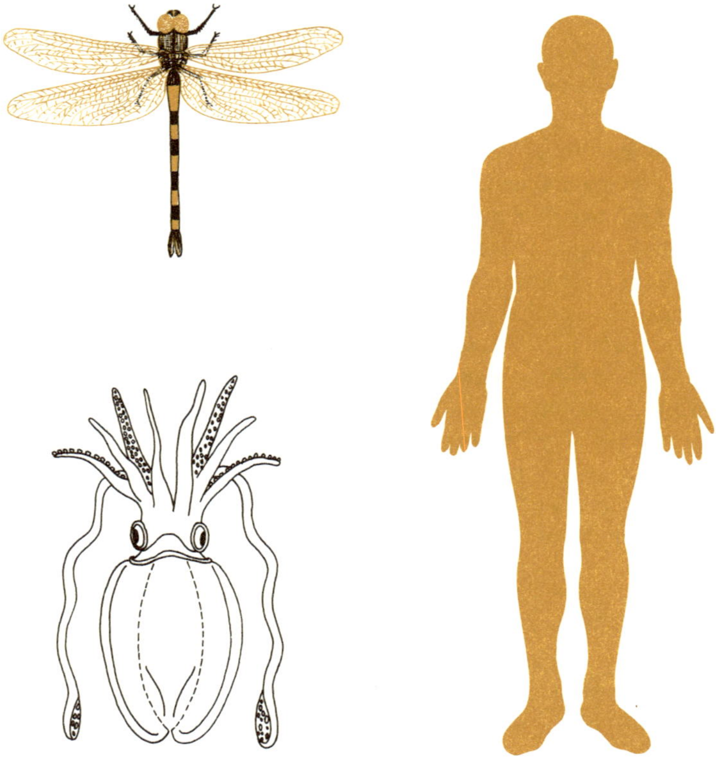

Figure 2-1 Most animals possess bilateral symmetry; that is, their right and left sides are mirror images of each other. The dragonfly, the squid, and man are just three examples.

It is also clear that objects such as the flower of a daisy, a starfish, and evergreen trees are symmetrical:

We can easily recognize that some of these objects possess greater symmetry than others; later we shall be concerned with exactly how symmetry properties are described in precise terms.

As fully as the works of nature, those of man depend on symmetry properties. The architect is vitally concerned with symmetry and balance in the design of buildings, both in the structures themselves and in their relationships to their surroundings. In sculpture and painting, the elements of symmetry are less definite but no less important: In both cases, the artist strives for balance in shapes, colors, and intensities. We may fail in efforts to describe quantitatively symmetry in a da Vinci or van Gogh, but the eye and mind recognize it instantly. There is symmetry, too, in the less exalted works of man—in fire hydrants, picket fences, and breakfast tables. Even our system of counting is symmetrical. If we consider the complete set of numbers, we see that for every positive number there exists a negative number of equal magnitude. Hence, the numbers are symmetrical about zero. A little reflection brings many additional cases of symmetry in everyday life to mind. But let us move on to attain a more precise way of talking about symmetry.

Symmetry properties are defined in terms of **symmetry operations.** Basically, to define one type of symmetry operation we ask what simple spatial rearrangements can be performed so that a structure equivalent to (i.e., superimposable on) the original structure will result. We may restate this in a slightly different way and say that symmetry operations are those for which, if we did not actually observe the operation, we would not be able to tell that any operation had actually taken place. To develop this point, let us imagine driving an imaginary spike through the center of one of the snowflakes shown in the photograph. If we pick up the snowflake by the spike and rotate it by 60°, by 120°, or by 180° and then return it to its resting place, no observer would be able to tell that rotation had occurred unless he had actually watched us perform the operation. Hence, these rotations are examples of symmetry operations. Another example can be developed for objects possessing bilateral symmetry (Figure 2-1). If we imagine placing a two-sided mirror through these objects so as to divide the objects into a left side and a right side and then reflect both sides of the object through the mirror, we would obtain structures indistinguishable from those with which we started. Hence, such a reflection is another example of a symmetry operation.

One geometric object that is particularly important for understanding the structure, and hence the chemistry, of carbon is the *tetrahedron:*

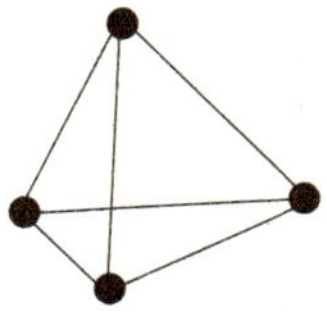

A tetrahedron is a three-dimensional solid with four vertices and four sides, each of which is a triangle. A regular tetrahedron, that is, one having all

These elegantly sculptured snow crystals provide a beautiful example of symmetry in nature. Although the details of the structure of each crystal are different, all of the crystals have a sixfold rotation axis, among other symmetry elements, perpendicular to the crystal plane. (Courtesy of NOAA.)

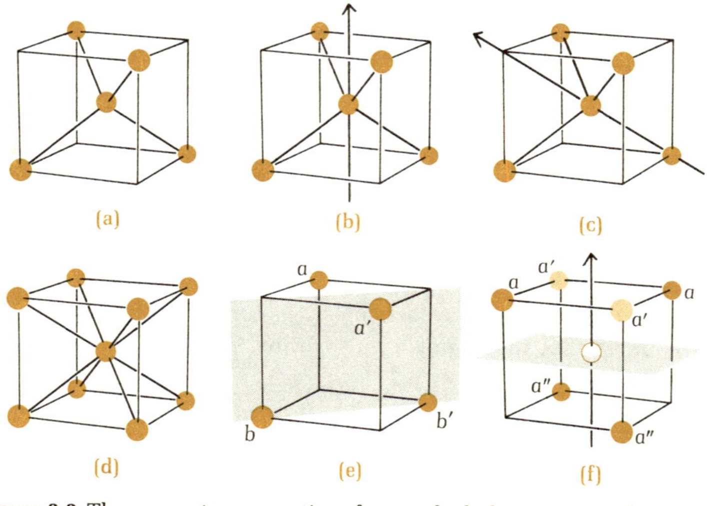

Figure 2-2 The symmetry properties of a tetrahedral structure and one with full cubic symmetry.

sides the same length, is a geometric figure that can be inscribed in a cube. Figure 2-2(a) shows the tetrahedral distribution of four identical atoms (i.e., each atom is a vertex of the tetrahedron) about an atom located at the center of the cube. We use atoms in our example here because eventually we shall wish to apply this geometry to the structure of molecules composed of atoms. In particular, we shall discover that carbon has a particular affinity for the tetrahedral structure. Let us consider symmetry operations that can be applied to this geometric object. Suppose we could drive our imaginary spike through the atom at the center of the cube as shown in Figure 2-2(b). As in the case of the snowflake, if we pick up the cube by the spike and rotate it by 180° and then return it to its resting place, no observer would be able to tell that any operation had taken place. Thus, this rotation is a symmetry operation. We speak of the imaginary axis as a **rotation axis,** and since two such operations return the system to its original position, this particular axis is said to be a *twofold axis*. The order of a rotation axis can be calculated by dividing 360° by the number of degrees of rotation of the symmetry operation. A few moments' study should convince us that such an axis could have been driven through the center of any of the three pairs of parallel faces of the cube and that there are thus three twofold rotational axes for a tetrahedral structure.

Figure 2-2(c) indicates a different way in which an axis can intersect a cube, this time through opposite corners. For this axis, a rotation of 120° will move the cube into a new but indistinguishable position. Since three such rotations ($360° \div 120° = 3$) are required to return each atom to its original position, this axis is said to be a *threefold axis*. Clearly, there are four such axes through the four sets of opposite points of the cube corners. For a tetrahedron, only these two types of rotational axes exist. In other structures, higher-fold axes may exist. For example, if an identical atom were located at each of the eight corners of a cube, as in Figure 2-2(d), three fourfold axes would exist. We shall find that it usually requires some time before it is easy to perceive symmetry operations and related symmetry elements by examining two-dimensional representations. It is easier to use a set of ball-and-stick models as a mechanism for understanding these operations, since they provide a three-dimensional representation.

Example *The snowflakes shown in the photograph are not only bilaterally symmetrical but possess rotation axes. Describe these. (After you have read the solution, can you analyze the rotation axes present in the flower of the daisy, the starfish, and the pine tree?)*

Solution *The snowflake has three kinds of rotation axes. We noted earlier that an axis passed through the center of the snowflake. Rotation of 60° about this axis is a symmetry operation; hence, this axis is sixfold (note that this axis is also threefold and twofold). A snowflake has a second*

Symmetry plays a strong role in most works of art. The bilateral symmetry of this mosaic, *Christ Enthroned, with Saints Vitale and Ecclesius,* from San Vitale, Ravenna, is evident. (Courtesy of MAS—Art Reference Library.)

kind of rotation axis, which passes through its opposite corners. Since a rotation of 180° about this axis is required to constitute a symmetry operation, this axis is twofold.

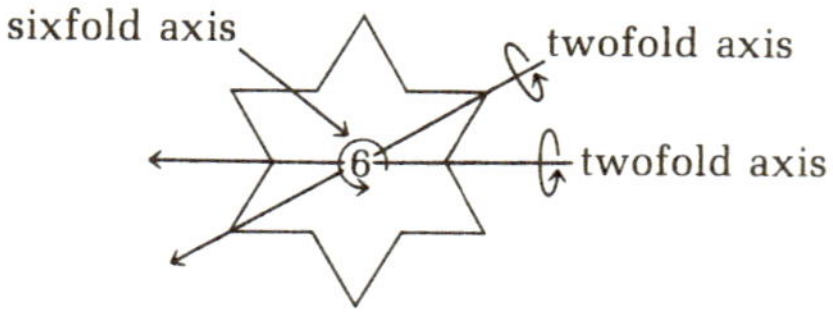

Clearly, there are three such axes in the snowflake. Finally, there are three more twofold rotation axes that pass midway between the corners of the snowflake. Thus, there are a total of seven rotation axes. The effect of the sixfold axis is to rotate the snowflake in the plane of the paper. The twofold axes, on the other hand, turn the snowflake over so that the bottom side is visible (but identical in appearance).

We can define another type of symmetry operation by examining the tetrahedral system. Let us imagine that we place a tiny, two-sided mirror in the position shown in Figure 2-2(e). The atom located at the position marked *a* is exactly where the reflection would appear for the atom marked *a′*. Note that the mirror bisects the atoms marked *b*, and thus reflection of half of the atom merely converts it to the other half. This symmetry operation is spoken of as a **mirror plane of symmetry.** If we consider the configuration carefully, it will show us that there are six different ways in which such mirrors can be lodged in the cube to reproduce the desired tetrahedral arrangement. The mirrors can be located along both diagonals of each of the three sets of parallel faces. Essentially, the mirror plane of symmetry operation starts with "half" of the configuration and "creates" the other "half" by reflection. Thus, it should be clear to us that any object having bilateral symmetry must have one or more mirror planes of symmetry.

Example *How many mirror planes of symmetry are present in the flower of the daisy?*

Solution *No less than eight mirror planes of symmetry are present in this object. Four of them are perpendicular to the plane of the flower and pass through the center of opposite petals. Four more are perpendicular to the plane of the flower but pass midway between the petals. If the petals of the flower looked the same on both sides, the plane of the flower would be an additional mirror plane, but experience tells us that we can distinguish between the front and back of a flower and this is not a true symmetry plane in a daisy. Clearly, any planar object has at least this one mirror plane.*

Yet another type of symmetry operation is, in principle, possible, but it is not applicable to the tetrahedral arrangement. Figure 2-2(d) shows eight identical atoms located at the corners of a cube arranged around an atom located at the center of the cube. If we consider the central atom and any of the atoms at a corner, a precisely identical atom is found at exactly the same distance from the center but in just the opposite direction. If this operation proves to be true for each atom in the arrangement, then the arrangement is said to possess a **center of symmetry.** Alternatively, the element around which the other elements are located may be called an **inversion center.**

Finally, we consider a more complex type of symmetry that exists in the tetrahedron. This symmetry operation is essentially a combination of the mirror plane of symmetry and rotational symmetry. Figure 2-2(f) shows a mirror plane through the central atom with a symmetry axis *perpendicular* to it that extends through the center of the cube. If we imagine that two successive operations are carried out, the tetrahedron reaches a position that cannot be distinguished from its original appearance. Let us rotate atom a about the axis by 90° to carry the atom temporarily to position a'. Reflection through the mirror plane would then carry the atom from a' to a'' and result in an arrangement identical to the original arrangement. Such a symmetry operation is called a **rotation-reflection** symmetry operation, and such a symmetry element is called a *fourfold rotation-reflection* (or fourfold alternating) *axis*. An examination will show that three such symmetry elements exist for this structure.

In summary, we note that an analysis of a regular tetrahedron reveals that this geometry has the following symmetry elements: three twofold rotational axes, four threefold rotational axes, six mirror planes of symmetry, and three fourfold rotation-reflection axes. This set of symmetry elements is unique to the tetrahedral geometry and, vice versa, the tetrahedral geometry may be described in terms of the symmetry elements. As we shall see, our study of the geometry of a tetrahedron, or any other configuration, will help us to understand the structures of molecules and hence their respective chemistries. The symmetry elements of a particular geometry determine a great deal about the chemistry of a molecule derived from a central atom. We should realize that the tetrahedral disposition of four identical atoms about a central atom is not the only geometry possible for a set of five atoms; we shall consider other possibilities. We should emphasize that for each geometry we might choose, a distinct set of symmetry elements exists, and these elements may in turn provide a basis for distinguishing between geometries.

MOLECULES AND STRUCTURE

Before we discuss the structure of individual molecules, it will be helpful to examine some features of chemical bonds that hold atoms together. Two particular parameters that describe bonds and their relationship to one another are **bond length,** the distance between the centers of the bonded atoms, and **bond angle,** the angle formed by the intersection of two chemical bonds at the center of the common atom. For a regular tetrahedron, the bond angle at the central atom formed between the bonds of any two of the others is by geometric arguments $109°28' \approx 109.5°$:

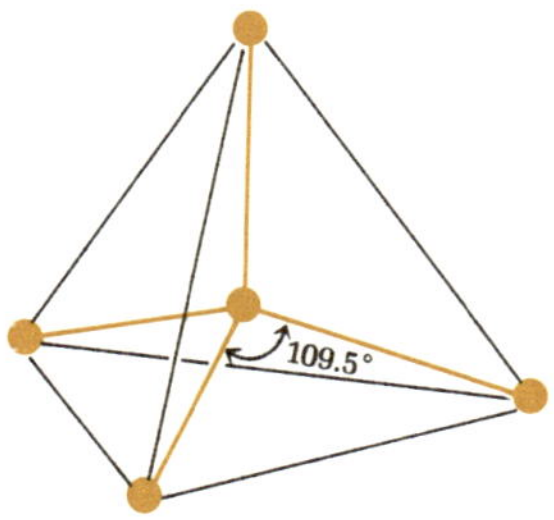

Bond lengths, however, are usually not determined by geometric arguments but depend very much on the nature of the bonded atoms.

It is worth pausing a moment to note the relationships between bond angles and lengths and some of the matters developed earlier. The size of a molecule having a specified number of bonds is determined largely by the lengths of these bonds. The molecular shape depends critically on bond angles and the relations between bond lengths. Symmetry properties are also determined by bond angles and relative bond lengths. To a large extent, the concepts of shape and size, symmetry properties, and bond lengths and angles are three ways of expressing what is basically the same set of information.

A very useful observation that will aid us in predicting the geometries of molecules is that bonds attached to a central element tend to spread out so that they are as far apart from each other in space as is possible. This can be illustrated by imagining that we tie four large balloons together. We would then observe that they naturally assume a shape such that the balloons are directed away from the center and are located at the four points of a tetrahedron. If one of the balloons were larger than the others, then the remaining three would crowd together in order to get farther away. Similar behavior occurs when atoms are bonded together. We shall explore this phenomenon later in the text.

METHANE

Although we shall be interested in the structures of molecules in general, including some of great complexity and elegance, let us begin our discussion of specific molecules with one of nature's simplest constructions, the *methane* molecule. This molecule will serve as a good illustration of many of the points just developed. Methane, a flammable gas ("marsh gas") which has a boiling point of $-161°C$, is composed of just five atoms: one carbon and four hydrogen. Thus, the molecular formula for methane is CH_4. Inherent in this simple formula are several very important questions that we may ask. Between which of the atoms do the chemical bonds exist? (This is equivalent to asking where the individual atoms are located.) Are all the chemical bonds linking the same types of atoms equivalent or not? What are the lengths of each of the chemical bonds? What are the angular relationships existing between the bonds?

To answer our first question, we learn that observation indicates that carbon atoms ordinarily form molecules where they are simultaneously bonded to four other atoms. In contrast, hydrogen atoms are ordinarily bonded to only one other atom. Thus, we may conjecture that the methane molecule is structured with a central carbon atom to which the four hydrogen atoms are linked. In such a structure, only carbon–hydrogen bonds exist (the answer to our first question).

One method of checking this assumption is to examine the methane molecule by one of several physical techniques available to modern chemistry. One such technique, applicable to our present problem, is *electron*

Figure 2-3 An electron diffraction photograph of the carbon disulfide molecule. From the information contained in this diffraction pattern, the trained chemist can deduce the structure of this molecule with great accuracy. (Courtesy of Prof. L. K. Montgomery.)

diffraction. This elegant but indirect and complicated technique uses an electron beam to produce a photographic record of the structure of a molecule. One example of an electron diffraction photograph is Figure 2-3. Examination of such photographs by specialists allows them to deduce molecular structures in great detail. Clearly, the electron diffraction pattern shown in Figure 2-3 does not provide information about molecular structure in as direct a way as an ordinary photograph provides information about structures of larger objects. To the trained eye of the specialist, however, the diffraction pattern does contain the necessary information to deduce the type of symmetry existing in the molecule's structure, its bond lengths, and its bond angles. An analysis of such photographs of methane gas molecules corroborates our deduction, namely, that the molecule consists of a central carbon atom surrounded by four atoms of hydrogen. Further, from the diffraction photograph we can deduce that the hydrogen atoms are in rapid motion.

Now let us try to answer our next questions, regarding the equivalence or nonequivalence of the carbon-hydrogen bonds and the spatial distribution of the atoms. Study of the electron diffraction pictures reveals a molecule with a carbon atom at its center with four equivalent hydrogen atoms attached by bonds of equal lengths and equal angular distribution from the central atom. This picture also agrees with experimental observations of simple derivatives of methane.

For example, if methane is treated with chlorine in the presence of ultraviolet radiation, a new molecule, *chloromethane*, is formed, in which one of the atoms of hydrogen has been replaced by a chlorine atom:

$$CH_4 + Cl_2 \longrightarrow CH_3Cl + HCl \qquad (2\text{-}1)$$

If we study the properties of this monosubstitution product, we discover that this reaction yields only one compound, corresponding to the molecular formula CH_3Cl. What are the properties of the compound that permit us to identify it? There are many properties, some simple and some sophisticated, that are useful identifying agents to the chemist. We may think of the properties of molecules that are used to identify molecules as being analogous to fingerprints, which are "properties" of people that are used to identify them. Electron diffraction pictures are one example of a molecular "fingerprint." Other examples are the melting and boiling points, color, and smell, all of which help to characterize molecules. If we return to our example, we discover that the "fingerprints" of all the chloromethane molecules are identical. Thus, we can tentatively conclude that each product molecule is the same and that only one chloromethane molecule exists. In fact, we can go further and say that for each simply substituted methane derivative CH_3X, one and only one chemical and physical entity can be obtained. This simple fact suggests a principal feature of the structural chemistry of methane: *All of the carbon-hydrogen bonds are equivalent.* Stated another way, we may say that each hydrogen atom occupies a physically and chemically equivalent position with respect to the central carbon atom. (If this conclusion were not true, then it ought to have been possible to construct two, three, or even four chloromethane molecules, each corresponding to the replacement of a particular hydrogen atom in a methane molecule in which these atoms are nonequivalent. We cannot even be certain about this latter statement, however, since it is quite possible that a further limitation may exist in that one of the hydrogen atoms may occupy a unique position with respect to the others and it may be that each time a hydrogen atom is replaced by an atom of chlorine or other element, it is that atom of hydrogen that is lost.)

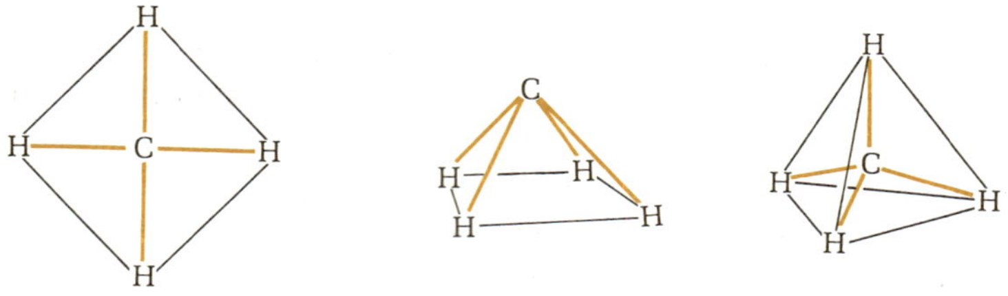

Figure 2-4 Three possible structures for the methane molecule: square-planar, pyramidal, and tetrahedral.

The equivalence of the four hydrogen atoms in the methane molecule clearly rules out several types of possible geometric structures but is consistent with three possibilities. The three possible alternatives, indicated in Figure 2-4, are the square, the square-based pyramid, and the tetrahedron. The choice between these structures may be deduced from the characteristics of the disubstitution products, just as the equivalence of the hydrogen atoms of the methane molecule was deduced logically from the study of its monosubstitution products. Thus, we treat the chloromethane molecule with additional chlorine and, in so doing, replace yet another atom of hydrogen with one of chlorine:

$$CH_3Cl + Cl_2 \longrightarrow CH_2Cl_2 + HCl \qquad (2\text{-}2)$$

By chemical analysis we identify the product as dichloromethane, CH_2Cl_2. Now we ask how many molecular constructions corresponding to the formula CH_2Cl_2 are possible based on each of the three structures in Figure 2-4? Clearly, in terms of symmetry there exist two distinct molecules that could be derived from a square-planar configuration,

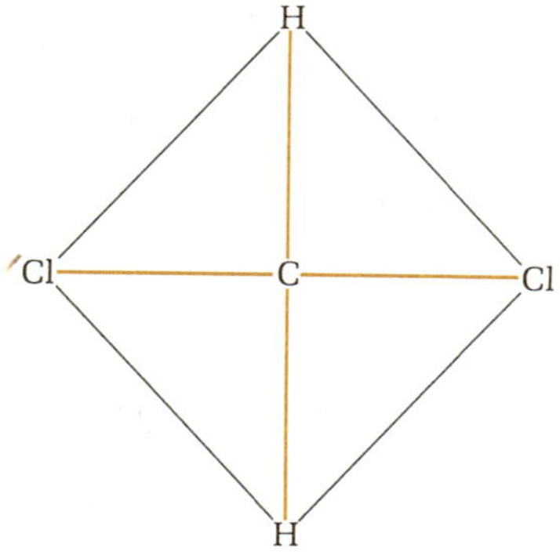

and

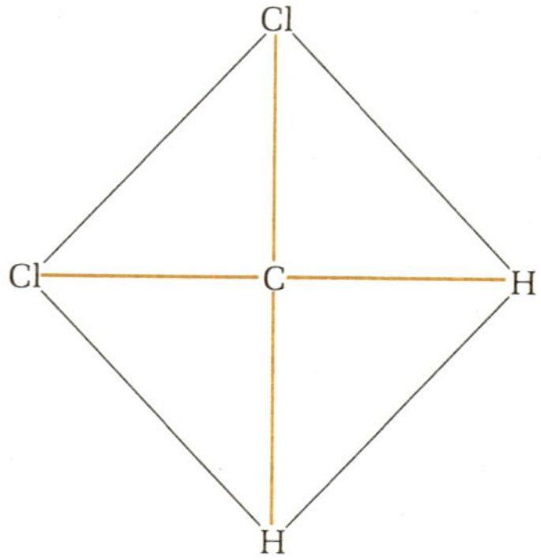

and two from the pyramidal structure,

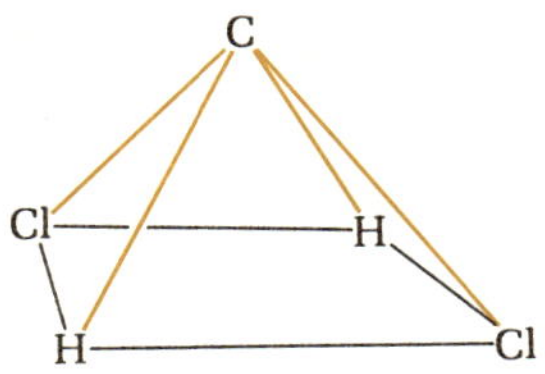

and

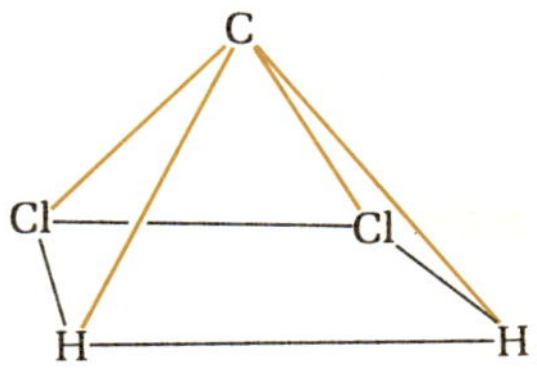

To understand clearly that the two representations of the square-planar dichloromethane molecule are, in fact, different species, let us examine their symmetry properties. We observe that for the molecule on the left there exists a twofold axis perpendicular to the plane through the carbon atom. For the molecule on the right, rotation about the same axis causes the chlorine and hydrogen atoms to change places. Thus, the molecule on the left has an element of symmetry that is lacking in the one on the right. We know that if two structures have different symmetry properties then they must represent different molecules. Thus, the square-planar configuration is ruled out as a possible geometric structure. A similar argument applies to our consideration and rejection of the two pyramidal structures. (In fact, if we view these from above, they look exactly the same as the square-planar representations.)

However, the tetrahedral structural representation of the dichloromethane molecule is unique:

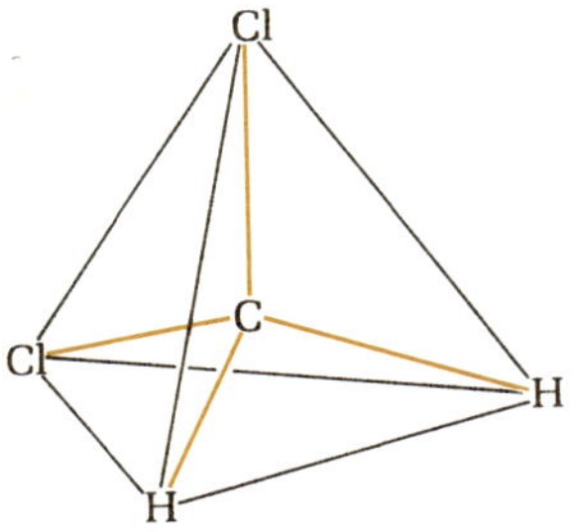

We note that each of the chlorine atoms is adjacent to both hydrogens and the other chlorine. In fact, in a tetrahedral structure, any atom is adjacent to each of the others. Thus, only one structure corresponding to CH_2Cl_2, or generally, CH_2X_2, can exist; this agrees with our original hypothesis. We should note that the symmetry properties of tetrahedral dichloromethane are quite different from those of tetrahedral methane. Of the three twofold rotational axes of the latter molecule, only one remains in the former, shown in Figure 2-5(a). None of the threefold axes remain, as inspection of this figure illustrates. Of the six mirror planes of methane, only two remain in dichloromethane: one that passes through the two atoms of hydrogen and midway between the two chlorine atoms, shown in Figure 2-5(b), and another that passes through the chlorines and midway between the hydro-

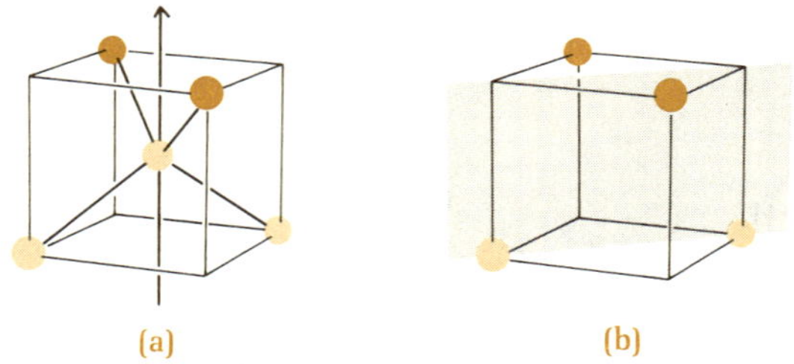

Figure 2-5 The symmetry properties of the dichloromethane molecule.

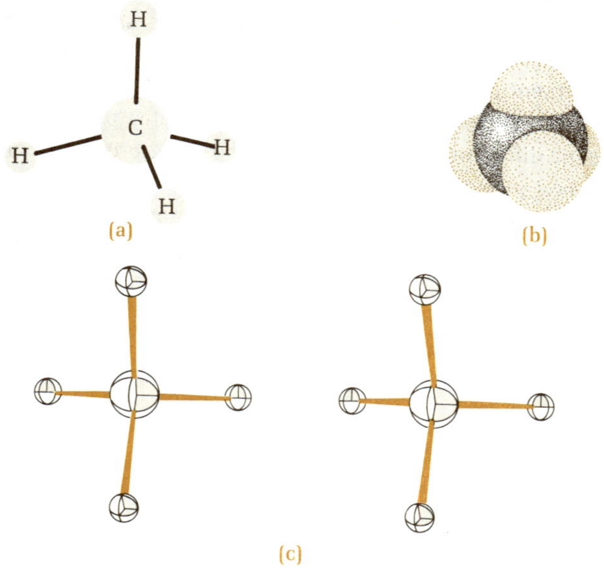

Figure 2-6 Three representations of the tetrahedral methane molecule: (a) ball-and-stick model, (b) space-filling model, and (c) stereo pairs.

gens. The fourfold inversion axes have been lost. Thus, we say that the dichloromethane molecule has lower symmetry than the methane molecule. In summary, we realize that we have discovered that, quite independent of source or method of preparation, there exists exactly one substance that corresponds to the formula CH_2Cl_2. This fact, plus the equivalence of the hydrogen atoms of methane, further confirms our earlier deduction that the methane molecule has a regular tetrahedral structure.

The fact that the methane molecule has a regular tetrahedral structure (along with other carbon compounds of the same general type) was first deduced by van't Hoff and Le Bel in 1874 on the basis of the arguments we have just followed. Since that time, modern experimentation has fully confirmed these results. Moreover, such experimentation has provided us with details of structure that cannot be deduced from indirect arguments. For example, electron diffraction studies indicate that the mean distance between the atom centers of the bonded carbon and hydrogens in methane is 1.107×10^{-8} centimeters (cm) or 1.107 angstroms (Å). We shall discover that one of the regularities of chemistry is that most C—H bonds, regardless of the molecule in which they appear, have approximately the same lengths as the C—H bonds in the methane molecule (this knowledge must be employed, however, with some discretion). The H—C—H bond angles of methane are all equivalent, of course, and are equal to the equivalent angles of the regular tetrahedron, 109.5°. Only one bond distance and one bond angle are necessary to describe completely the methane molecule.

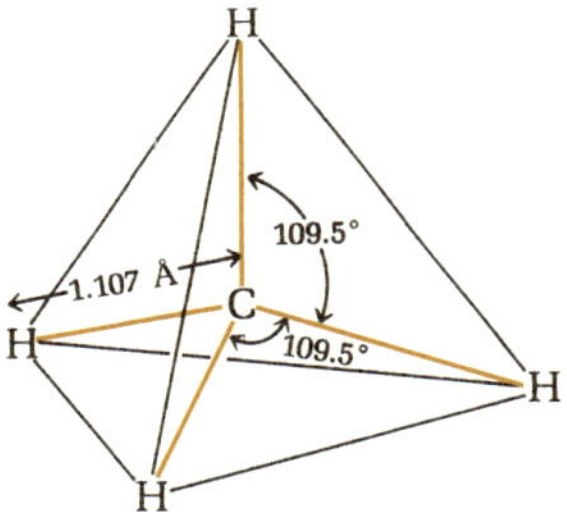

Let us look at the three different representations of the methane molecule shown in Figure 2-6. Each of these will help us to visualize properly the three-dimensional disposition of the atoms of methane. In the first representation [Figure 2-6(a)], the various atoms of the molecule are symbolized by spheres and the connecting bonds by rods. This *ball-and-stick model* is quite useful for illustrating which atoms are connected to which, for providing a basic impression of the three-dimensional organization of the atoms, and for permitting specification of bond distances and angles. A good representation of the actual shape of the molecule is better provided by the *space-filling model* shown in Figure 2-6(b). This visualization provides substantial insight into the actual physical shape of the molecule and the relative sizes of the various atoms. Finally, the *stereoscopic pairs* [Figure 2-6(c)] gives a view of the three-dimensional relationship of the atoms that cannot be obtained by simpler planar representations. (These pairs are best viewed through appropriate glasses, but these may prove unnecessary when the viewer becomes accustomed to examining them with an unaided eye. A piece of paper or other material held perpendicular to and between the stereo pair may aid in gaining the three-dimensional impression without the use of special glasses.) Thus, each of the representations provides us with special information about the methane molecule.

Our elementary knowledge about the structure of the methane molecule is useful in understanding some of its physical and chemical properties. As noted previously, methane has a boiling point of −161°C. This tendency to become a gas at such a low temperature is principally the result of being a very light molecule. Another contribution to the low boiling point is the nearly spherical shape of the molecule [Figure 2-6(b)], so that only atoms of one type are presented as part of the exterior surface. Molecules of this sort generally do not attract each other strongly, and therefore the individual molecules are able to escape into the gas phase rather easily. Hence, the physical makeup of methane explains its high volatility.

The melting point of methane is −183°C. Thus, methane exists in a liquid state in a range of only 22 degrees (by way of comparison, water is liquid over a range of 100 degrees). The symmetrical, nearly spherical, shape of methane permits these molecules to be readily packed into compact crystalline structures. Again, the physical makeup of methane explains its relatively high tendency to assume a different form, in this case going from a liquid form to a solid form. These points follow directly when we make the simple realization that molecules are a great deal more ordered with respect to each other in the solid phase than they are in the liquid or gas phases. Thus, any factor that makes the packing of molecules into compact and regular structures easier promotes the ease of formation of the solid phase, and thus a higher melting point results. To understand the significance of simple, symmetrical structures in this regard, consider the relative ease of packing marbles, cigars, and coat hangers into boxes. The most regular and most symmetrical of these structures is the marble, and no organizational effort is required at all by the packer; simply dumping the marbles into the box results in an ordered array. With the cigars, the packer has to work a little harder to ensure that all of them point the same way. With the coat hangers, the packing job becomes more complex, and the packer may have to expend a good deal of time and effort to achieve a neatly packed arrangement. We shall learn more about the influence of the relationship of molecular shape on the physical properties of molecules beyond the simple points we have just considered.

WEIGHT RELATIONSHIPS

We should emphasize that a particularly important fact embodied in the structural description of methane is that all molecules of this substance, regardless of their original source, are identical. The composition of methane is always the same. The reverse statement is also true. Namely, only a collection of one carbon atom and four hydrogen atoms arranged in a tetrahedral structure displays the properties of methane. We could logically describe methane as consisting of 20 percent (one in five) carbon atoms and 80 percent hydrogen. It is much more common, however, to consider not only the relative *number* of atoms of various types but also the *weight* of each type of atom. The actual weight of an individual carbon or hydrogen atom in familiar metric units is very small indeed, and a system of relative weights based on an arbitrary standard is more frequently used. The accepted standard currently used by chemists is to set carbon equal to a weight of exactly 12 atomic mass units (amu). On this scale, hydrogen has a relative weight of 1.0080 amu. Using these values, the weight of the methane molecule is 16.032 amu. Let us calculate the percentage makeup of the methane molecule on a weight basis. We calculate that it is

$$\frac{\text{weight of carbon atoms}}{\text{total weight of molecule}} = \frac{12.0}{16.032} \times 100\% \cong 75\%$$

$$\frac{\text{weight of hydrogen atoms}}{\text{total weight of molecule}} = \frac{4.032}{16.032} \times 100\% \cong 25\%$$

On a weight basis, the percentage makeup of methane is 75 percent carbon and 25 percent hydrogen. This result is strikingly different from the percentage makeup by number calculated above. Neither calculation is "more correct" than the other. Indeed, if we are discussing only molecular structure models, a percent description by number of a molecule makes a great deal of sense. In the laboratory, however, where we consider real molecules, it is much more difficult to count the various atoms in a particular sample. Since weight information can be used to calculate relative number information, we shall discover that it is much more common to describe the composition of a compound on a weight basis.

We continue with the development of these simple ideas about weight relationships, or *stoichiometry,* because we shall discover that they contribute and can be applied to the solution of problems with which we as chemists must deal in the laboratory. For example, suppose we wish to prepare the compound chloromethane according to the chemical reaction of Equation (2-1). How would we proceed? To begin with, we extract all of the pertinent information contained in this equation. This statement tells us that one molecule of methane reacts with one molecule of chlorine to yield one molecule of chloromethane and one of hydrogen chloride. Note that this is a *balanced* equation; that is, all of the atoms on the left-hand side of the equation also occur on the right-hand side. For ordinary chemical reactions, this must always be true, since we consider only processes that rearrange or redistribute atoms and not those that destroy or create them. Now, to perform the reaction we shall want to mix equal numbers of methane and chlorine molecules. We cannot weigh out individual molecules directly, but only very large collections of molecules. How, then, do we know how many grams of each reactant to weigh out? How many grams of product chloromethane will we expect? How much hydrogen chloride will be produced?

All of these questions, and a great many more related ones, can readily be answered by examining the relative weight relationships that exist in the reaction equation. It is helpful to state the following simple fact: Different atoms have different relative weights. It follows, then, that collections of the same number of atoms of different types will have different weights and that the weights of the collections will be in the ratio of the relative atomic weights. At this point, we shall find it helpful to develop some vocabulary of specific chemical terms. Let us remind ourselves that carbon has a relative weight of 12 amu on the atomic mass scale, and hydrogen has a weight of 1.008 amu. Thus, 12 grams (g) of carbon contain exactly the same number of atoms as 1.008 g of hydrogen. We define a *gram-atom* (g-atom) to be that number of atoms whose combined weight in grams is numerically equal to the atom's weight on the atomic scale. Let us extend this definition to molecules by noting that methane has a molecular weight of 16.032 [= (1 × 12) + (4 × 1.008)] and chlorine has a molecular weight of 70.906 (= 2 × 35.453). Thus, following our previous argument, the number of molecules in 16.032 g of methane is exactly equal to the number of molecules in 70.906 g of chlorine. We define a *mole* to be that number of molecules whose combined weight in grams is numerically equal to the molecule's molecular weight. We define **Avogadro's number** to be the number of atoms in a gram-atom or the number of molecules in a mole, and this number is equal to 6.023×10^{23}.

We are now in a position to continue to answer our questions. It is clear that our first answer is that in order to react a given number of methane molecules with an equal number of chlorine molecules, we conclude that for each 16.032 g of methane used we must employ 70.906 g of chlorine. Chloromethane has a molecular weight of 50.477 [= 12.00 + (3 × 1.008) + 35.453]. We repeat the same calculation for hydrogen chloride, which has a molecular weight of 36.461 (= 35.453 + 1.008).

$$
\begin{array}{ccc}
CH_4 + Cl_2 & \longrightarrow & CH_3Cl + HCl \\
16.032\ g + 70.906\ g & = & 50.477\ g + 36.461\ g \\
86.938\ g & = & 86.938\ g
\end{array}
$$

In general, in the laboratory it would be awkward and inconvenient to work always with just 1 mole of a material. Far more often, we shall wish to use some arbitrary amount of a particular compound, and it will be necessary to calculate what weight of each other reactant is required and what weight of each product will be formed. For example, if we have only 10 g of methane available, let us calculate the weights of each of the other compounds required. We observe from the calculation above that the weight of chlorine required always exceeds the weight of methane used; just 70.906/16.032 g of chlorine will be required for each gram of methane. We observe that not 1 mole but only 10.0/16.032 mole of methane is available, and therefore only (10.0/16.032) × 70.906 g = 44.2 g of chlorine will be needed. Amounts of each product produced will also be proportionately less. Only $(10.0/16.032) \times 50.477 = 31.5$ g of chloromethane and only $(10.0/16.032) \times 36.461 = 22.7$ g of hydrogen chloride will be produced. We can check our calculations by showing that the weight of products just equals the weight of reactants used:

$$
\begin{array}{lcccccc}
CH_4 + & Cl_2 & \longrightarrow & CH_3Cl & + & HCl \\
10\text{ g} + & \frac{10}{16.032} \times 70.906 & = & \frac{10}{16.032} \times 50.477 & + & \frac{10}{16.032} \times 36.461 \\
10\text{ g} + & 44.2\text{ g} & = & 31.5\text{ g} & + & 22.7\text{ g} \\
& 54.2\text{ g} & = & 54.2\text{ g} & &
\end{array}
$$

Thus, we have now answered all of our questions regarding the various weights of the reactants and products.

We reiterate the chemical principle involved in our study of the chloromethane reaction: Equal numbers of atoms, or molecules, of different types have weights that are in proportional ratio to their relative atomic or molecular weights. To test our understanding of this principle, let us ask ourselves how many grams of methane and chlorine must be reacted together to yield exactly 20 g of chloromethane? What can we conclude about the weight relationships for the reaction given in Equation (2-2)? (To solve this problem, we repeat the same series of calculations described in the previous paragraph.) Now we include two examples to show that we may use weight calculations of reactions in a variety of ways to obtain different sorts of information.

Example *The complete combustion of 10 g of methane according to the reaction*

methane + oxygen (O_2) $\longrightarrow$ carbon dioxide (CO_2) + water (H_2O)

yields 27.5 g of carbon dioxide and 22.5 g of water. What is the molecular composition of methane?

Solution *To solve this problem, let us approximate the atomic weights to be 12 for carbon, 16 for oxygen, and 1 for hydrogen. From this quantitative data, we can readily deduce a formula for methane. By examination of the reaction equation, we deduce that carbon dioxide comes from methane. The weight fraction of carbon in* CO_2 *is 0.273, and thus the weight of carbon in the original methane sample is* $27.5 \times 0.273 = 7.5$ *g. Likewise, the hydrogen present in water comes from the methane. The weight fraction of hydrogen in water is* $2/18 = 0.11$. *Hence, the weight of hydrogen present in the original methane sample is* $22.5 \times 0.11 = 2.5$ *g. Now let us convert the information about the weight composition into information about number composition. We divide through by the appropriate atomic weights: for carbon,* $7.5/12 = 0.625$; *for hydrogen,* $2.5/1 = 2.5$. *Thus, the ratio of carbon to hydrogen atoms in the methane molecule is* $0.625/2.5$ *or 1 to 4. This information is consistent with a molecular composition of* CH_4, C_2H_8,

Figure 2-7 Space-filling models for (a) chloromethane and (b) dichloromethane.

C_3H_{12}, *and so forth. To decide between these molecular compositions, additional information is required. In this particular case we can convince ourselves that* CH_4 *is the correct composition by realizing that carbon can form no more than four bonds and that none of the other compositions corresponds to a possible structure. The more general approach to making a choice between such compositions is to use additional experimental information to determine the molecular weight of the molecule. Several methods are available for determination of molecular weight. (This kind of data and reasoning was precisely what was used in the original determination of the molecular formula of methane.)*

Example *Analysis of the products of combustion of a hydrocarbon of an unknown structure (such as described in the preceding problem) reveals that the molecule of the hydrocarbon is 88.3 percent by weight carbon and 11.7 percent by weight hydrogen. Given that the molecular weight of this compound is 136, what is its molecular composition?*

Solution *Dividing the weight composition by the atomic weights yields the number composition: carbon,* $88.3/12 = 7.35$; *hydrogen,* $11.7/1 = 11.7$. *Therefore,* $C/H = 7.35/11.7 = 0.63$. *Alternatively, for each carbon atom there must be* $1/0.63 = 1.60$ *hydrogens. Of course, in any actual compound, only whole atoms can be present; therefore, for a* C:H *ratio of* 1:1.60, *the minimum formula with integral atoms is* C_5H_8. *However, the molecular weight of* C_5H_8 *is only 68, just one-half the given molecular weight. This ratio plus the molecular weight requires a composition of* $C_{10}H_{16}$.

Further opportunities to explore weight relationships in chemical reactions are provided in the study problems at the end of the chapter.

SUBSTITUTED METHANES

Before considering the structures of other molecules, let us consider those of chloromethane and dichloromethane to determine what modifications of the methane structure are necessary to describe these molecules. [We may wish to look again at chemical reactions Equations (2-1) and (2-2).] At the same time, we shall be looking at the symmetry properties of tetrahedral structures. Space-filling representations of chloro- and dichloromethane are shown in Figure 2-7. We notice immediately that the chlorine atom is substantially larger in size (and incidentally in weight) than the hydrogen atom; what modifications in molecular structure must occur to accommodate the replacement of smaller atoms by larger ones?

Since it is larger, the chlorine atom cannot approach a carbon atom as closely as a hydrogen atom does, so the C—Cl bond lengths must be longer than those for the C—H bonds; indeed, the bond length for C—Cl is 1.78 Å compared to 1.11 Å for the C—H bond. Furthermore, the disparity between the sizes of the chlorine and hydrogen atoms and their relative closeness to the carbon atom suggests that the basic molecular structure might be distorted somewhat by changes in the bond angles from those of a regular tetrahedron. In fact, the H—C—H angle in chloromethane is 111.9°, about 2.4° larger than in the case of methane. (At first glance, it would be reasonable to guess that if a chlorine atom is heavier and larger than a hydrogen, the H—C—H bond angle would be compressed and somewhat smaller than in methane. In the analogy that we made earlier with the orientation assumed by balloons, the chlorine "balloon" has a different shape, however; it is thinner near where the balloons are joined together and fatter at the outside. Therefore, the hydrogen balloons move a little closer to the chlorine and open the bond angle. To anticipate the contents of Chapter 4 for a moment, the "balloons" are bonds and the higher electronegativity of the chlorine draws the electrons away from the center. We shall examine this idea in greater detail in Chapter 4.)

A similar line of reasoning applies to the structure of the dichloromethane molecule. In this case, the two large chlorine atoms get in each other's way a little and hence force the Cl—C—Cl bond angle to be larger than the regular tetrahedral value. The angle turns out to be about 111°. Only one of the original four threefold axes of the regular tetrahedron are retained in chloromethane, and one of these is present in dichloromethane (see Figure 2-5). What other symmetry elements can we find for chloromethane and dichloromethane?

If we substitute yet one more chlorine atom for a hydrogen atom of methane, we obtain *chloroform* (trichloromethane), $CHCl_3$. Now, from a symmetry point of view there is great similarity here to the chloromethane case, since three atoms of one type and one of another exist around the central carbon atom. From the structural point of view, we would expect the Cl—C—Cl bond angle to be greater than 109.5° because of crowding of the large chlorine atoms into too little space. This expectation is borne out, since the actual bond angle is about 112°.

Finally, if we substitute chlorine for the last hydrogen atom, we obtain

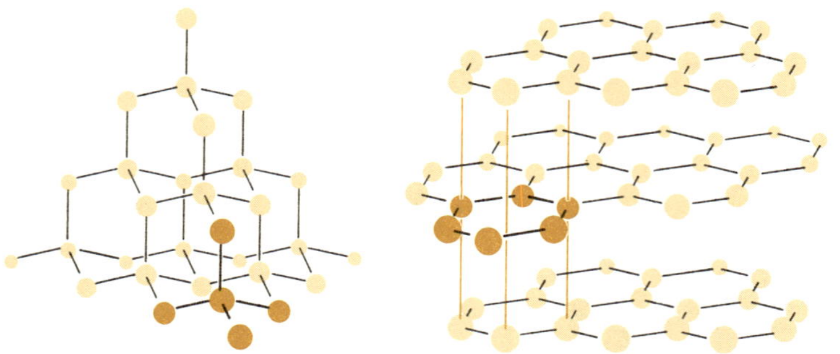

Figure 2-8 The structures for (a) diamond and (b) graphite.

carbon tetrachloride (tetrachloromethane), CCl_4, a molecule with the same symmetry properties as methane itself. The Cl—C—Cl bond angles must now be exactly 109.5° again (as indeed they are), just as in methane. The molecule is, of course, a great deal larger than methane, since the C—Cl bond distance is 1.77 Å.

The simple intuitive ideas about molecular structures derived from the substitution of atoms presented thus far suffice to help us predict or understand a variety of different structures or observations and their patterns.

Also, this brief discussion of methane and some of its chlorinated derivatives provides a basis for developing an understanding of the chemistry of the elements from which these molecules are derived—carbon, hydrogen, and chlorine. We shall now consider each of these elements in turn.

CARBON

Carbon is the fundamental building block for the construction of the molecules of all living things. Thus, we find it to be universally distributed throughout living nature. Pure carbon occurs principally in two forms: as *diamond* and as *graphite*. The structure of diamond [Figure 2-8(a)] is just what we might expect on the basis of our earlier observation that the carbon atom tends to form four chemical bonds simultaneously, with each bond directed to the corners of a tetrahedron. In diamond, each carbon atom forms four tetrahedrally arranged bonds with other carbon atoms; thus, the bonding tendencies of each carbon atom are remarkably well met in this structure, and the entire diamond crystal becomes essentially one molecule. Some of the physical properties of diamond, including its exceedingly low volatility and its extreme hardness, are due to this particular type of structure. Diamond is, in fact, the hardest material known. The carbon-carbon bond distance in diamond is 1.54 Å, which, as we shall see, is the common value for bonds of this type.

On the other hand, graphite is very different. As illustrated in Figure 2-8(b), the carbon atoms of this material are arranged in layers and within each layer in a hexagonal fashion. (This is, of course, not what we would intuitively expect on the basis of our discussion thus far.) However, given this structure, we may anticipate that graphite and diamond have very different properties; this, in fact, is true. Graphite is soft, black, and flaky. Furthermore, the carbon–carbon bond distance in graphite is only 1.42 Å. This short distance indicates that the structure of graphite probably has some complicating features; however, just what these may be we defer for the moment. The important point to realize is that two substances with quite different physical properties may have exactly the same chemical composition. We remind ourselves that both graphite and diamond consist only of carbon atoms. The distinctive properties of these two materials are derived from the ways in which the atoms are arranged in space. Thus, we conclude that not only the nature of the atoms but also their spatial arrangement is important in determining the physical and chemical properties of an element or compound. We shall encounter several additional examples emphasizing this point.

Carbon exhibits the ability to form long chains, or arrays, by linking up one carbon atom after another. This turns out to be the most characteristic and important feature of the chemistry of carbon. Both diamond and graphite exist in this form. The ability to form long chains in this manner is called *catenation tendency*. Carbon has the greatest catenation tendency of all the elements. As a consequence, molecules of great size and complexity can be formed that are based on carbon. Such molecules are useful for a variety of purposes; they include plastics, films, fabrics, and oils. Some of these compounds are discussed in Chapter 7. Other large molecules based on carbon are indispensable to life as we know it. These include proteins, nucleic acids, and polysaccharides. A discussion of these, too, is reserved for Chapter 7. We can note in passing that carbon is so intimately associated with living processes that great importance was attached to the analysis of lunar soil samples containing this element. Analysis shows that lunar soil contains exceedingly little carbon, about 100 parts per million. The carbon found was in forms that are not characteristic of those compounds occurring in living systems. Early examinations of lunar samples for characteristic molecules based on carbon have failed to provide evidence that life as we know it ever existed on the moon.

The essence of the chemistry of carbon lies in its compounds with hydrogen and their derivatives. Those compounds that contain only carbon and hydrogen, such as methane, are called **hydrocarbons.** A very large number of such compounds is known, and we encounter several of the simpler ones later in this chapter. Reserves of petroleum are the most important source of hydrocarbons. They are useful as solvents and fuels, among other things. The most characteristic feature of the chemistry of the hydrocarbons is their inertness; they are among the least reactive compounds known. This lack of reactivity can be taken advantage of in several ways. Just as great interest is attached to the possibility of finding evidence of life on the moon, there is much current interest in deducing the history of the development of life on earth. Much of our information concerning early life forms comes from fossils found in rock samples. Hydrocarbons, because of their stability over geological periods of time, are also useful as a kind of fossil, a *molecular fossil.* Just as we may use skeletal fossils to try and understand more about the development of life, so we may use the skeletons of molecules in the same way. We look for characteristic structural features of hydrocarbons, and sometimes other molecules, that we now recognize as being of biological origin. Studies similar to those done on lunar samples have been carried out on meteorites and other extraterrestrial samples. Hydrocarbons of apparent biological origin have been isolated from geological samples at least 2 billion years old, establishing a minimum age for the antiquity of life. We shall discuss this further in Chapter 17.

Many of the compounds derived from hydrocarbons by replacement of hydrogen atoms with halogen atoms are useful in a variety of ways. We have already encountered two important examples: Carbon tetrachloride is widely employed as a household cleaning agent, and chloroform was one of the earliest general anesthetics. Freon, the most generally used gas for refrigerators, is a simple compound of carbon, chlorine, and fluorine. Other simple compounds form the basis for many carbon-based polymers (Chapter 7).

Carbon forms two *oxides* of interest. The most common one is carbon dioxide, CO_2, familiar to almost everyone as Dry Ice. Carbon dioxide is a colorless, odorless gas and is a minor, but important, constituent of the atmosphere. Plants and other photosynthetic species use atmospheric carbon dioxide as the raw material from which they synthesize the complex carbon compounds that serve as food for animals and other living things (Chapter 19). Solution of atmospheric carbon dioxide in the oceans helps to maintain marine plants in a form suitable to support marine organisms.

When carbon dioxide is dissolved in water, part of it reacts with water to form carbonic acid: $CO_2 + H_2O \rightleftharpoons H_2CO_3$. Derivatives of carbonic acid, the *carbonates,* are important minerals. For example, limestone is calcium carbonate. Some aspects of the chemistry of carbon dioxide and carbonates are developed in Chapters 11 and 12.

The other important oxide of carbon, carbon monoxide (CO), is also colorless and odorless. It is a highly toxic gas, since it combines tightly with hemoglobin, the oxygen-transporting pigment of the blood, preventing the individual from acquiring sufficient oxygen to live. As described in Chapter 18, atmospheric contamination with carbon monoxide resulting from incomplete combustion of hydrocarbon fuels is an increasingly severe environmental problem.

Example *A standard U.S. automobile traveling 30 mph emits about 150 g of carbon monoxide per mile. How many moles of carbon monoxide are liberated by such a car for each mile traveled? How many moles are liberated per hour of travel at this speed?*

Solution *Carbon monoxide has a molecular weight of approximately 28. Therefore, the number of moles of this gas in 150 g is* $150/28 = 5.35$ *moles per mile traveled. In an hour, the car will travel 30 miles; hence the total amount of carbon monoxide liberated will be* $5.35 \times 30 = 160$ *moles.*

Carbon forms a variety of compounds with metallic elements. These are called *carbides* and are usually high-melting solids. The bulk of carbon occurring in solid form on the lunar surface is in the form of various metallic carbides.

The chemistry of the compounds of carbon is considered in more detail in several of the chapters to come, particularly in dealing with polymers, so we shall pick up later some of the lines of development begun here.

HYDROGEN

Elemental hydrogen at ordinary temperatures exists as a colorless diatomic gas, H_2. This is the lightest of molecules, molecular weight 2.0016, and is therefore, very volatile: Its boiling point is $-252.8°C$. Furthermore, molecular hydrogen is a highly symmetrical structure that packs readily into a crystalline lattice. Consequently, it has a very narrow liquid range of just 6°. Reflecting the small size of the atoms from which it is constructed, H_2 has a bond length of only 0.74 Å.

Elemental hydrogen is almost unknown in the atmosphere of the earth. There is considerable evidence that the primitive atmosphere of the earth consisted largely of hydrogen (see Chapter 17), but the gravitational attraction of the earth for this very light gas is quite weak. Consequently, essentially all of the H_2 escaped into space long ago, perhaps a few billion years ago. In contrast to our own atmosphere, hydrogen is the most abundant element in the universe. A reasonable approximation is to say that the universe is an ocean of hydrogen in which a few specks of nonhydrogen material can be found.

Hydrogen forms a large variety of compounds with many different elements. As a result, there is quite a bit of hydrogen, as a component of more complex molecules, in the crust of the earth. Considering all the elements, hydrogen ranks tenth in abundance on earth. **Isotopes** are forms of the same element that differ in their nuclear structure, a topic we shall consider in more detail in Chapter 16. Three isotopes of hydrogen exist naturally. An atom of the most common hydrogen isotope has one proton in its nucleus and one orbital electron. We note that for many elements, the number of neutrons in the nucleus is variable within rather narrow limits. Most of the hydrogen on earth contains no neutrons; this isotope is sometimes called *protium*. About 0.015 percent of terrestrial hydrogen contains one neutron in the nucleus; this isotope is called *deuterium*. Finally, about 10^{-15} percent of the hydrogen exists as *tritium*, having two neutrons in the nucleus. Of the three isotopes of hydrogen, only tritium is radioactive: It decomposes spontaneously to form an isotope of helium, emitting an electron in the process.

The chemical behavior of different isotopes of an element is very similar. This must be true, since chemical behavior largely reflects the charge on the nucleus (the number of protons) and the number of orbital electrons. Consequently, protium, deuterium, and tritium form analogous series of compounds. The properties of these compounds do differ in detail, since the weight of the isotopes does vary. For example, deuterium oxide, D_2O, and ordinary water, H_2O, have slightly different physical properties: Deuterium oxide melts at 3.82°C and boils at 101.4°C. For living organisms, the small differences in physical properties and chemical behavior between the isotopes of hydrogen can be crucial. Humans, for example, find that deuterium oxide is toxic if taken in substantial quantities. Some algae and bacteria, in contrast, can live perfectly well in pure deuterium oxide.

The chemical behavior of hydrogen is unique among all the elements. It is, in fact, a member of none of the usual chemical families. On the one hand, we might consider that hydrogen, with only one orbital electron, might act like the alkali metals, which have just one electron in their outermost electron shells. But the electron of hydrogen is held much more tightly by its nucleus, since it is very close to it, than are the outer electrons of the alkali metals. Since the chemical behavior of the alkali metals is dominated by their tendency to donate the outer electron to appropriate acceptors, the chemistry of hydrogen is necessarily distinct. On the other hand, we might argue that hydrogen should be a member of the halogen family since, like the halogens, addition of just one more electron will complete an electron shell. But the electron affinity of hydrogen is very much less than that for the halogens. Consequently, the species derived from hydrogen with the filled electron shell, the *hydride ion*, H^-, is much less stable than those derived from the halogens, F^-, Cl^-, Br^-, and I^-. The chemistry of hydrogen mimics neither that of the alkali metals nor that of the halogens nor, for that matter, that of any other of the chemical families.

Hydrogen forms of vast array of *hydrides* with nonmetals. Methane is one example of a carbon hydride. We shall encounter several more hydrides of carbon later in this chapter and, in the subsequent one, hydrides of several other nonmetals are discussed. In addition to these common and well-understood compounds, hydrogen forms hydrides with many metals. These are usually not as stable as the hydrides of nonmetals and show a variety of properties. For example, the hydrides of calcium and barium, CaH_2 and BaH_2, are good conductors of electricity, and are very powerful reducing agents. They are both colorless, crystalline compounds. These observations suggest that these hydrides must have substantial *ionic* character, deriving from the presence of cations and hydride ions in the crystal. Thus, to a fair approximation, these hydrides may be compared to an ionic crystal such as sodium chloride. The reducing power clearly indicates the presence of the hydride ion, a potent reducing agent. In contrast, the hydrides of lithium and beryllium, among others, conduct electricity less well and are less reactive. These solids are more nearly correctly described as *molecular* rather than ionic. Metal hydrides with intermediate properties are also known.

Of particular note are those metal hydrides formed from the adsorption of hydrogen gas onto pure metals. Titanium, for example, has the capacity to adsorb about 1600 times its own volume of hydrogen. Thus, hydrogen fits into spaces between the larger titanium atoms: Compounds of this type are called *interstitial compounds* and are frequently nonstoichiometric. The titanium hydride formed by adsorption of hydrogen into the interstices of the titanium lattice has the formula $TiH_{1.7}$. Other metals behave in a similar way. Palladium, platinum, and uranium form interstitial hydrides; these are useful reducing agents. It is not correct to think of these compounds as simply a physical association between a metal and a gas: The properties of both hydrogen and the metal change as a result, heat is liberated upon association, and substances having definite composition result. All of these are usually taken to indicate the formation of molecules.

CHLORINE

Like hydrogen, elemental chlorine exists as a diatomic gas, Cl_2. This gas is yellow-green and toxic. Elemental chlorine has a Cl—Cl bond distance of 1.99 Å; this indicates that the size of the chlorine atom is relatively larger than that of the hydrogen atom. Elemental chlorine weighs about 35 times as much as elemental hydrogen and is much less volatile than the latter. The boiling point is $-34.6°C$.

Very little chlorine exists on earth in the elemental form because of the high reactivity of this element. Chlorine is able to combine readily with most of the other elements as well as with a great many compounds.

Chlorine is a member of the group of elements known as the **halogens.** These elements are characterized by an outermost electron shell that needs

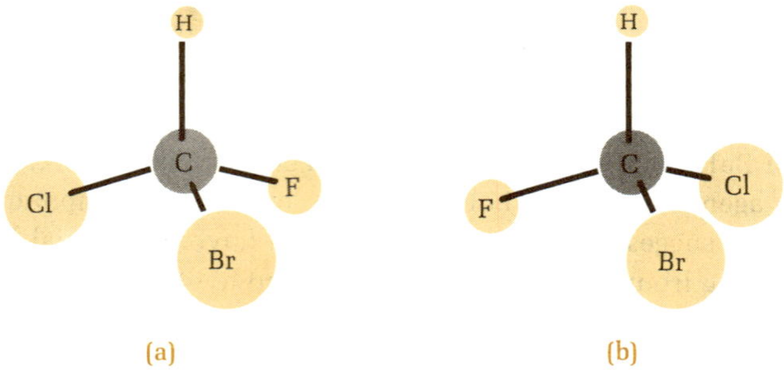

Figure 2-9 Two ball-and-stick models corresponding to the formula CHFClBr.

just one electron to obtain a noble gas electronic structure. Consequently, the halogens readily add one electron to form the corresponding anions. Indeed, the great majority of chlorine on earth is found in the oceans in the form of dissolved *chloride* ions. Salts of chlorides are also abundant in soil and rocks.

Although most chlorine occurs in the form of chloride, this element exists in a variety of other oxidation states. For example, there are four types of salts derived from combination of chlorine with oxygen. At the lowest level of oxidation is the *hypochlorite* ion, ClO^-, and the corresponding acid, hypochlorous acid. Hypochlorite ion is a weak oxidizing agent and finds much use as a bleaching agent. At higher oxidation states we find the *chlorous* ion, ClO_2^-, and chlorous acid, $HClO_2$, as well as the *chlorate* ion, ClO_3^-, and chloric acid, $HClO_3$. Finally, the highest oxidation state in which chlorine is found occurs in the *perchlorate* ion, ClO_4^-, and the associated acid, perchloric acid, $HClO_4$. This is the strongest common acid, having a great tendency to donate its proton to potential acceptors. Put another way, the perchlorate ion is an exceedingly weak base, having little tendency to associate with protons. For that matter, perchlorate ion has very little tendency to associate with cations generally. Its tendency to form complexes with dissolved metal ions in aqueous solution is one of the weakest known. Perchlorate is a very powerful oxidizing agent, and mixtures of this anion with oxidizable substances are potentially explosive. Indeed, perchlorate salts are used in certain solid fuels as rocket propellants. One such fuel consists of ammonium perchlorate and aluminum; oxidation of ammonium ion and aluminum by perchlorate liberates exceptionally large quantities of heat, which may be used to propel a rocket. This mixture is, as we can well imagine, quite treacherous.

MORE SUBSTITUTED METHANES

We have already noted that the symmetry properties of the tetrahedron, whether regular or not, require that there exist just one molecule corresponding to the formula CH_3X and just one corresponding to CH_2X_2. Let us examine the case of a molecule of the type CH_2XY in which, say, one hydrogen is replaced by chlorine and another by bromine:

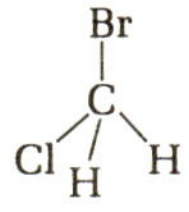

Study of the tetrahedral molecular structure of the chlorobromomethane molecule reveals that its structure is also unique. Of the original symmetry elements in a tetrahedron, only one mirror plane remains.

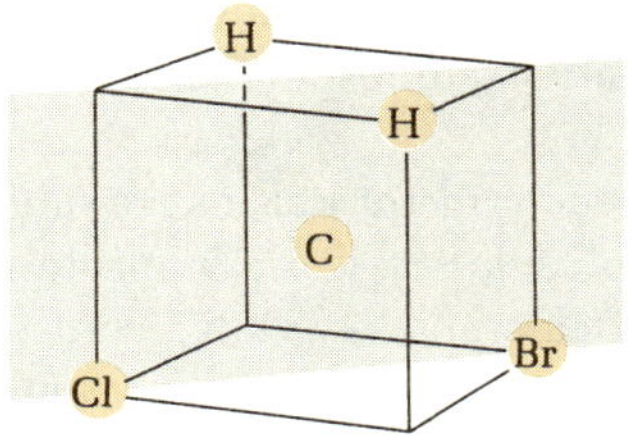

Regardless of how we dispose the two hydrogens, the chlorine, and the bromine around the central carbon atom, a molecule results in which the relationship of all the substituents to each other is identical. This is really the same as saying that if we place one representation on the table in a fixed position, we can always rotate any other representation in such a way that all atoms of the second representation occupy identical positions to those of the first. Such molecules are said to be *superimposable.*

CHIRAL MOLECULES

A different result is obtained when we consider a molecule of the type CHXYZ, fluorochlorobromomethane (CHFClBr) for example, in which the last symmetry element has vanished. Two ball-and-stick models of this molecule are given in Figure 2-9. Careful examination reveals that these molecules are, in fact, different. That is, there exists no way of superimposing them in space. As a result of having four different substituents on the central carbon atom, the molecule has acquired a definite spatial orientation. Let us imagine ourselves standing on the bromine atom facing the central carbon atom, with the hydrogen above us in the two structures of Figure 2-9. In structure (a), chlorine will be on our left side and fluorine on the right, and in structure (b), chlorine will be on the right and fluorine on the left. There is no way to turn one of these molecules in such a way that it can be superimposed on the other molecule. Thus, the asymmetry of each representation is unique to itself. For example, if we rotate structure (b) by 180° about the C—Br bond, the chlorine turns out to be on the right and the fluorine on the left as in structure (a), but now the hydrogen atom is below instead of above. Thus, trying to superimpose these structures is like trying to fit a right-handed glove into a left-handed one; short of ripping them apart, there just is no way to combine a left-handed item with a right-handed item. In fact, the relationship of the two structures of Figure 2-9 to each other can be considered to be analogous to that of a pair of gloves: Each glove is composed of exactly the same structural elements, but they are arranged differently in space. Like right- and left-handed gloves, the two fluorochlorobromomethane molecules are mirror images of each other. This is the first example of **stereoisomerism** that we have encountered. This term is applied to those situations in which molecules having the same elemental formulas (or atomic compositions) differ in their spatial arrangement of the atoms.

All molecules that are *not* superimposable on their mirror images are said to be **chiral.** The property of chirality in molecules is very important, particularly in the chemistry of living things. In general, observation shows that nature makes important distinctions between "left-handed" and "right-handed" molecules. Several examples of such distinctions will follow throughout this text, and a couple of examples are included below. We wish to consider how we can tell if a particular molecule is chiral or not.

It is frequently difficult to tell if a molecule and its mirror image are superimposable or not by inspecting two-dimensional models of compounds. Construction of a set of models permits us to decide this question, but this process is frequently time consuming and inconvenient. Determination of superimposability of a molecule and its mirror image can be readily made by considering symmetry elements. *All molecules that possess neither a plane nor a center of symmetry are chiral.* An equivalent test is to search for inversion axes of any order: If no such axis is present, then the molecule is chiral. If a plane or center of symmetry (or an inversion axis of any fold) is present, the molecule is said to be achiral.

Example *The property of chirality (like symmetry properties) is not restricted to molecules but may be possessed by any object. Which of the following objects is chiral? A starfish, an ear, a boot, a beer mug, a chair, a golf club, and a pair of scissors.*

Solution *We proceed by searching for symmetry elements in these objects. The starfish possesses several mirror planes, and a beer mug and a chair* *each possess one. Hence, these objects are achiral. Neither planes nor centers of symmetry are present in the ear, boot, golf club, or pair of scissors. These objects are chiral.*

Example *Which of the following molecules are chiral?*

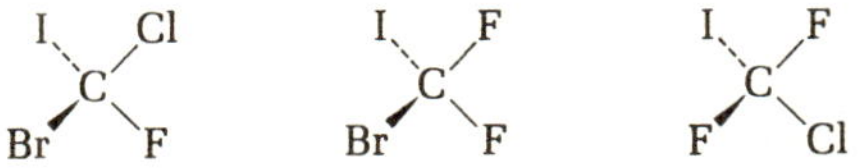

(a) (b) (c)

Solution *Molecule (a) possesses no plane or center of symmetry and so is chiral. Both (b) and (c) possess a plane of symmetry, which passes through the* I, Br, *or* Cl, *and* C *atoms, and, therefore, are achiral.*

OPTICAL ACTIVITY

In what ways does the difference in stereochemistry between the right-handed and left-handed members of a pair of chiral molecules manifest itself? If we measure many of the typical quantities used to characterize and identify molecules, no differences would be revealed at all. For example, melting point, boiling point, color, density, and index of refraction for the two stereoisomers would be identical. The differences between them do show up in two ways: the way in which they interact with light and in chemical reactions involving other molecules that are also chiral. Let us consider these two aspects of the behavior of chiral molecules one at a time.

To begin with, it is important to recognize that there are several types of stereoisomerism; the type encountered in the example of chiral molecules is only one among these. This particular kind of stereoisomerism (or just isomerism for short) is called **optical isomerism,** and the right-handed and left-handed members of the pair are referred to as *optical isomers.* These names are derived from the behavior of light as it passes through a medium containing chiral molecules. Ordinary light may be filtered so that the electromagnetic waves occupy a single plane; such light is termed *plane-polarized* (the properties of light are discussed in more detail in Chapter 14). When plane-polarized light passes through a medium containing chiral molecules, the plane of the light is rotated to a variable degree either to the left or right, depending on the nature and handedness of the chiral molecule. This phenomenon is shown schematically in Figure 2-10. For this reason, chiral molecules are frequently said to be *optically active.*

A pair of optical isomers differ in their ability to rotate the plane of polarized light only in terms of direction of rotation. Thus, if a solution at a specific concentration of one member of a pair of optical isomers rotates

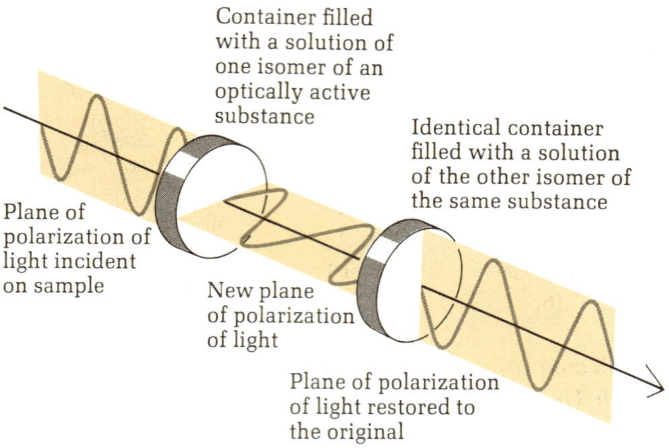

Figure 2-10 Rotation of plane-polarized light by an optically active material. As the polarized light passes through a solution containing one isomer of this material, the plane of polarization is rotated by an amount that depends on the nature and concentration of the optically active species and on the distance that the light travels through it. When this light passes through a solution identical to the first except that it is a solution of the other isomer, the light is rotated back to the original plane. Thus, we can conclude that the two isomers rotate the plane of polarized light to equal extents but in opposite directions.

the plane of the light by 100° to the right, then a solution at the same concentration of the other isomer must rotate the plane 100° to the left. The capacity of chiral molecules to rotate the plane of polarized light is frequently employed in their identification and characterization. Molecules that rotate the plane of polarized light to the right (clockwise to an observer facing the light source) are termed *dextrorotatory* and are designated (+); those that rotate the plane to the left are termed *levorotatory* and designated (−).

Example *Suppose that a particular solution of (+)-lactic acid,*

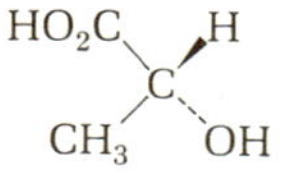

rotates the plane of polarized light x *degrees to the right. What predictions can we make about the rotation of the plane of polarized light by equally concentrated solutions of the following three molecules?*

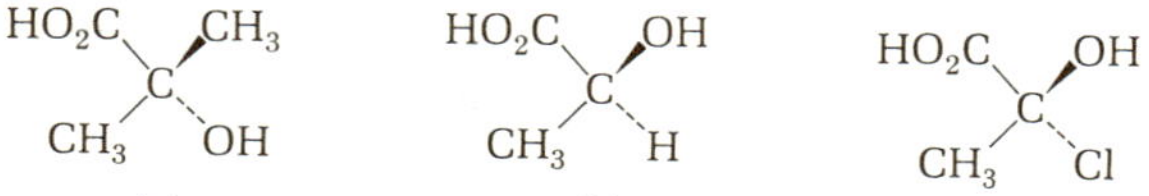

(a) (b) (c)

Solution *Molecule (a) is not chiral (it contains a mirror plane of symmetry) and therefore will not rotate the plane of polarized light. Molecule (b) is the mirror image of (+)-lactic acid and will rotate the plane of polarized light* x *degrees to the* left. *Molecule (c) is optically active but is not isomeric with lactic acid. Thus, we know that it will rotate the plane of polarized light but we can make no prediction about how much.*

As noted above, two optical isomers may differ in the nature and consequences of their reactions with other chiral molecules. Such differences are particularly important in biochemistry, the chemistry of life processes; several examples will be encountered in Chapters 7 and 22. At this point, we pause just long enough to note that a pair of optical isomers usually differ in how they taste and smell, in their effectiveness as drugs, in their toxicity, and so forth. As we shall see, these things are true because the molecules of living systems that are responsible for physiological responses are themselves chiral.

ETHANE

Let us continue our development of the structural chemistry of simple molecules based on the tetrahedral carbon atom by considering a special example of a monosubstituted methane: that derived by replacing one of the hydrogens of methane by a methyl (CH_3—) group. The resultant molecule, *ethane*, has the formula C_2H_6 or, more revealing, CH_3—CH_3. The new feature of this molecule is the existence of a carbon-carbon bond. The length of this bond in the ethane molecule, and in most other molecules possessing similar carbon-carbon bonds, is 1.543 Å. That this bond is somewhat longer than the carbon-hydrogen bond is the expected consequence of the fact that the carbon atom is larger than the hydrogen atom. Our knowledge of the structure of methane permits us to construct a very adequate structure for ethane: We simply remove one hydrogen atom from each tetrahedron and join the two tetrahedra at the vacant corner of each:

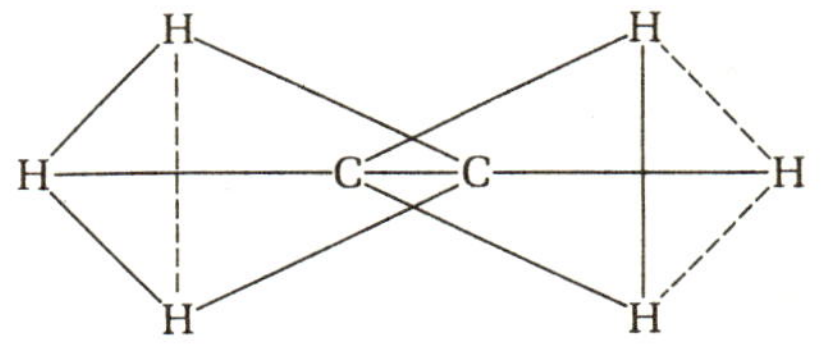

Now let us note a very important additional consideration. The structure is not completely specified until we establish the relative orientation of the two tetrahedra with respect to each other. That is, holding one tetrahedron fixed in space, we may rotate the other about the carbon-carbon bond. This

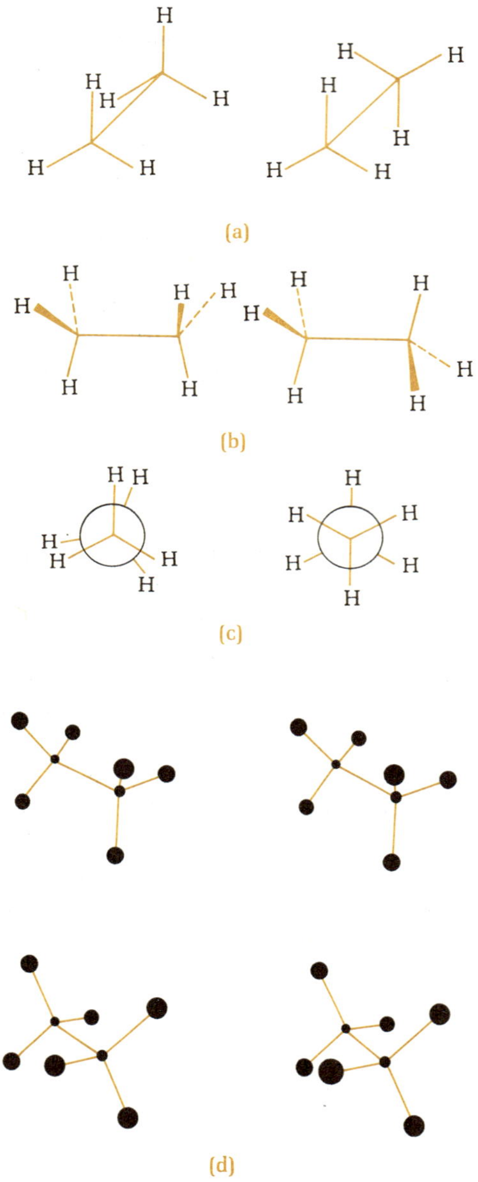

Figure 2-11 Several representations for the eclipsed and staggered conformations for the ethane molecule. (a) In the "sawhorse" structures the lower left-hand carbon is taken to be in the front. (b) In the dashed line-wedge convention, the dashed lines are for bonds receding from the plane of the paper, and the wedges are for bonds coming out of the paper. (c) In the Newman convention the structures are viewed along the axis of the carbon–carbon bond with the hydrogen atoms of the foremost carbon atom joined to it by full lines and those of the rearward one by partial lines. (d) Stereoscopic drawings for the eclipsed and staggered conformations for the ethane molecule.

changes the position of the hydrogen atoms of one methyl group relative to those of the other. Thus, the ethane molecule has available to it a variety of **conformations** and in this respect is more complex than the methane molecule, for which only one conformation can exist. Among the infinite different orientations that we can imagine, we have the two extreme **conformers** of the ethane molecule: the *eclipsed* one, in which the hydrogen atoms of one carbon are lined up directly with those of the other, and the *staggered* one, in which the hydrogen atoms of one carbon occupy positions between those of the other (Figure 2-11). Now, the different possible structures for the ethane molecule fit into our definition for stereoisomers, since the molecules differ only in the arrangement of the constituent atoms in space. Nevertheless, as we shall see, the various conformations of ethane are interconverted much too rapidly to permit their isolation as individual species. (The term "stereoisomers" is ordinarily reserved for forms stable enough to be capable of separation.) A variety of representations of the eclipsed and staggered conformations of the ethane molecule are illustrated in Figure 2-11. Of these representations, the stereo pair of structures perhaps provides the most informative picture of the three-dimensional molecular structure. Unhappily, we cannot usually rely on the availability of these structures; it is therefore necessary that we develop the ability to deduce from the two-dimensional representation a three-dimensional visualization. All of the planar representations illustrated are useful: Of them, the "sawhorse" method has the advantage of simplicity in construction and is the most generally employed.

What, then, is the structure of the ethane molecule? Neither the eclipsed nor the staggered conformer, nor an intermediate one, is adequate to describe this molecule. We shall discover that the rotation of the methyl groups with respect to each other around the carbon–carbon bond, while not completely free, is a very rapid process: At room temperature, rotation around this bond occurs a few billion times per second! Nevertheless, the staggered conformation, in which the atoms are as far removed from each other as possible, is more stable than the eclipsed one, and the ethane molecule spends more time in the former conformation than in the latter.

Clearly, the ethane molecule is not as symmetrical as methane: Ethane approximates a cylinder, while methane more closely resembles a sphere. This difference is reflected in the physical properties of the two molecules. Ethane has a boiling point of $-88.6°C$, more than 70° higher than that of methane. While this difference reflects principally the greater weight of the ethane molecule, it is also a consequence of the change in symmetry properties. The latter point is also of importance in determining the temperature range over which the materials are liquid. Recall that methane has a liquid range of only 22°. Ethane, in contrast, is liquid over a range of 83°.

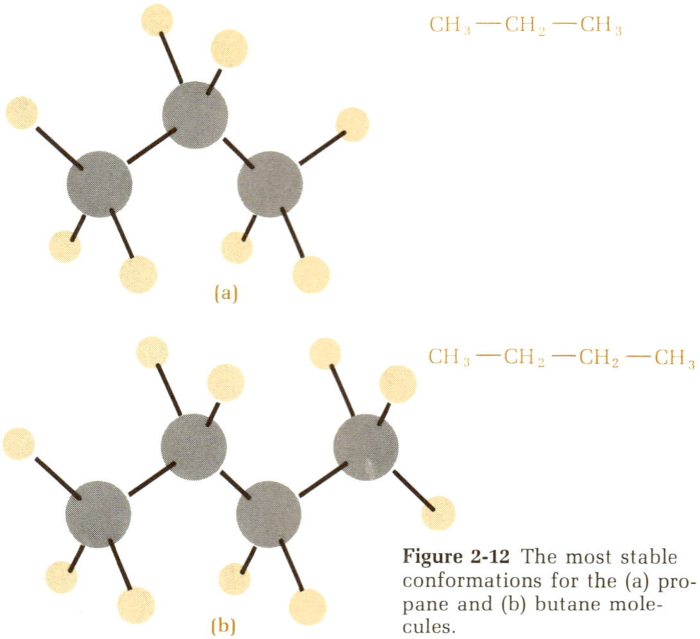

Figure 2-12 The most stable conformations for the (a) propane and (b) butane molecules.

PROPANE, BUTANE, AND PENTANE

The next two members of our hydrocarbon series, propane (CH_3—CH_2—CH_3) and butane (CH_3—CH_2—CH_2—CH_3), have structures that are just what we might expect on the basis of our previous discussion. Each carbon atom has a tetrahedral structure, with bond angles whose values are near those for the regular tetrahedron, and carbon-carbon and carbon-hydrogen bond lengths near those for methane and ethane. The most stable conformations for the propane and butane molecules, with all hydrogen atoms in the staggered configuration, are shown in Figure 2-12. Note the zigzag carbon chain, the necessary consequence of the tetrahedral carbon atom.

The butane molecule provides us with a new type of isomerism. For the above molecules that contain only one, two, or three carbon atoms, there exists only one way in which these can be associated (without making rings) by carbon-carbon bonds. With four carbon atoms, on the other hand, two possibilities exist. First, they may be arranged in a straight chain, as in the butane noted above; the resulting compound is called normal butane or *n*-butane. Second, they may be arranged in a branched structure; the resulting compound is called isobutane:

$$\underset{n\text{-butane}}{CH_3{-}CH_2{-}CH_2{-}CH_3} \qquad \underset{\text{isobutane}}{CH_3{-}\overset{\overset{\displaystyle CH_3}{|}}{C}H{-}CH_3}$$

Note that both compounds have the same molecular formula, C_4H_{10}, but that the structures are actually quite different. Isomerism of this type is called **structural isomerism.** Unlike optical isomers, structural isomers have different physical properties. Thus, *n*-butane melts at $-0.5°C$, while isobutane melts at $-12°C$.

Now let us consider the next family in the hydrocarbon series. For the five-carbon hydrocarbons, the *pentanes*, there exist three structural isomers:

$$CH_3{-}CH_2{-}CH_2{-}CH_2{-}CH_3 \qquad CH_3{-}\overset{\overset{\displaystyle CH_3}{|}}{C}H{-}CH_2{-}CH_3 \qquad CH_3{-}\underset{\underset{\displaystyle CH_3}{|}}{\overset{\overset{\displaystyle CH_3}{|}}{C}}{-}CH_3$$

n-pentane	isopentane	neopentane
bp = 36.1°C	bp = 27.9°C	bp = 9.5°C
mp = −129.7°C	mp = −159.6°C	mp = −16.6°C

Note that the symmetrical neopentane molecule has a very high melting point and short liquid range relative to the other two isomers.

The number of structural isomers possible increases very rapidly with an increasing number of carbon atoms. Thus, there are 5 isomers of the hexanes (6 carbon atoms), 9 isomers of the heptanes (7 carbon atoms), 75 isomers of the decanes (10 carbon atoms), and 366,319 isomers of the eicosanes (20 carbon atoms). Properties of some of these compounds are presented in Table 2-1. A particularly important point to reiterate is that in spite of the frighteningly large number of simple hydrocarbons that exist, their chemical physical properties are regularly related.

Example *How many structural isomers are there corresponding to the composition* C_3H_6ClBr?

Solution *We must search for all of the possible ways to distribute one chlorine atom, one bromine atom, and six hydrogen atoms along a chain composed of the three carbons. A little reflection will indicate that there are five ways of doing this. The five structural isomers are*

a. $CH_3{-}CH_2{-}CHClBr$
b. $CH_3{-}CHCl{-}CH_2Br$
c. $CH_3{-}CHBr{-}CH_2Cl$
d. $CH_2Cl{-}CH_2{-}CH_2Br$
e. $CH_3{-}CBrCl{-}CH_3$

Note that all other apparent possibilities are identical to one of those indicated here. Construction of molecular models may help us to understand this point.

Example *Which of the structural isomers of* C_3H_6ClBr *are chiral molecules?*

Table 2-1 Structures and physical properties for simple hydrocarbons

Name	Formula	mp (°C)	bp (°C)	Number of isomers
Methane	CH_4	−182.6	−161.7	1
Ethane	C_2H_6	−172.0	−88.6	1
Propane	C_3H_8	−187.1	−42.2	1
n-Butane	C_4H_{10}	−135.0	−0.5	2
n-Pentane	C_5H_{12}	−129.7	36.1	3
n-Hexane	C_6H_{14}	−94.0	68.7	5
n-Heptane	C_7H_{16}	−90.5	98.4	9
n-Octane	C_8H_{18}	−56.8	125.6	18
n-Nonane	C_9H_{20}	−53.7	150.7	35
n-Decane	$C_{10}H_{22}$	−29.7	174.0	75
n-Undecane	$C_{11}H_{24}$	−25.6	195.8	159
n-Dodecane	$C_{12}H_{26}$	−9.6	216.3	355
n-Tridecane	$C_{13}H_{28}$	−6	(230)	802
n-Tetradecane	$C_{14}H_{30}$	5.5	251	1,858
n-Pentadecane	$C_{15}H_{32}$	10	268	4,347
n-Hexadecane	$C_{16}H_{34}$	18.1	280	
n-Heptadecane	$C_{17}H_{36}$	22.0	303	
n-Octadecane	$C_{18}H_{38}$	28.0	308	
n-Nonadecane	$C_{19}H_{40}$	32	330	
n-Eicosane	$C_{20}H_{42}$	36.4		366,319
n-Triacontane	$C_{30}H_{62}$	66		4.1×10^9
n-Tetracontane	$C_{40}H_{82}$	81		6.2×10^{13}
n-Pentacontane	$C_{50}H_{102}$	92		
n-Hexacontane	$C_{60}H_{122}$	99		

Solution *Each of the structural isomers (a) through (e) must be examined for a plane or center of symmetry. Again, construction of molecular models or drawings intended to show three-dimensional aspects of these molecules may help us to see the answer. Compounds (a), (b), and (c) are chiral; (d) and (e) are not. Hence, the total number of isomers, structural plus optical, of* C_3H_6ClBr *is eight.*

CYCLIC MOLECULES

Let us quickly review what we have learned about the molecular structures of the hydrocarbons. We know that methane has a unique structure. With ethane, we first encountered the possibility of the existence of rotational isomers (conformers). When we considered longer chains, such as those of propane and *n*-butane, we discovered that the number of possible three-dimensional dispositions of the constituent atoms can become quite large. Thus, we may observe that these molecules begin to behave like flexible chains. If the chain is lengthened still further, as in the pentanes and hexanes, the carbon chain may adopt a conformation in which the molecule is approximately circular. That is, the last carbon atom may loop around and become positioned near the first atom. Indeed, there exists a series of molecules that actually contain cyclic structures. One of these is the hexane molecule in a cyclic conformation in which a chemical bond is formed between the first and last carbon atoms.

Cyclohexane

In two-dimensional form, we can write the structure of the *cyclohexane* molecule as

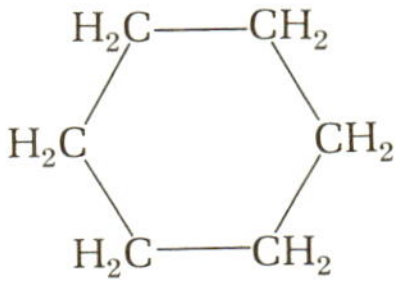

We can deduce a great deal about the intricacies of the structure of cyclohexane from our previous conclusions. We should expect the bonds about each carbon atom to be as nearly tetrahedral as possible. This simple statement tends to rule out the planar structure for cyclohexane shown in Figure 2-13. In the planar structure, each C—C—C bond angle would be 120°, the angle at each corner of a regular hexagon, but we already know that the ideal tetrahedral angle is 109.5°. Thus, to force cyclohexane to assume a planar structure would involve substantial distortion of the normal structure for quadruply bonded carbon. Furthermore, we have already seen that the (straight-chain) hydrocarbons have, as a result of the tetrahedral carbon atom, zigzag chains. Since a chain of six atoms is long enough to permit easy closing of the chain, we shall expect the cyclohexane molecule to retain this feature of the noncyclic molecules. This argument also tends to eliminate a planar structure.

If we adopt, then, the assumption of bond angles near those observed for simpler hydrocarbons, we can construct two models for cyclohexane. These are shown in Figure 2-14. Because of their superficial resemblance to the ordinary objects, these conformations of cyclohexane are called "chair" and "boat" forms. The student should search for the symmetry elements present in each structure. The threefold rotational axis of the chair form is particularly important to find. On what basis can we decide which of these is more likely to represent accurately the cyclohexane structure? We have already noted, in the cases of ethane, propane, and butane, that simply bonding two tetrahedral carbon atoms together can result in structures in which the hydrogen atoms on adjacent carbons can exist in either eclipsed or staggered conformations, the staggered form generally being the more stable. Inspection of the two structures of cyclohexane reveals that

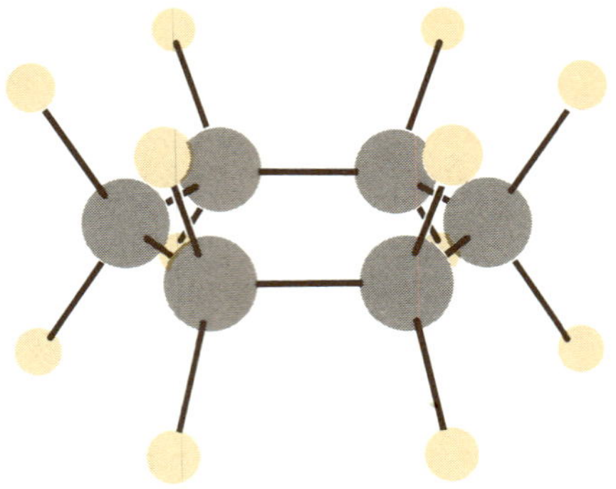

Figure 2-13 The planar structure for cyclohexane.

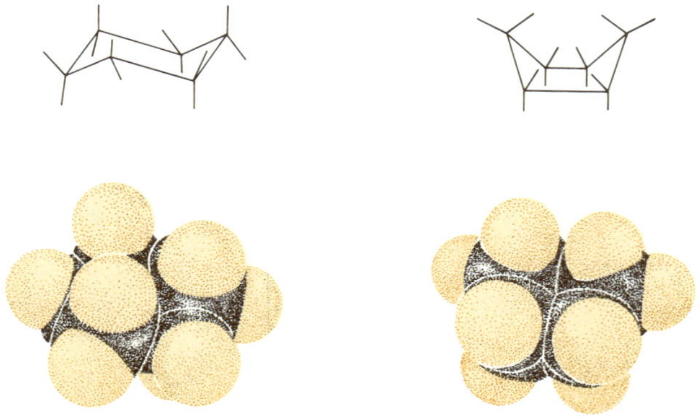

Figure 2-14 Two representations of the chair and boat forms of cyclohexane.

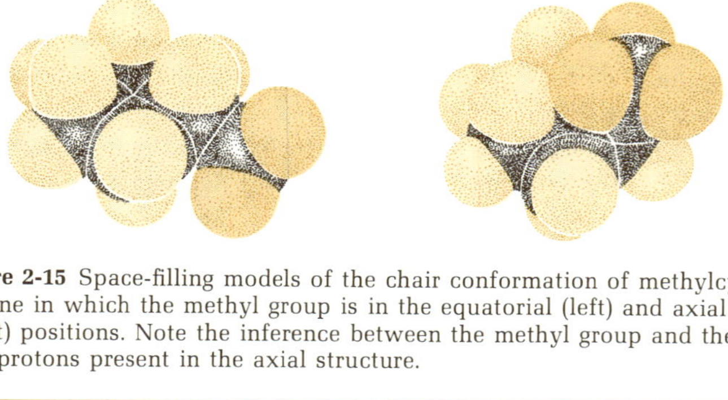

Figure 2-15 Space-filling models of the chair conformation of methylcyclohexane in which the methyl group is in the equatorial (left) and axial (right) positions. Note the inference between the methyl group and the ring protons present in the axial structure.

all hydrogens exist in the staggered conformation in the chair form, but that 8 of the 12 hydrogens exist in the eclipsed conformation in the boat structure. Hence, our straightforward deductions would suggest that the best structure for the cyclohexane molecule is the chair form; this is most certainly the case. As before, we can rely on physical methods to confirm our arguments.

Another analogy to the ethane molecule is found in cyclohexane: Just as the ethane molecule rapidly interconverts between the staggered and the eclipsed conformations but spends more time in the former, so the cyclohexane molecule flips back and forth from the chair to the boat form but spends about 99 percent of its time in the chair form.

There are two important points concerning the chemistry of cyclohexane that we have not previously encountered. First, in the boat form of this molecule the two hydrogen atoms on the out-of-plane carbon atoms across the ring from each other must be so close together that substantial repulsions exist between them (see Figure 2-14). The phenomenon of the repulsive interaction of two or more atoms in space is called **steric hindrance.** The steric hindrance that is present in the boat form of the cyclohexane molecule is less important in the chair form. This strengthens our conclusion about the relative stabilities of the two forms that we drew on the basis of other considerations.

Second, if we look at the chair conformation of the molecule and simply flip one end of the chair up and the other down, *ring inversion* occurs. That is, one chair conformation is converted into another. For cyclohexane, no net change occurs. On the other hand, if we look at a substituted cyclohexane, say one in which one hydrogen atom is replaced by a methyl group, ring inversion leads to a detectable change in structure:

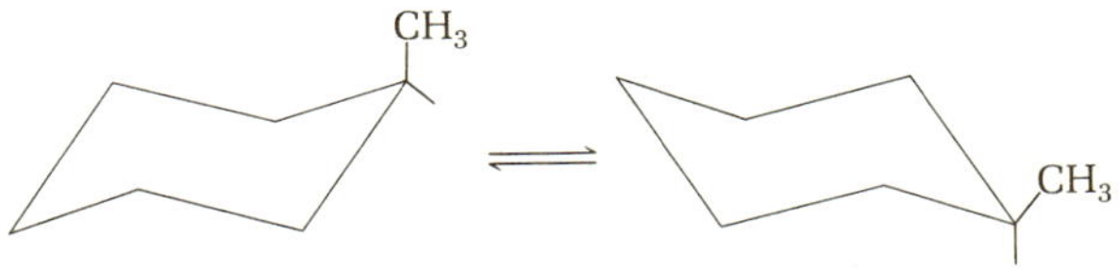

(2-3)

In the conformer on the left, the large methyl group occupies a position roughly perpendicular to the main plane of the ring (lying parallel to the threefold axis and thus called an *axial position*). In the form on the right, it occupies a position roughly in the main plane (*equatorial position*). In Figure 2-15, scale models of the chair forms of methylcyclohexane in which the methyl is in an axial and an equatorial position are provided. Inspection shows that less steric hindrance exists in the latter structure than in the former. Therefore, the methylcyclohexane molecule will spend more time in the latter conformation; interconversion between the two is again very rapid and occurs about a million times a minute at room temperature.

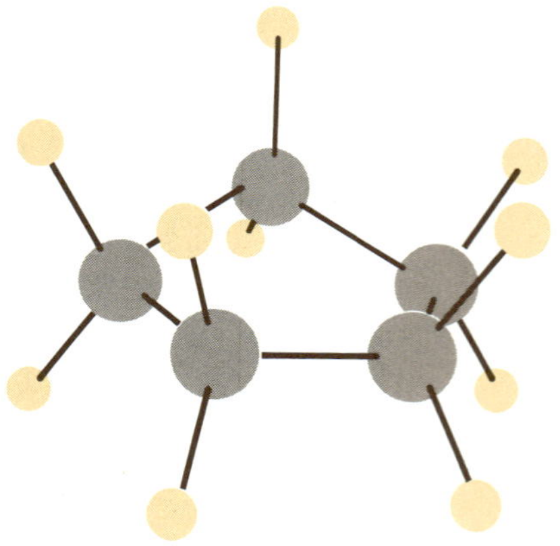

Figure 2-16 One conformation of the cyclopentane molecule.

Example *An expanded view of the chair conformation of the cyclohexane molecule, in which each hydrogen atom has been numbered, is provided below.*

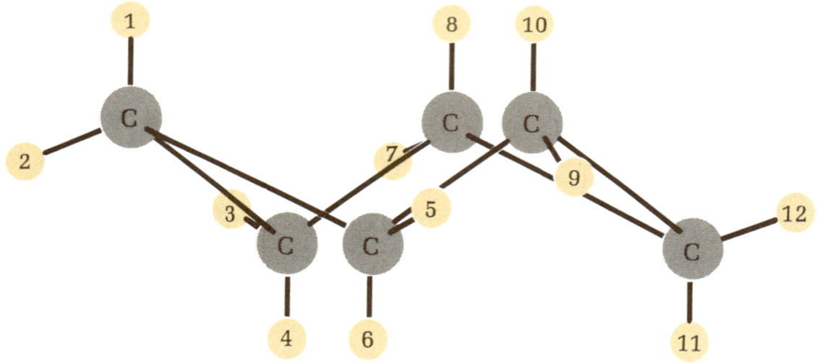

Which hydrogen atoms are equivalent to each other?

Solution *Only two classes of hydrogen atoms may be distinguished: those that are axial and those that are equatorial. Hence, hydrogens 1, 8, 10, 4, 6, and 11, which are axial, are all equivalent. Atoms 2, 3, 5, 7, 9, and 12, which are equatorial, are all equivalent.*

It is clear that the cyclohexane molecule is a good deal more symmetrical than the *n*-hexane molecule. It is not, therefore, unexpected that the melting points of these compounds are substantially different. The straight-chain molecule melts at $-95°C$, whereas the cyclic compound is a solid until 7°C. As before, this large difference reflects the relative ease with which molecules possessing a high degree of symmetry may be packed into crystalline structures. The boiling points, which depend more strongly on molecular weight than on symmetry properties, are not very different: 69°C for the straight-chain structure and 81°C for the cyclic.

Cyclopentane

Much of what we have said about cyclohexane also applies to *cyclopentane*, the cyclic hydrocarbon having five carbon atoms in the ring. Let us begin by again focusing attention on two principal points. First, the structure ought to have C—C—C bond angles as near 109.5° as possible and, second, the hydrogen atoms on adjacent carbon atoms ought to be in the staggered configuration when possible. A planar structure for cyclopentane is more attractive than the equivalent structure for cyclohexane. The internal angles for the regular pentagon are 108°, near the ideal value for tetrahedral carbon. However, a planar configuration requires that all hydrogens on adjacent carbon atoms be eclipsed. By puckering the cyclopentane molecule by bringing one carbon atom out of the plane of the other four, it is possible to retain all C—C—C bond angles near 108° while achieving the staggered configuration about four of the five carbon atoms. Such a structure is, therefore, the best compromise for the molecule in terms of avoiding both distortion of bond angles and eclipsing of hydrogens. Any of the five carbons may be the one that is out of plane, and any given molecule undergoes a rapid change between the five conformers. One such conformer for cyclopentane is shown in Figure 2-16.

Cyclobutane

For the construction of both the cyclohexane and cyclopentane rings, enough carbon atoms were available so that the ring could be closed without difficulty. In both cases, the best structures have bond angles and bond distances near those of the corresponding straight-chain structures; that is, near the ideal values for undistorted, unstrained systems. For the case of the *cyclobutane* molecule, in which the ring is formed from only four carbon atoms, the situation is different. The largest bond angles that can be achieved are 90° and this, of course, requires that the molecule be a planar structure with all hydrogen atoms eclipsed. Even bond angles this large are 19.5° smaller than the ideal tetrahedral value. Hence, it is clear that formation of the cyclobutane molecule will require substantial distortions of the normal structure of carbon atoms. Actually, even though it requires formation of bond angles even less than 90°, the cyclobutane molecule exists in a nonplanar conformation in which the unfavorable interactions due to the eclipsed hydrogen atoms are somewhat relaxed:

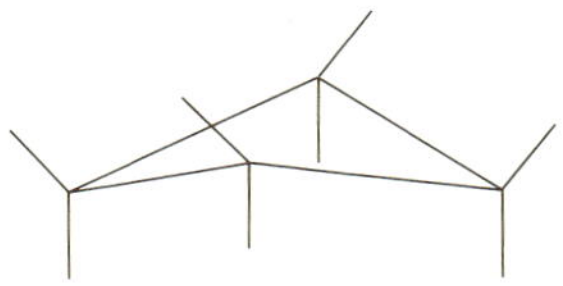

Cyclopropane

Finally, what must the situation be for the *cyclopropane* molecule, in which the cyclic structure is composed of just three carbon atoms? This molecule must be planar, with C—C—C bond angles of just 60°. Clearly, the normal bond angles of the carbon atom must be badly distorted to form this structure. Furthermore, the hydrogen atoms will all be eclipsed unless some distortion of the C—C—H and H—C—H bond angles occurs. Finally, the carbon–carbon bond distances in the cyclopropane molecule are only 1.52 Å, shorter than the normal value of 1.54 Å found in the straight-chain molecules and in cyclohexane and cyclopentane. In view of these considerations, perhaps the best way to visualize the structure of the cyclopropane molecule is to view the carbon–carbon bonds as springs rather than as straight rigid segments. If some flexibility is assigned to the carbon–carbon bond, then the cyclopropane molecule may be constructed with bent bonds:

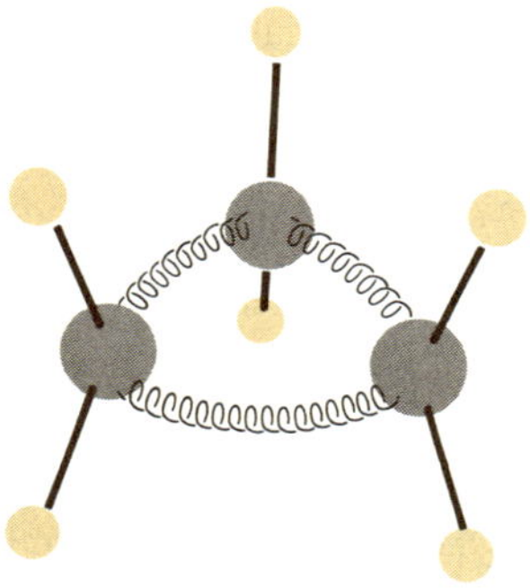

In this structure the C—C—C bond angles are widened significantly, and the distances between carbon atoms measured along the bond are 1.54 Å, which is considered to be the bond length in this molecule. However the cyclopropane molecule (or cyclobutane for that matter) is viewed, it is clear that it is a distorted, strained structure.

Propylene

We have noted that molecules such as pentane can be written as a variety of isomeric forms depending on the degree and nature of branching in the assembly of the carbon atoms. In a related sense, an isomeric form of cyclopropane exists in which we close the ring between adjacent carbon atoms instead of between the first and the last. This isomeric cyclopropane is called *propylene;* the structure may be viewed in the following manner, employing the bent bond representation:

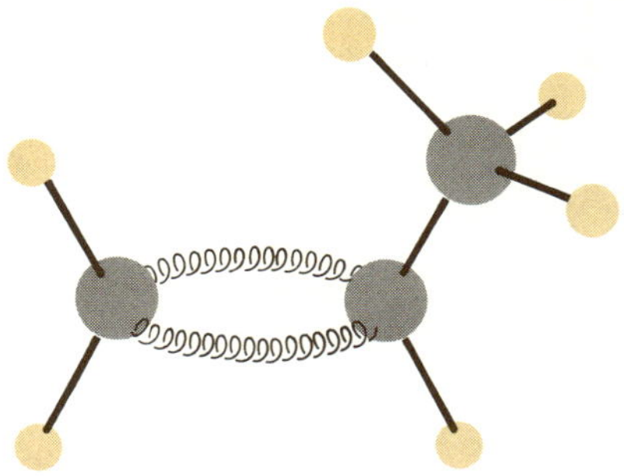

Thus, the propylene molecule has a "ring" containing only two members: The bond angles of the ring must be considered to be 0° if the bonds are considered rigid or to have small values if the bent bond view is adopted. The distance between the carbon atoms involved in the propylene "ring" is only 1.35 Å, a full 0.2 Å shorter than that found in ordinary unstrained structures. More about the structure of the propylene molecule can be readily deduced from what we already know about other molecules. We have learned that for cyclic molecules a planar structure requires that all hydrogen atoms be in the eclipsed conformation. Since the "ring" in propylene is planar in the simplest possible sense, this realization requires that the two carbon atoms of the "ring" and the four atoms, three of hydrogen and one of carbon, must all be in the same plane. Put another way, the equivalent of the eclipsed conformation for propylene is

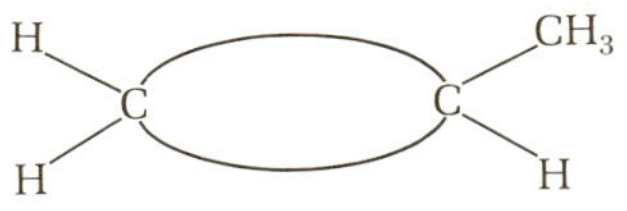

in which all atoms attached to the ring are in the plane of the paper (although the ring itself is not). The equivalent of the staggered conformation for propylene is

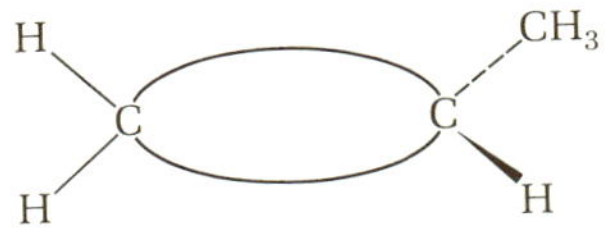

with two hydrogen atoms in the plane, the CH_3 group in back of the plane, and a hydrogen in front. Our conclusions require then that the structure of propylene be the eclipsed form. This fact has very important consequences for understanding the isomeric properties of related molecules, as we shall see shortly.

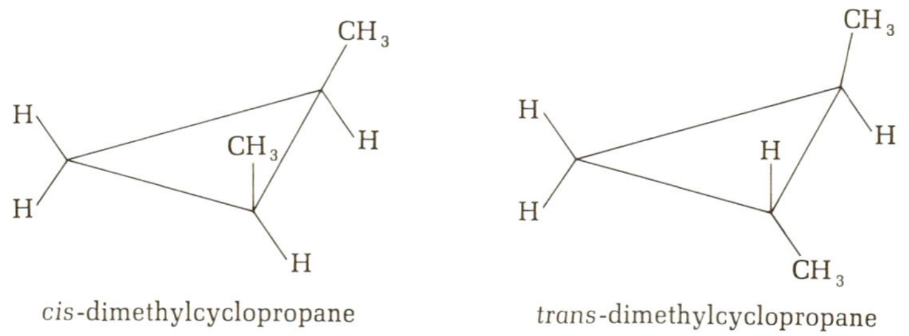

Figure 2-17 The two isomeric dimethylcyclopropanes.

SUBSTITUTED CYCLIC MOLECULES

Thus far, we have encountered several examples of isomerism. The forms of the ethane molecule, staggered and eclipsed, of the cyclohexane molecule, chair and boat, and of the methylcyclohexane molecule, axial and equatorial, are interconvertible through simple rotations about carbon–carbon bonds and, hence, these *conformers* (*stereoisomers* in the general sense) are not sufficiently stable to permit isolation of the individual members. On the other hand, the *structural* isomers of the butane molecule, *n*-butane and isobutane, are interconvertible only through the breakage and reformation of stable carbon–carbon bonds. These isomers are therefore isolable and may be characterized individually. (Let us recognize that, for example, the *n*-butane molecule has several rotational isomers similar to those of the ethane molecule; these, of course, are no more readily isolated than in the ethane case.) A type of stereoisomerism, **geometric isomerism,** which is related to, although distinct from, that encountered with the straight-chain compounds, exists in substituted cyclic hydrocarbons. Consider, for example, a cyclopropane molecule in which two of the hydrogen atoms on adjacent carbons have been replaced by methyl groups. There are two possible structures for the dimethylcyclopropane molecule: one in which both methyl groups are placed above (or below) the plane of the ring and one in which the methyl groups occupy opposite sides of this plane (Figure 2-17). Since existence of the cyclopropane ring imposes a rigidity on the molecule that prevents ready rotation about the carbon–carbon bonds, these geometric isomers are stable, isolatable species. The same considerations apply not only to other molecules such as cyclobutane and cyclopentane but to "cyclic" molecules such as propylene as well. The situation here is best understood if we refer to *ethylene*, the parent compound of propylene, which has the simpler structure

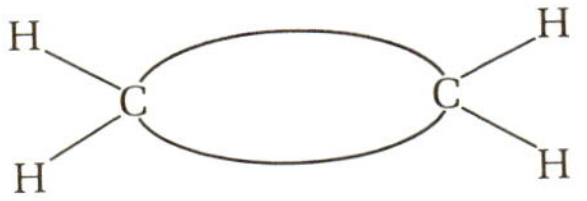

If we now construct a molecule in which one hydrogen on each of the carbon atoms is replaced by a methyl group, we have two structures corresponding to those of the dimethylcyclopropanes:

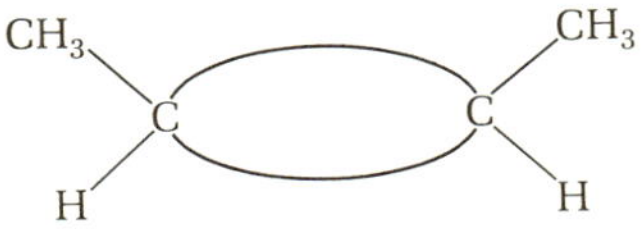

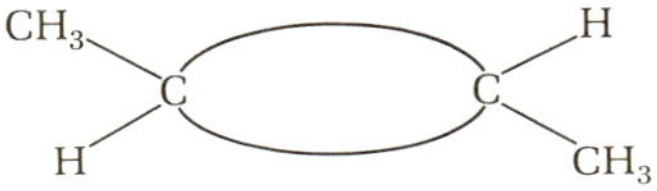

That structure that has both methyl groups on the same side of the "ring" is related to the cyclopropane molecule with both methyl groups above the "ring" and that with the methyl groups across the "ring" to that with the methyl groups on opposite sides of the "ring." This type of isomerism is sometimes referred to as *cis–trans* isomerism. The *cis* designation refers to those isomers in which the two substituents exist on the same side and the *trans* designation to those in which the substituents exist on the opposite sides of a ring structure.

Example *How many geometric isomers does the cyclic form of* $C_4H_5Cl_3$ *have (exclude all structures not having a four-membered ring)?*

Solution *We must ask how many distinct ways there are to dispose three chlorine and five hydrogen atoms on a cyclobutane nucleus, paying particular attention to both structural and geometric isomerism. There prove to be five possibilities:*

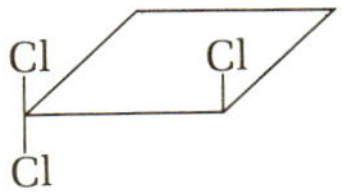

(a)

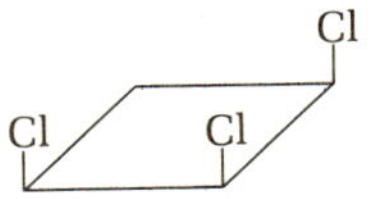

(c)

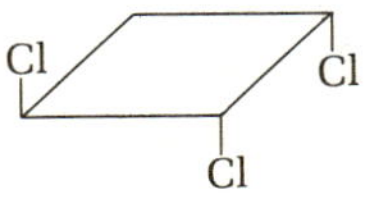

(e)

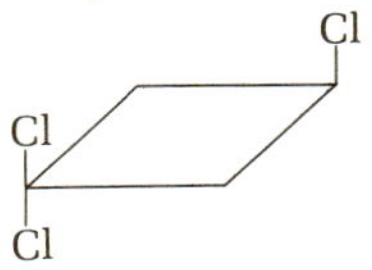

(b)

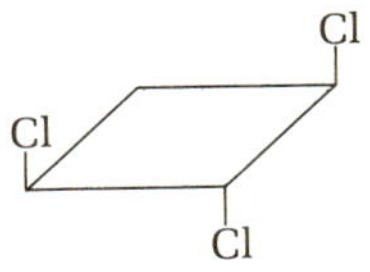

(d)

Figure 2-18 Space-filling models of the *cis* and *trans* isomers of ethylene substituted symmetrically with two methyl groups.

In these representations the carbon atoms are not explicitly shown; they are simply recognized as existing at the intersections of chemical bonds. We shall use this notation frequently from now on. Note that there are three structural types: (1) two chlorines on the same carbon with the third chlorine on an adjacent carbon, molecule (a); (2) two chlorines on the same carbon with the third chlorine on the opposite carbon, molecule (b); and (3) not more than one chlorine on any carbon atom, molecules (c), (d), and (e). The latter three are geometric isomers.

Example *Which of the above five isomers of* $C_4H_5Cl_3$ *is chiral?*

Solution *Molecules (a) and (e) possess neither a plane nor a center of symmetry and are chiral. Thus, there are a total of seven isomers of cyclobutanes of the composition* $C_4H_5Cl_3$, *including structural, geometric, and optical isomers.*

The existence of *cis–trans* isomers provides an example of the importance of steric hindrance in determining stabilities of molecules. In Figure 2-18, space-filling models of the *cis* and *trans* isomers of ethylene substituted with two methyl groups are shown. Clearly, the hydrogen atoms of the methyl groups are very much closer together in the *cis* than in the *trans* isomers. Actually, they are sufficiently close together in the *cis* isomer so that repulsive interactions occur between them. Although steric hindrance is not particularly important in this case, it is clear that in compounds of this type, in which the substituents are larger than the methyl group, steric hindrance will become very important indeed. Consider, for example, the *cis* isomer of the cyclopropane molecule substituted with two CH_3 groups:

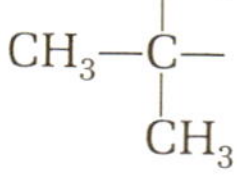

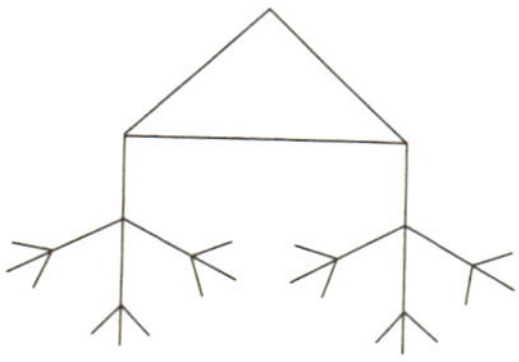

These two substituents get in each other's way, and steric hindrance results. The resulting greater stability of *trans* isomers as compared to *cis* isomers is reflected in the lower reactivity of the former species compared to the latter in reactions that destroy the ring structure, and hence also the steric hindrance. Consider, for example, the hydrogenation of the *cis* and *trans* isomers of dimethylcyclopropane:

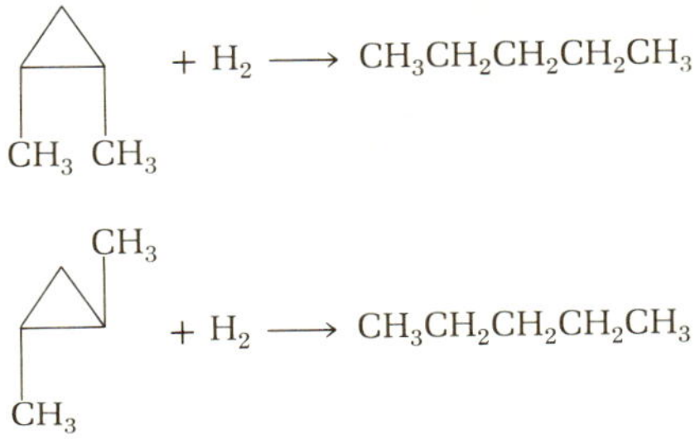

(2-4)

Since the *cis* isomer is less stable than the *trans* isomer and since the product molecules are identical due to breakage of the ring (it was the cyclic structure that permitted the separate existence of the isomers), we expect that hydrogenation of the *cis* isomer will occur more readily than it will for the *trans* isomer. This is precisely the result found. *Trans*-dimethylcyclopropane hydrogenates at a rate about equal to that for cyclopropane itself, whereas the *cis* isomer reacts faster under the same conditions.

Example *Which of the following two molecules would be more stable?*

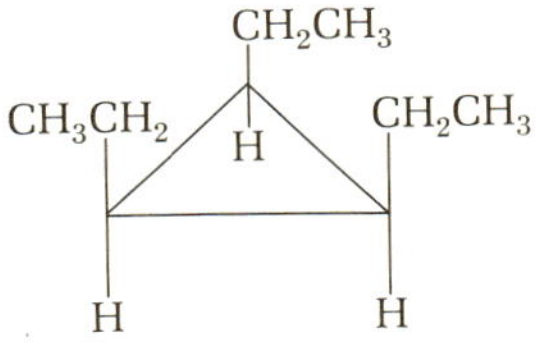

(a)

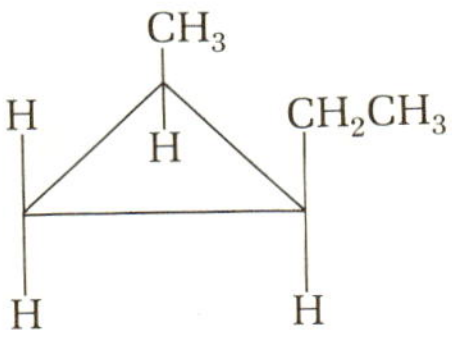

(b)

Solution *Both molecules are based on the cyclopropane skeleton, so we must look beyond this structural feature and focus attention on the nature and disposition of the substituent on the ring. Structure (a) has three ethyl groups cis to each other; but structure (b) has just one ethyl and one methyl group in the cis relationship. Hence, steric hindrance in (a) should be more severe than in (b). It follows that (b) is the more stable molecule.*

MACROCYCLIC MOLECULES

We may also encounter interesting structural chemistry in cyclic structures having a large number of carbon atoms in the ring. We have already seen

Figure 2-19 A space-filling model of a cyclic hydrocarbon molecule in which the chain is composed of 24 carbon atoms.

that cyclohexane can assume a very stable structure in which all of the carbon atoms have nearly ideal tetrahedral bond angles and in which all of the hydrogen atoms have the staggered conformation. By adding more carbon atoms to the cyclic structure, we would not expect any difficulties in arriving at unstrained structures. As the size of the ring increases, an increasing number of satisfactory conformations are available to the molecule, whereas only one is available to the cyclohexane molecule.

In Figure 2-19, a space-filling model of the cyclic structure formed from 24 CH_2 groups is shown. The molecule adopts a conformation that resembles a doughnut. Molecules of this kind pose an interesting question: Is it possible to form chainlike structures in which one cyclic molecule is threaded through the hole in the center of another? Diagrammatically, such structures would look like

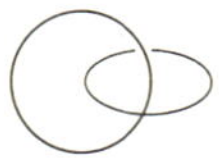

In recent years it has proven possible to answer this question in the affirmative. Ingenious synthetic approaches have yielded chainlike structures called *catenanes* after the Latin *catena*, a chain. Thus far, it has been possible to obtain only the simplest types of compounds in this class, which are like that indicated above.

Quite aside from catenanes, cyclic molecules have quite an interesting chemistry. For example, several classes of antibiotics possess cyclic structures. One such class, of which valinomycin is an example, is composed of molecules that associate with biological membranes and, by virtue of the holes within the molecules, permit the passage of ions to which the membranes are ordinarily impermeable. This forms the basis of their antibiotic action. More about this follows in Chapter 10.

ISOMERS: A SUMMARY

In the discussion preceding, several classes of isomers have been encountered. Let us summarize the distinctive features of each class now and amplify some of the concepts earlier introduced.

Stereoisomerism is the phenomenon of the different arrangement in space of the same collection of atoms linked together by chemical bonds. Put another way, we can frequently construct different molecules from the same collection of atoms. These are then called *stereoisomers*. Note that different arrangements of atoms in space that arise simply by rotations around chemical bonds are called *conformations*, not stereoisomers. We have encountered three types of stereoisomerism thus far. These include optical, structural, and geometric isomerism.

Optical isomerism occurs in those cases in which a molecule can exist in two stereoisomeric forms that are nonsuperimposable mirror images of each other. Thus, if the mirror image of a molecule is not identical with that molecule, this pair of molecules are *optical isomers*. A molecule that has an optical isomer is said to be *optically active*. Provided a molecule has no inversion axes, it will be optically active. If a molecule has one of these symmetry elements, it will not be optically active. Search for an inversion axis can be simplified by realizing that the presence of such an axis requires that the molecule have either a mirror plane or a center of symmetry. Hence, if either of these elements is present, we can be sure that an inversion axis also exists. Some examples:

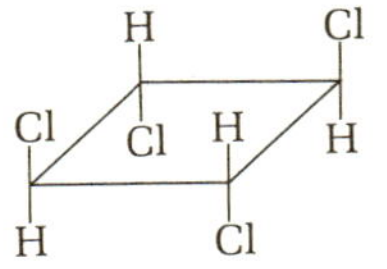

is not optically active, since it contains two mirror planes. These bisect the cyclobutane ring through carbon atoms at opposite corners and are perpendicular to it. This molecule contains no center of symmetry.

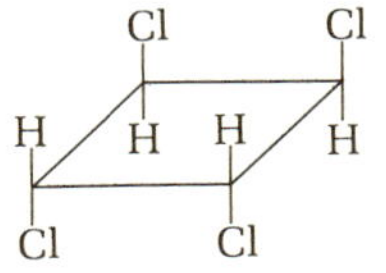

is not optically active either, even though both of the mirror planes in the above molecule are absent in this one. Note that this molecule has a center of symmetry at the center of the cyclobutane ring and a new mirror plane (which bisects the carbon–carbon bonds).

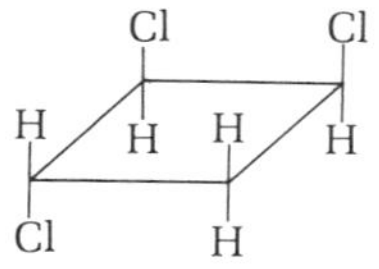

is, in contrast to those above, optically active, since it contains neither a mirror plane nor a center of symmetry (nor an inversion axis of any order).

Structural isomerism refers to the arrangement of the same collection of atoms disposed differently in space by virtue of different *bonding patterns.* That is, all of the atoms of two structurally isomeric molecules will not be identical in terms of the nature of the groups to which they are bound. Note that this is not true of optical isomers,,for which the bonding patterns are always identical.

Some examples: Suppose that we consider all of the structural isomers of the composition $C_4H_8Cl_2$. These include the following:

1. $CH_3CH_2CH_2CHCl_2$
2. $CH_3CH_2CHCl{-}CH_2Cl$
3. $CH_3CHCl{-}CH_2CH_2Cl$
4. $CH_2Cl{-}CH_2{-}CH_2{-}CH_2Cl$
5. $CH_3CHCl{-}CHCl{-}CH_3$
6. $CH_3CH_2{-}CCl_2{-}CH_3$
7. $CH_3{-}CH{-}CHCl_2$ (with CH_3 on the central CH)
8. $CH_3{-}CCl{-}CH_2Cl$ (with CH_3 on the central CCl)
9. $CH_2Cl{-}CH{-}CH_2Cl$ (with CH_3 on the central CH)

Note that the first six of these show the various ways in which we can distribute the two chlorine atoms among the carbon atoms arranged in the *n*-butane configuration and that the last three give the same information for the carbon atoms arranged in the isobutane configuration. If we examine these structures atom by atom, we will find that they do differ in terms of what is bonded to what; that is, the bonding patterns are different. Any other arrangement that we might draw is equivalent to one of those above, as we can see by analyzing the bonding patterns.

Geometric isomerism refers to stereoisomerism that arises from the disposition of equivalent groups on the same side of (*cis* isomers) or on opposite sides of (*trans* isomers) ring structures. Note that the bonding patterns are identical in isomerism of this type and that geometric isomers may or may not have optical isomers as well.

Some examples:

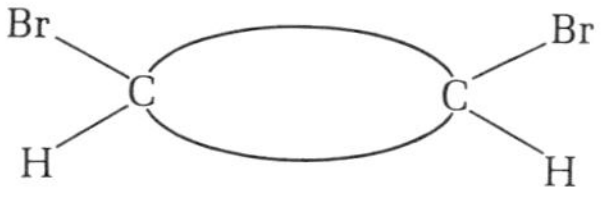

cis

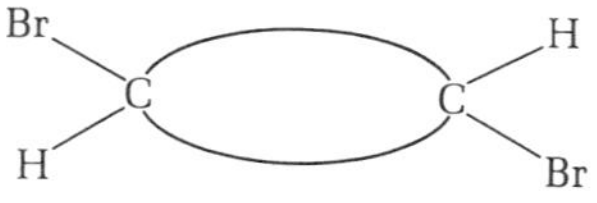

trans

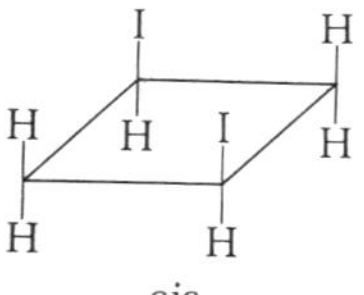

cis

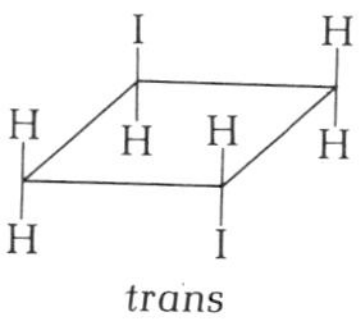

trans

Glossary

Avogadro's number the number of atoms in a gram-atom of an element or the number of molecules in a mole of a compound. Numerically, it is equal to 6.0235×10^{23}.

Bond angle the angle formed by a collection of three atoms, two of which are chemically bonded to the third.

Bond length the distance between the centers of chemically bonded atoms. Bond lengths are ordinarily expressed in angstroms.

Center of symmetry (inversion center) one of the types of symmetry element. Specifically, a point such that all lines drawn through it meet corresponding parts of the structure at equal distances on either side of it.

Chirality a property of a molecule lacking both a plane and center of symmetry. Chiral molecules have the property of rotating the plane of polarized light.

Conformation generally, the spatial arrangement of atoms in a molecule, but usually restricted to those arrangements that are in mobile equilibrium with other arrangements or conformations. Spatial arrangements not in such mobile equilibrium are usually called *configurations*.

Conformer one of two or more conformations of a molecule that are in mobile equilibrium with each other.

Geometric isomerism (*cis-trans* isomerism) a type of isomerism in which the isomers are distinguished by the nature of disposition of substituents on the two sides of a ring structure or double-bond structure.

Hydrocarbons those compounds derived entirely from carbon and hydrogen.

Interstitial compounds those compounds formed by the insertion of a molecule or atom into available sites in the crystalline lattice of another substance.

Isotopes variants on elements in which the atoms have the same atomic number but different atomic mass. That is, the number of nuclear protons is constant, but the number of nuclear neutrons varies.

Optical isomerism the phenomenon of nonidentical molecules having nonidentical mirror images related to each other as are right-handed and left-handed gloves.

Plane of symmetry (mirror plane) a plane passing through a structure in such a way as to divide the structure into two parts that are mirror images of each other.

Rotation axis an axis passing through a structure so that rotation of the structure about that axis by some number of degrees results in a geometry indistinguishable from the original. One of the types of symmetry element.

Stereoisomerism (isomerism) the phenomenon of two or more physically and chemically distinct molecules that have the same composition in terms of the atoms of which they are composed.

Steric hindrance (steric crowding) the interference of atoms or groups in the same molecule through their being crowded together in space.

Structural isomerism distinct arrangement of the same atoms in space which occurs by existence of different bonding patterns. For example, contrast *n*-butane and isobutane.

Symmetry operations those operations performed on a structure that leave the appearance of the structure unchanged. They include rotation axes and reflections through planes and centers of symmetry and through inversion centers.

Tetrahedral geometry geometry based on the tetrahedron, a solid figure having four triangular faces.

Study problems

Problems and questions about symmetry and molecular geometry

1. You can develop your ability to "see" symmetry elements by looking at common objects. What elements of symmetry are present in
   a. a chair
   b. a study table
   c. a flower vase
   d. a dog
2. Take four pennies and arrange them on your desk so that
   a. they lie in a straight line and have
      (1) a twofold axis perpendicular to your desk top (How many ways can you do this?)
      (2) a twofold axis parallel to your desk top
   b. Place them in a square arrangement heads up.
      (1) Can you arrange them so that a fourfold axis is present?
      (2) Is a fourfold axis always present?
      (3) What other elements of symmetry can be present?
3. Identify the elements of symmetry in
   a. the cyclohexane molecule in the boat form
   b. the cyclohexane molecule in the chair form
   c. the cyclohexane molecule in the (hypothetical) planar form
4. There exists an organic molecule, benzene, which has the composition C_6H_6. Replacement of one H by Cl yields just one chlorobenzene species, C_6H_5Cl. Replacement of a second H by Cl yields three distinct dichlorobenzene molecules, $C_6H_4Cl_2$. What conclusions can you draw about the geometry of benzene?
5. Suppose that you are an astronaut and have just landed on a strange planet. Being the only chemist on this interplanetary mission, you must investigate the nature of the molecules found in this new world. After conducting a few investigations, you find (to your amazement) that the hydride of carbon, enahtem, on this planet has the composition CH_5. Furthermore, you are able to prepare two separate and distinct chloroenahtem molecules, CH_4Cl, and three separate and distinct dichloroenahtem molecules, CH_3Cl_2. From these observations, draw all the conclusions that you can about the geometry of the enahtem molecule. How many molecules would exist corresponding to the composition CH_2Cl_3?
6. A center of symmetry means that if we rotate x degrees and reflect

through the mirror plane perpendicular to the rotation axis, we obtain a structure similar to the original one. How many degrees equals x?

7. Which has more elements of symmetry, $CHCl_3$ or CH_3Cl? Why?
8. The molecule ClF_3 has a twofold rotation axis and two mutually perpendicular mirror planes of symmetry. From this data, what can you conclude about the structure of this molecule?
9. The ion $[PtCl_2Br_2]^{2-}$ has three twofold rotation axes, three mirror planes, and an inversion center, but no other elements of symmetry. What do you conclude about the structure of this ion from these data?
10. By means of rotation around C—C bonds, the structures of hydrocarbons may be adjusted until the highest possible symmetry results. One isomer of pentane, C_5H_{12}, contains exactly three twofold axes of rotation (as well as other symmetry elements). What is the structure of this isomer?

Problems and questions about isomers

11. Draw all the structural isomers of chlorobromopropane, C_3H_6ClBr. For which of these will optical isomers exist? Defend your answers.
12. For each of the compounds below, decide whether or not an optical isomer will exist. Defend your decisions.

a. 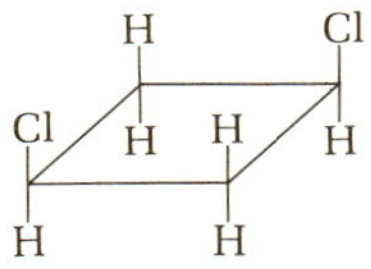

b. 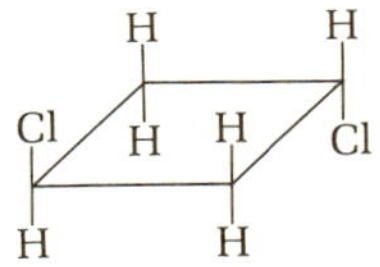

c. 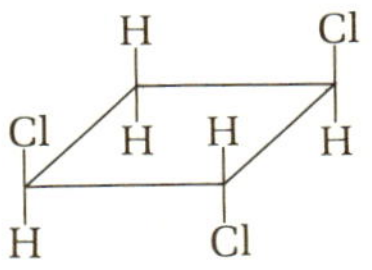

d. 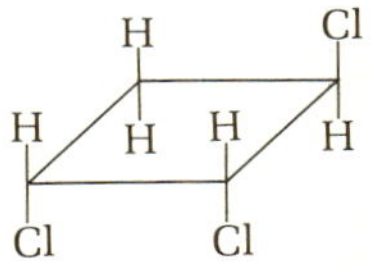

13. Which of the following molecules is chiral?

a. 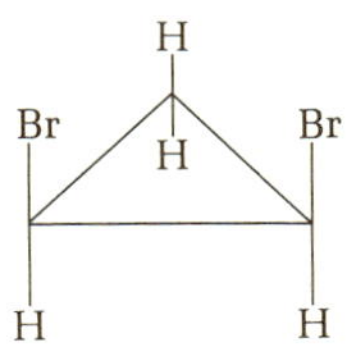

b. 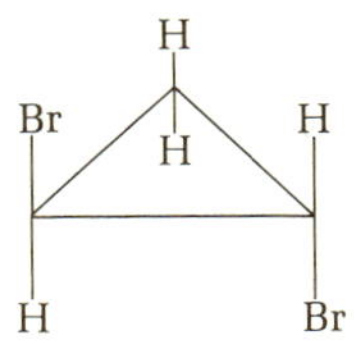

c. 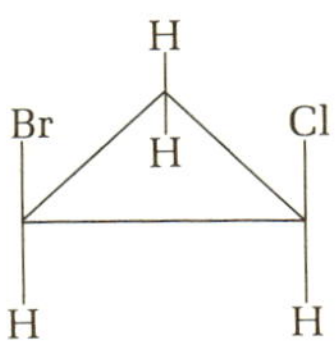

d. 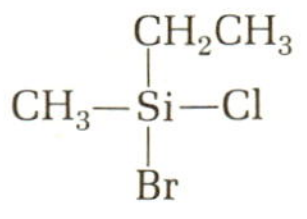

e. 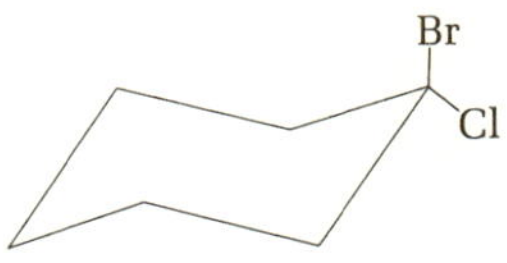

14. Are there optical isomers of

a. 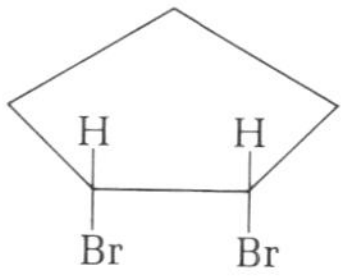

c. 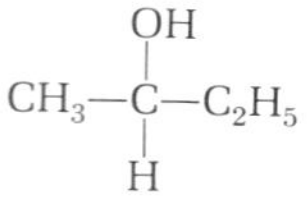

b. 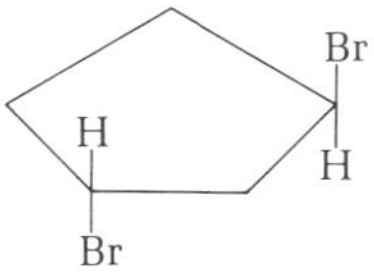

d. 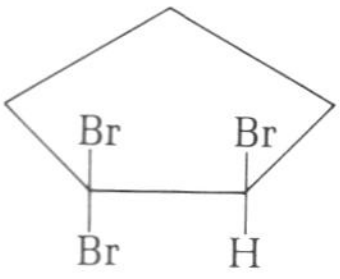

15. Draw all the structural isomers for the molecules corresponding to the following compositions:

a. $C_2H_4Cl_2$ b. $C_3H_6Cl_2$ c. C_6H_{14} d. $C_4H_6Cl_2$

16. How many isomers does the compound $C_5H_{11}F$ have? Include optical as well as structural isomers.

17. Below are provided several pairs of structures. For each pair, decide if they are optical isomers, structural isomers, geometric isomers, or not isomeric.

a. 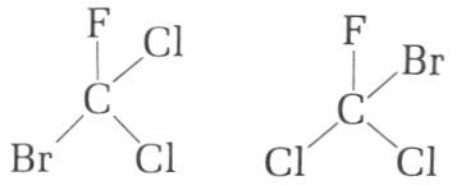

b. 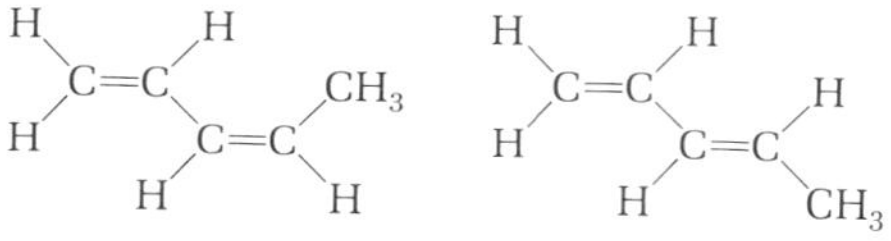

c. $CH_3CH_2CH_2{-}OH$ $CH_3{-}O{-}CH_2{-}CH_3$

d. 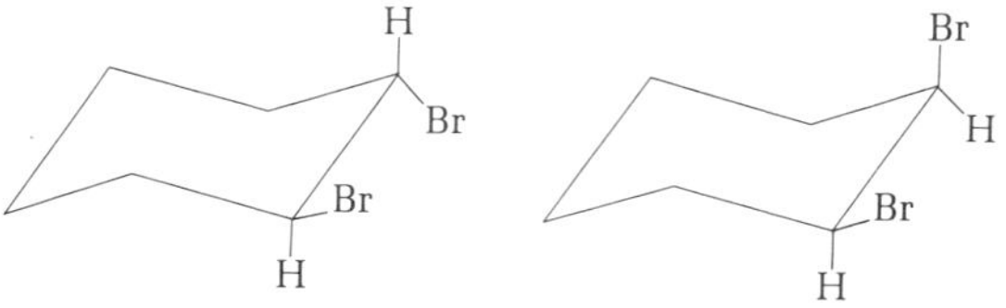

Some questions about molecular stability

18. Choose the structure in each of the following pairs that you expect to be more stable. Justify your choice in each case.

a. cyclobutane and cyclooctane

b. 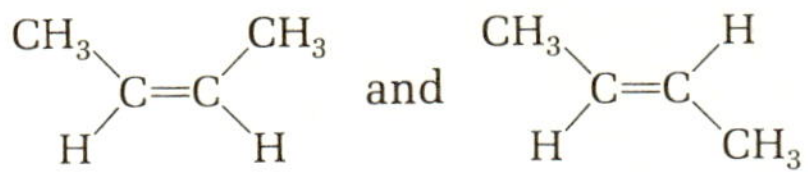

c. 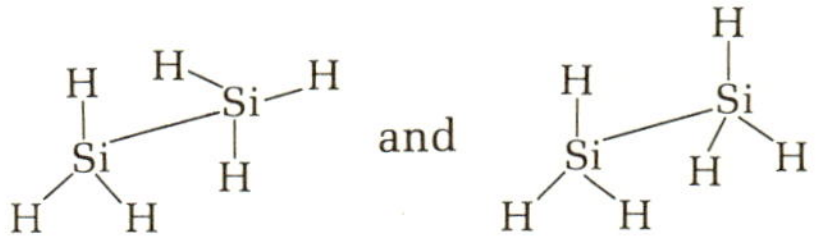

d. 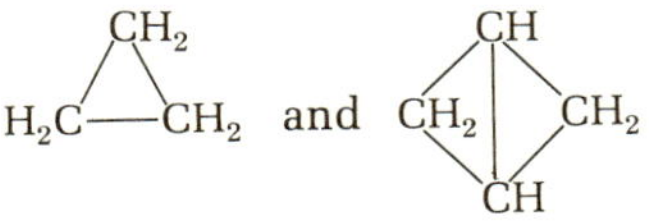

e. 

f. 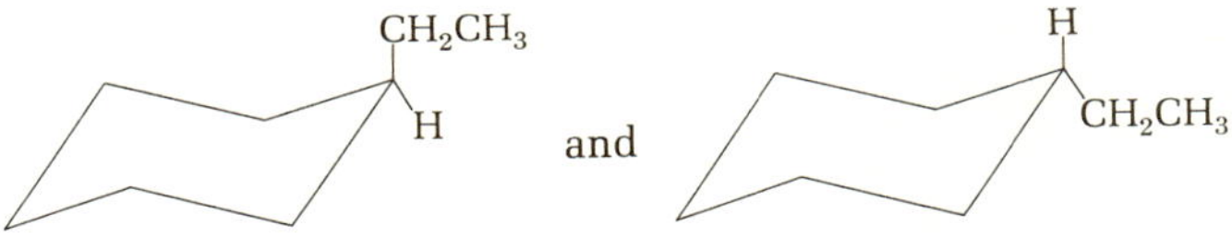

g. 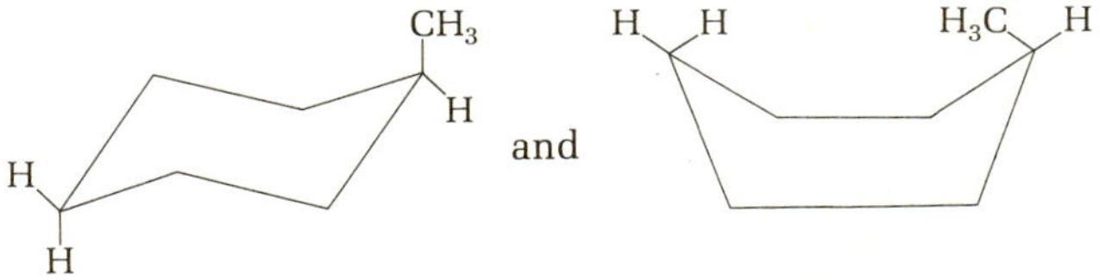

19. Draw structures for the most stable conformation of each of the following molecules:
    a. ethylcyclohexane
    b. methylcyclopentane
    c. chloroethane
    d. dimethylcyclobutane (with the methyl groups occupying positions on adjacent carbon atoms)
20. There exists a rather complicated hydrocarbon called *adamantane*, the structure of which is

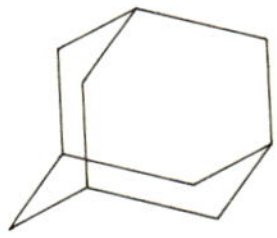

What would you predict about the stability of this molecule? Why?

21. Several years ago, a chemist succeeded in synthesizing a simple but unusual hydrocarbon called *cubane*. It has the composition C_8H_8 and the geometry of a cube:

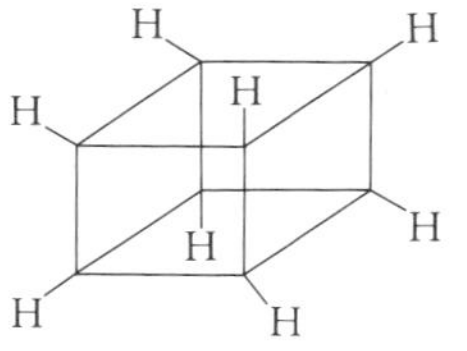

What would you predict about the stability of this molecule? Why? Cubane is a crystalline solid at room temperature. Can you guess what the shape of the crystals are? Do you think that it would be more challenging to synthesize cubane or *tetrahedrane*,

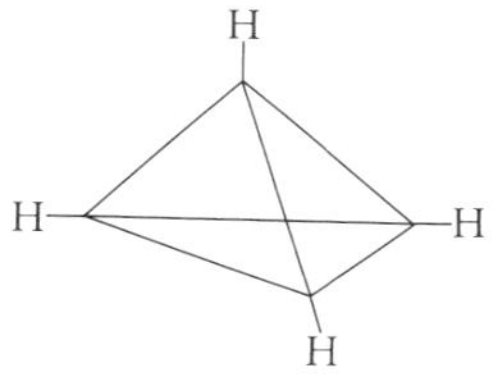

Some quantitative problems

22. Some simple hydrocarbons are listed below. For each of them, calculate the *number* and *weight* percentage of carbon and hydrogen.
   a. hexane
   b. cyclopentane
   c. cubane (see Problem 21)
   d. adamantane (see Problem 20)
23. The structures of some complex molecules based on carbon are provided below. Calculate the weight percentage of carbon in each of them. Incidentally, each of these molecules has some particular interest associated with it. How many can you identify?

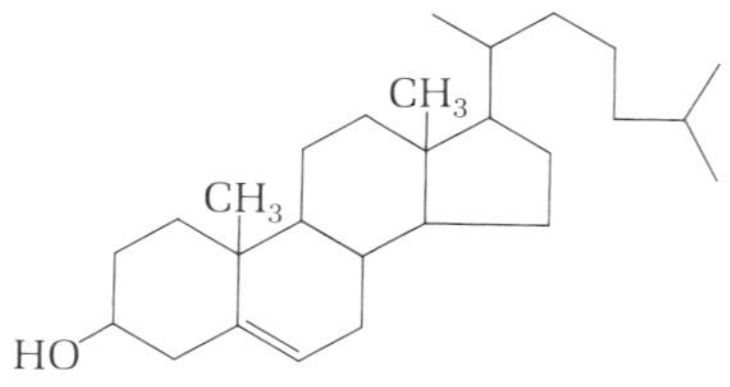

a. cholesterol

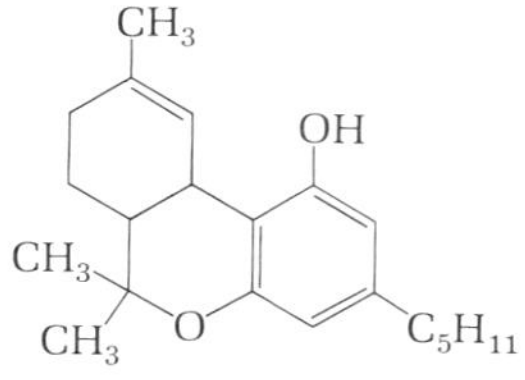

b. Δ^1-tetrahydrocannabinol

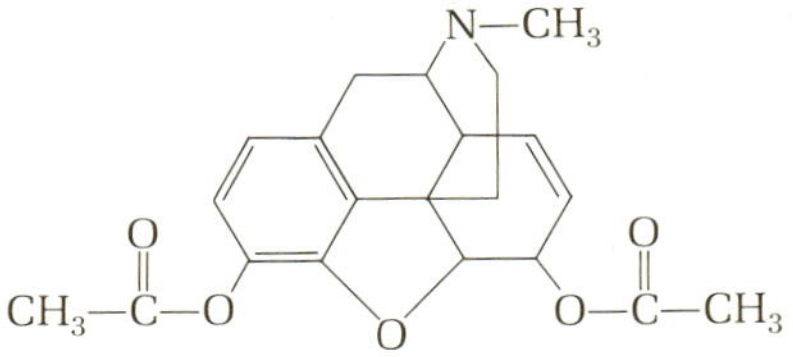

c. diacetyl morphine

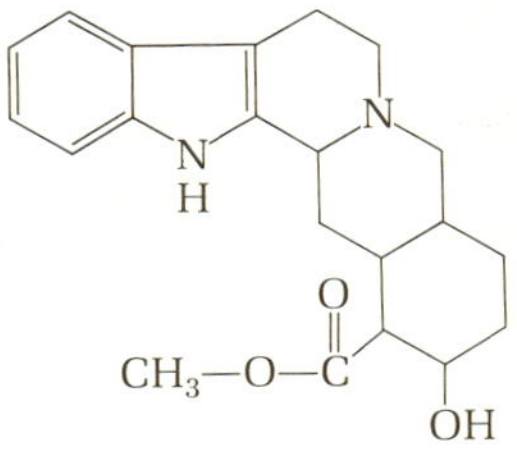

d. yohimbine

24. The combustion of butane gas to yield carbon dioxide and water is a process frequently employed to heat homes and other structures:

$$C_4H_{10} + \tfrac{13}{2}O_2 \longrightarrow 4CO_2 + 5H_2O$$

How many grams of carbon dioxide are produced in the combustion of 100 g of butane gas? How many grams of oxygen would be required to combust this quantity of butane?

25. Calcium carbide can be produced in the following process:

$$CaO + 3C \xrightarrow{2000^\circ C} CaC_2 + CO$$

How much elemental carbon would be consumed in the production of 100 g of calcium carbide? Calcium carbide is important because its reaction with water yields *acetylene,* the simplest member of those compounds having a *triple bond:*

$$CaC_2 + 2H_2O \longrightarrow H{-}C{\equiv}C{-}H + Ca(OH)_2$$

How many grams of acetylene could you produce with the 100 g of calcium carbide you just synthesized?

26. Carbon tetrachloride can be prepared by the chlorination of carbon disulfide:

$$CS_2 + 3Cl_2 \longrightarrow CCl_4 + S_2Cl_2$$

How many grams of carbon tetrachloride could be produced from chlorination of 50 g of carbon disulfide? How much chlorine would be required to effect this conversion?

Suggestions for further reading

I. Asimov *The World of Carbon,* Collier, New York, 1958.

R. Barker *Organic Chemistry of Biological Compounds,* Prentice-Hall, Englewood Cliffs, N.J., 1971, Chap. 3.

J. E. Fernandez *Modern Chemical Science,* Macmillan, New York, 1971, Chap. 11.

W. Herz *The Shape of Carbon Compounds,* Benjamin, Menlo Park, Calif., 1964, Chaps. 1, 2, 3, 11.

J. B. Lambert "The Shapes of Organic Molecules," *Sci. Am.*, **222,** 58 (January, 1970).

K. Mislow *Introduction to Stereochemistry,* Benjamin, Menlo Park, Calif., 1965.

H. Nechamkin *The Chemistry of the Elements,* McGraw-Hill, New York, 1968, Chaps. 1, 5.

L. Pauling and R. Hayward *The Architecture of Molecules,* Freeman, San Francisco, 1964.

E. Wasserman "Chemical Topology," *Sci. Am.*, **207,** 94 (November, 1962).

3

Other tetrahedral atoms

On chemistry in the home for fun and profit:
Then José Arcadio Buendía threw three doubloons into a pan and fused them with copper filings, orpiment, brimstone, and lead. He put it all to a boil in a pot of castor oil until he got a thick and pestilential syrup which was more like common caramel than valuable gold. In risky and desperate process of distillation, melted with the seven planetary metals, mixed with hermetic mercury and vitriol of Cyprus, and put back to cook in hot fat for lack of any radish oil, Ursula's precious inheritance was reduced to a large piece of burnt hog cracklings that was firmly stuck to the bottom of the pan.

Gabriel García Márquez
One Hundred Years of Solitude, 1970

In the last chapter we established the predisposition of the carbon atom to combine in a tetrahedral structure of four atoms or groups of atoms with the carbon atom in the middle, and we explored some of the consequences of this geometry for the chemistry of carbon compounds. What we have learned about the influence of structure on the chemistry of carbon compounds provides us with direct insight into the relation between structure and chemistry for compounds based on a number of other elements. In this chapter we investigate just what these insights are, pointing out both the structural similarities and distinctive differences that exist between compounds based on carbon and on other atoms. Silicon, germanium, tin, and lead form a series of elements chemically related to carbon and provide a starting point for our discussion. Subsequently, we shall discuss the structural chemistry of nitrogen and its relatives, oxygen and its relatives, the halogens, and noble gases. To help us organize our thoughts about chemistry, we first discuss the periodic table; then we shall return to structural chemistry.

THE IDEA OF PERIODICITY

One of the simplifying features for the understanding of chemistry of the elements is the concept of **periodicity.** The essence of this concept is that it is possible to organize the elements in such a way as to indicate simultaneously similarities in their behavior as well as their behavioral trends. Organization of the elements in this way is termed the **periodic table,** as shown in Figure 3-1. Some history follows later.

We can derive information from the periodic table both by reading down the columns and by reading across the rows. The former is simpler, and we consider many examples of the utility of doing so later in this chapter.

For the moment, let us note that each of the columns contains elements of related properties. Thus, the first row (Group IA) of elements is the **alkali metal family,** Li, Na, K, Rb, and Cs; the second row (Group IIA) is the **alkaline earth family,** Be, Mg, Ca, Sr, Ba, and Ra; the next to last (Group VIIA) is the **halogen family,** F, Cl, Br, I, and As; and the very last (Group 0, inert gases) is the **noble gas family,** He, Ne, Ar, Kr, Xe, and Rn. From our previous knowledge of chemistry, we are able to recognize quickly the similarities in the chemical behavior within each of these families. For example, the alkali metals form a complete series of stable, crystalline, high-melting salts of general composition MX with the halogens—NaCl, LiF, KI, and so forth. The noble gases are characterized by a great reluctance to form any compounds at all (with a few notable exceptions mentioned below). So regularly related are the properties of elements that lie in a single column of the periodic table that it has proved possible to predict the properties of many elements prior to their discovery. An example of this comes from the work of Dmitri Mendeleev, the father of the modern periodic table. In 1871, 100 years ago as these words are written, Mendeleev predicted many of the properties of germanium, an element that was isolated and characterized in 1886, 15 years later. How well Mendeleev's predictions were borne out is indicated in Table 3-1.

As an example of information that can be derived from reading the periodic table from left to right, let us consider the variation in those properties distinguishing *metals* from *nonmetals*. These include electrical conductivity and thermal conductivity, among others. Metals usually have high thermal and electrical conductivities. Nonmetals are just the opposite. In addition, there are several elements called semimetals or *metalloids*, and these exhibit intermediate values of these parameters.

Values of electrical conductivities for the elements of the first two rows of the periodic table are collected in Table 3-2. Note that those elements on the left have large values and hence are classified as metals, those on the right have very low values, and so must be nonmetals. Graphite and silicon have intermediate conductivities and are representatives of a class of elements called **semiconductors.** Properties of semiconductors are discussed in Chapter 19.

Other typical properties that distinguish metals and nonmetals include electron affinity, ionization energy, and electronegativity. These matters are not discussed until Chapter 4, but you may wish to look ahead to Tables 4-2, 4-3, and 4-4 at this time to note how these values change as a function of position in the periodic table.

Understanding how and why the periodic table provides the information it does follows from an understanding of the electronic structure of the elements. This topic is discussed in Chapter 4, and we shall return to ideas

Figure 3-1
Long form of the periodic table.

Transition elements

Period	IA	IIA	IIIB	IVB	VB	VIB	VIIB	VIIIB	
1	1 H 1.008								
2	3 Li 6.939	4 Be 9.012							
3	11 Na 22.990	12 Mg 24.31							
4	19 K 39.102	20 Ca 40.08	21 Sc 44.956	22 Ti 47.90	23 V 50.94	24 Cr 51.996	25 Mn 54.94	26 Fe 55.85	27 Co 58.93
5	37 Rb 85.47	38 Sr 87.62	39 Y 88.91	40 Zr 91.22	41 Nb 92.91	42 Mo 95.94	43 Tc 99	44 Ru 101.07	45 Rh 102.91
6	55 Cs 132.91	56 Ba 137.34	57–71 La–Lu Rare earths	72 Hf 178.49	73 Ta 180.95	74 W 183.85	75 Re 186.2	76 Os 190.2	77 Ir 192.2
7	87 Fr 223	88 Ra 226.05	89–103 Ac–Lw Actinides						

Rare earths	57 La 138.91	58 Ce 140.12	59 Pr 140.91	60 Nd 144.24	61 Pm 145	62 Sm 150.35	63 Eu 151.96
Actinides	89 Ac 227	90 Th 232.04	91 Pa 231	92 U 238.03	93 Np 237	94 Pu 242	95 Am 243

	IB	IIB	IIIA	IVA	VA	VIA	VIIA	0 Inert gases
								2 He 4.003
			5 B 10.81	6 C 12.011	7 N 14.007	8 O 15.999	9 F 19.00	10 Ne 20.183
			13 Al 26.98	14 Si 28.086	15 P 30.974	16 S 32.064	17 Cl 35.453	18 Ar 39.948
28 Ni 58.71	29 Cu 63.54	30 Zn 65.37	31 Ga 69.72	32 Ge 72.59	33 As 74.92	34 Se 78.96	35 Br 79.909	36 Kr 83.80
46 Pd 106.4	47 Ag 107.870	48 Cd 112.40	49 In 114.82	50 Sn 118.69	51 Sb 121.75	52 Te 127.60	53 I 126.90	54 Xe 131.30
78 Pt 195.09	79 Au 196.97	80 Hg 200.59	81 Tl 204.37	82 Pb 207.19	83 Bi 208.98	84 Po 210	85 At 210	86 Rn 222
64 Gd 157.25	65 Tb 158.92	66 Dy 162.50	67 Ho 164.93	68 Er 167.26	69 Tm 168.93	70 Yb 173.04	71 Lu 174.97	
96 Cm 247	97 Bk 249	98 Cf 251	99 Es 254	100 Fm 253	101 Md 256	102 No 253	103 Lw 257	

Table 3-1 A comparison of the observed properties of germanium and those predicted by Mendeleev in 1871

Property	Predicted	Observed
Atomic weight	72	72.59
Color	Dirty gray	Grayish white
Density	5.5 g/cm^3	5.47 g/cm^3
Product with air	GeO_2	GeO_2
Reactivity with acids	Slight	None with most
Density of GeO_2	4.7 g/cm^3	4.703 g/cm^3
Boiling point of $GeCl_4$	100°C	86°C

Table 3-2 Relative values of electrical conductivity for several elements as a function of position in the periodic table

Element	Li	Be	B	C	N	O	F	Ne
Conductivity	30	40	10^{-10}	0.1	Gas	Gas	Gas	Gas
Element	Na	Mg	Al	Si	P	S	Cl	Ar
Conductivity	60	60	100	10^{-9}	10^{-15}	10^{-20}	Gas	Gas

about periodicity then. However, we may use the periodic table in the meantime to expand our knowledge of the chemistry of elements based on the line of reasoning that we gained from our study of carbon. The periodic table will help us to group elements of similar chemical behavior together.

SILICON

Inspection of the periodic table reveals that silicon, germanium, tin, and lead lie directly beneath carbon in Group IV in order of increasing atomic weight. Some physical properties of these elements are collected in Table 3-3. Not all of the properties are of immediate interest to us, but we shall have occasion to refer back to them in later chapters. As noted above, elements that lie in the same vertical file of the periodic table generally form a related series of chemical compounds. Thus, we would expect that silicon, germanium, tin, and lead, like carbon, will tend to form compounds in which these atoms are simultaneously bonded to four substituents and that these will be disposed about the central atom in a tetrahedral array. This is found to be true by observation: Silane (SiH_4), germane (GeH_4), stannane (SnH_4), and plumbane (PbH_4) are compounds that correspond directly to methane (CH_4). They do all exist and are all believed to be regular tetrahedra. Nevertheless, each of these elements exhibits its own distinctive behavior in the physical and chemical properties of the com-

Table 3-3 Some physical properties of carbon, silicon, germanium, tin, and lead

	C	Si	Ge	Sn	Pb
Atomic number	6	14	32	50	82
Atomic weight	12.011	28.09	72.59	118.69	207.19
Density (g/ml)	3.51 (diamond) 2.0 (graphite)	2.33	5.35	7.28	11.34
Crystal structure	Diamond	Diamond	Diamond	(White, distorted; gray, diamond)	fcc[a]
Metallic radius (Å)	0.771	1.173	1.223	1.412 (white)	1.538
Atomic volume (ml)	5.3	11.7	13.5	16.2	18.2
Electrical resistivity	1375 (graphite)	10^5	89×10^3	11.5	21
Melting point	—	1410, 1423	937.2	231.8 (white)	327.4
Heat of fusion (kcal/mole)	—	11.1	7.6	1.72	1.14
Boiling point	3850[b]	2680	2830	2687	1751
Heat of vaporization (kcal/mole)	170	(105)	79.9	69.4	42.88

[a] Face-centered cubic structure.
[b] Sublimes

Source: R. T. Sanderson, *Chemical Periodicity*, Van Nostrand Reinhold, New York, 1960.

Figure 3-2 A schematic representation of the relative sizes of the carbon, silicon, germanium, tin, and lead atoms.

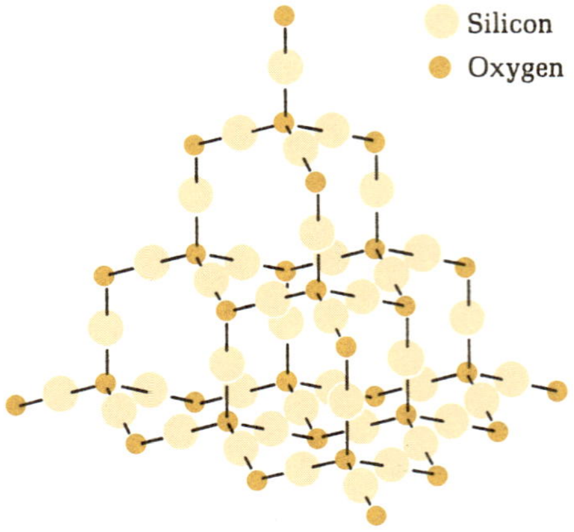

Figure 3-3 The crystal structure of cristobalite.

pounds it forms. Many of these may be understood on the basis of the relative sizes of the atoms involved (Figure 3-2). Clearly, very large differences exist in the sizes of these atoms; 6.023×10^{23} (Avogadro's number) atoms of carbon closely packed together occupy only 5.3 cubic centimeters (cm^3), the *atomic volume;* the same number of atoms of silicon, germanium, tin, and lead occupy 11.7, 13.55, 16.23, and 18.27 cm^3, respectively.

Silicon is the second most abundant element, next to oxygen, in the earth's crust. Although the pure element is not a generally familiar substance, it is a rather common one. Silicon occurs in a tetrahedral form that is similar to that of diamond (see Figure 2-8). As noted above (Table 3-2), elemental silicon is neither clearly a nonmetal, such as carbon (diamond, graphite, charcoal), nor clearly a metal, such as tin or lead, but exhibits some features of both types of materials. Physically, pure crystalline silicon is a hard, brittle, lustrous, gray substance. As a semiconductor, it finds wide industrial use, particularly in the manufacture of transistors, for which extremely pure silicon is required. In addition, silicon is employed as an additive in the production of alloys of steel. Elemental silicon can be produced by a variety of methods. Reduction of silicon dioxide (see the next section) with magnesium, aluminum, or carbon at elevated temperatures provides one method:

$$SiO_2 + 2Mg \longrightarrow Si + 2MgO$$

This method is not well suited to the production of pure silicon required for many purposes. Such silicon can be prepared by the reduction of K_2SiF_6 with aluminum according to the equation

$$3K_2SiF_6 + 4Al \longrightarrow 3Si + 2KAlF_4 + 2K_2AlF_5$$

Silicon oxides

Just as carbon forms the basis for the organic (living) world, silicon forms the basis for much of the mineral world. The silicon-containing minerals reflect the extensive chemistry that results from the tendency of silicon to form chemical bonds with oxygen. In this respect, the essence of the chemistry of carbon, which derives from the hydrides of carbon, is distinct from that of silicon, which is based on its oxides. As an example, consider silicon dioxide, SiO_2, which is commonly called *silica*. The name silicon dioxide and the formula SiO_2 are not particularly informative and may be misleading concerning the structure of the crystalline materials that have this stoichiometry, two oxygen atoms for each silicon atom. There are three common forms of SiO_2 that occur in nature: *quartz, tridymite,* and *cristobalite.* Given enough time, both tridymite and cristobalite would be converted to quartz, although these reactions are exceedingly slow at room temperature. In each of these forms, the silicon atom is surrounded by four oxygen atoms distributed tetrahedrally. The simplest of these, cristobalite, has a structure in which the silicon atoms occupy the same positions as the carbon atoms in diamond (Figure 2-8), and the oxygen atoms occur between each pair of silicon atoms. This structure is diagrammed in Figure 3-3. The stoichiometry is SiO_2, but the structure is tetrahedral. The structures of quartz and tridymite are related but more complicated, since the chains of SiO_4 tetrahedra have a helical character.

Example *Which of the following properties would you expect to be identical and which different in quartz, tridymite, and cristobalite?*

a. *density*
b. *weight percentage silicon*
c. *number percentage silicon*
d. *melting point*
e. *color*

Solution *The number and weight percentage of silicon in these crystals depends only on their composition. Hence, these are identical for the three silicon dioxides. The other quantities depend on both composition and organization of the atoms in space. Thus, these will be different for these three substances.*

The silicon oxides in combination with other elements, particularly sodium, potassium, aluminum, and iron, account for most rocks, clays, and

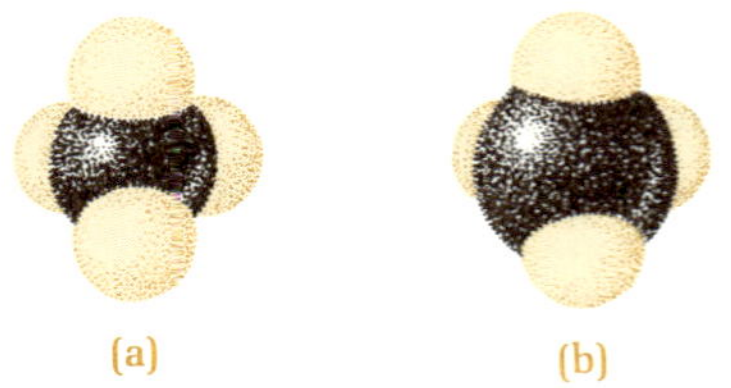

Figure 3-4 Space-filling models of the (a) methane and (b) silane molecules.

soils. Important minerals, which are the basic building blocks from which the rocks, clays, and soils are constructed, include orthoclase, mica, lazulite, talc, and kaolinite. Finally, a very important group of polymeric materials, the silicones, are related to the oxides of silicon. These will be discussed in Chapter 7.

Structures of the *silicates*, which are derived from the orthosilicate ion SiO_4^{4-}, and which include the minerals based on silicon, can be understood in more detail by reconsidering the structure of the silicon dioxides. When all the oxygen atoms in the SiO_4 tetrahedra are shared between two silicon atoms, the structures of the silicon dioxides are formed. But, clearly, intermediate stages may exist between the orthosilicate ion and the silicon dioxides in which a variable number of oxygen atoms are shared between two silicons. At one extreme, we have minerals such as *zircon*, which is a simple zirconium silicate, $ZrSiO_4$. Intermediate stages of condensation of silicates include *wollastonite*, $Ca_3(Si_3O_9)$, and *talc*, $Mg_3(Si_4O_{10})(OH)_2$.

Example *Below is provided a list of several minerals based on silicon. Arrange these in increasing order of condensation of orthosilicate tetrahedra.*

a. *beryl*, $Be_2Al_2(Si_6O_{18})$
b. *mica*, $KAl(Si_3AlO_{10})(OH)_2$
c. *olivine*, $(Mg)_2SiO_4$ [*mixed with* $(Fe)_2SiO_4$]
d. *quartz*, SiO_2

Solution *The crucial point is the ratio of silicon to oxygen atoms in the minerals. The highest ratio is present in olivine, next comes beryl, then mica (note that* Al *has replaced one* Si*) and finally, quartz.*

Silicon hydrides

Silicon forms a series of hydrides analogous to those of carbon and called **silanes.** The simplest of these, *silane*, has the formula SiH_4 and a regular tetrahedral structure. It is thus the silicon analog of methane. For the sake of comparison, space-filling models of silane and methane are shown in Figure 3-4. The silicon–hydrogen bond distance in silane is 1.48 Å, about 0.37 Å longer than the carbon–hydrogen bond in methane, reflecting the greater size of the silicon atom. The melting point of silane, −185 °C, is very close to that of methane, as might be expected from the similar symmetries of the two compounds. Silane boils at −112 °C, about 50° higher than methane, reflecting the increased molecular weight. Like methane, silane forms a series of derivatives in which one or more of the hydrogen atoms is replaced by a variety of substituents. Thus, for example, we have chlorosilane ($SiClH_3$), dichlorosilane ($SiCl_2H_2$), and so on, just as in the case of the chloromethanes. The symmetry properties of these molecules are identical to those of the corresponding carbon compounds. Furthermore, the two series of compounds are closely related in terms of bond angles and the deformation of bond angles caused by substituents of different sizes.

Replacing one of the hydrogen atoms of silane with a methyl group yields an analog of ethane, *methyl silane,* containing one atom of silicon. Replacement of one of the hydrogens with a silyl group yields another analog of ethane, *disilane,* in which both carbon atoms have been replaced by silicons. As indicated below, the structures of these analogs are similar, and all tend to exist in the staggered conformation.

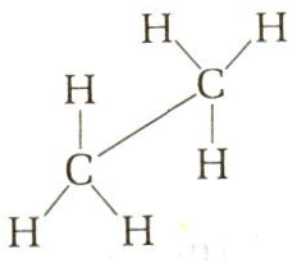

ethane

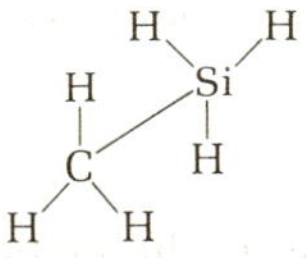

methyl silane

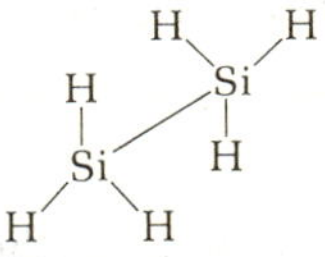

disilane

The distance between the central atoms will, of course, increase with increasing substitution of silicon for carbon:

$$\text{C—C}_{\text{ethane}} = 1.54\ \text{Å}$$
$$\text{C—Si}_{\text{methyl silane}} = 1.86\ \text{Å}$$
$$\text{Si—Si}_{\text{disilane}} = 2.30\ \text{Å}$$

Because of the greater distance between the hydrogens, the preference for the staggered conformation is less for disilane than for ethane.

It is possible to obtain the silicon analogs for propane, butane, pentane, and hexane, and other structural isomers are known for the butane analogs. Some of the physical properties of these two series of compounds, together with some related species, are collected in Table 3-4. A central distinction between the structural chemistry of compounds based on carbon and those based on silicon lies in the fact that compounds of silicon become less stable as the chains become longer. This behavior is in marked contrast to that of carbon. Thus, one may state that the **catenation tendency** of silicon is

Table 3-4 Some physical properties of the hydrides of carbon, silicon, germanium, tin, and lead

Type and property	M: C	Si	Ge	Sn	Pb
MH_4					
Melting point (°C)	−182.7	−185	−165	−150	
Boiling point (°C)	−161.3	−111.9	−88.1	−52	*ca.* −13
Decomposition temperature (°C)	800	450	285	150	0
M_2H_6					
Melting point (°C)	−172	−132.5	−109		
Boiling point (°C)	−88.7	−14.5	29		
M_3H_8					
Melting point (°C)	−189.9	−117.4	−105.6		
Boiling point (°C)	−44.5	52.9	110.5		
M_4H_{10}					
Melting point (°C)	−135	−84.3			
Boiling point (°C)	−0.5	107.4			
M_5H_{12}					
Boiling point (°C)	36.2	>100			
M_6H_{14}					
Boiling point (°C)	69.0	>100			

Source: T. Moeller, *Inorganic Chemistry,* Wiley, New York, 1952.

markedly less than that of carbon. Furthermore, the analogs of the cyclic carbon compounds (cyclohexane, cyclobutane, cyclopropane) have not yet been identified, and silicon analogs of compounds such as ethylene probably are quite unstable. Perhaps this fact can be understood in part simply in terms of the greater size of silicon compared to carbon. It may be increasingly difficult to force larger atoms into compact cyclic structures, although increased atomic size should be offset at least in part by increasing interatomic distance. Electronic factors are also important, but these must be considered later. Cyclic silicon compounds do not seem to be stable and, as a result, the enormous possibilities for structural variation that exist in carbon compounds are markedly reduced for those of silicon.

To summarize, we may note that both carbon and silicon have a tendency to form bonds to four substituents arranged tetrahedrally around the central atom. Silicon has less catenation ability than carbon and consequently forms a more restricted group of binary compounds—those that are constructed from only two elements. Finally, the chemistry of carbon centers around its combination with hydrogen, whereas the chemistry of silicon centers around its combination with oxygen. Consequently, we have many structural similarities, although the thrust of the chemistries differs very substantially.

The framework of these statements is a bit too limited. For example, the structural chemistry of silicon differs from that for carbon in that the former sometimes forms chemical bonds to five or, more commonly, six substituents. One example is the SiF_6^{2-} dianion.

Example *The silicon hexafluoride dianion,* SiF_6^{2-}, *contains, among other elements of symmetry, three fourfold axes of rotation and a center of symmetry. What is the structure of this ion?*

Solution *The six fluorides are disposed at equal distances from the central silicon at the corners of an octahedron:*

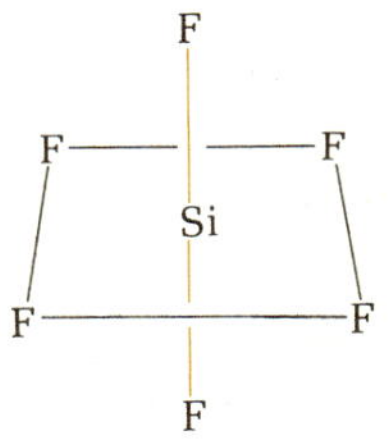

This geometry is considered in detail in Chapter 6.

GERMANIUM, TIN, AND LEAD

Germanium, the least abundant of the Group IV elements, is similar in many respects to silicon. In one modification, the elemental material has the diamond structure (Figure 2-8). Like silicon, germanium finds great use, particularly in slightly impure form, for the construction of transistors (Chapter 19). Tin and lead are also rather rare elements; each accounts for only about 0.001 percent by weight of the earth's crust. These elements are clearly metallic and have, of course, great industrial significance.

The differences between carbon and silicon structural chemistry are accentuated as we consider the properties of germanium, tin, and lead. As the atomic size increases, the catenation abilities of the elements decrease. Thus, germanium forms a series of hydrides somewhat more limited than silicon, whereas for tin only stannane, SnH_4, and distannane, Sn_2H_6, have as yet been prepared. Lead forms only plumbane, PbH_4, and this molecule exhibits only transient stability. Some physical properties of these hydrides are included in Table 3-4. The decreasing stability of the hydrides of this series of elements is paralleled in the behavior of the halide derivatives as well. Thus, the tetrahalides of carbon, silicon, and germanium are stable

Table 3-5 Bond distances (Å) for the hydrides and tetrahalides of carbon, silicon, germanium, tin, and lead

	M				
	C	Si	Ge	Sn	Pb
MH_4	1.11	1.48	1.52	1.70	*ca.* 1.84
MF_4	1.32	1.56	1.68	—	—
MCl_4	1.76	2.00	2.08	2.30	2.43
MBr_4	1.94	2.14	2.29	2.44	—
MI_4	2.15	2.43	2.50	2.64	—

Source: T. Moeller, *Inorganic Chemistry*, Wiley, New York, 1952.

(although the stability decreases markedly in this series); those of tin are not particularly stable; those of lead have only marginal stability and decompose on mild heating. Bond distances for the hydrides and tetrahalides of these elements are collected in Table 3-5.

Once again, we need to recognize that the correlations we are developing between the chemistries of different elements, while perfectly correct, are not all-inclusive. While it is certainly true that, for example, tin and lead form many compounds having tetrahedral geometry, we must not overlook the fact that both these elements also form compounds having just two substituents. For example, the chlorides of tin include both $SnCl_2$ and $SnCl_4$; those of lead include both $PbCl_2$ and $PbCl_4$.

NITROGEN

Just to the right of carbon in the first row of the periodic table is the element nitrogen. This element occurs principally in the form of the diatomic molecule N_2, which forms about 78 percent of the atmosphere (Chapter 18). The nitrogen molecule has the same symmetry properties as the hydrogen molecule, but it is a good deal heavier (a molecular weight of 28 compared to 2 for H_2). Thus, molecular nitrogen is a highly volatile species (bp $-195.8°C$), although not so volatile as hydrogen (bp $-259°C$). Molecular nitrogen is a very stable molecule indeed, and strenuous conditions are required for its conversion to the hydride, ammonia (NH_3), in the laboratory. This reaction is of great importance in nature, since the chemistry of nitrogen is intimately concerned with the chemistry of all life processes. Nitrogen is incorporated into living systems principally in the form of ammonia (NO_3^- is also important), although N_2 is the basic source of this element. Thus, living systems are faced with accomplishing the conversion of nitrogen, N_2, to ammonia, NH_3, the process of **nitrogen fixation.** Despite the fact that this is such a difficult laboratory process, nitrogen-fixing organisms, which include both plants and certain bacteria, manage to fix about 5×10^{12} tons of nitrogen per year under the mildest of conditions. We shall have more to say about how living systems are able to promote chemical reactions in Chapters 7 and 22.

The discovery of nitrogen is the result of an ingenious, but not very kind, experiment carried out by Daniel Rutherford, a Scottish physician. Rutherford took a bottle of air, put a mouse inside, closed the bottle, and waited until the unfortunate beast suffocated. The mouse, of course, used up all the oxygen in the air, leaving only nitrogen, carbon dioxide, and a few trace gases. Absorption of the carbon dioxide with lime left an essentially pure sample of nitrogen.

Further analyses have shown that nitrogen is composed almost exclusively of two isotopes, neither of which is radioactive. The predominant one is nitrogen 14, containing seven protons and seven neutrons in the nucleus. This isotope accounts for 99.635 percent of all nitrogen. Almost all the remainder is nitrogen 15, which has one more neutron in the nucleus. As is usual with isotopes, the chemistry of these two is very much the same.

Now we discuss the characteristics of those compounds based on the combination of hydrogen and nitrogen in some detail. Before we get into this topic, it is worthwhile to note that nitrogen forms a number of compounds with oxygen. Several of these are important for an understanding of the chemistry of the atmosphere, which is developed in Chapters 17 and 18. The principal oxides of nitrogen are collected together with some of their properties in Table 3-6. Of these oxides, three are of particular importance: NO_2 and NO because of their role in photochemical smog formation (Chapter 18), and NO_3^- because of its importance in agriculture and biology. Next to molecular nitrogen, nitrate ion is the most important form of nitrogen on earth. In addition to the fixation of molecular nitrogen noted above, plants also have the capacity to assimilate nitrate. This is accomplished in two steps. First, nitrate is reduced to nitrite and, second, the nitrite is reduced to ammonia: $NO_3^- \longrightarrow NO_2^- \longrightarrow NH_3$. Ammonia is the form actually involved in the synthesis of complex molecules. Since plants assimilate nitrate so readily, it has become an important fertilizer. Chile has very rich deposits of nitrate salts, and the production of fertilizers from these is economically important for Chile. The importance of fertilizers and related agricultural chemicals cannot be overestimated. Eighty-six percent of the world's people are born in the underdeveloped countries. Most of these people will pass their lives in a continual state of malnutrition. Malnutrition in childhood is now clearly recognized as leading to mental retardation. About 4 million people a year still die of starvation. Solution

Table 3-6 Some nitrogen oxides and oxyacids

Oxidation state	Oxide	Name	Remarks	Oxyacid	Name	Remarks
+1	N_2O	Nitrous oxide	Relatively unreactive, colorless gas	HONNOH	Hyponitrous acid	Unstable, weak acid
+2	NO	Nitric oxide	Reactive colorless gas			
+3	N_2O_3	Dinitrogen trioxide	Dissociates to NO and NO_2	HONO	Nitrous acid	Unstable, salts are more stable
+4	N_2O_4	Dinitrogen tetroxide	Colorless, dissociates to NO_2			
	NO_2	Nitrogen dioxide	Brown reactive gas			
+5	N_2O_5	Dinitrogen pentoxide	Colorless solid, contains NO_2^+ and NO_3^- ions	$HONO_2$	Nitric acid	Strong acid and oxidizing agent
"+6"	NO_3	Nitrogen trioxide	Unstable, never isolated			

Source: R. C. Johnson, *Introductory Descriptive Chemistry*, Benjamin, Menlo Park, Calif., 1966.

to these problems is partly, but not completely, a matter of proper food distribution. But even if the world's food supply were evenly divided among all its people, each of us would exist on the edge of malnutrition. In addition to growing more food, great effort is being directed to producing different foods.

At first glance, the relationship between the structural chemistry of carbon and nitrogen compounds is not so clear as that between, for example, carbon and silicon compounds. Whereas carbon and the elements directly below it in the periodic table tend to form bonds with four substituents, nitrogen tends to form bonds with only *three* substituents. Thus, the hydride of carbon, methane, has the formula CH_4, whereas that of nitrogen, ammonia, has the formula NH_3. What relationship can we find between the structural chemistry of methane and ammonia? To what extent can we employ our knowledge of the structures of more complicated carbon compounds to deduce information about those based on nitrogen? In point of fact, the structures of analogous molecules based on carbon and nitrogen are very closely related, provided we consider one important fact that we have not yet considered about nitrogen-containing molecules. In general: *Both carbon and nitrogen compounds have eight electrons disposed about the central atoms.* In methane these eight electrons form four two-electron bonds to four hydrogen atoms, whereas in ammonia these electrons form three two-electron bonds to three hydrogen atoms, and one pair of electrons is not involved in bond formation (an *unshared pair*). If, then, we regard the unshared electron pair of nitrogen as a substituent, it is clear that both carbon and nitrogen compounds have four substituents and, hence, that we may compare the geometries of these molecules directly. Not only are the geometries directly comparable, but they are very similar. That is, the ammonia molecule, like methane, has a tetrahedral structure:

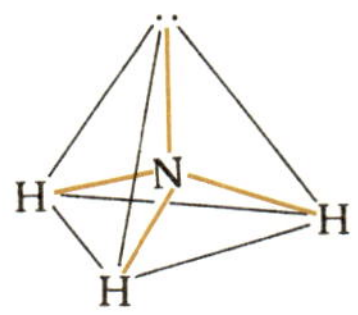

However, we note that the disposition of the atoms is not quite the same in that there is a missing atom that is replaced by an electron pair which is indicated by two dots.

There is a basic difference between the methane and ammonia molecules: In methane all the substituents are identical, whereas in ammonia we have three identical substituents and one, the unshared pair, that is distinct from the others. Thus, ammonia is really more closely analogous, for example, to the chloromethane molecule, $CClH_3$, than to methane itself. Remembering some of the consequences of substituting the chlorine atom for the hydrogen

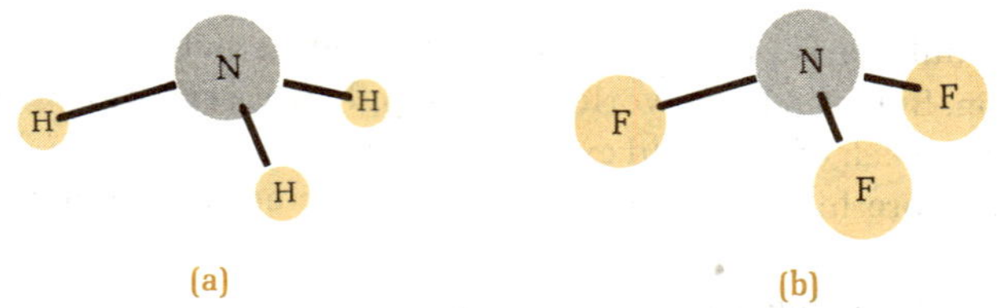

Figure 3-5 Ball-and-stick models for the (a) ammonia and (b) nitrogen trifluoride molecules drawn to scale.

atom in terms of the resultant change in geometry of the methane molecule, we can make educated guesses about the structure of ammonia. The unshared pair of electrons in ammonia actually behaves just like a large substituent: By occupying more space than the hydrogen atom, it forces the hydrogens closer together and hence causes the H—N—H bond angle to decrease. The H—N—H bond angle in ammonia is 107.3°.

Let us examine the symmetry properties of ammonia. We have noted above that ammonia is similar in this respect to chloromethane (or to chloroform, $CHCl_3$) and therefore lacks most of the symmetry properties of methane. Suppose that we drive an imaginary axis into the central nitrogen atom and through the unshared electron pair. Now, if we pick up the molecule by this axis, rotate it through 120°, and place it down again, the product appears unchanged from the original. Since three such rotations are required to return the molecule to its original position and since no other such axis exists, we know that ammonia has one threefold rotation axis:

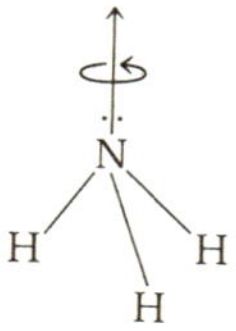

In addition, the ammonia molecule has three mirror planes of symmetry, all of which contain the axis in the plane:

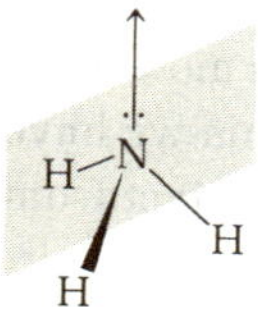

Like methane, ammonia forms a series of derivatives in which the hydrogen atoms are replaced by halogens or other atoms. Thus, the molecules

NH_3, NH_2Cl, $NHCl_2$, and NCl_3 are all known and form a series analogous to the chlorinated methanes. Of some interest is nitrogen trifluoride, NF_3, in which the F—N—F bond angle is just 102.1°. This value is substantially smaller than that for ammonia, a surprising finding at first glance. Basically, the fluorine atoms in NF_3 seem to have been crowded together rather more closely than the hydrogen atoms in NH_3. We know that fluorine is substantially larger than hydrogen. Why, then, does fluorine appear to behave as if the opposite were true? The important point that we have neglected thus far is that the N—H bond distance in ammonia is 1.015 Å, while that for the N—F bond in nitrogen trifluoride is 1.37 Å. Thus, the fluorine atoms are farther from the central nitrogen atom, and from each other, than are the hydrogen atoms. Clearly, the longer the bond distance from the central atom to the substituent, the larger the volume of space available to each substituent atom. Hence, the fact that fluorine seems to behave as if it were smaller than hydrogen when we compare the bond angles of ammonia and nitrogen trifluoride reflects principally the differences in bond lengths. This point is brought out graphically in Figure 3-5, in which models for these molecules are shown.

Alternatively, we can return to our picture of balloons tied at the central atom. The highly electronegative (a term defined in detail in Chapter 4) fluorine atom "draws" the electrons in the bond out from the center, thus narrowing these three at the crucial point close to the central atom. The "balloon" corresponding to the unshared electron pair then forces the F—N—F angle to decrease below the normal tetrahedral angle.

Replacing one of the hydrogen atoms in ammonia with a methyl function yields a new molecule, *methylamine*, $CH_3—NH_2$, which is structurally related to ethane. Replacement of such a hydrogen atom with the amino group, NH_2, yields the ethane analog *hydrazine*, $NH_2—NH_2$, in which both atoms of carbon have been replaced by nitrogen:

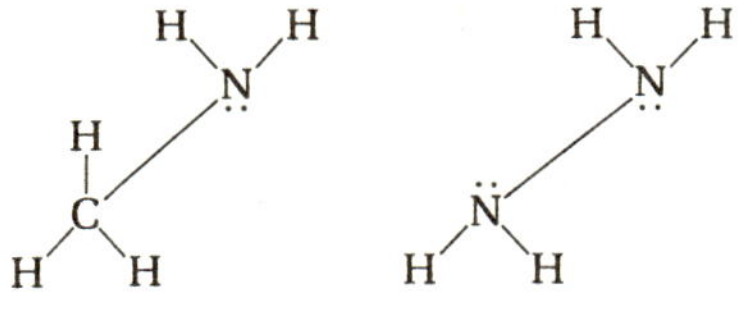

Notice that for the hydrazine, there exist two conformations of the staggered type: one in which the unshared electron pairs are as far apart as possible (a) and one in which they occupy adjacent positions (b):

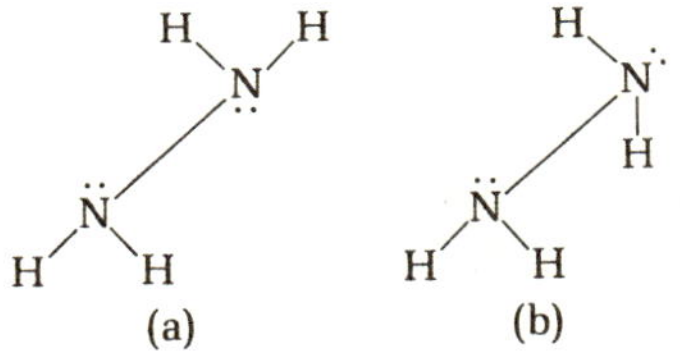

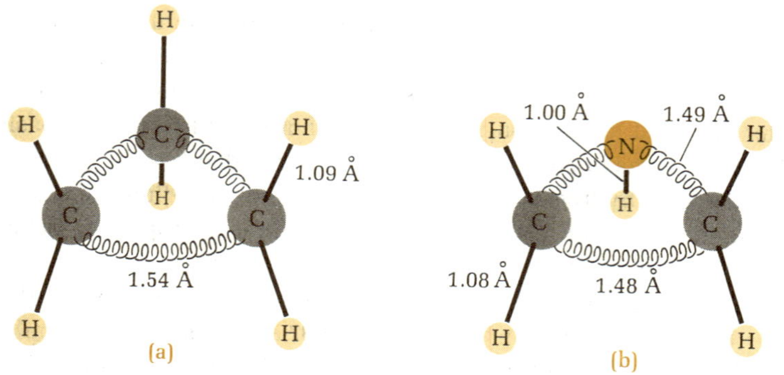

Figure 3-6 The molecular structures of (a) cyclopropane and (b) ethyleneimine.

The best evidence available at the moment suggests that the most stable form for the hydrazine molecule is (b). The most stable structures of the ethane, methylamine, and hydrazine molecules are then

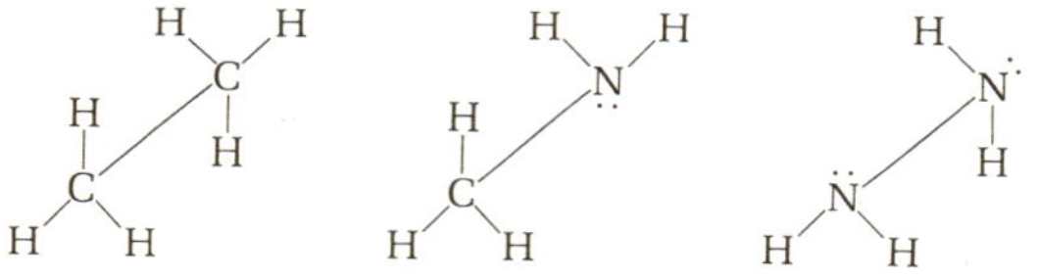

Example *What symmetry elements are present in the most stable conformation of hydrazine?*

Solution *Examination of this structure reveals that just one symmetry element exists; this one is quite difficult to find without actually building a model of the molecule. It is a twofold rotation axis that bisects and is perpendicular to the* N—N *bond and bisects the dihedral angle formed by the two unshared electron pairs on nitrogen. This symmetry element is present regardless of the conformation of the hydrazine molecule. If we think about the hydrazine molecule itself rather than one single conformation, other elements of symmetry are present. Remember, there is nearly free rotation about the* N—N *bond; therefore, we consider the symmetry properties of the most symmetrical form of hydrazine that we can obtain and analyze the symmetry elements it possesses. In the symmetrical form (a), there is a center of symmetry at the midpoint of the* N—N *bond. A plane of symmetry bisects the two* H—N—H *angles. Finally (and perhaps most difficult to see), this form has a twofold axis through the center of symmetry and perpendicular to the symmetry plane.*

The tendency of nitrogen toward catenation is very much less than that of carbon. Indeed, the hydrazine molecule is the only catenated molecule in this series that has yet been made, although the analogs of propane and butane, NH_2—NH—NH_2 and NH_2—NH—NH—NH_2, do occur in substituted form. On the other hand, nitrogen-containing analogs of the cyclic carbon compounds do occur rather widely. For example, ethyleneimine, C_2H_5N, has the same basic structure as cyclopropane except that an NH group replaces one of the CH_2 members of the ring:

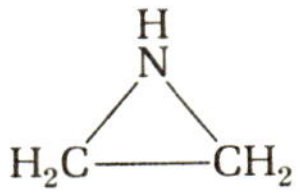

To emphasize the relationship between cyclopropane and ethyleneimine, the appropriate structures together with their molecular dimensions are shown in Figure 3-6. Likewise, analogs of cyclobutane, cyclopentane, and cyclohexane, in which the ring contains one nitrogen atom, are known, stable structures:

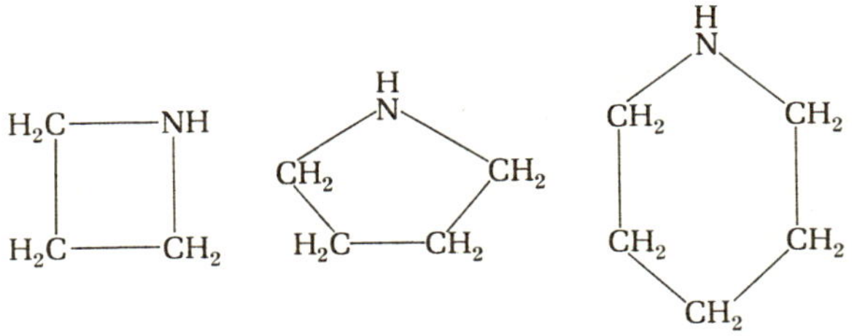

We have viewed the compounds of nitrogen in terms of four substituents: the three bonded atoms or groups and the unshared electron pair. If we choose the three bonded atoms or groups to be different species, such as in the methylsilylamine molecule, CH_3NHSiH_3, or generally in molecules of the type $NR_1R_2R_3$, we have a situation that is quite analogous to that encountered in derivatives of methane in which the four substituents are all different. We have already seen that, in this case, such compounds exist as optical isomers. Therefore, we expect that derivatives of nitrogen having four different groups, including the unshared electron pair, ought to exist as optical isomers as well:

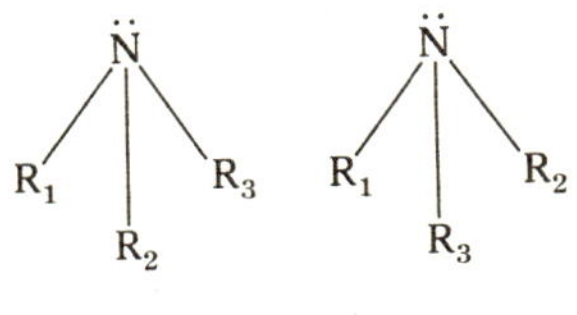

While all our reasoning thus far has been quite correct, it has proved possible only in exceptional cases to isolate optical isomers of nitrogen

Table 3-7 Some physical properties of nitrogen, phosphorus, arsenic, antimony, and bismuth

	N	P	As	Sb	Bi
Atomic number	7	15	33	51	83
Atomic weight	14.007	30.974	74.92	121.75	208.98
Density (g/ml)	0.96	1.82 (white)	5.72 (gray)	6.69	9.80
Crystal structure	(Several forms)	(White, P_4 tetrahedra; violet, chains; black, double layers of parallel chains)	(Yellow, As_4 tetrahedra; black, ?; gray, pyramidal)	(Yellow, ?; black, Sb_4 tetrahedra; vitreous; metallic, pyramidal)	Metallic, pyramidal
Metallic radius (Å)	0.74	1.10	1.21	1.41	1.52
Atomic volume (ml)	—	17.0 (white)	13.1	18.2	21.3
Electrical resistivity	—	10^{17} (white)	35	39	110
Melting point	−210.10	44.1 (white) 597 (red)	817 (gray) (36 atm)	630.5	271.3
Heat of fusion (kcal/mole)	0.086	0.15	6.62	4.74	2.6
Boiling point	−195.8	280 (white) 431 (red)[a]	613[a]	1637	1559
Heat of vaporization (kcal/mole)	0.67	2.97 7.2 (red)[a]	34.5[a]	16.23	36.2

[a] Sublimes.

Source: R. T. Sanderson, *Chemical Periodicity*, Van Nostrand Reinhold, New York, 1960. Copyright © 1960 by Litton Educational Publishing, Inc.

compounds. The reason for this difficulty is that the optical isomers are interconverted rapidly by a process in which the molecule inverts itself:

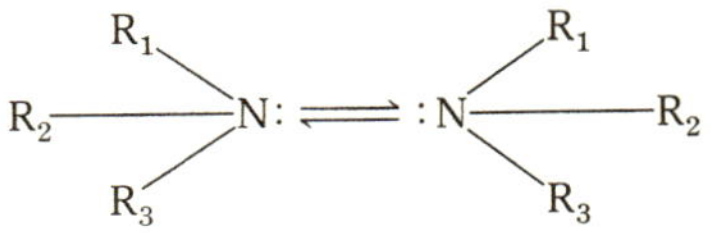

Processes of this type are very rapid indeed. For ammonia, the inversion takes place about 4×10^{10} times per second at room temperature. Other amines generally invert a great deal more slowly but still several thousands or tens of thousands of times per second.

The unshared pair of electrons of the ammonia molecule, or of derivatives of ammonia, have the capacity to form a chemical bond with species that contribute no electrons to the bond. One such electron-deficient bond-former is the proton, H^+. The molecule formed from the association of ammonia with a proton, the *ammonium ion*, NH_4^+, has a perfectly regular tetrahedral structure just like the methane molecule. Once the new proton enters into combination, it cannot be distinguished from those present earlier. Similarly, the molecule formed from the association of two protons with the hydrazine molecule, $^+NH_3—^+NH_3$, has a structure analogous to that of ethane. These simple concepts may be extended to include a variety of structures that are related in a most direct fashion to the corresponding carbon compounds.

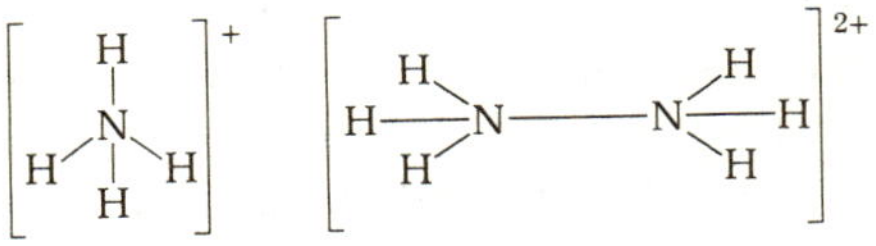

PHOSPHORUS, ARSENIC, ANTIMONY, AND BISMUTH

Just as silicon, germanium, tin, and lead lie directly below carbon in the periodic table, and hence form compounds with structures analogous in many ways to those of carbon, the elements phosphorus, arsenic, antimony, and bismuth lie directly below nitrogen. Some physical properties of these elements are collected in Table 3-7. What we have just learned about the structures of nitrogen compounds should therefore be useful in understanding the structures of simple compounds formed from the latter elements. We expect to find hydrides of these elements corresponding to ammonia, NH_3, with the three atoms of hydrogen and the unshared electron pair directed toward the four corners of a tetrahedron. This expectation is largely borne out. Thus, we have *phosphine* (PH_3), *arsine* (AsH_3), and *stibine* (SbH_3), and the highly unstable *bismuthine* (BiH_3). The trend toward

Table 3-8 Bond angles and bond distances for the hydrides and halides of phosphorus, arsenic, antimony, and bismuth

Compound	Bond angles (°)	Bond distance (Å)
NH_3	107.3	1.015
PH_3	93.8	1.42
AsH_3	91.6	1.52
SbH_3	91.5	1.71
PF_3	92.8	1.57
PCl_3	100.3	2.04
PBr_3	101.0	2.22
PI_3	100	2.46
AsF_3	96.2	1.71
$AsCl_3$	98.4	2.16
$AsBr_3$	99.7	2.36
AsI_3	100.2	2.54
SbF_3	88	2.03
$SbCl_3$	99.5	2.32
$SbBr_3$	98	2.52
SbI_3	98	2.75
$BiCl_3$	100	2.48
$BiBr_3$	100	2.63

decreasing stability in this series parallels the trend in the C to Pb group. Each of these compounds is an extremely toxic gas. While the gross structures of these hydrides are similar to ammonia, they do differ in detail. Thus, the H—P—H bond angle in phosphine is only 93.8°, a value much smaller than that for the H—N—H bond in ammonia, whereas the P—H bond distance is 1.42 Å, larger than that of the N—H bond for ammonia as expected, since the phosphorus atom is substantially larger than that of nitrogen. Bond angles and bond distances for all the hydrides in this series are collected in Table 3-8.

Of the Group V elements, only phosphorus and arsenic form stable hydrides analogous to hydrazine (P_2H_4 and As_2H_4). Phosphorus also forms both linear and cyclic compounds with as many as five linked P atoms. Thus, the catenation tendency of this group of elements is very limited indeed. Phosphorus, arsenic, antimony, and bismuth all form a complete series of trihalides. Bond angles for these compounds, collected in Table 3-8, are all near 100°, substantially less than the value for a regular tetrahedron but greater than those for the hydrides. The corresponding bond distances (Table 3-8) reflect the sizes of the atoms involved.

Note particularly in Table 3-8 the regular increase in bond angle for both the phosphorus and arsenic halide series as the halogen is changed from fluorine to iodine (the slight decrease at PI_3 suggests that the structure should be redetermined!). This is precisely what we expect from the "distorted balloon repulsion" picture. Fluorine, the most electronegative halogen, pulls electrons toward itself and allows the bond angle to decrease considerably from tetrahedral. Each halogen in turn is somewhat less electronegative than the preceding one, the "balloons" become progressively less distorted toward the halogen, and the bond angles consequently become gradually greater as we move through the series.

Elemental phosphorus

The structures of the elements themselves within this group are interesting and informative. Elemental phosphorus occurs in a variety of forms, including three principal *modifications:* white, red, and black phosphorus. White phosphorus, which is the most reactive form and inflames spontaneously on exposure to air, contains P_4 molecules that are regular tetrahedra:

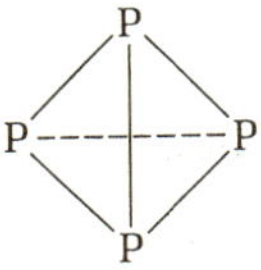

Phosphorus molecules in the gas phase are also regular tetrahedra. It is interesting to note that the symmetry properties of the P_4 tetrahedra are just the same as those of the methane molecule. The structure of red phosphorus, which is less reactive than the white modification and is stable in air although rather easy to ignite, has not been established with certainty. However, it seems likely that it is composed of P_4 tetrahedra linked by chemical bonds between the tetrahedra in the following fashion:

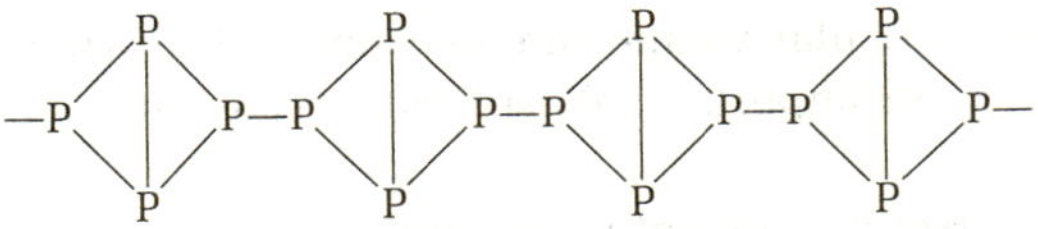

Finally, the structure of black phosphorus, which is the least reactive form and is difficult to ignite, is basically different. It is composed of corrugated sheets in which each phosphorus atom is bonded to three neighbors:

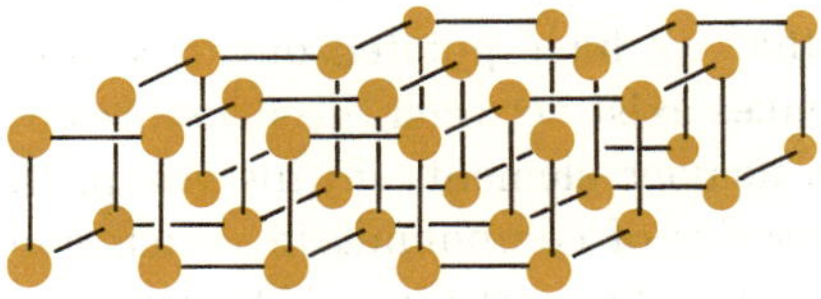

Black phosphorus is a flaky substance, reflecting the fact that it is rather easy to cleave the chemical bonds between the sheets. The distinct colors,

Table 3-9 Some oxides and oxyacids of phosphorus

Compound	Name	Remarks
P_2O_5	Phosphorous(V) oxide, commonly called phosphorous pentoxide	White solid; very powerful dehydrating agent
P_4O_6	Phosphorous(III) oxide	Form of phosphorous oxide that exists in the vapor phase
H_3PO_4	Phosphoric acid	Strong acid, the parent of the *phosphate* salts
H_3PO_3	Phosphorous acid	Parent of the *phosphite* salts
$H_4P_2O_6$	Hypophosphoric acid	Has the structure 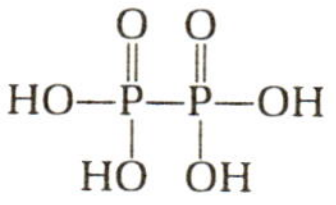
H_3PO_2	Hypophosphorous acid	

Table 3-10 Bond angles and bond distances for certain simple derivatives of oxygen

Compound	Bond angle (°)	Bond distance (Å)
OH_2	104.45	0.958
OF_2	103.2	1.41
OCl_2	111	1.68

physical properties, and chemical reactivities of these three forms of phosphorus emphasize the importance of the manner in which the atoms are arranged in space to the nature of the material.

A number of the oxides and oxyacids of phosphorus are important chemicals. The phosphates are particularly important as chemical fertilizers. Some information about the more important phosphorus oxides are collected in Table 3-9.

The stable forms of both arsenic and antimony are similar to black phosphorus. Both of these elements, however, do have unstable yellow forms that contain As_4 and Sb_4 tetrahedra, analogous to white phosphorus.

To this point, the structural chemistry of phosphorus, arsenic, antimony, and bismuth compounds has been directly related to that of nitrogen derivatives. Just as these similarities that we have pointed out exist and are important, so do distinctive differences exist between the structures of nitrogen compounds and those of the heavier elements in this group. In particular, the latter elements form a series of compounds in which *five* groups are bonded to the central atom instead of just three. Nitrogen does not form such compounds. For example, P, As, Sb, and Bi all form pentafluorides, MF_5. In addition, PCl_5, PBr_5, and $SbCl_5$ are known, as are some more complicated molecules. The structures of these and related molecules will be described in Chapter 6.

OXYGEN

Just to the right of nitrogen in the first row of the periodic table is the element oxygen. This element is one of the most abundant on earth. The oxygen molecule, O_2, forms about 20 percent of the atmosphere; the oxygen hydride, water (H_2O), covers about 80 percent of the earth's surface; and substantially all metals occur in nature in combination with this element. Indeed, all elements with the exception of helium, neon, and argon form compounds with oxygen, and it is possible that stable compounds with argon may eventually be obtained. So pervasive is the chemistry of oxygen that it is usually treated along with descriptions of the chemistry of other elements rather than being considered by itself. In addition to diatomic oxygen, this element also occurs to a limited extent in the form of *ozone*, O_3. Appreciable quantities of this gas, which is blue, do not exist in the atmosphere except at very high altitudes. It is produced through the passage of an electrical discharge through air or oxygen and accounts for the characteristic odor that accompanies such discharges.

Carbon forms four chemical bonds in its derivatives, and nitrogen forms three. It is hardly surprising, then, that oxygen should form only two. Nevertheless, the structural chemistry of oxygen compounds can be directly related to that of carbon just as was that of nitrogen. The four substituents in derivatives of oxygen consist of the two bonded atoms or groups and two unshared electron pairs. Thus, the hydride of oxygen, H_2O, has a tetrahedral structure:

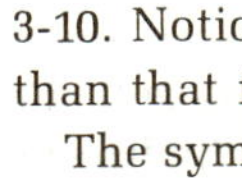

The tetrahedral structure is not quite regular: the H—O—H bond angle is just 104.45°. Like methane and ammonia, water forms a series of derivatives in which the hydrogen atoms are replaced by halogen or other atoms. The bond distances and bond angles for certain of these are collected in Table 3-10. Notice that in the OCl_2 molecule the bond angle is actually greater than that for a regular tetrahedron.

The symmetry properties of the water molecule are easy to identify and are shown diagrammatically in Figure 3-7. Clearly, there is one twofold rotational axis, which passes through the oxygen atom and between the two hydrogens. Furthermore, there exist two mirror planes: One is defined

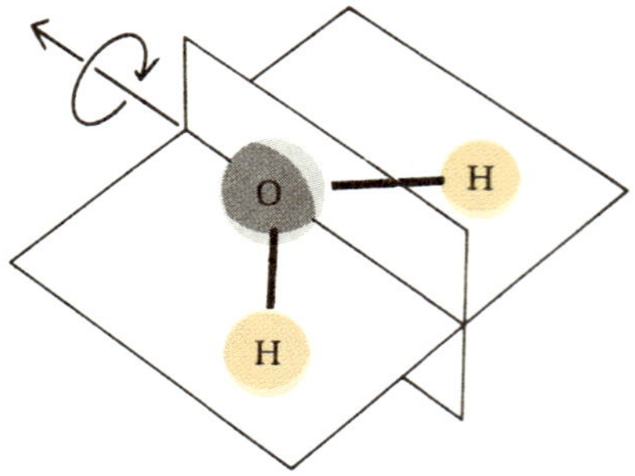

Figure 3-7 The symmetry elements of the water molecule.

Table 3-11 Some physical properties of oxygen, sulfur, selenium, and tellurium

	O	S	Se	Te
Atomic number	8	16	34	52
Atomic weight	15.999	32.064	78.96	127.60
Density (g/ml)	1.27 (at mp)	2.06 (rhombic)	4.82 (gray)	6.25
Metallic radius (Å)	0.74	1.04	1.17	1.37
Atomic volume (ml)	—	15.5	16.5	20.4
Electrical resistivity	—	2×10^{23}	—	—
Melting point	−218.9	119.0 (Monoclinic)	220.2 (gray) 217.4	450
Heat of fusion (kcal/mole)	0.053	0.34	1.25	4.28
Boiling point	−182.96	440.60	688	1390
Heat of vaporization (kcal/mole)	0.81	2.3	6.29	12.1

Source: R. T. Sanderson, *Chemical Periodicity*, Van Nostrand Reinhold, New York, 1960. Copyright © 1960 by Litton Educational Publishing, Inc.

by the plane of the molecule (such mirror planes occur, of course, for all planar molecules), and the other is a plane perpendicular to that of the molecule passing through the oxygen atom and midway between the two atoms of hydrogen.

Replacement of one of the hydrogen atoms of water with a methyl group yields a molecule of *methanol*, CH_3OH, which is structurally analogous to ethane, to methylamine, and to hydrazine. Replacing one of the hydrogen atoms by an NH_2 function yields *hydroxylamine*, NH_2OH, another structurally analogous molecule. Finally, replacement by an OH group yields *hydrogen peroxide*, H_2O_2, the last member of this series. Thus, the series of molecules CH_3CH_3, CH_3NH_2, CH_3OH, NH_2NH_2, NH_2OH, and HOOH all have a similar structure based on two tetrahedra joined along a line between the two central atoms.

Oxygen has a very limited tendency toward catenation; hydrogen peroxide is the only stable catenated hydride of oxygen. The catenated oxygen fluorides are somewhat more stable. Oxygen difluoride, OF_2, is a stable compound at room temperature. Dioxygen difluoride, F—O—O—F, is a yellow-orange solid which has marginal stability at $-50°C$. Trioxygen difluoride, F—O—O—O—F, is a dark red liquid which is stable only at temperatures near $-160°C$ or below. Finally, tetraoxygen difluoride, F—O—O—O—O—F, is stable only near $-200°C$. Compounds in which there are chains of more than four atoms of oxygen may be expected to be almost impossible to prepare.

SULFUR, SELENIUM, TELLURIUM, AND POLONIUM

The structures of many sulfur, selenium, tellurium, and polonium compounds are analogous to those of oxygen. These elements, of course, lie just below oxygen in the periodic table. Physical properties for these elements are collected in Table 3-11. All four form hydrides of the type MH_2. The H—M—H bond angles in these compounds differ substantially from the value for a regular tetrahedron. In H_2S the value is 92.25°, and in H_2Se the value is 91°. The compounds H_2Te and H_2Po are unstable, and their structures are not accurately known. The heavier Group VI elements also form halogen derivatives similar to those of oxygen. However, there is little to be added to our knowledge of molecular structure from a detailed consideration of these.

These elements do exhibit a tendency toward catenation which decreases with increasing atomic size. Indeed, sulfur is second only to carbon in its ability to form chainlike molecules. In the *sulfanes*, H—S—S—S ··· S—H, for example, the number of linked sulfur atoms may be very large. Both sulfur and selenium have the capacity to form ring structures, and tellurium exists as long chains in its elemental form. The structure of elemental sulfur is particularly interesting. In the solid state, elemental sulfur can assume several crystalline forms, two of which, rhombic and monoclinic, are common. In both modifications the sulfur occurs in the form of circular chains containing eight members arranged in a crownlike conformation:

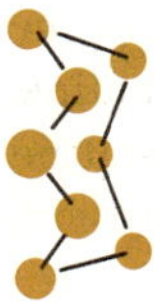

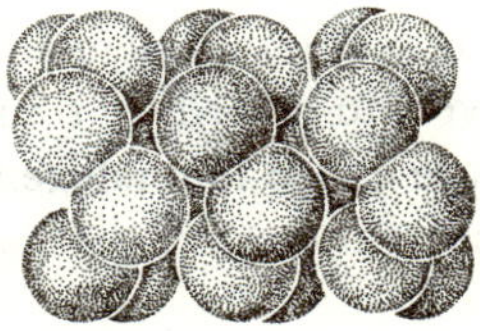

Table 3-12. Some physical properties of fluorine, chlorine, bromine, and iodine

	F	Cl	Br	I
Atomic number	9	17	35	53
Atomic weight	19.00	35.453	79.909	126.90
Density (g/ml)	1.108 (liquid at bp)	1.57 (liquid at bp)	3.14	4.942
Color of gas	Pale yellow	Yellow-green	Brown-red	Violet
Molecular volume of liquid at bp (ml)	34	45	54	68
Melting point	−219.62	−102.4	−7.2	113.6
Heat of fusion (kcal/mole)	0.061	0.77	1.26	1.87
Boiling point	−187.9	−34.6	58.2	184.5
Heat of vaporization (kcal/mole)	0.782	2.439	3.59	4.99

Source: R. T. Sanderson, *Chemical Periodicity,* Van Nostrand Reinhold, New York, 1960. Copyright © 1960 by Litton Educational Publishing, Inc.

Elemental selenium also occurs as cyclic Se_8 molecules. The oxides of sulfur are important molecules, as air pollutants among other things, and are discussed in Chapter 18.

We have already learned that heavier elements within the same group of the periodic table may exhibit some chemical features not found in the lighter elements. Thus, nitrogen forms only one chloride, NCl_3, whereas phosphorus forms two, PCl_3 and PCl_5. Similarly, sulfur, selenium, tellurium, and polonium all form halides that are structurally distinct from those of oxygen. Sulfur forms a variety of fluorides, the most prominent of which are SF_4 and SF_6, rather than SF_2, which corresponds to OF_2. Again, to recognize that differences do occur will suffice for our purposes at the moment. The geometries of these molecules will be described and examined in Chapter 6.

THE HALOGENS

Continuing across the first row of the periodic table, we encounter the element fluorine just to the right of oxygen. Below this element in this, the next to last column of the periodic table, we find chlorine, bromine, iodine, and astatine. Collectively, these elements are called the **halogens.** Astatine is a radioactive element which decomposes quite rapidly. It is, therefore, a difficult element to study and little is known of its chemistry. We shall not be further concerned with it. The other halogens occur as diatomic molecules in the elemental state: F_2, Cl_2, Br_2, and I_2. Some of their properties are collected in Table 3-12.

Fluorine, a green gas, boiling point −188°C, is exceptional among the elements in terms of its reactivity. With the exception of helium, neon, argon, and oxygen, fluorine reacts directly with all of the elements. Of particular note are the reactions of fluorine with the heavier noble gases discussed below. Because of its high reactivity, fluorine is an exceptionally corrosive and toxic substance.

Chlorine, like fluorine a pale green gas, boiling point −34.6°C, is found in nature principally as sodium chloride in sea water and in many solid deposits from ancient seas. It is a constituent of a good many minerals as well. The element is a reactive and toxic substance, although not so reactive as fluorine. The heavier element *bromine,* is a dark red liquid at room temperature, boiling point 58.8°C. It is much less abundant in nature than either fluorine or chlorine, although it occurs distributed in about the same manner. As a gas, bromine is a choking and toxic substance. The liquid causes painful skin burns, which heal very slowly. Of all the halogens, only *iodine* is easy to handle. This element occurs as a black solid at room temperature. It does not exhibit a liquid phase under ordinary conditions but sublimes directly to the gaseous state, which has an attractive violet color. Iodine is the least reactive of the halogens and has the fewest commercial uses. It is important in the human diet. Iodine deficiency in the diet produces an abnormality called *goiter.* To prevent this condition, very small quantities of iodine, about 1 part in 10^9, are frequently added to table salt (iodized salt).

Most of the chemistry of the halogens is just what we would expect on the basis of the discussion thus far. That is, each halogen tends to form chemical bonds to just one substituent. Thus, the corresponding hydrides are HF, HCl, HBr, and HI. These are acidic substances, and HF is one of the few compounds known that will etch glass. Consequently, it must be stored in wax-lined containers. If we wish to press our structural analogy of the elemental hydrides to the hydride of carbon, methane, to its absolute limit, we would be entitled to conclude that HF, HCl, HBr, and HI are tetrahedral molecules as well. To do this, however, we would have to have three positions of the tetrahedron occupied by unshared electron pairs and the fourth position occupied by hydrogen. Clearly, considerations about bond angles, symmetry, isomers, and the like that were important for more complicated molecules are not pertinent to these simple structures. The H—X bond distances do increase markedly with increasing size of the halogen as expected: H—F, 0.92 Å; H—Cl, 1.28 Å; H—Br, 1.42 Å; and H—I,

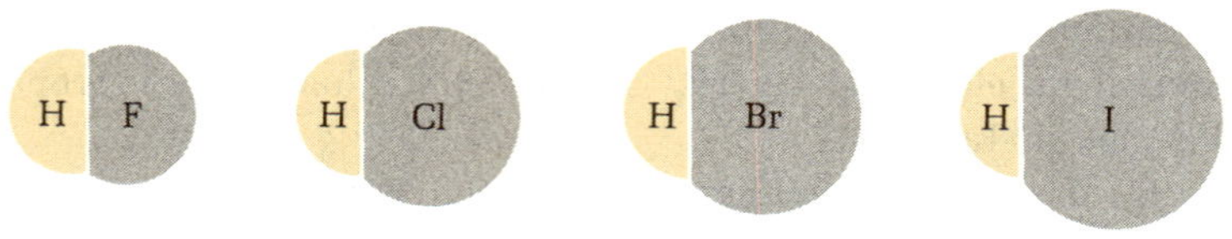

Figure 3-8 Space-filling models of the halogen hydrides.

Table 3-13 Some physical properties of helium, neon, argon, krypton, xenon, and radon

	He	Ne	Ar	Kr	Xe	Rn
Atomic number	2	10	18	36	54	86
Atomic weight	4.003	20.183	39.948	83.80	131.3	222
Density (g/ml) (liquid)	0.126	1.204	1.65 (solid)	2.6	3.06	4.4
Melting point	−272.1 (25 atm)	−248.6	−189.4	−156.6	−111.5	−71
Heat of fusion near mp (kcal/mole)	0.0033	0.078	0.265	0.36	0.49	0.8
Boiling point	−268.98	−246.03	−185.87	−152.9	−107.1	−65
Heat of vaporization near bp (kcal/mole)	0.022	0.44	1.50	2.31	3.27	43

Source: R. T. Sanderson, *Chemical Periodicity,* Van Nostrand Reinhold, New York, 1960. Copyright © 1960 by Litton Educational Publishing, Inc.

1.62 Å. The relative sizes of the halogen atoms can be appreciated from an examination of Figure 3-8, in which a space-filling model of each of the halogen hydrides is shown.

Just as the halides of carbon, nitrogen, oxygen, and their heavier analogs are important to the chemistry of these elements, so are the halides of the halogens themselves important to their chemistry. These constitute the *interhalogen compounds.* Many of the interhalogen compounds are structurally simple and are like the hydrides. These include BrCl, ICl, IBr, ClF, BrF, and others. There are also three additional series of compounds: (1) XX'_3 compounds such as ClF_3 and ICl_3, (2) XX'_5 compounds such as IF_5

and ClF_5, and (3) XX'_7 compounds such as IF_7. The structures of these compounds are interesting and complex.

THE NOBLE GASES

Finally, the last column of the periodic table contains the *noble gases:* helium, neon, argon, krypton, xenon, and radon. All of these elements occur in the atomic state, and all are low-boiling gases. The boiling points increase systematically with increasing size as we would expect; see Table 3-13. Classically, the noble gases were considered to be the least interesting of all the elements to the chemist simply because they seemed to form no chemical compounds at all. Two discoveries have dramatically changed this situation and have provided a good deal of excitement within the physical sciences. The first of these discoveries has to do with the absolutely remarkable properties of liquid helium. Helium liquefies at 4.18°K, and this material, He(I), has ordinary properties of viscosity, conductivity, and the like. However, if He(I) is cooled below 2.718°K, a second form, He(II), also liquid, is obtained which has properties that clearly distinguish it from all other known substances. Indeed, He(II) has frequently been referred to as the fourth state of matter. Its special properties include the following. (1) It has no viscosity at all (or at least none that can be detected by the most sensitive instruments available). It forms exceedingly thin films, which appear to flow without friction. In addition, (2) He(II) is a remarkable conductor of heat. It is the least-insulating substance known by several orders of magnitude.

The chemistry of the noble gases has been enlivened by the discovery that the heavier of these gases, particularly xenon, do indeed form chemical compounds, particularly with fluorine. For example, treatment of xenon with elemental fluorine yields, depending on the reaction conditions, XeF_2, XeF_4, and XeF_6. These are all solid crystalline materials at room temperature and have varying stabilities, the tetra- and hexafluorides being the most stable. In addition, xenon forms two compounds with oxygen, XeO_3 and XeO_4; the former is crystalline, the latter is a gas, and both are quite explosive. More complicated compounds exist as well: For example, $XeOF_4$ is a stable liquid substance. Less extensive information is available about the other noble gases. It is clear that krypton reacts with fluorine to form KrF_2 at least and that radon, which is radioactive and difficult to work with, does form a fluoride. Thus far, chemical compounds with helium, neon, and argon have not been produced. It seems likely that it will prove possible to obtain compounds with at least transient existence in the case of argon. Helium and neon will probably continue to deserve the appelation "noble gas" for the foreseeable future.

This scanning electron microscope located at The University of Chicago has a very high resolving power: about 5 angstroms. The electron source, scanning mechanism, and specimen for study are housed in the central structure. The signals recorded by electron detectors are displayed on an oscilloscope, where the image is obtained. (Courtesy of Prof. A. V. Crewe.)

SEEING IS BELIEVING

So far we have discussed the structures of a number of substances. In Chapters 6 and 7, attention is directed to the structures of many other substances, including simple compounds based on a variety of geometries and compounds of great size and complexity. Thus far, we have said relatively little about how such structures are established experimentally. As indicated in Chapter 2, the earliest workers in this area were forced to deduce molecular structures from indirect evidence: molecular weight, composition, behavior of other compounds based on the same element, number of isomers, optical activity, and the like. Although the use of these methods resulted in very important advances in our understanding of chemistry, their utility is limited. In the first place, such arguments never suffice to *establish* a structure. They only reveal that a particular structure is consistent with all available data. It is always possible that there exist other equally acceptable structures that simply have not occurred to chemists. In the second place, such arguments could never cope with the three-dimensional structures of the complex molecules typical of biological systems. In these cases literally millions of geometries may be possible, and indirect arguments are inadequate to distinguish between them.

More recent workers have had at their disposal many physical methods for the determination of structure. These vary in utility and complexity and in the type of information that they supply. We have seen one such method: electron diffraction. This method yields exceedingly precise information concerning the structures of molecules in the gas phase based on the way in which these molecules scatter electrons that strike them. This method is capable of establishing bond distances within about ± 0.001 Å. Although this method is suitable only for molecules in the gas phase, a related one, X-ray diffraction, which is mentioned in Chapter 7, provides almost as precise information concerning the structure of crystalline compounds. These and other methods available to the chemist, properly employed, have established the correctness of the structures we have discussed thus far and those that we shall encounter later. Nevertheless, there is a certain satisfaction in actually "seeing" something rather than relying on structural information cranked out by a computer into which was fed information about the scattering of electrons or X-rays. In this area too, great advances have been made in recent years.

The human eye is a marvelous instrument for seeing objects of the size that we normally encounter in our day-to-day living (see Chapter 15 for a brief discussion of the chemistry of vision). But at the atomic or molecular level, the eye becomes hopelessly inadequate to the task. The limitation is a fundamental one. For any instrument for seeing, be it the eye, a photographic plate, or more complicated recording devices, the limit of resolution

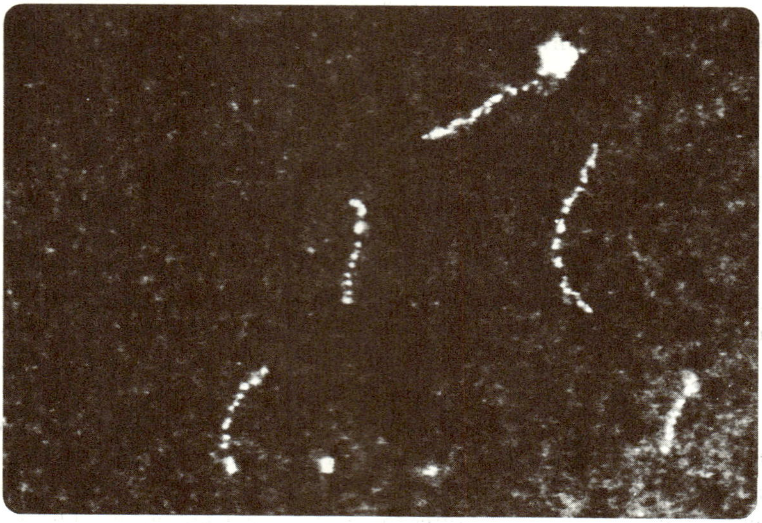

Single atoms and multiples of single atoms of thorium in the polymeric organic salt of 1,2,4,5-benzene tetracarboxylic acid are photographed onto a 25-angstroms-thick evaporated carbon film. The photograph is approximately 2100 angstroms wide and was taken with the high-resolution scanning electron microscope in the laboratory of Prof. A. V. Crewe. (Courtesy of Prof. A. V. Crewe.)

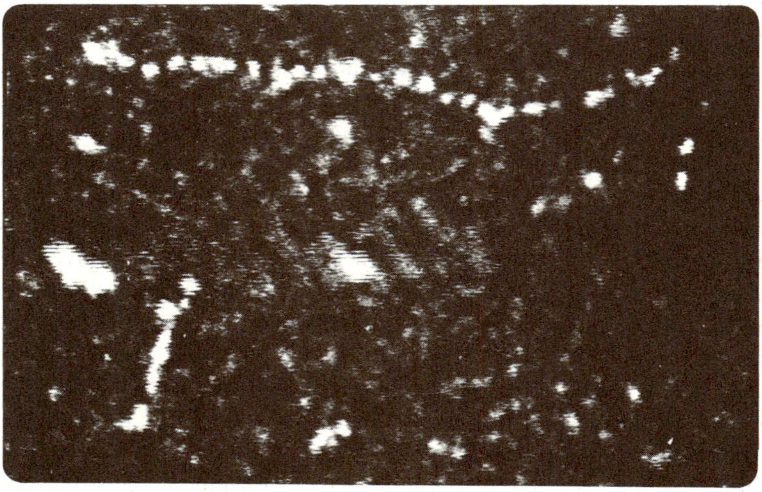

Thorium-benzene tetracarboxylic acid chain that has been sprayed onto a 25-angstroms-thick evaporated carbon film. The small spots along the chain are single atoms of thorium. The photograph is approximately 670 angstroms wide and was taken with the high-resolution scanning electron microscope in the laboratory of Prof. A. V. Crewe. (Courtesy of Prof. A. V. Crewe.)

is established by the wavelength of light employed for observation. The human eye is sensitive to wavelengths of light that vary from about 3500 to 8000 Å. It is, therefore, theoretically impossible to see objects smaller than 3500 Å in their smallest dimension, even employing the best light microscope. Remember that atomic diameters are just a few angstroms and that very large macromolecules usually have a diameter less than 100 Å. Hence, to get down to the molecular level we must use some device to "see" that is sensitive to radiation having very short wavelengths.

One such device is the transmission *electron microscope*. This instrument functions much like a normal light microscope except that electrons having very short wavelengths are employed in place of light. The best such microscopes have a resolving power near 5 Å. This is more than adequate to reveal many features of large molecules; a number of examples are provided in Chapter 7. A further recent advance has been the development of *scanning* electron microscopy. In this case a flying spot of electrons is used to scan an object, and a televisionlike image is produced. Using this instrument, it has been possible to *detect individual atoms* of high molecular weight and to reveal something about the shapes of larger molecules. Several examples of the use of this technique are provided in these pages. Seeing is believing!

Glossary

Alkali metal family a family of elements consisting of lithium, sodium, potassium, rubidium, cesium, and francium.

Alkaline earth family a family of elements consisting of beryllium, magnesium, calcium, strontium, and barium.

Catenation tendency a measure of the capacity of an element to form stable chemical bonds with itself.

Halogen family a family of elements consisting of fluorine, chlorine, bromine, iodine, and astatine.

Interhalogen compounds those compounds formed by combination of one halogen with another in any of several possible ratios.

Nitrogen fixation the reduction of molecular nitrogen to ammonia that is carried out by a variety of living organisms.

Noble gas family a family of particularly unreactive elements including helium, neon, argon, krypton, xenon, and radon.

Periodic table an arrangement of the elements that indicates which of them fall into each class (see Figure 3-1).

Periodicity this term refers to the fact that the elements fall into several distinct classes in terms of their physical and chemical behavior.

Semiconductors (metalloids) materials that are neither clearly metals nor nonmetals but that have intermediate properties, especially that of electrical conductivity.

Silanes the silicon hydrides.
Silicon oxides this class includes a large number of species derived from the combination of silicon and oxygen, forming the basis for the mineral world.
Sulfanes catenated sulfur hydrides, H—S—S—S ··· S—H.

Study problems

Periodic properties

1. Describe how electrical resistivity of the elements changes as we move across the periodic table from left to right.
2. Below are listed a few properties of elements. In each case, predict whether or not the property would be expected to change in a uniform and consistent way as we move down a column in the periodic table. Defend carefully your responses.
   a. atomic radius
   b. volume occupied by 1 g-atom
   c. boiling point
   d. melting point
   e. density
   f. electrical conductivity
3. As we move down a column of the periodic table, does the chemistry of the elements successively encountered become simpler or more complex? Or can any reliable prediction be made?
4. Would you expect the physical and chemical properties of $CaCl_2$, $SrBr_2$, and BaF_2 to be similar or quite distinct? Why?

Quantitative problems

5. Calculate the weight percentage of hydrogen in the hydrides of the Group IV elements: methane, silane, germane, stannane, and plumbane.
6. Treatment of xenon with elemental fluorine yields a compound that was found to contain 63 percent xenon and 37 percent fluorine by weight. What compositions for this xenon fluoride are consistent with this analysis?
7. Treatment of radon with elemental fluorine yields a compound containing only one radon atom and having a molecular weight of 260. What is the formula for this radon fluoride?
8. Plants contain an enormous variety of complex substances containing nitrogen that are collectively called *alkaloids*. Many of these compounds have interesting physiological properties. One of the most important is *reserpine*, widely used in the treatment of mental disturbances. The weight percentage composition of reserpine is 66 percent carbon, 6.6 percent hydrogen, 6.2 percent nitrogen, and 21.2 percent oxygen. What is the simplest elemental composition that is consistent with these data?
9. Space ships are powered by a variety of mixtures of fuel and oxidizing agents. The reaction between these liberates the energy necessary for propulsion of the vehicle. The fuel used to propel the Apollo 17 lunar module consisted of a mixture of hydrazine and dimethylhydrazine, $(CH_3)_2NNH_2$; the oxidizing agent was nitrogen tetroxide. The reactions involved are

$$2N_2H_4 + N_2O_4 \longrightarrow 3N_2 + 4H_2O$$
$$(CH_3)_2NNH_2 + 2N_2O_4 \longrightarrow 2CO_2 + 3N_2 + 4H_2O$$

Calculate the quantity of N_2O_4 required to oxidize completely a mixture consisting of 1 mole each of hydrazine and dimethylhydrazine. How much water would be produced in the process?

Symmetry and isomer problems

10. Define the elements of symmetry present in F_2O_2, SiO_4^{2-}, $NHCl_2$, PH_3, and S_8.
11. Xenon tetrafluoride has the following symmetry elements: five mirror planes of symmetry, one fourfold axis of rotation, four twofold axes of rotation (not counting the fourfold axis), and a center of symmetry. What is the geometry of this molecule?
12. The molecules considered in Chapters 2 and 3 have tetrahedral geometries. Can you think of any reasons why this geometry occurs so frequently in chemistry?

Suggestions for further reading

I. Asimov *The World of Nitrogen,* Collier, New York, 1962.

B. Bertman and R. A. Guyer "Solid Helium," *Sci. Am.,* **217,** 84 (August, 1967).

C. C. Delwiche "The Nitrogen Cycle," *Sci. Am.,* **223,** 136 (September, 1970).

J. T. Hall and D. A. Keyworth *Brief Chemistry of the Elements,* Benjamin, Menlo Park, Calif., 1971.

R. C. Johnson *Introductory Descriptive Chemistry,* Benjamin, Menlo Park, Calif., 1966.

W. L. Jolly *The Chemistry of Non-Metals,* Prentice-Hall, Englewood Cliffs, N.J., 1966.

H. Nechamkin *The Chemistry of the Elements,* McGraw-Hill, New York, 1968.

G. L. Pollack "Solid Noble Gases," *Sci. Am.,* **215,** 64 (October, 1966).

C. J. Pratt "Sulfur," *Sci. Am.,* **222,** 62 (May, 1970).

R. T. Sanderson *Chemical Periodicity,* Van Nostrand Reinhold, New York, 1960.

4

The structure of the atom

An early alternative to electronic structure as the basis for chemical bonding: The [atoms] move in the void and catching each other up jostle together, and some recoil in any direction that may chance, and others become entangled with one another in various degrees according to the symmetry of their shapes and sizes and positions and order and they remain together and thus the coming into being of complete things is effected.

Simplicius on Leucippus
in H. Diels and W. Kranz, *Die Fragmente der Vorsokratiker*

In the previous two chapters, we have discussed the structures of quite a number of molecules in terms of the spatial disposition of the constituent atoms with respect to one another. In so doing, we have made use of one of the most generally useful concepts in all of chemistry: "—" (the bond). A single straight line is usually drawn between atoms as a symbol that they are linked by a chemical bond. Without worrying about the nature of this bond, or the nature of the atoms themselves, we have been able to give a rather satisfactory description of the geometry of many molecules. More examples follow in Chapters 6 and 7. Indeed, until well into the twentieth century, chemistry was almost wholly concerned with exactly the same type of description that we have studied in Chapters 2 and 3. But this leaves two fundamental questions unanswered: What is a chemical bond? Why does each type of atom form a definite number of bonds with a specific geometry? In this chapter and the next, we search for answers to these questions, first by considering the electronic structure of the atom and then by considering the interactions between atoms.

SOME EARLY THOUGHTS ABOUT ATOMS

Since chemistry is largely concerned with the ways in which molecules are held together by chemical bonds between atoms and with the making and breaking of such bonds, the theory of the atom is the cornerstone of chemistry. The better the structure and properties of the atom are understood, the better the understanding of molecular structure and chemical reactions. Consequently, it is not surprising that a great many chemists have been and are vitally concerned with atomic theory. If we understood this matter in sufficient detail, chemistry would take on more of the character of physics, a science in which theory is relatively more important than experiment in comparison to chemistry. But we do want to remember that in both physics and chemistry, as in other sciences, experimental results are the final arbiter of truth.

Because of the crucial importance of atomic theory for chemistry, it is not surprising that important milestones in this area are regarded as great

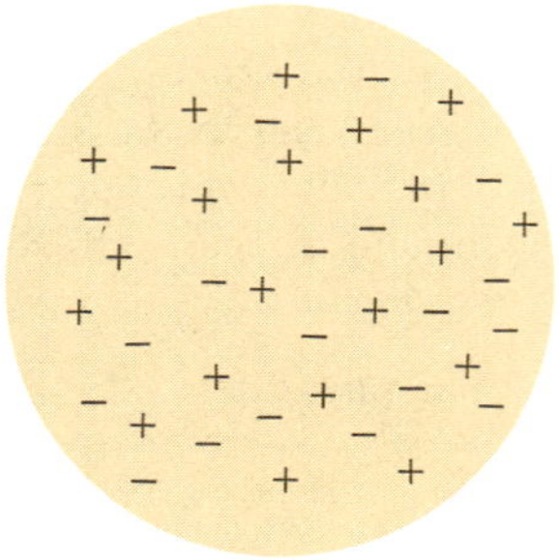

Figure 4-1 The 1910 model of an atom. The total number of positive and negative charges are equal, but the positive charge is spread uniformly whereas the negative charge is concentrated at the sites of electrons.

contributions to chemistry in particular and science generally. The history of atomic theory is long and complex, having its origins long before the birth of Christ and continuing to this day. We shall neither recount this history nor describe many of the crucial experimental observations at this time. Much of this information is supplied throughout the remainder of this book. For the moment, we shall move almost directly to a qualitative description of the atom as it is currently understood. This will provide us with the necessary basis for the description of chemical bonding that follows in Chapter 5.

The basis for descriptive chemistry was made with advancement of the proposal of John Dalton and the modifications and extensions that followed during the latter half of the nineteenth century. By 1910, it had been recognized that atoms were not indivisible units but were themselves formed from more elementary particles. At that time, the atom was thought to be a uniform, positively charged sphere, approximately 10^{-8} cm in diameter, with electrons embedded in the sphere to balance the positive charges. A schematic model is provided in Figure 4-1. A beautiful experiment, described in Chapter 16, carried out near the end of the nineteenth century by Ernest Rutherford clearly established that the atom was actually composed of negatively charged electrons moving about a small, but very dense, nucleus. (The chemistry of the nucleus itself is also considered in Chapter 16.) This model of the atom was refined in an important way by the Danish physicist Niels Bohr in 1913, when he incorporated the first ideas of the modern quantum theory of the atom into the planetary model of Rutherford (the Bohr atom is discussed at the end of Chapter 14). Important as this theory was, it fell almost as soon as it was proposed under the weight of new understanding derived from the development of **quantum mechanics** by Schrödinger and Heisenberg. This theory has provided the basis for the modern description of the atom and of chemical bonding. Perhaps quantum mechanics may one day be overthrown by yet a more satisfactory theory, but it is the best theory that we have at the moment. It is to the principal conclusions of this theory for the understanding of atoms that we now turn our attention. Related considerations deriving from quantum mechanics are treated in Chapter 14.

ATOMIC DESCRIPTION

The present description of atoms differs from the planetary model principally in one respect: We do not now believe that it is possible to follow the motion of a single electron within an atom as we are able to follow the motion of the earth about the sun. It appears meaningless to attempt to describe the pathway or orbit followed by an electron (see also Chapter 14). Rather, there are certain regions in space, called **orbitals,** occupied by electrons. According to quantum theory, it is not possible to know where an electron is in an orbital at any instant or how electrons move from one orbital to another. Atoms may be compared to a modern apartment house of rather free-form design with a central lobby (the nucleus) around which apartments (orbitals) are distributed. Without detailed study, it would be difficult for the casual passerby to say which apartments are occupied and which are not. Our first concern, then, is to describe the shape and size of the "apartments" available to the electrons and subsequently to see if there are simple rules to tell us which are occupied. It is most important to recognize that the "apartments" exist whether or not sufficient electrons can be found to "lease" them all.

s orbitals

Although a large number of shapes of orbitals exist, three basic shapes suffice to describe the most common and interesting ones. The simplest and most symmetrical orbitals available to electrons are called s orbitals and are spherically disposed about the nucleus as depicted in Figure 4-2. Numerous s orbitals are available within the atomic framework (seven are commonly used), but all, fortunately, have identical symmetry properties and differ principally in their size. Figure 4-2 also shows, in appropriate sizes, the first three orbitals of s type. The orbitals are numbered in order of increasing size and the number, usually referred to as ***n*,** is called the **principal quantum number.** This designation comes from the detailed theory of the structure of matter (the quantum theory) and its fundamental significance can concern us much later. For the moment it will suffice to say that each orbital is described by a number n, which is related to the orbital size.

Now that we know that the s orbitals are spherically symmetrical, let

This photograph shows a portion of the interior of a large and powerful particle accelerator. This instrument and related ones have been of great value in understanding details of atomic and nuclear structure as well as the nature of transformation of the elementary particles from which atoms are constructed. (Courtesy of Lawrence Radiation Laboratory.)

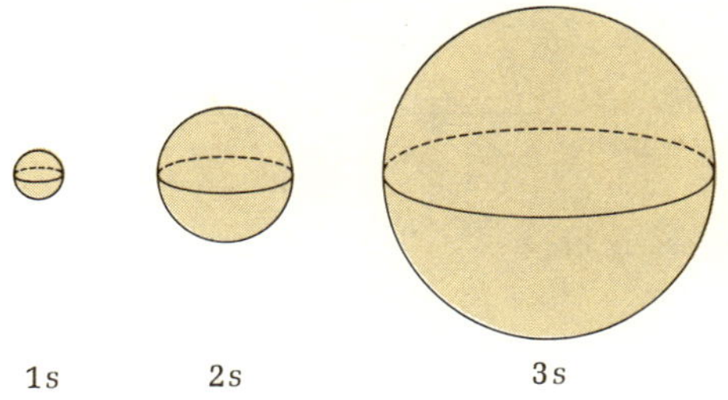

Figure 4-2 Shapes and sizes of 1s, 2s, and 3s orbitals.

us inquire more closely into where within an s orbital the electrons are to be found. Since modern atomic theory forbids us to ask how electrons move from one location in space to another, we can only ask: *Where in space are we most likely to find an electron at any given time?* This question can be rephrased in the following way: If it were possible to take stop-motion photography of electrons within each orbital and count the number of times an electron is found in a particular position, what statements could we eventually make about the probability of the electron being found at a specific spot? This would be analogous to describing a use pattern for an apartment: During the day the bedroom is occupied 35 percent of the time, the kitchen 10 percent, and so on. At any particular instant, the inquirer outside the apartment cannot go beyond a statistical answer as to the present location of the inhabitants. Similarly, we can only tell you what percentage of the time an electron spends in a particular region of space. We never know when it is actually there.

Fortunately, since s orbitals spherically surround the nucleus, the probability of finding an electron depends only on the distance from the nucleus and not on the direction. We may need to convince ourselves that this follows from our discussions of the symmetry of various shapes as applied to a spherical object. Any line passing through the center of the sphere is an infinite axis of rotation, and any plane through the center is a mirror plane. If we can specify the probability of finding the electron as a function of distance from the nucleus, no further information is necessary. If an electron is in a particular orbital, the probability of finding it somewhere in that orbital is unity (it *must* be there somewhere). We might expect considerable variation in probability as distance varies from the nucleus, but the sum of all probabilities must be unity. There might very well be (and indeed are) regions where the probability of finding the electron is low, or where it is high. But the description of orbitals in terms of probabilities of finding an electron as a function of position in space is just a

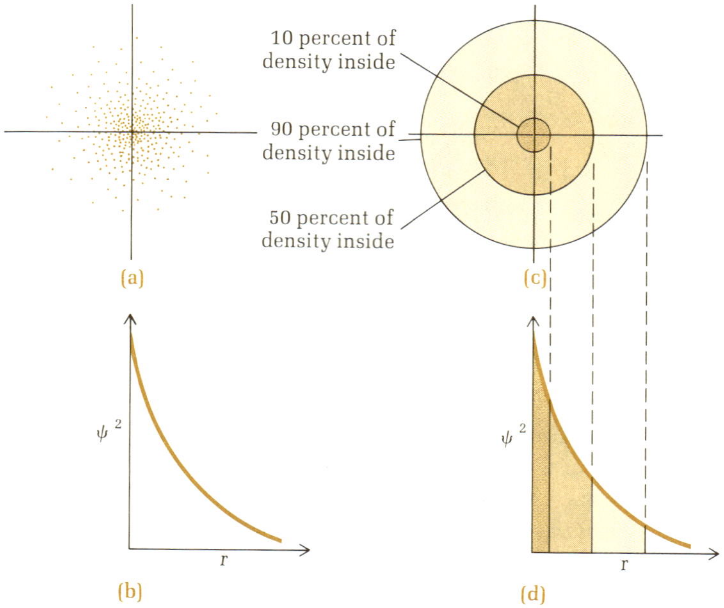

Figure 4-3 Four ways of representing the probability distribution for an electron in a 1s oribtal. In (a), electron density as a function of position in space is shown by the density of dots; equivalent information is contained in (b), in which the probability function ψ^2 is plotted as a function of distance from the center of the atom, *r*. In (c), contour lines have been drawn about the atomic center so as to enclose specified amounts of the total electron probability in the smallest possible volumes; this information is translated into a ψ^2 versus *r* plot in (d). (From G. C. Pimentel and R. D. Spratley, *Understanding Chemistry*, Holden-Day, San Francisco, 1971.)

little more complicated than it might seem to be at first glance. Let us return to the simplest of orbitals, the 1s case, for a moment and see the various ways in which this can be done.

To begin with, if we study quantum mechanics we discover that its theory describes electrons in terms of a mathematical quantity called the **wave function,** which is designated by ψ. Wave functions are almost always very complicated mathematical functions, but their exact description need not concern us. The important point is that the square of the wave function, ψ^2, measures the probability of finding an electron within a given volume. What does ψ^2 look like for an electron in a 1s orbital? We can use several pictures to try to answer this question. One quantity of interest is the *probability density distribution,* that is, how ψ^2 varies with distance from the center of the atom. Suppose that we take a planar cut through the nucleus of an atom and represent the electron density for a 1s orbital in terms of a density of dots: The greater the density of dots, the greater the value of ψ^2, the probability function. For this case, the picture in Figure 4-3(a) would be generated. Note that the electron density is greatest near the center of the atom and decreases with increasing distance from the center. This same information can also be provided as a plot of ψ^2 versus r, the distance from the atomic nucleus; this is shown in Figure 4-3(b). Perhaps a more generally useful picture conveying the same information is to draw contour lines about the center enclosing certain percentages of the probability within the smallest possible volume. Such a picture and its translation into a ψ^2 versus r plot are provided as Figures 4-3(c) and 4-3(d). It is precisely this type of representation that has already been used in Figure 4-2, where the lines used to indicate orbital size have been drawn to include an arbitrary fraction of the electron probability. Thus, each line encloses, for example, 90 percent of the probability. A crucial point that follows from this is that *the atom has no boundaries;* for an electron in a 1s orbital, the value of ψ^2 is everywhere finite, even at great distances from the nucleus.

There is another way entirely to show distribution of electron density as a function of position. To understand this, we need to realize that the actual probability of finding an electron within a given volume element at some distance from the nucleus is given by the product of the electron *density* and the *size* of the volume element. As shown in Figure 4-3, the size of the volume element increases with increasing distance from the nucleus; quantitatively, it increases with r^2. Hence, a plot of $r^2\psi^2$ versus r yields a picture of probability (not density) as a function of distance from the nucleus. Figure 4-4 shows such plots for an electron in 1s, 2s, and 3s orbitals of the hydrogen atom. Observe that for the 1s orbital the probability of finding an electron very near the nucleus is low, even though the electron density is maximal there, but rises rapidly farther from the nucleus. The probability curve reaches a maximum and then falls sharply once more. The probability of finding the electron within 2 Å of the nucleus is very high in a 1s orbital, although the electron may well be farther away at times. The most probable distance (the maximum in the probability curve) from the nucleus for a 1s electron is 0.53 Å (it is interesting to note that this is exactly the radius of the planetary orbit required by the older theory).

The probability distributions for 2s and 3s orbitals are more complex. Note that for the 2s orbital, the probability distribution has two maxima, and for the 3s, three maxima occur. For the 2s orbital, the most probable place to find an electron would be about 2.5 Å from the nucleus (once more with no preference for a particular direction). There is a significant probability of finding the electron only 0.5 Å away, but practically none at all

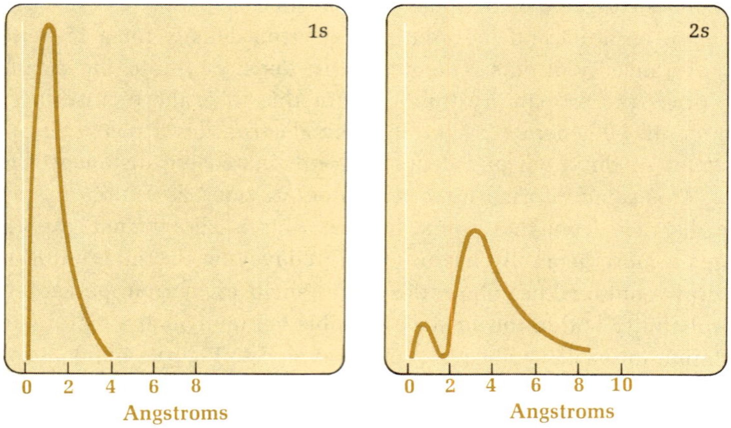

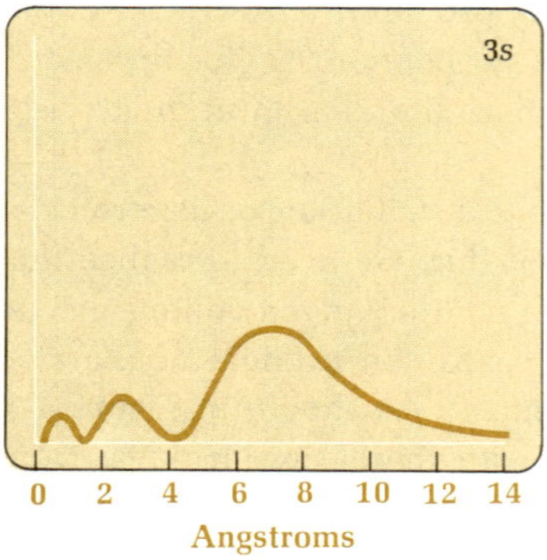

Figure 4-4 Probability distribution of electrons in hydrogen 1*s*, 2*s*, and 3*s* orbitals.

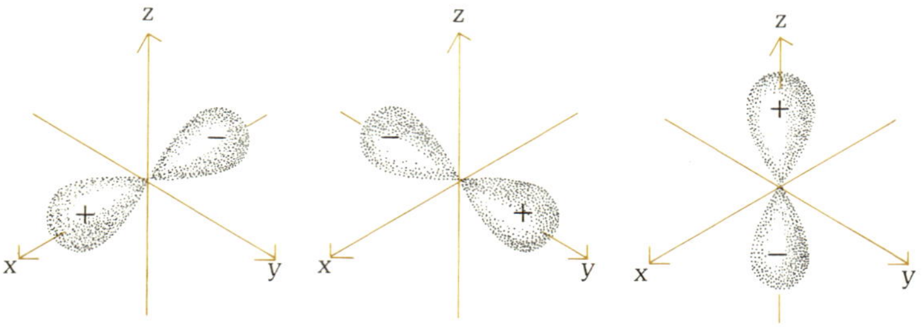

Figure 4-5 The three *p* orbitals.

of finding it 1.2 Å away. There are regions in space that the electron totally avoids, and in this case a spherical surface results called a **nodal surface.** It is indeed a curious "apartment" in which we might find a person in the living room or bedroom but never between: The futility of inquiring into the pathway followed by an electron is clearly demonstrated in this instance if the electron can be in one position or another but not in an intervening spot. It is worth noting that the number of nodal surfaces is always one less than the principal quantum number *n*.

In Figure 4-4, the probability curves appear to fall to zero a few angstroms from the nucleus, but this is a result only of the inadequacy of the drawing. The probabilities in fact approach zero asymptotically. The nonvanishing probability of finding electrons at large distances from the nucleus makes it impossible to draw an absolutely accurate representation of the distribution of an electron within an orbital. Pictures such as those in Figure 4-2 can be drawn only if we agree on some common convention for describing the limits of the atom. It is convenient to draw a boundary surface such that the electron will most likely be found within the bounds 90 percent of the time.

p orbitals

The second type of orbital is called a *p* orbital. Figure 4-5 shows the general shape associated with *p* orbitals. In contrast to s orbitals, the characteristic most immediately obvious for *p* orbitals is that the direction from the nucleus is important. In fact, there are three mutually perpendicular directions from the nucleus where the probability of finding a *p* electron is high. It is convenient to draw arbitrary x, y, and z axes through the centers and refer to these *p* orbitals as p_x, p_y, and p_z. As with s orbitals, there are many *p* orbitals in each atom. Each has the shape shown, and the size is again related to the principal quantum number (always given first). We can perhaps describe this most vividly by showing probability graphs for *p*-type orbitals. Figure 4-6 shows the distribution in 2*p* and 3*p* orbitals (note that no orbital of 1*p* type exists; this follows from the detailed mathematical theory, and we provide a qualitative rationalization below.)

d orbitals

Finally, let us describe orbitals of the type called *d*. As with *p* orbitals, it is necessary to discuss both size and shape. Again, 3*d*, 4*d*, 5*d*, . . . orbitals exist, and the shapes of all *d* orbitals are similar; there are only differences in size and probability distributions of electrons within the orbital. Figure 4-7 shows the shapes commonly assigned to the five different *d* orbitals.

Description of the shapes of *d* orbitals is arbitrary and can be confusing

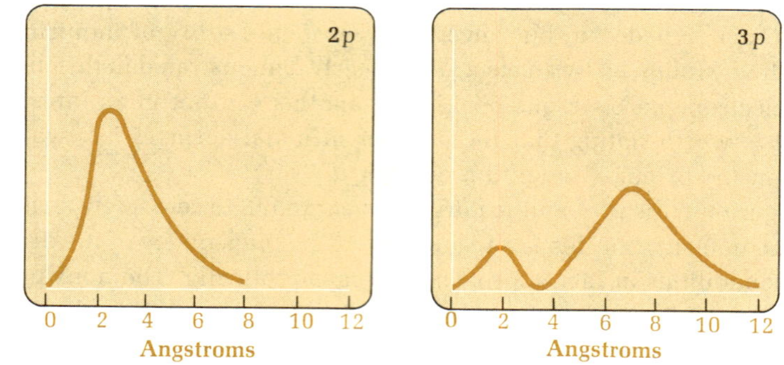

Figure 4-6 Probability distribution of electrons in 2*p* and 3*p* orbitals.

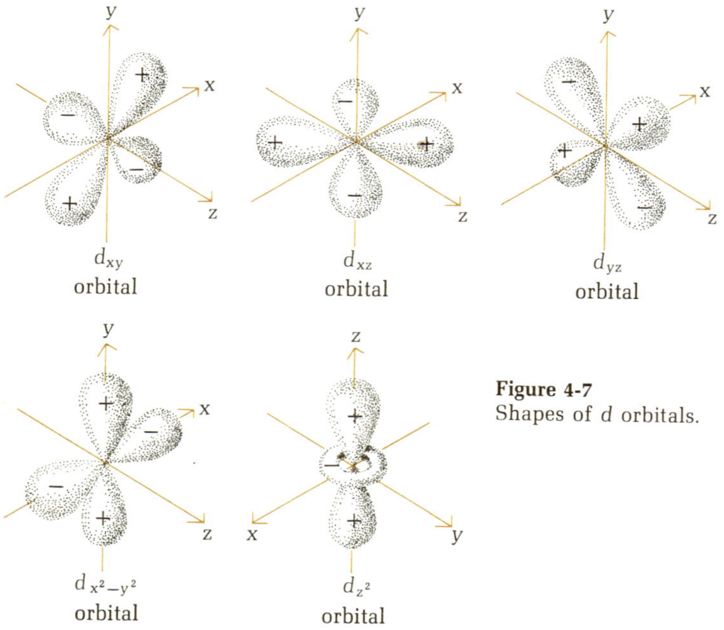

Figure 4-7
Shapes of *d* orbitals.

to the student. The problem is simply this: There are only five *d* orbitals, but objects in three dimensions are commonly referred to three mutually perpendicular axes. The *d* orbitals are basically shaped like four-leaf clovers, and it would be simplest to describe six of them. The first four shapes of Figure 4-7 are relatively simple to describe and remember. One *d* orbital has the xy plane as a plane of symmetry, and the regions of high probability lie along the two diagonals bisecting the axes. This orbital is referred to as the d_{xy} orbital, since the xy plane contains the four maxima. The d_{xz} orbital has the xz plane as the plane of symmetry, and the d_{yz} orbital is similar. The orbital described for reasons that need not concern us as $d_{x^2-y^2}$ also has the xy plane of symmetry, but regions of high probability lie along the axes rather than along the diagonals between.

Clearly, it should be possible to draw two more orbitals like the $d_{x^2-y^2}$ orbital but lying in the yz and xz planes, respectively. We now have a set of six orbitals that can be described d_{xy}, d_{yz}, d_{xz}, $d_{x^2-y^2}$, $d_{x^2-z^2}$, $d_{y^2-z^2}$. Unfortunately, the atomic design allows for only five, but at the same time insists that the same symmetry be kept as though all six were possible. With these restrictions, we adopt a solution agreed to by convention and combine the last two of the set of six to form a single orbital. Note that each of those to be combined has a z component; the combined single orbital, called d_{z^2}, shows this in its shape. Regions of high probability lie along the z axis, with a second doughnut-shaped region of high probability surrounding the z axis in the xy plane. Note the signs of the mathematical descriptions of the *d* orbitals as indicated in each figure.

For the first *d* orbital (3*d*), the probability distribution has a single maximum near 5 Å and two nodal planes. Higher *d* orbitals have interior regions of zero probability, as do the higher orbitals of the *s* and *p* type. In summary, observe that the overall size of any orbital is related to its principal quantum number. Moreover, the number of nodal surfaces is always one less than the principal quantum number regardless of the orbital type (and thus shape).

It is important to recognize that the orbital framework described above exists for every atom and is independent of the number of electrons present in that atom. Even for hydrogen, with only a single electron to house, there exist an infinite number of orbitals of *s*, *p*, *d*, and more complex shapes. In fact, if energy is added to the hydrogen atom (by passing a spark through it, for example) the single electron may move from one orbital to another (see Chapter 15). The "apartment" structure is very similar for all atoms, and this greatly simplifies the problem of describing the structure of any particular atom. We need decide only two things: (1) how many electrons are present in a particular atom of interest, and (2) what rules govern the order in which various orbitals are occupied.

The number of electrons present in any atom is immediately available

Table 4-1 The quantum numbers

Name	Symbol	Allowed values	Comments
Principal quantum number	n	1, 2, 3, 4, . . .	Largely determines the energy of the orbital and its size
Azimuthal quantum number	l	0, 1, . . . , $(n - 1)$	Determines shape and symmetry properties of the orbital
Magnetic quantum number	m	$-l, -l + 1, \ldots, 0, \ldots, l - 1, l$	Determines the orientation of the orbital in space
Spin quantum number	s	$\frac{1}{2}, -\frac{1}{2}$	Determines the spin of the electron in a given orbital

in a periodic table (Figure 3-1). The atomic number shown for each element corresponds exactly to the positive charge on the nucleus of the atom. In an electrically neutral atom, exactly that same number of electrons must be present. If the atoms gain or lose electrons to become charged ions, the number of charges carried by the ion shows the number of electrons that must be added to or subtracted from the atomic number. Remembering how many electrons are associated with any given type of atom is of no more importance than remembering the telephone numbers of all one's own friends. We tend to remember the most frequently used phone numbers, however, and we shall probably soon learn the number of electrons in the most common atoms.

ELECTRON SPIN

Each orbital may be occupied by two electrons, just as most apartments are occupied by husband and wife. The two electrons occupying a particular orbital must be of opposite types. The difference between electrons resides in a property we refer to as *spin*. The negatively charged and extremely small electrons behave as though they were spinning on an axis. [It should be clear that the description given here is based on the picture of an electron as a particle, not as a type of very short wavelength "light" (see Chapter 14).] Since a moving electrical charge (even spinning) generates a magnetic field, the difference in spin between electrons may be detected by taking advantage of this fact. If a beam of hydrogen atoms, each with a single electron, is passed through a magnetic field, it is split into two components. Equal numbers of electrons appear to be spinning in opposite directions. It is not difficult to flip the direction of spin of a particular electron. The only important rule to us at the moment is that *when two electrons occupy a single orbital, they must have opposite spin.*

THE COMPLETE SET OF QUANTUM NUMBERS

If we think back over the material just covered, we can conclude that four pieces of information would suffice to characterize any electron occupying an atomic orbital. First, we must say something about the size of the orbital by specifying the principal quantum number. Second, the shape and symmetry properties of the orbital must be given by indicating whether the orbital is *s*, *p*, *d*, and so on. Third, the orientation of the orbital in space must be defined (i.e., p_x, p_y, etc.) and, finally, the spin of the electron must be specified. In fact, all of this information is contained in a completely specified wave function for the electron and appears as four **quantum numbers,** of which the principal quantum number is one. Since these numbers may aid in remembering, for example, the number of each type of orbital, we describe them in Table 4-1.

Note carefully how these quantum numbers relate to what we have already developed. Thus, for principal quantum number 1, the rules collected in Table 4-1, tell us that only orbitals with an **azimuthal quantum number l** = zero are allowed. Thus, a value of $l = 0$ must correspond to an *s* orbital. No *p* orbitals having principal quantum number 1 are possible. For $n = 2$, orbitals having $l = 0$ and $l = 1$ are allowed. These correspond to *s* and *p* orbitals, respectively. For the *p* orbitals, $l = 1$, we note that the allowed values of the **magnetic quantum number m** are -1, 0, and 1. These must correspond then to p_x, p_y, and p_z orbitals. Further analysis of this type will generate all the types of *d* orbitals, $l = 2$. For a principal quantum number greater than 3, these rules tell us that yet other kinds of orbitals exist which we have not yet described and, indeed, have no need to describe. Finally, there is the **spin quantum number s,** which takes on either the value $+\frac{1}{2}$ or $-\frac{1}{2}$. This quantum number determines the spin of the electron.

ORDER OF ORBITAL FILLING

We must next consider the preferences that electrons have for various orbitals. Fortunately, the atoms long ago decided on the relative desirability of the various apartments and never change their minds about this subject. In any apartment house, we might expect the first inhabitant to occupy the best apartment even though all of the others are empty. If the 1*s* orbital is vacant, it is always the orbital preferred by the first occupant. When a second electron becomes available, it finds it most convenient to become

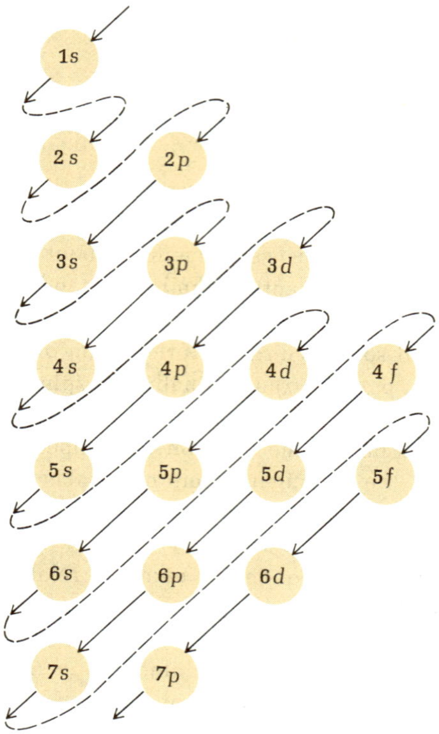

Figure 4-8 Order of filling orbitals. (From T. Moeller, *Inorganic Chemistry, An Advanced Textbook*, Wiley, New York, 1952.)

of opposite spin and occupy the 1s orbital jointly with the first inhabitant. This orbital is now fully occupied, and if a third electron wishes to join the atom it must make do with the next most desirable apartment.

The factor determining desirability of orbitals is not a difficult one to understand. Bear in mind that the nucleus of the atom is positively charged and that each electron bears a negative charge. Elementary physics tells us of the attraction between opposite charges, and we might expect that the most favorable orbital would be the one that maximizes the possibility for interaction between the nucleus and the electron. When this orbital is occupied, the next most electrostatically favorable orbital is chosen, and so forth.

We can build up the structure of any particular atom by sequentially filling the most favorable orbitals until we have exhausted the supply of available electrons. The regular order with which orbitals are filled leads to close similarity in properties between atoms of similar atomic number. As the atomic number increases, each atom has essentially the electronic structure of the atom immediately preceding it with only the additional problem of finding a home for one more electron. The order in which orbitals are filled is as follows: 1s, 2s, 2*p*, 3s, 3*p*, 4s, 3*d*, 4*p*, ... It is relatively difficult to remember a string of numbers such as that given above, and perhaps the diagram shown in Figure 4-8 will help you to derive the sequence of orbitals when you need it. Note that the diagram contains a type of orbital not discussed, the *f* type. We shall not describe this orbital further than to say that the shapes of *f* orbitals are more complex than those of *d* orbitals and that there are seven of them compared to the five shapes of *d* orbitals, three shapes of *p* orbitals, and only one type of s orbital. The first *f* orbitals available are the 4*f* type; altogether 14 electrons can be held in these seven orbitals. The order of filling orbitals may be obtained from the diagram in simple fashion: We start at the top right and move diagonally down to the left. When we have passed through the entire array of orbitals, we return once more to the top right and repeat the left and down motion. We could continue the list with the eighth, ninth, and higher levels, but these will not prove useful to us in our present study of chemistry.

To see how the above ideas may be applied, let us describe the electronic structure of the silicon atom. Silicon has atomic number 14, and consequently 14 electrons must be placed in orbitals about a single silicon nucleus. The 1s orbital is the most favorable available in any atom and 2 electrons can be placed in it, leaving 12 electrons to be housed elsewhere. The next most favorable orbital is the 2s orbital, and it also can account for 2 electrons, leaving 10 not accounted for. The 2*p* orbitals are the next most favorable and each of the three can hold 2 electrons; 2 electrons go into the p_x orbital, 2 into the p_y, and 2 into the p_z; there remain 4 electrons to be accounted for. The 3s orbital can hold 2, and the last 2 gather into 3*p* orbitals. Electrons by nature prefer to occupy orbitals singly if they can. The reason for this preference can be readily understood: The electrons in the different *p* orbitals are farther apart than they would be in a single orbital. Thus, the repulsion between the negatively charged electrons is minimized. Since the 3*p* orbitals are identical except with respect to their direction along the coordinate axes, 1 electron goes into each of two orbitals, say the $3p_x$ and the $3p_y$. All 14 electrons have now been accounted for. It is convenient to describe this structure explicitly in the following way: $1s^2 2s^2 2p^6 3s^2 3p^2$. Table 4-2 collects the electronic structures of all 103 elements known at the present time. Note that we have abbreviated the descriptions somewhat, since the inner structure of each atom is similar to those preceding it in atomic number. Since the elements at the extreme right of the periodic table form relatively few compounds and are similar

Table 4-2 The electronic configuration of atoms

Atomic number	Symbol	Electronic structure	Atomic number	Symbol	Electronic structure
1	H	$1s^1$	47	Ag	(Kr)$5s^1 4d^{10}$
2	He	$1s^2$	48	Cd	(Kr)$5s^2 4d^{10}$
3	Li	(He)$2s^1$	49	In	(Kr)$5s^2 4d^{10} 5p^1$
4	Be	(He)$2s^2$	50	Sn	(Kr)$5s^2 4d^{10} 5p^2$
5	B	(He)$2s^2 2p^1$	51	Sb	(Kr)$5s^2 4d^{10} 5p^3$
6	C	(He)$2s^2 2p^2$	52	Te	(Kr)$5s^2 4d^{10} 5p^4$
7	N	(He)$2s^2 2p^3$	53	I	(Kr)$5s^2 4d^{10} 5p^5$
8	O	(He)$2s^2 2p^4$	54	Xe	(Kr)$5s^2 4d^{10} 5p^6$
9	F	(He)$2s^2 2p^5$	55	Cs	(Xe)$6s^1$
10	Ne	(He)$2s^2 2p^6$	56	Ba	(Xe)$6s^2$
11	Na	(Ne)$3s^1$	57	La	(Xe)$6s^2 5d^1$
12	Mg	(Ne)$3s^2$	58	Ce	(Xe)$6s^2 4f^2$
13	Al	(Ne)$3s^2 3p^1$	59	Pr	(Xe)$6s^2 4f^3$
14	Si	(Ne)$3s^2 3p^2$	60	Nd	(Xe)$6s^2 4f^4$
15	P	(Ne)$3s^2 3p^3$	61	Pm	(Xe)$6s^2 4f^5$
16	S	(Ne)$3s^2 3p^4$	62	Sm	(Xe)$6s^2 4f^6$
17	Cl	(Ne)$3s^2 3p^5$	63	Eu	(Xe)$6s^2 4f^7$
18	Ar	(Ne)$3s^2 3p^6$	64	Gd	(Xe)$6s^2 4f^2$
19	K	(Ar)$4s^1$	65	Tb	(Xe)$6s^2 4f^9$
20	Ca	(Ar)$4s^2$	66	Dy	(Xe)$6s^2 4f^{10}$
21	Sc	(Ar)$4s^2 3d^1$	67	Ho	(Xe)$6s^2 4f^{11}$
22	Ti	(Ar)$4s^2 3d^2$	68	Er	(Xe)$6s^2 4f^{12}$
23	V	(Ar)$4s^2 3d^3$	69	Tm	(Xe)$6s^2 4f^{13}$
24	Cr	(Ar)$4s^1 3d^5$	70	Yb	(Xe)$6s^2 4f^{14}$
25	Mn	(Ar)$4s^2 3d^5$	71	Lu	(Xe)$6s^2 4f^{14} 5d^1$
26	Fe	(Ar)$4s^2 3d^6$	72	Hf	(Xe)$6s^2 4f^{14} 5d^2$
27	Co	(Ar)$4s^2 3d^7$	73	Ta	(Xe)$6s^2 4f^{14} 5d^3$
28	Ni	(Ar)$4s^2 3d^8$	74	W	(Xe)$6s^2 4f^{14} 5d^4$
29	Cu	(Ar)$4s^1 3d^{10}$	75	Re	(Xe)$6s^2 4f^{14} 5d^5$
30	Zn	(Ar)$4s^2 3d^{10}$	76	Os	(Xe)$6s^2 4f^{14} 5d^6$
31	Ga	(Ar)$4s^2 3d^{10} 4p^1$	77	Ir	(Xe)$6s^2 4f^{14} 5d^7$
32	Ge	(Ar)$4s^2 3d^{10} 4p^2$	78	Pt	(Xe)$6s^1 4f^{14} 5d^9$
33	As	(Ar)$4s^2 3d^{10} 4p^3$	79	Au	(Xe)$6s^1 4f^{14} 5d^{10}$
34	Se	(Ar)$4s^2 3d^{10} 4p^4$	80	Hg	(Xe)$6s^2 4f^{14} 5d^{10}$
35	Br	(Ar)$4s^2 3d^{10} 4p^5$	81	Tl	(Xe)$6s^2 4f^{14} 5d^{10} 6p^1$
36	Kr	(Ar)$4s^2 3d^{10} 4p^6$	82	Pb	(Xe)$6s^2 4f^{14} 5d^{10} 6p^2$
37	Rb	(Kr)$5s^1$	83	Bi	(Xe)$6s^2 4f^{14} 5d^{10} 6p^3$
38	Sr	(Kr)$5s^2$	84	Po	(Xe)$6s^2 4f^{14} 5d^{10} 6p^4$
39	Y	(Kr)$5s^2 4d^1$	85	At	(Xe)$6s^2 4f^{14} 5d^{10} 6p^5$
40	Zr	(Kr)$5s^2 4d^2$	86	Rn	(Xe)$6s^2 4f^{14} 5d^{10} 6p^6$
41	Nb	(Kr)$5s^1 4d^4$	87	Fr	(Rn)$7s^1$
42	Mo	(Kr)$5s^1 4d^5$	88	Ra	(Rn)$7s^2$
43	Tc	(Kr)$5s^1 4d^6$	89	Ac	(Rn)$7s^2 6d^1$
44	Ru	(Kr)$5s^1 4d^7$	90	Th	(Rn)$7s^2 6d^2$
45	Rh	(Kr)$5s^1 4d^8$	91	Pa	(Rn)$7s^2 5f^2 6d^1$
46	Pd	(Kr)$4d^{10}$	92	U	(Rn)$7s^2 5f^3 6d^1$

Table 4-2 *(Continued)*

Atomic number	Symbol	Electronic structure	Atomic number	Symbol	Electronic structure
93	Np	$(Rn)7s^25f^46d^1$	99	Es	$(Rn)7s^25f^{11}$
94	Pu	$(Rn)7s^25f^6$	100	Fm	$(Rn)7s^25f^{12}$
95	Am	$(Rn)7s^25f^7$	101	Md	$(Rn)7s^25f^{13}$
96	Cm	$(Rn)7s^25f^76d^1$	102	No	$(Rn)7s^25f^{14}$
97	Bk	$(Rn)7s^25f^86d^1$	103	Lw	$(Rn)7s^25f^{14}6d^1$
98	Cf	$(Rn)7s^25f^{10}$			

in electron structure, we describe any element as having a *core* similar to the preceding *noble gas* structure and list only the electrons beyond the core.

A close examination of Table 4-2 will show some slight irregularities not covered by our rules discussed above. For example, chromium (atomic number 24) would be expected to have the structure $(Ar)4s^23d^4$ but actually has the structure $(Ar)4s^13d^5$. Nature appears to have a preference for half-filled orbitals (not well understood but related to the argument given above that leads to single occupancy of an orbital before double occupancy occurs). Close examination of the table will show several such examples: See Cu (29), Nb (41), Pd (46), Ce (58), and others. In the main, however, our generalizations are remarkably accurate and easy to use.

IONIZATION ENERGIES

The removal of an electron from one of the orbitals of an atom is called *ionization*. Ions may also be formed by addition of electrons to an atom or group of atoms. The forces that hold electrons close to the nuclei of atoms are electrical in nature. Furthermore, the attractive force between a positively charged nucleus and the negative electrons decreases rapidly as the distance between the two increases. Coulomb's law tells us that the attractive force between particles of different charge varies inversely as the square of the distance between them. Removal of an electron from its position in an orbital near the nucleus of an atom requires us to expend energy to overcome the attractive forces. To move the electron to an orbital only slightly less favorable to form an *excited state* may require only relatively small amounts of energy. Complete removal of an electron from the atom generally requires the expenditure of substantial amounts of energy. Moreover, if an atom contains several electrons, we expect that the energy

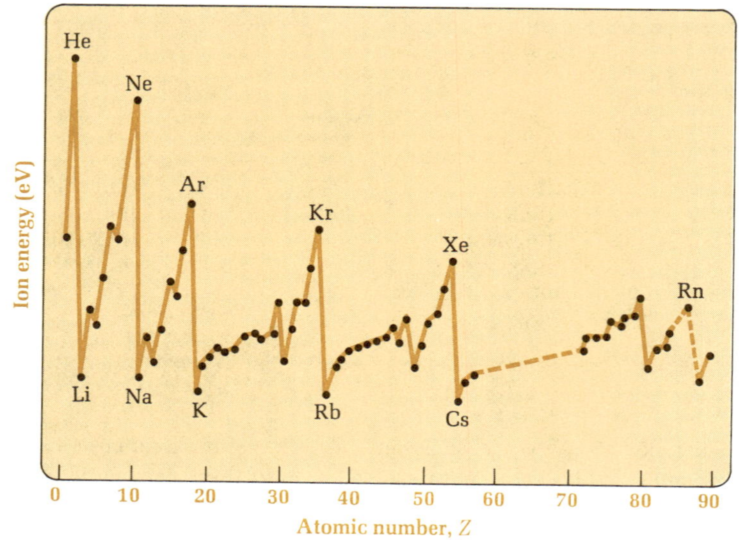

Figure 4-9 Ionization potential versus atomic number.

Table 4-3 Selected values of ionization potentials (eV)

H 13.6							He 24.6
Li 5.4	Be 9.3	B 8.3	C 11.3	N 14.5	O 13.6	F 17.4	Ne 21.6
Na 5.1	Mg 7.6	Al 6.0	Si 8.1	P 11.0	S 10.4	Cl 13.0	Ar 15.8

required to remove the second electron will always be substantially greater than that required to remove the first electron, since the second and succeeding electrons must be removed from atoms already carrying an excess of positive charge.

We might at first believe that the removal of a single electron from an atom would be relatively independent of, or would at least vary smoothly with, atomic number, but examination of Figure 4-9 shows that there are marked variations in the energy required. Compare this figure with the periodic table in Chapter 3 (Figure 3-1). Some systematic relationship appears to hold between atomic number and **ionization energy** that is reflected as a sawtoothed curve. Note that to remove the first electron from the helium atom requires nearly twice as much energy as is required to remove the electron from hydrogen. However, it is substantially easier to remove the electron from a lithium atom than from hydrogen. As we proceed to the right in the first period to elements of higher atomic number, it becomes progressively more difficult. Table 4-3 collects the first ionization energies for a number of the elements at the beginning of the periodic table. The ionization energies given in this table are expressed in units of electron volts (eV). This energy unit is the amount of energy acquired by an electron when it falls through a potential difference of 1 volt (V).

Let us turn our attention to Table 4-3 to see if some simple explanations for the variation in ionization energies can be found. The most obvious difference between helium and hydrogen is that the former contains two electrons and its nucleus bears two positive charges. To remove the negative electron from a doubly charged helium nucleus might be expected to require about twice as much energy as is required for hydrogen ionization ($2 \times 13.6 = 27.2$). If we consider only the problem of overcoming attractive forces between the electron and the nucleus, we might expect the ionization energy of helium to be 2.6 eV higher than it actually is. How can we account for the lower value? Bear in mind that both the first and second electrons are in the 1s orbital. Although two electrons of opposite spin are allowed to occupy any orbital, some repulsion between the like-charged electrons must be expected. The presence of the second electron in a helium atom then makes it slightly easier to remove the first electron than we might otherwise have expected. The third element, lithium, shows a substantial drop in ionization energy. It requires only a little more than a third as much energy to remove the first electron from the lithium atom as from hydrogen, although the nucleus of the lithium atom bears three positive charges. The large drop in ionization energy reflects the fact that the electron removed is from the 2s orbital. Reference to Figure 4-2 shows that an electron in a 2s orbital spends a majority of its time substantially farther away from the nucleus than does a 1s electron. In fact, the maximum in the probability curve occurs only slightly less than 3 Å from the nucleus, whereas the maximum in the 1s orbital occurs at only 0.53 Å from the nucleus. (Strictly speaking, the distances quoted are those appropriate to the hydrogen atom itself, but the principle applies to heavier atoms as well). A second and somewhat more subtle factor also is important. The 2s orbital extends substantially farther from the nucleus than the 1s orbital, and both 1s electrons are interposed between an electron in this orbital and the nucleus; the effect of the nuclear charge is thus partly *screened* by the charge of the 1s electrons. An electron in the 2s orbital "feels" less effectively the positive charge on the nucleus than it would if no such screening were

In contrast to the minuteness of atoms and subatomic particles, the physical apparatus needed to detect and determine the properties of these particles is usually enormous in size. This photograph shows the huge hydrogen bubble chamber at the Argonne National Laboratory into which a beam of particles from a synchrotron enters from the left. As the particles pass through the bubble chamber, which contains 6500 gallons of liquid hydrogen, they leave "footprints" in the form of minute bubbles that permit the physicist to deduce some of their basic properties. (Courtesy of Argonne National Laboratory.)

Table 4-4 Electron affinities of some elements (eV)

H 0.7							He 0
Li 0.54	Be 0	B 0.3	C 1.13	N 0.2	O 1.48	F 3.62	Ne 0
Na 0.73	Mg 0	Al 0.4	Si 1.90	P 0.80	S 2.07	Cl 3.82	Ar 0
					Se —	Br 3.54	Kr 0
						I 3.24	Xe 0
							Rn 0

present. Thus, both screening by inner electrons and greater distance contribute to the sharp drop in ionization potential. For the same reasons that apply in the hydrogen versus helium comparison, the ionization potential of beryllium is slightly less than twice that of the lithium atom. Note that a drop in ionization energy occurs with the boron atom when the 2*p* orbitals begin to fill. The ionization energies rise once more as successive electrons are put into the p_y and p_z orbitals (carbon and nitrogen). When the second electron is added to the p_x orbital (with oxygen), the curve dips slightly (owing to electron-electron repulsion) and rises once more until the shell of electrons is completed with the neon atom. Figure 4-9 shows the general variation that occurs, then, as successive energy levels are occupied in the heavier elements.

Three principal factors can be seen to be operative: As we proceed from atoms of lower to atoms of higher atomic number, the increase in nuclear charge tends to cause a general increase in ionization energy. But this is offset both by screening of inner electrons and by the successive use of orbitals of greater radius. The net result of the three factors is to cause a gradually declining ionization energy for any series of similar electronic structure. In Figure 4-9, we observe that a gradual decrease of ionization energy occurs with the noble gases from helium through radon.

ELECTRON AFFINITIES

We should also consider briefly a quantity that is essentially the opposite of ionization energy: **electron affinity.** We have emphasized that orbitals exist for every atom whether or not electrons are present to place into those orbitals. If electrons are added to such an unoccupied orbital, ions bearing negative charge are produced. If we add an electron to a neutral atom we expect that energy will be evolved, since the negative electron is brought much closer to the positively charged nucleus: On simple Coulombic grounds, we expect energy release from this process. Table 4-4 shows the energy, in electron volts, released when single electrons are added to some of the lighter elements. Again, note the periodicity of the changes observed.

ELECTRONEGATIVITY

Those elements that most readily accept electrons would be expected to lose electrons least readily, since the same factors affect both quantities. A high electron affinity will generally, therefore, be associated with a high ionization potential. Probably the simplest way to summarize this informa-

Table 4-5 Electronegativities of some elements

H																
2.1																
Li	Be											B	C	N	O	F
1.0	1.5											2.0	2.5	3.0	3.5	4.0
Na	Mg											Al	Si	P	S	Cl
0.9	1.2											1.5	1.8	2.1	2.5	3.0
K	Ca	Se	Ti	V	Cr	Mn	Fe	Co	Ni	Cu	Zn	Ga	Ge	As	Se	Br
0.8	1.0	1.3	1.5	1.6	1.6	1.5	1.8	1.8	1.8	1.9	1.6	1.6	1.8	2.0	2.4	2.8
Rb	Sr	Y	Zr	Nb	Mo	Tc	Ru	Rh	Pd	Ag	Cd	In	Sn	Sb	Te	I
0.8	1.0	1.2	1.4	1.6	1.8	1.9	2.2	2.2	2.2	1.9	1.7	1.7	1.8	1.9	2.1	2.5

tion in a single quantity is to average the ionization energy and the electron affinity. Such an average, properly scaled, is called the **electronegativity** of the atom. (We should observe that electronegativity has been defined in a number of ways. Various definitions lead to slightly different numerical values of electronegativity, but in that the relative values are of most interest to us we need not be concerned at this time with these small differences.) More complex procedures can also be used to obtain electronegativities of atoms, and Table 4-5 collects the numerical values for a number of the simpler elements. The simple concept of electronegativity can be used to give many qualitative explanations of the properties of bonds within molecules, and we shall use this concept in Chapter 5 and later. We shall also examine there the effect of electronegativity which deforms the "balloons" of Chapters 2 and 3.

PERIODICITY IN TERMS OF ATOMIC STRUCTURE

In the previous chapter we noted that a number of properties of atoms and molecules vary in a predictable way according to the periodic table of the elements (Figure 3-1). We are now in a position to understand more about why this must be so. The discussion in this chapter has established that certain regularities exist in terms of electronic structure. For example, all of the elements in a given column have essentially the same electronic structure in terms of the outermost electron shell (refer to Table 4-2). It is, therefore, not surprising that these elements form a series of structurally related compounds.

If we look more closely at our information concerning periodic properties, we may see that the first element in each column has distinctive properties. That is, the first element is less like the others in its column than the others are like themselves. For example, among the Group IV elements, carbon alone forms compounds containing just four bonds. We have already noted that silicon and the heavier elements of this group also form compounds having five or six bonds. Similarly, among the Group V elements, only nitrogen is restricted to forming compounds with just three bonds. Phosphorus and the heavier elements also form compounds containing five chemical bonds to the central element. For another example, let us return to Table 4-5 and look again at the electronegativities of the elements. Note that Si, Ge, and Sn all have the same electronegativity, which is substantially different from that of carbon. Similarly, P, As, and Sb have similar electronegativities, less than that of nitrogen. Further examination of this and other tables of data will provide additional examples of unique characteristics of the lightest elements of each column of the periodic table. To what is this distinctive difference due?

We can cite three reasons for the above observations. Two of these are pertinent here, but we must reserve the third until we complete our discussion of chemical bonding in the next chapter. In the first place, the matter of atomic *size* is important. As we move down a column in the periodic table, the size of the atoms increases regularly, but the largest percentage difference occurs between the first and second elements of the column. That is, the difference in size between F and Cl is more dramatic than that between Cl and Br or between Br and I. The same is true for the other columns of the periodic table as well. Now, in part, the number of bonds that a particular element can form with other atoms is limited by its size. Clearly, the smaller the atom is, the more difficult it will be to crowd a large number of other atoms around it. On this basis, we can understand that nitrogen forms only one chloride. NCl_3, but that the larger phosphorus forms two, PCl_3 and PCl_5. Similarly, oxygen forms just OCl_2, whereas sulfur forms a chloride having the composition SCl_4.

The other important factor we can appreciate at this point is that first-row elements have *just four low-energy orbitals* available for bond formation. Thus, the outermost shell has just the 2*s* and the three 2*p* orbitals available. It is certainly true that the higher energy orbitals exist for these elements, but the increase in energy to reach them is so great that they are of little actual importance in understanding the chemistry of these elements. But for the second and subsequent rows of the periodic table, there is also a set of *d* orbitals available for bond formation at reasonable energies. This fact, too, helps account for the ability of the second- and subsequent-row elements to form compounds having a larger number of bonds to the central elements than is possible for the first-row elements.

Our rudimentary consideration of atomic structure has begun to yield insight into the structures of the compounds that we discussed previously. Much additional insight comes from a consideration of the nature of the chemical bond.

Glossary

Azimuthal quantum number l the quantum number most closely related to orbital shape and symmetry. It takes on the values 0, 1, 2, . . . , $(n - 1)$; these numbers correspond to the designations s, p, d, ...

Electron affinity the energy released upon the capture of an electron by a neutral molecule or atom.

Electronegativity a quantitative scale that measures the tendency of the elements to acquire electrons; thus, the difference in electronegativities of two atoms involved in the formation of a chemical bond measures the polarity of that bond (the degree of charge separation in the bond).

Ionization energy the energy required to remove an orbital electron from an atom or molecule.

Magnetic quantum number m the quantum number defining the disposition in space of an orbital. It takes on the values $-l, -(l+1), \ldots, 0, \ldots, l$; these values correspond to the subscripts in the designations p_x, p_y, d_{xy}, and so on.

Nodal surface a surface on which the probability of finding a particular electron is zero.

Orbitals the locus occupied by an electron associated with an atomic nucleus. The size, shape, symmetry properties, and spatial orientation of each orbital is defined by the quantum numbers n, l, and m.

Principal quantum number n the quantum number most closely associated with orbital size and energy. It takes on the values 1, 2, 3, 4, ...

Quantum mechanics a theory of the physical world that underlies our current concept of atomic and molecular structure, among other things.

Quantum numbers a set of four numbers that determines the properties of orbital electrons.

Spin quantum number s that quantum number, $\frac{1}{2}$ or $-\frac{1}{2}$, that specifies the spin of an electron.

Wave function a mathematical expression deriving from quantum mechanical calculations that completely defines the behavior of an electron.

Study problems

1. Draw diagrams showing the shape of the following orbitals:
   a. $2p_x$ b. 3s c. $4d_{z^2}$
2. Use Figure 4-8 to describe the electronic structure of the following atoms or ions (atomic numbers given in parentheses):
   a. B (5) c. Zr (40) e. Cl^- (17)
   b. As (33) d. Sn^{2+} (50)
3. Why do we describe the electron distribution within an orbital in terms of probabilities?
4. Briefly describe the four quantum numbers (and their meanings) that specify the wave function of an electron in an atom.
5. What does it mean to say that an atom is in an excited state? Has been ionized?
6. Why is the second ionization potential of an atom always greater than the first?

7. The first ionization potential of the rare gas atoms (He, Ne, Ar, etc.) is always much greater than that of the next alkali metal (Li, Na, K, etc.). Why?
8. Briefly summarize the factors that strongly influence the ionization potential of an atom.
9. Contrast the concepts of electron affinity and electronegativity.
10. Draw a diagram to illustrate the differences between d_{xy} and $d_{x^2-y^2}$ orbitals.
11. Describe the symmetry elements present in the following orbitals:
    a. $2p_x$ b. d_{xz} c. d_{z^2}
12. Suggest at least three additional atoms or ions whose electronic structures could be the same as that of argon.
13. How does electronegativity generally change as we move down a column in the periodic table? To the right along a row?

Suggestions for further reading

V. H. Booth *The Structure of Atoms,* Macmillan, New York, 1964.

A. L. Companion *Chemical Bonding,* McGraw-Hill, New York, 1964 (paperback).

H. B. Gray *Electrons and Chemical Bonding,* Benjamin, Menlo Park, Calif., 1964 (paperback).

S. Hecht *Explaining the Atom,* Compass, New York, 1954.

R. M. Hochstrasser *Behavior of Electrons in Atoms,* Benjamin, Menlo Park, Calif., 1964.

G. E. Ryschkewitsch *Chemical Bonding and the Geometry of Molecules,* Van Nostrand Reinhold, New York, 1963 (paperback).

G. T. Seaborg *Man-Made Transuranium Elements,* Prentice-Hall, Englewood Cliffs, N.J., 1963.

H. H. Sisler *Electronic Structure Properties and the Periodic Law,* Van Nostrand Reinhold, New York, 1963 (paperback).

5

The chemical bond

Could he have been discussing chemical bonding?
We must be grateful to Lord Russell for the unequaled skill with which he has left the vast darkness of the subject unobscured.

Alfred North Whitehead
commenting on a lecture by Bertrand Russell

In Chapters 2 and 3 we developed the structural chemistry of carbon and other elements that form tetrahedral molecules by beginning with a simple but crucial postulate: Atoms are held together in a stable geometry by **chemical bonds.** The groundwork for understanding chemical bonds was laid in the last chapter with the development of the electronic structure of atoms. But before plunging directly into how we as chemists describe bonding in molecules, let us reflect a little on our earlier observations so as to understand as clearly as possible what it is that we wish to explain. There are several important points.

Suppose we look back at methane, our starting point in structural chemistry. The first observation that we made was that methane has the composition CH_4. This raises the first question: Why does carbon form four bonds to hydrogen and not three or five or some other number? Until now, we accepted the tendency of carbon to combine with other atoms in groups of four based on our experience. We expect that any successful theory of chemical bonding should account for the *composition of molecules.* The second discovery we made concerning methane is that it is tetrahedral. Why is it tetrahedral and not square-planar, pyramidal, or some other geometry? Our theory of chemical bonding should provide an explanation for the why of the *geometries of molecules.* A third requirement of our theory can be realized if we employ a slightly more complicated molecule as an example: We know that the collection of atoms C_4H_{10} can be arranged in two different ways to yield two distinct, stable molecules, *n*-butane and isobutane. These structural isomers are stable indefinitely and are not interconvertable at any detectable rate. This requires that our theory account for the fact that chemical bonds (certain kinds at least) have a particular stability. That is, that appreciable energy is required to break and rearrange the bonds. In Chapter 8 we shall learn that each type of chemical bond has its own characteristic stability or bond energy. For the moment, we shall expect a theory of chemical bonding to provide a qualitative explanation for the *stability of molecules.* Finally, we may note that chemical bonds have other properties that can be measured and that must be explained. We have previously noted one of these, bond lengths. A complete theory of chemical bonding should account for bond lengths as well as the other parameters of bonds. To some extent, our discussion will stop short of a complete development of all of these matters. Some of them require the use of mathematical methods beyond our grasp, and most require that we focus attention on matters more detailed than are justifiably our concern. The important point is that we learn to understand the basis of the fundamental properties of chemical bonds, and that we learn to use this information in novel situations so as to expand our understanding of chemistry generally.

The importance of the electronic structure of atoms to the theory of chemical bonding was recognized as early as 1916. In that year W. Kossel, a German physicist, and G. N. Lewis, an American chemist, pointed out that stable ions have electronic structures corresponding to the noble gas configuration. Elements with slightly more electrons tend to lose them (potassium loses one electron to achieve the configuration of argon), whereas those with nearly completed shells tend to gain electrons (the sulfur atom gains two electrons to form the S^{2-} ion with the argon electronic structure). Lewis realized that many atoms could attain noble gas configuration by sharing electrons with other atoms and, on this basis, made the inspired suggestion that *covalent bonds* between atoms result from sharing two electrons between them. The Lewis concept was extensively developed in the next years, following the full development of the quantum theory of atomic structure.

ELECTRON DONOR-ACCEPTOR INTERACTIONS

Representation of molecules in terms of the *Lewis electron dot structures,* in which the valence (outermost, associated with bonding) shell of electrons for each atom is explicitly shown as dots, still remains a useful and commonly employed method to help understand chemistry. Three examples of such Lewis structures are

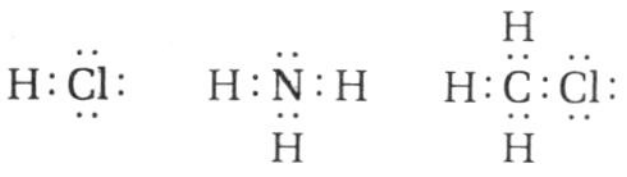

Note that here and in succeeding Lewis structures, we emphasize distributions of electrons and for clarity in that regard may distort the actual geometric structures as discussed in detail in Chapters 2 and 3. Note also that in each of these structures, each atom has a complete octet of electrons, with the exception of hydrogen which assumes a noble gas configuration (He) with only two electrons. Those electrons shown as occupying positions between atoms are considered to be shared between them and to form the chemical bond linking them.

Example *Construct Lewis electron dot representations for each of the following molecules:*

a. F_2 c. O_2 e. BF_3
b. H_2O_2 d. C_2H_4

Solution *In writing Lewis structures, there are two guiding considerations. First, we must provide the correct number of valence electrons in the structure, and second, each atom must assume,* as nearly as possible, *a noble gas configuration. Using these rules, we can arrive at satisfactory representations for each of the above cases. (a)* F_2: *Each fluorine atom contributes 7 valence electrons so that we have 14 in all. The structure*

has the necessary number of electrons, and each fluorine atom has a complete octet. Hence, this structure must be correct. (b) H_2O_2: *Each hydrogen contributes 1 valence electron, and each oxygen contributes 6. There are, therefore, 14 valence electrons in all. The structure*

meets both the criteria of the number of electrons and complete valence shells. (c) O_2: *In this case there are 12 valence electrons to be distributed between the two atoms. To arrive at a complete octet for both of them, it is necessary to share 4 electrons between them:*

The two oxygen atoms may be said to be linked by a double bond. *(d)* C_2H_4: *Each carbon atom contributes 4 valence electrons, and each hydrogen contributes 1. The 12 electrons may be arranged in the following way,*

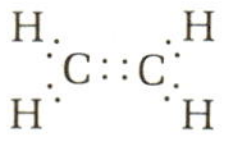

so that each atom has a complete valence shell. The two carbon atoms are linked by a double bond. (e) BF_3: *There are a total of 24 valence electrons, 7 from each fluorine atom and 3 from the boron. No matter how hard we try, it is not possible to distribute these so that each atom has a complete octet (note that fluorine does not form double bonds). A choice is therefore necessary as to which atom will be left with an incomplete octet. Since fluorine has the greater electronegativity, we fill its octet at the expense of boron. Note also that this results in a more symmetrical distribution. The best that we can do is*

The fact that the boron atom is associated with only six electrons is an important consequence that we develop in the text immediately below. As we shall see, modern theories of chemical bonding retain the central idea of the sharing of electrons between atoms. Before moving on to consider these theories, let us push the concept developed by Lewis a little further.

The example cited just above shows that the best representation for the BF_3 molecule assigns the central boron atom just six electrons. Consequently, we expect that BF_3 will try to complete the octet of electrons if the opportunity presents itself. Compounds (or ions) that are capable of accepting a pair of electrons are called **Lewis acids.** This is the most general definition of an acid that we have; a more restricted definition, due to Brønsted, is developed in detail in Chaper 11. Lewis acids include all cations and those uncharged molecules having the capacity to coordinate with a pair of electrons. Many such compounds are known. Since elements in a single column of the periodic table (Chapter 3) generally have related properties, we might search for additional Lewis acids in the boron family. Directly below boron, we find aluminum. The Lewis structure for $AlCl_3$, for example, is

```
:Cl:  :Cl:
   Al
  :Cl:
```

Like BF_3, $AlCl_3$ has a vacant molecular orbital and is a Lewis acid. Just to the left of boron in the periodic table is beryllium. Since this element has fewer electrons than boron, we might expect that beryllium compounds would also be Lewis acids. The Lewis structure for $BeCl_2$ confirms this supposition:

```
:Cl:Be:Cl:
```

$BeCl_2$ has, in fact, *two* empty orbitals.

To understand the importance of Lewis acid character, let us return to the ammonia molecule, discussed in some detail in Chapter 3. We remember that NH_3 has an unshared pair of electrons, $:NH_3$. Consequently, ammonia is capable of donating a pair of electrons to a suitable acceptor. Such compounds are known as **Lewis bases.** Other Lewis bases include water, carbon monoxide, dimethyl ether, and trimethylamine:

```
H:O:H     :C:::O:     CH3:O:CH3     CH3:N:CH3
                                        CH3
```

Ordinarily, Lewis bases have unshared electron pairs on oxygen, nitrogen, sulfur, and sometimes on carbon. Moreover, all anions have Lewis base character. But note that BF_3, which is uncharged and does have unshared electron pairs on fluorine, is not a Lewis base.

Example *From the following list of molecules and ions, select those that are Lewis acids, those that are Lewis bases, and those that have neither Lewis acid nor base character:* Fe^{2+}, $NH(CH_3)_2$, PH_3, CCl_4, CH_3^+, H_2NOH, CH_3SCH_3, C_4H_{10}, $Al(CH_3)_3$.

Solution *Of these species,* Fe^{2+}, CH_3^+, *and* $Al(CH_3)_3$ *can act as acceptors of electron pairs and are Lewis acids. The species* $NH(CH_3)_2$, PH_3, H_2NOH, *and* CH_3SCH_3 *all have unshared electron pairs that can be donated to*

suitable acceptors and are Lewis bases. The molecule C_4H_{10} is without Lewis acid or base character, and CCl_4 is such a weak Lewis base (unshared electron pairs on uncharged halogen) that it might well be classified with C_4H_{10}.

It is not difficult to anticipate the next stage in our argument. On the one hand, we have compounds capable of accepting a pair of electrons and, on the other, compounds capable of donating a pair of electrons. Hence, Lewis acids and bases can interact with each other. One typical example is

$$BF_3 + :NH_3 \longrightarrow \begin{array}{c} \mathrm{H}\;\;\mathrm{F} \\ \mathrm{H}:\ddot{\underset{..}{\mathrm{N}}}:\ddot{\underset{..}{\mathrm{B}}}:\mathrm{F} \\ \mathrm{H}\;\;\mathrm{F} \end{array} \tag{5-1}$$

In the formation of this compound, boron has accepted a pair of electrons from nitrogen. The bond so formed is called an **electron donor-acceptor bond** (or a **dative bond**). Bonds of this type are generally not as strong as ordinary bonds, but they are stronger than many types of intermolecular interactions (see Chapter 7). Consequently, compounds formed by making electron donor-acceptor interactions may be isolated and characterized. Such compounds are frequently called **molecular complexes,** suggesting that the individual molecules forming the complex retain some of their individual character in the complex.

Example *Predict a structure for $BF_3 \cdot NH_3$.*

Solution *From our discussion in Chapter 3, we know that the geometry of four substituents around nitrogen is tetrahedral. In this compound, boron has now four substituents as well and we might expect these to be oriented tetrahedrally. Consequently, this molecule may be structurally analogous to ethane. In fact, the structure has been determined very precisely by X-ray diffraction methods (Chapter 7) and is*

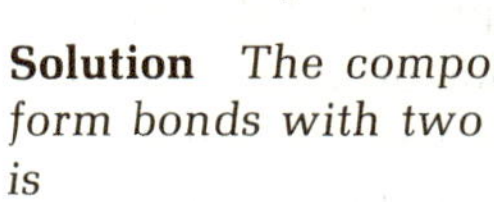

in agreement with our qualitative conclusions.

Example *The compound $BeCl_2$ forms an electron donor-acceptor bond with dimethyl ether. Predict the structure of this molecular complex.*

Solution *The compound $BeCl_2$ has two vacant orbitals, and hence can form bonds with two molecules of CH_3—O—CH_3. The resulting structure is*

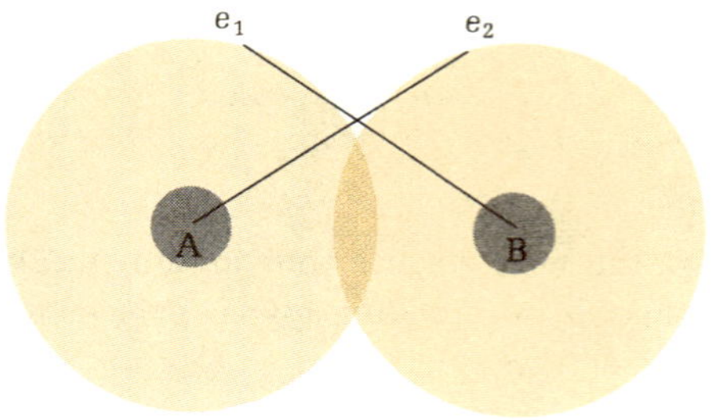

Figure 5-1 Two hydrogen atoms shown at bonding distance without disturbance of the atomic orbitals.

With this background deriving from the considerations of Lewis theory, let us move on to consideration of two more modern theories of chemical bonding. But remember that the concept of a covalent bond as the sharing of two electrons between atoms will remain essentially unchanged.

CONTRAST OF VALENCE BOND AND MOLECULAR ORBITAL METHODS

At present there are two widely used and nearly equivalent theories to account for chemical bonding. The first of these, called the **valence bond** (VB) **theory,** preserves in many respects the simplicity of the Lewis view. The VB method concentrates on describing the bonds that hold together each pair of atoms in a molecule. Complex molecules are then viewed as a collection of individual bonds that hold together all of the atoms of the molecule. On occasion, we find it impossible to write a satisfactory *single* structure in the VB method and are forced to describe a molecule in terms of a type of average of several structures. We shall develop the VB method in more detail later in this chapter.

A second approach to bonding develops a picture of the bonding in the molecule as a whole rather than as a collection of individual bonds between nearest neighbors. It often turns out that in such a scheme we can identify a portion of the total description that nicely corresponds to bonds between adjacent atoms very similar to the bonds of the VB approach. At other times, the picture that results is that of a bond extending over many atoms and perhaps even over the entire molecule. The basic approach of this method is to develop a system of orbitals similar to the atomic orbitals (AO's) but taking into account the presence of many or all of the atomic nuclei. The derivation of the name **molecular orbital** (MO) **theory** would seem apparent. The MO method in many instances provides the simpler picture, and we shall describe it first.

THE MOLECULAR ORBITAL METHOD

The molecular orbital approach involves the following conceptual stages: (1) The nuclei of the elements present in a molecule are arranged into the geometric pattern corresponding to its structure; (2) the atomic orbitals of the various nuclei are combined to form new molecular orbitals; and (3) the appropriate number of electrons are placed into the molecular orbitals. Let us emphasize that molecular orbitals resemble atomic orbitals in the sense that they describe the regions of space in a molecule where electrons are to be found. In the same way that the specification 1s orbital brings to mind a certain picture of electron distribution, so similar names are assigned to the molecular orbitals. The only essential difference between atomic orbitals and molecular orbitals is that the former are confined to individual atoms and the latter may extend over the complete molecule. The shapes of the molecular orbitals are the result of the location of the positively charged nuclei and the total electron distribution. It is a choice of the scientist to describe them by combining individual atomic orbitals to closely approximate what nature establishes directly.

The rules of combination of atomic orbitals are simple: The extent of formation of a molecular orbital depends on the degree to which the atomic orbitals extend into the same space (*overlap* in space). If two atomic orbitals are combined, two new molecular orbitals result. Any atomic orbitals that are relatively far from the orbitals on other atoms appear essentially unchanged in the molecule. Thus, in general, the molecular orbitals that house the inner or core electrons of each atom are essentially similar in shape to those of each individual atom. The electrons in the core tend to remain near the nucleus, much as though no bond had been formed. However, if the orbitals on one atom overlap in space with those on another atom, the combination of orbitals changes the picture considerably.

The simplest case, the hydrogen molecule, consists of just two electrons and two protons 0.74 Å apart. Let us first suppose that two ground-state hydrogen atoms could be brought this close without effect on the individual atomic orbitals. One result of this supposition can be seen in Figure 5-1. In this figure we show the two nuclei at A and B and draw a circle around the nucleus at a distance corresponding to the most probable distance from the nucleus for the electron, 0.53 Å. Note that the circles about the two atoms intersect. If the electron belonging to atom A happens to lie in the general direction of B, there is a considerable region in space in which it will actually lie closer to nucleus B than to A. Since the attraction between an electron and a nucleus rises sharply as the distance between them decreases, the electron belonging to A will be at this moment more strongly attracted to nucleus B than to its own. In fact, it no longer would be appropriate

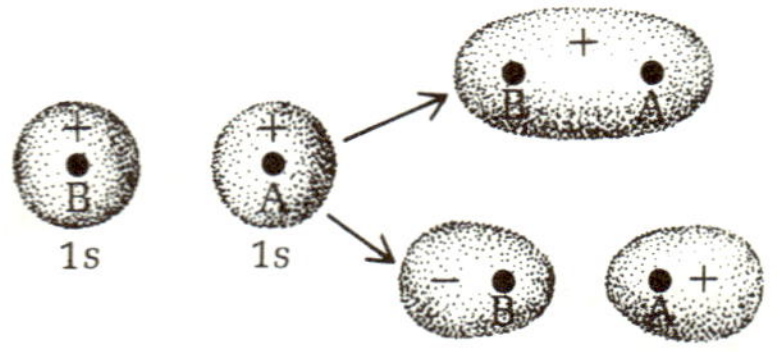

Figure 5-2 Formation of sigma bonding and antibonding molecular orbitals from 1s atomic orbitals of two hydrogen atoms. (From W. J. Moore, *Physical Chemistry*, 3rd ed., Prentice-Hall, Englewood Cliffs, N.J., 1962.)

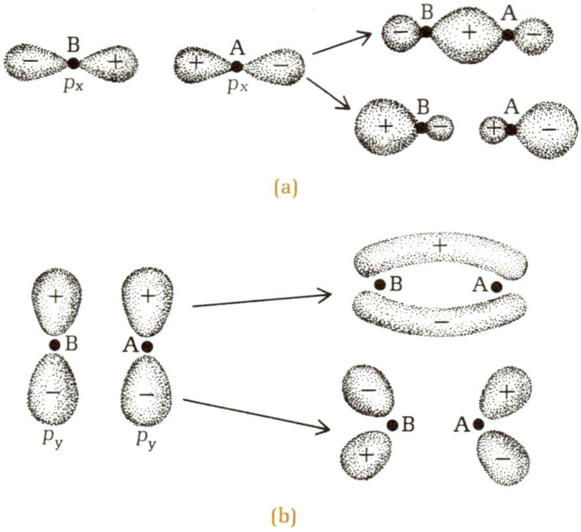

Figure 5-3 (a) Formation of sigma-type bonding and antibonding molecular orbitals from *p* orbitals. (b) Formation of pi-type bonding and antibonding molecular orbitals from *p* orbitals. (From W. J. Moore, *Physical Chemistry*, 3rd ed., Prentice-Hall, Englewood Cliffs, N.J., 1962.)

to think of this electron as belonging to A. Under such a circumstance it is no surprise to find that our picture of the isolated atoms must be substantially altered when they are within bonding distance of one another. The strongest interactions can be expected when the electrons are near or *between* the two protons. In such a position the negative charge serves to bond together the two protons. The total energy of the system is at a minimum.

The description of the mathematics and physics underlying formation of molecular orbitals from atomic orbitals is far beyond the scope of this book.

There are some simple rules that we can examine that will be qualitatively useful in our own studies.

As noted above, when atomic orbitals are combined to produce molecular orbitals, one molecular orbital is formed for each atomic orbital used. In the simplest terms, this reflects the fact that each atomic orbital can accommodate two electrons, so as many as four electrons may reside in the two overlapping atomic orbitals. Each molecular orbital can also accept only two electrons, so two molecular orbitals must be formed to accommodate all four electrons. As two hydrogen atoms are gradually brought closer together, the two 1s atomic orbitals fuse gradually and form two molecular orbitals involving both nuclei. These are shown in Figure 5-2. Note the two distinctive shapes. Electrons in the molecular orbital drawn above spend the majority of their time between the two atoms, the most stable possibility. In the lower molecular orbital, quite the opposite is true. Observe that the more stable molecular orbital resulting from coalescence of the two atomic orbitals is the one in which the electrons are found between the two nuclei a substantial fraction of the time and thus provide the "glue" to hold the two atoms together. This orbital may be occupied by one or two electrons, resulting in bonding of the two atoms to one another; consequently, we speak of this orbital as a **bonding orbital.** In the second case the electrons of the individual atoms can be seen to spend a majority of their time on opposite sides of the nuclei. In fact, they are most unlikely to be found between two atoms. The picture is just the reverse of the bonding situation, and we speak of this type of orbital as an **antibonding orbital.** Note that the bonding molecular orbital results when the signs of the two atomic orbitals are alike and the antibonding molecular orbital results when they are different.

Under the proper circumstances, any two atomic orbitals can combine to form one bonding and one antibonding molecular orbital provided the individual atomic orbitals overlap in space. The combination of two *p* orbitals to form a single molecular orbital is shown in Figure 5-3(a). In this instance we have combined two *p* orbitals lying along a single axis. The result of this combination is a bonding molecular orbital that is cylindrically symmetrical (the cross section of the orbital at any point along the axis is circular). It is conventional to refer to such an orbital as a **sigma**-type molecular orbital. (The molecular orbitals formed from the 1s atomic orbitals of hydrogen are also sigma orbitals.) Figure 5-3(a) also shows the antibonding orbital formed by two *p* orbitals overlapping along a single axis. Observe that a node occurs midway between the two atoms. Again we note that electrons in this molecular orbital would spend the majority of their time on opposite sides of the two nuclei. It should be apparent that no bonding would result from use of this orbital. The cross sections of this molecular orbital are also all circular. It is an *antibonding sigma orbital.*

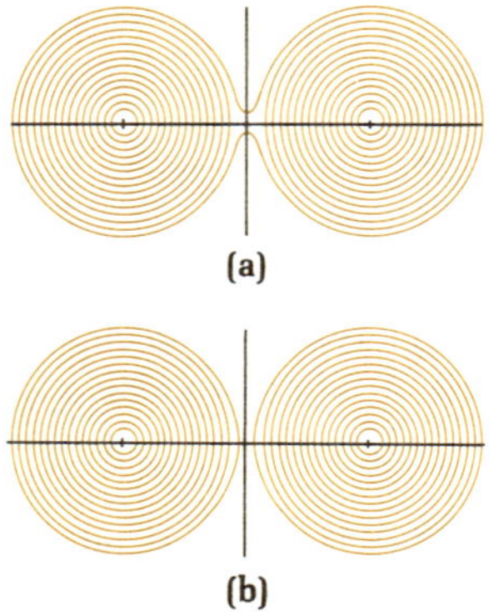

Figure 5-4 (a) The bonding molecular orbital formed from two lithium 1s orbitals. (b) The antibonding molecular orbital formed from two lithium 1s orbitals. (From A. C. Wahl, *Science*, **151,** 961 (1966). Copyright © 1966 by the American Association for the Advancement of Science.)

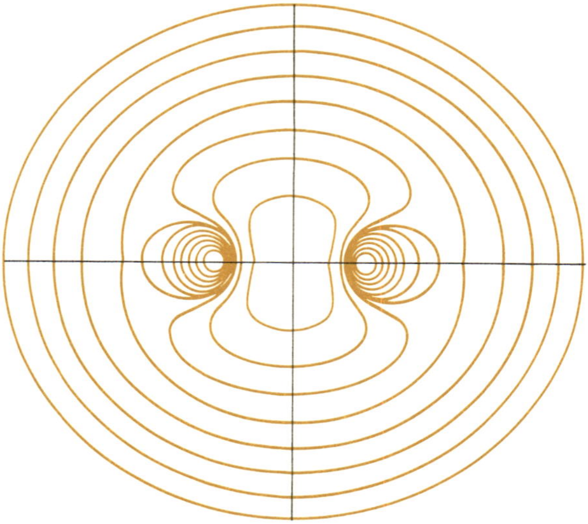

Figure 5-5 The bonding orbital formed from 2s orbitals of two lithium atoms. (From A. C. Wahl, *Science*, **151,** 961 (1966). Copyright © 1966 by the American Association for the Advancement of Science.)

Overlap between two *p* orbitals along a common line gives a sigma orbital. A second type of combination of *p* orbitals can also occur in contrast to the single manner of combination of s orbitals. Figure 5-3(b) shows two *p* orbitals overlapping side by side rather than end to end. The bonding orbital in this case tends to place the electrons above and below the internuclear line, but again the general region between the two nuclei is occupied by the electrons a substantial fraction of the time. In contrast, the antibonding orbital shown at the bottom of the figure again clearly indicates that the electrons strongly prefer not to be found between the nuclei. In fact, the plane bisecting the A–B axis is a nodal plane. Orbitals of the type depicted in Figure 5-3(b) are referred to as **pi** orbitals. A cross section perpendicular to the bond axis is not circular and in fact has two separate regions where electrons are most likely to be. This is the characteristic of all pi orbitals.

HOMONUCLEAR DIATOMIC MOLECULES

To explore more fully the meaning of bond formation in molecular orbital terms, let us examine the simple series of two-atom molecules beginning with Li_2 and running through Ne_2. For simplicity's sake, we first consider only those molecules in which both atoms are the same chemical element; they are therefore called *homonuclear diatomic* molecules.

The Li_2 molecule

Remember that the isolated lithium atom has the structure $1s^2 2s^1$; in the diatomic molecule there will be six electrons for which we must account. To a very good approximation, the bonding between the two lithium nuclei is the result of utilization of the 2s orbitals to form a bonding molecular orbital and the occupation of this orbital by two electrons. The remaining four electrons may, to a first approximation, be thought of as occupying nearly undisturbed 1s atomic orbitals on each atom. For the sake of completeness, however, we shall examine the disposition of all six orbitals. For this purpose we adopt a new type of figure, in which the electron probability is shown by a contour diagram; in this diagram we may think of the outermost contour as enclosing the electron 90 percent of the time, the next perhaps 85 percent, and so forth. Figure 5-4 shows such a contour diagram for the lithium bonding and antibonding molecular orbitals formed from the pair of 1s orbitals. Observe that the electron density concentration between the two nuclei is not greatly different in either orbital from the density that would result if the two orbitals were totally independent of one another. Bear in mind that each molecular orbital described in Figure 5-4 can accommodate two and only two electrons. Both the bonding and the antibonding orbitals in these buried molecular orbitals show that the electrons spend the majority of their time in the immediate vicinity of the nuclei. Electrons in the bonding molecular orbital contribute very little to binding the two nuclei together. Those in the antibonding orbital, correspondingly, provide a very weak force tending to drive them apart. In any event, the bonding and antibonding effects just balance one another.

In contrast, combination of the 2s orbitals produces a bonding and an antibonding molecular orbital quite different from the atomic orbitals. Since only two electrons remain to be placed in molecular orbitals, the antibonding orbital is not occupied, and for the moment we shall not concern ourselves with it. The diagram for the bonding orbital is shown in Figure 5-5; this

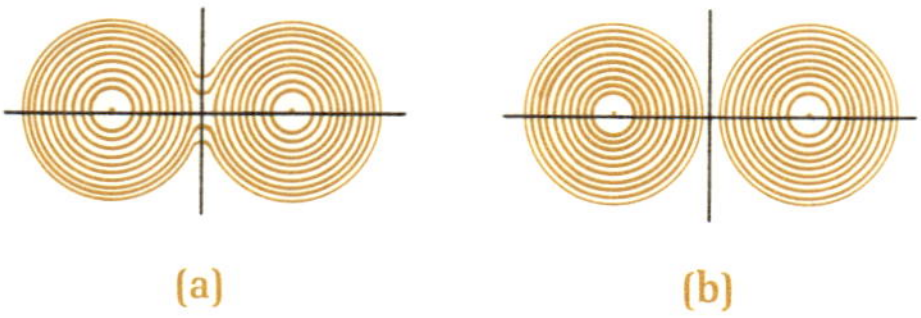

Figure 5-6 Bonding and antibonding orbitals from 1s orbitals of two boron atoms. (a) Bonding. (b) Antibonding. (From A. C. Wahl, *Science*, **151,** 961 (1966). Copyright © 1966 by the American Association for the Advancement of Science.)

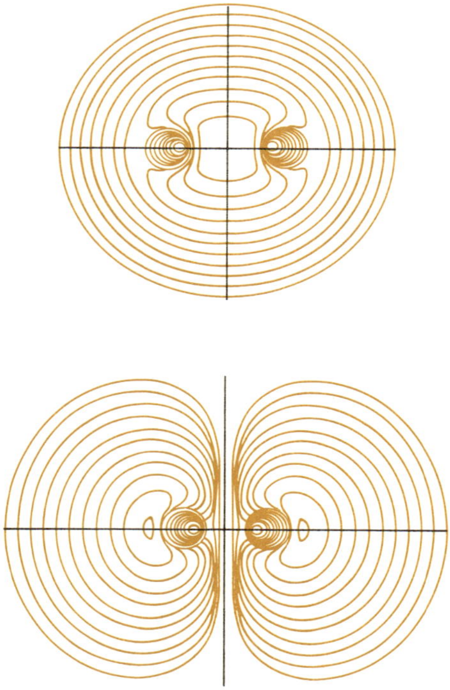

Figure 5-7 Bonding and antibonding orbitals from 2*s* orbitals of two boron atoms. (From A. C. Wahl, *Science*, **151,** 961 (1966). Copyright © 1966 by the American Association for the Advancement of Science.)

should be compared with the bonding orbital formed from the 1s orbitals. Note that here too the contour density around the individual nuclei is large (in part this reflects the inner "hump" in the probability curve for the atomic orbital—see Figure 4-2), but electrons spend a significant proportion of their time between the two nuclei; bonding results.

The Be_2 molecule

In the Be_2 molecule, combination of the pair of 1s orbitals leads to a bonding and an antibonding molecular orbital and four electrons occupy these.

Combination of the pair of 2s orbitals also leads to a bonding and an antibonding orbital, and in the Be_2 molecule four more electrons are available to fill both orbitals. As in the case of the first-level filled bonding and antibonding orbitals in the Li_2 molecule, the amount of bonding is just cancelled by the amount of antibonding. There is no apparent reason for such a molecule to be held together and, in fact, the Be_2 molecule has not been observed. Molecular orbital theory suggests that such a molecule, if prepared, would be of low stability. It decidedly does not state that no such molecule can be prepared!

The B_2 molecule

In the B_2 molecule, a total of ten electrons must be accommodated. Bonding and antibonding orbitals are formed from the 1s combination and from a combination of 2s orbitals as shown in Figures 5-6 and 5-7; eight electrons are required to fill these orbitals.

If the axis of the molecule is defined as the z axis, both the p_x and the p_y orbitals are in the correct spatial orientation to produce pi molecular orbitals, and the p_z orbitals overlap to form sigma bonding and antibonding orbitals. Both the pi and sigma bonding orbitals concentrate electrons between the nuclei, and there is no simple way for us to decide which orbitals will lie at the lowest energy and be the first to be used by the molecule. Chemists therefore followed the direct procedure of studying the B_2 molecule to find which it has used. Solution of the problem depends on the fact that the p_x and p_y orbitals are identical in the molecule in every respect. The p_z orbital is unique in that it lies along the internuclear line; no such feature serves to distinguish the other two. If two orbitals cannot be distinguished on the basis of energy, they are said to be *degenerate*. Either the pi orbital formed from p_x-p_x or that formed from p_y-p_y overlap is an equally satisfactory place for the first electron of the remaining two. However, when the second electron is added, it will prefer to go to the unoccupied orbital, since then there is no restriction on its spin and it will be farther in space from the other electron (desirable because of the repulsion of like charges). Thus, if the electrons enter the two pi orbitals, they will do so with no spin restrictions. Of course, if the sigma orbital from p_z-p_z overlap were used, the two electrons would of necessity have paired spins, since only a single bonding orbital is available.

We have observed in Chapter 4 that an individual electron behaves in some respects like a tiny magnet. If electrons are spin-paired, however, the magnetic effects of one electron are just cancelled by the other. When we subject the B_2 molecule to study of its magnetic properties, we find that it behaves just as we expect for a molecule with two spin-unpaired electrons; we conclude that the last two electrons have gone into the two

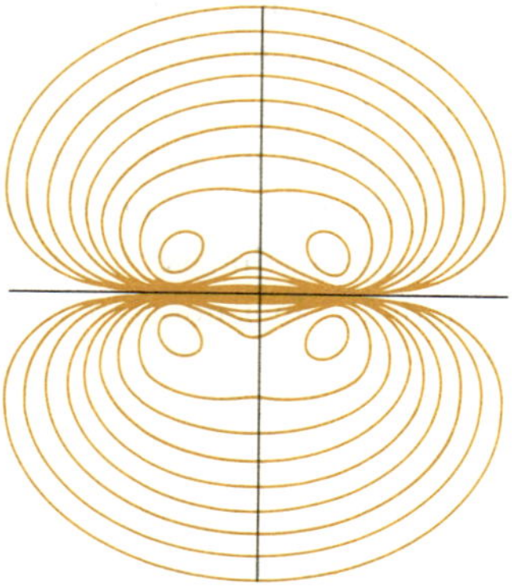

Figure 5-8 A bonding pi orbital from overlap of p_x or p_y orbitals of two boron atoms. (From A. C. Wahl, *Science,* **151,** 961 (1966). Copyright © 1966 by the American Association for the Advancement of Science.)

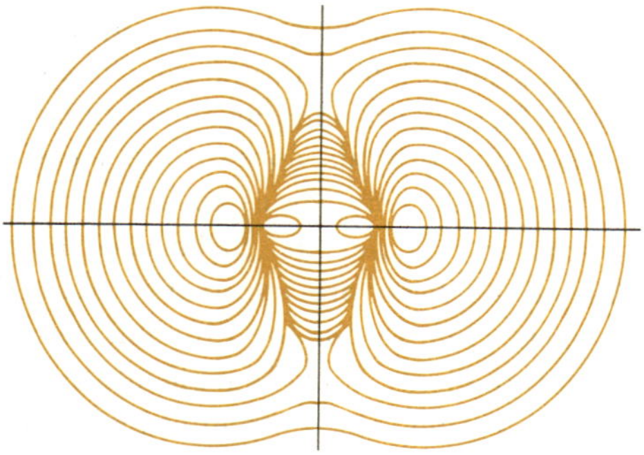

Figure 5-9 A sigma orbital formed by overlap of two p_z orbitals from two nitrogen atoms. (From A. C. Wahl, *Science,* **151,** 961 (1966). Copyright © 1966 by the American Association for the Advancement of Science.)

Figure 5-10 An antibonding orbital formed from the p_x or p_y orbitals of two oxygen atoms. (From A. C. Wahl, *Science,* **151,** 961 (1966). Copyright © 1966 by the American Association for the Advancement of Science.)

degenerate pi orbitals. The resulting bonding orbital is shown in Figure 5-8.

Let us pause a moment to note that the molecule has no knowledge of our choice of x and y axes. The only significant fact to the molecule is that the two orbitals are perpendicular to one another. The consequence of this molecular unawareness is of some importance: It means that *wherever* we choose to draw the x axis, we shall obtain a pi orbital of the type shown in Figure 5-8 and one just like it at right angles. This can be true only if the two pi orbitals *together* have the symmetry of a cylinder centered on the z axis.

The C_2 molecule

The molecular orbitals required to accommodate the 12 electrons in C_2 can be built from the proper atomic orbitals just as in the case of B_2. The first 10 electrons fill these just as they did for B_2, and the last 2 electrons also enter those two bonding molecular orbitals from p_x-p_x and p_y-p_y overlap. Four electrons now occupy the two degenerate pi orbitals and are spin-paired. The C_2 molecule is a well-known molecule in chemistry at high temperatures. It is a principal component of the gas formed when carbon is evaporated at 5000°C.

The N_2 molecule

In the nitrogen molecule, N_2, two more electrons must be accommodated than in the C_2 molecule. We might guess at first that these electrons would half-fill the antibonding pi orbital. However, it turns out that the bonding sigma orbital formed from the two p_z orbitals is filled first. This orbital is shown in Figure 5-9. Note the very effective use that electrons make of the internuclear region in this orbital. Note that the bonding molecular orbitals formed from the 2*p* atomic orbitals are now completely occupied.

The O_2 molecule

We now turn to the oxygen molecule. With two more electrons to accommodate than the nitrogen molecule, we must seek an additional orbital. The only orbitals still available are the two antibonding pi orbitals shown in Figure 5-10 (which are each capable of accepting two electrons) and the antibonding sigma orbital. The two electrons available in oxygen are found in the degenerate pi orbitals. As with B_2, the orbitals are only half-filled, and there is no restriction on the electron spins. The magnetic behavior of oxygen confirms the presence of two unpaired electrons in the molecule.

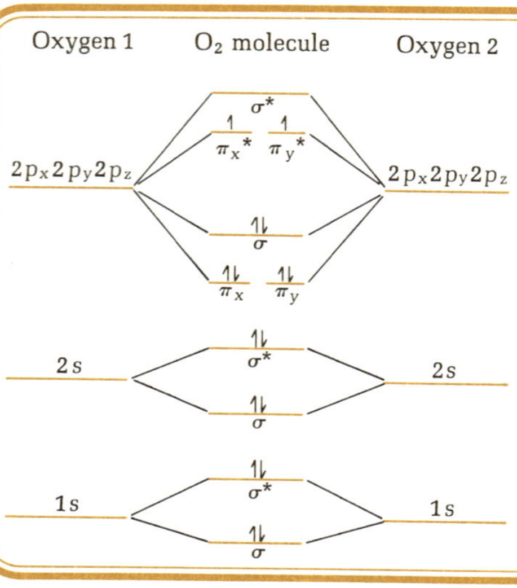

Figure 5-11 Molecular orbital diagram for the O_2 molecule (the two spin states are symbolized by arrows).

Table 5-1 Comparison of molecular orbital populations

	Li_2		Be_2		B_2		C_2		N_2		O_2		F_2		Ne_2	
	B	A	B	A	B	A	B	A	B	A	B	A	B	A	B	A
1 sigma	2	2	2	2	2	2	2	2	2	2	2	2	2	2	2	2
2 sigma (2s)	2	—	2	2	2	2	2	2	2	2	2	2	2	2	2	2
2 pi ($2p_x$ and $2p_y$)	—	—	—	—	2	—	4	—	4	—	4	2	4	4	4	4
2 sigma ($2p_z$)	—	—	—	—	—	—	—	—	2	—	2	—	2	—	2	2
Sum	4	2	4	4	6	4	8	4	10	4	10	6	10	8	10	10

The F_2 molecule

In the fluorine molecule, the two additional electrons enter the antibonding pi orbitals; since four electrons are now in those orbitals, the spins must be paired.

The Ne_2 molecule

Finally, in Ne_2, the extra two electrons would enter the antibonding sigma orbital formed from the two $2p_z$ orbitals. Since the total number of electrons in antibonding orbitals now equals the number in bonding ones, no net bonding occurs. Ne_2 is unknown.

ORBITAL ENERGIES

What factor determines the preference of an electron for a particular orbital? Or, what is equivalent, what factors establish the relative energies of different orbitals? In the main, the most important factor is Coulombic forces: An electron is best suited where it can interact most strongly with the positive nuclei. It is this fact that distinguishes between bonding and antibonding orbitals. In bonding orbitals the space between the nuclei is preferentially occupied and the electron is thus near both nuclei. In antibonding orbitals the electron is near only one or the other nucleus. You may wish to review Figures 5-4 through 5-10 to understand this point as clearly as possible. Electrons occupy the orbitals of lowest Coulombic energy first. We can use this fact to construct a simple diagram to describe the electron distribution for any molecule. Figure 5-11 shows how the structure of the oxygen molecule is described by such a diagram.

A close study of Figure 5-11 will be profitable. Note that it shows schematically the atomic orbitals of one oxygen atom on the left and a second oxygen atom on the right arranged in such a fashion that energy increases toward the top of the diagram. The molecular orbitals formed from the atomic orbitals are shown in the center, and the brown lines serve to draw attention to the origins of the molecular orbital. The Greek letters σ (sigma) and π (pi) identify the symmetry of the molecular orbitals, and the antibonding orbitals (always of higher energy than the bonding) are identified as σ^*, π^*, and so on.

Table 5-1 collects our conclusions for a comparison of the entire series of homonuclear diatomics. Note that Be_2 and Ne_2 would both have equal numbers of bonding and antibonding electrons; there is nothing to hold the molecules together and neither is known. In Li_2, B_2, and F_2, the number of bonding electrons exceeds the number of antibonding ones by two. (In the Lewis sense, these two electrons correspond to one bond between the atoms.) In both C_2 and O_2 there are four more electrons in bonding orbitals than antibonding, and in N_2 there are six more bonding electrons. In terms of the Lewis picture, we would describe these as double and triple bonds. The results of the two descriptions are very similar, although the methods used to arrive at them are vastly different.

In summary, in application of the molecular orbital theory we start with the nuclei of each of the atoms fixed in the stereochemical arrangement corresponding to the established structure. Through a combination of the proper atomic orbitals, molecular orbitals characterizing that structure are established, and then all the available electrons are placed in the preferred molecular orbitals. If we were faced with the problem of describing the bonding in a molecule with unknown structure, obvious difficulties would arise that have not been important in our discussion of diatomic molecules.

Table 5-2 Directional characteristics of some hybrid bonds

sp	Linear
sp^2	Planar trigonal
sp^3	Tetrahedral
dsp^2	Square planar ($d_{x^2-y^2}$)
dsp^3	Square-based pyramid (d_{z^2})
d^2sp^3	Octahedral (d_{z^2} and $d_{x^2-y^2}$)
$(sp^2 + d_{z^2}p_z)$	Trigonal bipyramid

The most direct approach would be to try a variety of structures, derive appropriate molecular orbitals for each structure, and then evaluate them. Fortunately, chemical experience helps us to limit the number of structures by providing knowledge about preferred molecular geometries. This, of course, is the basis of classical stereochemical theory.

THE VALENCE BOND METHOD

The valence bond approach seeks to explain chemical bonding in terms that are more directly related to the observations of classical chemistry. For example, it would be helpful to understand the basis for the tetrahedral disposition of bonds about the carbon atom. Molecular orbital theory can account for this observation, but the necessary calculations are long and difficult. Valence bond theory, on the other hand, provides us with a simple yet reasonably accurate explanation. Let us begin with the example of the carbon atom to examine how this theory can be applied to a real chemical problem.

Let us pause to remember the electronic structure of atomic carbon: $1s^22s^22p^2$. Application of the principles discussed earlier in this chapter could well lead us into error unless we have thought the problem through in careful fashion. We might guess that the 2s electrons occupy a filled subshell and are consequently relatively undisturbed in the formation of a molecule incorporating the carbon atom. We would then be concerned principally with the *p* orbitals and anticipate formation of molecules such as CH_2. Best use of the *p* orbitals of carbon and the 1s orbitals of the hydrogen atoms to form molecular orbitals would occur if the H—C—H bond angle were 90°. Although there is abundant evidence for the existence of CH_2 as a highly reactive molecule (see p. 298), we recognize the much stronger tendency of carbon to form four bonds. One simple way to achieve a capability of forming four bonds would be to take one electron from the 2s orbital of the carbon atom and place it in the remaining, and yet unoccupied, 2*p* orbital. An atom in an *electronic excited* state is thus first formed from the ground-state atom; that is, we have supplied energy to rearrange the electronic configuration. The energy necessary to form the excited-state from the ground-state atom is referred to as **promotion energy.** Now each of the four orbitals in the second shell contains a single electron.

Although the carbon atom now has an apparent ability to form four bonds to hydrogen by overlap with the atomic orbitals of carbon to form molecular orbitals, a serious problem still remains: The three 2*p* orbitals are directed at right angles to one another, whereas the s orbital spherically surrounds the nucleus. If each hydrogen were in the most favorable position to interact with each atomic orbital, three of the hydrogens would be at the ends of the three *p* orbitals. Thus, each H—C—H bond angle would be 90° for this part of the structure. The fourth hydrogen would no doubt be somewhere on the far side of the carbon from the other three hydrogens and would interact with the carbon 2s orbital. We know, of course, that this does not represent the real structure of the methane molecule. We therefore ask how our theory can lead to suitably directed (and stronger) bonds.

The valence bond method suggests that the atomic orbitals present on the carbon atom be examined to see if possible recombination of them can be achieved to provide a better description of four equivalent orbitals. We say that the individual atomic orbitals are **hybridized** to form a new set of orbitals characteristic of the **valence** state of the atom. The mathematics of this operation is beyond our interests. It is sufficient for our purposes to observe that an equal mixture of each of the atomic orbitals is mathematically equivalent to the formation of four new identical orbitals directed towards the corners of a tetrahedron. These new orbitals have some of the character of s orbitals and some of the character of *p* orbitals, and thus it is appropriate to speak of them as **hybrid orbitals.** On a simple numerical basis, the hybrid orbitals have one-fourth s character and three-fourths *p* character. We speak of this as sp^3 hybridization. It is then a simple matter to form methane from the valence-state carbon atom. We simply bring four hydrogens into appropriate location with respect to the four hybrid orbitals and form bonds by allowing the overlap of the different orbitals.

Hybridization of orbitals can be carried out in any number of ways depending on the number of bonds to be formed, but there are a few hybrid orbitals that are particularly useful in the valence bond method. Table 5-2 shows the directional characteristics associated with certain types of hybridization. Figure 5-12 shows the individual shapes of sp^3 hybrid orbitals as well as the tetrahedral disposition of these four hybrid orbitals. In this particular figure the small tail of the individual orbital has been omitted for clarity of the drawing. Figures 5-13 and 5-14 show somewhat more explicit diagrams of the bonding orbitals that can be formed by mixing an s atomic orbital with one or two *p* orbitals. If only one *p* orbital is used, two hybrid orbitals directed at 180° to each other are formed; if two *p* orbitals are used, a triangular disposition of sp^2 hybrid orbitals is attained.

Figure 5-12 An individual sp^3 hybrid orbital, and the collection of four orbitals showing tetrahedral orientation.

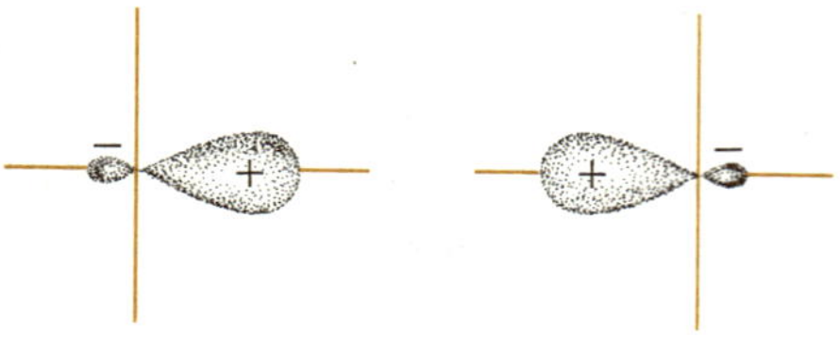

Figure 5-13 Two *sp* hybrid orbitals.

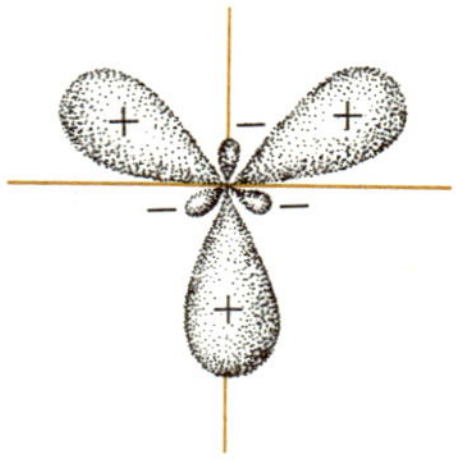

Figure 5-14 Three sp^2 hybrid orbitals.

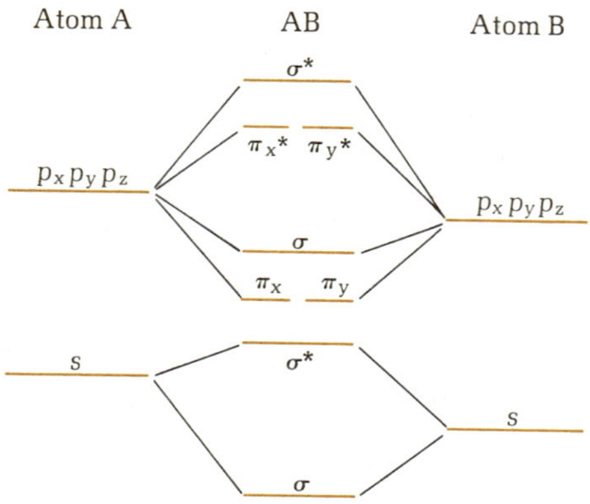

Figure 5-15 Molecular orbitals for diatomics of unequal electronegativities.

For elements in the third complete row of the periodic table and beyond, there are also *d* orbitals of energies sufficiently close to s and *p* orbitals so that hybrid orbitals may involve mixtures with *d* character as well.

Table 5-2 includes configurations adopted by three such hybrid orbitals. Note that either a trigonal bipyramid or a square-based pyramid needs five atomic orbitals and can be achieved by use of only a single *d* orbital. If the unique axis of the pyramid is defined as the z axis, use of the d_{z^2} orbital produces a trigonal bipyramid, whereas use of the $d_{x^2-y^2}$ orbital yields the square-based pyramid. Use of both the d_{z^2} and the $d_{x^2-y^2}$ orbitals results in an octahedral disposition of the hybrid bonds. These hybridizations will be of importance to us in Chapter 6.

HETERONUCLEAR MOLECULES

We have previously examined the molecular orbitals used in diatomic molecules composed of identical atoms. The great diversity of chemical compounds arises from the many types of combinations that can be formed between unlike atoms, and we must now extend our theories to this situation.

In Chapter 4, we have seen that isolated atoms differ considerably in the strength of interaction with electrons; for example, systematic variation in both ionization potential and electron affinity occurs in a series of atoms with increasing atomic number. These two properties simply reflect the fact that some atoms hold their electrons more strongly than do others. We might expect that when a chemical bond is formed between two atoms with markedly different electron-holding power, that bond must in some way reflect the preference of the individual atoms. Figure 5-15 indicates the way in which atomic orbitals combine to form molecular orbitals in the general case of unlike nuclei. For simplicity's sake, this diagram shows only the orbitals principally involved in bonding, the **valence orbitals.** One or more complete shells of underlying atomic orbitals may be filled, but in general the molecular orbitals do not disturb the energy or the electron distribution in these orbitals significantly.

Let us concentrate first on the sigma orbital formed by overlap of the individual atomic s orbitals. Note that as in the *homonuclear* case, both a bonding and an antibonding sigma orbital are formed. Atom B, however, corresponds to a more stable home for electrons than atom A (an electron in the s orbital of B is of lower energy); the bonding orbital is closer to atom B in energy, whereas the antibonding orbital is closer to that of A. The importance of this observation is perhaps most easily seen by examining electron distribution in these orbitals as shown in Figure 5-16. Note that in the bonding sigma orbital the electrons tend to spend slightly more

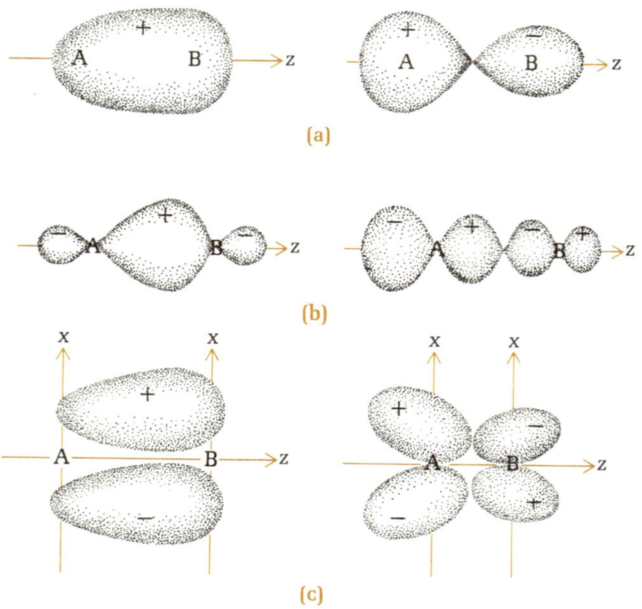

Figure 5-16 Electron distribution of molecular orbitals of atoms with B more electronegative than A. (a) Sigma orbitals formed from two s orbitals. (b) Sigma orbitals formed from two *p* atomic orbitals lying on a common axis. (c) Pi-type orbitals formed from atomic p_x orbitals. (From H. B. Gray, *Electrons and Chemical Bonding*, Benjamin, Menlo Park, Calif., 1964. Copyright, © 1964, W. A. Benjamin, Inc.)

Table 5-3 Some properties of bonds in diatomic molecules and ions

	Bond order	Bond length (Å)	Bond strength (kcal/mole)
CC	2	1.31	150
BN	2	1.28	92
BO	$2\frac{1}{2}$	1.20	185
CN	$2\frac{1}{2}$	1.17	188
CO^+	$2\frac{1}{2}$	1.12	
CO	3	1.13	256
NO^+	3	1.06	
CN^-	3	1.14	
N_2	3	1.10	225

of their time near atom B than near atom A, whereas just the reverse is true in the antibonding orbital. Figure 5-16 also shows that the corresponding statement applies to the sigma orbital formed by overlap of the p_z orbitals along the interatomic line and to the pi orbital formed by overlap of either the p_x or the p_y orbitals. It is often important to remember and note explicitly that the electron distribution within a bond is not symmetrical, but is polarized to place greater electron density near the more electronegative element. A convenient way of indicating this is by use of a symbol to indicate a separation of charge, $\overset{\delta+\delta-}{\text{A—B}}$. It is, of course, just this distortion of electron density in the bonds that result in the "distorted balloon" pictures of Chapters 2 and 3.

It will be instructive to compare a series of heteronuclear molecules with the homonuclear cases studied earlier. Table 5-3 collects some properties of interest. The net number of bonds formed between two atoms, the **bond order,** can readily be deduced from the number of electrons in bonding orbitals in excess, divided by two, of those in antibonding orbitals. The most significant observation to be made is the following: A direct correlation exists between the number of bonds thus calculated and the bond length or bond energy. For the moment we shall simply note that bond energy is a measure of how tightly two bonded atoms are held together. The greater the energy, the stronger the bond. Detailed discussion of this is provided in Chapter 8.

In general, the greater the number of bonds between the two atoms, the closer the atoms approach one another and the greater the bond strength. A series of molecules or ions having essentially the same electronic structure, such as BO, CN, and CO^+, are said to be **isoelectronic.** This is an important concept, since it permits us to recognize similarities in geometry, bond order, and other parameters of bonding for what might appear to be quite distinct species.

Triatomic molecules and other more complex species can be treated essentially in the same fashion. Figure 5-17 shows the molecular orbitals of the linear molecule BeH_2. Note that the most stable bonding sigma orbital places a heavier concentration of electrons near the hydrogens than the beryllium, and the antibonding orbital reverses this order. This polarization follows directly (as previously explained) from the fact that hydrogen is a substantially more electronegative element than beryllium. Note that a second bonding molecular orbital of sigma type is also present, which concentrates electron density to an even greater extent on the hydrogen atoms. In fact, just as the atomic $2p_z$ orbital has a node at the beryllium atom, so does the molecular orbital formed from it shown in Figure 5-17(b). Figure 5-18 shows a diagrammatic comparison of the orbital energy levels, and we may deduce that the four bonding electrons present in beryllium

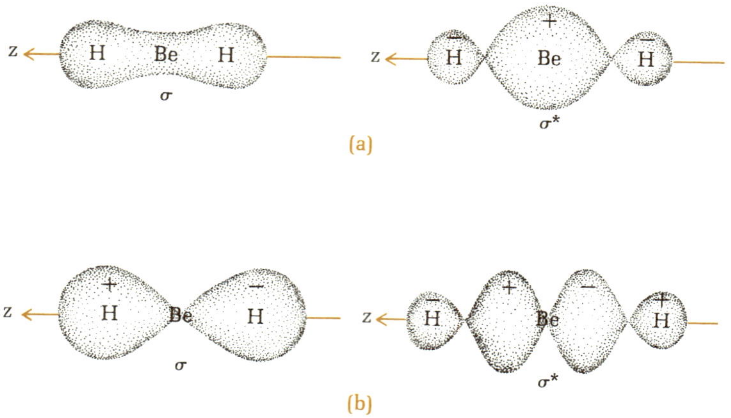

Figure 5-17 Electron distributions in molecular orbitals for BeH_2 (H more electronegative than Be). The orbitals in (a) are formed using the 1*s* hydrogen orbital and the 2*s* orbital of beryllium, whereas the $2p_z$ orbital of beryllium is used for the molecular orbitals shown in (b). (From H. B. Gray, *Electrons and Chemical Bonding,* Benjamin, Menlo Park, Calif., 1964. Copyright, © 1964, W. A. Benjamin, Inc.)

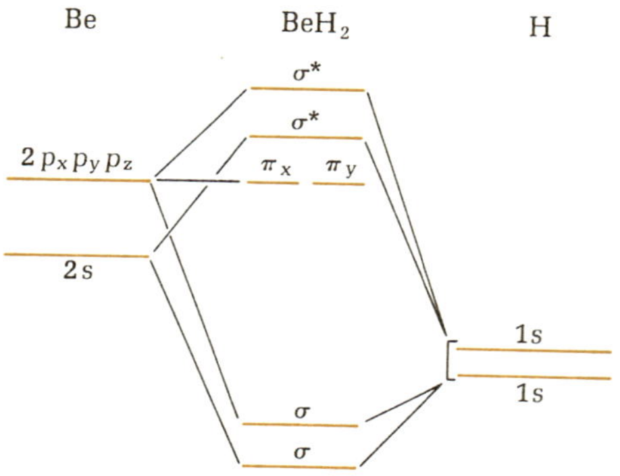

Figure 5-18 Molecular orbital energy levels for BeH_2.

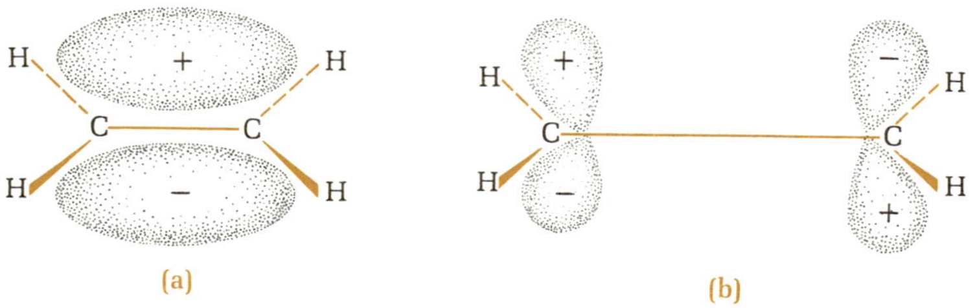

Figure 5-19 Pi bonding (a) and antibonding (b) molecular orbitals from $2p_z$ atomic orbitals in ethylene.

hydride will occupy the two bonding sigma orbitals. The remaining 2*p* orbitals on the beryllium atom are essentially unaffected by bond formation.

Let us now turn to two simple examples of the application of molecular orbital theory to carbon compounds. Specifically, we wish to examine the electronic structures of the ethylene molecule C_2H_4 and benzene molecule C_6H_6. The structure of the ethylene molecule was discussed in Chapter 2. All six of the atoms are in the same plane, and the observed H—C—H angle is 117°. The electronic structure of ethylene is most commonly discussed by a combination of a valence bond with a molecular orbital description. (We should note that mixture of the theories is not essential. If the atomic nuclei are fixed in the proper places and all orbitals are used to form molecular orbitals, some will be very nearly localized; these are considered to be valence bonds in the treatment we give here.) On each carbon, 2*s*, $2p_x$, and $2p_y$ orbitals are first mixed to form a set of hybrid sp^2 orbitals. One of the hybrid bonds from each carbon points towards the other carbon, and the remaining two point toward the hydrogen nuclei. Overlap of each hybrid orbital with the 1*s* orbital on the hydrogen results in bond formation similar to the bonds discussed above for BeH_2. Since the electronegativity of hydrogen and carbon are very nearly identical, the bond is not markedly polarized in either direction. Similarly, the hybrid orbitals of the carbons that point toward one another overlap to form a sigma bond substantially equivalent to that which would be formed in a homonuclear diatomic molecule. A total of 12 electrons are available in the valence shell (4 from each of the two carbons and 1 from each of the four hydrogens). We have thus far described five sigma bonds holding the framework of the molecule in place, and these require 10 electrons. There remain 2 electrons and only two orbitals, the p_z atomic orbitals of the two carbon atoms. These are properly oriented to overlap and allow formation of a pi bond of the type shown in Figure 5-19. A corresponding antibonding orbital is, of course, formed simultaneously, but electrons are not available to fill this orbital. In this view, the two halves of the molecule are held together by a direct sigma bond between the carbon atoms and a pi bond. Both of these bonding orbitals are filled with two electrons, and thus a *double bond* exists between the halves.

The benzene molecule may be similarly described, but a new feature emerges. The benzene molecule consists of six carbons arranged at the corners of a regular hexagon, with each hydrogen bonded to a carbon by a bond directed away from the center of the hexagon. A description of the bonding in the molecule is suggested by the geometry. The angles formed at the points of the hexagon are 120°, and we are thus strongly led to an sp^2 hybridization for each carbon atom. Two of the hybrid orbitals on each carbon are directed toward adjacent carbons, and the third is directed toward the attached hydrogen. As with ethylene, overlap of these orbitals

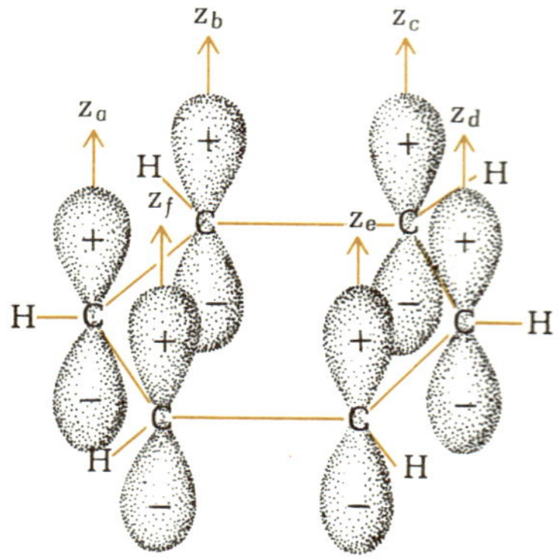

Figure 5-20 $2p_z$ orbitals on carbon of benzene used to form delocalized molecular orbitals. (From H. B. Gray, *Electrons and Chemical Bonding,* Benjamin, Menlo Park, Calif., 1964. Copyright, © 1964, W. A. Benjamin, Inc.)

Figure 5-21 Pi bonding and antibonding molecular orbitals of benzene. (From H. B. Gray, *Electrons and Chemical Bonding,* Benjamin, Menlo Park, Calif., 1964. Copyright, © 1964, W. A. Benjamin, Inc.)

then leads to three sigma bonds about each carbon. A count of the electrons will show that of the 30 electrons available in the valence shell, the bonding pattern thus far described accounts for 24. Six electrons remain to be accounted for, and the six p_z orbitals centered on the six carbons are available to house them.

Two approaches can be used at this stage. Overlap of the p_z orbitals on two adjacent carbons could lead to a pattern similar to that in ethylene, with the formation of alternate single and double bonds. It is at once clear that there are two possible patterns for the formation of such a bonding arrangement, and the choice between them would seem most arbitrary indeed. In fact, a careful study of the structure of benzene shows that all six bonds are identical in length and suggests that the true electronic structure must be some average of the limiting structures obtained in this method. In valence bond terms, we speak of **resonance** involving the limiting structures and call the "average" structure a **resonance hybrid.** The molecular orbital approach is somewhat more complex but avoids this unsatisfying feature of the valence bond method. Figure 5-20 shows the six available orbitals, which are to be combined to form six molecular orbitals. The resulting molecular orbitals are shown in diagrammatic form in Figure 5-21. Three bonding and three antibonding orbitals are formed, and each is capable of accepting one electron pair. Clearly, the six available electrons just suffice to fill the bonding pi orbitals. From the point of view of the number of bonds connecting the carbon atoms, these six electrons are equivalent to three bonds spread over the six atoms. In addition, there are six sigma bonds or a total of nine bonds connecting the six atoms. On the average, each carbon is attached to the adjacent one by one-and-a-half bonds, but the molecular bonding picture does not attempt to average these in the way necessary in the valence bond approach. We shall see other examples of similarly delocalized bonds in the final section of this chapter.

PERIODIC PROPERTIES REVISITED

At the end of Chapter 4, we noted that three factors sufficed to account for the unusual characteristics of the first row of elements in the periodic table compared to others in the same chemical family. Two of these related to size and the number of low-energy orbitals available for bonding. The final factor is *the ability of these elements to form pi bonds with themselves or with other first-row elements*. We have encountered two examples: ethylene and benzene. In these molecules, the formation of pi bonds accounts for the existence of multiple bonds. Such pi bonding is not very important for heavier elements, perhaps because these atoms are too large to permit effective overlap of the *p* orbitals. At any event, silicon, phosphorus, and other elements do not readily form double or triple bonds to themselves or to other elements. This simple fact has enormous implications for the understanding of the descriptive chemistry of the first-row elements compared to heavier ones. Let us consider one or two examples.

First, let us consider the elements themselves. Oxygen can form multiple bonds with itself and so exists as a gaseous diatomic molecule: O=O. But sulfur cannot, and this element exists as a solid under the same conditions, with a complex ring structure (see Chapter 3). Similarly, nitrogen forms multiple bonds with itself and the element exists as N≡N, a volatile gas. But the element immediately below nitrogen in the periodic table, phosphorus, does not form such bonds and, as a consequence, exists as a polymeric solid in the elemental form (Chapter 3).

The oxides of carbon and silicon provide other examples. Since carbon forms multiple bonds to oxygen, its oxide has the simple structure O=C=O, carbon dioxide. But silicon does not form multiple bonds to oxygen and so exists as a polymeric SiO_2 structure (Chapter 3) in several modifications. Other examples of this type of behavior occur to us with a little thought. This is a good time to review some of the structural and descriptive chemistry presented in the earliest chapters.

FURTHER CONSIDERATION OF MOLECULAR GEOMETRY

Up to now, we have been concerned with how theories of chemical bonding can account for what we know about molecular geometries; this line of argument is continued in the next chapter as well. But we can turn the direction of the argument around if we wish and ask what the bonding theories will tell us about the geometries of structures with which we are not familiar. For example, is CO_2 a linear molecule? Is O_3 a linear molecule? Is NO_3^- a planar ion? These are not questions with which we have dealt previously, but it turns out that we can use valence bond and molecular orbital theory to develop reasonably accurate pictures of bonding in these structures and, hence, their geometries. In terms of these examples, let us see how this is done.

For each case, it will prove convenient to follow this procedure:

1. Draw a Lewis electron dot structure for the species in question.
2. All orbitals involved in the formation of sigma bonds or occupied by unshared electron pairs are considered to be valence (localized) orbitals. The geometry of these is deduced in terms of the hybridization about each atom necessary to provide the requisite number of orbitals.
3. Multiple bonds are considered to be formed from one localized sigma bond and one or more localized or delocalized pi bonds.

Using the procedure outlined in these three parts, let us consider the structure of carbon dioxide, CO_2. To begin with, we recognize that there are 16 valence electrons to be accounted for. The best Lewis structure that we can draw, remembering the octet rule, is

:Ö::C::Ö:

Note that this picture accounts for the 16 valence electrons and that each atom has a complete octet. Now notice that the central carbon atom makes two sigma bonds, one to each of the oxygens, but possesses no unshared electron pairs. It follows that the hybridization of this carbon atom will be *sp*. The remaining two *p* orbitals will form pi orbitals extending over the whole molecule and providing for the double bonds to oxygen. We have already answered our question about the geometry of CO_2: The *sp* hybridization at carbon requires that the molecule be linear. But we can continue with the argument and develop a fuller picture of the bonding in CO_2. Turning now to oxygen, we note that each oxygen makes one sigma bond (to carbon) and has two unshared pairs of electrons. To accommodate these, hybridization at oxygen must be sp^2. The remaining *p* orbital, then, is involved in pi bond formation. Overall, the structure of CO_2 then is

Suppose that we turn to the slightly more complicated example of sulfur dioxide, SO_2. In this case there are 18 valence electrons to worry about. The best Lewis structures that can be drawn are

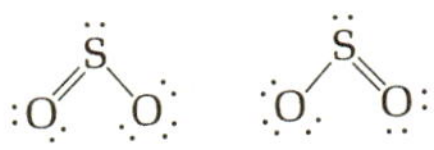

Note that provision is made for all 18 electrons and that all atoms satisfy the octet rule. The two Lewis structures are quite equivalent, just as they are in the case of benzene, so we know that the actual structure of SO_2 will be a resonance hybrid of these two. Note that the central sulfur atom must form two sigma bonds and accommodate an unshared electron pair. This requires a total of three valence orbitals, and it follows that hybridization at sulfur will be sp^2. We conclude that this molecule, in contrast to CO_2, is bent. The remaining *p* orbital on sulfur will be involved in the formation of a pi molecular orbital extending over the complete molecule. The same type of analysis reveals that each oxygen atom must also be

This painting by Georges-Pierre Seurat, *Port-en-Bessin, Entrance to the Harbor,* is an excellent example of work by the pointillist school of painters. Note how units of dark and light are combined to form an image with properties not present in any of the elements that form it. Similarly, combination of atoms in specific ways leads to formation of molecules having new and unique properties. [Seurat, *Port-en-Bessin, Entrance to the Harbor* (1888), oil on canvas, $21\frac{5}{8} \times 25\frac{5}{8}''$, Collection, The Museum of Modern Art, New York. Lillie P. Bliss Collection.]

hybridized sp^2. Thus, a picture of bonding in SO_2 is

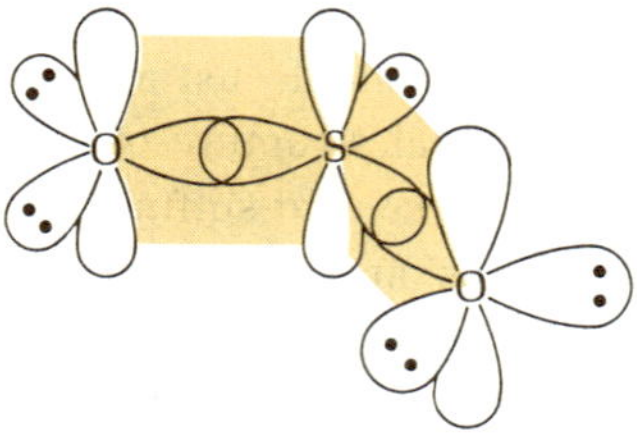

Example *Is ozone, O_3, a linear or a bent molecule?*

Solution *This case is very similar to that for SO_2. There are again 18 electrons to be accounted for, so that the best Lewis structures are*

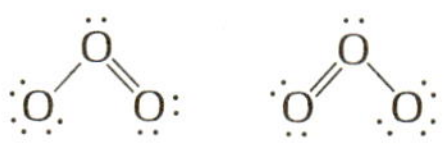

Note that ozone, too, is a resonance hybrid of these two valence bond pictures. To form two sigma bonds and accommodate the unshared electron pair, the central oxygen atom must be hybridized sp^2. It follows that ozone is bent.

Let us consider one final example of deducing geometries from bonding theories: NO_3^-. In this case there are a total of 24 electrons, 5 from the nitrogen, 6 from each oxygen, and 1 to account for the net unit negative charge. Three Lewis structures are equally acceptable for this ion:

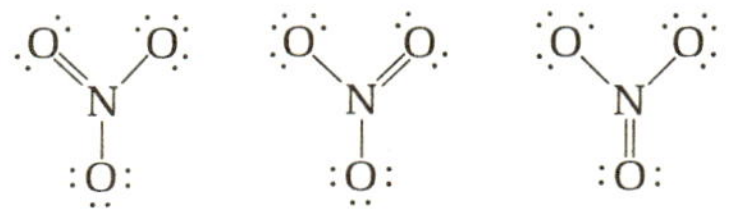

Since these structures are equivalent, we once again recognize that the actual structure of the NO_3^- ion is an average of the three. Thus, each of the nitrogen–oxygen bonds has a bond order near 1.33. The central nitrogen must make three sigma bonds, one to each oxygen. Therefore, the hybridization must be sp^2. It follows that NO_3^- is planar and has O—N—O bond angles of 120°. The *p* orbital on nitrogen and a *p* orbital on each of the oxygens form a pi molecular orbital that is occupied by the pair of electrons involved in formation of the multiple bonds:

The simple ideas developed here can be extended to arrive at the geometries for a great many species. Some additional examples are provided in the study problems at the end of the chapter.

Theories of chemical bonding and their relation to molecular structure have been developed to a point adequate for our purposes. In the next chapter we proceed to examine some of the consequences of these considerations for molecules having new and more complex geometries.

Glossary

Bond order the number of chemical bonds linking two bonded atoms; may be calculated by subtracting the number of electrons in antibonding orbitals from the number in bonding orbitals and dividing the difference by two.

Electron acceptor–donor bond (dative bond) a chemical bond formed in the interaction of a Lewis acid with a Lewis base. The members of the structure created through formation of this type of bond retain more of their individual character than would be the case for covalent bond formation.

Hybrid orbitals those atomic orbitals derived by the process of hybridization; sp^2, sp^3, dsp^3, and so forth.

Hybridization the construction of new sets of atomic orbitals by combining various sets of *s*, *p*, and *d* orbitals.

Isoelectronic structures those chemical species, atoms, molecules, or ions that have closely related electronic structures.

Lewis acid any chemical species having a vacant orbital and capable of accepting a pair of electrons.

Lewis base any chemical species having appreciable capacity to donate a pair of electrons to a Lewis acid.

Molecular complex the class of structures derived from formation of an electron acceptor–donor bond.

Molecular orbital theory a theory of chemical bonding in which both bonding and nonbonding electrons are placed in orbitals that extend over the entire molecule and that are constructed from the orbitals of the atoms involved.

Promotion energy the energy required to form an electronically excited state from the ground state.

Resonance a molecule or ion for which more than one equivalent valence bond representation may be written is said to exhibit resonance; that is, it has a structure that is an average of the individual valence bond pictures.

Resonance hybrid the structure of any molecule or ion for which resonance exists is said to be a resonance hybrid.

Valence bond theory a theory of chemical bonding in which electrons occupy localized orbitals that are either the atomic orbitals themselves or orbitals derived from them by hybridization.

Valence orbitals those orbitals involved in chemical bonding.

Study problems

1. Reasoning from molecular orbital theory, indicate how many chemical bonds ought to connect the two carbon atoms in C_2. Would you expect any electrons to be unpaired? Indicate clearly your line of reasoning used in reaching these conclusions.
2. The species N_2 and NO^+ both have ten electrons in their valence shells. Use a molecular orbital energy diagram to compare bonding in the two molecules (oxygen is more electronegative than nitrogen) and compare

the electron distribution in the two sigma bonding levels of the two molecules.

3. Show how electrons are distributed in the bonding sigma orbital of the B—N molecule (overlap of the two $2p_z$ orbitals of the atoms). Information:

	Atomic number	Atomic weight	Ionization potential	Electronegativity
B	5	10.8	8.3	2.0
N	7	14.0	14.5	3.0

4. Explain in terms of molecular orbital theory why chemists do not consider it a worthwhile task to try to prepare Be_2.
5. Consider the diatomic molecule B_2 from a molecular orbital point of view. Show what orbitals are used and predict the B—B bond order.
6. Show how electrons are distributed in the bonding sigma orbital of the NO molecule (overlap of the two $2p_z$ orbitals of the atoms). Information:

	Atomic number	Atomic weight	Ionization potential	Electronegativity
N	7	14.0	14.5	3.0
O	8	16.0	13.6	3.5

7. Using the molecular orbital scheme derived for homonuclear diatomic molecules, predict whether the following molecules will be stable:
   a. HH^+ b. BeB
   Atomic numbers are H = 1, Be = 4, and B = 5.

Suggestions for further reading

A. L. Companion *Chemical Bonding,* McGraw-Hill, New York, 1964.

H. B. Gray *Electrons and Chemical Bonding,* Benjamin, Menlo Park, Calif., 1964.

E. Griswold *Chemical Bonding and Structure,* Heath, Lexington, Mass., 1968.

P. F. Lynch *Orbitals and Chemical Bonding,* Houghton Mifflin, Boston, 1966.

H. H. Sisler *Electronic Structure, Properties, and the Periodic Law,* Van Nostrand Reinhold, New York, 1963.

J. C. Speakman *Molecules,* McGraw-Hill, New York, 1966.

6

More about molecular geometry

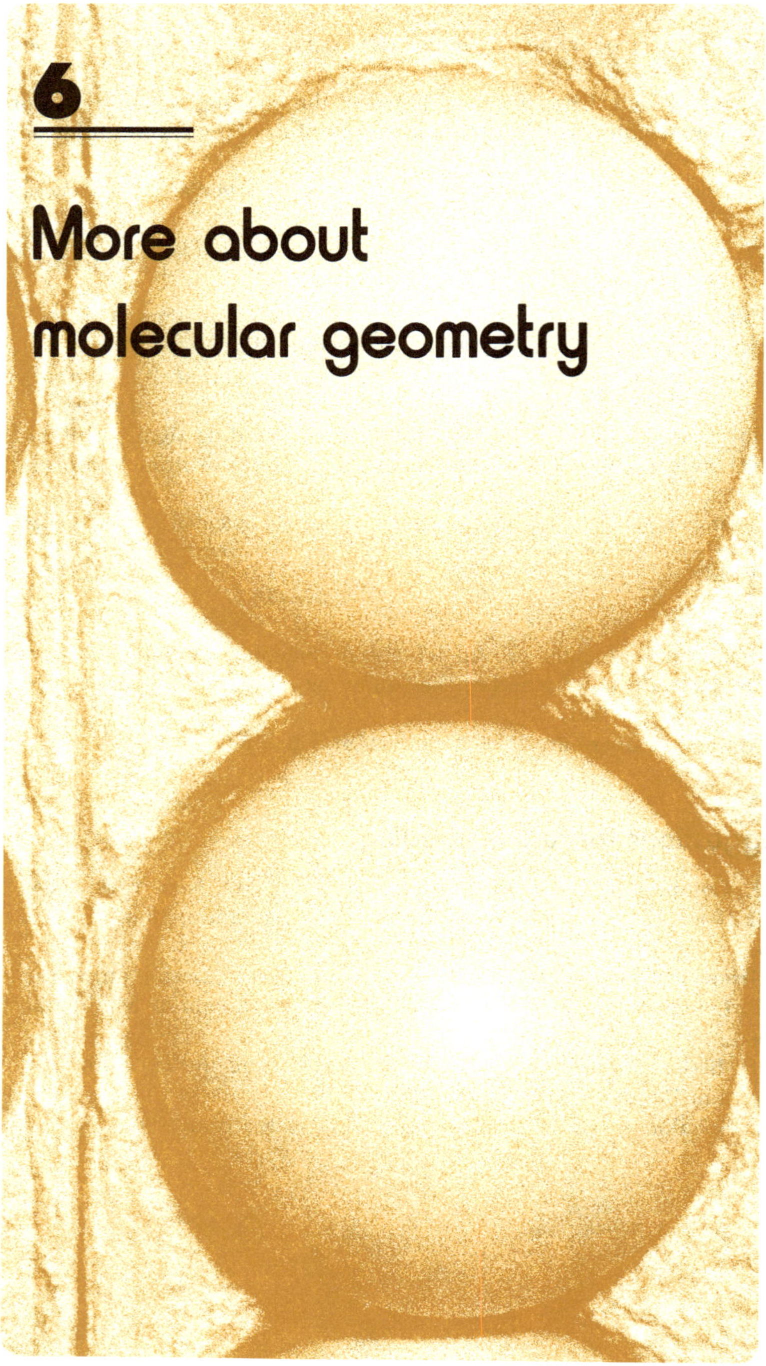

Philosophy is written in this grand book—I mean the universe—which stands continually open to our gaze, but it cannot be understood unless one first learns to comprehend the language and interpret the characters in which it is written. It is written in the language of mathematics, and its characters are triangles, circles, and other geometrical figures, without which it is humanly impossible to understand a single word of it; without these, one is wandering about in a dark labyrinth.

Galileo, 1623

In Chapters 2 and 3 we concentrated our attention on molecules of tetrahedral geometry. Although tetrahedral disposition of bonds is of prime importance in the chemistry of carbon compounds and finds ample illustration in the chemistry of other elements as well, several other geometric arrangements are important in the study of many of the elements. We shall examine examples of several nontetrahedral geometries, and we shall both review and apply the concepts of symmetry as applied to these structures. We shall also examine the electronic factors that lead to various geometries.

Many of the examples we shall use in this chapter are molecules (or ions) formed by metals acting as *Lewis acids* in combination with groups called **ligands** acting as *Lewis bases*. When compounds of this type were first prepared, the nature of the bonding was not at all well understood and they were said to be **metal complexes.** More commonly, they are now called **coordination compounds.** The metals are characterized by low electron affinities and low ionization potentials. In reactions with more electronegative elements, metals tend to become positive ions by loss of one or more electrons from their outermost shell. Both the orbitals vacated when these electrons are lost and empty orbitals of similar energy are available to take part in electron sharing with a suitable ligand.

Ligands are simply ions or molecules that have electron pairs not used in bonding. They may be as simple as the chloride ion, which has filled 3s and 3*p* orbitals, or they may be more complex molecules, such as ammonia, that have one unshared pair of electrons. Table 6-1 lists some typical ligands that are often found joined to metal atoms. Note that the ligands may be either electrically neutral or negatively charged.

Analysis of the structure of a metal complex can be simply carried out in stages. Let us consider a complex formed from a metal M and several ligand molecules L bearing a charge *c*: $ML_m{}^c$. The charge on the complex may be either positive, negative, or zero. Thus, typical coordination compounds have the formulas $Co(NH_3)_6{}^{3+}$, $FeCl_4{}^-$, and $Ni(CO)_4$. To understand the structure of such complexes, we wish first to derive the electronic structure of the central atom M. This can be readily done by studying the number and type of ligand as well as the net charge on the complex and observing that the overall charge must be accounted for by the sum of the

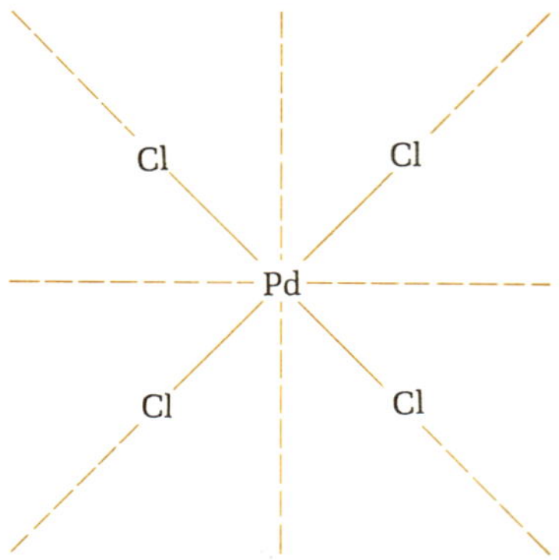

Figure 6-1 The structure of the $PdCl_4^{2-}$ ion with indicated symmetry elements. The dashed lines show locations of mirror planes perpendicular to the plane of the paper and also represent twofold axes in the plane.

Table 6-1 Electron dot structures for several common ligands

$:\ddot{F}:^-$	$:C\equiv N:^-$	$:NH_3$ (H—N(—H)—H)
$:\ddot{Cl}:^-$	$:\dot{N}=\ddot{O}:$	$:PF_3$ (F—P(—F)—F)
$:\ddot{Br}:^-$	$:C\equiv O:$	$:\ddot{S}(CH_3)—CH_3$
$:\ddot{I}:^-$	$:H^-$	$H_2\ddot{N}—CH_2—CH_2—\ddot{N}H_2$

individual charges. In $Co(NH_3)_6^{3+}$, for example, the three positive charges must be the sum of the charges on the cobalt atom and the six molecules of ammonia. Since we know that the ammonia molecule is uncharged, the cobalt ion must be the source of the three positive charges in this case. In this compound, then, the electronic structure of cobalt must account for the 27 electrons present in the neutral atom less the 3 electrons responsible for the charge.

The $FeCl_4^-$ ion, in contrast, bears a net negative charge. As Table 6-1 shows, each chloride ion carries a single negative charge as a ligand. If four chlorides in the complex carry a total of four negative charges yet the complex as a whole bears only a single charge, the iron atom must cancel three of them and is itself assigned a +3 charge; it now has 26 − 3 or 23 electrons about the nucleus. Finally, in the compound $Ni(CO)_4$ we recognize that each CO is uncharged (see Table 6-1). Since the complex as a whole is uncharged, the Ni atom carries just the 28 electrons associated with the neutral atom.

Example *What is the electrical charge of the metal atom in each of the following coordination complexes?*
a. $CoCl_3(NH_3)_3^{-1}$ b. $Fe(CN)_6^{3-}$ c. $Co(NO)(CO)_3$

a. *The three chloride ions contribute one negative charge each; the ammonias are neutral. It follows that the cobalt bears a net charge of +2.*
b. *The cyanides contribute one negative charge each. Hence, the iron must have a charge of +3.*
c. *The NO and CO ligands are neutral. Since the complex is electrically neutral too, the cobalt must be present as the neutral atom.*

Let us now examine a number of examples to see how the molecular geometry of a large number of elements may be correlated by valence bond theory.

SQUARE-PLANAR COMPOUNDS

The metals of Group VIII of the periodic table (and particularly the compounds of platinum and palladium) provide the most common examples of square-planar structures. Figure 6-1 shows the structure of the $PdCl_4^{2-}$ ion and indicates some of the symmetry elements present. Perhaps the most obvious symmetry elements are the fourfold axis perpendicular to the plane of the paper and the mirror plane represented by the paper itself. Indicated in the figure by dashed lines are four mirror planes perpendicular to the plane of the paper. Can you see that the fact that there are four such planes necessarily follows from the presence of the fourfold axis? Note that the same dashed lines are also twofold axes of rotation in the plane of the paper. The central atom lies at a center of symmetry in the molecule as well. Finally, we should recognize that if a molecule has a fourfold axis, that axis can also be viewed as a twofold axis. Since this will always be true we rarely call attention to it specifically, although a truly exhaustive list of symmetry elements would include both the twofold and the fourfold axes. We can see then that the square-planar structure is one of very high symmetry indeed.

How does the electronic structure of the palladium atom lead to the square-planar geometry for the $PdCl_4^{2-}$ ion? Let us first investigate the electronic structure of the atom in the complex. The four chloride ligands

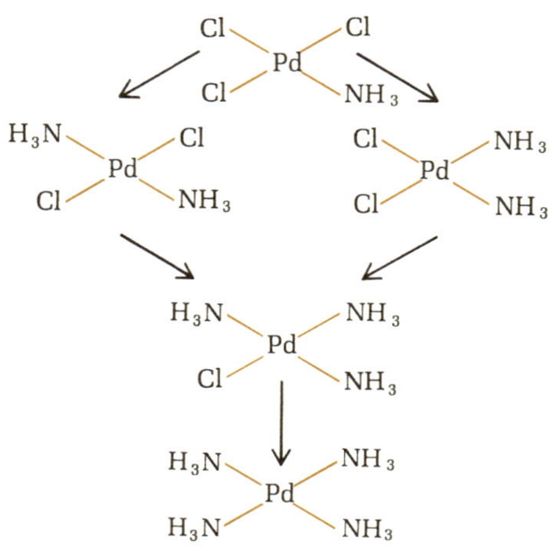

Figure 6-2 Structures derived from the $PdCl_4^{2-}$ ion by successive replacement of Cl^- ions by NH_3 molecules.

contribute four negative charges to the complex, yet the net charge is only -2; clearly, the palladium atom must have a $+2$ charge to accomplish this change. The electronic structure of the neutral palladium atom is given in Table 4-2 as (Kr)$4d^{10}$. The two electrons lost must then be removed from the 4*d* shell, leaving one of the *d* orbitals vacant. To allow formation of bonds to the four chlorines, four orbitals will be required from the metal. The lowest additional orbitals still unfilled in palladium are the 5*s* and 5*p* orbitals. The four orbitals used are then the 4*d*, 5*s*, and two 5*p*'s. If the molecular symmetry axis is the *z* axis, the p_x and p_y orbitals as well as the $d_{x^2-y^2}$ and *s* orbitals are all in the proper plane for bonding. Hybridization to form dsp^2 orbitals then allows bonding of the four chlorines in the square-planar arrangement.

In the treatment given above, we have first partitioned the electrons to give electron-rich ligands and electron-poor metals, hybridized the metal orbitals, and then formed covalent bonds using the electrons from the ligand only. An alternative, molecular orbital method is possible, which reaches the same net description of bonding without the somewhat artificial partition of electrons. The valence bond procedure will be used, however, because of its elegant simplicity, for the remainder of our discussion.

Example *Using the valence bond method, predict the shape of* $Pt(PF_3)_2Cl_2$. *Draw all the isomers of this coordination complex.*

Solution *The* PF_3 *ligands carry no charge; therefore, the platinum atom must have a charge of* $+2$ *to balance the negative charges contributed by the chlorides, since the complex itself is electrically neutral. Referring to Table 4-2, we can deduce that the electronic structure for* Pt^{2+} *is* (Xe)$4f^{14}5d^{8}$.

Hence, the four bonds to the ligands will be formed from one d, one s, and two p orbitals: dsp² hybrids are square-planar. *There are just two isomers:*

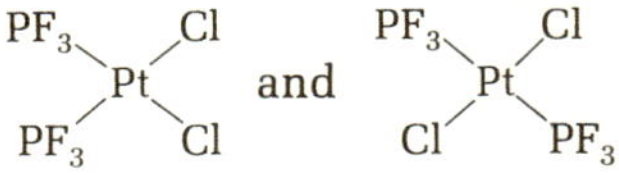

It is instructive to follow the changes that occur in the symmetry of the molecules as we perform successive substitutions, just as we did earlier for the methane molecule. Figure 6-2 shows the structure of the $PdCl_3NH_3^-$ ion and subsequent substitution products. Note that the single substitution has destroyed a number of symmetry elements present in the parent structure. The center of symmetry, the fourfold axis (and the twofold axis coincident with it), three of the four mirror planes perpendicular to the plane of the paper, and three of the four twofold axes in the plane of the molecule have all disappeared. In fact, there now remain only the mirror plane of the paper, one mirror plane perpendicular to it, and the twofold axis at their intersection.

It is easy to observe that no structure can ever be devised for these planar molecules that would destroy the mirror plane that includes all atoms. Therefore, every planar molecule has this minimum symmetry element, and this is sufficient to prevent optical activity for any substitution pattern. Other types of isomerism are possible, however. Figure 6-2 shows the *cis* and *trans* modifications possible for the $PdCl_2(NH_3)_2$molecule. *Cis-trans* isomerism is very common in square-planar molecules, and a large number of related pairs has been prepared.

TETRAHEDRAL METAL COMPLEXES

Let us examine two additional coordination compounds of metals to see examples of tetrahedral complexes. The complex $CdCl_4^{2-}$ is an ion that appears to be very similar to the $PdCl_4^{2-}$ ion just discussed. However, the chlorides are positioned about the central metal in tetrahedral fashion. Although the cadmium atom (atomic number 48) has just two more electrons than palladium, the structure is decidedly different. Why do two more electrons have such a dramatic effect?

Let us carry through our analysis just as we did for the $PdCl_4^{2-}$ ion. First, we observe the four negative charges of the four chloride ions must be compensated by two positive charges on the cadmium to account for the net -2 charge on the ion. The cadmium atom has the electronic structure $(Kr)5s^24d^{10}$ according to Table 4-2, and two of the electrons must be removed to form the ion. Should we remove them from the 5s or from the 4*d* shell?

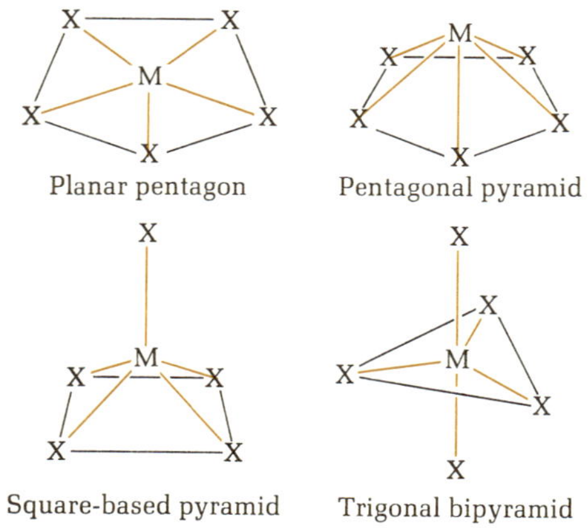

Figure 6-3 Possible geometries for distribution of five groups about a central atom.

Fortunately, nature is quite systematic in its operation, and when such a question arises it always turns out that the electrons are taken from the shell with the largest principal quantum number, the 5s level in this case. The cadmium ion is then left with a filled 4*d* shell, and both *s* and *p* levels unoccupied in the fifth shell. Consequently, when four atomic orbitals are hybridized to form bonds to the four chloride ions, the 5s and three 5*p* orbitals are used; tetrahedral geometry results.

The compound $Ni(CO)_4$ provides us with an illustration of one final point. Here the complex carries no charge, and further, the ligands are uncharged CO molecules; consequently, the nickel atom has the same number of electrons as the neutral atom, 26. The structure of the element itself is $(Ar)4s^2 3d^8$, but prior to hybridization *all electrons are packed into the shells of lowest principal quantum number* to give $(Ar)3d^{10}4s^0$. Now, hybridization of four orbitals to free four orbitals for bonding to CO finds the 4*s* and 4*p* orbitals available, and again a tetrahedral molecule results from the sp^3 hybridization.

MX_5 MOLECULES

It is obvious that nontetrahedral structures must be written for atoms in which five or more groups are bonded to the central atom; we have already seen (in Chapter 3) a few such cases. Generally, nature seems to prefer structures of at least some degree of symmetry, and in the case of five groups distributed around a central atom there are at least four highly symmetrical possibilities that warrant our attention. These are illustrated in Figure 6-3. It will be convenient to refer to these four structures as **planar-pentagonal,**

pentagonal-based pyramidal, square-based pyramidal, and **trigonal-bipyramidal,** respectively. In the square-based pyramidal case, the M atom might be in, above, or below the plane of the X groups, but the symmetry properties are the same for the three cases.

Examples of molecules that display a planar-pentagonal structure are rare if not nonexistent, as are those with pentagonal-pyramidal structure. Examples of both other pyramidal structures are known. PF_5 and related compounds provide many examples of the trigonal bipyramid. Those molecules that adopt square-based pyramidal geometry are often more complex, however. Thus, BrF_5 has this geometry, but an unshared pair of electrons is also present (similar to those in ammonia) and it is probably better to treat this as an example of an MX_6 molecule.

Let us examine the symmetry characteristics of the important square-based pyramid and trigonal bipyramid. The first clear difference is readily seen in that the *principal axes* are fourfold and threefold, respectively. Observe that for each model there are twofold axes present as well. Only one twofold axis is present in the square-based pyramid, and it wholly coincides with the fourfold axis. Three twofold axes are present in the trigonal pyramid, each in the plane of three X atoms and running through one of the X atoms. Note that for both structures the principal axes are unchanged in *order* by moving the M atom above or below the plane of the appropriate X atoms, although the number of twofold axes is reduced to zero in the trigonal bipyramid case. The trigonal bipyramid possesses a mirror plane perpendicular to the principal axis, but the square-based pyramid does not. On the other hand, both structures possess several mirror planes that *include* the principal axis. Three such mirror planes are present for the trigonal bipyramid and four for the square-based pyramid, reflecting the order of its principal axis. Finally, let us observe that neither of the structures possesses a center of symmetry.

The bonding in the PF_5 molecule may be analyzed in a very similar fashion to that employed earlier for the metal complexes. In this case, the formal partition into five negative fluoride ions and a phosphorus atom with five positive charges is even less realistic than before. Nevertheless, the approach still leads to an understanding of the stereochemistry, and we shall continue in this manner. We do note in passing, however, that a treatment such as was used for methane in Chapter 5 could be used in which the compound is analyzed in terms of five neutral fluorine atoms and a neutral phosphorus atom hybridized into a valence state with five electrons in the hybrid orbitals. In the ionic approach, the electronic structure of the phosphorus atom (Ne)$3s^2 3p^3$ is stripped to the neon core. To form five bonds it is necessary to use the *s*, all three *p* orbitals, and one orbital from the 3*d* level to achieve dsp^3 hybridization. Trigonal-bipyramidal stereochemistry is achieved if the d_{z^2} orbital is used.

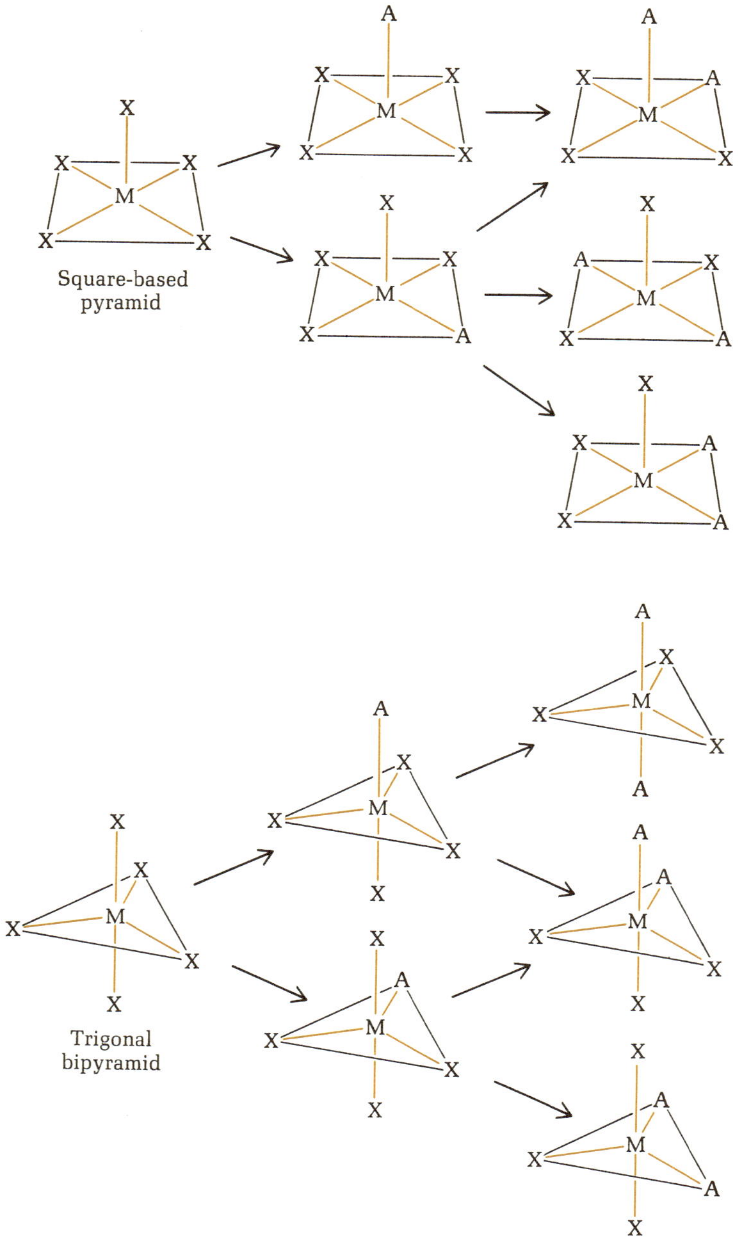

Figure 6-4 Structural isomers formed by successive substitution in square-based and trigonal-bipyramidal structures.

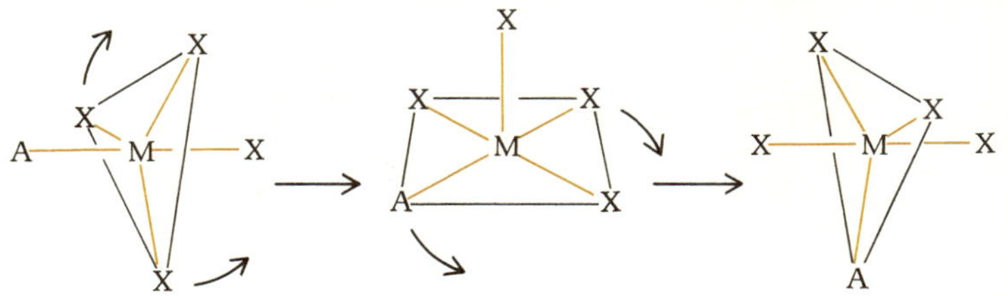

Figure 6-5 Structural changes that can interconvert the square-based pyramidal and trigonal-bipyramidal forms.

We have seen in Chapter 2 that arguments based on the number of isomers possible on substitution is a useful qualitative way in many instances of deducing which of several possible geometries is adopted by a particular molecule. Figure 6-4 compares the number of isomers possible for each structure for molecules of composition MAX_4 and MA_2X_3. The number of isomers is identical for substitution in either structure. Thus, in contrast to the tetrahedral versus square-planar problem, where the number of isomers may be used to establish structure, no such selection between alternatives may be made for MX_5 molecules.

Example *Draw all of the structural isomers of a molecule* MX_3AB *on the basis of a trigonal-bipyramidal geometry. If any isomer is optically active, indicate it.*

Solution *The structural isomers are*

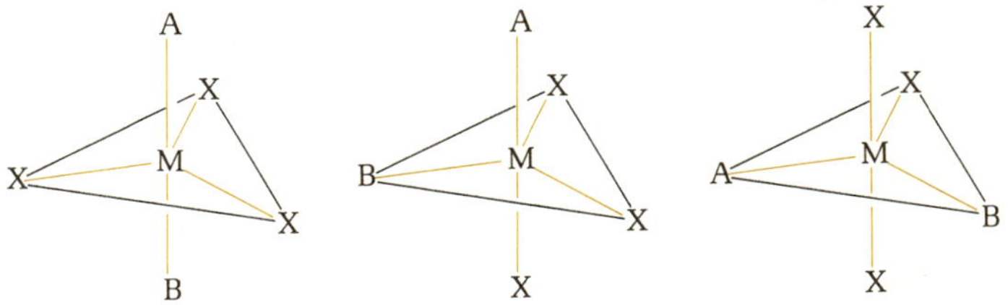

Each structure contains a mirror plane, and therefore none is optically active.

Let us examine somewhat more closely the relationship between the pyramidal and bipyramidal structures. At first glance, these structures appear quite different but, in fact, the differences are smaller than appears on the surface. In Figure 6-5 the bipyramidal molecule has been drawn from a different perspective, with one of the three equivalent positions pointed upward. Arrows indicate a suggested direction for moving the two remain-

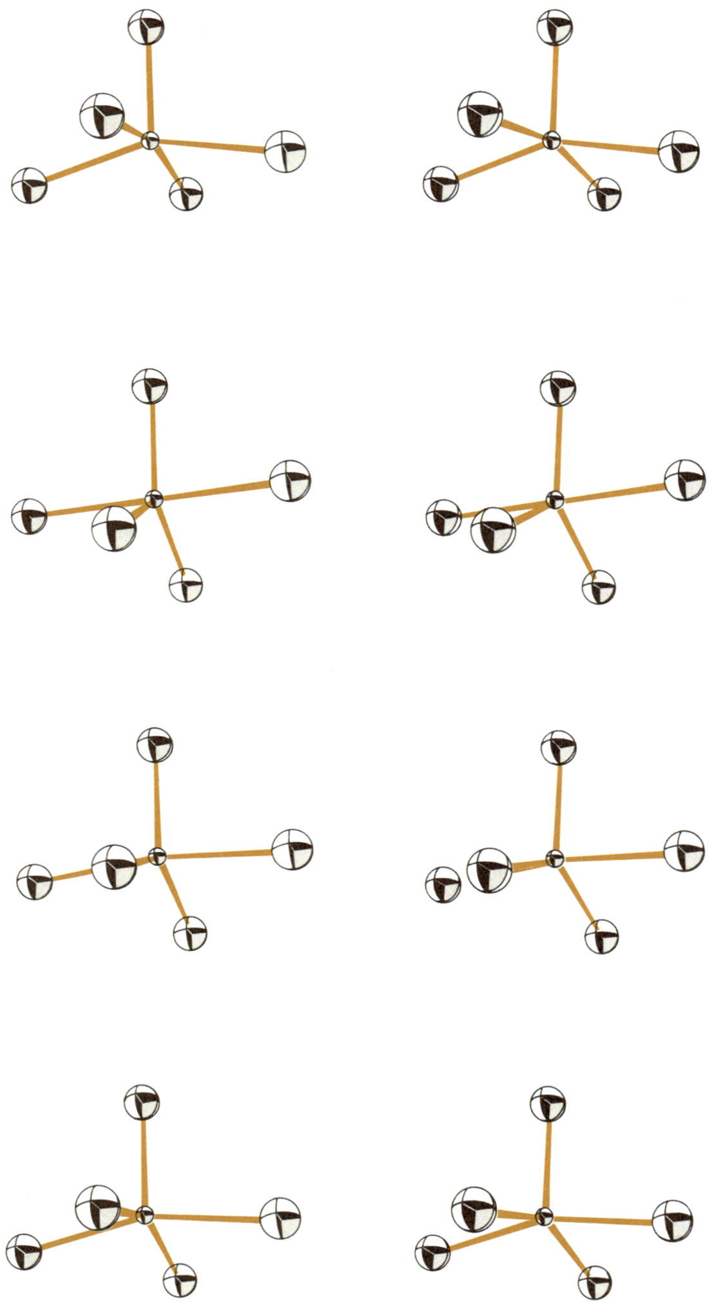

Figure 6-6 Stereoscopic views of the trigonal-bipyramid to square-based pyramid conversion.

ing equatorial atoms upwards toward the third. When this motion has been carried far enough, the two equatorial atoms reach a common plane with two former axial atoms and, in fact, the structure has assumed square-based pyramidal geometry with the M atom in the basal plane. In a trigonal bipyramid, the angle formed from an equatorial atom to the central atom to a second equatorial atom is 120°. The deformation indicated requires opening this angle to 180°. Lesser bending of these bonds accompanied by a simultaneous bending down of the axial atoms produces a square-based arrangement with M *above* the plane. Figure 6-6 displays this transition with stereoscopic pairs.

Finally, we should note that once the square-based pyramidal structure has been achieved, it makes no difference at all to the molecule which of the two opposite points on the square is bent back to form again a trigonal-bipyramidal arrangement. As shown in the figure, this molecular deformation can provide a means for interchanging the axial and equatorial positions for either pyramidal structure. Such extensive changes might appear difficult, but several lines of evidence suggest that in some cases just such molecular deformations do occur.

With this introduction to the possible stereochemistry of MX_5 molecules, let us return to the specific cases mentioned in Chapter 3. In contrast to the results derived from a study of isomer numbers for the methane molecule in Chapter 2, we have just seen that we shall be unable to distinguish from such a study between the square-based pyramidal and trigonal-bipyramidal for such molecules as PF_5. Actually, the structures of PF_5 and PCl_5 were among the first determined by the electron diffraction technique mentioned in Chapter 2 and have been known for over 30 years. Both substances adopt the trigonal-bipyramidal arrangement, and it is interesting to observe that the axial and equatorial distances are somewhat different. In PCl_5 the axial chlorines are 2.04 Å from the phosphorus, whereas the equatorial halogens are 2.19 Å away.

Careful study of the molecular structure of pentahalides of the Group V elements illustrates the care that must be used in reaching conclusions about structure based on overall formulas of compounds. As stated above, there is no doubt that the structure of PCl_5 in the gas state is trigonal-bipyramidal. However, structures can also be studied in the solid state by a technique using diffraction of X-rays rather than electrons. When this technique is applied to solid PCl_5, we unexpectedly find that no PCl_5 molecules exist. Instead, one of the phosphorus–chlorine bonds is broken, and a chloride ion is transferred from one molecule to another to form an ionic compound $PCl_4^+PCl_6^-$. In the PCl_4^+ ion the four chlorine atoms surround the phosphorus atom in a purely tetrahedral arrangement (the P—Cl bond distance of 1.98 Å is changed somewhat from the distance in the uncharged PCl_5 molecule). The tetrahedral shape is a consequence of the fact that only four

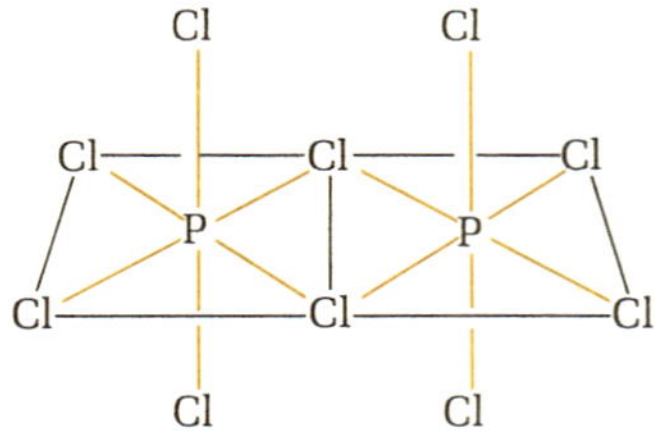

Figure 6-7 The probable structure of PCl_5 in the liquid state.

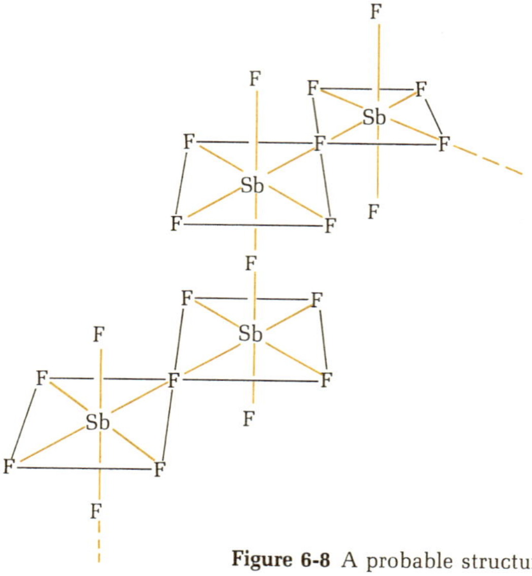

Figure 6-8 A probable structure for liquid SbF_5.

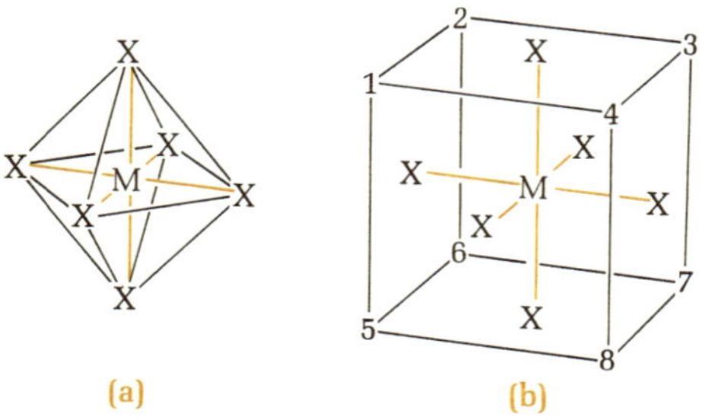

Figure 6-9 (a) An octahedral MX_6 compound. Yellow lines represent bonds; brown lines are included to sketch the octahedral shape. (b) The relation of an octahedron to a cube centered on the M atom.

chlorides are now bound to phosphorus, and sp^3 hybridization suffices without the need for a *d* orbital. The PCl_6^- ion supplies our first example of a sixfold substituted central atom and we consider its structure in more detail below. Recent studies show that, in solution, PCl_5 probably has an intermediate structure shown in Figure 6-7.

Similarly, antimony pentachloride, $SbCl_5$, and presumably antimony pentafluoride have the trigonal-bipyramidal structure in the gaseous state. Antimony pentachloride is also known to have this structure in the solid state, but the situation is more complex for SbF_5. This material melts at 8.3°C and boils at 103°C and in these two properties closely resembles water. Most liquids, including water, are highly fluid substances resembling gasoline in their properties, but liquid antimony pentafluoride is highly *viscous* and much more resembles molasses in its properties. It is believed that this property results from strong interactions between the molecules forming extended structures of the type shown in Figure 6-8.

There are many known compounds in which one or more atoms in the pentagonal-pyramidal arrangement are substituted. For example, the mixed compound PCl_2F_3 is known. Examination of Figure 6-2 shows that one might expect three different geometric isomers for a compound with such formula. In fact, chemists have thus far been able to prepare only one molecule of this formula, and it has been shown to have the two chlorines in axial position with the three fluorines lying in the equator of the molecule.

Before leaving MX_5 molecules, we might examine one more case. Compounds of the type $(NH_4)_2SbCl_5$ can be prepared and are found to contain the $SbCl_5^{2-}$ ion. In this ion, X-ray studies have shown that the five chlorines assume the square-based pyramid arrangement with the antimony slightly below the plane of the pyramid base. Just as the structure of ammonia may be viewed as a tetrahedral arrangement with an electron pair occupying one of the tetrahedral sites, we view the case of the $SbCl_5^{2-}$ as an octahedral arrangement in which the sixth site is occupied by an electron pair.

OCTAHEDRAL SYMMETRY

Figure 6-9(a) shows the **octahedral geometry** for an MX_6 compound. The symmetry of the octahedron may be seen most readily by examining Figure 6-9(b), which shows the relation of the octahedron to the cube (compare the symmetry analysis of the tetrahedron shown in Figure 2-1). Perhaps we should first observe that since only one M atom is present at the center of the cube, each symmetry operation must carry the M atom into itself; this is possible only if every symmetry element intersects the central M atom. The cube has fourfold axes (and consequently twofold axes as well) through the centers of each pair of opposed faces. These centers

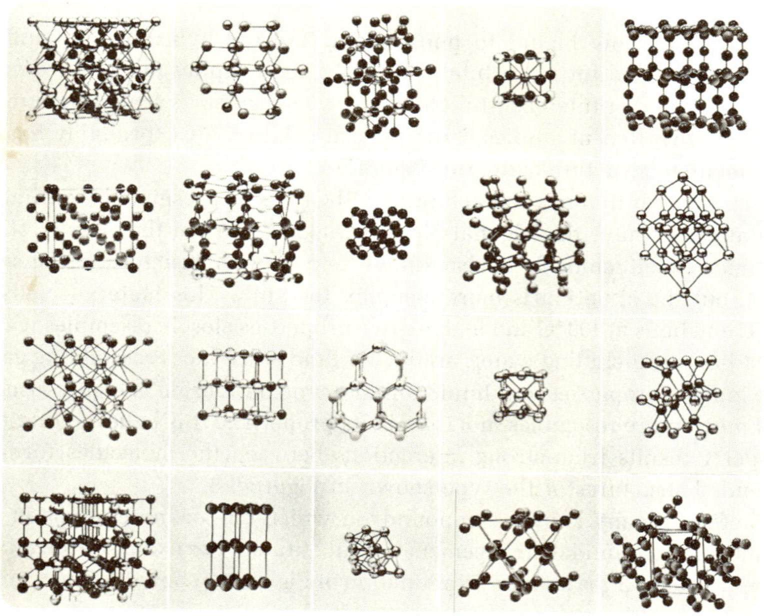

The arrangements of atoms in crystal lattices of many common minerals exhibit a variety of symmetries. A number of examples of crystal structure models are shown above. Top row, left to right: cuprite, zinc blende, rutile, perovskite, tridymite. Second row: cristobalite, potassium dihydrogen phosphate, diamond, pyrite, arsenic. Third row: cesium chloride, sodium chloride, wurtzite, copper, niccolite. Fourth row: spinel, graphite, beryllium, carbon dioxide, alpha quartz. The symmetry of the atomic arrangement is frequently reflected in that of the macroscopic crystal. Sodium chloride, for example, forms cubic crystals, just as one might expect from an examination of the atomic arrangement. (Courtesy of Bell Telephone Laboratories.)

are now occupied by X groups, but the fourfold axes remain unchanged. Numerous mirror planes also are present. Examine first two parallel edges on opposite sides of the central M atom (the 2–6 and 4–8 pair, for example). The plane defined by these lines is a mirror plane. Since the cube has 12 edges and these are needed in pairs to define planes, six such mirror planes are present. Furthermore, three planes are present parallel to the faces of the cube and passing through the central atom. We can also readily see that the octahedral geometry possesses a center of symmetry.

Two classes of rotational axes remain to be found that require closer inspection. Each line of intersection of mirror planes is a twofold axis; one such line passes between any two adjacent X groups and the central M atom. Finally, threefold axes pass through each pair of opposed points of the cube (such as 1 and 7).

OCTAHEDRAL COMPLEXES

Let us pause to examine the PCl_6^- ion mentioned above from an electronic point of view. First, note that our earlier analysis of the electronic structure of the central atom need not be changed. An analysis in which all bonding electrons are assigned to the chlorines still leads to a phosphorus atom with the neon shell and all succeeding orbitals unoccupied. Now, however, we must plan to hybridize a sufficient number of orbitals to accommodate six chloride ions. Four atomic orbitals are supplied by the 3s and 3*p* levels, and thus two of the 3*d* orbitals are required. Table 5-2 shows that the resulting d^2sp^3 hybridization leads to an octahedral shape.

Analysis of the $SbCl_5^{2-}$ ion will be very similar, since the electronic structure of antimony parallels that of phosphorus with two additional levels filled. Now, however, the antimony must allow places for five chlorides and two extra electrons, one pair. Six orbitals are still required, and hence d^2sp^3 bonding again results. Observe that the electron pair is found on the fourfold axis of the ion and that the equatorial chlorides are somewhat displaced away from the lone pair. This is another case where the unshared pair of electrons exerts the steric effect of a large atom.

Perhaps a more typical example of an octahedral complex is presented in the $Co(NH_3)_6^{3+}$ ion. We know at once that the NH_3 groups need be assigned no charge, and hence the cobalt carries the full three positive charges. The neutral cobalt atom has the $(Ar)4s^23d^7$ structure according to Table 4-2, and by stripping electrons first from the outermost shell we arrive at $(Ar)3d^64s^0$ for the +3 ion. To allow bonding to six ammonia molecules, six atomic orbitals must be provided and hybridized. Use of the two remaining 3*d*, the 4*s*, and the three 4*p* orbitals leads to d^2sp^3 hybridization and a simple explanation of the octahedral stereochemistry.

As a final example, consider the neutral molecule $Cr(CO)_6$. The carbon monoxide groups are neutral, and hence the central chromium has the same number of electrons as the element, 24. In the element, however, the structure is $(Ar)4s^13d^5$ (note the use of half-filled shells), and before hybridization the six electrons are first compressed into the 3*d* shell. Again d^2sp^3 bonding is possible, and chromium hexacarbonyl is octahedral.

Substitution in octahedral complexes

The symmetry of the octahedron requires that replacement of a single ligand by a different one will result in only a single isomer. For example, only one ion of formula $Co(NH_3)_5Cl^{2+}$ is known. This single substitution lowers the symmetry of the ion just as does the first chlorination of methane to give methyl chloride as discussed in Chapter 2. Now, in the complex only a single fourfold axis remains, and the four ammonias related by it are,

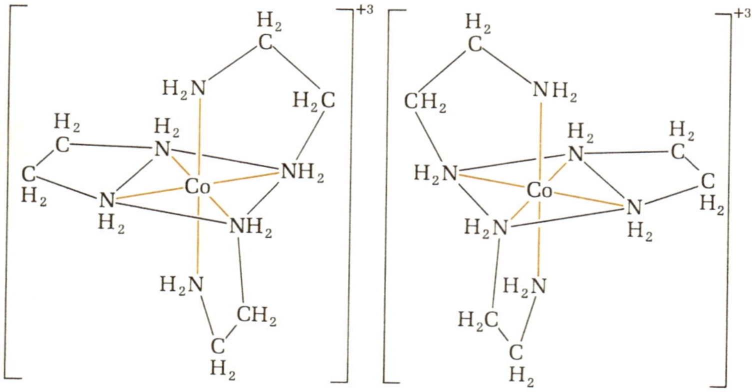

Figure 6-10 Optical isomers of the $Co(H_2NCH_2CH_2NH_2)_3^{3+}$ ion.

of course, identical. The single ammonia on the remaining fourfold axis is different by symmetry, however. Therefore, when a second ligand is replaced to give ions of formula $Co(NH_3)_4Cl_2^+$, two isomers are formed:

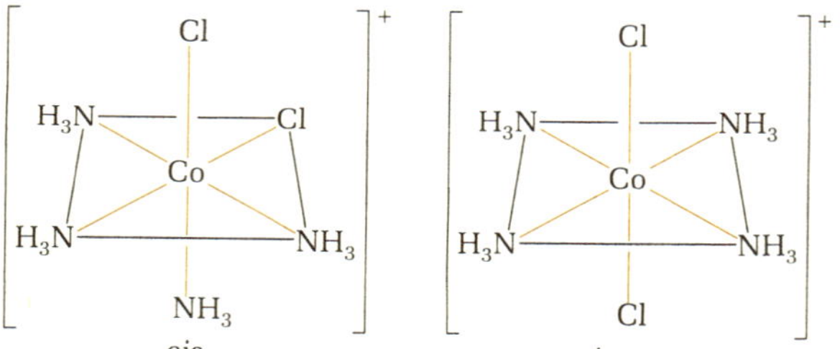

cis *trans*

Similarly, a third substitution produces the two isomers

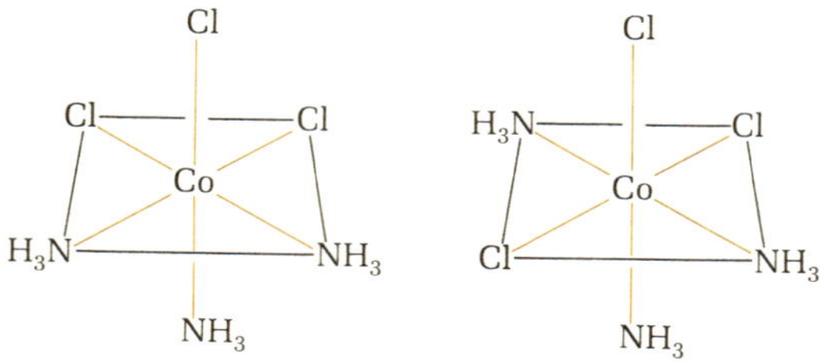

Finally, optical isomerism is possible in properly substituted octahedral complexes. A common example is provided by the cobalt complex ion formed by replacement of the six ammonias with six nitrogens linked together in pairs. The reagent $H_2N—CH_2—CH_2—NH_2$ (called ethylenediamine and often abbreviated to en) has two nitrogens that can function as ligands provided they occupy adjacent sites in the complex. Such a reagent is said to be a **chelate,** a word derived from the Greek word for claw. The two optical isomers of $Co(en_3)_3^{3+}$ are illustrated in Figure 6-10. Note that these structures have neither mirror planes nor centers of symmetry.

Example *List all the symmetry elements for a trisubstituted octahedral molecule of the type*

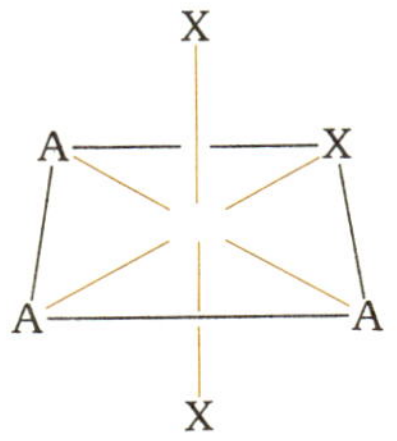

Solution *There are just one twofold rotation axis and two mirror planes, both of which include this axis. No center of symmetry is present. Can you draw an isomer of this structure that possesses a threefold rotation axis?*

We have now briefly explored the molecular geometries that occur most frequently in chemistry, The electronic basis for these geometries, in terms of hybridization of the central atom, has also been developed. In later chapters we shall have occasion to refer again to the chemistry of coordination complexes.

Glossary

Chelate a reagent possessing more than one site capable of donating an electron pair to a metal ion or atom.
Coordination compound a compound formed by combining a metal ion or atom with one or more molecules of a Lewis base(s).
Ligand generally, any Lewis base; usually, a Lewis base coordinated to a metal ion or atom.
Metal complex synonymous with coordination compound.
Octahedral geometry an important molecular geometry in which a central ion or atom is surrounded by six ligands arranged at the corners of a regular octahedron (equivalent to having the ligands at the center of each face of a cube).
Planar-pentagonal geometry a rare molecular geometry in which a central atom is surrounded by five substituents that occupy the corners of a regular pentagon, the plane of which includes the central atom.
Square-based pyramidal geometry a rare molecular geometry in which a central atom is surrounded by five substituents organized at the corners of a square-based pyramid.
Trigonal-bipyramidal geometry an important molecular geometry in which a central atom is surrounded by five substituents located at the corners of a trigonal bipyramid.

Study problems

1. Assume that an atom has pentagonal-pyramidal geometry (see Figure 6-3). How many isomers of formula MX_3H_2 can exist? Specify the symmetry elements present in each isomer and identify any optical isomers.
2. Explain why metal atoms in general are Lewis acids.
3. The compound PF_3Cl_2 has been extensively studied. What structures can you suggest for this composition? Apparently, only a single isomer exists. Can you suggest any explanation for this fact?
4. All isomers of $Co(NH_3)_3Cl_3$ are based on an octahedral shape. Draw each possible structure and list the symmetry elements in each.
5. Is it possible to have optical isomers of a coordination compound with square-planar geometry? Why?
6. Draw *d* and *l* forms of $Co(NH_3)_2Cl_2Br_2^-$.
7. What single electronic property is characteristic of all ligands?
8. If the coordination compound MX_2A_2 is square-planar, how many isomers are expected?
9. Draw the structure of *cis*-$Pt(NH_3)_2Br_2$.
10. Use the valence bond method to predict the geometry of the following compounds:

a. $B(NH_2)_3$	c. $FeCl_4^-$	e. $Ag(NH_3)_2^+$
b. $Fe(CO)_5$	d. $Rh(NH_3)_6^{+3}$	f. $SnCl_4$

Suggestions for further reading

F. Basolo and R. C. Johnson *Coordination Chemistry,* Benjamin, Menlo Park, Calif., 1964.

D. F. Martin and B. B. Martin *Coordination Compounds,* McGraw-Hill, New York, 1964.

R. K. Murmann *Inorganic Complex Compounds,* Van Nostrand Reinhold, New York, 1964.

J. F. Quagliano and L. M. Vallarino *Coordination Chemistry,* Heath, Lexington, Mass., 1969.

7

The structure of polymers

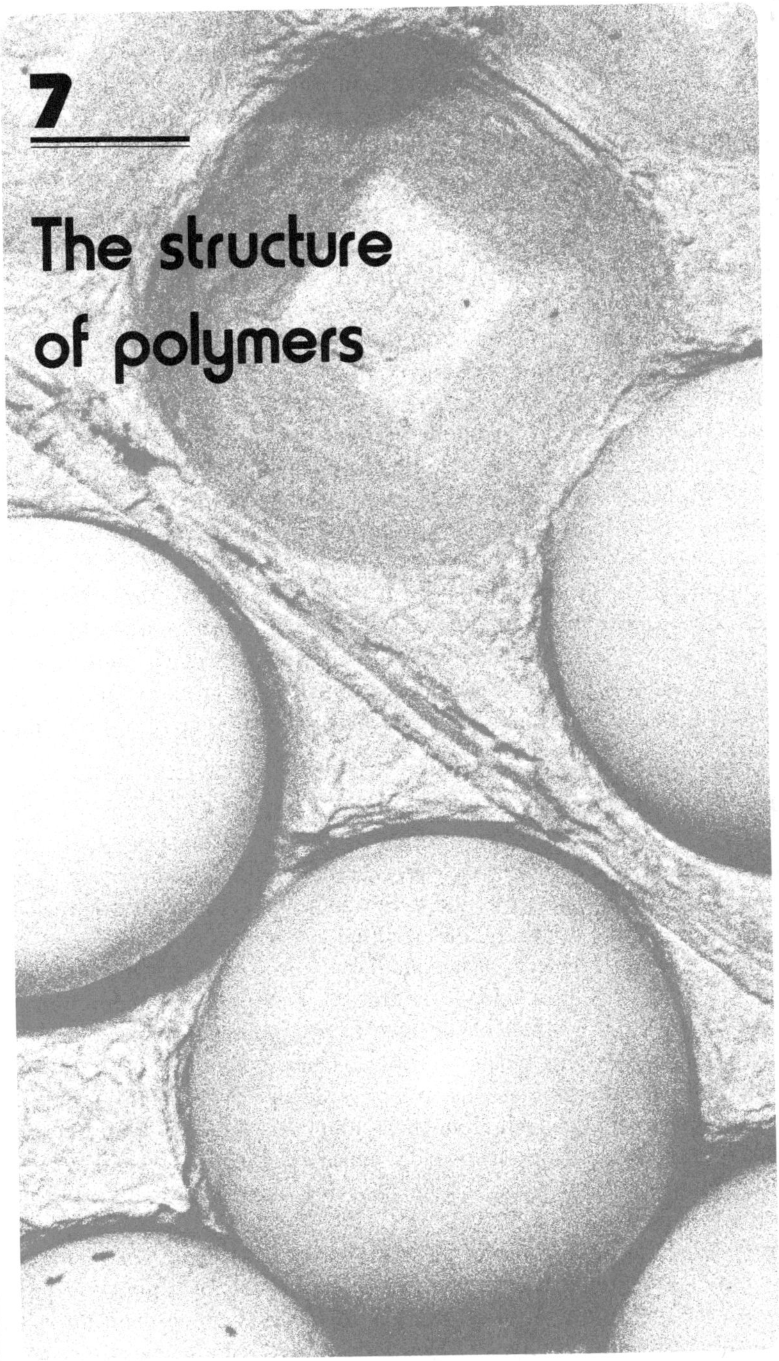

The structures of the compounds that we have considered thus far have been simple, for the most part. In the case of methane, for example, we were able to specify the molecular structure clearly by stating that the atoms of hydrogen are arranged about the carbon atom at the corners of a regular tetrahedron and that the C—H bond distance is 1.11 Å. Similarly, the structure of phosphorus trichloride can be specified by stating that the three chlorine atoms occupy three corners of a tetrahedron (or the base of a trigonal pyramid if we ignore the unshared electron pair) with Cl—P—Cl bond angles of 100.3° and a P—Cl bond length of 2.04 Å. Thus, for compounds of these types, the structure is determined by specifying (1) which atoms are connected to which, (2) the geometry of the molecule (or what is equivalent, the bond angles), and (3) the distances between bonded atoms, or the bond lengths.

Structures for some of the molecules that we have encountered have been less easy to deal with. Thus, cyclohexane has a kind of molecular flexibility that is not found in methane or phosphorus trichloride. This molecule can potentially exist in many conformations and does in fact spend appreciable time in two, the chair and boat forms. Intuitively, it is clear that the larger a molecule is, the more difficult it will become to define a precise structure for it as a consequence of its potential to adopt a variety of conformations in space. In very large molecules, frequently termed **macromolecules,** this problem becomes formidable indeed. Almost all macromolecules are **polymers.** We treat some aspects of the structures of polymers in this chapter.

POLYMERS: BASIC CONCEPTS

Polymers are classified into two groups, depending on the nature of the element on which they are based. *Organic* polymers are based on carbon; *inorganic* polymers are based on anything else. Of these, the most familiar are the organic polymers, and we shall spend most of our time discussing them. Man has depended fundamentally on organic polymers since the dawn of civilization. Materials derived from one or more kind of polymer are capable of satisfying most of our basic needs. These include cotton, silk, fur, leather, wood, rubber, oils, paper and parchment, paint, rope, and waxes. Objects synthesized from these materials have been and continue to be of the greatest importance to man; dwellings, clothing, weapons, writing material, lubricants, and musical instruments, to mention some of the more important.

In view of the importance of organic polymers, it is not surprising that, once man knew the basis of their structure, many attempts have been made to manufacture synthetic ones. Studies in this area have been underway since the 1920s and, today, the synthetic polymer industry is one of the largest of any kind. We come into contact every day with many synthetic polymers and the products derived from them: fabrics, adhesives, synthetic rubber, films and packaging material, and plastics of all types. Attempts are being made to manufacture polymers with the necessary properties to serve as structural elements in bridges, large buildings, and homes. With the importance of polymers in mind, let us begin to learn something about their chemistry.

The word "polymer" derives from Greek words that mean "many member." And indeed, the characteristic feature of polymers is the generation of a complete structure by the repetition of one or a few basic structural units. These structural units are called **monomers.** Thus, the monomers bear the same relationship to the polymer that the links of a chain do to the chain itself. This can be simply illustrated in the case of the well-known polymer Teflon:

$$F_2C{=}CF_2 \longrightarrow -(-CF_2-CF_2-)_n-$$

In the formation of this polymer, nothing more is involved than adding one monomer to another, time after time. No atoms or molecules are lost or gained in the process. Such polymers are called **addition polymers.**

In many cases the formation of polymer from monomers takes place with loss of some simple molecule such as water. Graphically, we can illustrate this relationship between monomer and derived polymer by considering the process of linking together molecules of glycine, H_2N-CH_2-COOH, one after another:

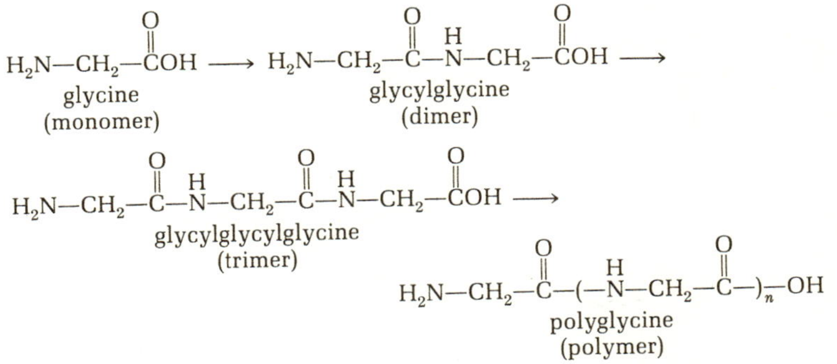

(7-1)

Note that the repeating unit enclosed in parentheses is missing the elements of water, a hydrogen from the nitrogen and an OH group from the carbon. One molecule of water is produced as each link is added to the chain. Polymers in which the addition of one monomer to the other results in the production of some molecule are called **condensation polymers.** Both addition and condensation polymers are of great importance, but only the latter kind is found in nature. With a fundamental idea of a polymer in hand, let us consider the properties of some actual examples, beginning with synthetic polymers.

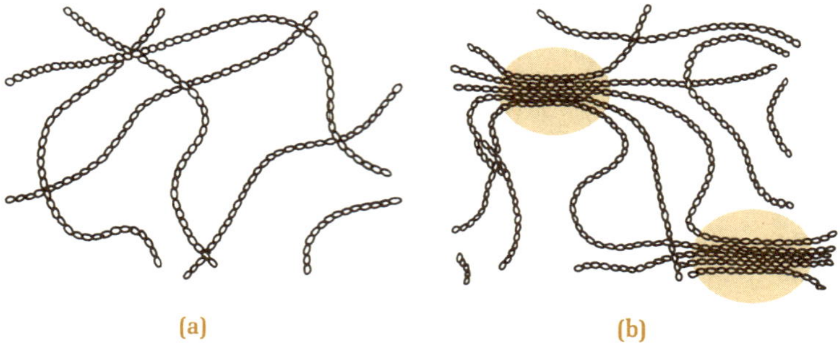

Figure 7-1 Two ways in which polymer chains may exist. In (a) the structure is amorphous, with the chains in a random coil region. In (b) the chains have acquired more order, with regions of crystalline structure shown in yellow.

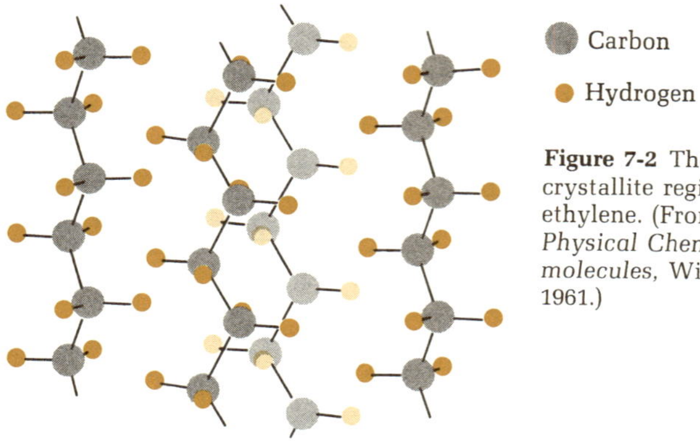

Figure 7-2 The structure of crystallite regions of polyethylene. (From C. Tanford, *Physical Chemistry of Macromolecules*, Wiley, New York, 1961.)

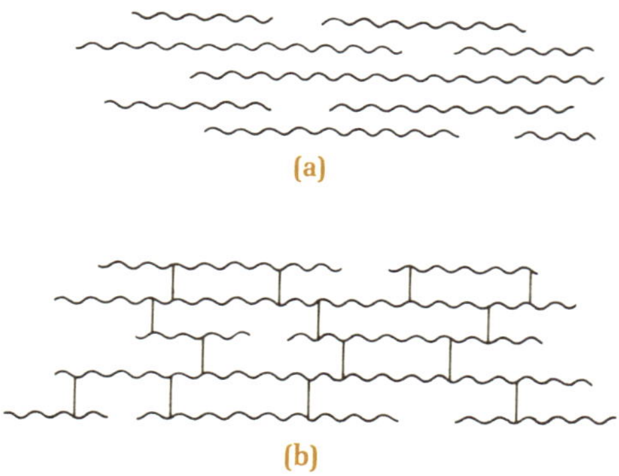

Figure 7-3 Comparison of crystalline and cross-linked polymers. (a) A crystalline network of polymer chains. (b) A cross-linked network of polymer chains.

SYNTHETIC POLYMERS

Synthetic polymers are named as derivatives of the monomers from which they are constructed, although many will be more familiar by their trade names (Lucite, Teflon, etc.). Thus, polyethylene is prepared from ethylene, polystyrene from styrene, and polyglycine from glycine. At present we wish to study two things about polymer chemistry: What are the structures of the polymeric materials, and how are these related to the physical properties of the polymers?

Let us begin with one of the simplest possible polymers, *polyethylene:*

$$—CH_2—CH_2—CH_2—CH_2—CH_2—CH_2—CH_2—CH_2—CH_2—CH_2— \quad \text{or} \quad (CH_2)_n$$

Polyethylene is a plastic familiar to almost everyone, since it has many industrial uses. Principal uses include packaging films, wire insulation, and blown and molded articles. This simple fact requires us to answer an important question: What determines the properties of polyethylene so that it can be used for manufacture of both transparent films and wire insulation, for example? We can define at least three variables that will determine the properties of polyethylene. First, we have the *degree of polymerization,* that is, the number of monomeric units that comprise the polymer, or the molecular weight of the polymer. Certainly, we would expect that a sample of polyethylene in which each polymer chain contained just 10 carbon atoms would be quite different from a sample in which each contained 10,000 carbon atoms. We might expect that polymers with long chains would be stronger and more fibrous than those with short chains.

The second variable is the *degree of crystallinity.* If we consider just one polyethylene molecule, it will be able to assume a very large number of conformations in space. It is more or less like a strand of well-cooked spaghetti, flexible and able to adopt a variety of complicated, twisted, irregular conformations. Such chains are referred to as *random coils.* But the polyethylene chain is a compact regular structure, and we can well imagine packing them together in a highly organized way. Thus, if we can get many of the chains to line up in a parallel fashion, we can form small crystalline regions, or *crystallites.* Schematically, we contrast an amorphous polyethylene and a partially crystalline polyethylene in Figure 7-1. Even though the forces that hold the polyethylene chains together in the crystalline regions are weak (see the discussion later in the chapter), crystalline polyethylene is a strong, tough, abrasion-resistant material with a melting point of 130°C. In Figure 7-2, some of the details concerning the way that polyethylene chains pack together are shown.

The third variable at our disposal is the *degree of cross-linking.* As mentioned above, polyethylene chains are held together by weak intermolecular forces. But it is also possible to introduce strong covalent linkages

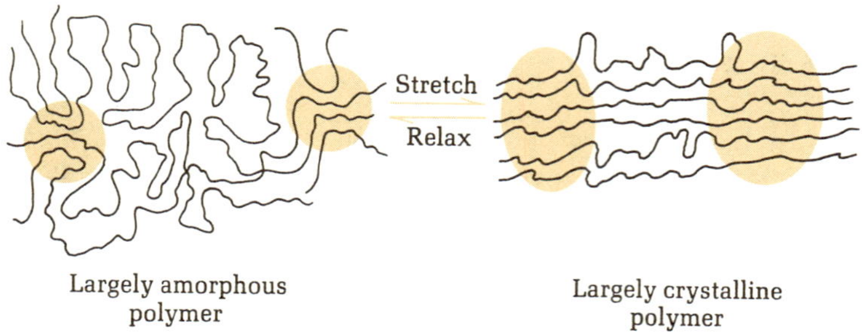

Figure 7-4 A representation of the change in organization of a polymer in the stressed and unstressed states, which underlies the phenomenon of elasticity. (From J. D. Roberts and M. Caserio, *Basic Principles of Organic Chemistry*, Benjamin, Menlo Park, Calif., 1964. Copyright, © 1964, W. A. Benjamin, Inc.)

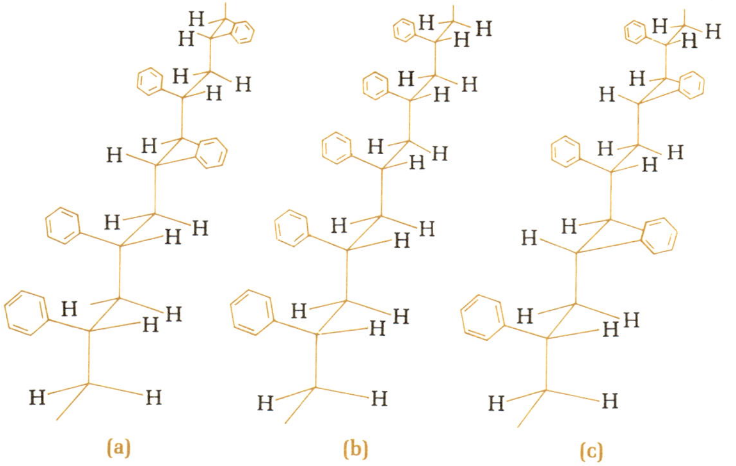

Figure 7-5 Configuration of (a) atactic, (b) isotactic, and (c) syndiotactic polystyrene.

between the chains in the course of the synthetic process. These linkages serve to give the polymer great strength and rigidity: the more cross-links, the more rigid the structure. The distinction between a crystalline polymer and a cross-linked one is made in Figure 7-3. The cross-links serve to maintain the crystalline character under conditions in which it might otherwise be lost. To understand more about the importance of cross-linking, let us consider the case of a more complicated polymer.

Polyisoprene is a derivative of isoprene. The relationship between monomer and polymer structure is

$$\underset{\text{isoprene}}{CH_2{=}\underset{\displaystyle CH_3}{\underset{|}{C}}{-}CH{=}CH_2} \longrightarrow \underset{\text{polyisoprene}}{+CH_2{-}\underset{\displaystyle CH_3}{\underset{|}{C}}{=}CH{-}CH_2{+}CH_2{-}\underset{\displaystyle CH_3}{\underset{|}{C}}{=}CH{-}CH_2{+}CH_2{-}} \tag{7-2}$$

Polyisoprene, which can be prepared synthetically in the laboratory, also occurs in nature as *rubber*. The structure of polyisoprene is markedly different from that of polyethylene. The interactions between the various chains are quite weak, and most of the material is not organized into crystallites. Perhaps the interactions are weakened largely as a consequence of steric hindrance between the methyl groups along the chain. Polymers in which crystallites are completely absent are said to be **amorphous.** Such polymers have low tensile strength and are subject to plastic flow, consequences of the weakness of interchain interactions. Actually, the properties of synthetic polyisoprene are similar to those of natural rubber, and the synthetic material finds many of the same uses. We know that natural and artificial rubbers are not readily subject to plastic flow but rather exhibit the interesting property of **elasticity.** The basis for this behavior is shown in Figure 7-4. Although polyisoprene is largely amorphous, it does have a sufficient tendency toward organization so that when tension is applied to the polymer it assumes an ordered structure. When tension is relaxed, the polymer shrinks back into its relatively nonordered state. Polymers that behave in this fashion are called **elastomers.**

Polyisoprene provides an excellent example of the importance of cross-linking for the physical properties of polymers. By adding more and more cross-links to the sample, the material loses elasticity and becomes progressively stiffer. Adding sufficient cross-links finally results in the formation of a very hard substance called *ebonite* which, among other things, is used in the manufacture of bowling balls. Clearly, cross-linking the polymer chains has an enormous effect on the properties of artificial rubber. There remains one additional way of varying the properties of polymers that does not show up in the chemistry of polyethylene or polyisoprene. This is the property of making the chains themselves stiffer. One of the ways to accomplish the stiffening of a polymer chain is to add large bulky groups to it that tend to prevent bending of the chains. An excellent example is provided by the case of polystyrene.

The structure of *polystyrene* exhibits some features that we have not encountered thus far. The monomer-polymer relationship is

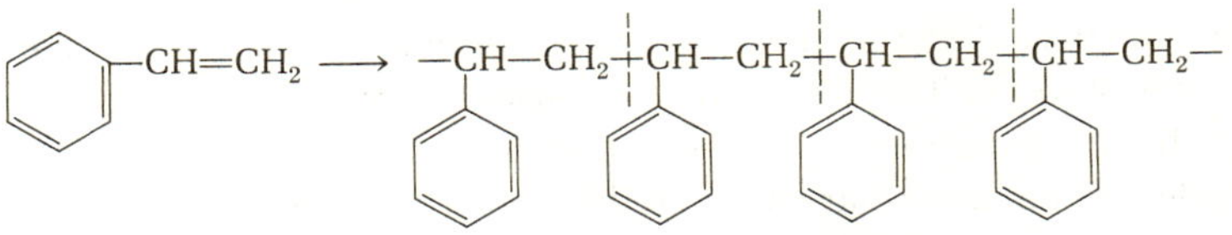

styrene polystyrene

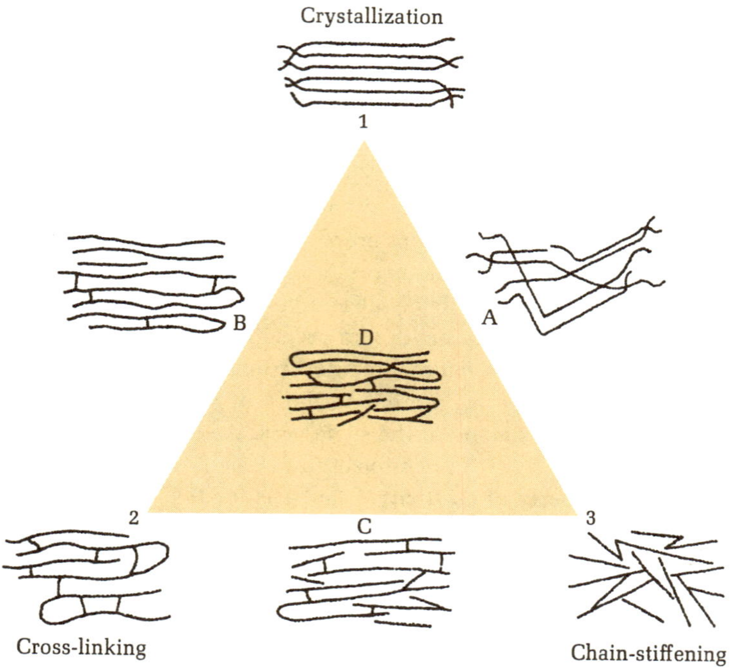

Figure 7-6 Combined features make it possible to achieve various properties with polymers. Each corner of the triangle represents one of the three basic principles for making a polymer rigid and temperature-resistant; the sides of the center of the triangle indicate various combinations of the principles. The chart on page 203 gives examples of the possibilities. [From H. F. Mark, "The Nature of Polymeric Materials," *Sci. Am.*, **217**, 148 (September, 1967). Copyright © 1967 by Scientific American, Inc. All rights reserved.]

It is possible to prepare a number of structurally distinct polystyrenes that differ with respect to the disposition of the phenyl groups, 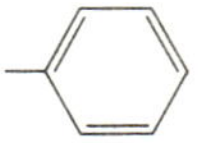, about the molecular axis. Three of these are called *atactic, isotactic,* and *syndiotactic* polystyrene. The structures of these polymers are shown in Figure 7-5. In atactic polystyrene the phenyl groups are randomly disposed on both sides of the molecular chain; in the isotactic polymer, on one side only; and in the syndiotactic polymer, on both sides with a perfectly

Location	Polymer characteristics	Examples	Uses
1	Flexible and crystallizable chains	Polyethylene	Pails, pipes, thin films
		Polypropylene	Steering wheels
		Polyvinyl chloride	Plastic pipes and sidings
		Nylon	Stockings, shirts, dresses, coats
2	Cross-linked, amorphous networks of flexible chains	Phenol, formaldehyde	Television casings, telephone receivers
		Cured rubber	Tires, transport belts, hoses
		Styrenated polyester	Finish on automobiles and appliances
3	Rigid chains	Polyimides	High-temperature insulation
		Ladder molecules	Heat shields
A	Crystalline domains in a viscous network	Terylene (Dacron)	Fibers and films
		Cellulose acetate	Fibers and films
B	Moderate cross-linking with some crystallinity	Neoprene	Oil-resistant rubber goods
		Polyisoprene	Particularly resilient rubber goods
C	Rigid chains, partly cross-linked	Heat-resistant materials	Jet and rocket engines and plasma technology
D	Crystalline domains with rigid chains between them and cross-linking between chains	Materials of high strength and temperature resistance	Buildings and vehicles

alternating regularity. The physical properties of the polymers reflect these structures. As might have been anticipated from our comments about the polyisoprene molecule, the atactic species is amorphous because of the difficulty of packing the bulky phenyl groups into any sort of ordered pattern. The isotactic polymer, on the other hand, in which the phenyl groups are all oriented in the same direction, permits the formation of crystalline regions in which each polystyrene chain is in the form of a helix. The crystallinity of the syndiotactic polymer is intermediate between the other two. Isotactic polystyrene has much greater tensile strength and heat

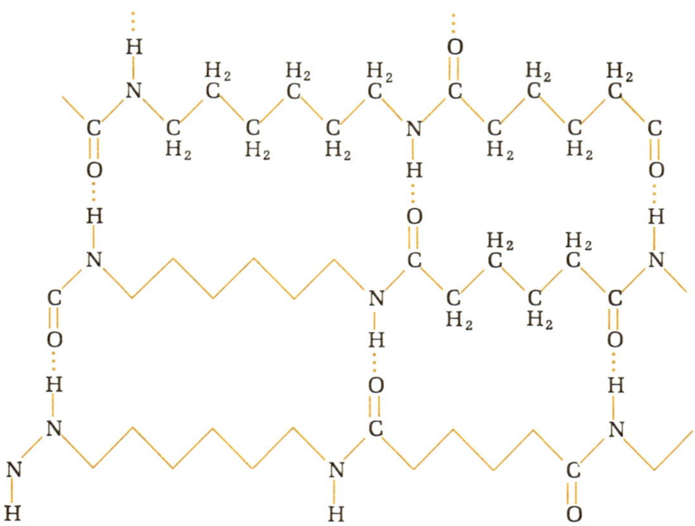

Figure 7-7 The structure of nylon 66.

resistance than the atactic structure, as expected on the basis of the extent of crystallinity. But in all of the cases the presence of the bulky phenyl groups lends stiffness to the polymer chains.

This brings us to one final point about polymer properties—the *stiffness* of the polymer chain. This is not something we can change for polyethylene, since it is largely determined by the structure of the monomer and the type of bond that holds the monomers together. But by choosing these correctly, we can construct individual polymer chains that are very flexible, such as polyethylene, only moderately flexible, such as the polystyrenes, or quite stiff and inflexible, such as the polyimides. A typical polyimide chain is

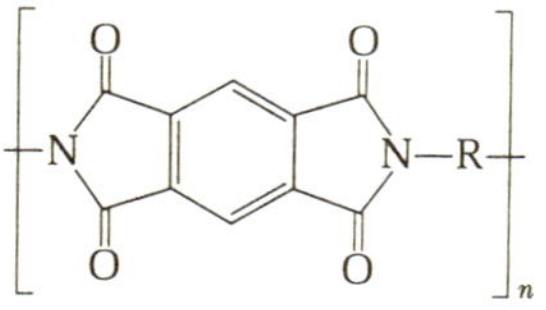

Before proceeding to consider other kinds of polymers, let us summarize what we have said so far in terms of the qualities of polymers as they relate to crystallinity, cross-linking, and chain stiffening. This is done in Figure 7-6 and in Table 7-1.

Table 7-1 The relationship between polymer structure and polymer properties

Physical nature of polymer	Molecular requirements
Hard (difficult to scratch)	High molecular weight High crystallinity Cross-linking or network structure *Example:* phenol-formaldehyde copolymer (Bakelite)
Strong (cannot easily be pulled apart)	High molecular weight Cross-linking or network structure Partly crystalline—crystallites embedded in amorphous matrix that acts like a cement *Example:* polymethylmethacrylate (Lucite, Plexiglas) reinforced with glass fiber
Fibrous	High molecular weight Linear macromolecules Long parallel arrangements of crystalline and amorphous regions *Example:* Polyacrylonitrile (Acrilan, Orlon)
Leathery	High molecular weight Linear macromolecules with slight degree of cross-linking; fragments of the chains are free to move under stress Low crystallinity *Example:* vinyl chloride-vinyl acetate copolymer (vinyl floor covering)
Rubbery	Linear macromolecules with little cross-linking; entire chains are free to move under stress High molecular weight Low crystallinity (but crystallinity increases with elongation) *Example:* polybutadiene (Buna rubber)
Soft, waxy	Low molecular weight ($<$10,000) Low crystallinity *Example:* polyvinyl acetate chewing gum
Thermoplastic (able to soften and assume new shapes by application of heat and pressure)	Linear macromolecules Little cross-linking Relatively low molecular weight *Example:* cellulose acetate
Thermosetting (not able to melt or flow without decomposition)	High molecular weight Highly cross-linked or network structure *Example:* urea-formaldehyde resin

Table 7-2 A collection of representative synthetic polymers and their uses

Monomer(s)	Formula	Physical type	Trade names	Uses
Ethylene	$CH_2{=}CH_2$	Semicrystalline	Polyethylene, Alathon	Film, containers, piping
Vinyl chloride	$CH_2{=}CHCl$	Atactic, semicrystalline	Polyvinyl chloride, Geon	Film, insulation, piping
Vinyl fluoride	$CH_2{=}CHF$	Atactic, semicrystalline	Tedlar	Coatings
Vinyl chloride	$CH_2{=}CHCl$	Crystalline	Saran	Tubing, fibers, film
Chlorotrifluoroethylene	$CF_2{=}CFCl$	Atactic, semicrystalline	Kel-F	Gaskets, insulation
Tetrafluoroethylene	$CF_2{=}CF_2$	Crystalline	Teflon	Gaskets, valves, insulation, filter felts, coatings
Propylene	$CH_2{=}CHCH_3$	Isotactic, crystalline		Fibers, molded articles
Ethylene	$CH_2{=}CH_2$	Amorphous	Nordel	Rubber articles
Hexafluoropropylene	$CF_2{=}CFCF_3$	Amorphous	Viton	Rubber articles
Isobutylene	$CH_2{=}C\ (CH_3)_2$	Amorphous	Vistanex, Oppanol	Pressure-sensitive adhesives
Isobutylene	$CH_2{=}C\ (CH_3)_2$	Amorphous	Butyl rubber	Inner tubes
Chloroprene	$CH_2{=}C\ (Cl)CH{=}CH_2$	Amorphous	Neoprene	Rubber articles
Isoprene	$CH_2{=}C\ (CH_3)CH{=}CH_2$	Amorphous	Natural rubber, Ameripol, Coral rubber	Rubber articles
Styrene	$CH_2{=}CHC_6H_5$	Atactic, semicrystalline	Styron, Lustron	Molded articles, foam
Styrene	$CH_2{=}CHC_6H_5$	Isotactic		
Vinyl acetate	$CH_2{=}CHO_2CCH_3$	Amorphous	Polyvinyl acetate	Adhesives
Vinyl alcohol	$(CH_2{=}CHOH)$	Crystalline	Polyvinyl alcohol	Water-soluble adhesives, paper sizing
Formaldehyde	$CH_2{=}O$	Crystalline	Delrin	Molded articles
Acrylonitrile	$CH_2{=}CHCN$	Crystalline	Orlon	Fiber
Methyl methacrylate	$CH_2{=}C\ (CH_3)CO_2CH_3$	Atactic amorphous	Lucite, Plexiglas	Coatings, molded articles
Ethylene terephthalate	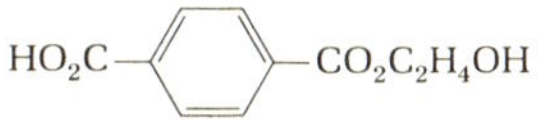	Crystalline	Dacron, Mylar, Cronar, Terylene	Fiber, film
Hexamethylenediamine adipic acid salt	$NH(CH_2)_6NH_2$ / $CO\ (CH_2)_4CO_2H$	Crystalline	Nylon, Zytel	Fibers, molded articles

Source: Modified from J. D. Roberts and M. Caserio, *Basic Principles of Organic Chemistry*, Benjamin, Menlo Park, Calif., 1964. Copyright, © 1964, W. A. Benjamin, Inc.

Nylon is a polymer distinct from those considered above, since it is formed from two complementary molecules rather than a single type of monomer. Actually, the term nylon refers to a class of polymers of related structure. We shall consider specifically nylon 66. This polymer is derived from adipic acid and hexamethylenediamine:

$$\underset{\text{adipic acid}}{HO_2C\text{—}CH_2CH_2CH_2CH_2\text{—}CO_2H} + \underset{\text{hexamethylenediamine}}{H_2N\text{—}CH_2CH_2CH_2CH_2CH_2CH_2\text{—}NH_2} \longrightarrow$$

$$\underset{\text{nylon 66}}{\left[\overset{O}{\overset{\|}{C}}\text{—}(CH_2)_4\text{—}\overset{O}{\overset{\|}{C}}\text{—}\overset{H}{\overset{|}{N}}\text{—}(CH_2)_6\text{—}\overset{H}{\overset{|}{N}} \right]_n} \qquad (7\text{-}3)$$

Nylons are named on the basis of the number of carbon atoms in the monomers; thus, nylon 66 reflects the fact that hexamethylenediamine has six carbon atoms and adipic acid has six carbon atoms. Nylon 66 is an exceptionally strong material as a result of its ability to form strong bonds between chains. These are formed by the sharing of a hydrogen atom between two groups as shown in Figure 7-7. Linkages of this type are called *hydrogen bonds* and are discussed in more detail later.

A summary of many types of polymers, together with some indication of their physical properties and, when applicable, their trade names, is provided in Table 7-2.

It is instructive to consider some of the polymers listed in Table 7-2 in terms of their impact on daily living. In one way or another, most of us make use of polyethylene, Saran, Teflon, Neoprene, Styron, Dacron, Orlon, Lucite, Bakelite, nylon, and other polymers daily. This is not only the space age, jet age, and atomic age, but the polymer age as well.

Example *For each of the following polymers, deduce the structure of the monomer(s) from which they are constructed:*

a.
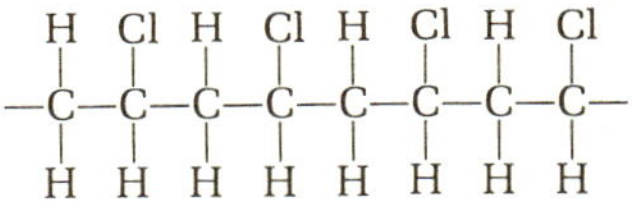

b.
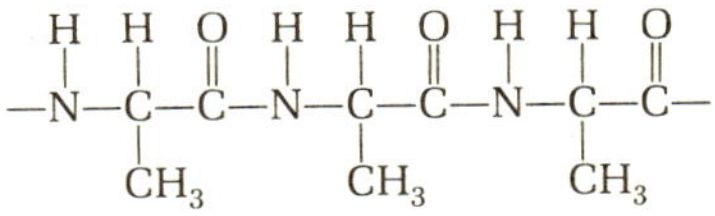

c.
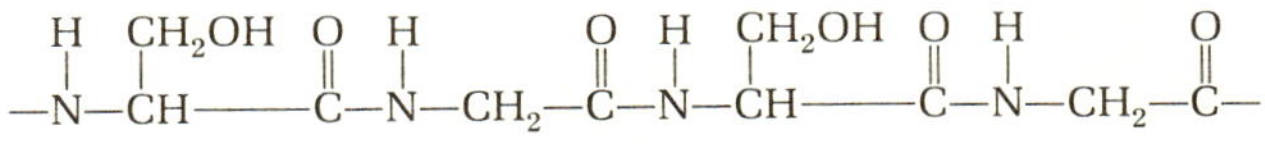

Solution

a. *The monomer is*
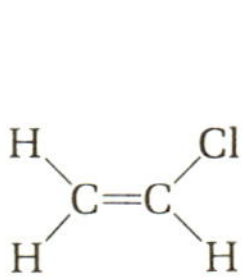

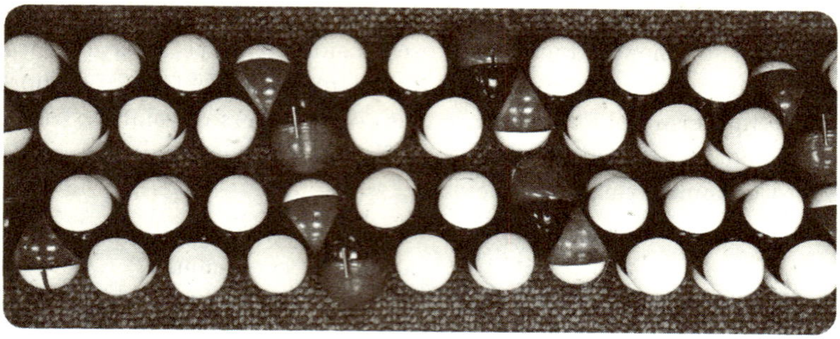

This photograph of a space-filling model of nylon shows how several independent polymer chains line up parallel to each other to form a strong fibrous product. The individual chains are held together to form sheet structures by hydrogen bonds between oxygen atoms on one chain and hydrogen-nitrogen groups on another. (Courtesy of Du Pont Company.)

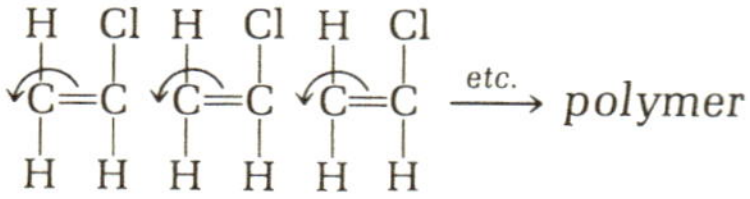

b. *The monomer is* 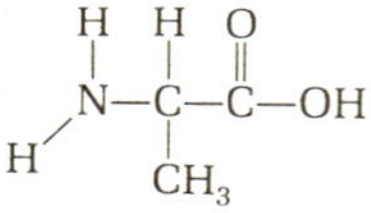

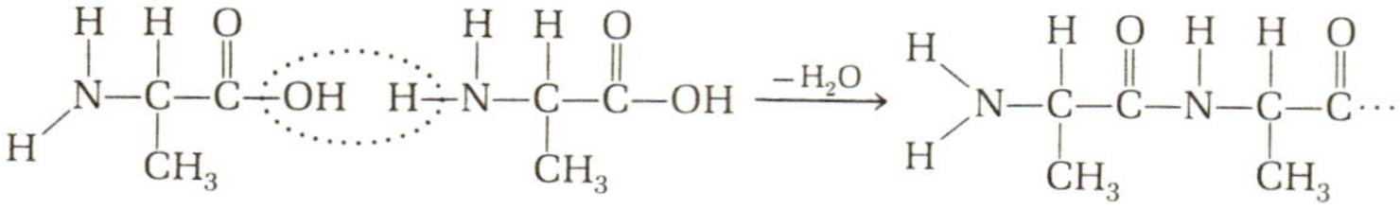

c. *The monomers are* 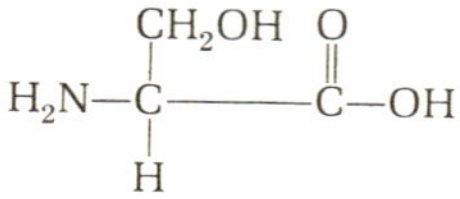 *and* 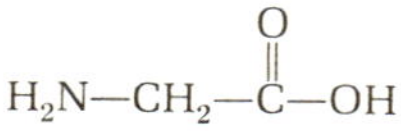

They are linked together as shown in the preceding case.

Example *Which of the following three polymers would you expect to show the highest degree of crystallinity?*

a. 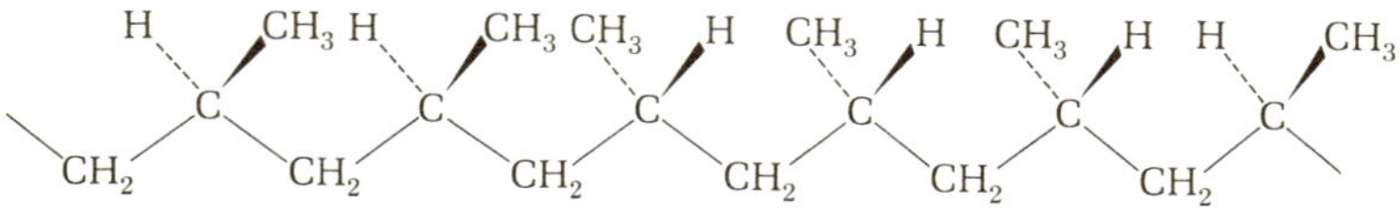

b. 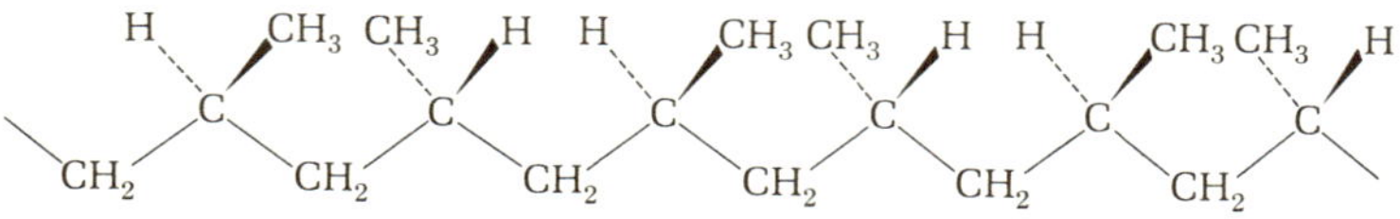

c.
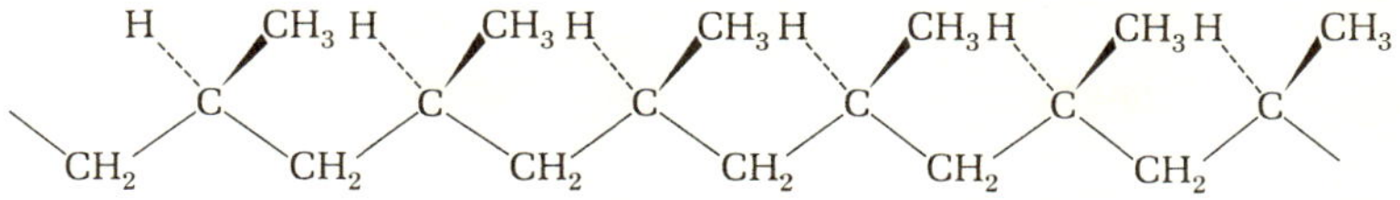

Solution *Polymer (c), the isotactic one, with all the methyl groups on the same side of the chain, will be the most crystalline one, just as in the case of the polystyrenes. The atactic polymer (a) will be the least crystalline one.*

THE SYNTHESIS OF POLYMERS

There are a large number of ways of synthesizing organic polymers. One of the most ingenious was discovered by Carl Zeigler, who won the Nobel Prize in chemistry as a result of his efforts. Since the method that he developed is beautiful in concept and result and since it is based on some of the considerations of the last chapter, we shall describe it briefly.

The point of the method is to find a convenient and inexpensive way to synthesize polyethylene. Thus, we need some way to speed up, **catalyze,** the addition of one molecule of ethylene to another, and so on, until a long chain is formed. To accomplish this task, Zeigler synthesized a complicated complex of titanium:

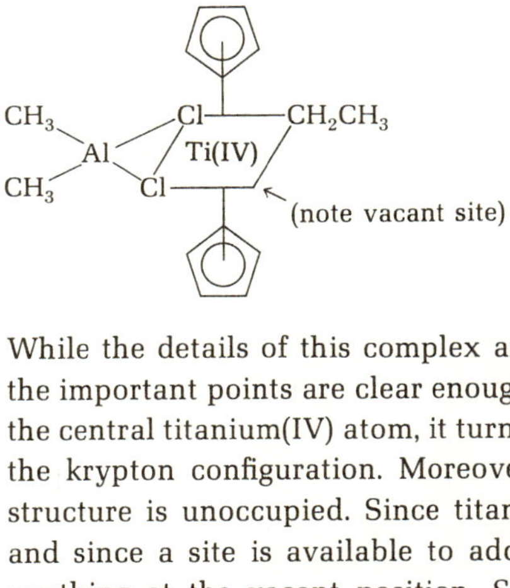

While the details of this complex are beyond the scope of our treatment, the important points are clear enough. If we count up the electrons around the central titanium(IV) atom, it turns out that it lacks just two of achieving the krypton configuration. Moreover, one of the sites of the octahedral structure is unoccupied. Since titanium badly wants two more electrons and since a site is available to add them, this complex will add almost anything at the vacant position. Specifically, it will add a molecule of ethylene:

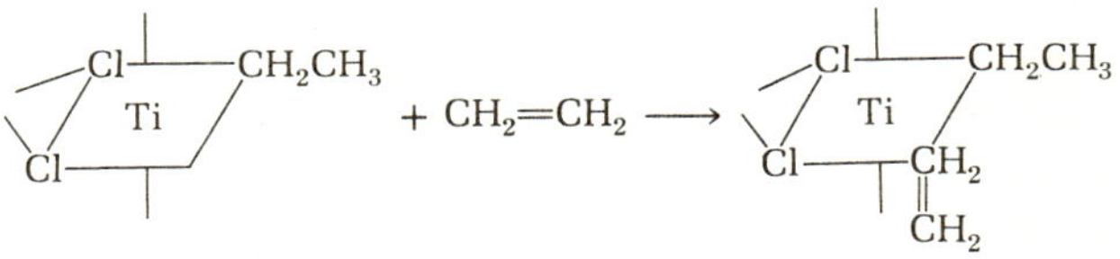

Note now that the resident ethyl group is adjacent to the newly complexed ethylene molecule. Moreover, the newly complexed ethylene is made more reactive as a result of the complexation reaction. Consequently, a facile reaction takes place in which the ethyl group adds to the ethylene group:

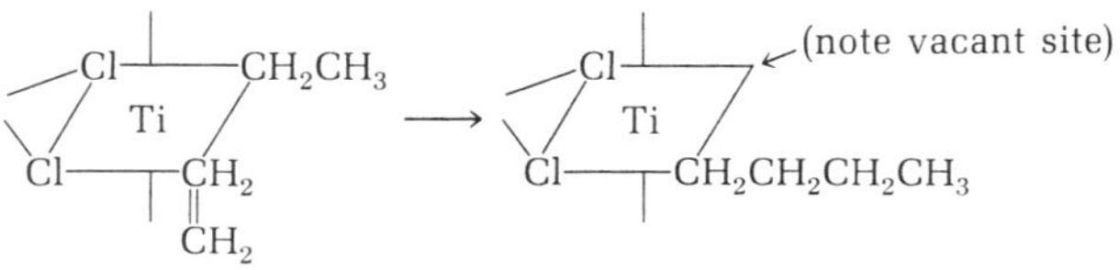

But note that this reaction leaves us back where we started except that the ethyl group has been converted to a butyl one. Addition of another molecule of ethylene follows, followed by another condensation, and so on. Eventually, a very long chain can be formed.

The catalyst is ideally suited to serve its purpose. First, the electronic structure of titanium is crucial to providing a driving force for addition of a molecule of ethylene. Second, the geometry of the complex not only provides a site for ethylene adsorption but also places the ethylene as close as possible to the ethyl group, facilitating the reaction in this way. Finally, the ethylene is made more reactive as a result of complexation. All of these things are crucial to the construction of an ideal catalyst for this very important reaction.

POLYMERS NOT BASED ON CARBON

All of the examples of polymers considered thus far have carbon as the most important structural element of the backbone. While such artificial polymers constitute the largest single group, they are by no means the only ones that are important in daily life or interest for the chemist. Let us continue by considering a few polymers based on elements other than carbon.

First, let us return to something we discussed in Chapter 3: The catenation tendency of all other elements is much less than that of carbon and frequently is quite limited.

Only sulfur exhibits a catenation tendency sufficient to permit the formation of polymers constructed on a backbone of it alone. Thus, most attention has been directed to finding those sequences of two or more elements that can be joined together in a regular manner to construct long chains.

Sulfur does have a catenation tendency sufficiently great to allow the formation of reasonably long chains. While this fact proves of little importance practically, it does result in an interesting chemical behavior for liquid sulfur. As the temperature of liquid sulfur is raised from about 160°C to 200°C, the **viscosity** of the liquid increases markedly—about 10,000-fold. This behavior is quite unusual. Viscosity is a quantitative measure of the tendency of a liquid to flow: The higher the viscosity, the less the flow tendency. Thus, we say that syrup has a higher viscosity than water. Generally, increase in temperature decreases the viscosity of liquids; syrup flows more easily in July than in January. The unusual viscosity behavior of liquid sulfur is the consequence of the opening of the S_8 rings characteristic of elemental sulfur with the formation of long chains:

$$S_8(\text{cyclic}) \longrightarrow S_8(\text{linear}) \longrightarrow S_{16}(\text{linear}) \longrightarrow \text{etc.} \tag{7-4}$$

The formation of the long chains results in an increased viscosity, since the chains become entangled with one another and this impedes their ability to move and tumble independently in the flow process.

Turning attention now to those polymers that have more than one type of atom along the backbone, those constructed from chains of oxygen and silicon are probably the most common and useful. Actually, we have already seen one structure similar in some respects. The crystal structure of cristobalite (Figure 3-3) does have a three-dimensional network of Si—O bonds. The corresponding polymers have either one-dimensional or two-dimensional networks, depending on whether the polymer is linear or cross-linked. The simplest such polymers, **siloxanes,** are made by the condensation of *silanols:*

$$n\mathrm{HO}-\underset{\underset{\mathrm{R}}{|}}{\overset{\overset{\mathrm{R}}{|}}{\mathrm{Si}}}-\mathrm{OH} \longrightarrow \mathrm{HO}-(\underset{\underset{\mathrm{R}}{|}}{\overset{\overset{\mathrm{R}}{|}}{\mathrm{Si}}}\mathrm{O})_n\mathrm{H} + (n-1)\mathrm{H_2O} \tag{7-5}$$

The silanols are produced by hydrolysis of the halide:

$$\mathrm{Cl}-\underset{\underset{\mathrm{R}}{|}}{\overset{\overset{\mathrm{R}}{|}}{\mathrm{Si}}}-\mathrm{Cl} + 2\mathrm{H_2O} \longrightarrow \mathrm{HO}-\underset{\underset{\mathrm{R}}{|}}{\overset{\overset{\mathrm{R}}{|}}{\mathrm{Si}}}-\mathrm{OH} + 2\mathrm{HCl} \tag{7-6}$$

The properties of the siloxanes can be varied by changing the degree of polymerization or the nature of the R group. If R is capable of forming chemical bonds to itself, cross-linked polymers can be obtained. Perhaps the major use of siloxanes is as *silicone oils,* which find many industrial uses since they are resistant to thermal degradation and to chemical attack. The R groups are methyl in the silicones.

Many related cases might be cited, including polymers based on backbones of boron and nitrogen, carbon and nitrogen (the epoxy resins, for example), phosphorus and nitrogen, and so on. Most of these are insufficiently stable to be industrially useful. In addition, *coordination polymers* frequently occur in nature. An example of such a polymer is palladium chloride, which has the structure

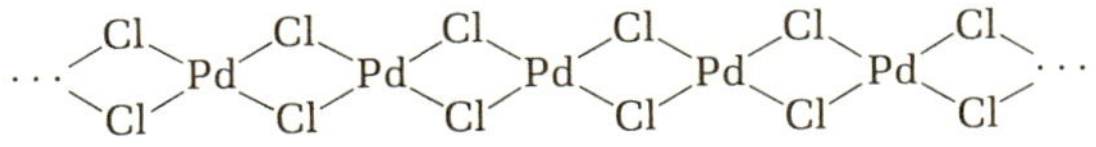

WEAK INTERACTIONS BETWEEN MOLECULES

Our discussion of polymers this far has indicated the importance of structures that result from noncovalent interactions between independent polymer chains. In the discussion that follows, we shall learn that such interactions are also important for biological polymers, as are interactions of one part of a polymer chain with another part of that same chain. Let us, then, pause in our treatment of polymers long enough to indicate briefly the principal types of weak interactions that exist between molecules and atoms.

One of the most important weak interactions, particularly for the understanding of the structure and stability of biological macromolecules, is the **hydrogen bond.**

Schematically, a hydrogen bond may be visualized as

$$X{-}H \cdots Y$$

As indicated in this representation, the hydrogen atom is shared between two atoms, X and Y, although it is more strongly bonded to one, X in this representation, than to the other, Y. Hydrogen bonds of appreciable strength are only formed when X and Y are the first-row elements of substantial electronegativity: specifically, nitrogen, oxygen, and fluorine. There is good evidence that the hydrogen atom itself can move back and forth between its two bonding partners very rapidly. In fact, the proton may oscillate across the hydrogen bond about 10^{13} times per second.

In comparison with most chemical bonds, the hydrogen bond is a very weak one. As we shall see in Chapter 8, most covalent bonds possess bond energies of 50 kcal/mole to about 100 kcal/mole. In contrast, a typical hydrogen bond has a bond energy in the vicinity of 5 kcal/mole. Despite the apparent weakness of this interaction, the formation of such hydrogen bonds in substantial numbers has profound consequences for the behavior of chemical systems. Many physical properties of water, hydrogen fluoride, and ammonia reflect hydrogen bonding (Chapter 10).

A chemical situation in which a hydrogen atom is shared between two other atoms is not a typical one. We are accustomed to thinking of hydrogen as bonding to only one atom at a time. Consequently, it is not simple to characterize the hydrogen bond in a convenient way and, in fact, discussion in the chemical literature concerning the nature of the hydrogen bond continues. It is probably fair to conclude that the hydrogen bond is partially ionic in character and partially covalent in character. From our discussion above, we note that with molecules such as water the oxygen atom is partially negatively charged, while the hydrogen atom has a partial positive charge. Consequently, we may view a hydrogen bond as an electrostatic interaction in the following way:

$$\overset{\delta+}{X{-}H} \cdots \overset{\delta-}{Y}$$

At the same time, we can provide a molecular orbital description of a hydrogen bond in terms of the p orbital of the X atom to which the hydrogen is most firmly bound, in terms of the p orbital largely occupied by the unshared pair on the Y atom, and in terms of the 1s orbital of the hydrogen atom itself. Using these three orbitals, we can construct a set of molecular orbitals in the usual way into which the pair of electrons involved in hydrogen bond formation may be placed. This procedure clearly differs somewhat from our usual one, since ordinarily we have a pair of electrons distributed between two atoms. In the case of the hydrogen bond, the pair of electrons occupies a molecular orbital that extends over three atoms:

A common and important weak interaction involves the formation of **ionic bonds** between groups of opposite charge:

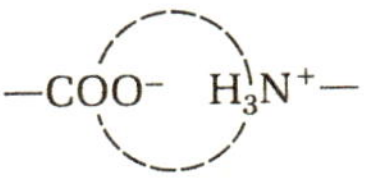

Details of the energetics of ionic bond formation in aqueous solution are complicated by the role of solvent water as it interacts with dissolved substances (see Chapter 10). But two or three points are worth bearing in mind. First, electrostatic forces are relatively long range: The electrostatic force varies as the inverse of the second power of the distance between the two charges:

$$e_1 \underset{d}{\longleftrightarrow} e_2 \qquad F_{\text{electrostatic}} = k\,\frac{e_1 e_2}{d^2} \tag{7-7}$$

Many other forces between molecules or ions fall off much more rapidly with distance. Second, the side chains of several of the amino acids that constitute proteins (p. 221 ff.) carry a net positive or negative charge under physiological conditions. These amino acids include aspartic and glutamic acids, histidine, arginine, and lysine. Hence, we encounter a number of charged groups on the protein surface, and we may expect that electrostatic forces will play an important role in the adsorption of molecules and ions on the protein. Finally, note that there are a number of other types of interaction that have an electrostatic basis. For example, if a molecule has a permanent dipole moment, that dipole can interact with a charged group to yield a net binding force:

an *ion-dipole* interaction: $\leftarrow\!\!\!+\!\!\!\longrightarrow \ \oplus$ (7-8)

In a related manner, two molecules each having a dipole moment can interact in a favorable way:

a *dipole-dipole* interaction: $\begin{matrix}\longleftarrow\!\!\!+\!\!\!\longrightarrow\\ \longleftarrow\!\!\!+\end{matrix}$ (7-9)

As you may have guessed, ion–dipole interactions and dipole–dipole interactions are not as strong as ion–ion interactions. Moreover, the interaction energy decreases more rapidly with distance between the two interacting species than it does for ion–ion interaction.

One of the subtlest of intermolecular forces is also based on electrostatic considerations, even though the interacting molecules or atoms may have no net charge or net dipole moment. Suppose we focus attention on two argon atoms in the gas phase and ask how these can interact with each other. The argon atom is spherically symmetrical in terms of its charge distribution. But at any given moment the charge distribution about the nucleus is not likely to be perfectly uniform, and the atom may have a *transient* dipole moment. If this atom now encounters another argon atom, the transient dipole moment in the first will tend to induce a transient dipole moment in the second. This new dipole will stabilize the first. At some moment an instant later, we may find that the two dipoles have reversed; we shall find that the two continue to oscillate back and forth, generating a net attractive force called *London dispersion forces*. Schematically, we can view the situation as

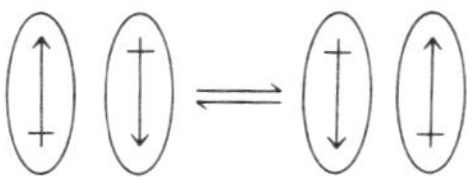

(7-10)

The strength of dispersion forces depends crucially on two quantities: the *polarizability* and the *distance* between the oscillating dipoles. The polarizability is a measure of the deformability of the electron cloud associated with an atom or molecule. Clearly, the greater this deformability, the easier it will be to induce a larger oscillating dipole in the electron cloud. Dispersion forces fall off very rapidly with increasing distance between the interacting atoms or molecules; the energy is inversely proportional to the *sixth* power of the distance. Hence, if the interacting dipoles retreat from each other by a factor of two in distance, the interaction energy between them decreases by a factor of 64 (2^6).

Finally, there is a tendency for nonpolar molecules in aqueous solution to bind to each other with what has been termed **hydrophobic** or **apolar bond formation:**

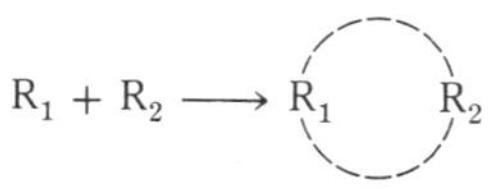

(7-11)

R_1, R_2 are hydrophobic species in water

Hydrophobic bond formation is thought to be quite important in aqueous solution and may, in fact, largely determine the exact three-dimensional structure of protein molecules. Measurements have indicated that hydrophobic bond formation is driven by a favorable entropy change. How can we account for this observation? What does it mean in terms of the nature of the forces responsible for hydrophobic bond formation?

As is detailed in Chapter 10, hydrophobic molecules in aqueous solution have the capacity to organize solvent water molecules about them into a kind of "icelike" structure. This organization is accompanied by a loss in the freedom of these water molecules to move about or rotate, and hence is accompanied by a loss of stability of the system. Association of the hydrocarbonlike structures with themselves frees these molecules of water. Hence, the driving force for hydrophobic bond formation is largely the increase in freedom (positive entropy change) resulting from the liberation of solvent molecules organized about the hydrophobic entities (see Chapter 8).

This brief list does not exhaust, by any means, the ways in which atoms, molecules, and ions can interact with one another. It does include, however, those forces that are likely to be most important in determining the three-dimensional structures of polymers. With these considerations in mind, let us return to the matter of biological polymers and explore their structural organization.

NATURAL POLYMERS

There are many naturally occurring polymers. We have already noted that natural polymers have well served the interests of mankind in many ways. But in addition to their usefulness in daily life, natural polymers also are intimately involved in the most basic aspects of life itself. Indeed, no matter how simple the life form may be, it will be found to contain at least two types of natural polymers: proteins and nucleic acids. Even in the simplest viruses, which are on the borderline between life and nonlife, there is nothing but nucleic acid and proteins. More complicated systems, which can be more realistically called living, inevitably contain these kinds of molecules but have additional ones as well. Any definition of life as we know it must contain the specification that proteins and nucleic acids play crucial roles in it. Beginning with the nucleic acids, let us consider the chemistry of these fascinating molecules together with the relation between their structure and biological function.

NUCLEIC ACIDS

The **nucleic acids,** which include the largest molecules yet discovered with molecular weights up to 40 billion, occur as two types: **deoxyribonucleic acid** and **ribonucleic acid.** These are generally referred to by their abbreviations, DNA and RNA, respectively. The structure of the DNA molecule is well established, while that for RNA is not. Therefore, we shall concentrate our attention on the former. The basic structural features of the DNA molecule are the following: First, it is composed of four types of monomeric

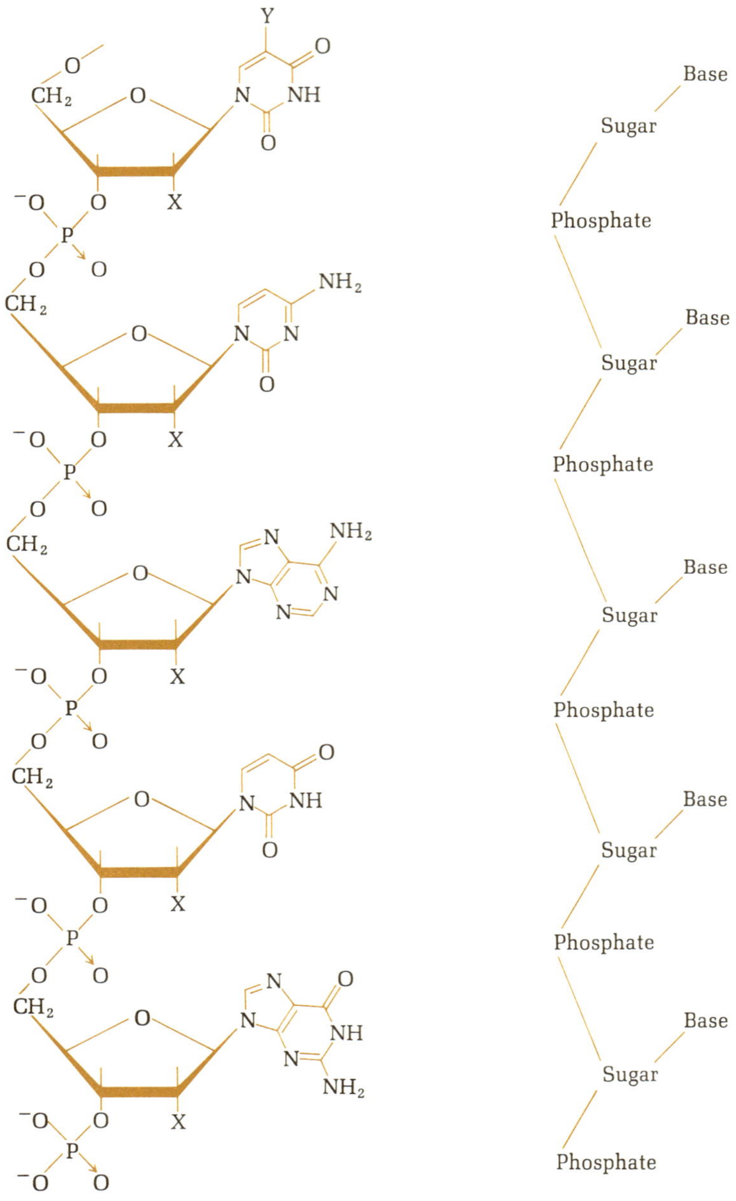

Figure 7-8 A segment of a hypothetical DNA chain.

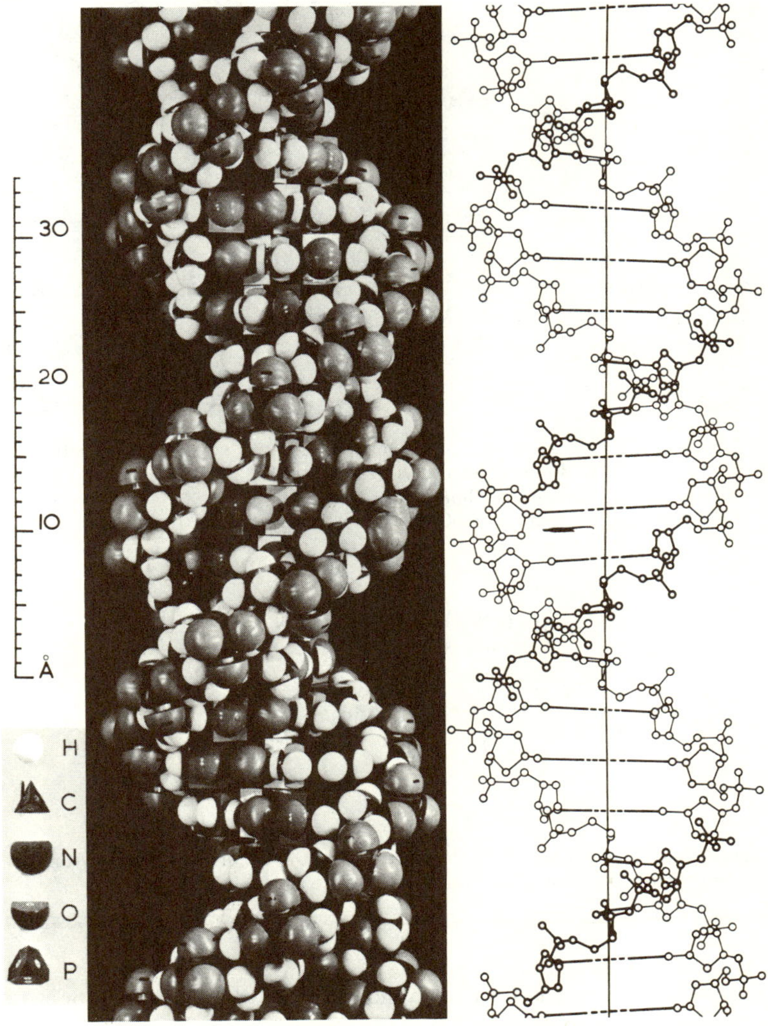

Figure 7-9 Ball-and-stick and space-filling models of the double-stranded helical DNA molecule. (Courtesy of Prof. M. H. F. Wilkins.)

An electron micrograph of DNA from the bacterial virus T2. A careful examination of this long molecule will reveal that it is circular. (From H. R. Mahler and E. H. Cordes, *Biological Chemistry*, 2nd ed., Harper & Row, 1971, New York. Courtesy of C. A. Thomas, Jr., and L. MacHattie.)

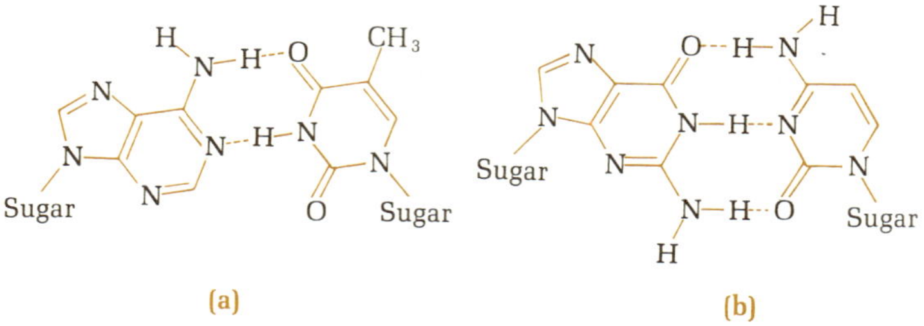

Figure 7-10 The structures of the base pairs that occur in DNA. (a) Adenine–thymine base pair. (b) Guanine–cytosine base pair.

units. Each monomer is itself complex structurally and is composed of a base, a sugar, and phosphate. The monomers can be characterized by the nature of the base present, since the sugar and phosphate parts are the same for all. The four types of base that occur are named *a*denine, *g*uanine, *c*ytosine, and *t*hymine; hence, the designations A, G, C, and T.

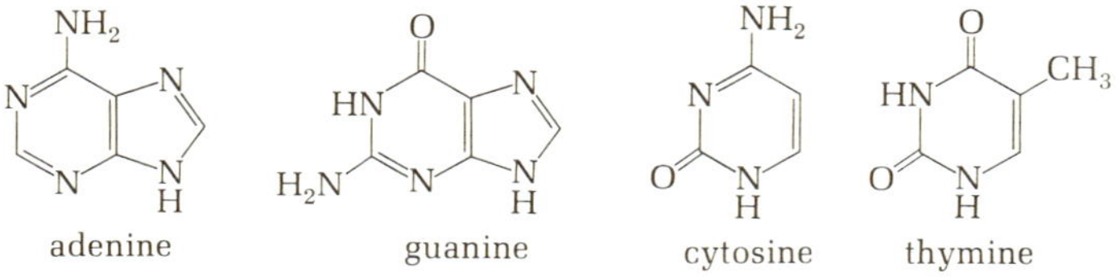

The monomers are joined together through the phosphate units as shown in Figure 7-8. What we have then is a backbone composed of sugar and phosphate with the bases hanging onto it.

A central distinction between the nucleic acids and the other polymers thus far considered is that the individual molecules of the nucleic acids are themselves highly structured entities. Each molecule, as it occurs in nature, is composed of not one but two chains, which are wound about each other in a helical fashion. In Figure 7-9, both ball-and-stick and space-filling models of the double-stranded helical DNA molecule are shown. These molecules can be enormous indeed. Molecular weights in the several billions are known. Such molecules are so large that the largest dimension would be easily visible to the unaided eye. Examination of the structures in Figure 7-9 will reveal that the bases of the nucleic acid molecule are on the inside of the helical structure. Furthermore, they are paired. That is, each base on one chain is paired with one on the opposite chain through weak but specific interactions. This gives rise to two important features of the molecule. First, the base pairing along the molecular backbone severely limits the flexibility of this molecule. Consequently, it behaves in solution much as a rigid rod and on this account solutions of DNA are quite viscous, since the large rods interfere with the freedom of motion of the solvent molecules. For example, a solution of 1 g of a high molecular weight DNA in 100 milliliters (ml) of water is so viscous that the containing vessel may be turned upside down without fear that the solution will spill. Second, the base pairing along the backbone is specific. That is, each A is paired with a T, each T with an A, each G with a C, and each C with a G. The structures of the paired bases, A–T, and G–C, are shown in Figure 7-10. This view of the base pairs would be obtained by looking along the long axis of the DNA molecule; that is, the plane of the bases is perpendicular to this axis. Thus, in the representations in Figure 7-9, the base pairs appear as lines since we are looking at them edge-on. Note that the specificity in formation of pairs of bases is due to the formation of hydrogen bonds between them. The structures of the bases are such that, within the limitations imposed by the double-helical structure of DNA, hydrogen bonding can be obtained only between A and T and between G and C. Thus, one chain is the complement of the other. All of these considerations have a great deal to do with the function of the nucleic acids in living systems.

INFORMATION STORAGE BY DNA

Any living entity, cellular or not, must contain a certain reservoir of *information*. This information provides for orderly cellular development, metabolism, and reproduction. It specifies the structures of all cellular constit-

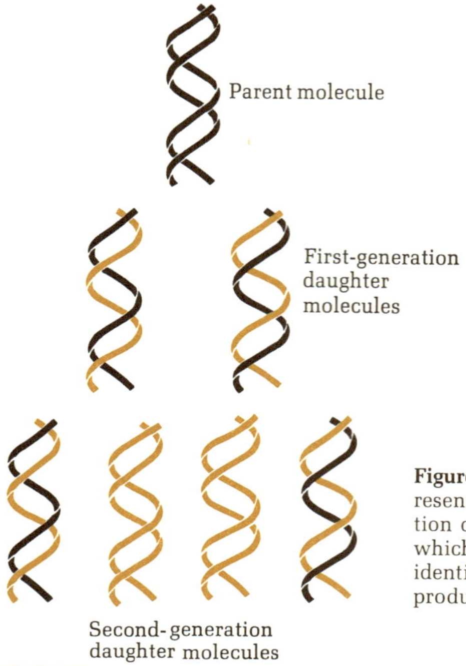

Figure 7-11 A schematic representation of the replication of a DNA molecule in which daughter molecules identical to the parent are produced.

uents. Upon cell division, this information must be passed from parent cell to progeny cell; upon reproduction, from parent to progeny. What is the nature of this central information storehouse, and how is the information employed for maintenance of life? Let us return to our molecule of DNA and make a basic observation: *The DNA molecule is capable of storing information.* Indeed, any polymer that contains more than one type of monomeric unit may be employed as an information storage unit. Let us suppose that the sequence of bases along one of the two chains of the DNA duplex is

AGATTGGGAAAACTAGGATCTGAATTGTCTAATGGTTGA

This may prove to be a perfectly intelligible message (if we know the language), just as a string of letters on this printed page may constitute an intelligible message. The alphabet of the DNA language is simple, since it contains just four letters: A, G, C, and T. Furthermore, it turns out that all words in the DNA language contain just three letters. Finally, the DNA language has one punctuation mark, the period, which is also a three-letter word, TGA. With this information in mind, we can rewrite the above message:

AGA TTG GGA AAA CTA GGA TCT GAA TTG TCT AAT GGT

Now, if we know what these words mean or have a means for translating them into some language that we do understand, this message would make sense to us.

To understand a bit more about this, we need to know an additional fact: Protein molecules are responsible for virtually all the structure and functions of the living cell. Proteins are themselves polymers, as we shall see directly. The important point at the moment is that the information contained in DNA is employed for the maintenance of life through its use to *specify the structures of proteins*. Thus, to use the information possessed by the DNA molecule for the purposes of life, we must translate the message into a new language, that of the protein molecules. There exists complex translation machinery in all living cells that converts the DNA information into the structure of proteins (see Chapter 21). This is simply accomplished conceptually, since each of the words in the message above, for example, designates one monomeric unit of a protein molecule. More about this will follow when we come to the discussion of protein structures. For now, we wish to emphasize one additional point: *The DNA molecule is capable of transmitting information*. Clearly, if DNA is going to be the central information storehouse of the cell, some means must be found to transmit this information to the daughter cells upon cell division. This means is inherent in the structure of the DNA molecule itself as a consequence of its complementary double-stranded structure. Thus, if the DNA molecule unwinds and the strands come apart, the synthesis of two new strands—one complementary to each of the original strands—can occur, since A always lines up opposite to T and so forth; we shall end up with two double-stranded molecules, each identical to the original. Each daughter cell could then receive one of these new molecules. The process is shown schematically in Figure 7-11.

It may be readily appreciated then, that the structure and function of the DNA molecule are intimately associated. If we wished to invent a molecule capable of the storage and transmission of information, it would be difficult to improve on the molecule nature has chosen to perform this function.

PROTEINS

Proteins are ubiquitous components of living material and perform several functions. These include crucial roles in cellular architecture, catalysis, metabolic regulation, contractile processes, and defense mechanisms. Among the polymers that occur in living systems or that have been synthesized in the laboratory, the proteins are perhaps the most complicated. They contain no less than 20 types of monomeric units. These are called **amino**

acids and are structurally related to glycine, which in fact is an amino acid that occurs in proteins:

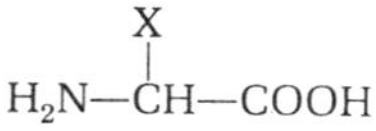

glycine

The different amino acids differ in the nature of the group X in this structure. Clearly, in glycine X is H. The chain of a protein molecule then looks like

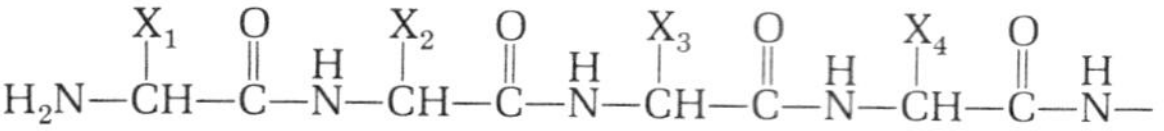

This is just the same as the basic structure of polyglycine, except that a variety of different groups may be substituted for one of the hydrogen atoms in each $—CH_2—$ group. Note also the similarity of the nylon and protein structures [Equation (7-3)] in that the monomers of each are joined by a $-\overset{\overset{O}{\|}}{C}-\overset{\overset{H}{|}}{N}-$ link. The structures of the amino acids that commonly occur in proteins are collected in Table 7-3. Note the use of three-letter abbreviations.

Both the portion of the amino acids that is common to all and the portion that varies from one to another are of interest to us. The amino and carboxyl groups are the moieties through which one amino acid is linked to another. Hence, these two groups, plus the α-carbon atom between them, form the backbone of the molecule. The side chains, on which our classification in Table 7-3 is based, are appended to this backbone and are responsible for the individual characteristics of protein molecules. Clearly, the nature and arrangement of these side chains serve to differentiate one protein from another at the most basic level of structural organization. It will prove of much interest later to understand the types of interactions in which each of the groups participate. Let us, therefore, consider the properties of the amino acids in terms of the nature of their side chains.

We may distinguish three classes of amino acids: *nonpolar, uncharged polar,* and *charged polar.* Into the nonpolar class we may place Gly, Ala, Val, Leu, Ile, Phe, Met, and Pro. In addition, Tyr and Trp have substantial nonpolar characteristics but also have the capacity to interact with water in a favorable way. These side chains are related to compounds that themselves have little or no solubility in water. Placing such groups in an aqueous environment has the effect of inducing the water molecules in the immediate vicinity to adopt an ordered structure (Chapter 10). These groups tend to form hydrophobic bonds.

The individual amino acids differ of course, in their degree of hydrophobicity: The larger the nonpolar side chain, the more hydrophobic the amino acid. Space-filling models of the pertinent side chains are shown in

Figure 7-12. Clearly, Phe is the bulkiest of the group and Gly is the smallest. Two of the amino acids in this category have special properties: Phenylalanine can interact favorably with other aromatic systems by virtue of overlap between the two pi electron systems, and methionine can interact with metal ions because the sulfur atom possesses unshared electron pairs. Nature appears to have chosen the nonpolar amino acids both on the basis of the variety of hydrophobic bulk and molecular shape and on the basis of ability to participate in specialized interactions. Note, specifically, that the amino acid having a hydrocarbon structure corresponding to the side chain of methionine does not occur in proteins; methionine is capable of providing the necessary bulk and yields the metal-binding capacity as a bonus.

The uncharged polar group of amino acids contains a variety of structures, being comprised of Ser, Thr, Asn, Gln, Trp, and Tyr. Each of these side chains is capable of interacting with water molecules through formation of hydrogen bonds. Similar interactions with other polar molecules are possible as well and, in fact, occur regularly in proteins, as we shall see below. Here again, a variety of sizes and shapes is present in the side chains, and both Tyr and Trp can interact with other aromatic systems, as noted above for the case of Phe. Moreover, the hydroxylic amino acids can act as nucleophilic reagents, and all members of the groups can interact with metal ions.

Finally, several amino acids bear a full charge at neutral pH and fall into the charged polar category: Asp and Glu are negatively charged, and His, Lys, and Arg bear a positive charge. These side chains can be involved in electrostatic interactions as well as hydrogen bond formation. They will strongly prefer an aqueous environment and, insofar as is possible, will occupy positions in which they are fully solvated by the available water molecules. We cannot expect to find a charged polar side chain surrounded on all sides by nonpolar ones. Later we shall see how these general considerations are translated into specific protein structures.

Proteins are very large molecules. The largest have molecular weights of a few million. It is therefore clear that opportunity for enormous structural variation exists in these molecules. If a protein has 20 different amino acids as the monomeric units, we can construct 400 different molecules having just 2 monomer units (dipeptides), 8000 different ones having 3 units (tripeptides), and 20^{100} different ones having 100 amino acid units. This variability accounts for the existence of protein molecules that have many different physical and chemical properties, which in turn accounts for their use in so many different cellular functions. Let us consider some of the aspects of the structure of just two proteins: myoglobin and carboxypeptidase.

To understand protein structures, it is first necessary to determine the

Table 7-3 Structures for the amino acids that occur in proteins

Name	Structure
I. Aliphatic amino acids	
1. Glycine (Gly)	$H_2N—CH_2—CO_2H$
2. Alanine (Ala)	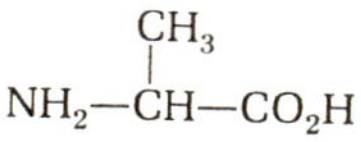
3. Valine (Val)	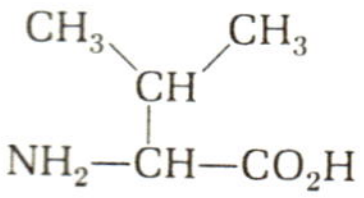
4. Leucine (Leu)	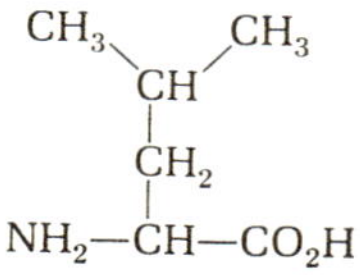
5. Isoleucine (Ile)	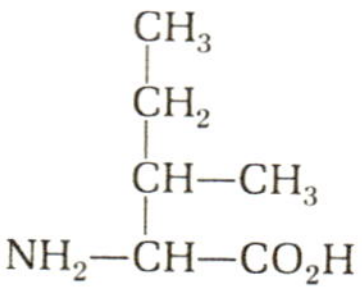
II. Hydroxyamino acids	
6. Serine (Ser)	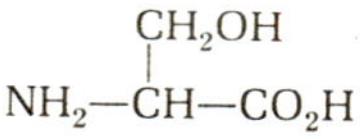
7. Threonine (Thr)	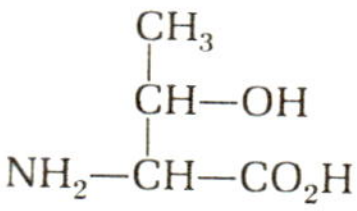
III. Dicarboxylic amino acids and their amides	
8. Aspartic acid (Asp)	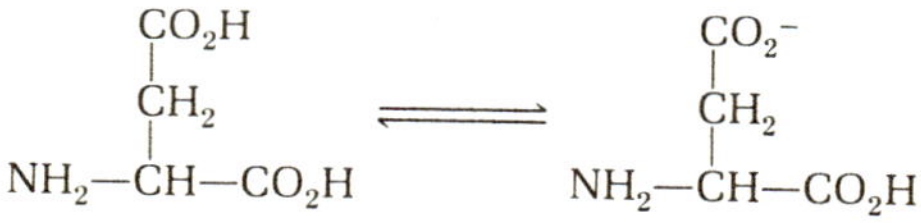
9. Asparagine ($AspNH_2$) (or Asn)	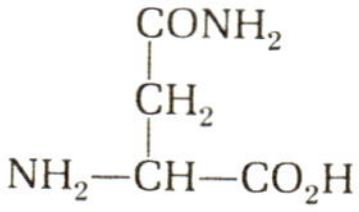
10. Glutamic acid (Glu)	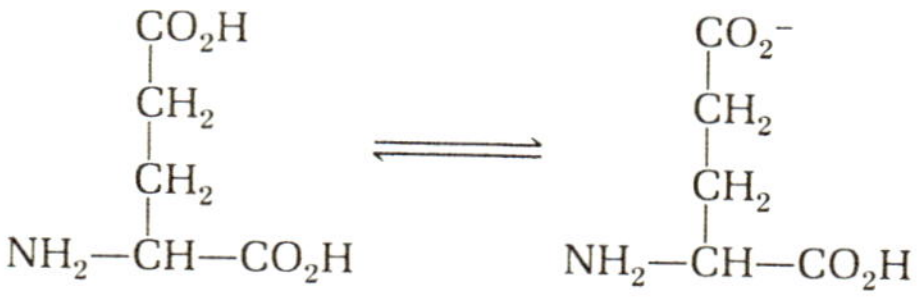
11. Glutamine ($GluNH_2$) (or Gln)	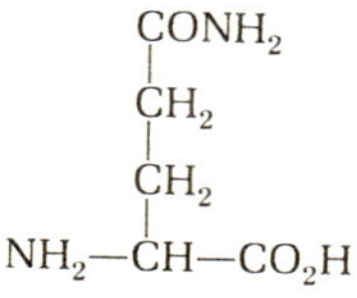

Table 7-3 *(Continued)*

Name	Structure
IV. Amino acids having basic functions	
12. Lysine (Lys)	
13. Histidine (His)	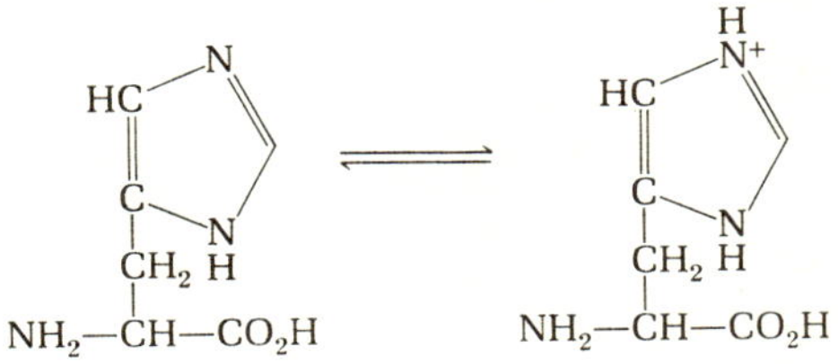
14. Arginine (Arg)	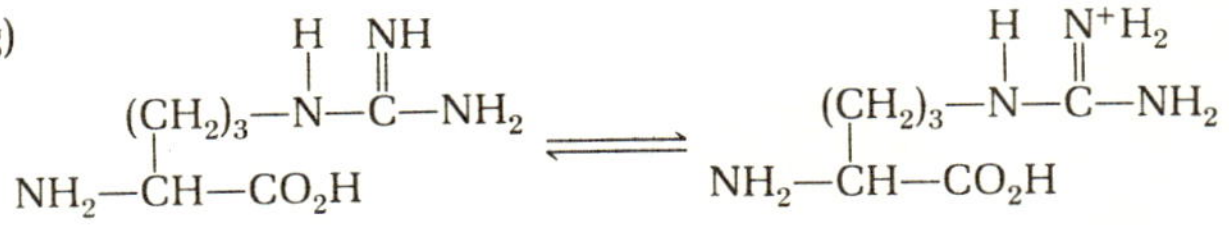
V. Aromatic amino acids (histidine included in category IV)	
15. Phenylalanine (Phe)	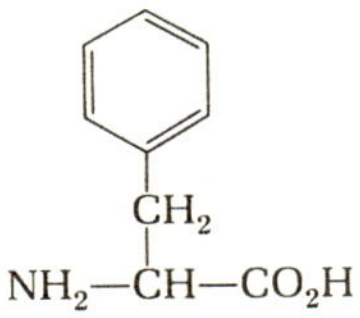
16. Tyrosine (Tyr)	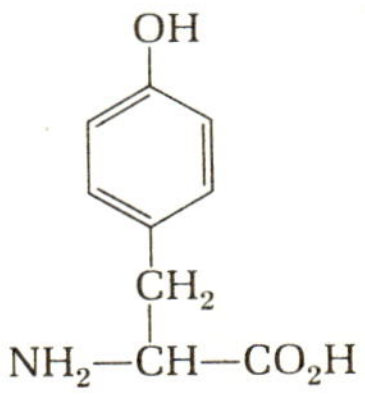
17. Tryptophan (Trp)	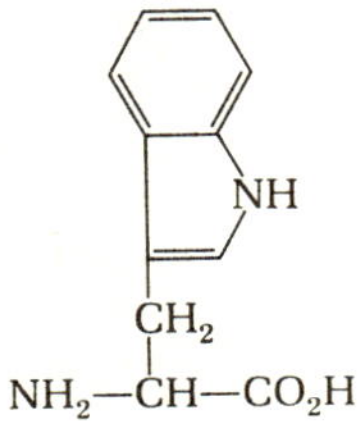
VI. Sulfur-containing amino acids	
18. Cysteine (CySH)	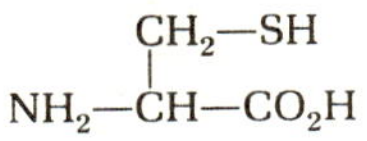
19. Cystine (CyS-SCy)	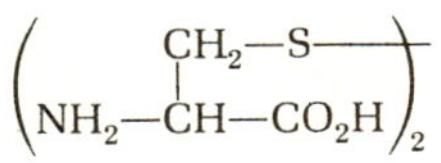
20. Methionine (Met)	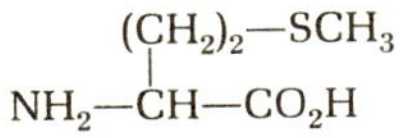
VII. Imino acids	
21. Proline (Pro)	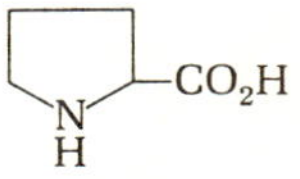

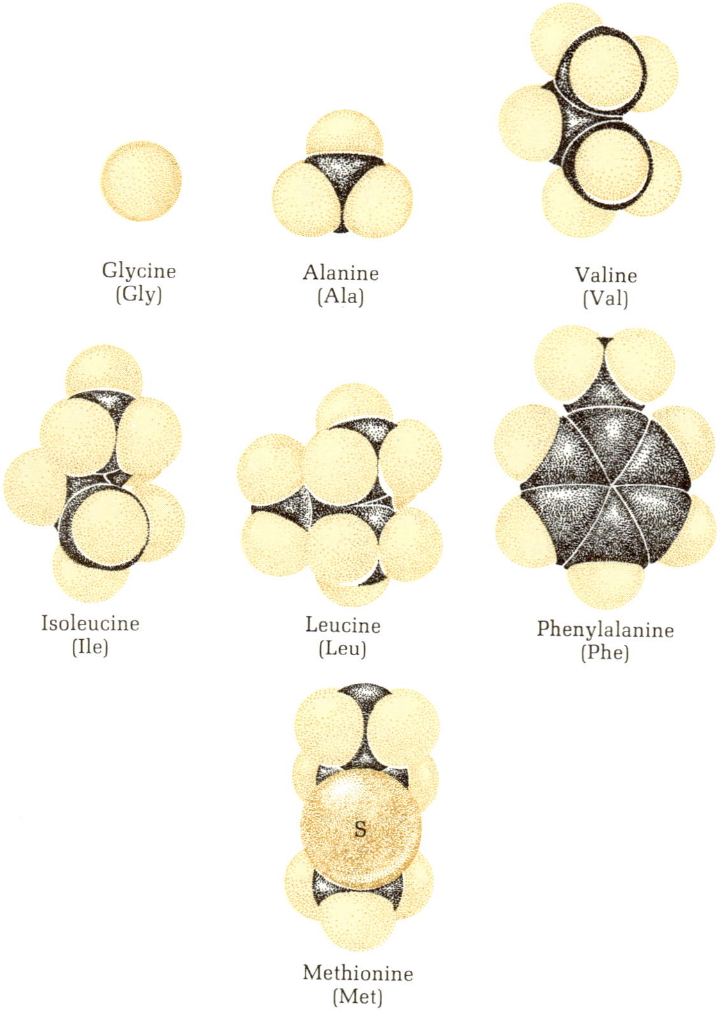

Figure 7-12 Space-filling drawings of the side chains of the nonpolar amino acids. The drawings give an accurate representation of the relative sizes of the atoms.

sequence of amino acids along the polypeptide chain. It was not until 1955 that the sequence of amino acids for a protein was determined. In that year F. Sanger, a British biochemist, completed his work on the structure of insulin, a small protein that functions as a hormone in regulating the metabolism of carbohydrates. Since then, the amino acid sequences for a great many proteins have been determined. As an example, let us consider the protein *myoglobin*, which is an oxygen-storage protein in mammalian muscle. This protein is composed of a linear chain of 153 amino acids; the sequence for myoglobins from six species are collected in Figure 7-13. The abbreviations employed in this figure have been given in Table 7-3. Note that the structures are related but not identical in the three cases. This is generally found to be true: Proteins that perform the same function in different organisms have related but distinct sequences of amino acids.

The nature of the amino acid substitutions in the myoglobins in Figure 7-13 is worth noting. In almost every case, hydrophobic amino acids replace hydrophobic ones or polar amino acids replace polar ones. That is, although the details of the structure of the amino acid at a particular site changes, the essential nature of the amino acid does not. This is precisely the result that we would have expected. We might readily expect that more drastic changes, substitution of a polar amino acid for a hydrophobic one, for example, might so change the structure of the protein that it would have altered biological properties. We shall, in fact, encounter one such example later.

How do we go about determining the amino acid sequence for a protein? This is a multistep process which principally consists of (1) specifically cleaving the protein into a number of smaller pieces; (2) determining the sequence of these pieces by available chemical means; and (3) putting all the information together to deduce the complete structure. An example of how this is done is shown in Figure 7-14, in which the structures of individual parts of one segment of the insulin molecule are arranged to show how they relate to the whole structure.

Example *Suppose that partial hydrolysis of a small polypeptide yields products having the following sequence of amino acids:*

a. Ala · Ser
b. Ile · His · Leu · His · Glu · Ala
c. Ala · Ser · Phe
d. Phe · Gly · Glu

What is the sequence of the intact polypeptide?

Solution *The only possibility is*

Ile·His·Leu·His·Glu·Ala·Ser·Phe·Gly·Glu

Our next problem is to define just how the polypeptide chain of a protein molecule is arranged in space. To examine this matter, let us turn our attention to a single protein, *carboxypeptidase*, for which such a study is complete.

	1	2	3	4	5	6	7	8	9	10	11
Sperm whale	Val	Leu	Ser	Glu	Gly	Glu	Trp	Gln	Leu	Val	Leu
Harbor seal	Gly			Asp				His			
Porpoise	Gly										
Cow	Gly			Asp					Ala		
Horse	Gly			Asp					Gln		
Man	Gly			Asp					—		

	26	27	28	29	30	31	32	33	34	35	36
Sperm whale	Gln	Asp	Ile	Leu	Ile	Arg	Leu	Phe	Lys	Ser	His
Harbor seal		Glu	Val								
Porpoise			Val							Gly	
Cow		Glu	Val						Thr	Gly	
Horse		Glu	Val						Thr	Gly	
Man		Glu	Val							Gly	

	51	52	53	54	55	56	57	58	59	60	61
Sperm whale	Thr	Glu	Ala	Glu	Met	Lys	Ala	Ser	Glu	Asp	Leu
Harbor seal	Ser		Asp	Asp		Arg	Arg				
Porpoise											
Cow											
Horse											
Man	Ser		Asp								

	76	77	78	79	80	81	82	83	84	85	86
Sperm whale	Leu	Lys	Lys	Lys	Gly	His	His	Glu	Ala	Glu	Leu
Harbor seal											
Porpoise										Asn	
Cow											Val
Horse											
Man											Ile

	101	102	103	104	105	106	107	108	109	110	111
Sperm whale	Ile	Lys	Tyr	Leu	Glu	Phe	Ile	Ser	Glu	Ala	Ile
Harbor seal											
Porpoise											
Cow									Asp		
Horse									Asp		
Man										Ser	

	126	127	128	129	130	131	132	133	134	135	136
Sperm whale	Asp	Ala	Gln	Gly	Ala	Met	Asn	Lys	Ala	Leu	Glu
Harbor seal				Ala			Lys				
Porpoise											
Cow							Ser				
Horse							Thr				
Man							Asp				

	151	152	153
Sperm whale	Tyr	Gln	Gly
Harbor seal	Phe	His	
Porpoise	Phe	His	
Cow	Phe	His	
Horse	Phe		
Man	Phe		

12	13	14	15	16	17	18	19	20	21	22	23	24	25
His	Val	Trp	Ala	Lys	Val	Glu	Ala	Asp	Val	Ala	Gly	His	Gly
Asn			Gly				Thr						
Asn			Gly										
Asn	Ala		Gly										
Asn			Gly						Phe				
Asn			Gly				Pro		Ile				

37	38	39	40	41	42	43	44	45	46	47	48	49	50
Pro	Glu	Thr	Leu	Glu	Lys	Phe	Asp	Arg	Phe	Lys	His	Leu	Lys
								Lys					
								Lys					
								Lys					
								Lys					
								Lys					

62	63	64	65	66	67	68	69	70	71	72	73	74	75
Lys	Lys	His	Gly	Val	Thr	Val	Leu	Thr	Ala	Leu	Gly	Ala	Ile
Arg				Asn								Gly	
				Asn								Gly	
				Asn								Gly	
				Thr	Ala							Gly	
				Ala								Gly	

87	88	89	90	91	92	93	94	95	96	97	98	99	100
Lys	Pro	Leu	Ala	Gln	Ser	His	Ala	Thr	Lys	His	Lys	Ile	Pro
	His							Asn				Val	
												Val	

112	113	114	115	116	117	118	119	120	121	122	123	124	125
Ile	His	Val	Leu	His	Ser	Arg	His	Pro	Gly	Asn	Phe	Gly	Ala
						Lys			Ala	Glu			
									Ala	Glu			
					Ala	Lys			Ser				Ala
						Lys							Ala
Val	Asp			Gln		Lys							Ala

137	138	139	140	141	142	143	144	145	146	147	148	149	150
Leu	Phe	Arg	Lys	Asp	Ile	Ala	Ala	Lys	Tyr	Lys	Glu	Leu	Gly
			Asn										
							Thr						
			Asn		Ala		Glu				Val		
			Asn										
					Met		Ser	Asp					

Figure 7-13 A comparison of the amino acid sequences of sperm whale, harbor seal, porpoise, cow, horse, and man myoglobins. Differences from sperm whale proteins are indicated by the appropriate substitution. Where no difference occurs, the entry is omitted. (Courtesy of Prof. F. R. N. Gurd.)

Complete sequence	1 2 3 4 5 6 7 8 9 10 Phe · Val · Asp · Glu · His · Leu · Cys · Gly · Ser · His ·
Sequences of individual parts of the chain	Phe · Val · Asp · Glu · His · Leu · Cys · Gly Ser · His · His · Leu · Cys · Gly · Ser · His ·
Complete sequence	11 12 13 14 15 16 17 18 19 Leu · Val · Glu · Ala · Leu · Tyr · Leu · Val · Cys ·
Sequences of individual parts of the chain	Tyr · Leu · Val · Cys · Leu · Val · Glu · Ala Ala · Leu Ala · Leu · Tyr Leu Val · Glu · Ala · Leu
Complete sequence	20 21 22 23 24 25 26 27 28 Gly · Glu · Arg · Gly · Phe · Phe · Tyr · Thr · Pro ·
Sequences of individual parts of the chain	Gly · Thr · Pro · Gly · Glu · Arg · Gly Gly · Phe Gly · Phe · Phe · Tyr · Thr · Pro ·

Figure 7-14 The structure of one of the two chains of the insulin molecule. Note how the overlapping sequences of the individual parts of the molecule help to provide the complete sequence.

CARBOXYPEPTIDASE

Carboxypeptidase is one of the proteins found in the pancreatic juice of man and other mammals. It is excreted into the small intestine, where it aids in the digestive process by catalyzing the hydrolysis of other proteins. The hydrolysis products, the amino acids, are then employed for the synthesis of those proteins the organism may require. We shall consider some of the aspects of the promotion of chemical reactions by enzymes in Chapter 22. For the moment, let us focus our attention first on the structure of carboxypeptidase as a representative of proteins generally and then on the means for the specification of such structures employing the information of DNA.

Carboxypeptidase consists of a chain of 309 amino acids linked sequentially in a single chain as in the case of myoglobin. The sequence of amino acids in carboxypeptidase provides us with a great deal of information concerning the structure of this molecule but is far from telling the whole story. Proteins represent the highest degree of structural organization within an individual molecule. The polyethylene molecule, for example, exists as a random coil by itself but may be organized to some extent through interactions with its neighbors. DNA molecules have a more fixed structure because of their rigid double-stranded helical structure. Many protein molecules, carboxypeptidase included, have even more complex structures. That is, the polymer exists essentially in only *one* of the infinity of possible conformations available to it. Thus, each atom has its position fixed with respect to other atoms of the molecule. Since we are dealing with a molecule containing several thousand atoms, it is clear that specifications of the complete structure will be no easy task. How do we go about it?

The complete specification of three-dimensional structures for protein molecules has only been realized in recent years. At the moment, detailed information concerning protein structures is available for about a dozen proteins. The first protein for which such detailed three-dimensional information was available was myoglobin from the sperm whale. This structural information was the culmination of almost three decades of work by two British biochemists, J. C. Kendrew and M. F. Perutz. This most important series of investigations led to a sharing of the Nobel Prize in chemistry by these scientists in 1965. The information was obtained by employing another fingerprinting method available to the chemist: X-*ray diffraction*. In some ways this technique is similar to electron diffraction (Chapter 2). In both cases the chemist bombards the molecule of interest with particles or radiation, X-rays or electrons in these two cases, and obtains a scattering pattern, that is, a picture of how the particles or radiation is scattered by the molecule. This, in turn, permits him to go back and understand a great deal about the arrangement of the atoms that did the scattering. Electron diffraction is generally used for compounds composed of a few atoms only, 10 or 20, in the gas phase (although it does find uses for condensed phases as well). X-Ray diffraction methods can be adapted to use with very large molecules but require a more ordered organization of molecules with respect to each other than is present in the gas phase. The most detailed information is obtained through the use of crystalline preparations, and the necessity of using a crystalline protein is a significant limitation on the application of this technique. An X-ray diffraction pattern obtained from carboxypeptidase is shown in Figure 7-15. The light spots on the film are called *diffraction maxima*. The intensity of each diffraction maximum is a function of the location and nature of each atom in the molecule. That is, all the atoms contribute to each maximum. Clearly, as in the case of electron

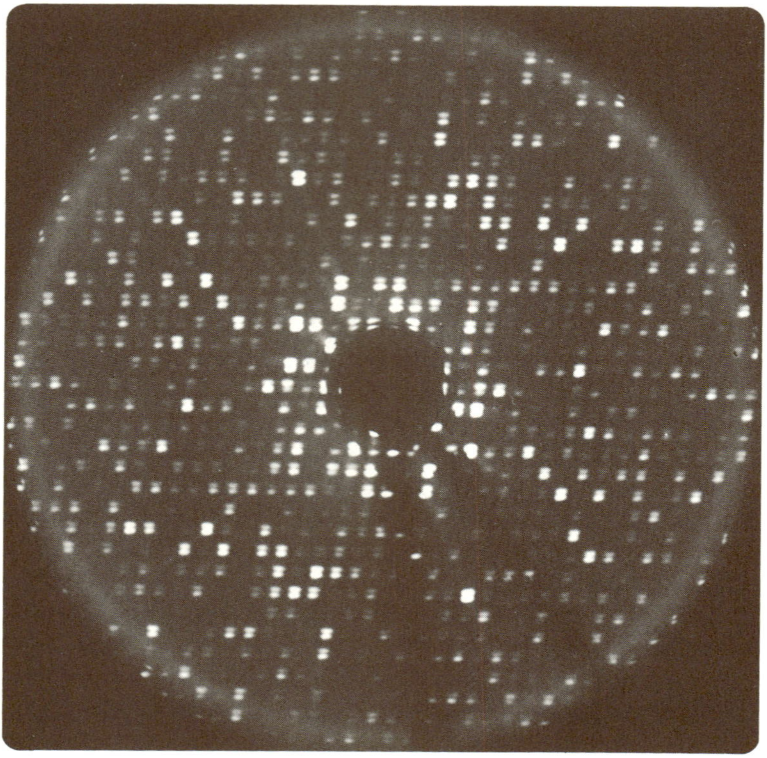

Figure 7-15 An X-ray diffraction pattern for carboxypeptidase. (Courtesy of Prof. W. N. Lipscomb.)

diffraction, X-ray diffraction provides only an indirect picture of the molecule. Nevertheless, the picture is an extremely detailed one. It just requires informed interpretation. Diffraction patterns of the sort of Figure 7-15 can be translated into **electron density maps.** In such maps (an example is shown in Figure 7-16), electron densities are plotted as contour lines. Thus, each line traces out a region of uniform electron density. The regions of highest electron density correspond to atoms. Positions of atoms correspond to places on the map where there are many concentrated lines. Since each type of atom has its own characteristic electron density, the nature of the atom as well as its location can be determined. Thus, if we simply collect enough data, analyze it with sufficient rigor, and construct enough electron density maps, the complete structure of the protein can, in principle, be deduced.

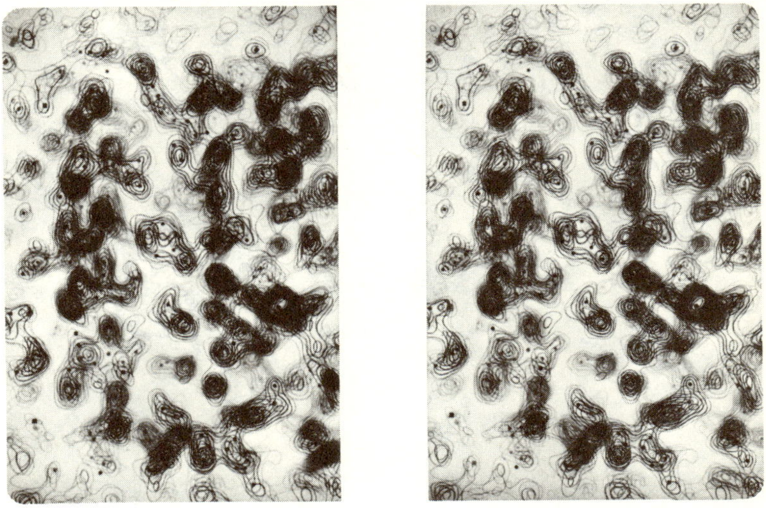

Figure 7-16 An electron density map for a portion of the carboxypeptidase molecule. Lines correspond to contours of constant electron density. Concentrations of lines correspond to centers of high electron density which, in turn, define positions of atoms, •. (Courtesy of Prof. W. N. Lipscomb.)

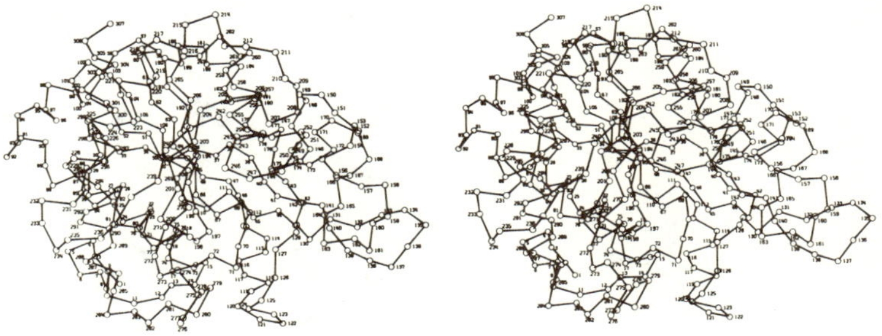

Figure 7-17 A stereo pair showing the complete three-dimensional structure of carboxypeptidase. (Courtesy of Prof. W. N. Lipscomb.)

It is easier to say it than do it. Determination of the structure of carboxypeptidase, the work of Prof. W. N. Lipscomb and his coworkers at Harvard University, required a battery of X-ray cameras together with equipment for automatic data collection, use of high-speed computers, and perhaps 30 man-years of time. The total investment of money necessary to determine the complete structure ran into several hundreds of thousands of dollars. The result of it all is shown in Figure 7-17, in which the backbone configuration of the molecule is shown as a stereo pair. Clearly, it is complex. A view of an important part of the whole molecule is shown in Figure 7-18.

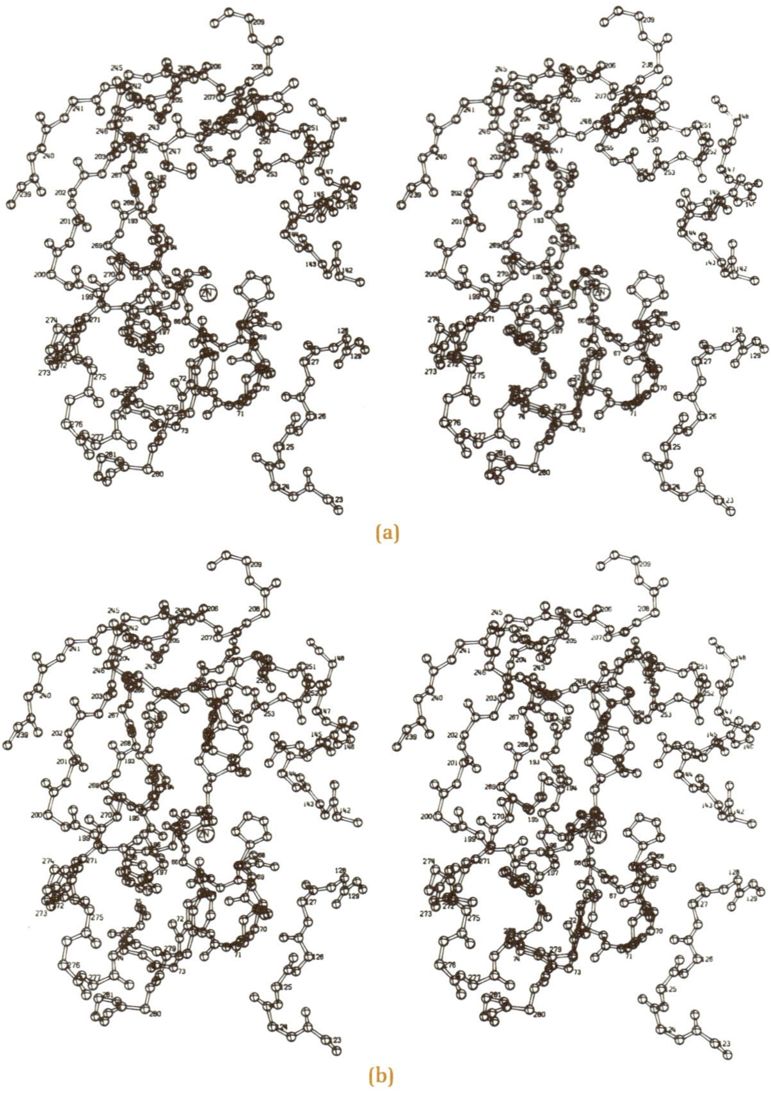

Figure 7-18 Two sets of stereo pairs showing a restricted portion of the carboxypeptidase molecule. Attention is focused on just that region of the molecule that is actively involved in the catalytic process. This region contains one atom of zinc, the only metallic atom in the molecule, which is clearly visible. In the upper pair (a), this portion of carboxypeptidase is shown in the form that it assumes in the absence of substrate molecules, that is, in a resting state. Note the position of the ring of the tyrosine in position 248. In the lower pair (b), the same site is shown in the form that it assumes in the process of catalysis, that is, in an active state. Note that the ring of tyrosine 248 has swung down 180° and is now rather near the zinc. (Courtesy of Prof. W. N. Lipscomb.)

Here the three-dimensional relationships involved for those parts of the molecule that come together in one region in space can be seen more clearly. Structures of the type shown in these figures are one of the great triumphs of modern chemistry. The consequences of knowing them will be explored more fully in Chapter 22, where the relationship of structure to function is explored.

An important point about the three-dimensional structure of carboxypeptidase is the following: *It is a unique structure and required for the biological function* of the molecule. If it is destroyed or even slightly altered by mistreatment, carboxypeptidase becomes a biologically impotent molecule. This is true in general for protein molecules whatever their function may be. How is this structure related to the sequence of amino acids along the polypeptide chain? That is, given that we have a linear chain of amino acids, how do we complete the synthetic process by arranging the chain properly in space? The point here is simple: Nature does it for us. The structure of a protein in the biologically active form is precisely that which the molecule normally assumes. Thus, if we could hold onto a carboxypeptidase molecule by the two ends of the chain and stretch it out into a linear structure and then let go of the ends, it would assume that structure shown in Figure 7-17.

This brings up the next point: What determines the three-dimensional conformation of a protein molecule? We know that there are a large number of forces that combine to determine the most stable conformation (see the discussion of forces above). Although a complete analysis of this question is not yet available, we can point out some of the important features. Let us return to the structures of the amino acids for a moment and consider the nature of the side chains. It is only reasonable that a protein should adopt a conformation that has most of the hydrophobic amino acids on the inside protected from water and able to create their own hydrophobic environment. At the same time, all of the charged and other hydrophilic amino acids should appear on the protein surface, where the favorable interactions with water can occur. This is exactly what is found. In substantially every case investigated so far, the interior of the protein is hydrophobic and the outside hydrophilic.

What would happen if we changed the nature of just one amino acid along the entire polypeptide chain? As noted above, it depends on the type of change we make. If we substitute a hydrophobic amino acid for another hydrophobic one, probably nothing much will happen to the overall structure or biological function, as shown by the various amino acid sequences of the myoglobins (Figure 7-13). On the other hand, if we substitute a hydrophilic one for a hydrophobic one, we have replaced an amino acid that occupied a position on the inside with one that would much prefer to be on the outside. As a result, the whole structure of the protein molecule

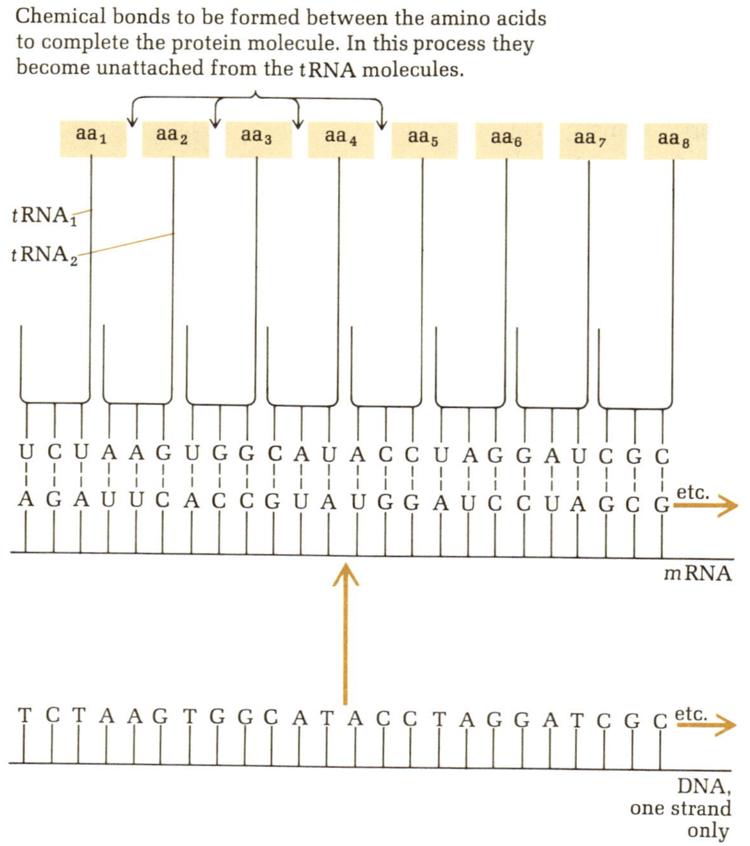

Figure 7-19 A schematic visualization of how the sequence of bases in DNA can be translated into a sequence of amino acids in a protein. The *t*RNA molecules act as adaptors. They have the capacity to recognize both the amino acid and the code word for the amino acid in the *m*RNA molecule. Hence, they are a translator of one language into the other.

may change appreciably in arriving at a new energy minimum. Since the biological activity of the protein depends crucially on its total structure, we might expect a single amino acid interchange of this type to have rather serious consequences. And indeed it may. An excellent example of this is associated with the disease *sickle cell anemia*. People who have this disease have abnormal hemoglobin, an oxygen-carrying protein of blood, which functions poorly compared with normal hemoglobin. As a result, the disease is quite a serious one. It is known that the hemoglobin molecules of normal people and those who are afflicted with sickle cell anemia differ in just one respect: A position normally occupied by a valine (hydrophobic) is replaced by a glutamic acid (hydrophilic). Other than this one change, the molecules are identical. Clearly, the change of just one amino acid has changed the structure of the molecule sufficiently so that it can no longer function efficiently. What in fact happens is that the abnormal hemoglobin becomes insoluble when combined with oxygen and actually precipitates within the red blood cell. This gives the cells their characteristic "sickle" shape, from which the disease takes its name. There is a marked increase in the viscosity of the blood and its consequence is a decreased ability to transport oxygen, resulting in anemia.

SPECIFICATION OF PROTEIN STRUCTURE

Finally, let us return for a moment to the information contained in the DNA molecule and its translation into protein structure. *Each three-letter word in the DNA language corresponds to one amino acid.* Thus, the sequence of words along the DNA molecule specifies a sequence of amino acids along the protein molecule. The message indicated on p. 220 may be translated into the structure of a protein in the following hypothetical manner:

AGA	TTG	GGA	AAA	CTA	GGA	TCT	GAA	TTG	TCT	AAT	GGT
Arg	Leu	Gly	Lys	Leu	Gly	Ser	Val	Leu	Ser	Asn	Gly

The meaning of the period is that the structure of the protein molecule is complete, and synthesis of the amino acid chain may be terminated. Thus, there is a one-to-one mapping of the structure of the DNA molecule onto the structure of the protein molecule as the result of a translational process.

Just how the translation is accomplished is considered in more detail in Chapter 21. However, we can understand the basis of the process at this time. There are two important points. First, the individual amino acids are attached to small nucleic acid molecules, **transfer RNA** or *t*RNA. This attachment is specific in the sense that each type of *t*RNA recognizes only one of the various amino acids. Thus, glycine is attached to a glycine *t*RNA, alanine to an alanine *t*RNA, and so forth. This recognition involves the sequence of bases along the *t*RNA molecules. Second, a messenger molecule, called **messenger RNA** or *m*RNA, is synthesized from the DNA molecule by taking advantage of the complementary base-pairing property. Thus, the *m*RNA molecule has a base sequence complementary to that of the DNA from which it is synthesized. Third, the amino acid-*t*RNA complex can recognize one of the three-letter words of *m*RNA. It does this in the manner suggested by the structure of the DNA molecule itself: complementary base pairing. Thus, if the three-letter word in the *m*RNA is AGA, this would correspond to the sequence UCU of the *t*RNA chain. By a sequence of such interactions, the amino acids can be lined up in the proper order and then zipped off with the formation of the complete protein. A schematic visualization is provided in Figure 7-19.

What happens if one of the letters in the message in the DNA molecule is wrong? For example, suppose a DNA word that ought to read AGG reads AGT instead. In all probability, AGG is the code word for one amino acid and AGT is the code word for another. Thus, when a protein is synthesized from the DNA molecule, it will have a mistake in it. It is precisely this kind of mistake that results in the production of the abnormal hemoglobin of sickle cell anemia. That is, a single mistake in the information center results in a disease. Actually, many such diseases are known: hemophilia, albinism, and phenylketonuria are examples. Errors in the DNA molecules of cells are called *mutations*. Most such errors are harmful. Mutations are induced by radiation and a number of chemicals. It is precisely because of errors of this and related types that people are concerned about nuclear fallout and the like.

Glossary

Addition polymers those polymers whose synthesis involves only the addition of one monomer to another without the elimination of any molecule.

Amino acids compounds having the general structure $H_2N\text{—}\overset{\displaystyle R}{\overset{|}{C}}H\text{—}COOH$; these are the building blocks (monomers) from which proteins are made.

Amorphous lacking structure; more particularly, lacking crystallinity.

Catalysis acceleration of a chemical reaction through the intervention of an outside agent (the catalyst) in a process in which the catalyst is neither consumed nor produced and in which the overall thermodynamics of the reaction is unchanged.

Condensation polymers those polymers whose synthesis involves the addition of one monomer to another with the elimination of a new molecule in the process.

Degree of cross-linking a measure of the number of cross-linked regions existing in a polymer preparation.

Deoxyribonucleic acid (DNA) a double-stranded polynucleotide containing the bases adenine, guanine, thymine, and cytosine and the sugar deoxyribose; associated with the biological functions of information storage and transfer.

Elasticity the capacity of a substance to reform its original shape following mechanical distortion.

Elastomers polymers having elastic properties.

Electron density map a plot of electron density as a function of position in a crystal; derived from X-ray diffraction studies and employed to aid in structure elucidation.

Hydrogen bonds weak chemical bonds in which a single hydrogen atom is shared between atoms of two electronegative elements.

Hydrophobic or apolar bonds the tendency of two or more hydrophobic molecules or groups present in an aqueous environment to associate with each other.

Ionic bonds chemical bonds formed from electrostatic interaction between two charged groups.
London dispersion forces weak, short-range interactions resulting from mutually induced dipole moments in two molecules or atoms having no permanent dipole moment.
Macromolecule any very large molecule, but almost invariably a polymer of high molecular weight.
Messenger RNA (*m*RNA) a species of RNA molecule that acts as a carrier of genetic information from the DNA to the site of protein synthesis.
Monomer the individual unit from which polymers are constructed.
Nucleic acids polynucleotides; includes RNA and DNA.
Polymer any molecule made by the consecutive linking together of many monomer units.
Protein the class of polymers based on the α-amino acids as monomers; these polymers play several essential functions in living systems.
Ribonucleic acid (RNA) a single-stranded polynucleotide containing the bases adenine, guanine, uracil, and cytosine and the sugar ribose; involved in the process of protein synthesis.
Siloxanes polymers based on silanols as monomeric units.
Transfer RNA (*t*RNA) small RNA molecules that act as adaptors between the amino acids and *m*RNA in the process of protein synthesis.
Viscosity the tendency of a solution to resist flow.

Study problems

1. For each of the following polymers, identify the monomer(s) from which it was derived:

a. 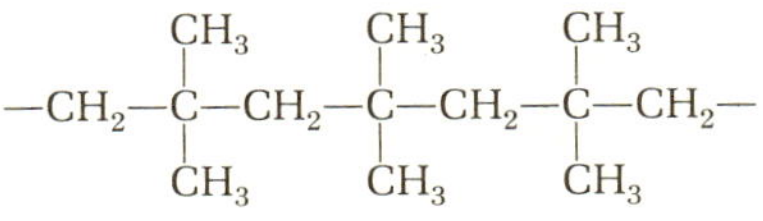

b. $—CH_2—O—CH_2—O—CH_2—O—CH_2—O—CH_2—$

c. 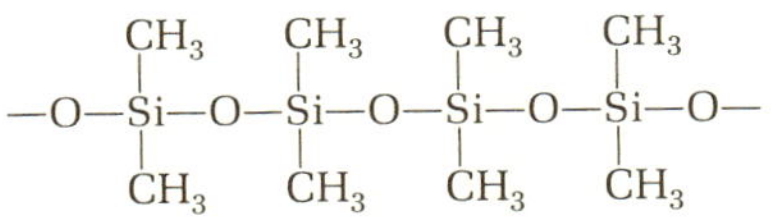

d. 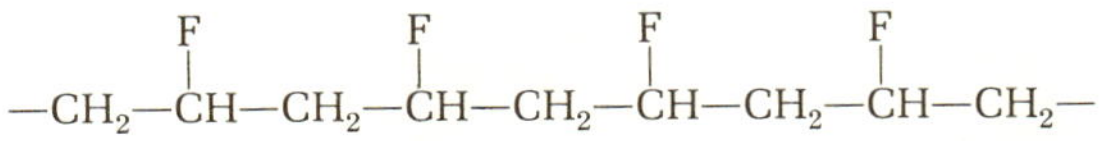

e. 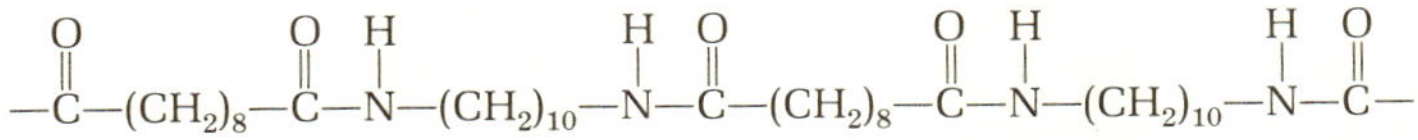

f.

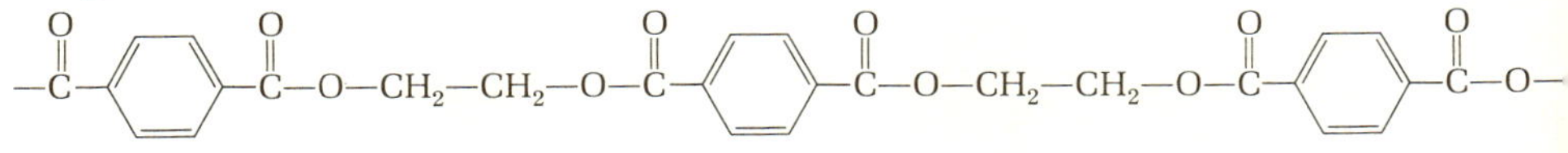

g. $-CH_2-C(Cl)=CH-CH_2-C(Cl)=CH-CH_2-C(Cl)=CH-CH_2-$

2. For each of the following polymers, calculate the molecular weight:

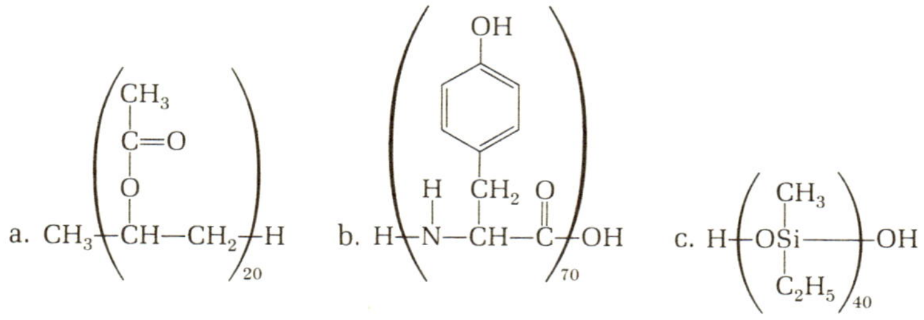

3. Design strategies to synthesize polymers of each of the following types (i.e., given a bottle of monomer A and one of monomer B, how would you proceed to make the indicated copolymers?):
   a. AABBABAAABAABBABBBABBABA . . .
   b. ABABABABABABABABABAB . . .
   c. AAAAAAAAAAAAAAAAAAAAA . . .

```
    B       B       B
    B       B       B
    B       B       B
    B       B       B
    ⋮       ⋮       ⋮
```

4. If polyethylene is irradiated with X-rays, cross-links are introduced into it. How would you expect the properties of this polymer to change as a result of such irradiation? Why?
5. If a sample of atactic and a sample of isotactic polystyrene are slowly heated and the temperature of the samples is measured as a function of the amount of heat added, the following two curves are generated. How do you explain, in molecular terms, the difference observed?

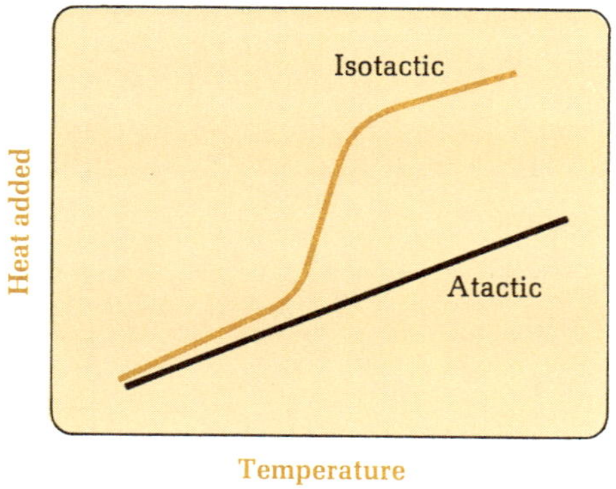

6. Poly-*p*-phenylene is a polymer composed of a linear sequence of benzene rings linked together:

Indicate the properties that you might expect for a high molecular weight sample of this polymer.

7. Which of the following polymers would you expect to have the best elastic properties?
   a. polyethylene
   b. isotactic polypropylene
   c. a copolymer of ethylene and propylene
   Defend your answer clearly.
8. Explain what might happen to the three-dimensional structure of a protein such as carboxypeptidase if it were placed in an organic solvent such as ethanol, CH_3CH_2OH. Justify your predictions.
9. Aqueous solutions of double-stranded DNA are usually quite viscous. If such a solution is gradually heated, the viscosity drops dramatically over a small, characteristic temperature range. How can you account for this observation? Note: Formaldehyde, a reagent that reacts with amino groups, $—NH_2$, does not react with DNA at ordinary temperatures but does react strongly at temperatures above that at which the decrease in solution viscosity occurs.
10. Suppose that the following peptide fragments were obtained from partial degradation of a polypeptide having 13 amino acids in a linear chain:

    Glu · Ala · Leu · Tyr · Leu · Val
    Cys · Gly · Glu · Arg
    His · Leu · Val · Glu
    Leu · Val · Cys

    What is the structure of the original polypeptide?
11. Suppose that a molecule of DNA contains the following sequence of bases along the polymer chain: . . . TGCATCTCGATGTGA . . . . Let this DNA fragment specify the following order of amino acids along a polypeptide chain: . . . Ala · Gly · Val · Lys . . . . What sequence of amino acids would be specified by a portion of a DNA molecule having the base sequence . . . ATGATCTGCATCTCGTGA . . . ?
12. For polynucleotides (i.e., DNA or RNA) of the following types, indicate what types of proteins (polypeptides) would be synthesized using each of them as information source:
    a. AAAAAAAAAAAAAAAAAAAAAAAAA . . .
    b. ABABABABABABABABABABABAB . . .
    c. ABCABCABCABCABCABCABCABCABC . . .
13. Construct a clear picture of the DNA molecule, explicitly indicating its principal structural features. What characteristic of the structure of this molecule accounts for its biological function of information storage? What characteristic of the structure of this molecule accounts for its biological function of information transfer (during cell division)?

Suggestions for further reading

K. F. O. Driscoll *The Nature and Chemistry of High Polymers,* Van Nostrand Reinhold, New York, 1964.
J. D. Watson *Molecular Biology of the Gene,* Benjamin, Menlo Park, Calif., 1965.
J. D. Watson *The Double Helix,* Atheneum, New York, 1968.

Scientific American has published a number of excellent articles concerning the structure of biological macromolecules. A few of these are

F. H. C. Crick, "The Genetic Code" (October, 1966).
D. C. Phillips, "The Three Dimensional Structure of an Enzyme Molecule" (November, 1966).
C. Yanofsky, "Gene Structure and Protein Structure" (May, 1967).
A. Mirsky, "The Discovery of DNA" (June, 1968).
L. Bragg, "X-ray Crystallography" (July, 1968).
A. Kornberg, "The Synthesis of DNA" (October, 1968).
N. Sharon, "The Bacterial Cell Wall" (May, 1969).

8

The energetics of chemical reactions

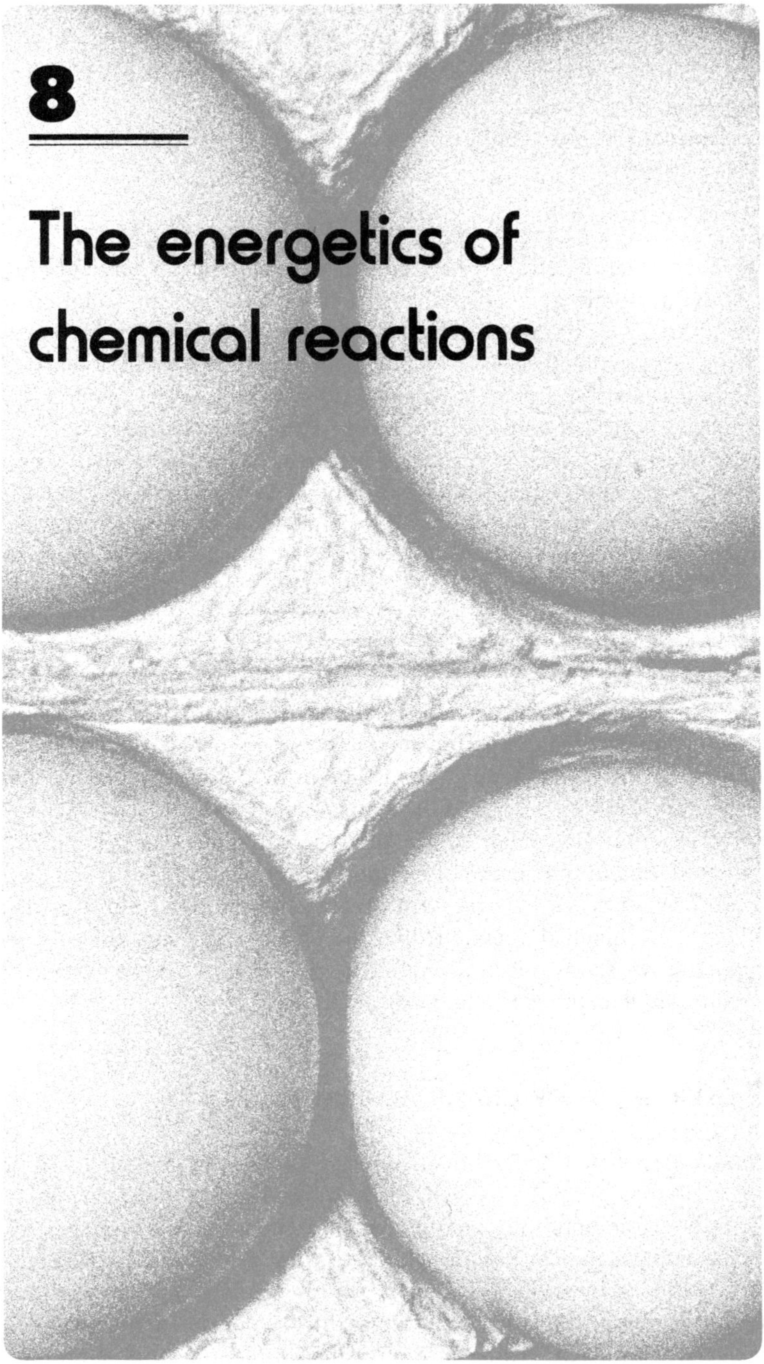

Herein is taken a dim view of equilibrium constants:
I look upon amity and enmity as affections of human beings, and I have not yet found it explained by any, how such appetites can be placed in bodies inanimate and devoid of knowledge or of so much as sense.

Robert Boyle
The Sceptical Chymyst, 1661

Our consideration of chemistry so far has dealt with *chemical statics,* that is, with atomic and molecular structures and with the relationships between them. Our interest has been centered about several questions: Can a particular architecture exist for a given molecule? Will a particular architecture exist for a given molecule? How do molecular structures vary with subtle changes in molecular makeup? How can we understand molecular structures in terms of atomic structures? What types of forces are involved in maintaining molecular structures? With some capacity to provide answers for these questions now in hand, we are in a position to approach the field of *chemical dynamics,* that is, chemical reactions and chemical reactivity. Previously, we have treated questions such as these only in passing (see, for example, Figure 6-5 or 6-6 as an example of a change of structure).

We shall center our consideration of chemical dynamics about a set of questions related to those for chemical statics, but this time focused not on the structures themselves but on the changes that can occur in these structures: What types of reactions occur in or between molecules? Can a particular reaction occur? Will a particular reaction occur? How do chemical reactions depend on subtle differences in molecular makeup? How can we understand the course of chemical reactions in terms of molecular structure? What types of forces are involved in the determination of the features of a chemical reaction? In this chapter we consider the most fundamental of these questions. That is, we examine those factors that determine the energetics of chemical reactions.

THE ENERGETICS OF CHEMICAL REACTIONS

Understanding something about the energy changes in chemical reactions is a very important matter indeed. Such understanding will provide information about the conditions under which a chemical reaction will occur if it can be made to occur at all, whether or not a hypothetical compound is likely to be stable, whether a given compound might prove useful as a fuel, and the like. Basically, what we want to know at this stage is: *Can a given reaction occur?* To be sure, this qualitative question is too limited to serve all our purposes. Quantitatively, the question becomes: To what *extent* will a particular chemical reaction occur? But the qualitative point of view will suffice for the moment. Before we delve into its various aspects, we may explain that matters relating to the energy changes in chemical reactions fall into the broad field of **thermodynamics.** Thermodynamics is a complex subject, broadly interdisciplinary in nature and mathematical in formulation. Nevertheless, we can acquire knowledge of the basic aspects of the subject and learn to make judgments about pertinent questions in chemistry without treating thermodynamics in detail.

A chemical reaction is just a particular kind of physical process: one in which chemical bonds are made, broken, or rearranged. Thus, we can understand the energetics of chemical reactions in the same terms that we understand the energetics for any physical process. Before considering chemical reactions as such, let us think about more familiar processes to gain insight into those factors that determine the course of all physical events. Basically, there are two such factors: an *energetic* one related directly to physical forces and an *organizational* one related to information. Pushing an automobile down the street is a physical process ruled principally by the energetic factor, that is, the necessity of overcoming physical forces, such as the frictional force between the tires and the street, that resist this physical process. It is more difficult to push a Cadillac than a Volkswagen, since the opposing physical forces are greater for the former. On the other hand, assembling an automobile from the constituent parts is a process described not only by the energetic factor but by the organizational one as well. That is, in addition to the physical forces that must be overcome in putting the automobile together, it is necessary to put some effort into organizing the parts in the proper fashion. In assembling a wristwatch the organizational factor will be rather more important relative to the energetic one compared to the matter of assembling an automobile. The organizational problem may be just as formidable with the wristwatch, but little real physical effort will be required for this case. Let us note that in order to assemble either a wristwatch or an automobile, the organizational problem must be overcome by the input of information.

Intuitively, we know that all physical processes tend to occur spontaneously in the directions of *minimum energy* and *maximum disorder*. A ball placed on the side of a hill will roll down the hill to a position of lower energy, not up the hill to a position of higher energy. The molecules of a glass of water poured into a lake will distribute themselves throughout the lake, a state of maximum disorder, rather than remaining together, a more organized condition. Some physical processes, such as those just noted, are dominated by either the energetic or organizational factor, while for others, both factors are important.

To understand the distinction between the energetic factor and organizational factor more clearly, let us consider the relative difficulty of

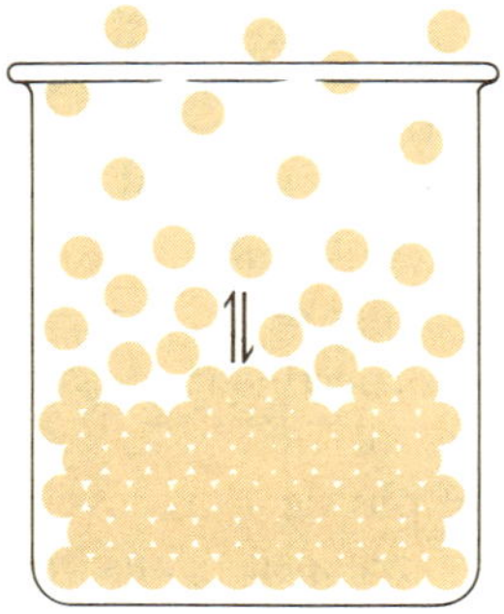

Gaseous state: Molecules are relatively far apart and experience little mutual attractive interaction. On the other hand, the molecules have very little organization.

Evaporation: Interconversion between the two states is controlled by the balance between energetic and organizational factors.

Liquid state: Molecules can experience favorable interactions with each other, tending to stabilize this state, but are organized into a lattice-type structure.

Figure 8-1 A diagrammatic representation of the liquid and gaseous phases and the interconversion between them.

picking up, card by card, a deck of cards spread out on a table: first, in a random order, and second, in a specific order. Clearly, the former task is simple and may be accomplished more quickly. Nevertheless, the amount of energy needed to pick up 52 cards from a table does not depend on the order in which they are picked up. Thus, the energetic factor is the same for the two processes, and the latter is the more difficult because of the organizational factor; that is, aside from the energy input, this process requires the input of information to permit the necessary organization to be introduced into the final product.

As an example of a process that involves a balance between the tendency toward minimum energy and maximum disorder, let us consider the evaporation of a liquid, shown diagrammatically in Figure 8-1. In the liquid state the rapidly moving individual molecules enjoy favorable energetic interactions with one another. The nature and strength of these interactions vary markedly, of course, with the nature of the molecules, but even atoms of helium do have mutually attractive interactions in the liquid state. In the gas phase, in which the distances between the molecules are, on the average, much greater than in the liquid phase, these favorable interactions are largely lost. Hence, the purely energetic factor will tend to maintain the system in a condensed state. In contrast, molecules in a liquid state are highly organized relative to the same collection of molecules in a gaseous state. Hence, the organizational factor tends to maintain the system in a dispersed state. The balance between these, which is clearly temperature-dependent, determines the actual state of the system under any specified set of conditions. We shall return to this example below to explore its consequences further.

With these simple ideas in mind, let us examine each of these factors independently in more detail.

ENERGY AND ENTHALPY

There are three things that we need to know about the nature of thermodynamics before proceeding very much further with our arguments. In the first place, thermodynamics deals only with **macroscopic properties.** These are the properties exhibited by large collections of atoms or molecules rather than the properties of the individual particles themselves **(microscopic properties).** Such macroscopic properties include pressure, temperature, and energy. Pressure and temperature are properties that owe their existence to collections of molecules; on the other hand, we may speak of the energy of a single molecule. Nevertheless, thermodynamics is concerned with the energy of collections of molecules only. Indeed, we need never make reference to the molecular nature of substances in any thermodynamic argument, although it sometimes proves helpful to do so. At the same time, as we shall see, thermodynamics does provide insight into physical events at the molecular level. In the second place, thermodynamics most commonly deals with **equilibrium states,** that is, states in which the temperature and pressure are uniform throughout and in which no *net* chemical change is occurring. Finally, thermodynamics divides the universe in two parts: the *system* is that part under scrutiny and the *surroundings* is everything else.

Example *Which of the following are macroscopic properties?*

a. *density (weight per unit volume)*
b. *hardness*
c. *physical state (gas, liquid, solid)*
d. *color*

Solution *Each of the above is properly considered a macroscopic property; none are aspects possessed by individual atoms or molecules. In the last analysis, of course, all properties of matter do reflect the nature of the constituent atoms or molecules and the interactions among them.*

What is energy? Or, what is more important for our consideration, how can we describe energy changes? There are two processes by means of which energy may be transferred between system and surroundings: *work* and *heat*. These are, of course, readily interchangeable; work done against frictional forces, as in braking an automobile, generates heat; heat, as generated in an internal combustion engine, may be employed for the production of work, such as that necessary to propel an automobile along the highway. One of the most fundamental statements of thermodynamics deals with the relationship between work, heat, and energy changes and is known as the **first law of thermodynamics:**

$$\Delta E = q + w \tag{8-1}$$

Stated in words, Equation (8-1) says: "The change in energy of a system, ΔE, is equal to the heat added *to* the system, q, plus the work, w, done

A powerful *Saturn V* rocket blasts the Apollo 11 moon mission off the launching pad. The amount of thrust developed by rocket engines has been determined beforehand from a knowledge of the energy change involved in the chemical reaction providing the driving force. (Courtesy of NASA.)

on the system." The first law of thermodynamics is, then, just a precise way of saying that, in any ordinary physical process, energy is conserved. The law is a fundamental statement of belief and is not derivable from any more basic considerations. It simply sums up all of our experiences about the way physical systems behave. The first law of thermodynamics may be applied to an immense variety of physical processes: the blowing up of a balloon, the operation of an engine, the melting of ice, and so forth, but let us concentrate on the application of the law to chemical reactions. For example, how can we evaluate the energy change for the chlorination of methane?

$$CH_4 + Cl_2 \longrightarrow CH_3Cl + HCl \qquad (8\text{-}2)$$

If we carry out this reaction in a closed vessel of specified volume, then the chemical reaction can do no work on the surroundings. This is true because chemical reactions generally do only **pressure–volume work,** that is, change volume against a pressure, $w = P\,\Delta V$. (We ignore here and in subsequent discussions the rather specialized cases where chemical reactions do mechanical work or the more common cases where electrical work is done.) The operation of the internal combustion engine provides a common example of a chemical reaction performing pressure–volume work. When the sparkplug fires, the fuel in the combustion chamber ignites and burns: hydrocarbons + $O_2 \longrightarrow CO_2 + H_2O$. The resulting hot gases expand against the piston, forcing it down and increasing the volume of the combustion chamber:

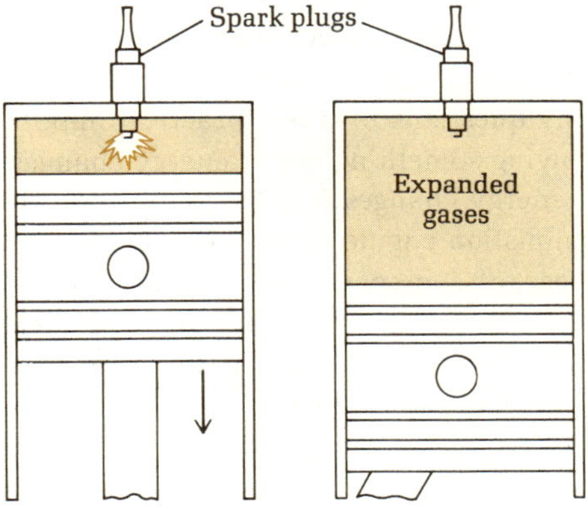

This increase in volume against the pressure exerted by the piston, reflecting the energy required to propel the car and to overcome frictional forces, constitutes the work done by the chemical reaction. Similarly, if we carry out a chemical reaction in a balloon and the balloon expands during the reaction process, then the reaction system will do work on the surroundings by virtue of increasing its volume against atmospheric pressure. Conversely, if the balloon contracts during a reaction, the surroundings do work on the system by expanding its volume against the pressure inside the balloon. At any event, if there is no volume change, $\Delta V = 0$, then w must be equal to zero as well for such chemical reactions, and the first law of thermodynamics reads:

$$\Delta E = q_v \qquad (8\text{-}3)$$

in which we have denoted the heat change with a subscript to emphasize that this statement is true only for those chemical reactions occurring at constant volume. Thus, measurement of the amount of heat released or absorbed during the chemical reaction at constant volume yields directly the change of energy for the chemical reaction. These energy changes are usually expressed in units of calories (cal), the amount of heat needed to raise 1 g of water 1°C. The chlorination of methane [equation (8-2)] at constant volume results in the evolution of heat. Thus, according to our definitions above, q for this reaction must be a *negative* quantity, and hence ΔE must be negative as well. Such reactions are said to be **exothermic.** Chemical reactions at constant volume that absorb heat from the surroundings, and therefore have *positive* values of ΔE, are said to be **endothermic.**

Once we have obtained a value of ΔE for a chemical reaction, what use can we make of it? What useful information does it supply? Provided that we measure the energy change for our reaction under conditions in which the reactants and products are present at the same temperature, so that the kinetic energies of these species are not different, the energy change measures the difference in internal energy, that is, *the difference in stabilities,* between reactants and products. Thus, we may derive information relevant to one of the central questions in chemistry: *What are those factors that contribute to the stabilities of molecules?* Quite aside from this fundamental consideration, there are many questions of direct practical importance that may be answered by knowing something about energy changes for chemical reactions. Thus, such energy changes for the combustion of ordinary gasoline in an internal combustion engine tell us how much heat is liberated in the process (in this case, of course, we must correct for the work done by the system, since the reaction is not at constant volume), and hence determine the specifications for the cooling system for that engine. Similarly, the chemist at the laboratory bench is interested in knowing about the nature of the energy change for his chemical reaction, since it provides a guideline as to whether he must provide heat to make the reaction go or provide for the dissipation of heat generated by the reaction to keep it under control.

Clearly, many chemical reactions are not conducted at constant volume: witness the operation of an internal combustion engine as an example. Furthermore, in the laboratory, it is common to run chemical reactions in systems that are open to the atmosphere, and that therefore occur at constant atmospheric *pressure,* rather than in closed systems at constant *volume.* How can we deal with this situation? For chemical reactions at constant pressure, it will prove helpful to speak about a new quantity, the **enthalpy,** H, rather than energy and to deal with changes in enthalpy rather than changes in energy. The enthalpy is a quantity closely related to energy and is defined as

$$\boldsymbol{H = E + PV} \tag{8-4}$$

Since the product of pressure and volume has the units of energy, enthalpy has these units as well. We may write the change in enthalpy as follows:

$$\Delta H = \Delta E + \Delta(PV) \tag{8-5}$$

But if, and only if, the pressure is constant, this equation simplifies to

$$\Delta H = \Delta E + P\,\Delta V \tag{8-6}$$

But $P\,\Delta V$ is just the work done by the system on the surroundings: $P\,\Delta V = -w$. Since $\Delta E = q + w$, it follows, if only PV work is considered, that

$$\Delta H = \Delta E - w = q_p + w - w = q_p \tag{8-7}$$

in which we have again employed a subscript to emphasize that the relationship is true only for a reaction at constant pressure. Thus, we have demonstrated that, for a reaction at constant pressure, the enthalpy is equal to the heat change. We have, in summary, the twin relationships

$$\begin{aligned} \Delta E &= q_v \\ \Delta H &= q_p \end{aligned} \tag{8-8}$$

Clearly, values of ΔH for chemical reactions may be obtained directly from the heat change for the reaction at constant pressure or from the heat change and pressure change for the reaction at constant volume, employing Equation (8-5) and realizing that $\Delta V = 0$.

For many chemical reactions, ΔE and ΔH are very nearly the same. Thus, if a liquid is converted into another liquid or a solid into another solid at constant pressure, very little volume change will occur, so that the $P\,\Delta V$ term will be small and we shall have the relationship $\Delta H \cong \Delta E$. On the other hand, if a liquid is converted into a gas, for example, then a very large volume change may occur, and the $P\,\Delta V$ term may not be negligibly small. However, provided that the volume change occurs against a reasonably small pressure and that the ΔH is not too near zero, it may still be true that $P\,\Delta V \ll \Delta H$ and therefore that $\Delta H \cong \Delta E$. Experience has indicated that it is more convenient for the chemist to deal with changes in enthalpy than with changes in energy. We shall, therefore, use enthalpy values rather than energy values in the succeeding discussion. We shall wish to keep in mind the relationship between them, however.

Example *For which of the following would you expect $\Delta E \cong \Delta H$?*
a. *dissolution of salt in water*
b. *melting of ice*
c. *burning of gasoline in an internal combustion engine*

Solution *When salt is dissolved in water or when ice melts, the total volume of the system changes very little. Therefore, it is likely that $P\,\Delta V \ll \Delta H$, so that ΔE and ΔH are approximately equal. On the other hand, combustion of gasoline in a cylinder of a piston engine involves a sizable change of the volume of the system against a large pressure. It follows that $P\,\Delta V$ is probably large enough so that ΔE and ΔH are significantly different.*

The units of heat and work

The first law of thermodynamics describes energy changes in terms of changes in heat content and work done. To facilitate computations of energy

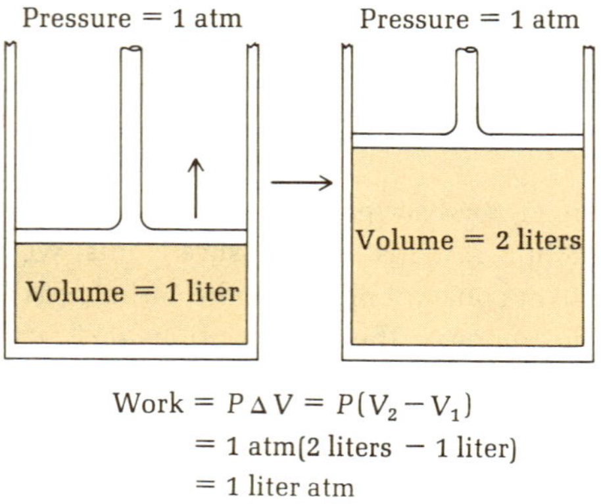

Figure 8-2 The calculation of pressure-volume work.

changes using the first law, it is important to understand something about the units employed to measure these quantities. As noted above, heat is usually measured in units of calories (cal), the amount of heat necessary to raise 1 g of water 1°C, or in kilocalories (kcal), equal to 1000 cal. Work is ordinarily expressed in units other than calories. For example, if we wish to calculate the pressure-volume work involved in the expansion of a gas from a volume of 1 liter to a volume of 2 liters against an external pressure of 1 atmosphere (atm) as shown in Figure 8-2, the work is simply given by

$$w = P\,\Delta V = 1 \text{ atm} \times (2 - 1) \text{ liters} = 1 \text{ liter atm}$$

Clearly, calories and liter atmospheres are both units of energy and must be interconvertible. Another usual unit of energy is the *erg*, the amount of energy necessary to raise a 1-g weight 1 cm. The following set of conversion factors will prove useful in making calculations.

1 liter atm = 24.2 cal
1 cal = 0.0413 liter atm
1 liter atm = 1.01×10^{9} ergs
1 erg = 0.99×10^{-9} liter atm
1 cal = 4.17×10^{7} ergs
1 erg = 2.40×10^{-8} cal

Calculations of enthalpy changes

To proceed with our examination of the energetic factor in chemical reactions, we must state explicitly one central fact that, until now, has been implicitly assumed, and examine the consequences of that fact: *The energy change or enthalpy change for a process, under a specified set of conditions, is a unique quantity independent of how that change is brought about.* Such quantities are called **state functions** since they depend only on the nature of the initial and final states and not on the nature of the pathway between them. Let us take a simple example. Suppose that we wish to evaluate the enthalpy change for a reaction in which ice at −20°C is converted into water at 30°C. There are any number of ways in which this transformation might be accomplished. For example, we could heat the ice so that all of the ice remains at the same temperature until it reaches 0°C, then melt it so that the water formed is all at 0°C, and then heat the water uniformly until the temperature of 30°C is reached. On the other hand, we might heat the ice to a temperature of 30°C (melting can be prevented by raising the pressure sufficiently), and then permit the ice to melt at that temperature. These alternative routes may be shown diagrammatically as follows:

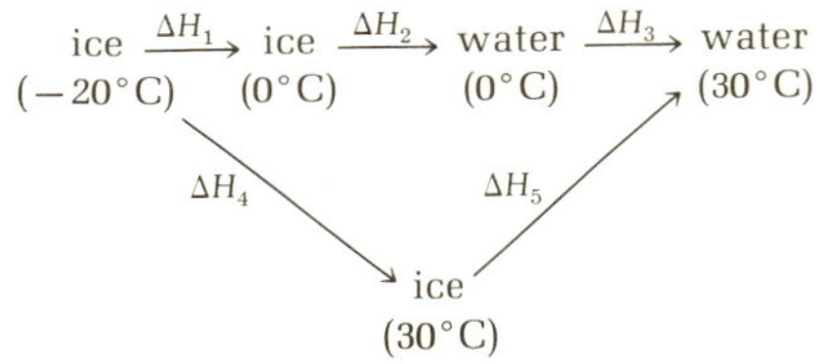

The important point is that the energy change and the enthalpy change for the overall process is the same regardless of the pathway. Stated another way: $\Delta H_1 + \Delta H_2 + \Delta H_3 = \Delta H_4 + \Delta H_5$. Thus, energy and enthalpy are state functions. Clearly, not all quantities are of this type. Thus, the quantity that is the distance traveled between New York and Los Angeles depends not only on the straight-line separation between these two cities but also on how one chooses to make the trip: The distance traveled on a direct jet flight is substantially different from that on a steamer that sails through the Panama Canal.

One very important point follows from the definition of a state function: *The changes in the value of such a function for any process in the forward and reverse directions are equal in magnitude and opposite in sign.* Thus, in our example just above, the enthalpy change for the transformation of ice at −20°C to water at 30°C is $\Delta H_{\text{forward}} = \Delta H_1 + \Delta H_2 + \Delta H_3$: It follows for the transformation of water at 30°C to ice at −20°C, $\Delta H_{\text{reverse}} = -\Delta H_{\text{forward}} = -\Delta H_1 - \Delta H_2 - \Delta H_3$. Once we have determined the change in any state function for any process, we immediately know the change for that function in the opposite direction.

Now, the fact that enthalpy changes, for example, are independent of pathway has a very important consequence for the evaluation of these changes. We are now at liberty to choose any convenient pathway for the reaction and to obtain the overall enthalpy change by summing those for the individual steps. Let us illustrate this point with an example. Suppose that we wish to know ΔH for the combustion of carbon (in the form of graphite) to carbon monoxide:

$$\text{C(graphite)} + \tfrac{1}{2}\text{O}_2\text{(g)} \longrightarrow \text{CO(g)} \qquad \Delta H = ? \tag{8-9}$$

The value of ΔH for this reaction is almost impossible to measure directly. It is difficult to carry out this reaction quantitatively, since combustion of graphite tends to yield carbon dioxide rather than carbon monoxide and, even under carefully selected and controlled conditions, such combustion always yields some carbon dioxide. On the other hand, it is rather easy to evaluate ΔH for reaction (8-9) indirectly by measuring the values of ΔH for two reactions that can be easily studied:

$$\text{C(graphite)} + \text{O}_2\text{(g)} \longrightarrow \text{CO}_2\text{(g)} \qquad \Delta H' \tag{8-10}$$

$$\text{CO(g)} + \tfrac{1}{2}\text{O}_2\text{(g)} \longrightarrow \text{CO}_2\text{(g)} \qquad \Delta H'' \tag{8-11}$$

Now, if we simply subtract (8-11) from (8-10), the stoichiometry is just that of (8-9), and hence $\Delta H = \Delta H' - \Delta H''$. Diagrammatically, we can picture the situation as follows:

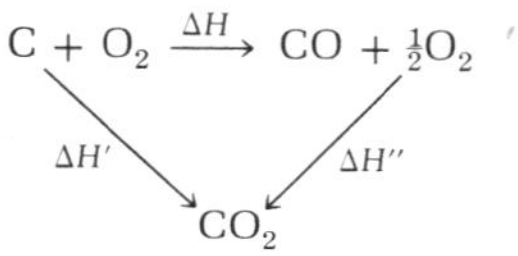

$$\Delta H + \Delta H'' = \Delta H'$$
or
$$\Delta H = \Delta H' - \Delta H''$$

Example *Let us suppose that we have measured enthalpy changes for the following reactions:*

(1) $N_2 + \frac{1}{2}O_2 \longrightarrow N_2O \qquad \Delta H_1$
(2) $\frac{1}{2}N_2 + \frac{1}{2}O_2 \longrightarrow NO \qquad \Delta H_2$
(3) $\frac{1}{2}N_2 + O_2 \longrightarrow NO_2 \qquad \Delta H_3$

Using these values, calculate the enthalpy changes for

(a) $N_2O + \frac{1}{2}O_2 \longrightarrow 2NO$
(b) $NO + \frac{1}{2}O_2 \longrightarrow NO_2$
(c) $N_2O + \frac{3}{2}O_2 \longrightarrow 2NO_2$

Solution *To solve these problems, we must find ways of combining the three equations of known ΔH to obtain the new ones.*

(a) *We can arrive at Equation (a) by multiplying Equation (2) by 2 and subtracting Equation (1):*

$$\begin{array}{rlll} & N_2 + O_2 & \longrightarrow 2NO & 2\,\Delta H_2 \\ (-) & N_2 + \frac{1}{2}O_2 & \longrightarrow N_2O & \Delta H_1 \\ \hline & \frac{1}{2}O + N_2O & \longrightarrow 2NO & 2\,\Delta H_2 - \Delta H_1 = \Delta H_a \end{array}$$

(b) *Equation (b) is obtained by subtracting Equation (2) from Equation (3):*

$$\begin{array}{rlll} & \frac{1}{2}N_2 + O_2 & \longrightarrow NO_2 & \Delta H_3 \\ (-) & \frac{1}{2}N_2 + \frac{1}{2}O_2 & \longrightarrow NO & \Delta H_2 \\ \hline & \frac{1}{2}O + NO & \longrightarrow NO_2 & \Delta H_3 - \Delta H_2 = \Delta H_b \end{array}$$

(c) *Finally, Equation (c) is just twice Equation (3) minus Equation (1):*

$$\begin{array}{lll} N_2 + 2O_2 \longrightarrow 2NO_2 & & 2\,\Delta H_3 \\ (-)\ N_2 + \tfrac{1}{2}O_2 \longrightarrow N_2O & & \Delta H_1 \\ \hline \tfrac{3}{2}O_2 + N_2O \longrightarrow 2NO_2 & & 2\,\Delta H_3 - \Delta H_1 = \Delta H_c \end{array}$$

Additional examples of calculations of this kind are collected in the study problems at the end of the chapter. Clearly, the use of such calculations permits us to obtain values of ΔH that we cannot conveniently measure and, moreover, may permit us to avoid making a great many measurements unnecessarily. However, the actual values of enthalpy changes depend, of course, on the precise initial and final conditions. To realize the advantages noted above and to deal with all of the information in the most efficient manner, it has proved convenient to tabulate those enthalpy changes for the formation of pure compounds from their elements under conditions in which the entire system experiences a uniform temperature of 25°C, a pressure of 1 atm, and in which all substances are present in their stable forms under these conditions. Such enthalpy changes are termed **standard enthalpies of formation** and are denoted ΔH_f^0. Some representative enthalpies of formation are collected in Table 8-1.

The values of Table 8-1 may be employed to calculate the ΔH values for those reactions that involve the substances listed. In general, these calculations can be made by employing the following equation, which is simply a quantitative statement of the considerations developed above:

$$\boldsymbol{\Delta H^0 = \Sigma\, \Delta H_f^0(\text{products}) - \Sigma\, \Delta H_f^0(\text{reactants})} \tag{8-12}$$

Thus, the enthalpy change for the oxidation of water to hydrogen peroxide,

$$H_2O + \tfrac{1}{2}O_2 \longrightarrow H_2O_2 \tag{8-13}$$

is given by

$$\begin{aligned} \Delta H^0 &= \Delta H_f^0(H_2O_2) - \Delta H_f^0(H_2O) \\ &= -32.53 - (-68.32) = 35.79 \text{ kcal/mole} \end{aligned}$$

Note that, according to our definition given above, the standard enthalpy of formation for all elements must be zero. Our calculation tells us that formation of hydrogen peroxide according to Equation (8-13) is an endothermic process; that is, it occurs with absorption of heat by the system. The reverse reaction, decomposition of hydrogen peroxide, is exothermic. This observation is in accord with and a partial explanation for what we know from experience: Hydrogen peroxide is not a very stable substance and tends to decompose to oxygen and water.

A related example is provided by the isomerization of *cis*-2-butene to *trans*-2-butene:

Table 8-1 Enthalpies of formation, ΔH_f^0, at 298.16°K, for a series of common substances (kcal/mole)

Inorganic compounds

$H_2O(g)$	−57.79	$CaSO_4(s)$	−315.6
$H_2O(l)$	−68.32	$BaO(s)$	−133.5
$H_2O_2(g)$	−32.53	$BaCO_3(s)$	−291.3
$O_3(g)$	34.0	$BaSO_4(s)$	−345.3
$HCl(g)$	−22.06	$Ba(NO_3)_2(s)$	−237.1
$SO_2(g)$	−70.96	$BaCl_2(s)$	−205.6
$SO_3(g)$	−94.45	$Fe_2O_3(s)$	−196.5
$H_2S(g)$	−4.81	$FeCl_2(s)$	−81.5
$H_2SO_4(l)$	−194.5	$FeCl_3(s)$	−96.8
$SF_6(g)$	−289.0	$FeS(s)$	−22.7
$N_2O(g)$	19.49	$FeSO_4(s)$	−220.5
$NO(g)$	21.60	$FeCO_3(s)$	−178.7
$NO_2(g)$	8.09	$Al_2O_3(s)$	−399.1
$NH_3(g)$	−11.04	$AlCl_3(s)$	−168.3
$HNO_3(l)$	−41.6	$Al_2(SO_4)_3(s)$	−822.4
$(NH_4)_2SO_4(s)$	−282.2	$CuO(s)$	−37.6
P(s) (white)	0	$CuCl_2(s)$	−49.2
P(s) (red)	−4.2	$CuS(s)$	−11.6
$P_4(g)$	14.1	$CuSO_4(s)$	−184.0
$PH_3(g)$	1.3	$CuCO_3(s)$	−142.2
$H_3PO_4(s)$	−305.7	ZnO	−83.2
C(s) (graphite)	0.0	$ZnCl_2(s)$	−99.2
C(s) (diamond)	0.45	$ZnS(s)$	−49.2
$CO(g)$	−26.41	$ZnSO_4(s)$	−234.9
$CO_2(g)$	−94.05	$ZnCO_3(s)$	−194.3
$CaO(s)$	−151.8	$NiO(s)$	−58.4
$Ca(OH)_2(s)$	−235.6	$NiCl_2(s)$	−73.0
$CaCO_3(s)$	−288.4	$NiSO_4(s)$	−213.0
$CaCl_2(s)$	−190.0	$Ni(CO)_4(g)$	−144.7

Organic compounds

Gases		Liquids	
Methane, CH_4	−17.89	Methanol, CH_3OH	−57.02
Ethane, C_2H_6	−20.24	Ethanol, C_2H_5OH	−66.35
Propane, C_3H_8	−24.82	Acetic acid, CH_3COOH	−116.4
n-Butane, C_4H_{10}	−29.81	Benzene	11.72
Isobutane, C_4H_{10}	−31.45	$CCl_4(l)$	−15.6
Acetylene, C_2H_2	54.19	$CCl_4(g)$	14.5
Ethylene, C_2H_4	12.50	$CH_3Cl(g)$	−13.7
Propylene, C_3H_6	4.88	$CH_2Cl_2(g)$	−15.8
1-Butene, C_4H_8	0.28	$CHCl_3(g)$	−16.8
cis-2-Butene, C_4H_8	−1.36		
trans-2-Butene, C_4H_8	−2.40		
Isobutene, C_4H_8	−3.34		

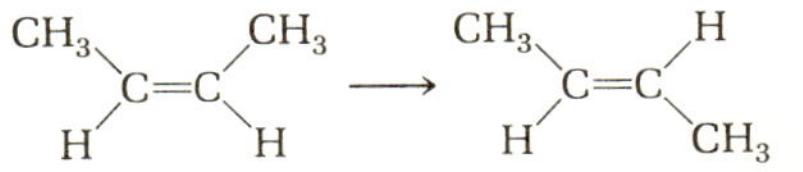

(8-14)

Reference to Table 8-1 indicates that ΔH^0 for this reaction is given by

$$\Delta H^0 = \Delta H_f^0(\textit{trans}\text{-2-butene}) - \Delta H_f^0(\textit{cis}\text{-2-butene})$$
$$= -2.40 - (-1.36) = -1.04 \text{ kcal/mole} \qquad (8\text{-}15)$$

Thus, this reaction is exothermic and proceeds with the evolution of heat. This result also confirms our conclusion drawn in Chapter 2 that the *trans* isomer of butene is more stable than the *cis* isomer.

Example *One common method for the synthesis of ethanol is the hydration of ethylene:*

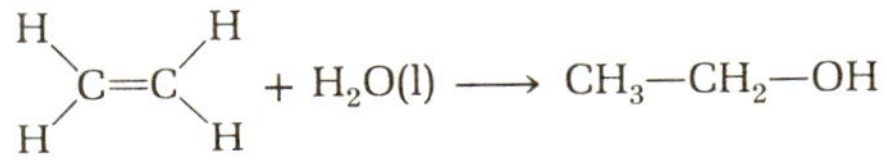

Calculate the enthalpy change for this reaction.

Solution *We need to insert values for the standard enthalpies of formation (Table 8-1) into Equation (8-12):*

$$\Delta H^0 = \Sigma\, \Delta H_f^0(\text{products}) - \Sigma\, \Delta H_f^0(\text{reactants})$$
$$= -66.35 - (12.50 - 68.32) = -66.35 - (-55.82)$$
$$\boldsymbol{\Delta H^0 = -10.53 \text{ kcal/mole}}$$

Thus, this reaction is mildly exothermic.

BOND ENERGIES

Calculations of the type just presented are very useful and permit the evaluation of values for ΔH for a large number of chemical reactions. They are, nevertheless, limited in scope. To perform these calculations, it is necessary to know the enthalpies of formation for each of the reactants and products involved. Since there are an enormous number of compounds known, more than a million of which are derivatives of carbon alone, it is clear that calculation of enthalpy changes for chemical reactions in general starting with enthalpies of formation would be exceedingly cumbersome. Indeed, insufficient values of ΔH_f^0 are known to permit such calculations for more than a small fraction of the possible chemical reactions. Calculating enthalpy changes can be greatly simplified by focusing our attention not on the entire molecules involved but just on those chemical bonds that actually undergo change in the reaction. The discussion just concluded suggests how this can be done. Let us consider a very simple chemical reaction involving the formation of gaseous atoms from the elements. For example, we may consider the formation of hydrogen atoms from

Table 8-2 Some bond energies (kcal/mole)

Diatomic molecules		Polyatomic molecules		Polyatomic molecules (*Continued*)	
H—H	104.2	C—H	98.7	C—O	85.5
O═O	119.1	N—H	93.4	C═O[b]	192.0
N≡N	225.8	O—H	110.6	C═O[c]	166
C═O[a]	255.8	S—H	83	C═O[d]	176
F—F	36.6	P—H	76	C═O[e]	179
Cl—Cl	58.0	N—N	39	C—F	116
Br—Br	46.1	N═N	100	C—Cl	81
I—I	36.1	O—O	35	C—Br	68
H—F	134.6	S—S	54	C—I	51
H—Cl	103.2	N—O	53	C—S	65
H—Br	87.5	N═O	145	C═S[f]	128
H—I	71.4	C—C	82.6	N—F	65
		C═C	145.8	N—Cl	46
		C≡C	199.6	O—F	45
		C—N	72.8	O—Cl	52
		C═N	147	O—Br	48
		C≡N	212.6		

[a] Carbon monoxide.
[b] For carbon dioxide.
[c] For formaldehyde.
[d] Other aldehydes.
[e] Ketones.
[f] For carbon disulfide.

Source: J. D. Roberts and M. Caserio, *Basic Principles of Organic Chemistry*, Benjamin, Menlo Park, Calif., 1964. Copyright, © 1964, W. A. Benjamin, Inc.

elemental hydrogen and of oxygen atoms from elemental oxygen:

$$\tfrac{1}{2}H_2 \longrightarrow H\cdot \qquad \Delta H_f^0 = 52.1 \text{ kcal/mole} \tag{8-16}$$

$$\tfrac{1}{2}O_2 \longrightarrow O\cdot \qquad \Delta H_f^0 = 59.6 \text{ kcal/mole} \tag{8-17}$$

These values of ΔH_f^0 measure the enthalpy change for the rupture of H—H bonds and O—O bonds, respectively. Can we generalize this idea? That is, can we define similar values for the enthalpy change for rupture of various types of chemical bonds in addition to those for decomposition of the elements? Within certain limitations noted below, the answer to this question is "yes." These values are referred to as **bond energies.**

In order to see, in a simple case, how these can prove useful, let us calculate the enthalpy of formation of HI:

$$H_2 + I_2 \longrightarrow 2HI \tag{8-18}$$

We can visualize this overall reaction as the sum of several separate hypothetical steps:

$$I_2 \longrightarrow 2I\cdot \qquad \Delta H_{I_2} \tag{8-18a}$$

$$H_2 \longrightarrow 2H\cdot \qquad \Delta H_{H_2} \tag{8-18b}$$

$$2(I\cdot + H\cdot \longrightarrow HI) \qquad \Delta H_{HI} \tag{8-18c}$$

$$\text{sum} \quad H_2 + I_2 \longrightarrow 2HI \qquad \Delta H = \Delta H_{I_2} + \Delta H_{H_2} - 2(\Delta H_{HI}) \tag{8-18d}$$

In reactions (8-18a) and (8-18b), we have simply broken each reactant molecule into suitable parts so that their recombination in (8-18c) leads to the observed products. Dividing the reaction up in this manner permits us to calculate the overall enthalpy change from the bond energies, which are collected in Table 8-2.

$$\begin{aligned} \Delta H &= \Delta H_{H_2} + \Delta H_{I_2} - 2(\Delta H_{HI}) \\ &= 104.2 + 36.1 - 2(71.4) = 2.5 \text{ kcal/mole} \end{aligned} \tag{8-19}$$

Note that calculations of enthalpy changes from standard enthalpies of formation, Equation (8-12), and from bond energies, Equation (8-19) for example, take different forms. In the former case the overall enthalpy change is obtained by subtracting the enthalpies of formation for reactants from those for products. In the latter case the enthalpy change is obtained by subtracting the bond energies of the products from those of the reactants. Each calculation is correct and reasonable; the apparent inconsistency in the forms of the calculations derives from the fact that enthalpies of formation and bond energies are quite different things. Enthalpies of formation may be either positive or negative depending on whether the compound in question has a higher or lower enthalpy content than the elements from which it is derived. Bond energies, in contrast, are always positive quantities, since energy must be expended to dissociate chemical bonds. The important point is to understand the basis for the method of calculation in each case and to avoid confusion of the two.

The calculation in Equation (8-19) is reasonably rigorous, and we may be confident that it provides us with a good estimate of the enthalpy change for the reaction. However, not all calculations carried out by this method are so reliable. To see this point, let us examine the case of the bond energy for dissociation of the C—H bond. We might evaluate this number by measuring the enthalpy change for the reaction

$$CH_4 \longrightarrow CH_3\cdot + H\cdot \qquad \Delta H = 102 \text{ kcal/mole} \tag{8-20}$$

Clearly, only a single C—H bond has been broken in this reaction and, as indicated, a value of ΔH of 102 kcal/mole is obtained. But what happens if we do a similar measurement and consider the energetics of removing a second hydrogen atom from methane:

$$CH_3\cdot \longrightarrow CH_2\colon + H\cdot \qquad \Delta H = 87 \text{ kcal/mole} \tag{8-21}$$

Clearly, this value is not the same as that previously obtained. Removal of the third and fourth atoms of hydrogen from methane has values of ΔH of 125 and 81 kcal/mole, respectively. Thus, not all C—H bonds have exactly the same bond energy. The values provided in Table 8-2 are typical values and are generally reliable to ± 3 or 4 kcal/mole. Thus, deviations as extreme as those just noted for the successive removal of hydrogen atoms from methane are the exception, not the rule. Nevertheless, they serve to introduce a necessary note of caution into the willy-nilly use of calculations of this type. This consideration is not important for the formation of HI, since there is only one kind of H—H bond (that in H_2), one kind of I—I bond, and one kind of H—I bond. Hence, we can have a substantial degree of confidence that this calculation at least is a reasonable one. Let us look at some other examples.

What are the energetics of the reaction between H_2 and Cl_2 in comparison with that for H_2 and I_2 just considered?

$$H_2 + Cl_2 \longrightarrow 2HCl \tag{8-22}$$

By reference to Table 8-2, we can calculate that this reaction is expected to be exothermic by about 44.2 kcal/mole. Indeed, if H_2 and Cl_2 are mixed, the reaction is quite violent once it starts. Note the necessity for considering both reactant and product states before making judgments about reactions. If we considered just reactant states, we would have concluded that the reaction involving I_2 would have been the more exothermic, since the energy needed to break the Cl—Cl bond is greater than that for the I—I bond. Looking at the product state as well, we see that the energy liberated in formation of the H—Cl bond is more than enough greater than that for the H—I bond to make up for this difference.

To further illustrate calculations of this type, let us consider the combustion of butane, the principal component of natural gas, which is frequently employed for the heating of homes. This suggests that the reaction

$$CH_3{-}CH_2{-}CH_2{-}CH_3 + 6.5O_2 \longrightarrow 4CO_2 + 5H_2O \tag{8-23}$$

must be very exothermic. To calculate the enthalpy change per mole for this reaction, we can proceed as follows. First break the bonds of the reactants:

$$C_4H_{10} \longrightarrow 4\cdot\dot{\underset{\cdot}{C}}\cdot + 10H\cdot \qquad \Delta H = (10 \times 98.7) + (3 \times 82.6) \text{ kcal} = 1235 \text{ kcal}$$
$$6.5O_2 \longrightarrow 13\cdot O\cdot \qquad \Delta H = 6.5 \times 119.1 \text{ kcal} = 774.2 \text{ kcal}$$

Then, form the bonds in the products:

$$4\cdot\dot{\underset{\cdot}{C}}\cdot + 8\cdot O\cdot \longrightarrow 4CO_2 \qquad \Delta H = -8 \times 192 \text{ kcal} = -1536 \text{ kcal}$$
$$10H\cdot + 5\cdot O\cdot \longrightarrow 5H_2O \qquad \Delta H = -10 \times 110.6 \text{ kcal} = -1106 \text{ kcal}$$

The sum of these changes is -633 kcal/mole. The reaction is very exothermic indeed, in accord with our expectations.

We can use bond energies to calculate standard enthalpies of formation if we wish. To provide an additional example of enthalpy calculations employing bond energies and to understand more about the relationship of bond energies and enthalpies of formation, let us calculate the enthalpy of formation of ammonia from the elements:

$$\tfrac{3}{2}H_2 + \tfrac{1}{2}N_2 \longrightarrow NH_3 \tag{8-24}$$

Referring to Table 8-2, we proceed as follows:

$$\tfrac{3}{2}H_2 \longrightarrow 3H\cdot \qquad \Delta H = 1.5 \times 104.2 = 156.3 \text{ kcal}$$
$$\tfrac{1}{2}N_2 \longrightarrow N\cdot \qquad \Delta H = 0.5 \times 225.8 = 112.9 \text{ kcal}$$
$$3H\cdot + N\cdot \longrightarrow NH_3 \qquad \Delta H = 3 \times 93.4 = -280.2 \text{ kcal}$$

Subtracting the enthalpy change for the formation of the bonds in the product ammonia molecule from those involved in decomposition of the reactants, we obtain an overall enthalpy of formation of ammonia of −11 kcal/mole. This value is in excellent agreement with the standard free enthalpy of formation of ammonia of −11.04 kcal/mole (Table 8-1). This agreement is a result of the fact that the average bond energies for the N—H bonds of ammonia are very close to the N—H bond energy given in Table 8-2. Had ammonia not been a typical example of a molecule in which N—H bonds occur, the agreement between the two numbers would have been less satisfactory.

Example *Higher animals derive the energy required to sustain the life process by the complete oxidation of sugars to carbon dioxide and water. The equation for such oxidation of glucose is*

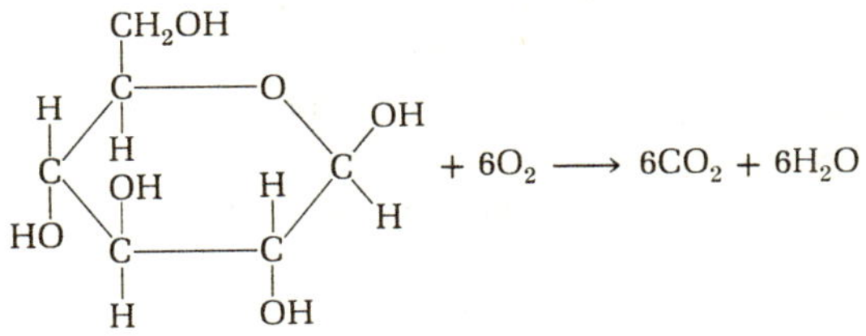

Using bond energies, calculate the enthalpy change for this reaction.

Solution *The total input of energy to rupture chemical bonds in the reactants is given by*

$$5 \times \Delta H_{C-C} = 5 \times 82.6 = 413.0$$
$$7 \times \Delta H_{C-O} = 7 \times 85.5 = 588.5$$
$$7 \times \Delta H_{C-H} = 7 \times 98.7 = 690.9$$
$$5 \times \Delta H_{O-H} = 5 \times 110.6 = 553.0$$
$$6 \times \Delta H_{O=O} = 6 \times 119.1 = 714.6$$
$$2960.0 \text{ kcal/mole}$$

The energy gained on formation of the bonds of the product molecules is given by

$$12 \times (-\Delta H_{C=O}) = 12 \times (-192.0) = -2304.0$$
$$12 \times (-\Delta H_{O-H}) = 12 \times (-110.6) = -1327.2$$
$$-3631.2 \text{ kcal/mole}$$

The sum of the enthalpy changes is

$$\begin{array}{rr} & 2920.0 \\ & -3631.2 \\ \hline \Delta H = & \mathbf{-671.2\ kcal/mole} \end{array}$$

This value is within a few percent of that measured experimentally for this reaction. Note that the five O—H bonds in glucose are conserved in the reaction process; we could, therefore, have made the calculation somewhat simpler by not calculating the enthalpy change for their destruction and reformation.

Example *Calculate the enthalpy change for conversion of cis-butene to trans-butene using bond energies.*

Solution *Since the type and number of bonds in the product and reactant are the same, ΔH is calculated to be zero. This illustrates one type of limitation of these calculations—no account is taken of steric hindrance.*

ENTROPY

At the outset of our discussion of the energetics of chemical reactions, we invoked two factors as being important: energy and organization. We have now explored the basis of the energy or enthalpy changes in chemical reactions and are able to proceed to a consideration of organizational changes. We have already noted that any physical process requiring organization also requires the input of energy (using work in its general sense rather than as it is defined in the first law of thermodynamics). The converse is certainly true as well: Real systems spontaneously tend to become disorganized. This leads us directly to the **second law of thermodynamics.** This law tells us that there exists some quantity called **entropy,** S, which, like energy and enthalpy, is a state function, and that the entropy of the universe, system plus surroundings, increases in any real physical process. It is not easy to formulate the second law of thermodynamics in a way that is clear to everyone. Indeed, there are about as many statements of this law as there are chemists. Perhaps the following list of statements of the second law will assist you in understanding the main points. But note clearly: All of these statements are correct and equivalent. In some of them you will note specification of processes as "reversible" or "irreversible." A "reversible" process is a useful idealization thought up by chemists and describes one that is carried out infinitely slowly and carefully so that it is in a state of balance at all times. Thus, the slightest diminution in the driving force for a reversible reaction would cause the reaction to proceed in the other direction. All real processes are irreversible, although the state of reversibility can be approached experimentally.

Statements of the second law of thermodynamics

1. "Every system which is left to itself will, on the average, change toward a condition of maximum probability" (G. N. Lewis).
2. "The state of maximum entropy is the most stable state for an isolated system" (Enrico Fermi).
3. "In an irreversible process, the total entropy of all bodies concerned is increased" (G. N. Lewis).
4. "When any actual process occurs it is impossible to invent a means of restoring *every* system concerned to its original condition" (G. N. Lewis).
5. "Every physical or chemical process in nature takes place in such a way as to increase the sum of the entropies of all the bodies taking part in the process. In the limit, i.e., for reversible processes, the sum of the entropies remains constant" (Max Planck).
6. "A transformation whose only final result is to transform into work heat extracted from a source which is at the same temperature throughout is impossible" (Max Planck).
7. "The energy of the universe is constant, the entropy increases toward a maximum" (R. J. E. Clausius on the first law and second law).
8. "Gain in information is loss in entropy" (G. N. Lewis).
9. "Entropy is time's arrow" (A. Eddington).

More sophisticated considerations than concern us here indicate that we can calculate entropy changes as the ratio of a change in heat content and temperature. Entropy has the units of calories per mole degree. Consequently, to understand the importance of entropy changes in the same terms that we use for energy or enthalpy changes, we must multiply the entropy change by the absolute temperature. Thus, the important quantity is not usually ΔS but $T\,\Delta S$. As one example, an entropy change of 100 cal/mole deg at 300°K corresponds to an energy change of 30 kcal/mole, while at 500°K this entropy change corresponds to 50 kcal/mole. The temperature dependence of the entropy contribution to the overall energetics of chemical reactions has most important consequences which we shall consider later.

We can gain a little better quantitative understanding of entropy by considering some simple examples of entropy content. To begin with, let us consider four pennies and the various ways of arranging them with respect to heads and tails. Overall, there are five possibilities: four heads, three heads and one tail, two heads and two tails, one head and three tails, and four tails. These possibilities are called *macrostates*. Let us label our pennies 1, 2, 3, and 4 so that we can distinguish them. For the macrostate in which all of the pennies are heads, there is only one *microstate:* penny 1 must be a head, penny 2 must be a head, and so forth. Similarly, there exists only one microstate for the case of four tails. But how about the

You may have to look more than once to realize it, but thermodynamics has something to say about this work, *The Waterfall,* by the Dutch artist M. C. Escher. The second law of thermodynamics says that what is shown here just can't be. (Courtesy of M. C. Escher, Escher Foundation—Haags Gemeentemuseum—The Hague, *Waterval.*)

macrostate in which we have three heads and one tail? We have four microstates that all yield this macrostate. These are

(a)	penny 1—head penny 2—head penny 3—head penny 4—tail	(c)	penny 1—head penny 2—tail penny 3—head penny 4—head
(b)	penny 1—head penny 2—head penny 3—tail penny 4—head	(d)	penny 1—tail penny 2—head penny 3—head penny 4—head

Similarly, the case of one head and three tails will have the corresponding sets of four microstates. Finally, the macrostate consisting of two heads and two tails will have a set of six microstates, as the student may verify for himself. Now, we can say that the macrostate that has the most associated microstates has the greatest entropy. Thus, the state of two heads and two tails has the greatest entropy content, while those composed of four heads or four tails have the least. Thus, entropy content corresponds to something that we know from our everyday experience: If you flip four pennies, it is more likely that you will end up with two heads and two tails than with, for example, four heads. Thus, this simple physical process, like all physical processes, tends to occur in such a way as to maximize entropy.

Clearly, there are many such examples. There is only one way (microstate) to arrange a deck of cards so that we have 1, 2, 3, . . . of spades followed by 1, 2, 3, . . . of hearts, and so on, a macrostate. In contrast, if we relax our restrictions and require only that all 13 spades precede all 13 hearts and so on, there are a great many ways of arranging the cards (microstates) that correspond to this macrostate. The latter has the greater entropy. The point is that the degree of organization in any system composed of a collection of things, whether molecules or baseballs, may be appreciated by asking how many ways you can arrange this collection so as to obtain that system—the more ways, the greater the entropy.

Example *The laws of poker are determined by the considerations of probability and entropy. To see this point clearly, let us consider the probability of obtaining some poker hands in a five-card stud game. Calculate the number of ways (microstates) that correspond to a royal flush (ace, king, queen, jack, and ten of the same suit), a straight flush (any sequence of five cards of the same suit), and four of a kind.*

Solution *There are only four possible sets of cards that correspond to a royal flush: one for each of the four suits. There are 40 possibilities corresponding to a straight flush (36 if we subtract those that are royal flushes), 10 possibilities for each suit (ace, 2, 3, 4, 5; 2, 3, 4, 5, 6; 3, 4, 5, 6, 7; etc.)*

Calculating the number of hands that contain four of a kind is more difficult. We approach this question by asking how many possibilities there are as each card is dealt. The first card may be any of the 52, and the nature of this card will be taken to determine the four. Therefore, the second card may be one of only three of the same value as the first, the third card must be one of only two, and the fourth card must be the last remaining of the determined face value. But the last card may be any of the remaining 48. Thus, the number of ways of obtaining four of a kind is $52 \times 3 \times 2 \times 1 \times 48$. *But we must note that the order of obtaining the four cards that determine the value of the hand is irrelevant. There are* $4 \times 3 \times 2 \times 1 = 24$ *ways of obtaining these four cards. It follows that the number of different ways of drawing four of a kind is*

$$N = \frac{52 \times 3 \times 2 \times 1 \times 48}{24} = 624$$

We can formulate our statements to this point in a precise mathematical way. For any system, the entropy is given by

$$S = k \ln W \tag{8-25}$$

In this expression, k is Boltzmann's constant, 1.38×10^{-16} erg deg^{-1} molecule^{-1}, and W is the number of microstates that correspond to the macrostate of interest. Hence, we can see that the entropy is logarithmically related to the number of microstates.

Example *The number of combinations of cards that correspond to each of the possible poker hands is collected in the list below. Calculate the entropy that corresponds to each hand.*

Hand	W
1. royal flush	4
2. straight flush	36
3. four of a kind (XXXX0)	624
4. full house (XXYYY)	3744
5. flush (all the same suit)	5108
6. straight (five-card sequence)	10200
7. three of a kind (XXX00)	54912
8. two pair (XXYY0)	123552
9. one pair (XX000)	1098240
10. bust hand	1302540

Solution *We can calculate the entropy corresponding to each hand directly from Equation (8-25) and the data provided above. The results are royal flush, 2.76; straight flush, 7.13; four of a kind, 12.81; full house, 16.40; flush, 17.02; straight, 18.40; three of a kind, 21.72; two pair, 23.30; one pair, 27.60; bust hand, 28.10.*

With this background in mind, let us see how these ideas apply to some chemical reactions. Remember: The greater the entropy increase in going from reactants to products, the more this factor will aid the reaction; and conversely, the greater the entropy decrease, the more this factor will oppose the reaction.

Those chemical reactions in which one molecule is cleaved into two, elimination reactions,

$$A \longrightarrow B + C \tag{8-26}$$

generally have a positive entropy change since the two product molecules are independent while the same collection of atoms was constrained to behave as a unit in the reactant. Similarly, chemical reactions in which two molecules unite to form one, addition reactions,

$$A + B \longrightarrow C \tag{8-27}$$

tend to have a negative entropy change. A slightly more complicated example is provided by the cyclization of 1-hexene to form cyclohexane:

$$CH_3{-}CH_2{-}CH_2{-}CH_2{-}CH{=}CH_2 \longrightarrow \tag{8-28}$$

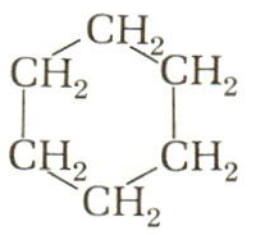

The cyclohexane molecule has less freedom of motion of the various atoms available to it than the straight-chain compound, and consequently, this reaction is hindered by the decrease in entropy in the system. In this case the effect amounts to about 6 kcal/mole at room temperature.

Some of the most striking examples of entropy changes are found in the reactions of highly organized polymeric molecules in nature. Let us return to the carboxypeptidase molecule for a moment. As may be appreciated from the diagram of this molecule on p. 233, this is a very highly organized molecule indeed. Hence, if we were to permit the molecule to unfold and exist as a random coil with literally thousands of conformations available to it instead of just one, we would expect that the entropy would increase quite substantially. Indeed, the entropy increase can amount to 100 kcal/mole or more at room temperature! Since this protein molecule is really rather stable at room temperature, it follows that the enthalpic contribution to its stability must be even greater than 100 kcal/mole, thus providing a fine illustration of the strength of weak forces when present in great numbers.

THE THIRD LAW OF THERMODYNAMICS

Our considerations up to this point permit us to answer another question: What is the entropy content of a perfectly ordered system? That is, what is the entropy content of a macrostate to which only one microstate

Table 8-3 A tabulation of absolute entropies, S^0, at 298°K (cal/mole deg)

Monoatomic gases		Diatomic gases		Polyatomic gases	
He	30.13	H_2	31.21	H_2O	45.1
Ne	34.95	D_2	34.6	CO_2	51.1
Ar	36.98	F_2	48.6	NO_2	57.5
Kr	39.19	Cl_2	53.3	NH_3	46.0
Xe	40.53	Br_2	58.6	CH_4	44.5
		N_2	45.7	C_2H_6	54.8
		O_2	49.0		

Liquids		Solid elements		Solid compounds	
Br_2	36.4	Ag	10.20	BaO	16.8
H_2O	16.73	B	1.7	CaO	9.5
Hg	18.17	C(graphite)	1.37	CuO	10.4
CH_3OH	30.3	C(diamond)	0.6	ZnO	10.5
C_2H_5OH	38.4	Ca	9.95		
		S(rhombic)	7.62		

Source: B. H. Mahan, *University Chemistry*, 2nd ed., Addison-Wesley, Reading, Mass., 1968.

corresponds? In chemistry, in which we ordinarily deal only with very large collections of molecules, the achievement of perfect order is necessarily an idealization, but one that can be approached experimentally. At exceedingly low temperatures, crystalline materials exhibit very little disorder indeed: each molecule occupies its own position in the crystal lattice and is essentially immobile in that lattice. Experimentally, it has proved possible to reach the exceedingly low temperature of 0.000001°K. These considerations generate the **third law of thermodynamics:** *The entropy of a perfect crystal of any element or compound is zero at a temperature of absolute zero.* Note that this result follows directly from Equation (8-25), since W is unity. This makes it possible to calculate absolute entropies for elements and compounds. A number of these are collected in Table 8-3. These absolute entropies may be used to calculate entropy changes for chemical reactions:

$$\Delta S^0 = \Sigma\, S^0(\text{products}) - \Sigma\, S^0(\text{reactants}) \qquad (8\text{-}29)$$

That is, the standard entropy change, ΔS^0, is given by the sum of the absolute entropies of the products less the sum of the absolute entropies of the reactants.

Note that the concept of absolute entropies is distinct from that for enthalpies or energies. Since entropy is associated with the number of microscopic states corresponding to a macroscopic state, we have a natural choice for the reference zero entropy state. In contrast, our values for enthalpies of formation are based on an arbitrary reference zero: that the enthalpy of formation of the elements in their standard states is zero. Thus, values for enthalpies are relative, those for entropies are absolute.

FREE ENERGY

We have now identified the two factors that determine the energetics of chemical reactions: changes in enthalpy and changes in entropy. Knowing the values for these changes, we can gain understanding about our original question applied to any chemical reaction: Can it occur? Nevertheless, our understanding as it stands now is not ideal, since for some chemical reactions a favorable enthalpy change may be opposed by an unfavorable entropy change, or vice versa, and we may not know which of these changes is the dominant one. Is there, then, some one quantity that combines both enthalpy changes and entropy changes and the value of which will permit us to make quantitative statements about the overall energy situation? The answer to this question is "yes," and the quantity of interest is the *free-energy change*. The **free energy,** G, is defined as

$$G = H - TS \tag{8-30}$$

Since the free energy depends only on quantities that are state functions, it is a state function as well. For processes occuring at constant temperature, we may rewrite Equation (8-30) to obtain an expression for the change in free energy:

$$\Delta G = \Delta H - T\Delta S \tag{8-31}$$

We frequently use the change in free energy to tell us to what extent a chemical reaction will proceed: If the free-energy change is negative, an **exergonic** reaction, the reaction will tend to proceed to completion and will more closely approach completion the greater the free-energy decrease; if the free-energy change is positive, an **endergonic** reaction, the reaction will have little tendency to occur and less of the reaction will occur the greater the free-energy increase. We shall see shortly just how we can make this statement more quantitative.

How can we obtain values of the free-energy change for chemical reactions? One method is immediately suggested by Equation (8-31). If we know the enthalpy and entropy changes, we can proceed to obtain the free-energy change at any temperature by using this equation. Let us remember that changes in the parameters that we are discussing are dependent on the temperature, pressure, physical state of the reactants and products, and the like. Therefore, it proves convenient to discuss *standard free-energy changes*, ΔG^0, for chemical reactions by which we understand that the

Table 8-4 Standard free energy of formation, ΔG_f^0, at 298°K (kcal/mole)

Inorganic compounds

H_2O	−54.64	C(graphite)	0.0
H_2O_2	−24.7	C(diamond)	0.69
O_3	39.06	CaO	−144.4
HCl	−22.77	$CaCO_3$	−269.8
HBr	−12.72	$Ca(OH)_2$	−214.3
HI	0.31	$CaCl_2$(s)	−179.3
SO_2	−71.79	$CaSO_4$(s)	−315.6
SO_3	−88.52	SiO_2	−192.4
H_2S	−7.89	Fe_2O_3	−177.1
H_2SO_4(l)	−194.5	$FeCl_2$(s)	−72.2
SF_6(g)	−289.0	$FeCl_3$(s)	−80.6
N_2O	24.9	FeS(s)	−23.3
NO	20.72	$FeSO_4$(s)	−194.8
NO_2	12.39	$FeCO_3$(s)	−161.1
NH_3	−3.97	Al_2O_3	−376.8
HNO_3(l)	−41.6	$AlCl_3$(s)	−150.3
$(NH_4)_2SO_4$(s)	−282.2	$Al_2(SO_4)_3$(s)	−741.0
CO	−32.81	CuO	−30.4
CO_2	−94.26	Cu_2O	−34.98
P(s) (white)	0	$CuCl_2$(s)	−39.0
P(s) (red)	−2.9	CuS(s)	−11.7
P_4(g)	5.8	$CuSO_4$(s)	−158.2
PH_3(g)	3.2	$CuCO_3$(s)	−123.8
H_3PO_4(s)	−267.5	ZnO	−76.05
AgCl	−26.22	$ZnCl_2$(s)	−88.3
AgBr	−22.39	ZnS(s)	−48.1
AgI	−15.81	$ZnSO_4$(s)	−209.0
BaO	−126.3	$ZnCO_3$(s)	−174.9
$BaSO_4$	−350.2	NiO(s)	−51.7
$BaCO_3$	−272.2	$NiCl_2$(s)	−61.9
$Ba(NO_3)_2$(s)	−190.1	$NiSO_4$(s)	−184.9
$BaCl_2$(s)	−193.8	$Ni(CO)_4$(g)	−140.3

Organic compounds

Gases		Liquids	
Methane, CH_4	−12.14	Methanol, CH_3OH	−39.73
Ethane, C_2H_6	−7.86	Ethanol, C_2H_5OH	−41.77
Propane, C_3H_8	−5.61	Acetic acid, CH_3COOH	−93.8
n-Butane, C_4H_{10}	−3.75	Benzene, C_6H_6	29.76
Isobutane, C_4H_{10}	−4.3	Carbon tetrachloride, CCl_4(l)	−15.6
n-Pentane, C_5H_{12}	−2.0	Carbon tetrachloride, CCl_4(g)	−14.5
Isopentane, C_5H_{12}	−3.5	CH_3Cl(g)	−13.7
Neopentane, C_5H_{12}	−3.6	CH_2Cl_2(g)	−15.8
Ethylene, C_2H_4	16.28	$CHCl_3$(g)	−16.8
Acetylene, C_2H_2	50.00		
1-Butene, C_4H_8	17.09		
cis-2-Butene, C_4H_8	15.74		
trans-2-Butene, C_4H_8	15.05		
Isobutene, C_4H_8	13.88		
1,3-Butadiene, C_4H_6	36.01		
Methyl chloride, CH_3Cl	−14.0		

reaction occurs at 25°C, at a pressure of 1 atm, and with the reactants and products in their stable physical forms under these conditions. Since a large number of enthalpies and entropies of formation have been tabulated (see Table 8-1, for example), these values can be employed to calculate **standard free energies of formation;** a representative collection is provided in Table 8-4. These values may then be employed to calculate standard free-energy changes just as we have employed related values to calculate enthalpy changes above, that is, by use of an equation of the form

$$\Delta G^0 = \Sigma\, \Delta G_f^0(\text{products}) - \Sigma\, \Delta G_f^0(\text{reactants}) \tag{8-32}$$

Using the appropriate values from Table 8-4, we can calculate the standard free-energy change for the oxidation of water to hydrogen peroxide [Equation (8-13)]:

$$\begin{aligned}\Delta G^0 &= \Delta G_f^0(H_2O_2) - \Delta G_f^0(H_2O)\\ &= -24.7 - (-54.64) = 29.94 \text{ kcal/mole}\end{aligned} \tag{8-33}$$

Thus, this reaction is endergonic by nearly 30 kcal/mole. This value tells us that water and oxygen in their standard states have very little tendency to form hydrogen peroxide in its standard state. Note that the free-energy change is not quite so positive as the standard enthalpy change calculated earlier for this reaction [Equation (8-13)]; thus, this reaction must proceed with a slightly favorable entropy change. Similarly, the isomerization of *cis*-2-butene to *trans*-2-butene [Equation (8-14)] has a standard free-energy change of

$$\begin{aligned}\Delta G^0 &= \Delta G_f^0(\textit{trans}\text{-2-butene}) - \Delta G_f^0(\textit{cis}\text{-2-butene})\\ &= 15.05 - 15.74 = -0.69 \text{ kcal/mole}\end{aligned} \tag{8-34}$$

Thus, this reaction is exergonic by 0.69 kcal/mole, in accord with our knowledge that the *trans* isomer is more stable than the *cis* isomer. Note that this value is less negative than that obtained for the standard enthalpy change, −1.04 kcal/mole [Equation (8-15)]; hence, this reaction must have a slightly negative entropy change. On the whole, we can conclude that *cis*-2-butene in its standard state will tend to form the *trans* isomer spontaneously in its standard state. Note that we have said nothing about whether this reaction *will* actually occur in any given time period. That is a question of reaction *kinetics*, a topic that we shall consider in the next chapter. All thermodynamics tells us is that the reaction *can* occur.

Example *What is the standard free-energy change for the conversion of diamond to graphite?*

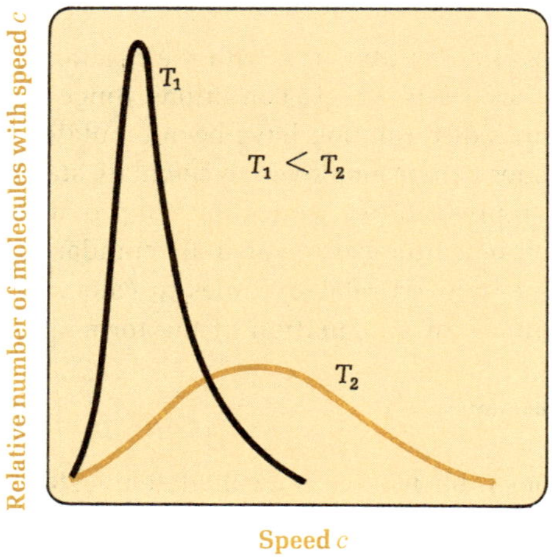

Figure 8-3 Molecular speed distribution for a gas at two temperatures.

Solution *Employing Equation (8-32) and the data in Table 8-4, we can readily make the calculation*

$$\begin{aligned}\Delta G^0 &= \Sigma\, \Delta G_f^{\,0}(\text{products}) - \Sigma\, \Delta G_f^{\,0}(\text{reactants})\\ &= 0.0 - (0.69)\\ &= -0.69 \text{ kcal/mole}\end{aligned}$$

This tells us that graphite, not diamond, is the stable form of carbon. What does it tell us about the rate *at which diamond is converted into graphite?*

Temperature and free-energy changes

Free-energy changes are generally a function of temperature. Thus, an exergonic reaction may become endergonic as the temperature is changed. One clear example is provided by the evaporation of a liquid: At temperatures below the boiling point, the process is endergonic and does not tend to proceed spontaneously, while at temperatures above that point, the reaction becomes exergonic. What can we understand about the temperature dependence of free-energy changes from our preceding discussion?

Example *For the vaporization of a liquid,*

$$\text{X(liquid)} \rightleftharpoons \text{X(gas)}$$

what must be the sign of the free-energy change, ΔG, above the boiling point? Below the boiling point? Given that, near the boiling point, $\Delta H = 9590$ cal/mole and that $\Delta S = 25.7$ cal/mole deg for the vaporization of a particular liquid, calculate the boiling point for that liquid.

Solution *Above the boiling point, the process is exergonic and ΔG is negative; below the boiling point, ΔG must be positive. At the boiling point,*

ΔG is zero. This permits us to calculate the boiling point using the data given and Equation (8-31):

$$\Delta G = \Delta H - T\,\Delta S$$

or

$$T = \frac{\Delta H - \Delta G}{\Delta S} = \frac{(9590 - 0)\ \text{cal/mole}}{25.7\ \text{cal/mole deg}}$$

$$= 373°\text{K} = 100°\text{C}$$

An important point can be understood by considering one example: the distribution of velocities of gas molecules as a function of temperature. In Figure 8-3, the number of molecules having a given velocity is plotted against the velocity for a gas at two temperatures. At the higher temperature, we can note two changes. First, the average velocity has increased, and second, the distribution of velocities has broadened. The latter point is the important one for our consideration. It is equivalent to saying that the extent of disorder in the system has increased with increasing temperature. This is a generally true observation: *Temperature is the weighting factor between order and disorder.* As the temperature of the system increases, the extent of disorder increases as well. Actually, this point is made explicitly in the definition of the free energy [Equation (8-30)], in which the entropy contribution enters as *TS*. Now let us return to our case of the evaporation of a liquid. We have already concluded that evaporation is opposed by enthalpic considerations, which are relatively temperature-independent but favored by entropic changes. Thus, at sufficiently low temperatures, the enthalpic considerations are predominant and the liquid phase will be the stable state. As the temperature is increased, the order–disorder transition becomes more important, entropic considerations become dominant, and the liquid-to-gas transition becomes spontaneous. The increase in temperature coupled with a positive entropy change has caused the free-energy change to become negative.

The equilibrium constant

By realizing one final point, we can provide ourselves with another means of calculating standard free-energy changes and of using these values to obtain information about the extent to which chemical reactions will occur. For any chemical reaction, we can define an **equilibrium constant** that specifies relative concentrations of reactants and products in their standard states under conditions in which no net chemical reaction occurs, that is, when the reaction system is in a state of equilibrium. In general, we can write a chemical reaction in the following form:

$$a\text{A} + b\text{B} \rightleftharpoons c\text{C} + d\text{D} \qquad (8\text{-}35)$$

The associated equilibrium constant is defined as

$$\mathbf{K} = \frac{(\mathbf{C})^c(\mathbf{D})^d}{(\mathbf{A})^a(\mathbf{B})^b} \tag{8-36}$$

in which the parentheses refer to concentrations of the indicated species. In formulating equilibrium constant expressions, note that: (1) concentrations of products appear in the numerator and those of reactants in the denominator; (2) concentration terms are multiplied together, not added; and (3) coefficients denoting stoichiometry in the chemical equation appear as exponents in the equilibrium constant expression. Some examples: For the reaction of Equation (8-13), the equilibrium constant is given by

$$K = \frac{(H_2O_2)}{(H_2O)(O_2)^{1/2}} \tag{8-37}$$

for that of Equation (8-14), by

$$K = \frac{(\textit{trans}\text{-2-butene})}{(\textit{cis}\text{-2-butene})} \tag{8-38}$$

and so forth.

Example *Formulate expressions for the equilibrium constant for each of the following reactions:*

a. $CO_2 + H_2O \rightleftharpoons H_2CO_3$

b. $CH_3{-}Cl + OH^- \rightleftharpoons CH_3{-}OH + Cl^-$

c. $NiCl_2 + 2Cl^- \rightleftharpoons NiCl_4^{2-}$

d. $C_6H_{12}O_6 + 6O_2 \rightleftharpoons 6CO_2 + 6H_2O$

Solution *Remembering that coefficients in the equations appear as exponents in equilibrium constant expressions, we have*

a. $K = \dfrac{(H_2CO_3)}{(CO_2)(H_2O)}$

b. $K = \dfrac{(CH_3OH)(Cl^-)}{(CH_3Cl)(OH^-)}$

c. $K = \dfrac{(NiCl_4^{2-})}{(NiCl_2)(Cl^-)^2}$

d. $K = \dfrac{(CO_2)^6(H_2O)^6}{(C_6H_{12}O_6)(O_2)^6}$

We may view the statement in Equation (8-36) in two ways. In the first place, it is an experimental fact that, regardless of the initial concentration employed, the ratio of reactants and products at equilibrium will be those specified by the equilibrium constants generated as indicated. In the second place, we can justify the indicated formulation of equilibrium constants on the basis of the rates of the reactions in the forward and reverse directions. While the topic of reaction rates is reserved for the next chapter, we can note at this point that a state of equilibrium must be one in which the rate of the reaction in the forward direction is equal to the rate of the reaction in the reverse direction. Now, for the rate of the forward reaction in Equation (8-35), we have the expression $\text{rate}_f = k_f(A)^a(B)^b$ and for the rate in the reverse direction, $\text{rate}_r = k_r(C)^c(D)^d$. In these expressions, k_f and k_r are constants, called rate constants, that relate rates to concentrations.

If at equilibrium these rates are equal, it follows that

$$k_f(A)^a(B)^b = k_r(C)^c(D)^d \tag{8-39}$$

Thus, we have

$$\frac{(C)^c(D)^d}{(A)^a(B)^b} = \frac{k_f}{k_r} \tag{8-40}$$

Thus, the equilibrium constant is just the ratio of the rate constants for the reaction in the forward and reverse directions.

Equilibrium constants and standard free-energy changes both measure the tendency of a chemical reaction to occur, and hence some relationship must exist between them. The relationship is

$$\Delta G^0 = -RT \ln K \tag{8-41}$$

In this expression T is the absolute temperature, K is the equilibrium constant, and R is called the *gas constant* and has a value of 1.98 cal/deg. Clearly, if we know the standard free-energy change, Equation (8-41) will permit us to obtain the equilibrium constant for the reaction and vice versa. For example, suppose we wish to know the ratio of *trans*- to *cis*-2-butene at equilibrium. Our previous calculation tells us that the standard free-energy change is −0.69 kcal/mole = −690 cal/mole [Equation (8-34)]. It follows, then, that the equilibrium constant at 298°K defined in Equation (8-38) must be given by

$$\begin{aligned} -690 &= -1.98 \times 298 \ln K \\ -690 &= -1.98 \times 298 \times 2.3 \log K \\ \log K &= 0.51 \\ K &= 3.2 \end{aligned} \tag{8-42}$$

Thus, Equation (8-38) tells us that under these conditions and at a state of equilibrium, the concentration of *trans*-2-butene will be 3.2 times that of *cis*-2-butene. Put another way, the equilibrium mixture will consist of 76 percent of the *trans* isomer and 24 percent of the *cis* isomer.

Example *At ordinary temperatures, methylcyclohexane exists as an equilibrium mixture of two conformations: one in which the methyl group is axial and one in which the methyl group is equatorial:*

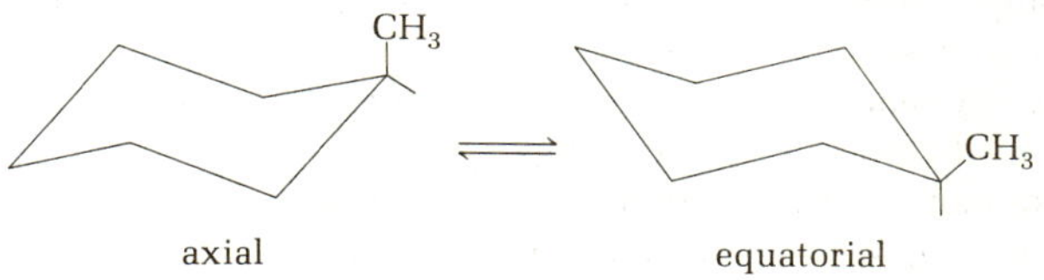

At 25°, methylcyclohexane adopts the equatorial conformation about 90 percent of the time and the axial one about 10 percent of the time. What is the standard free-energy change, ΔG^0, for this ring inversion reaction?

Solution *To solve this problem, we make use of Equation (8-41), which relates free-energy changes to equilibrium constants:*

$$\Delta G^0 = -RT \ln K = -RT \ln \frac{(\text{equatorial})}{(\text{axial})}$$
$$= -1.98 \text{ cal deg}^{-1} \times 298 \text{ deg} \times 2.303 \log 9$$
$$= -1360 \times 0.955 \text{ cal}$$
$$= -1295 \text{ cal}$$

We shall deal later in this book with many examples of systems at equilibrium. Two areas of considerable importance are equilibria involving ionizations of weak acids or bases and equilibria between slightly soluble solids and their solutions. The principles we have discussed above will serve to explain many phenomena that we shall study later.

Glossary

Bond energy the enthalpy change for the homolytic cleavage of a chemical bond: A—B $\longrightarrow$ A· + B·

Endergonic describes any reaction having a positive free-energy change, that is, one that does not tend to occur spontaneously.

Endothermic describes any reaction having a positive enthalpy change, that is, one that absorbs heat from the surroundings.

Energy, *E*, a state function that is conserved in ordinary chemical reactions. Changes in this quantity are expressed in terms of work or heat.

Enthalpy, *H*, a state function related to energy and defined by the expression $H = E + PV$

Entropy, *S*, a state function that describes the randomness of a system in terms of its individual elements. Mathematically, it can be defined in terms of the number of microscopically distinguishable states, W, corresponding to a macrostate: $S = k \ln W$.

Equilibrium constant a number that specifies relative concentrations of reactants and products in their standard states under conditions in which no net chemical reaction occurs.

Equilibrium state a state in which the system of interest has uniform temperature and pressure throughout and in which no net chemical transformations occur.

Exergonic describes any reaction having a negative free energy change, that is, one that tends to occur spontaneously.

Exothermic describes any reaction having a negative enthalpy change, that is, one that liberates heat into the surroundings.

First law of thermodynamics the mathematical expression of the conservation of energy: $\Delta E = q + w$, in which q is heat added to the system and w is work done on the system.

Free energy, G, a state function that defines quantitatively the tendency of a chemical reaction to proceed spontaneously; defined in terms of enthalpy and entropy: $G = H - TS$.

Macroscopic properties those properties that owe their existence to collections of atoms or molecules.

Microscopic properties those properties that can be defined for individual atoms or molecules.

Pressure-volume work the work done in expanding a system by a specified volume against a certain pressure: $w = P\,\Delta V$.

Second law of thermodynamics the statement that the entropy of an isolated system tends to increase. That is, isolated systems tend toward maximum disorder.

Standard enthalpy of formation the enthalpy change for synthesis of a compound from its elements at 25°C and 1 atm pressure, all substances being present in their stable physical forms.

Standard free energy of formation the free-energy change for synthesis of a compound from its elements at 25°C and 1 atm pressure, all substances being present in their stable physical forms.

State function any function whose change is determined solely by the nature of the initial and final states of the system, that is, whose change is independent of pathway.

Third law of thermodynamics the entropy of a perfect crystal at a temperature of absolute zero is zero.

Thermodynamics the science that treats energy changes in chemical and physical processes.

Study problems

Some problems related to the first law of thermodynamics

1. When 1 mole of ice melts at 0°C and a constant pressure of 1 atm, 1440 cal of heat is absorbed by the system. One mole of ice occupies 0.0196 liters, whereas 1 mole of water occupies 0.0180 liters. Calculate ΔH and ΔE for this process. Note: 1 liter atm is equal to 24.2 cal.
2. Using the bond energies collected in Table 8-2, calculate values of ΔH^0 for each of the following reactions:
   a. $H_2 + Cl_2 \longrightarrow 2HCl$
   b. $H_2 + CCl_4 \longrightarrow CH_2Cl_2 + Cl_2$
   c. $H_2C{=}O + H_2 \longrightarrow CH_3OH$
   d. $CH_3{-}CH_3 + O_2 \longrightarrow CH_3{-}CHO + H_2O$
   e. $N_2 + 3Cl_2 \longrightarrow 2NCl_3$
   f. $2CO + O_2 \longrightarrow 2CO_2$
3. Using the standard enthalpies of formation collected in Table 8-1, calculate values of ΔH^0 for each of the following reactions:
   a. $2SO_2 + O_2 \longrightarrow 2SO_3$
   b. $2C(\text{graphite}) + O_2 \longrightarrow 2CO$
   c. $2C(\text{diamond}) + O_2 \longrightarrow 2CO$
   d. $2CuS + 3O_2 \longrightarrow 2CuO + 2SO_2$
   e. $2H_2S + 3O_2 \longrightarrow 2H_2O + 2SO_2$
   f. $P_4 + 6H_2 \longrightarrow 4PH_3$
   g. $BaO + 2HNO_3 \longrightarrow Ba(NO_3)_2 + H_2O$

4. Which of the following would be considered a macroscopic property of a sample of lithium?
   a. ionization potential
   b. electron affinity
   c. temperature
5. Calculate the enthalpy of formation of solid $Ca(OH)_2$ from the following data:

$$H_2(g) + \tfrac{1}{2}O_2(g) \longrightarrow H_2O(l) \qquad \Delta H = -68.3 \text{ kcal}$$
$$CaO(s) + H_2O(l) \longrightarrow Ca(OH)_2(s) \qquad \Delta H = -15.3 \text{ kcal}$$
$$Ca(s) + \tfrac{1}{2}O_2(g) \longrightarrow CaO(s) \qquad \Delta H = -151.8 \text{ kcal}$$

6. Suppose that a liquid is placed in a cylinder with a movable piston at a temperature of 300°K. The pressure on the piston, 1.5 atm, is equal to the vapor pressure of the liquid at this temperature. Heat is now added to the system so as to vaporize the liquid at a constant temperature and pressure. When 100 cal of heat has been added, the volume of the system is increased by 0.50 liters. From this data, calculate each of the following:
   a. the number of moles of liquid vaporized (use the perfect gas law)
   b. the work performed in the expansion process
   c. ΔE
   d. ΔH

Some questions and problems about the second law of thermodynamics

7. Suppose that a gas is contained in a cylinder and constrained to a certain volume with a piston. If the piston is now withdrawn so that the available volume increases, the gas will expand so as to occupy all of the available volume. Why?
8. Indicate how the enthalpy and entropy change (increase, decrease, or remain constant) for each of the following processes. Briefly justify your answer.
   a. the isothermal expansion of an ideal gas into an evacuated chamber
   b. the melting of ice
   c. contraction of a stretched rubber band (rubber bands cool off when they contract)
9. From the data in Table 8-3, calculate the entropy change for each of the following reactions:

   a. $C(graphite) + O_2(g) \longrightarrow CO_2(g)$
   b. $CH_4(g) + 2O_2(g) \longrightarrow CO_2(g) + 2H_2O(g)$
   c. $Ca(s) + \tfrac{1}{2}O_2(g) \longrightarrow CaO(s)$

10. Listed below are several pairs of substances. In each case, choose that member of the pair expected to have the greater absolute entropy. Unless specified otherwise, assume 1 mole of each substance at the same temperature and pressure.

    a. C(graphite), Ag(s)
    b. B (300°K), B (400°K)
    c. $Cl_2(g)$, $2Cl(g)$
    d. Kr (1 atm), Kr (0.01 atm)

Some questions about equilibria and free energy

11. For each of the reactions in Problems 2 and 3, formulate the equilibrium constant expression.
12. Using the free energies of formation collected in Table 8-4, calculate the values of ΔG^0 for each of the following reactions:
   a. $ZnS(s) + O_2(g) \longrightarrow Zn(s) + SO_2(g)$
   b. $2CuS(s) + 3O_2(g) \longrightarrow 2CuO(s) + 2SO_2(g)$
   c. $4FeS(s) + 7O_2(g) \longrightarrow 2Fe_2O_3(s) + 4SO_2(g)$
   d. $2C(graphite) + O_2(g) \longrightarrow CO_2(g)$
13. In the reaction

   $$PCl_3(g) + Cl_2(g) \longrightarrow PCl_5(g)$$

   at equilibrium at 460°K, the pressures are $PCl_3 = 0.25$ atm, $Cl_2 = 0.25$ atm, and $PCl_5 = 0.5$ atm. What is the equilibrium constant for the reaction and what is ΔG^0?
14. For the dissociation of hydrogen iodide,

   $$2HI \rightleftharpoons H_2 + I_2$$

   at equilibrium at a temperature of 450°K, we have the following concentrations: $5.62 \times 10^{-3}M\ H_2$, $5.93 \times 10^{-4}M\ I_2$, and $1.27 \times 10^{-2}M$ HI. Calculate the equilibrium constant for this reaction.
15. The equilibrium constant for the dissociation of dinitrogen tetroxide according to the equation

   $$N_2O_4 \rightleftharpoons 2NO_2$$

   is $1.07 \times 10^{-5}M$. If a solution that contains 0.129M dinitrogen tetroxide is in a state of equilibrium with NO_2, what is the concentration of the latter species?
16. From the data of Table 8-4, calculate equilibrium constants for the following reactions:
   a. $CH_3CH_2CH_2CH_3 \rightleftharpoons CH_3CH(CH_3)CH_3$
   b. $CaCO_3 \rightleftharpoons CaO + O_2$
   c. $CH_2{=}CH_2 + H_2 \rightleftharpoons CH_3CH_3$
17. Consider the reactions

   $$2C(s) + H_2(g) \longrightarrow C_2H_2(g) \quad (1)$$
   $$6C(s) + 3H_2(g) \longrightarrow C_6H_6(l) \quad (2)$$
   $$3C_2H_2(g) \longrightarrow C_6H_6(l) \quad (3)$$

   a. Do you expect ΔS for (1) to be positive or negative? Why?
   b. Is reaction (2) exergonic or endergonic?
   c. What is the numerical value for the equilibrium constant for reaction (3)?

Suggestions for further reading

H. A. Bent "Haste Makes Waste: Pollution and Entropy," *Chemistry,* **44,** 6 (1971).

J. A. Campbell *Why Do Chemical Reactions Occur?,* Prentice-Hall, Englewood Cliffs, N.J., 1965.

B. H. Mahan *Elementary Chemical Thermodynamics,* Benjamin, Menlo Park, Calif., 1963.

K. B. Morris *Principles of Chemical Equilibrium,* Van Nostrand Reinhold, New York, 1965.

C. J. Nyman and R. E. Hamm *Chemical Equilibrium,* Heath, Lexington, Mass., 1968.

G. C. Pimentel and R. D. Sprately *Understanding Chemical Thermodynamics,* Holden-Day, San Francisco, 1969.

E. Schrödinger *What Is Life?,* Cambridge University Press, London, 1967.

9

Chemical reactivity

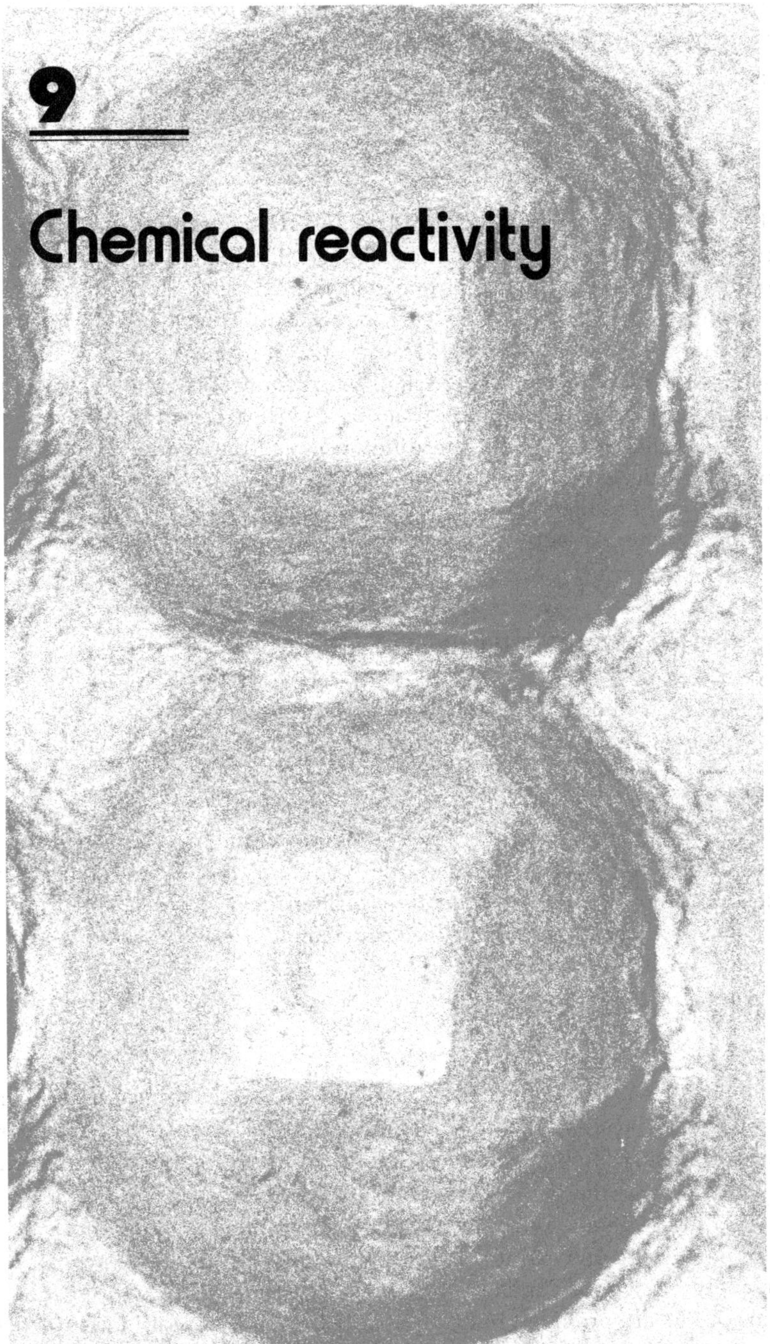

The course of a chemical reaction:
Bursts of helpless laughter and a labored altercation also played an intermittent part in the proceedings.

L. Durrell, *Bitter Lemons*

In the last chapter we examined those factors that determine to what extent a given chemical reaction can proceed. Basically, this involves the relative stabilities, or, exactly, the differences in free energies, between reactants and products. At this point we turn our attention to a related question: Given that a chemical reaction *can* occur, *will* it occur? Or, more precisely, at what rate does it occur? As we shall see, this question can be considered in a manner closely related to that adopted earlier. But in this instance we shall focus our attention on the differences in free energies between reactants and the **transition state,** a structure that is intermediate between reactants and products. In so doing, we hope to provide additional insight into the structure–reactivity relationships developed in Chapter 2.

TRANSITION STATE THEORY

To understand more about the relationship between energy changes and rates of chemical reactions, let us consider the interconversion of axial and equatorial fluorine atoms in the PF_5 molecule. As detailed in Chapter 6, (p. 186 and Figures 6-5 and 6-6), distortion of a trigonal-bipyramidal molecule, the stable geometry of PF_5, into a square-pyramidal geometry provides a route for the interchange of axial and equatorial positions. Thus, if we focus our attention on a specific atom of fluorine, F*, which, for example, may occupy an axial position in the original molecule, we have the following sequence of processes:

$$PF_5\text{(bipyramidal, F* axial)} \rightleftharpoons PF_5\text{ (pyramidal)} \rightleftharpoons PF_5\text{ (bipyramidal, F* equatorial)}$$

How can we describe the energy changes involved in terms of the geometry of the PF_5 system? To do so requires that we construct a **free-energy diagram.** In such diagrams the free energy of the system is plotted against the reaction coordinate, a quantity that bears a more-or-less direct relationship to the geometry of the system. Thus, we move along the reaction coordinate in the process of converting reactants to products. Two possible free-energy diagrams for the interchange of axial and equatorial fluorine atoms in PF_5 are provided in Figure 9-1. Let us look at that in Figure 9-1(a) first. In this diagram the bipyramidal PF_5 molecule is shown lying in a free-energy well whose geometry corresponds to the normal PF_5 geometry.

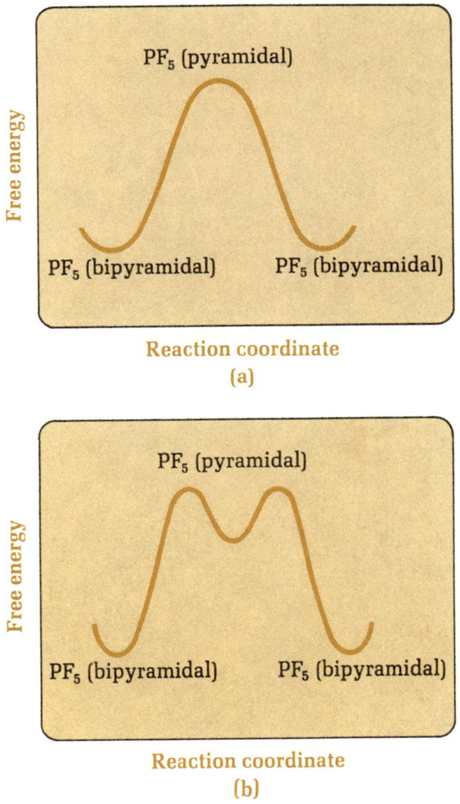

Figure 9-1 A free-energy diagram for the interchange of axial and equatorial fluorine atoms in the PF_5 molecule. (a) The square-pyramidal geometry is considered to be the geometry of highest energy, the transition state. (b) This geometry is shown as an energy minimum, an intermediate.

Movement along the reaction coordinate in either direction results in an energy increase since we are distorting the most stable bipyramidal geometry. As we move to the right along the reaction coordinate, we progress from the bipyramidal geometry to the square-pyramidal one. Eventually, we pass through this geometry and return to the original one, which must have the same free energy as initially. Note that in this representation, the square-pyramidal geometry is considered to be the geometry of highest energy in the reaction process. This geometry is called the *transition state*. The transition state is always the point of highest energy in these diagrams. In Figure 9-1(b) the square-pyramidal geometry is shown as an energy minimum. In this case the square-pyramidal form of PF_5 is said to be an

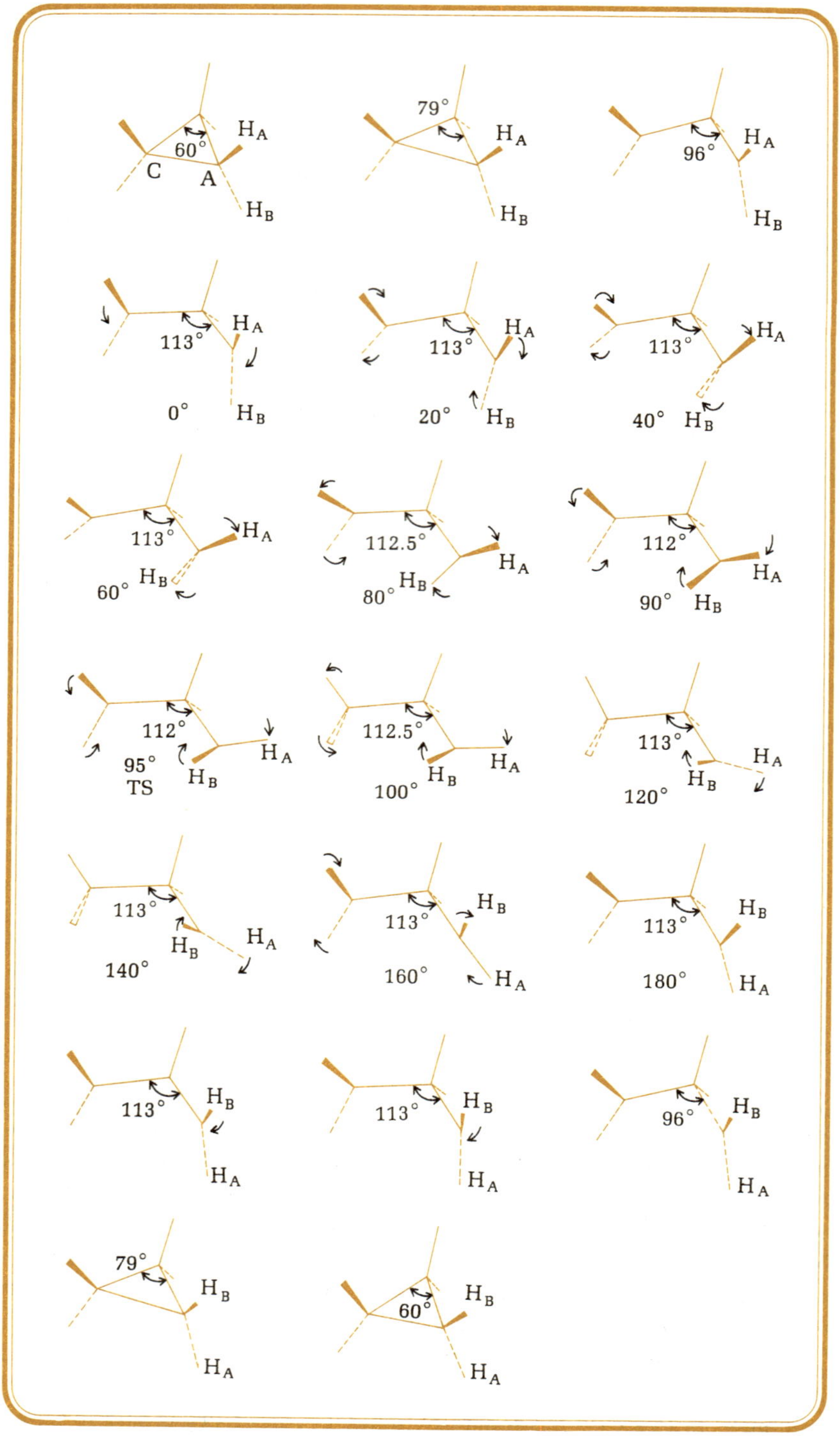

60°
C
A
H_A
H_B
79°
H_A
H_B
96°
H_A
H_B
113°
H_A
0°
H_B
113°
H_A
20°
H_B
113°
H_A
40°
H_B
113°
H_A
60°
H_B
112.5°
H_A
80°
H_B
112°
H_A
90°
H_B
112°
95°
TS
H_B
H_A
112.5°
H_B
H_A
100°
113°
H_A
H_B
120°
113°
H_B
H_A
140°
113°
H_B
160°
H_A
113°
H_B
180°
H_A
113°
H_B
H_A
113°
H_B
H_A
96°
H_B
H_A
79°
H_B
H_A
60°
H_B
H_A

intermediate in the reaction process. Intermediates always correspond to energy minima between the minima of reactants and products in free-energy diagrams. In this case the transition state for our reaction has a geometry intermediate between those of bipyramidal and pyramidal PF_5.

Note that, *a priori*, we have no way of knowing if the square-pyramidal form of PF_5 is an intermediate, having a transient existence, or a transition state, having a lifetime of only about 10^{-13} sec. A properly designed experiment might permit us to decide, although questions of this type are frequently difficult to answer. In some relatively simple cases, it is possible to calculate a reaction pathway and approximate structure of the transition state from knowledge of molecular structures and associated parameters. One such example, for the isomerization of a cyclopropane, is shown in Figure 9-2. This figure shows a detailed reaction pathway and is worthy of careful consideration. For most reactions, calculations of this kind are not yet possible (even in this case we cannot be certain of the validity of the results), and we approximate the transition state by choosing that which is most consistent with experimental data.

In the free-energy diagram of Figure 9-1, there are two free-energy differences that determine the behavior of the system. The first of these is between the reactant and product states, ΔG. This difference is, of course, the overall free-energy change for the reaction and determines the extent to which the reaction will occur as developed in Chapter 8 (see pp. 269–276). In this case the free-energy diagram has been drawn with the reactant state at the same free energy as the product state. Thus, this reaction is *isoenergetic*. The second of these is the difference between the reactant state and the transition state, $\Delta G^{\ddagger}$, the **free energy of activation.** This difference determines, in part, the *rate* of the reaction. Qualitatively, if this difference is large, the reaction will occur slowly, whereas if the difference is small, the reaction will occur rapidly. Quantitatively, we can develop this idea by stating that there exists a direct relationship between the **rate constant,** k, for the reaction (the larger the rate constant, the faster the reaction) and $\Delta G^{\ddagger}$:

$$k = \text{constant} \times e^{-\Delta G^{\ddagger}/RT} \qquad (9\text{-}1)$$

Figure 9-2 A calculated reaction pathway for geometric isomerization of cyclopropane. The overall process is simply to interchange the positions of H_A and H_B with respect to the plane of the cyclopropane ring. Note that the process begins with a stretching and then rupture of a C—C bond in the ring. The rotation of the principal methylene (CH_2) group, interchanging the positions of H_A and H_B, is accompanied by first a clockwise, then a counter-clockwise rotation of the adjacent methylene group. These rotations occur so as to minimize the repulsions between these groups. When the rotation of the methylene group is complete, the C—C bond is reformed, terminating the reaction process. The structure corresponding to the transition state is labeled TS. (From J. A. Horsley et al., *J. Amer. Chem. Soc.*, **94**, 279 (1972). Copyright © 1972 by the American Chemical Society.)

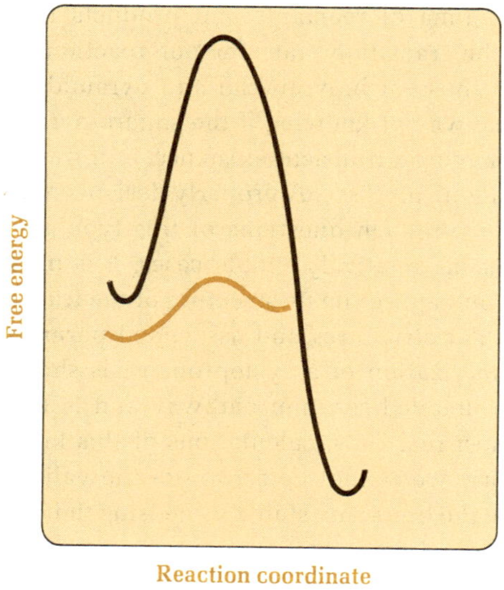

Figure 9-3 Free-energy diagrams for slow exergonic reactions (brown) and rapid endergonic reactions (yellow).

While we shall focus our attention almost exclusively on the free energy of activation in determining reaction rates, we ought to recognize that this does not completely determine the reaction rate. Basically, the free energy of activation is a measure of the equilibrium constant for formation of the transition state. To complete the picture, we also must be concerned with the rate of passage over the barrier, that is, with the rate at which the transition state decomposes to products. As noted above, this time is generally on the order of 10^{-13} sec, but it does depend on the temperature. This dependence is hidden in the "constant" of Equation (9-1). It is perfectly legitimate to discuss the changes in reaction rates that result from changes in molecular structure and the like in terms of the free energy of activation.

The free energy of activation may be expressed in terms of the *enthalpy of activation*, $\Delta H^\ddagger$, and *entropy of activation*, $\Delta S^\ddagger$, since $\Delta G^\ddagger = \Delta H^\ddagger - T\,\Delta S^\ddagger$. We may then rewrite Equation (9-1) in the form

$$k = \text{constant} \times e^{\Delta S^\ddagger/R}\, e^{-\Delta H^\ddagger/RT} \qquad (9\text{-}2)$$

Hence, we may analyze rates of reactions in terms of the contributions of enthalpic and entropic factors just as we analyze extents of reactions in these terms. Before we turn to these considerations, let us pause to consider the relationship between extents and rates for chemical reactions.

In what we have just said, we can see that rates and extents of chemical reactions depend on different free-energy changes, $\Delta G^{\ddagger}$ and ΔG, and nothing that we have said suggests that there is a *necessary* relationship between them. This is, in point of fact, true. It is quite possible to have reactions that have a large negative ΔG, so that the reaction will proceed nearly to completion, and that also have a very large $\Delta G^{\ddagger}$, so that the reaction rate is exceedingly slow. Similarly, it is possible to have endergonic reactions that proceed rapidly (although not very far, as is required by the overall free-energy change). Free-energy diagrams that apply to these cases are shown in Figure 9-3. There are many examples of reactions that fall into these categories. For example, simple hydrocarbons such as hexane, octane, and decane are thermodynamically unstable with respect to their elements, carbon and hydrogen. That is, their free energies of formation are positive. Nevertheless, these compounds are stable over geological periods of time and are, in fact, isolable from ancient organic deposits. Similarly, the reaction of hydrogen gas with oxygen gas to yield water is violently exothermic (burning liquid hydrogen with liquid oxygen is a good means for rocket propulsion); nevertheless, we can mix the gases and let them stand as long as desired without detectable reaction. For reactions in this class, there exists no pathway for the reaction that is low enough in energy to permit it to occur at a reasonable rate. That is, the distance up the energy barrier is simply too great for a molecule to get to the top very often. The problem for a molecule is akin to that of a man who would walk from one valley to another at a lower altitude but by a pathway that involves scaling the Matterhorn. It can be done but it does not happen very often.

The examples of thermodynamically allowed but kinetically forbidden reactions are, however, not the usual case. Generally, reactions that are "downhill" thermodynamically are also rapid, because those factors responsible for stabilizing the product relative to the reactant usually also stabilize the transition state. When this is the case, a parallel between kinetic and thermodynamic behavior is observed. As a specific example, let us consider the simple dissociation of butane and tetramethylbutane. Reasonable transition states for these reactions may be formulated as involving some stretching and weakening of the central C—C bond that is cleaved in the product:

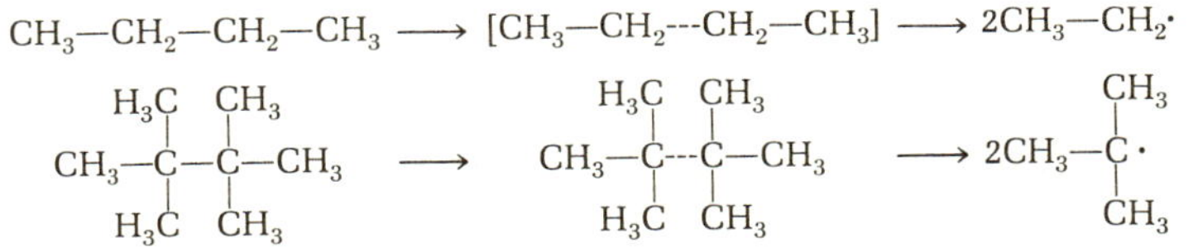

(9-3)

One reason (not the only one) that the latter reaction is favored thermodynamically over the former is that the steric crowding of the methyl groups is relieved in the products. But this is also true in the transition state. Since

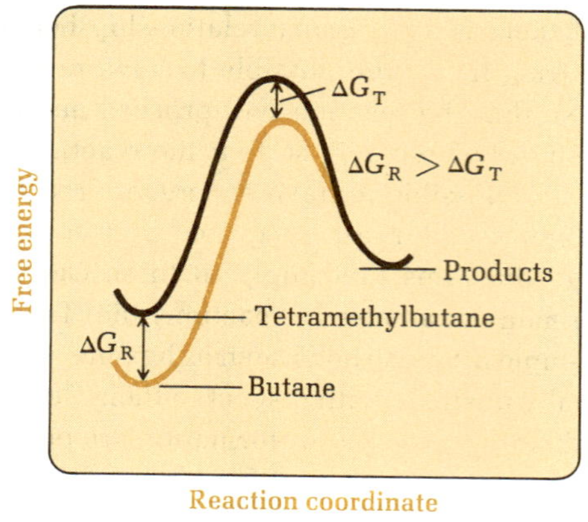

Figure 9-4 Free-energy diagrams for the dissociation of *n*-butane and its tetramethyl derivative.

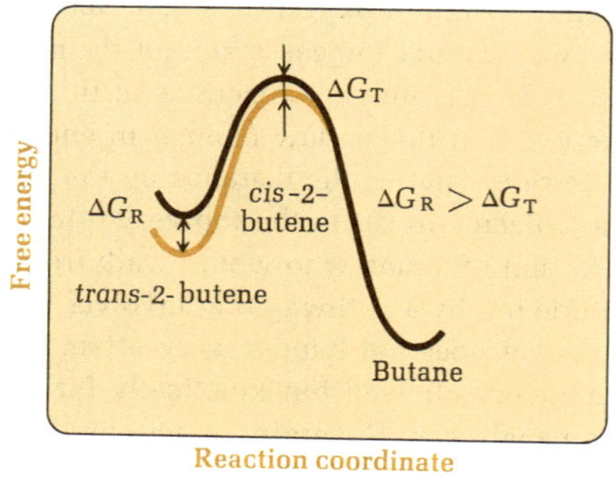

Figure 9-5 Free-energy diagrams for the hydrogenation of the *cis* and *trans* isomers of 2-butene.

the C—C bond is being stretched, the methyl groups will have relatively more room in the transition state than in the reactant. Thus, one factor that stabilizes the product state with respect to the reactant also stabilizes the transition state with respect to the reactant. Thus, a more favorable reaction pathway exists for the latter reaction, and we shall expect it to proceed more rapidly than the former as well as more nearly to completion. These points are shown diagrammatically in Figure 9-4. By focusing attention only on steric factors, we have simplified a complex matter. Nevertheless, the point developed is a valid and correct one.

A related case is provided by the hydrogenation of the *cis* and *trans* isomers of 2-butene:

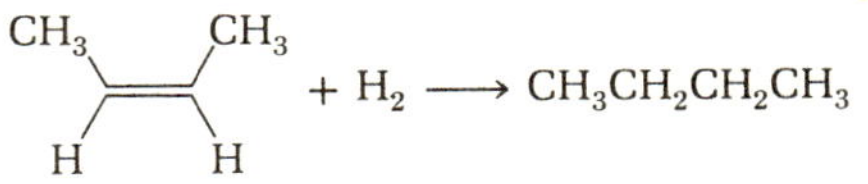

(9-4)

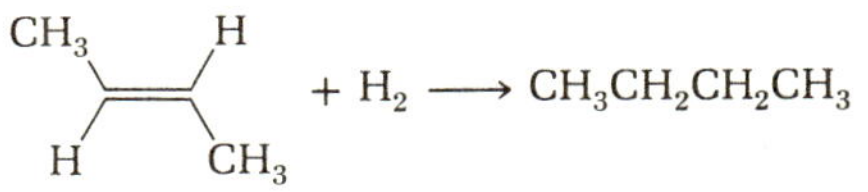

(9-5)

In this case the products are identical, so any differences in reactivity must depend on the reactants. We have already demonstrated that the *trans* isomer is the more stable of the two, since it avoids the steric hindrance between the methyl groups that is present in the *cis* isomer (see Chapter 2, pp. 64–66 and Chapter 8, pp. 257, 271). As we move from reactant state to product state in the hydrogenation reaction, we are moving from a state in which a double bond exists between the two central carbon atoms to one in which only a single bond is present. Consequently, in the transition state, which must have a structure intermediate between these two, the double-bond character must be partially lost. Thus, we shall expect to find added freedom of rotation around this bond and, consequently, a diminution of the steric hindrance in the *cis* isomer. The net effect of all this is to make the energy difference between the reactant and transition states less for the *cis* isomer than for the *trans* isomer. Thus, the former is the more reactive. The energetic situation for this reaction is shown in Figure 9-5. In this case too, the kinetic behavior parallels the thermodynamic behavior: that reaction that proceeds to the greater extent also proceeds more rapidly.

Example *A common reaction in chemistry is the formation of acetals from the addition of alcohols to aldehydes according to the process*

$$CH_3CHO + 2ROH \rightleftharpoons CH_3{-}\underset{\displaystyle OR}{\overset{\displaystyle OR}{\overset{|}{\underset{|}{C}}}}{-}H + H_2O$$

The position of the equilibrium for this reaction has been measured for a number of alcohols in a system containing 1 mole of acetaldehyde to 5 moles of alcohol. With ethanol, 78 percent of the aldehyde is converted to acetal at equilibrium; with cyclohexanol, 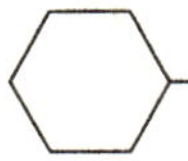—OH, *56 percent is converted to acetal at equilibrium; and with t-butyl alcohol,* $(CH_3)_3$—COH, *only 23 percent is acetal at equilibrium. Account for these observations. What prediction would you make about the relative rate of these three reactions?*

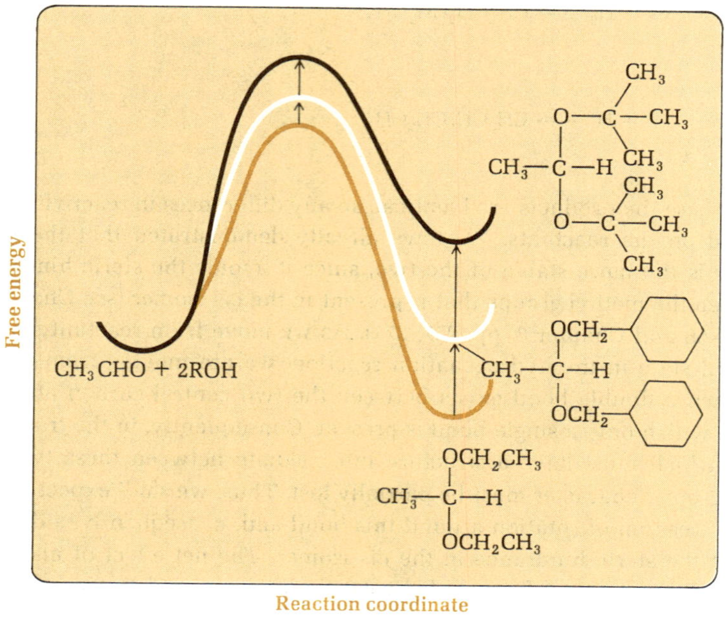

Figure 9-6 Free-energy diagrams for reaction of acetaldehyde with three alcohols. Note the striking effect of steric factors on both transition state and product energies.

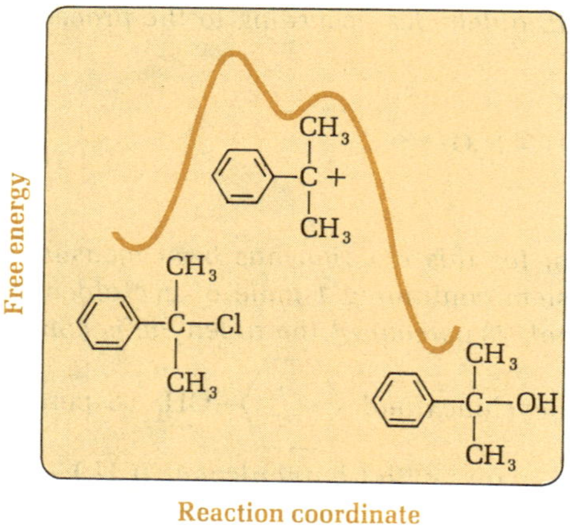

Figure 9-7 Free-energy diagram for the hydrolysis of an organic halide involving one intermediate.

Solution *We can understand these results on the basis of steric crowding in the acetal. The structures of the three acetals are*

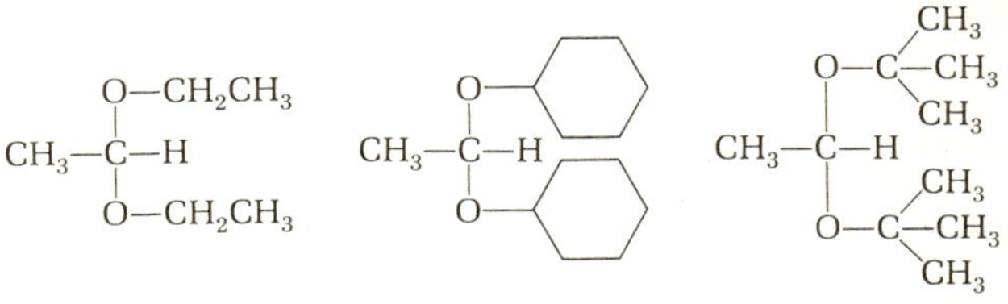

Note the increasing steric crowding as we go from the acetal derived from ethanol to that derived from cyclohexanol to that derived from t-butyl alcohol. Since no such steric crowding exists in the reactants, we shall expect that the products will be specifically destabilized relative to reactants by this steric crowding. This accounts for the qualitative trend of the equilibrium positions. We would expect that the rates would follow the trend of equilibrium constants: The amount of steric crowding in the transition state will not be as great as that in the product acetals, but it will exist to some extent. Thus, we would expect ethanol to react most rapidly and t-butyl alcohol to react least rapidly. These points are summarized in the free-energy diagram of Figure 9-6.

REACTION MECHANISMS

Having examined the relationship between thermodynamic and kinetic behavior, it is worth keeping in mind the central distinction between them. In investigating the energetics of a reaction, we ignored the *pathway* of the reaction; with kinetics we are centrally concerned with the pathway. A description of the pathway of a chemical reaction, including the *structure of any intermediates* and the *structure of the transition state*, is said to constitute the mechanism of the reaction. As an example of a **reaction mechanism,** let us consider the hydrolysis of an organic halide according to the pathway

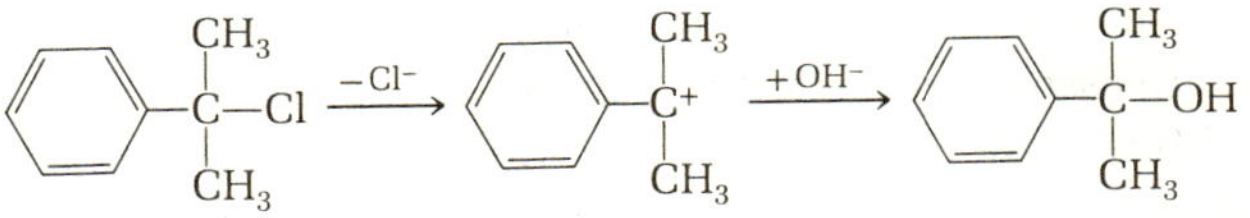

(9-6)

This reaction proceeds through one intermediate: the **carbonium ion** (an ion possessing a positively charged carbon atom) derived from the substrate by loss of the chloride ion. Reaction of the carbonium ion with hydroxide ion completes the reaction. The free-energy diagram that corresponds to this pathway is shown in Figure 9-7. Note two points. First, the intermediate corresponds to a minimum in free energy. Second, the highest free-energy

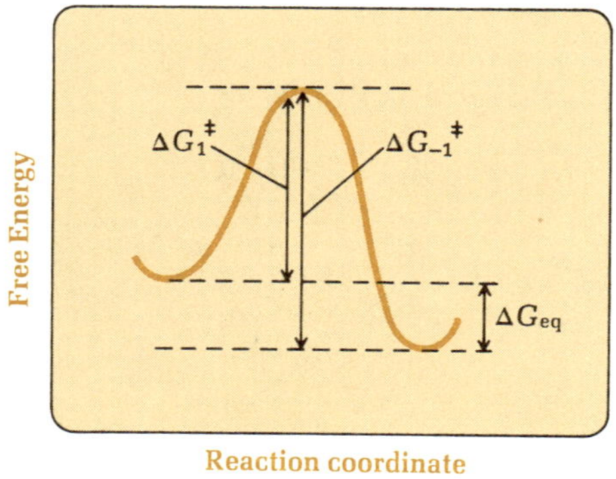

Figure 9-8 Free-energy relations for a generalized, exergonic reaction.

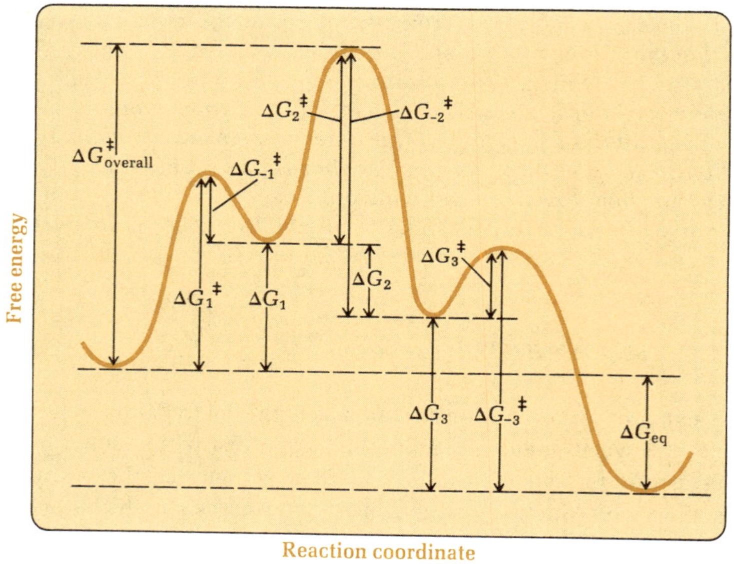

Figure 9-9 One possible free-energy diagram for a reaction pathway involving two intermediates.

point in the diagram is associated with the formation of the carbonium ion. It is, then, the structure associated with this step that is the transition state for the reaction, and furthermore, it is the rate of this step that determines the rate of the overall reaction. *The rate constant for all chemical reactions is largely determined by the free-energy difference between reactant and transition states* (although we may be concerned with the rate of passage across the energy barrier as well).

Prior to moving on to related matters, it may be well to review the relationship between rate constants, equilibrium constants, and free-energy changes. We consider the following two examples. For a simple reaction that proceeds without the formation of intermediates,

$$A \underset{k_{-1}}{\overset{k_1}{\rightleftharpoons}} B \qquad K_{eq} = \frac{(B)}{(A)}$$

the thermodynamics and kinetics of the system are described by one equilibrium constant and two rate constants, k_1 and k_{-1}, as noted above. These quantities are determined, respectively, by the free-energy differences between products and reactants, transition state and reactants, and products and transition state. We picture this statement in Figure 9-8, in which the subscripts on the free-energy changes relate these to the associated rate or equilibrium constant. These basic ideas may be generalized directly to more complicated reactions that involve intermediates; we simply consider each successive reaction individually and, additionally, consider the overall rate and equilibrium constants.

Thus, for the reaction pathway involving two intermediates,

$$A \underset{k_{-1}}{\overset{k_1}{\rightleftharpoons}} I_1 \underset{k_{-2}}{\overset{k_2}{\rightleftharpoons}} I_2 \underset{k_{-3}}{\overset{k_3}{\rightleftharpoons}} B \qquad K_{eq} = \frac{(B)}{(A)}$$

$$K_1 = \frac{(I_1)}{(A)} \qquad K_2 = \frac{(I_2)}{(I_1)} \qquad K_3 = \frac{(B)}{(I_2)}$$

we can formulate a free-energy diagram, clearly not the only possibility, of the sort shown in Figure 9-9, in which the free-energy changes are again related to the associated rate and equilibrium constants by subscripts. Note that free-energy differences between successive energy minima determine equilibrium constants for successive steps in the overall reaction. As always, the free-energy difference between product and reactant states determines the overall equilibrium constant. Note also that free-energy differences between free-energy minima and the following or preceding maxima are associated, respectively, with rate constants in the forward and reverse directions. The free-energy difference between the transition state, the free-energy maximum, and the reactant state largely determines the overall rate of the reaction.

Among other points, our considerations thus far establish that there is an intimate relationship between reactivities and mechanism. At this point, we wish to discuss the nature of this relationship in some detail, at the same time emphasizing the relationship between molecular structure, electronic structure, and reactivity.

Example *Construct a free-energy diagram for a reaction that (a) is endergonic, (b) occurs through one intermediate, (c) the decomposition of which is the rate-determining step.*

Solution *In constructing the free-energy diagram, the conditions set down can be met as follows. First, the free energy of the product state must be greater than that of the reactant state to account for the endergonic character of the reaction. Second, there must be one free-energy minimum between reactant and product states to account for the formation of one* *intermediate. Finally, the maximum free energy reached along the reaction coordinate, the transition state, must occur between the intermediate and product states to accord with rate-determining intermediate decomposition. Thus, the free-energy diagram will take the form*

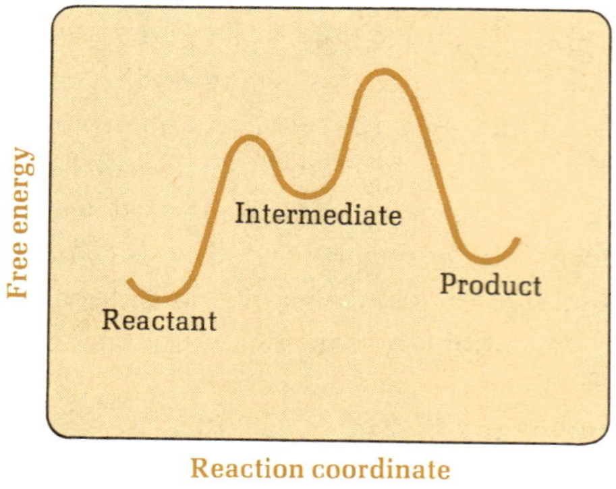

HYDROLYSIS OF HALIDES

Let us begin our discussion of reactivities (rates) and mechanisms by searching for explanations to account for the following observation: The reactivities of the four methyl halides toward anionic (negatively charged) reagents, such as hydroxide ion, are in the order $CH_3F < CH_3Cl < CH_3Br < CH_3I$.

$$CH_3X + OH^- \longrightarrow CH_3OH + X^- \tag{9-7}$$

To explain the order of reactivity in this series, we must compare the effect of the nature of the halogen on the stabilities of the reactant and transition states. We can understand a great deal about this matter with respect to the reactant state since we know the structures of the four methyl halides. But we must choose some reaction pathway in order to know the structure of the transition state. For reactions of the type of Equation (9-7), there exist several possible reaction pathways. We consider two examples at this point: dissociative and associative pathways. In the first of these, the reaction is viewed as involving the decomposition of the methyl halide, a *unimolecular* reaction, to yield a carbon compound bearing a positive charge, a *carbonium ion,* and the halogen anion, followed by association of the carbonium ion with the hydroxide ion [see also Equation (9-6) and Figure 9-7]:

$$CH_3X \longrightarrow [CH_3{}^{\delta+}\text{---}X^{\delta-}] \longrightarrow CH_3{}^+ + X^- \tag{9-8}$$

$$CH_3{}^+ + OH^- \longrightarrow CH_3\text{—}OH \tag{9-9}$$

$$\text{sum} \quad CH_3\text{—}X + OH^- \longrightarrow CH_3\text{—}OH + X^- \tag{9-10}$$

In this sequence of reactions we have chosen the first of the two steps to be the slow one. Hence, the transition state is associated with this step, and a structure for it has been enclosed in brackets in Equation (9-8). Since the reactant state has the C—X bond fully formed and the product state has it fully cleaved, it follows that this bond must be partially broken, or stretched, in the transition state, as depicted.

In the second possible pathway, an associative process, the reaction is viewed as a single step in which the hydroxide ion attacks the carbon atom from the side opposite to that occupied by the halogen atom:

$$OH^- + CH_3{-}X \longrightarrow HO{---}\underset{H}{\overset{H\quad H}{C}}{---}X^- \longrightarrow CH_3{-}OH + X^- \qquad (9\text{-}11)$$

In this case the transition state may be approximated to a structure in which the central carbon atom makes partial chemical bonds with the incoming and outgoing groups along a single line and three full bonds to the hydrogen atoms located in a plane perpendicular to this line. We can understand something about how such a reaction occurs by reviewing for a moment our knowledge of chemical bonding. In compounds such as methyl halides, in which carbon forms four single bonds to four substituents, the chemical bonds are of the sp^3 type (see pp. 160–162, Chapter 5). Such bonds are of the sigma type. Adopting the valence bond point of view, the orbital on carbon has one large lobe, which forms part of the bond to the halide atom, and a small lobe pointing away from this atom. As the hydroxide ion approaches the methyl halide from the back side, the unshared electron pair on oxygen can begin to form a bond with carbon by overlapping with this small lobe of the orbital. As the reaction progresses, bond formation between oxygen and carbon increases and that between carbon and halide decreases: The lobes of the orbital on carbon are essentially interconverted. Visually, we can picture the process in the following way:

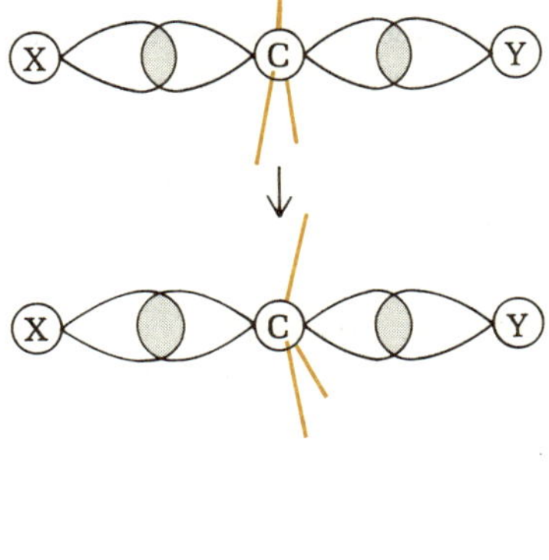

(9-12)

Although, depending on the nature of the halide, either mechanism may occur, experimentation has clearly established that the mechanism for these simple reactions [Equation (9-7)] is associative (9-11) rather than dissociative [(9-8) and (9-9)]. We might have guessed this result beforehand just by comparing the stabilities of the two transition states. Since the reactant states are the same, the reaction will proceed via the pathway that involves formation of the most stable transition state. In the dissociative mechanism, we form a CH_3^+ ion as an intermediate. This is a very unstable species indeed: Carbon does not tolerate a positive charge very well, and in this simple ion there is little opportunity for stabilization of this charge [this is not the case for the carbonium ion in Equation (9-6)]. Since the transition state for a reaction is the energy maximum, it must be even less stable than the CH_3^+ ion. Hence, the transition state, which involves partial formation of the carbonium ion, of Equation (9-8) must lie at very high energy. This pathway is, consequently, unfavorable. The associative mechanism, which avoids the formation of any unstable intermediates, thus intuitively seems the better of the two, as indeed it is.

With these considerations in mind, we can return to our question and try to evaluate the influence of the various halogen atoms on the stabilities of reactant and transition states. Beginning with the reactant state, we can understand one important cause of the order of reactivity by returning to the bond energies collected in Table 8-2. Clearly, the C—F bond is the strongest and the C—I bond is the weakest of the series. Thus, the strength of the bond cleaved in the reaction is inversely related to reactivity as expected; the substrate with the strongest bond reacts least rapidly. Thus, we may suppose that one factor in establishing the indicated order of reactivity is reactant state stabilities. Is this the only important factor? Or are differences in transition state stabilities important as well? We can gain useful information regarding this point by considering what happens in some related, and even simpler, reactions. These are the isotopic exchange reactions:

$$\begin{aligned}
CH_3\text{—}F + F^{*-} &\longrightarrow CH_3\text{—}F^{*} + F^{-}\\
CH_3\text{—}Cl + Cl^{*-} &\longrightarrow CH_3\text{—}Cl^{*} + Cl^{-}\\
CH_3\text{—}Br + Br^{*-} &\longrightarrow CH_3\text{—}Br^{*} + Br^{-}\\
CH_3\text{—}I + I^{*-} &\longrightarrow CH_3\text{—}I^{*} + I^{-}
\end{aligned} \qquad (9\text{-}13)$$

The starred atoms are considered chemically identical to the corresponding unstarred ones but may be distinguished on the basis of, for example, radioactivity. In these symmetrical reactions, the bonds being made and broken are the same, so that the overall thermodynamics is not a factor ($\Delta G^0 = 0$). Here, too, we find that the reaction of the fluoride is the slowest and that of the iodide is the fastest. Thus, it must be true that I^- is simply a better "leaving group" than Br^-, which is better than Cl^-, which is better than F^-. Stated another way, the transition state of Equation (9-11) is basically more stable when X is I than when it is F. The reason for this is rather complicated but basically reflects the fact that fluoride ion is a small, compact ball of negative charge while the iodide ion is a large, diffuse ball of negative charge. As a result, the iodide ion has a kind of flexibility, enabling it to accommodate itself to the demands of the reaction in a way not available to the fluoride ion. Iodide ion is more **polarizable** than fluoride. It is a little like the relative difficulties of shoving a 2-in. ball-bearing and a 2-ft balloon through a 1-in. hole. Even though the balloon is a great deal larger than the ball-bearing, its ability to accommodate itself to the task at hand makes it the easier object to deal with.

To sum up, the order of reactivity for the basic hydrolysis of the methyl halides is due both to the reactant state stabilities, as reflected in the bond energies of the bonds to be cleaved, and the transition state stabilities, as reflected in the relative abilities of the leaving ions to accommodate themselves to the demands of the reaction.

A related series of reactions introduces some new complications. This series is comprised of the chlorinated methanes: CCl_4, $HCCl_3$, H_2CCl_2, and H_3CCl. The facts to be dealt with are illustrated simply as

$$\begin{array}{llll} CCl_4 + OH^- \longrightarrow HO\text{—}CCl_3 + Cl^- & \text{inert} \\ HCCl_3 + OH^- \longrightarrow HO\text{—}CHCl_2 + Cl^- & \text{moderately reactive} \\ H_2CCl_2 + OH^- \longrightarrow HO\text{—}CH_2Cl + Cl^- & \text{slightly reactive} \\ H_3CCl + OH^- \longrightarrow HO\text{—}CH_3 + Cl^- & \text{very reactive} \end{array} \tag{9-14}$$

(Note: Only the last of these products is stable; the others undergo further reactions spontaneously. These need not concern us.) In this series of reactions, we have a discontinuity: Chloroform, $HCCl_3$, is out of order. Other than this case, we see that the reaction rate increases as the number of chlorine atoms decreases. Thus, we need to explain two things: the general trend of increasing reactivity with decreasing chlorine substitution and the exceptional behavior of $HCCl_3$.

In each of these reactions, we are cleaving a C—Cl bond. While it is certainly true that the strengths of the C—Cl bonds in each of the four compounds will not be exactly the same, it is very unlikely that the differences in reactant state stabilities as measured by these bond energies will be nearly as great as those in the reaction series just discussed. Thus, we expect that reactant state stabilities will be of less importance in establishing the reactivity order than in the preceding case. Viewing the reactions as proceeding through an associative reaction pathway [Equation (9-11)], we can understand the general order of reactivity in terms of transition state stabilities. All we need to remember is that the chlorine atom is a great deal larger than the hydrogen atom. Thus, in a transition state formed from the attack of hydroxide ion on CCl_4, there simply is not room for the hydroxide ion and the four chlorides to be bound simultaneously to the carbon atom. This transition state will, then, be very much less stable than that formed from hydroxide ion and CH_3Cl, in which such steric crowding does not occur. The reaction of H_2CCl_2 ought to involve a transition state more stable than that formed in the case of CCl_4 but less stable than that formed from CH_3Cl, and hence H_2CCl_2 ought to react at an intermediate rate, which it does.

Clearly, the explanation offered thus far cannot account for the abnormally great reactivity of $HCCl_3$. What kind of explanation must we search for in this case? Generally, it proves true that compounds that react abnormally within a given series whose relative reactivities may otherwise be readily explained react by different mechanisms. This is the case here. The pathway for reaction of $HCCl_3$ is dissociative:

$$HCCl_3 + OH^- \longrightarrow {}^-CCl_3 + H_2O \qquad (9\text{-}15)$$

$$CCl_3^- \longrightarrow :CCl_2 + Cl^- \qquad (9\text{-}16)$$

$$:CCl_2 + H_2O \longrightarrow HO\text{—}CHCl_2 \qquad (9\text{-}17)$$

One key intermediate in this reaction scheme is $:CCl_2$, a divalent carbon compound of a type that we have not previously encountered. Such divalent carbon compounds are called **carbenes;** these species undergo a variety of exceptionally interesting chemical reactions which are, unfortunately, beyond the scope of our current interests. The other key intermediate is the CCl_3^- ion. Ordinarily, carbon has a very low tolerance for negative charge. However, in cases in which delocalization of the negative charge is possible, the ions can be rather stable. This is just what happens here: The chlorine atoms are very good withdrawers of electrons. Thus, a large fraction of the negative charge will reside on the three chlorines, which have a high tolerance for such charge (electronegativity), and not on carbon. The combined electron affinity of the three chlorine atoms is sufficient to make this ion stable enough so that it has transient existence.

Example *Construct a free-energy diagram for the hydrolysis of* $HCCl_3$ *given that (a) formation of the carbene [Equation (9-16)] is the slowest step in the process; (b) the equilibrium constant for Equation (9-15) is less than 1; (c) the overall reaction is exergonic.*

Solution *The following free-energy diagram meets all these criteria, as you may justify.*

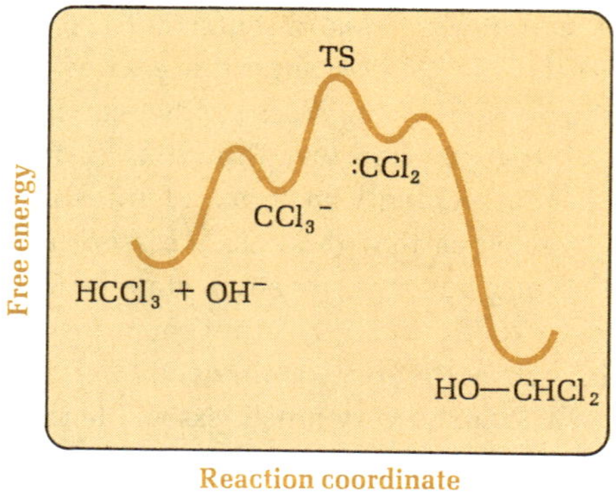

This diagram can be modified in one way and still accord with our conditions. Which way?

Let us leave the halides of carbon for the moment and turn to those of nitrogen, both to see what we can understand about the nitrogen halides themselves and to compare their behavior with those of the corresponding carbon compounds. The fluorides of nitrogen, NF_3, NF_2H, and NFH_2, make a remarkable series in terms of their reactivity. NF_3 is quite a stable compound, nearly inert to hydrolysis in water but susceptible to hydrolysis in acidic media. In marked contrast, NF_2H is extremely reactive with almost any **nucleophilic** (electron-rich) **reagent.** Rather little is known about the reactivity of NFH_2. The situation in this case is clearly related to that encountered just above for the carbon chlorides in which that species, $HCCl_3$, in which just one of the halide atoms has been replaced by hydrogen, exhibits extraordinary reactivity. In fact, the explanation for the reactivity of NF_2H is very similar to that for $HCCl_3$: A special reaction pathway exists. For the former compound, the reaction with hydroxide ion proceeds as follows:

$$:NF_2H + OH^- \longrightarrow :\ddot{N}F_2^- + H_2O \tag{9-18}$$

$$:\ddot{N}F_2^- \longrightarrow :\ddot{\underset{..}{N}}F + F^- \tag{9-19}$$

$$:\ddot{\underset{..}{N}}F + H_2O \xrightarrow[\text{many steps}]{} :NH_3 + F^- \tag{9-20}$$

Here again we have two crucial intermediates. One, the :NF species, is an example of a monovalent nitrogen compound, analogous to the divalent carbon compounds already encountered, the carbenes. The corresponding nitrogen species are termed **nitrenes.** The other, the NF_2^- ion, owes its stability to the electron-withdrawing power of the fluorine atoms, although we ought to recognize that nitrogen does tolerate a negative charge a great deal better than does carbon.

Example *List the following anions in increasing order of stability:* NF_2^-, NCl_2^-, $NHCl^-$, CCl_2H^-, CF_3^-, *and* CH_3^-.

Solution *There are two matters that concern us here: first, the tolerance of the central atom for negative charge and, second, the tolerance of the substituents for negative charge. These are measured approximately by the electronegativities of the elements (Table 4-5). Nitrogen, being more electronegative than carbon, tolerates a negative charge much better than carbon. Thus, we might expect the three carbon anions to be the least stable in this list. (Incidentally, it is* not *true that all nitrogen anions are more stable than all carbon anions.) Among these,* CH_3^- *is the least stable, having no electronegative substituents, and* CF_3^- *is the most stable, having three;* CCl_2H^- *is intermediate. The same consideration indicates that* $NHCl^-$ *is less stable than* NCl_2^-*, which is less stable than* NF_2^-*, fluorine being more electronegative than chlorine. Thus, the entire list reads:*

$CH_3^- < CCl_2H^- < CF_3^- < NHCl^- < NCl_2^- < NF_2^-$

What about the relative reactivity of nitrogen halides and carbon halides? In the first place, NF_3 is more reactive than CF_4. This fact can be readily understood on the basis of an associative mechanism. Two factors are important. First, the bond strength of an N—F bond, 65 kcal, is substantially less than that of a C—F bond, 102 kcal. Second, the three fluorine atoms of NF_3 provide far less steric interference to the attack of an anion than do the four fluorine atoms of CF_4. On the other hand, it is difficult to rationalize this behavior in terms of a dissociative mechanism. Thus, NF_3 would form a NF_2^+ ion and CF_4 would generate a CF_3^+ ion in the rate-determining steps. While neither of these species would be very stable, the nitrogen cation should be less stable than the carbon cation. As a second example, NCl_3 is very much more reactive than CCl_4. The difference in reactivity is quite remarkable: CCl_4 is stable indefinitely in water; NCl_3 reacts very rapidly. We can understand something about this behavior just by looking at the products of the reaction of NCl_3. Provided this compound reacts by nucleophilic attack of water or hydroxide ion at the nitrogen atom, in a fashion analogous to that already described for reactions of carbon, we ought to expect to find the following stoichiometry:

$$2H_2O + NCl_3 \longrightarrow 3HCl + HNO_2 \qquad (9\text{-}21)$$

Now, the actual products are observed to be hypochlorous acid and ammonia:

$$3H_2O + NCl_3 \longrightarrow NH_3 + 3HOCl \qquad (9\text{-}22)$$

These products can be understood only by recognizing that the first reaction must have occurred by attack of water at chlorine rather than at nitrogen:

$$\begin{array}{llll} H_2\ddot{O}\!: \rightarrow Cl\text{—}\ddot{N}Cl_2 \longrightarrow & {}^+H_2O\text{—}Cl & + & {}^{-}\!\!:\!\ddot{N}Cl_2 \\ & \quad\downarrow & & \quad\downarrow \text{many steps} \\ & HO\text{—}Cl & & :NH_3 + HOCl \end{array} \qquad (9\text{-}23)$$

Why does CCl_4 not react by an analogous pathway? We can understand this matter by again focusing our attention on an intermediate as a model for the transition state. The key intermediate generated from NCl_3 by attack of water on chlorine is NCl_2^-, while the corresponding species derived from CCl_4 is CCl_3^-. As a consequence of the greater capacity of nitrogen than carbon to tolerate a negative charge, the former intermediate is very much the more stable. Thus, not only is the reactant state of the carbon halide more stable than that for the nitrogen compound (by virtue of the greater bond strength of the former), but also the transition state for the carbon compound is much less stable than that for the nitrogen species. The net effect is a remarkable reactivity difference.

HALIDES OF SECOND-ROW ELEMENTS

Turning now from considerations regarding the reactivity of halides of the first-row elements, let us see what novel factors are introduced in comparing reactivities as we move down a column in the periodic table. To begin with, let us consider the relative reactivities of the halides of carbon and silicon. We have already emphasized that carbon tetrafluoride and tetrachloride are very unreactive toward hydrolysis; in contrast, the corresponding halides of silicon are rapidly and completely hydrolyzed under very mild conditions. These reactions have the stoichiometry

$$SiX_4 + 2H_2O \longrightarrow SiO_2 + 4HX \qquad (9\text{-}24)$$

To what do the halides of silicon owe their considerable reactivity? While part of this reactivity can be accounted for in terms of the lower strength of Si—X bonds compared to C—X bonds, a thorough understanding of this situation requires that we recall the following from our earlier discussions: Elements of the second row of the periodic table have the capacity to expand the outer shell of electrons beyond the usual octet, while those of the first row do not. This simple fact permits us to make a small but crucial distinction between the reaction mechanism for hydrolysis of carbon halides and silicon halides: Silicon compounds are capable of forming a stable adduct with water which may subsequently decompose with loss of halide, while carbon compounds are not. Schematically, we have

$$H_2O + CF_4 \longrightarrow \underset{\text{transition state}}{\left[H_2O\cdots \overset{F\ \ F}{\underset{F}{C}} \cdots F \right]} \longrightarrow H_2\overset{+}{O}{-}CF_3 + F^- \xrightarrow{\text{many steps}} \text{final products} \qquad (9\text{-}25)$$

$$H_2O + SiF_4 \longrightarrow \underset{\text{transition state}}{[H_2O\cdots SiF_4]} \xrightarrow{-H^+} HO{-}SiF_4^- \longrightarrow HO{-}SiF_3 + F^- \xrightarrow{\text{many steps}} \text{final products} \qquad (9\text{-}26)$$

The crucial distinction is the following: The transition state for the reaction of the carbon halide, which is an energy maximum, corresponds structurally to an intermediate, which is an energy minimum, for the reaction of the silicon halide (Figure 9-10). It follows directly that the silicon halide will have the lower energy reaction pathway since, at a time in which the carbon halide system is still laboring uphill toward an energy maximum, the silicon halide system will have passed through its energy maximum and be going downhill toward the intermediate. Looked at in a slightly different way, hydrolysis of the carbon halide involves both bond formation and bond breaking, while that for the silicon halide involves just bond formation in

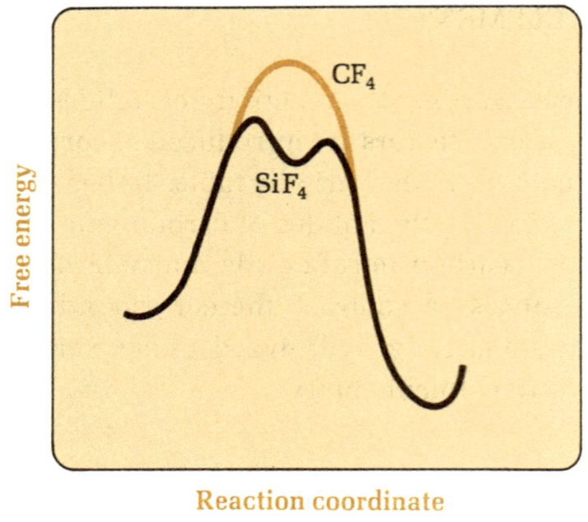

Figure 9-10 Free-energy diagrams for the hydrolysis of CF_4 and SiF_4.

Table 9-1 A summary of several factors that influence chemical reactivity

Factor	Examples
Nature of the reaction pathway	1. Hydrolysis of CF_4 and SiF_4 2. Hydrolysis of CH_2Cl_2 and $CHCl_3$ 3. Hydrolysis of PCl_3 and NCl_3 4. Hydrolysis of CCl_4 and NCl_3
Steric effects	1. Hydrogenation of *cis*- and *trans*-2-butene 2. Hydrolysis of CH_2Cl_2 and CH_3Cl 3. Hydrolysis of CF_4 and NF_3
Bond strengths	1. Hydrolysis of CH_3F and CH_3Cl 2. Hydrolysis of CCl_4 and $SiCl_4$ 3. Hydrolysis of CF_4 and NF_3
Delocalization of charge	1. Hydrolysis of $CHCl_3$ and CH_2Cl_2 2. Hydrolysis of NHF_2 and NH_2F
Expansion of octet of electrons (coupled with change in reaction pathway)	1. Hydrolysis of CF_4 and SiF_4 2. Hydrolysis of PCl_3 and NCl_3
Accommodation ability (polarizability)	1. Hydrolysis of CH_3Br and CH_3I 2. Isotopic exchanges of CH_3F and CH_3Cl

the transition state. Thus, the silicon system avoids the necessity of finding the energy needed to cleave partially a covalent bond in the transition state.

As a final example of those factors that influence reactivity, let us consider the relative reactivities of the halides of nitrogen and phosphorus, in particular NCl_3 and PCl_3. We already know that NCl_3 reacts very readily with water; PCl_3 reacts positively violently. The stoichiometry of the hydrolysis of PCl_3 is

$$PCl_3 + 3H_2O \longrightarrow H_3PO_3 + 3HCl \quad (9\text{-}27)$$

We can immediately realize that the reactions of PCl_3 and NCl_3 are different in one crucial respect: The products derived from the former must be derived via pathway involving the attack of water on the phosphorus atom, while we already know that water attacks NCl_3 on chlorine. Now, the latter owes its high reactivity to the fact that attack on chlorine does occur and generates the relatively stable anion NCl_2^-. Clearly, the high reactivity of PCl_3 must derive from different sources. In fact, PCl_3 hydrolyzes by a route related to that for the silicon halides and involves a stable tetracovalent intermediate:

$$PCl_3 + H_2O \longrightarrow \underset{\text{transition state}}{[H_2O\cdots PCl_3]} \xrightarrow{-H^+} HO\text{—}PCl_3^- \longrightarrow HO\text{—}PCl_2 + Cl^- \xrightarrow{\text{many steps}} H_3PO_3 + 2Cl^- \quad (9\text{-}28)$$

Hence, the same comments directed toward the explanation of the reactivity of the silicon halides also apply to this case.

In the preceding few pages, we have identified several reasons why reactivities of related compounds toward the same reagent may differ very substantially. These are summarized, together with appropriate examples, in Table 9-1. Let us remember that all of these factors serve to alter the relative stabilities of reactant and transition states, which is, after all, the factor that matters most in determining chemical reactivities.

KINETICS AND MECHANISM

The preceding discussion immediately has established that kinetics and mechanism for chemical reactions are intimately associated. In particular, we have recognized that the mechanism of the reaction determines the kinetics of the reaction. At the same time, we have pointed out that kinetic behavior may provide clues to mechanism: The abnormal reactivity of chloroform in comparison to other chlorides of methane suggests that chloroform reacts with this nucleophile by an independent mechanism. It is this latter aspect that we wish to develop further at this time. Explicitly, what can we learn from reaction kinetics about the mechanism of a particular reaction?

At the outset, we ought to recognize two points. In the first place, there are a number of means of investigating mechanisms for chemical reactions.

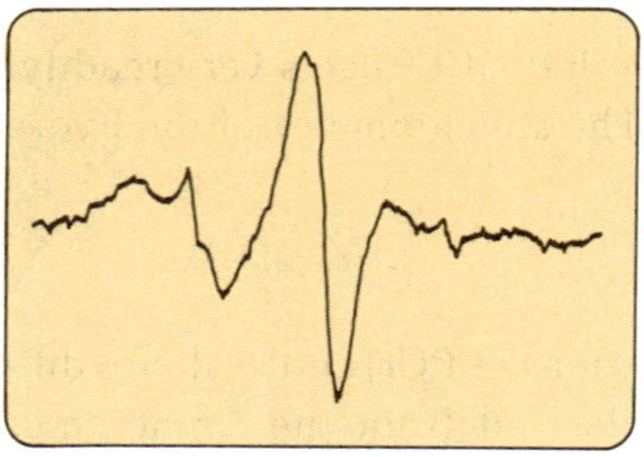

Electron spin resonance spectrum typical of a free radical partially immobilized through attachment to a large molecule such as the antibody.

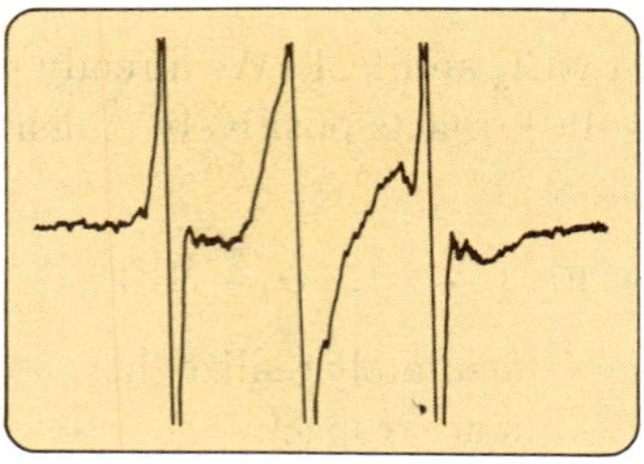

Electron spin resonance spectrum typical of a free radical tumbling rapidly in solution.

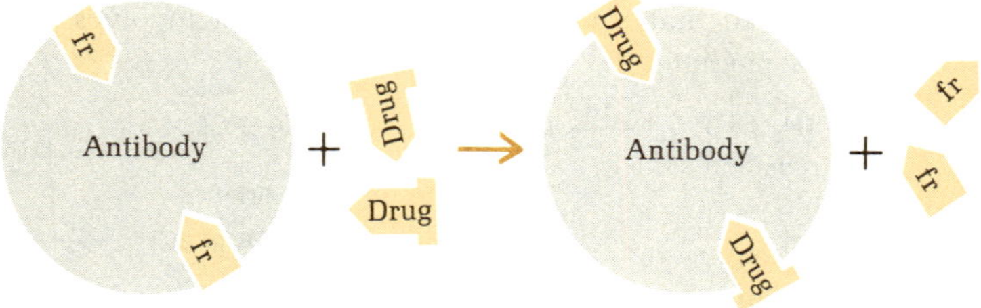

Schematic view of the displacement of free radicals associated with antibody molecules by molecules of a drug having a related size and shape.

Studies of the kinetic behavior of molecules can be useful in a great many connections. A recent and unusual use of kinetics is for the detection of heroin users among the population. This somewhat complex method works in the following way. Molecules that have an unpaired electron ("free radicals") possess a permanent magnetic field that gives rise to an "electron spin resonance spectrum." The way this spectrum looks depends on how fast the free radical tumbles in solution: The sharp three-line spectrum shown on the right of the upper part of this figure is characteristic of a free radical tumbling freely in solution; the more diffuse spectrum shown next to it results when the free radical becomes associated with a large protein molecule that rotates and tumbles much more slowly than the small molecule itself. Clearly, we can readily distinguish between the free radical bound to a large molecule and one that is not. In order to make this observation useful for detecting heroin use, it is necessary to prepare antibodies, a class of large protein molecules, that have the capacity to react specifically with morphine-type drugs (heroin is one example). When an analog of morphine that contains an attached free radical is added to such antibodies, it attaches to them and the resulting spectrum is broad and diffuse. If a few drops of urine from a person using heroin are now added, the heroin metabolities displace the morphine free radical from the antibodies through competition for available binding sites, as shown in the lower part of the figure. This results in a sharpening of the free radical, which is readily detected by a person using a relatively sophisticated type of equipment. The whole procedure requires just a few seconds and thus is useful for large-scale screening of possible heroin users, such as troops returning from Indochina.

These include product identification, the use of isotopic tracers, the identification of intermediates by spectroscopic means, reaction kinetics, and so on. Thus, kinetics is one tool in the chemist's bag for working on these problems but not the only one. In the second place, kinetic studies by themselves never suffice to establish a mechanism. We can only say that a certain mechanism is or is not consistent with the experimental facts derived from kinetics. Thus, careful work demands the use of ancillary techniques in addition to kinetic studies to investigate reaction mechanisms. Nevertheless, the study of reaction kinetics can frequently be of enormous use to the chemist.

In general, the rate of a chemical reaction depends on two things: the *rate constant,* which is determined by the free-energy difference between reactant and transition states, and the *concentrations* of the reactants. The description of reaction rate in terms of a rate constant and concentrations is called the **rate law** for the reaction. To begin with, let us consider rate laws for the simplest type of chemical reaction, the **elementary process,** defined as a reaction that cannot be subdivided into simpler processes. This is equivalent to saying that the free-energy diagram for an elementary process has no energy minima save those for reactants and products, that is, that no intermediates occur along the reaction pathway. The simplest type of elementary process is the decomposition of a single molecule:

$$\mathrm{A} \xrightarrow{k} \mathrm{P} \tag{9-29}$$

It is intuitively clear that the rate of this reaction, expressed in terms of moles of A consumed or moles of P produced per unit time, will be greatest when the concentration of A is the greatest. Thus, if we measure the concentrations of A and P as a function of time, we shall observe the curves shown in Figure 9-11. The rate of change of the concentrations of A and P decrease as the concentration of A decreases. Quantitatively, we can express this fact in the following way:

$$v = -\frac{d\mathrm{A}}{dt} = \frac{d\mathrm{P}}{dt} = k(\mathrm{A}) \tag{9-30}$$

Note that the velocity of the reaction, v, may be expressed as the rate of change of the concentration of A as a function of time, $d\mathrm{A}/dt$, or as the rate of change of the concentration of P with respect to time, $d\mathrm{P}/dt$. These must be equal in magnitude and opposite in sign, since each time we destroy one molecule of A, one molecule of P must be generated. The important point about the rate law for any elementary process is that there is a direct correspondence between the stoichiometry for the process and the concentration dependence of the rate. Thus, for the reaction of Equation (9-29), one molecule of A is involved in the reaction and the rate depends on the concentration of A to the first power. Such reactions are said to be *first-*

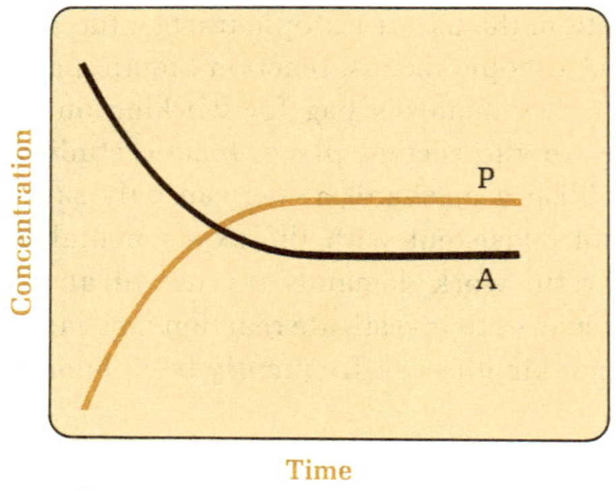

Figure 9-11 The time dependence of the concentrations of A and P for an elementary process A ⟶ P.

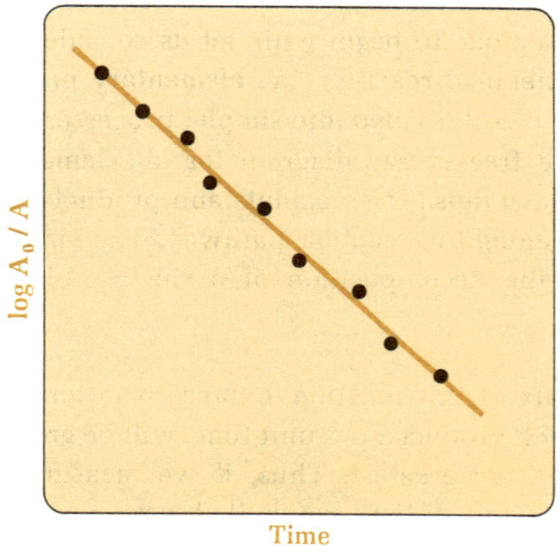

Figure 9-12 A typical plot for a chemical reaction showing first-order kinetics.

order. To understand the point more fully, let us consider the following two examples.

Case 1. A reaction *second-order* in A.

reaction: $2A \xrightarrow{k} P$

rate law: $v = k(A)^2$

Case 2. A *second-order* reaction, *first-order* in A and *first-order* in B.

reaction: $A + B \xrightarrow{k} P$

rate law: $v = k(A)(B)$

We can conclude: For any elementary process, the rate of the reaction is given by the product of the rate constant and the concentrations of all concerned species raised to the power equal to the number of molecules that actually participate in the process. The sum of all exponents on the concentration terms in the rate law, determines the **order** of the reaction.

The form of a rate law for a reaction is determined experimentally from the dependence of concentrations on time. Thus, we may write Equation (9-30) in an alternative form that expresses concentrations as a function of time rather than velocities as a function of concentrations:

$$A = A_0 e^{-kt} \quad \text{or} \quad \log A_0/A = kt \tag{9-31}$$

in which A_0 is the initial concentration of A. The form of Equation 9-31 suggests an experimental test of whether a given reaction is first-order in A or not: If it is, then a plot of log A_0/A against time must be linear (see Figure 9-12). If it is, then the reaction has been determined to be first-order in A, and the rate constant is given by the slope of the straight line. If not, then we must search for agreement between the data and predictions based on another rate law. At any event, the point here is not how to establish rate laws for chemical reactions, but the fact that we can establish them and what we can learn from them once they are known.

The central point of all kinetic considerations is this: *The composition of the transition state for a chemical reaction is given directly by the concentration terms in the rate law.* Thus, once we have established the rate law, we have established the composition of the transition state as well. This is obviously so from our statements about the rate laws for elementary processes. Since no intermediates occur in elementary processes, the composition of the transition state must correspond to the stoichiometry of the reaction which, in turn, provides the rate law. Quite aside from the fact that we usually do not know in advance if a particular reaction is an elementary process or not, the situation is generally not so simple. Let us consider some specific examples.

Returning to the hydrolysis of the halides of carbon for a moment, it is true that the rate law for the reaction of chloromethane with hydroxide ion has the form

$$v = k(CH_3Cl)(OH^-) \tag{9-32}$$

That is, the transition state must contain the atoms of chloromethane and hydroxide ion. This is consistent with the mechanism for this reaction depicted in Equation (9-11) but not with that of Equations (9-8) and (9-9). Were the dissociative mechanism in fact the correct one, then the rate law would have had to be

$$v = k(CH_3Cl)$$

since the hydroxide ion is not present in the transition state. The reaction of Equation (9-6), in which the first step is rate-determining (see Figure 9-7), has just such a rate law. Thus, kinetic studies suffice to distinguish between associative and dissociative mechanisms for halide hydrolysis to the extent of being inconsistent with one of the possibilities.

The limitation of kinetic studies in terms of establishing reaction mechanisms is that only the composition if the transition state is established, not the *arrangement* of the atoms within the transition state. To see this point clearly and understand the sort of difficulties that can arise, let us consider the hydrolysis of chloroform according to the pathway of Equations (9-15) through (9-17), in which the second step is rate-determining; that is, the transition state for the overall process is associated with the second in the series of three elementary processes. We can then write the rate law as

$$v = k(CCl_3^-) \tag{9-33}$$

While this expression is certainly correct, it is not very useful, since we cannot readily determine the concentration of CCl_3^- in solution. However, we can recognize that this species is in equilibrium with chloroform and hydroxide ion according to Equation (9-15), and we can therefore write

$$K_{eq} = \frac{(CCl_3^-)}{(HCCl_3)(OH^-)} \quad \text{or} \quad (CCl_3^-) = K_{eq}(HCCl_3)(OH^-) \tag{9-34}$$

(note that the concentration of water has been absorbed into the equilibrium constant). We can now rewrite Equation (9-33) in the form

$$v = kK_{eq}(HCCl_3)(OH^-) = k'(HCCl_3)(OH^-)$$

But the form of this rate law is just the same as that for the hydrolysis of chloromethane [Equation (9-32)], although the mechanisms are quite distinct. Thus, kinetic studies may eliminate some mechanisms definitively but can never establish mechanisms definitively. Mechanisms that have the same collections of atoms in the transition state (plus or minus molecules of solvent whose involvement in the reaction cannot be detected experimentally) are said to be *kinetically indistinguishable.*

Example *The simple molecule urea,* $H_2N{-}\overset{\overset{\displaystyle O}{\|}}{C}{-}NH_2$, *is the end product of nitrogen metabolism in man. Studies of the synthesis of this compound have been carried out a number of times in different laboratories, and two mechanisms have been suggested:*

$$NH_4^+ + NCO^- \longrightarrow H_2N{-}CO{-}NH_2$$
$$NH_3 + HNCO \longrightarrow H_2N{-}CO{-}NH_2$$

In the first reaction, synthesis is viewed as resulting from the reaction between ammonium ion and cyanate ion; in the second, ammonia and cyanic acid are considered to be the reactants. Explain clearly why kinetic experiments alone will never suffice to distinguish between these two possibilities.

Solution *Kinetics, the determination of the rate law for the reaction, suffices to determine only the composition of the transition state. Note that the total composition of the two sets of reactants are identical. The only difference is in the position of a single proton. But kinetics never tells us geometry for transition states, only compositions. If we want to confirm this qualitative argument, we can do so by writing down the rate expression for the two reactions and proving that these are equivalent in terms of the acid-base equilibria involved. We have two rate laws:*

$$v = k(NH_4^+)(NCO^-)$$
$$v = k'(NH_3)(HNCO)$$

and two equilibria:

$$NH_4^+ \rightleftharpoons NH_3 + H^+ \qquad K_1 = \frac{(NH_3)(H^+)}{(NH_4^+)}$$

$$HNCO \rightleftharpoons NCO^- + H^+ \qquad K_2 = \frac{(NCO^-)(H^+)}{(HNCO)}$$

Substituting the two equilibrium expressions in the first of the rate laws, we obtain

$$v = k\frac{(NH_3)(H^+)}{K_1}\frac{K_2(HNCO)}{(H^+)} = k\frac{K_2}{K_1}(NH_3)(HNCO)$$

But the resulting expression is identical to that for the second rate law in which $k' = k(K_2/K_1)$. Since the two rate laws have exactly the same form, it follows that the two mechanisms are kinetically indistinguishable.

Glossary

Carbenes divalent carbon compounds.
Carbonium ion a positively charged ion in which the charge is localized largely or completely on carbon.
Elementary process a chemical reaction that occurs in a single step without the formation of intermediates.
Free-energy diagram a plot of free-energy content of a reaction system against the reaction coordinate, a measure of the geometry of the system.
Free energy of activation the difference in free energy between the reactant state and transition state for a chemical reaction; this quantity largely determines the rate constant for the reaction.
Intermediate a molecular geometry corresponding to a free-energy minimum between those of reactants and products in a free-energy diagram.
Nitrenes monovalent nitrogen compounds.
Nucleophilic reagent (nucleophile) any reagent having an unshared pair of electrons that can be used to attack an electron-deficient site. Examples include all anions, plus amines, water, alcohols, and the like.

Polarizability a measure of the deformability of the electron cloud of an atom or molecule under the influence of electrical forces.

Rate constant a quantity linking the rate of a chemical reaction to the concentrations of the reacting species. The rate constant is largely determined by the free energy of activation for the reaction.

Rate law a quantitative expression in which the dependence of the rate of a chemical reaction on a rate constant and the concentrations of reacting species is provided.

Reaction mechanism the pathway of a chemical reaction, usually taken to include the specification of the structure of all intermediates that occur as well as specification of the structure of the transition state.

Reaction order the sum of all exponents of concentration terms that occur in the rate law for a chemical reaction.

Transition state the molecular geometry that corresponds to the point of maximum free energy in a free-energy diagram.

Study problems

1. Sketch a graph to show energy versus reaction coordinate for an *endergonic* reaction that proceeds through *two* intermediates. Label the graph to show ΔG^0 and appropriate $\Delta G^{\ddagger}$s.
2. A free-energy diagram for a chemical reaction is reproduced below. From an examination of this diagram, provide answers together with brief justifications to the following questions.

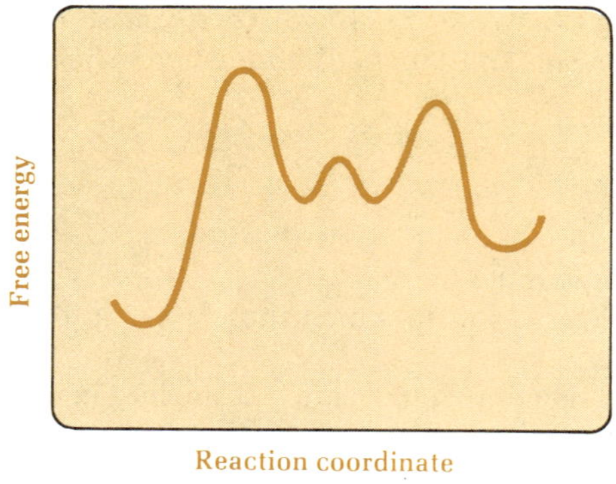

   a. Is the overall reaction exergonic or endergonic?
   b. Is the formation of the first intermediate from the reactants exergonic or endergonic?
   c. How many intermediates occur in the overall reaction process?
   d. What free-energy change determines the overall rate of the reaction? (Indicate it on the diagram.)
   e. Is the overall rate of the reaction associated with any single rate constant?
   f. If the answer to (e) is yes, which one? If no, why not?
3. The comparative free-energy diagrams for hydrolysis of CCl_4 and PCl_3 are shown on the next page.

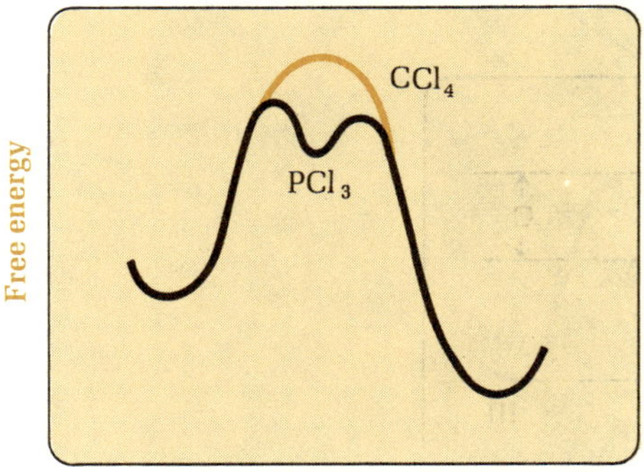

a. Label the transition state for each reaction.
b. Which hydrolysis will occur more rapidly?
c. Explain what feature of electronic structure is responsible for the difference between the two curves.
d. The hydrolysis of PCl_3 is second-order. What does this statement mean?

4. Consider a chemical reaction of the type

$$R{-}Cl + OH^- \longrightarrow R{-}OH + Cl^-$$

in which R is some organic (carbon-containing) group. Suppose that we measure the rate and extent of disappearance of R—Cl for this reaction and for two similar reactions in which hydroxide ion is replaced by methoxide ion (CH_3O^-) and bromide ion (Br^-) with the following results:

a. The reactions with hydroxide ion and methoxide ion proceed at the same rate and essentially to completion.
b. The reaction with bromide ion is slower than those with hydroxide or methoxide ion and does not proceed to completion.

Rationalize these results in terms of some reasonable reaction pathway. Construct free-energy diagrams for each reaction, showing how the differences between them account for the observations.

5. The kinetics of the reaction

$$F_2 + 2ClO_2 \longrightarrow 2FClO_2$$

has been studied in the gaseous state. The data at 250°K may be summarized as follows:

Rate of disappearance of F_2 (moles/liter sec)	Reactant concentrations (moles/liter) F_2	ClO_2
0.6×10^{-3}	0.05	0.010
2.4×10^{-3}	0.05	0.040
1.2×10^{-3}	0.10	0.010

Formulate the rate law for this reaction.

6. Interpret the following free-energy diagram as fully as possible, including specification of the significance of each of the indicated free-energy differences.

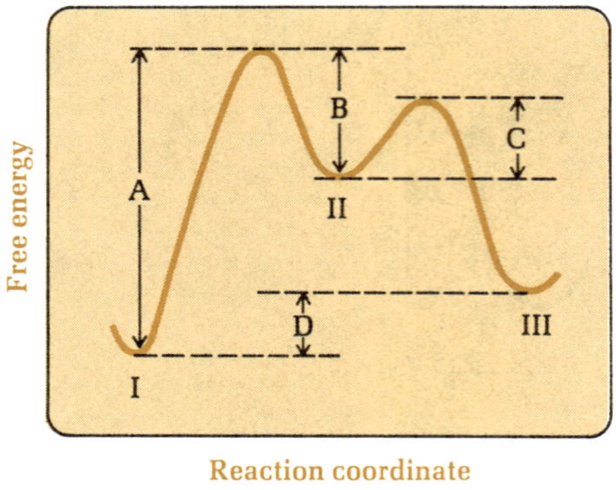

7. The hydrolysis of OCl_2 yields only HOCl as reaction product. In contrast, the hydrolysis of OF_2 yields molecular oxygen plus fluoride ions. From this information, which of the following follows?
   a. Hydrolysis of OCl_2 proceeds with attack on chlorine, not oxygen.
   b. OF_2 is more reactive than OCl_2 toward hydrolysis.
   c. Hydrolysis of OF_2 proceeds with attack on oxygen, not fluorine.
   d. OF_2 reacts by a dissociative mechanism.
   e. Both statements (a) and (c) are true.
   f. Both statements (c) and (d) are true.
   g. Statements (a), (c), and (d) are true.
   h. All of these statements are true.
   i. None of these statements is true.

8. NF_3 is more reactive toward alkaline hydrolysis than is CF_4. This principally reflects the fact that
   a. NF_3 reacts by a dissociative mechanism but CF_4 reacts by an associative one
   b. the N—F bond is weaker than the C—F bond
   c. CF_3^+ is more stable than NF_2^+
   d. the attack of hydroxide ion on NF_3 is less sterically hindered than the attack on CF_4
   e. both (a) and (b)
   f. both (b) and (c)
   g. both (b) and (d)
   h. all of the above

9. Which of the following statements are true? (1) $:CF_2$ is an example of a carbene; (2) NCl_2^- is more stable than CCl_3^-; (3) hydrolysis of NCl_3 occurs with attack of water on nitrogen, not on chlorine.

a. none	e. (1) and (2)
b. just (1)	f. (1) and (3)
c. just (2)	g. (2) and (3)
d. just (3)	h. all

Suggestions for further reading

J. A. Campbell *Why Do Chemical Reactions Occur?*, Prentice-Hall, Englewood Cliffs, N.J., 1965.

J. B. Dence, H. B. Gray, and G. S. Hammond *Chemical Dynamics*, Benjamin, Menlo Park, Calif., 1968.

J. O. Edwards, E. F. Greene, and J. Ross "From Stoichiometry and Rate Law to Mechanism," *J. Chem. Educ.*, **45**, 381 (1968).

H. Eyring and E. M. Eyring *Modern Chemical Kinetics*, Van Nostrand Reinhold, New York, 1963.

G. M. Harris *Chemical Kinetics*, Heath, Lexington, Mass., 1966.

Part two

WATER

Water is one of the most abundant and familiar substances on earth. It is also one of the most remarkable. Not only is the chemistry of water itself surprisingly complex and instructive, but the chemistry of those reactions that occur in water as solvent are of great importance in both living and nonliving systems. In this section of our investigation into chemistry, these matters are probed in modest detail. In Chapter 10 the chemical properties of pure water are developed, and in Chapters 11 and 12 we consider many of the reactions that characteristically take place in this solvent. Any discussion of water would be incomplete without consideration of one of our great natural resources—the oceans. As developed below, recovery of fresh water from the oceans is assuming ever greater importance for many countries, including our own. Thus, the final chapter of this part is devoted to describing the chemistry of desalination. Prior to entering directly into chemical matters, let us pause just long enough to begin to understand just how remarkably important the oceans and their resources are.

With each passing year, man is technologically utilizing, for a variety of purposes, the oceans to a greater extent. Everyone recognizes that the oceans are a valuable source of food and that recovery of oil and natural gas from the floor of the ocean are economically important processes. Utilization of other of the oceans' capacities is less well recognized, but it seems certain that ever-increasing advantage will be taken of the enormous stores of elements and minerals that are present there. Of all of the things we might extract from the ocean, the most abundant is certainly the most important. That most abundant resource is the water itself.

The ocean may provide for water requirements of two types. The first is the necessity for increasing quantities of fresh water for a variety of purposes, and the second is the utilization of the oceans' waters for industrial processes that do not require fresh water. Looking at the former point, the requirements of the United States in 1970 for fresh water are summarized in the following table.

Data for the utilization of fresh water in the United States in 1970

Major industries	1.9×10^{11} gallons/day
Irrigation	1.8×10^{11} gallons/day
Other uses (household)	0.3×10^{11} gallons/day
Total	4.0×10^{11} gallons/day

1970 population of the United States: 200,000,000
Per person consumption: 2,000 gallons/day

As detailed in this table, the utilization of fresh water for industrial, irrigating, and household purposes consumed 400 billion gallons of fresh water per day. By dividing that quantity by the population of the United States in 1970, we can calculate that each man, woman, and child utilized approximately 2,000 gallons of water per day. This corresponds to the use of about 16,000 lb of water per day per person at that time. Clearly, life as it is now conducted in this country requires phenomenal quantities of fresh water each day. There are currently about 700 desalination plants operating in the world, each of which has a capacity to generate 25,000 gallons or more per day of fresh water from sea water. One coastal community in the United States generates its entire water supply, some 2.6 million gallons of water per day, from desalination of the sea. Inasmuch as the oceans are by far the principal water reservoir on earth, and in view of the continually rising demand for supplies of fresh water, it seems certain that desalination processes will gain increasing importance in coming years.

Turning now to industrial utilization of ocean water without desalination, we note that currently about 20 million gallons of ocean water per minute are employed industrially. This vast volume of water serves three principal purposes. In the first place, it serves as a coolant; this is particularly important for those power plants that derive their power from nuclear reactors. Such reactors generate immense quantities of heat, which is ordinarily dissipated by the circulation of water through the plant. The resulting rise in temperature of the water used as coolant has some effects, but is probably not terribly important for those plants located in coastal areas which can utilize ocean water for this purpose; however, it is a serious problem when smaller bodies of water are employed. For example, the water temperature of a small river that was almost completely diverted to function as a nuclear power plant coolant recently rose to 140°F. Such thermal pollution of rivers, of course, destroys the ecological balance. In the second place, the waters of the ocean are very largely employed as a waste diluent. That is, the waters are mixed with the by-products of industrial activity and then returned to the ocean in a polluted form. Such pollution probably is more dangerous to the oceans at the moment than the thermal pollution noted above. Even considering the vast size of the oceans, it is already clear that widespread contamination is not only possible but has already occurred. Finally, about 1 million gallons of ocean water per minute are processed worldwide for the recovery of minerals. We shall have a bit more to say about this later.

These few comments about water and the oceans certainly suffice to establish their importance. We return to the chemistry of the ocean and to the extraction of fresh water therefrom later. But we begin our considerations with the chemistry of water itself.

10

The properties of water

On the chemistry of water:
And the children of Israel went into the midst of the sea upon the dry ground: and the waters were a wall unto them on their right hand, and on their left.

Exod. 14:22

In some respects the chemistry of liquids is more complicated than that of either solids or gases. Under ordinary conditions most of the former are highly organized substances and most of the latter are highly disorganized. Clearly, liquids occupy a position intermediate between the two, but themselves exhibit a remarkable diversity of behavior: This reflects the strength of mutual interaction of the atoms or molecules forming the liquid phase. If these are strong, the liquid may possess a surprising degree of organization; if they are weak, the liquid may be almost gaslike.

Liquid water is a fluid possessing a remarkable collection of physical and chemical properties. These suit it ideally for the functions it performs as one of the basic entities supporting living organisms. To understand the properties of liquid water and why they are different from those of most other liquids, let us begin by discussing some of these properties in comparison with those of corresponding hydrides of elements near oxygen in the periodic table.

MELTING POINTS, BOILING POINTS, AND HEATS OF VAPORIZATION

To begin to understand some of the curious properties of water, consider the data collected in Table 10-1, in which melting points, boiling points, and heats and entropies of vaporization of the hydrides of several of the first-row elements are collected. Since each molecule contains just ten electrons, these hydrides (and neon) are said to form an *isoelectronic series*. They all possess basically the same electronic structure, and therefore we expect the series of compounds to have related physical and chemical properties (see Chapters 2, 3, and 6).

Example *To review the concept of isoelectronic molecules, decide which of the following pairs of compounds are isoelectronic:*

a. *methane and ammonia*
b. *carbon monoxide and nitrogen*
c. *benzene and borazine,*

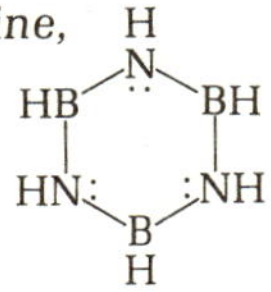

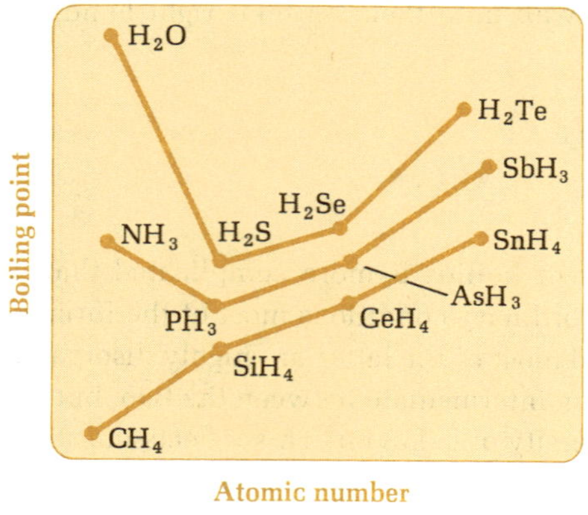

Figure 10-1 Boiling points for three isoelectronic series of hydrides as a function of atomic number of the element from which the hydrides are derived.

Table 10-1 Melting points, boiling points, and heats and entropies of vaporization for an isoelectronic series of first-row hydrides

Substance	Melting point (°C)	Boiling point (°C)	ΔH_v (cal/mole)	ΔS_v (cal/mole deg)
CH_4	−184	−161	2200	19.7
NH_3	−78	−33	5500	23.1
OH_2	0	100	9750	26.1
FH	−92	19	7220	24.7
Ne	−249	−246	415	15.4

Solution *As may be demonstrated either by drawing Lewis electron dot structures or by using molecular orbital theory, each of these pairs of molecules is isoelectronic. If you have any trouble appreciating this fact, refer back to Chapter 5 for appropriate review.*

The data in Table 10-1 clearly indicate that this expectation is not realized; quite the contrary, the melting points, boiling points, and heats and entropies of vaporization all pass through maxima with water. The total variation in these properties in this isoelectronic series is quite striking when we realize that the molecular weights and molecular sizes are quite similar. If, for the moment, we view methane and neon as "normal" members of this series, then ammonia, water, and hydrogen fluoride appear to have melting points, boiling points, and heats and entropies of vaporization much greater than would be expected. To see this point somewhat more clearly, refer to Figure 10-1, in which the boiling points of a series of hydrides for various columns in the periodic table are plotted against atomic number. Note that one series of these hydrides—methane, silane, germane, and stannane—exhibit boiling points that change uniformly with the molecular weight of the molecules. Thus, as we expect, the lightest member of the series, methane, has the lowest boiling point, and the heaviest member of the series, stannane, has the highest boiling point. For the other two series of elements, the same trend is evident with the exceptions of water and ammonia. Thus, for example, stibine has a higher boiling point than arsine, which has a higher boiling point than phosphine, which, however, has a boiling point much lower than that of ammonia. The same comments apply to water and the hydrides of the elements directly below oxygen in the periodic table. Clearly, our assumption that the behaviors of water and ammonia are abnormal is justified by the data in this figure.

Taken as a whole, the data in Table 10-1 and Figure 10-1 strongly suggest that substantial intermolecular forces exist between independent molecules in the liquid and solid states in the cases of water, ammonia, and hydrogen fluoride. To the extent that intermolecular forces are present in the solid state but are disrupted in the liquid state, such forces will tend to raise the melting point of the material. This follows directly from the fact that more kinetic energy must be supplied to the molecules to overcome the intermolecular forces that stabilize the solid state. Thus, water, ammonia, and hydrogen fluoride seem to possess stronger intermolecular interactions in the solid state than the comparison compounds. Similarly, the existence of such interactions in the liquid state, which are lost in the gas phase, will tend to raise the boiling point for these substances. Again, note the elevated values for water, ammonia, and hydrogen fluoride compared to those for methane and neon. The **heat of vaporization,** which measures the amount of heat energy that must be supplied to convert one mole of the substance of interest from the liquid to the gaseous state is also a measure of the intermolecular interactions that must be overcome in going from the liquid to the gaseous phase.

Entropies of vaporization for the hydrides of concern to us here are listed in the last column in Table 10-1. These values measure the difference between the entropies of the gases and the corresponding liquids:

$$S_{\text{gas}} - S_{\text{liquid}} = \Delta S_v \qquad (10\text{-}1)$$

A little reflection concerning the meaning of entropies (see Chapter 8) will reveal that these values also tell us something about the intermolecular interactions in the liquids. Two points are important. First, strong intermolecular interactions should lead to significant structural organization in the liquids. Values of entropy are a measure of such organization. Second, the entropy changes will largely measure the degree of organization in the liquids, since little organization exists for most gases (gaseous HF is some-

Table 10-2 Typical heats of fusion, heats of vaporization, and specific heats

Substance	ΔH_f (cal/g)	ΔH_v (cal/g)	Specific heat (cal/g deg)
Water	79.7	540	1.0
Ammonia	108.1	—	1.23
Methanol (CH_3OH)	16	289.2	0.6
Benzene	30	94	0.5
Acetone ($CH_3{-}\overset{O}{\overset{\|}{C}}{-}CH_3$)	23	125	—
Acetic acid ($CH_3{-}\overset{O}{\overset{\|}{C}}{-}OH$)	45	97	—

thing of an exception to this generalization). It follows that entropies of vaporization will reflect principally the degree of structural organization in the liquids, which in turn reflects the strength of intermolecular interactions. In accord with our previous conclusions, values of ΔS_v are greatest for water and smallest for methane and neon.

How are values of ΔS_v obtained? Calculation of entropy changes is frequently quite complicated, but in this case no problem is presented. The vaporization of a liquid meets two criteria. First, this change is *reversible* since the liquid and vapor are in equilibrium throughout the process. Second, the transformation is *isothermal;* that is, it occurs at the same temperature throughout. For reversible isothermal processes, the entropy change is calculated by dividing the enthalpy change by the absolute temperature at which the process occurs:

$$\Delta S_v = \frac{\Delta H_v}{T_b} \tag{10-2}$$

The values listed in Table 10-1 have been calculated using this equation.

Example *The heat of fusion, ΔH_f, for the melting of ice is given in Table 10-2, as 79.7 cal/g. What is the entropy of fusion for this process?*

Solution *Like the vaporization of a liquid, the melting of a solid is a reversible isothermal process. Hence, we may employ an equation analogous to (10-2) to make the required calculation:*

$$\Delta S_f = \frac{\Delta H_f}{T_m}$$

To begin with, we need the heat of fusion in the proper units, calories per mole. Since the molecular weight of water is 18, it follows that $\Delta H_f = 79.7$ cal/g $\times$ 18 g/mole = 1435 cal/mole/273.2 deg = 5.25 cal/mole deg.

Finally, let us note that the boiling point of a substance reflects both the entropy and heat of vaporization. Remembering the definition of free-energy changes, $\Delta G = \Delta H - T\,\Delta S$, and recognizing that $\Delta G = 0$ at equilibrium, it follows that $T_b = \Delta H_v/\Delta S_v$. These comments apply with equal force to the process of melting (fusion).

Taken together, these facts establish that the interactions of water molecules with other water molecules, of ammonia molecules with other ammonia molecules, and of hydrogen fluoride molecules with other hydrogen fluoride molecules are very much stronger than the comparable interactions between methane molecules and between neon atoms. We now wish to give consideration to the nature and strength of these interactions.

THE ELECTRONIC STRUCTURE OF WATER

A schematic representation of the water molecule is

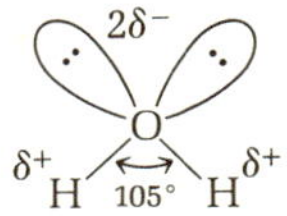

As discussed in Chapter 3, we view the two unshared pairs of electrons of the water molecule as substituents and the geometry around the central oxygen atom as tetrahedral, although not completely regular. The angle formed by the two atoms of hydrogen and the oxygen atom is a little less than that for a regular tetrahedron, while the angle between the two unshared pairs of electrons and the central oxygen atom is a little larger than that for a regular tetrahedral molecule. Of particular interest is the observation that the water molecule has substantial polar characteristics. The oxygen atom is quite electronegative (Chapter 4). That is, it has a strong affinity for electrons. Consequently, the oxygen atom tends to deplete the hydrogen atoms of electron density. Viewed in another way, the bonds that link the oxygen to the hydrogen atoms are polarized in such a way that the majority of the electron density resides near the central oxygen atom. Thus, the oxygen atom is negatively charged with respect to the hydrogens and conversely; this is the meaning of the symbols $\delta-$ and $\delta+$ in the above drawing. The ammonia and hydrogen fluoride molecules may be similarly described. Nitrogen and fluorine, like oxygen, are strongly electronegative elements and consequently form polar hydrides. Such molecules have the interesting capacity to form a special type of chemical bond called a hydrogen bond, the basic aspects of which have been described in Chapter 7.

As noted earlier, hydrogen bonds are relatively weak chemical bonds formed from the sharing of a single hydrogen atom between two electro-

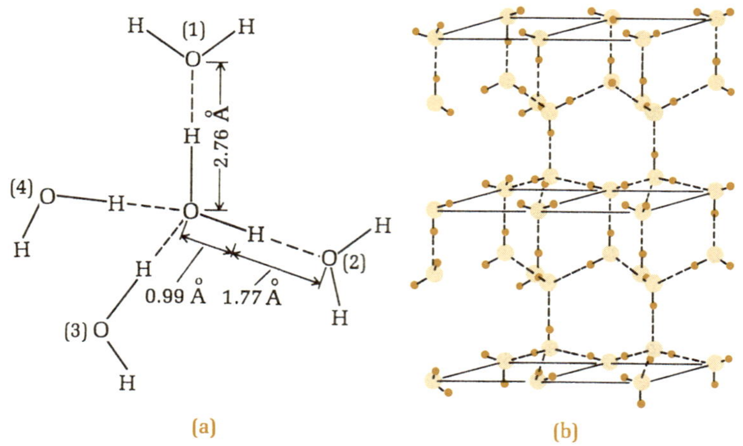

Figure 10-2 Tetrahedral coordination of water molecules in ice. (a) Molecules (1) and (2) as well as the central molecule lie entirely in the plane of the paper. Molecule (3) lies above this plane and molecule (4) lies below it, so that oxygens (1), (2), (3), and (4) lie at the corners of a regular tetrahedron. (b) The arrangement of water molecules in ice, showing a larger part of the crystal.

negative elements (pp. 212–213). Of the hydrogen bonds that exist in ammonia, water, and hydrogen fluoride, those in ammonia are the weakest and those in hydrogen fluoride are the strongest, although the difference is not a marked one:

N—H---N < O—H---O < F—H---F

What is more important in deducing the physical properties of these compounds than the strength of the hydrogen bonds formed is the ability of the molecules to form extended three-dimensional networks of hydrogen bonded structures. Note that ammonia, by virtue of having only a single unshared pair of electrons, and hydrogen fluoride, by virtue of having only a single proton, have limited abilities to form large arrays. Principally, such molecules form hydrogen bonded structures that are either rings or chains. Water, in contrast, has two protons to use in hydrogen bonds and two unshared pairs of electrons to act as acceptors of hydrogen atoms from other water molecules. Consequently, water is capable of forming extended three-dimensional hydrogen bonded arrays. This ability accounts for the fact that, of the three compounds under consideration, water has the greatest ability to interact with itself in the liquid and solid states. To understand the structural basis of such interaction, let us consider the structure of ice.

Ice occurs in a variety of structural modifications depending on the temperature and pressure of the system. The ice we are accustomed to encountering is called Ice I and has the structure shown in Figure 10-2. Note that each water molecule has associated with it four other water molecules distributed tetrahedrally about the central oxygen atom. On the average, each water molecule forms hydrogen bonds to two other water molecules. To what extent are structures of this type retained in liquid water?

THE STRUCTURE OF LIQUID WATER

To gain some understanding into the question just raised, let us consider the thermodynamics of the transition of water in the solid state to water in the liquid state.

$$H_2O(s) \longrightarrow H_2O(l) \qquad \Delta H^0 = 1.44 \text{ kcal/mole} \qquad (10\text{-}3)$$

Enthalpy changes accompanying the conversion of solids to liquids at the same temperature are called **heats of fusion** and are analogous to the heats of vaporization mentioned above. Suppose that in the melting of ice all of the hydrogen bonds that maintain the structure of ice (Figure 10-2) are broken. What enthalpy change would we expect? Since each water molecule makes hydrogen bonds to two other water molecules and since each hydrogen bond has a strength of approximately 5 kcal/mole, breaking all such hydrogen bonds should be accompanied by an energy input near 10 kcal/mole. The actual heat of fusion (1.44 kcal/mole) is only about 14 percent of this value, suggesting that only about 14 percent of the hydrogen bonds of ice are broken in the conversion of ice to liquid water. Consequently, water, in the liquid state, retains a marked degree of structure. Ultimately, in the gas phase at low pressures, all hydrogen bonds are broken and only isolated molecules exist. The retention of many hydrogen bonds in liquid water accords with the physical properties detailed above: specifically, the high heat of vaporization and high boiling point of liquid water.

The ordered structure of water also accounts for the unique roles of the hydrated proton and hydroxide ion in electrical conductivity through water. In the case of most ions, such as sodium ion or chloride ion, the conductance of electricity requires the physical migration of that ion through the aqueous medium. This permits the translocation of a charge from one point in the solution to another. In contrast, protons and hydroxide ions may be involved in chemical reactions that result in such charge translocation, but without the physical migration of any distinct species through the solution. This is possible since substantial chains of hydrogen bonded water molecules exist in the liquid state. The attack of a hydroxide ion on one end of this chain can generate hydroxide ion at the other, and conversely, the attack of a hydronium ion on one end of this chain can generate hydronium ion at the other. The mechanisms for these charge translocation reactions are shown below:

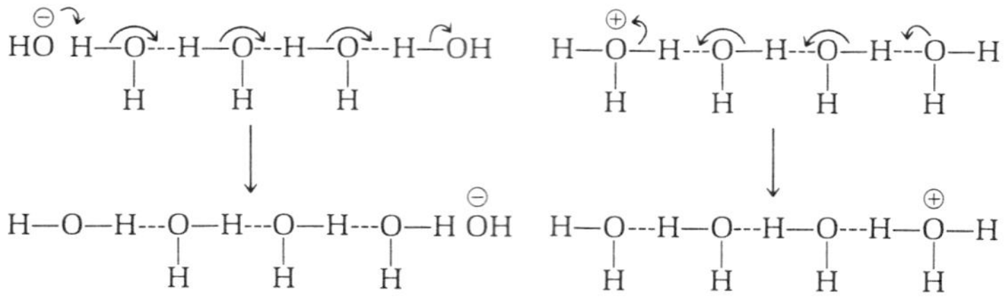

Clearly, charge transfer mechanisms related to these are impossible for ions other than those derived directly from the solvent itself.

Charge transfer reactions of the kind shown above ought to be even more facile in ice, which has a more highly ordered structure, than they are in water. In fact, proton mobility in ice is 1600 times greater than proton mobility in water, emphasizing the importance of the hydrogen bonded structure and of mechanisms of the type shown just above for proton translocation reactions.

HEATS OF VAPORIZATION AND FUSION; SPECIFIC HEAT

We have noted above that the heat of fusion of ice is 1.44 kcal/mole. This corresponds, in other units, to 79.7 cal/g. This heat of fusion is one of the largest known, emphasizing the major structural transition from the completely hydrogen bonded ice state to the partially hydrogen bonded liquid water state. In fact, only ammonia and a few salts such as sodium fluoride and sodium chloride are known to have higher heats of fusion than water.

We have also noted above that the heat of vaporization of water is 9.75 kcal/mole, which corresponds to 540 cal/g. This heat of vaporization is the largest such value known. This again emphasizes the extent of the structural organization in liquid water which is lost in the transition to the gaseous state.

The amount of heat required to raise 1 g of water 1°C (its **specific heat**) is used to define the calorie; 1 cal is the amount of heat needed to raise 1 g of water from 15°C to 16°C. Like the heats of fusion and vaporization, the heat capacity of water is very large in comparison with most other substances. In fact, only liquid ammonia has a greater specific heat. Representative values of heats of fusion, heats of vaporization, and specific heats are collected in Table 10-2. Note that the amount of heat required to raise 1 mole of a substance 1°C is called the **heat capacity** of that substance.

Example *Calculate the amount of heat in calories required to vaporize 100 g of methanol (bp 65°C) initially at 25°C.*

Solution *We must divide the overall process into two stages to make the required calculation: First, heat the methanol from 25°C to 65°C, the boiling point, and second, vaporize the methanol at 65°C. The first process requires*

$$100 \text{ g} \times 0.6 \text{ cal/g deg} \times (65 - 25) \text{ deg} = 2{,}400 \text{ cal}$$

The second requires

$$100 \text{ g} \times 289.2 \text{ cal/g} = 28{,}920 \text{ cal}$$

The sum of these is 31,320 cal.

The points made just above establish that water is a nearly ideal substance to function as a temperature-regulating device. Because of its high heat capacity, water can gain or lose relatively large amounts of heat while sustaining only modest temperature changes. Moreover, because of the high heat of fusion and the high heat of vaporization, extensive heat exchanges between water and its environment can be sustained with the melting of a relatively small amount of ice or the boiling of a relatively small amount of water. Consequently, both in terms of the environment generally and in terms of living organisms, water functions as an important temperature-regulating device.

EXPANSION OF WATER ON FREEZING

Water is nearly unique among known substances in that it expands rather than contracts upon freezing. Water has a maximum density at 4°C, and on cooling it contracts until this temperature is reached and then expands and, upon freezing, continues to expand until it occupies a volume some 11 percent greater than the liquid at the same temperature. This unusual behavior is more than a laboratory curiosity. In fact, maintenance of life on this planet is dependent on precisely this exotic quality of water. This point has been elegantly stated in part of a remarkable book by L. Henderson entitled *The Fitness of the Environment*, written in 1913. Henderson states,

> And so it would be with lakes, streams, and oceans were it not for the anomaly and the bouyance of ice. The coldest water would continually sink to the bottom and there freeze. The ice, once formed, could not be melted, because the warmer water would stay at the surface. Year after year the ice would increase in winter and persist through the summer, until eventually all or much of the body of water, according to the locality, would be turned to ice. As it is, the temperature of the bottom of a body of fresh water cannot be below the point of maximum density; on cooling further the water rises; and ice forms only on the surface. In this way the liquid water below is effectually protected from further cooling, and the body of water persists.

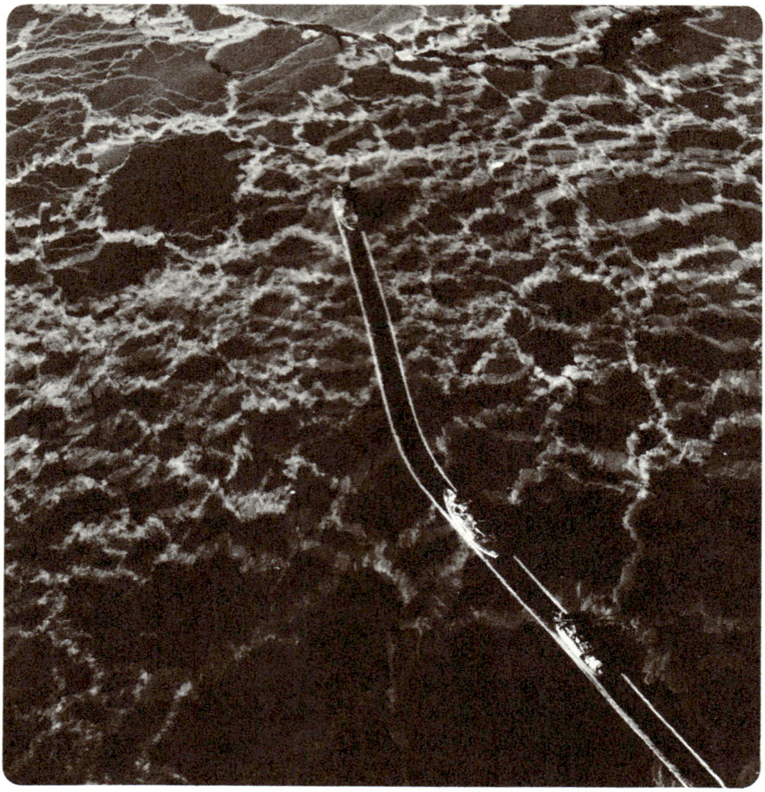

Water has the peculiar property of expanding when it freezes. This can create some slippery problems for getting about in the polar seas. Here three U.S. Navy icebreakers deal with the problem. (Courtesy of U.S. Navy.)

Table 10-3 Dielectric constants for selected liquids

Substance	Dielectric constant
Water	80.1
Ammonia	26.7
Sulfur dioxide	15.6
Hydrogen cyanide (HCN)	116
Ethanol (CH_3CH_2OH)	24.3
Benzene	2.28
Hexane	1.89

In the spring the first warm weather melts the ice, and at the earliest possible moment all ice vanishes.

Henderson's point can be demonstrated very nicely in the laboratory. If a block of ice is held at the bottom of a beaker of water, it is perfectly possible to boil that water by directing the flame of a bunsen burner at the surface without melting the ice at the bottom. The evaporation of water in the boiling process dissipates the heat, which then has no mechanism for reaching the ice at the bottom of the container. If water were to contract upon freezing, the ice would simply sink to the bottom and no mechanism would exist for the melting of that ice. Consequently, as Henderson states, in a period of a few years most of the bodies of water on the earth would be converted to the ice state.

DIELECTRIC CONSTANT

Like most of its other properties, the dielectric constant of water is unusual. The **dielectric constant** is a value that defines the forces between charged particles in the particular medium. Specifically, the equation relating dielectric constants to the force between charged particles is given in Equation (10-4), where D is the dielectric constant:

$$F = \frac{e_1 e_2}{Dr^2} \tag{10-4}$$

We note that the higher the dielectric constant, the smaller the force between the charged particles will be. In this equation e_1 and e_2 refer to the charges on the particles and r to the distance between them. Specifically, the dielectric constant of water at ordinary temperatures is 80. Contrast this with values of typical hydrocarbons, which range from 1 to perhaps 5 or 6. Dielectric constants for some liquids are collected in Table 10-3; these indicate the range of values encountered and will permit us to deduce something about the relationship between molecular structure and dielectric constant.

One of the consequences of the very high dielectric constant of water is that this solvent is a very good medium for the solution of ions. This is partially the result of water's ability to minimize the electrostatic interaction between oppositely charged groups; hence, ions (of sodium and chlorine for example) do not interact as strongly in a water medium as in a medium with a lower dielectric constant. If such ions were placed in a medium of low dielectric constant, such as benzene or carbon tetrachloride, the forces between the oppositely charged ions would be very strong indeed, and, in consequence, the ions would draw together and precipitate from the solution. Solvents of low dielectric constant are very poor solvents for ionic species.

Example *Calculate the electrostatic force between a sodium ion and a chloride ion (assume each ion to be a point charge) at a distance of 10 Å from each other in water, ammonia, and benzene.*

Solution *We must employ Equation (10-4) and the dielectric constants from Table 10-3. The charge on each ion is the electronic charge, 4×10^{-10} electrostatic units (esu). Thus, for water,*

$$F \text{ (dynes)} = \frac{(4.8 \times 10^{-10})^2 \text{ esu}^2}{80.1 \times (10^{-7})^2 \text{ cm}^2} = 2.9 \times 10^{-5} \text{ dynes}$$

Using the appropriate dielectric constants, we can calculate in the same way that $F = 8.6 \times 10^{-5}$ dynes in ammonia and 101×10^{-5} dynes in benzene. Note that the strengths of these interactions are inversely related to the solubility of sodium chloride in these solvents: NaCl is very soluble in water, sparingly soluble in ammonia, and essentially insoluble in benzene.

STRUCTURES OF DISSOLVED SUBSTANCES

Since liquid water is a highly organized substance, a question of substantial importance is the effect dissolved substances may have on water structure. We can approach this question by asking what the structure of water molecules is in the immediate vicinity of such substances, and how this compares with the structure of liquid water itself. Let us consider a series of cases beginning with what is perhaps the simplest one, the proton.

While some dispute still exists as to the structure of the hydrated proton in aqueous solution, there is universal agreement on one point: The proton itself is hydrated and does not exist free in aqueous solution. A good deal of evidence suggests that the structure of the hydrated proton is best written as $H_9O_4^+$. Structurally, this formulation corresponds to a hydronium ion surrounded by three additional solvent molecules:

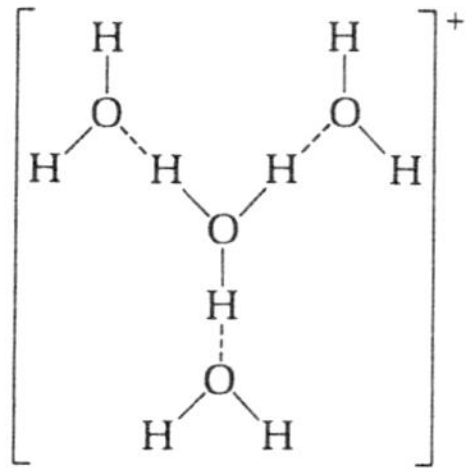

One of the lines of evidence suggesting this structure for the hydrated proton has to do with the rapidity of proton transfer reactions in aqueous solutions. We turn to this argument a little later in the discussion.

Let us now consider the solvated form of ions in solution. Here we can distinguish three cases: very small ions, larger ions, and very large ions, which include substantial hydrophobic character.

Let us consider first very small ions. These include the lithium cation and the fluoride anion, for example. These species, because of their small size, have a high charge density and a strong electrostatic field emanating from them. This electrostatic field tends to overwhelm the electrostatic interactions of water molecules with each other and has the effect of orienting solvent water molecules in a highly structured way about the dissolved ion similar to that present in the $H_9O_4^+$ ion. This highly structured water is called the **hydration sphere** of the ion, and the number of water molecules contained therein is called the *hydration number*. When discussing the behavior of dissolved ions in water, it is proper to think of the hydrated structure as the unit of function. That is, the ion with its hydration molecules moves as a unit and reacts as a unit in solution. A possible structure for the inner hydration shell of an ion such as lithium is

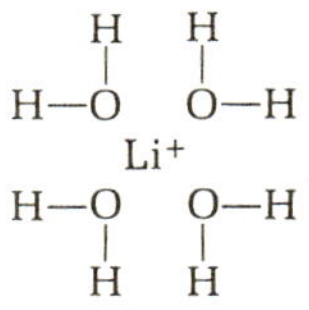

in which the negative ends of the water molecules are shown closely associated with the positive lithium atom. The influence of the hydration shell on the properties of dissolved ions can be understood if we realize that, in terms of hydrated volumes, lithium is larger than sodium, which is larger than potassium, which, in turn, is larger than the cesium ion. This is just the opposite order of the radii of the atoms themselves.

As the size of the ions of constant charge becomes larger, the charge density of the ions necessarily becomes smaller. Consequently, the electrical field associated with the ions is weaker at the distance of closest approach of a second ion or molecule. In such cases the electrical field of the dissolved ion is not sufficiently strong to dominate the electrostatic interactions of the water molecules with each other. In consequence, a relatively weak electrostatic field is imposed on the interactions of the water molecules with themselves. The result is that the solvent water molecules tend to become less organized than they are in the normal liquid state. The situation is a little like imposing the music of the Rolling Stones on top of that of the Boston Symphony Orchestra. Although both groups produce music, the result of the combination of the two is rather more chaotic than either alone.

For very large ions, particularly those that have substantial hydrophobic components, such as the tetrabutylammonium ion, the situation is more complicated. Let us approach this condition by asking what the structure of solvent water molecules is in the vicinity of dissolved hydrocarbon species. There is substantial evidence that the presence of hydrocarbons

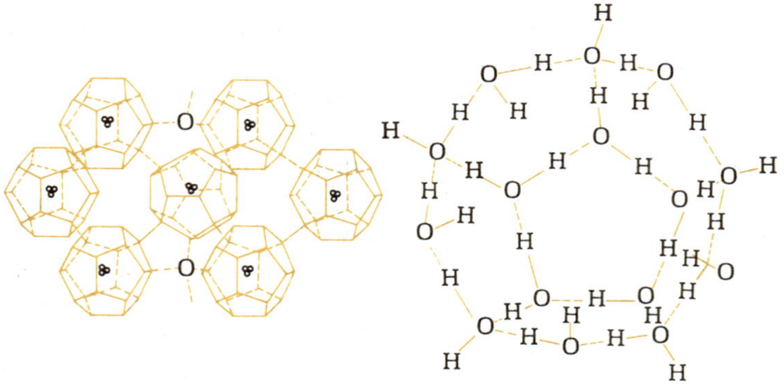

Figure 10-3 A possible structure for water clathrates. The enclosing water molecules are considered to form pentagonal dodecahedra having a water molecule at each corner. The center of each dodecahedron may be occupied by a solute molecule. As shown on the left, the dodecahedra may be hydrogen bonded together through water molecules between them. Such structures have been considered as a model for liquid water itself, in which water molecules occupy the positions indicated for solutes. [From L. Pauling, *Science*, **134**, 16 (1961). Copyright © 1961 by the American Association for the Advancement of Science.]

in aqueous solution has a strong structure-orienting influence. As an extreme example of such influence, we might cite the inclusion compounds or **clathrates**, which are formed by surrounding one molecule of a hydrocarbon by a cage of water molecules. One such example is provided by the hydrate of methane (see Figure 10-3). A collection of water molecules is able to form a highly organized structure into which one molecule of methane may be placed. The methane molecule is referred to as the guest molecule and the water molecules as the host. The physical properties of the hydrate of methane are remarkable. Despite the fact that the guest molecule is very volatile and boils at nearly $-200°C$, the crystalline hydrate of methane is stable even above the melting point of ice.

The situation of hydrophobic ions in aqueous solution resembles that of the crystalline hydrate of methane, although the organization of water about the exposed hydrophobic portion is not so extreme as in the previous case. Nevertheless, it is clear that very large hydrophobic ions behave as structure-forming agents in aqueous solution.

WATER AND BIOLOGICAL MEMBRANES

What we have just developed in terms of the interaction of water with dissolved substances helps us to understand the basis of several very interesting antibiotics and related compounds.

Biological membranes act as barriers to the free diffusion of ions. Indeed, this may be their primary role in living systems. Since these membranes are largely constructed from molecules that are hydrophobic in nature, they interact very poorly with charged species. As a result, entities such as potassium and sodium ions do not readily pass across biological membranes. This impermeability to ions permits cells to maintain large ionic gradients across short distances. This is quite important for the survival of cells and is related to a number of important phenomena. Among marine organisms, for example, are those that can "pump" certain ions across the cell membrane and then retain them at high concentrations since they cannot readily leak out again. Changes in membrane permeability underlie some of the most important physiological activities. For example, the transmission of an impulse along a nerve is the direct consequence of the loss of the impermeability of the membrane of the nerve cells to sodium and potassium ions. Since the maintenance of ionic gradients is so important to living systems, molecules that interfere with this function might be suspected of possessing *antibiotic* properties. This turns out to be the case.

Some representative molecules that associate with membranes and destroy their permeability properties include valinomycin, nystatin, and gramicidin S, all useful antibiotics. Although diverse structurally, all these molecules have one property in common: They each possess a cavity in the center which is lined with oxygen atoms, and each presents a hydrophobic exterior. A simple model for these cyclic antibiotics is provided by "crown" compounds, so-called because of their appearance when viewed in three dimensions. One example has the structure

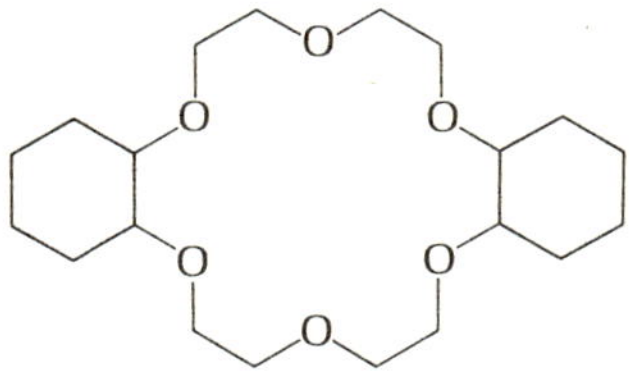

These oxygen atoms can interact with cations just as the oxygen atoms of water do. The hydrophobic exterior provides for solubility of the complex within the membrane. Hence, these cavities provide a suitable environment for cations, such as sodium and potassium. By virtue of their inclusion within the cavity, these molecules provide a means by which the ions can diffuse across biological membranes. Molecules such as valinomycin and nystatin essentially form passages through these membranes by which cations can travel from one side to the other. The resultant destruction of the permeability properties of the membrane leads to eventual cell death.

Finally, we ought to note that these molecules are rather specific for individual ions. This follows from the fact that cations of different sizes fit more or less well into the cavity: Those that are readily accommodated pass more freely through the membrane than do those that do not fit as well. Valinomycin, for example, increases the permeability of membranes to potassium ion about 40 times more than the permeability to sodium ion.

THE IONIZATION OF WATER

One of the simplest, yet most important, acid–base reactions that we shall consider is that which occurs in aqueous solution itself. Briefly, we refer to an acid as a molecule that can donate a proton and to a base as a molecule that can accept one. If we consider again the electronic structure of water, it is clear that the molecule may serve both as an acid and a base. In fact, it does so when it forms hydrogen bonds with itself. Each water molecule is capable of donating two protons to acceptors, and each water molecule is capable of accepting two protons as well. Consequently, we can have a reaction in aqueous solution between two water molecules with the transfer of a proton from one to the other:

$$H_2O + H_2O \rightleftharpoons H_3O^+ + OH^- \tag{10-5}$$

For this reaction it is possible to formulate an equilibrium constant in the usual way:

$$K_{eq} = \frac{(H_3O^+)(OH^-)}{(H_2O)^2} \tag{10-6}$$

This equilibrium constant can be simplified somewhat by recognizing that the concentration of water in an aqueous solution is essentially a constant (see Chapter 12). We can then absorb the water concentration term in the denominator of the above equation into the equilibrium constant and formulate the equilibrium constant as follows:

$$K_w = (H_3O^+)(OH^-) \tag{10-7}$$

At 25°C, the value of K_w is almost exactly $10^{-14}M^2$. It is customary to write the negative logarithm of equilibrium constants in the form p*K*. The p simply says "take the negative logarithm of the quantity that follows." Consequently, we can say that, at 25°C, pK_w is equal to 14.00. This quantity is remarkably temperature-dependent. At 0°C, pK_w is equal to 14.94, while at 37°C, the same quantity has a value of 13.59. Remembering that these numbers are logarithms of the equilibrium constants, it is clear that the self-ionization of water decreases by almost an order of magnitude between 25°C and 0°C.

The rate of the reaction between the hydrated proton and hydroxide ion is the greatest known for a bimolecular reaction in aqueous solution. By using special techniques, it has been possible to demonstrate that the second-order rate constant for this reaction is near $3 \times 10^{10}M^{-1}\ sec^{-1}$. This is a remarkable rate constant when we realize that it is about three times more rapid than the calculated rate for diffusion-controlled reactions in aqueous solutions. It is generally true that the most rapid reaction possible in solution is the reaction that occurs every time the two reactants bump into each other in solution. Such processes have rates that are called *diffusion-controlled*. How is it possible to have a reaction that proceeds even more rapidly than this apparent upper limit? To examine this situation we need to return to our formulation of the hydrated proton not as H_3O^+ but as $H_9O_4{}^+$. If the latter species is the important one in solution, then it ought to be possible to pull a proton off of that complex from any of the six exposed sites in the molecule. Thus, in fact, a collision between a hydronium ion and a hydroxide ion is not a requirement for the reaction, and it is possible to have proton transfer occurring through the intermediacy of one or more water molecules. Recall in this connection our earlier discussion of the electrical conductance through water by hydronium and hydroxide ions.

Glossary

Clathrate a type of compound in which one of the components is enclosed in a structure formed by another. Clathrates usually show a definite composition but a nonintegral stoichiometry.

Dielectric constant a property of gases, liquids, and solids that is inversely proportional to the force between charges embedded in the substance.

Heat capacity the quantity of heat required to raise the temperature of 1 mole of a substance 1°C.

Heat of fusion the quantity of heat required to convert 1 g of a solid to the liquid form at the melting point of the solid at a pressure of 1 atm.

Heat of vaporization the quantity of heat required to convert 1 g of a liquid to the vapor form at the boiling point of the liquid at a pressure of 1 atm.

Hydration sphere those molecules of water that are more or less tightly bound to a dissolved ion and that, with the ion, form a structural unit.

Ionization a chemical reaction involving the heterolytic cleavage of a chemical bond:

$$M{-}X \longrightarrow M^+ + X^-$$

Specific heat the quantity of heat required to raise the temperature of 1 g of a substance 1°C.

Study problems

Some quantitative problems

1. The specific heat of liquid ammonia is 1.23 cal/deg g. How much heat would be required to raise the temperature of 1 mole of ammonia 15°C?

2. The *molar* heat capacity of graphite is 2.06 cal/deg mole. How much heat would be required to increase the temperature of 100 g of graphite from 20°C to 150°C?
3. Experimentally, it is observed that 11.5 cal of heat are required to raise the temperature of 21 g of tin metal from 30°C to 40°C. What is the specific heat of tin?
4. One of the early methods employed to determine atomic weights was based on the observation, made in 1819 by Dulong and Petit, that the molar heat capacities of the elements then known (with the exception of carbon) were near 6 cal/deg mole. Experimentally, the specific heats of several elements have been shown to be the following:

Element	Heat capacity (cal/deg g)
Mg	0.235
K	0.178
Cu	0.0921
Pb	0.0314

On the basis of the rule of Dulong and Petit, use these data to calculate the atomic weights of these elements. How do your results compare with the currently accepted values?
5. The heat of fusion of methanol is 16 cal/g. How much heat would be required to melt 2 moles of methanol?
6. Given that the heat of fusion of some substance is 45 cal/g and that 2.7 kcal are required to melt 1 mole of it, calculate the molecular weight for this substance.
7. Given that 2.08 kcal of heat are required to melt 1 mole of solid chloroform, what is the heat of fusion of this substance in calories per gram?
8. The vaporization of water is frequently employed as a means of dissipating the heat generated in chemical reactions. The reaction between molecular hydrogen and chlorine to generate HCl is exothermic by 44.2 kcal/mole:

$$H_2 + Cl_2 \longrightarrow 2HCl \qquad \Delta H = -44.2 \text{ kcal/mole}$$

How many moles of HCl would be produced in this reaction when sufficient heat had been generated to completely vaporize 1000 g of water?
9. How many grams of ammonia could be vaporized by the heat generated by the condensation of 100 g of water?
10. Sublimation is the process of direct conversion of a solid to a gas without passing through the liquid phase. Under proper conditions, most substances undergo sublimation. The heat of sublimation of ice is 0.68 cal/g. About 80 percent of this value is considered to reflect the breakage of hydrogen bonds. What is the bond strength of the hydrogen bonds in ice?
11. From the data in Table 10-2, calculate the entropies of fusion for methanol and benzene.

12. For a solvent of dielectric constant 50, at what distance of separation would the force between a sodium ion and a chloride ion be 20×10^{-5} dynes?

Qualitative questions

13. The entropy of fusion of ice is less than the entropy of vaporization of liquid water. What conclusions can you draw from this fact concerning the structure of liquid water?
14. How can you reconcile the fact that water expands considerably upon freezing, although most other liquids contract upon solidification?
15. The dielectric constants for several substances are collected in Table 10-3. Can you see any relationship between molecular structure and dielectric constant? Estimate the dielectric constants for methanol, carbon tetrachloride, and formamide, $H{-}\overset{\overset{\displaystyle O}{\|}}{C}{-}NH_2$.
16. In water, dissolved HF reacts with fluoride ion to form a new ion, HF_2^-:

    $$HF + F^- \rightleftharpoons HF_2^-$$

    What structure can you suggest for HF_2^-?
17. Ammonium fluoride is the only substance known to have appreciable solubility in ice. How do you account for this fact? (Hint: It may prove worthwhile to consider what the crystal structure for ammonium fluoride might be.)

Suggestions for further reading

J. T. Edsall and J. Wyman *Biophysical Chemistry*, Academic, New York, 1958, Chap. 2.

"The Oceans," *Sci. Am.*, **221** (September, 1969). The whole issue.

L. K. Runnels "Ice," *Sci. Am.*, **215**, 118 (December, 1966).

11

Fundamentals of acid-base chemistry

Concerning solutions to acid-base equilibrium problems:
The answers to the riddle of life are not found in the back of the book.

Charlie Brown

Determination and regulation of the acidity of aqueous solutions are important in a great many instances. One example, considered in some detail at the end of this chapter, concerns the chemistry of blood. The maintenance of life requires that the acidity of blood be maintained as a constant over a very narrow range independent of the physiological state of the person. Another example, the technology of the desalination of sea water, is considered in Chapter 13. There are many other examples that we cannot treat in detail but that are important nevertheless. These include the efficiency of some drugs, the quality of wines, the shape and properties of proteins, the suitability of commercial detergents and other cleaning products, the technology of culturing tumor cells, the pollution of streams and rivers, and so forth. Acid-base reactions in water solutions are thus critical to the concerns of the pharmacologist, physiologist, biochemist, industrial chemist, vintner, cell biologist, and conservationist, among others.

The most interesting and important processes involving acid-base reactions are usually the most complicated. Thus, a superficial knowledge of this field will not serve to prepare us to deal with real problems. For example, the regulation of the acidity of the blood involves a complicated set of equilibria: oxygenation of hemoglobin, the hydration of carbon dioxide, the ionization of carbonic acid, and the ionization of the bicarbonate ion. Thus, we must deal with the following set of reactions:

$$Hb + O_2 \rightleftharpoons HbO_2 + H^+ \qquad (Hb = \text{hemoglobin})$$
$$H_2O + CO_2 \rightleftharpoons H_2CO_3$$
$$H_2CO_3 \rightleftharpoons HCO_3^- + H_3O^+$$
$$HCO_3^- \rightleftharpoons CO_3^{2-} + H_3O^+$$

Treatment of sets of equilibria such as this is not trivial. Consequently, we shall begin our discussion of equilibria in aqueous solution with much simpler cases. Specifically, solutions of strong acids and bases will be examined first, followed by treatment of solutions of weak acids and bases, and, finally, of the phenomenon of buffering. We shall then be in a position to turn to more complex cases.

THE LOWRY-BRØNSTED CONCEPT

Historically, the concepts of acidity and basicity have undergone a number of revisions. The earliest productive point of view was that of Arrhenius, who proposed in the late nineteenth century that all acids function by producing the hydrated proton in solution and that all bases produce the hydroxide ion in solution. There arose, in succeeding years, two difficulties with the Arrhenius concept: (1) It appeared more reasonable to think of acids as species capable of transferring protons rather than producing them through dissociation, and (2) it became clear that many species other than hydroxide ion had basic properties.

In consequence, a more general view, attributed to Lowry and Brønsted, was proposed in 1923. The **Lowry-Brønsted view** defines an **acid** as *a species having the capacity to donate or transfer a proton to an acceptor* and a **base** as *a species having the capacity to accept or add a proton from a donor.*

The Lowry-Brønsted definitions are not the most general ones currently employed. Later we shall have the opportunity to explore the Lewis concepts of acidity and basicity. But for the moment, the idea of acids and bases as compounds that donate and accept protons will suffice for our purposes. Let us proceed, then, to examine some acid-base equilibria.

ACID-BASE REACTIONS IN PURE WATER

As noted in Chapter 10, water molecules have both acidic and basic properties. Thus, one water molecule may donate a proton to the other:

$$H_2O + H_2O \rightleftharpoons H_3O^+ + OH^- \qquad (11\text{-}1)$$

Here one molecule functions as an acid and the other as a base. Viewing the reaction in the opposite direction, hydronium ion functions as an acid and hydroxide ion as a base. Hydronium ion and the water molecule derived from it by loss of a proton are called a **conjugate acid-base pair.** Similarly, the water molecule that acts as an acid and the hydroxide ion derived from it are called a conjugate acid-base pair. This terminology is quite general and applies to all acids and the bases derived from them by loss of a proton.

The concept of pH

Just as the negative logarithms of equilibrium constants are designated as pK, the negative logarithm of the concentration of the hydrated proton is designated as **pH:**

$$\mathbf{pH \equiv -\log(H^+)} \qquad (11\text{-}2)$$

Values of pH are commonly used to express the concentration of the hydrated proton in solution since it proves very convenient to do so. Thus, for a solution containing a concentration of hydronium ions (more exactly, hydrated protons) of $1.0 \times 10^{-6}M$, the pH is simply 6. It is clearly easier to describe the situation using pH rather than the concentration terms.

Example *Calculate the pH of solutions containing the following molar concentrations of the hydrated proton:*

a. 1×10^{-4} b. 2.0 c. 1.8×10^{-7} d. 3.9×10^{-11}

Solution *In each case we proceed by applying the definition of* pH *expressed in Equation (11-2). That is, we find the negative logarithm of each of the above numbers. The results are*

a. 4 b. -0.3 c. 6.75 d. 10.41

In reaching answers (c) and (d), you must recall that the mantissa of a logarithm (the numbers following the decimal) is always positive. Thus, for (c), $\log(1.8 \times 10^{-7})$ *is* $-7.00 + 0.25$ *or* -6.75.

Example *Calculate the concentration of the hydrated proton in solutions of the following values of* pH.

a. 5.0 b. 0.0 c. 8.64 d. 11.34

Solution *In this case we must again use Equation (11-2), but now we must find the antilogarithm of the negative of each value of* pH. *Thus, first change each* pH *to its negative value (more generally, multiply the* pH *by* -1*) and look up the antilogarithm of the resulting number in a log table or on your slide rule. The results are*

a. $1 \times 10^{-5}M$ b. $1.0M$ c. $2.29 \times 10^{-9}M$ d. $4.47 \times 10^{-12}M$

What is the pH of pure water? To answer this question we need to make use of two fundamental relationships that are general to all acid-base problems: dissociation constants and the electrical neutrality of solutions. We already know that for water at 25°C, $(H_3O^+)(OH^-) = K_w = 10^{-14}M^2$. The **principle of electrical neutrality** for solutions requires that the net charge of the solution be zero. That is, the number of positive charges must be exactly equal to the number of negative charges. In pure water the only positively charged species are hydrated protons, and the only negatively charged ones are hydroxide ions. It follows, then, that the concentration of these two species must be equal. We have, then, two equations that describe the situation in pure water:

$$(H_3O^+)(OH^-) = K_w = 10^{-14}M^2 \quad (11\text{-}3)$$

$$(H_3O^+) = (OH^-) \quad (11\text{-}4)$$

Putting the second of these equations into the first, we obtain

$$(H_3O^+)^2 = 10^{-14} \quad (11\text{-}5)$$

From this equation it follows, by taking the square root of both sides, that

$$(H_3O^+) = 10^{-7}M \quad \text{or} \quad \text{pH} = 7 \quad (11\text{-}6)$$

One of the functions of the system serving this pool is to ensure that the water is free of pathogenic organisms while maintaining the pH of the water near neutrality at the same time. (Courtesy of Seaboard Pools, Inc.)

Table 11-1 Criteria for acidic and basic aqueous solutions

Acidic solutions	Basic solutions
$(H_3O^+) > (OH^-)$	$(H_3O^+) < (OH^-)$
$(H_3O^+) > 10^{-7}$	$(H_3O^+) < 10^{-7}$
$(OH^-) < 10^{-7}$	$(OH^-) > 10^{-7}$
$pH < 7$	$pH > 7$
$pOH > 7$	$pOH < 7$

Table 11-2 Some examples of strong acids and bases

Acids	Bases
HCl	$NaOH$
HBr	$LiOH$
$HClO_4$	KOH
HNO_3	$CsOH$
H_2SO_4	
$HSCN$	

From our knowledge of the condition of electrical neutrality, it follows immediately that the concentration of hydroxide ion must be $10^{-7}M$ as well:

$$(OH^-) = 10^{-7}M \quad \text{or} \quad pOH = 7 \tag{11-7}$$

Example *As pointed out in Chapter 10, the* pK_w *for water at 0°C is 14.94. What is the concentration of hydrated protons in water at this temperature?*

Solution *From the value of* pK_w, *we can calculate the dissociation constant of water at this temperature:*

$$pK_w = -\log K_w = 14.94$$

Finding the antilogarithm of −14.94 reveals that $K_w = 1.15 \times 10^{-15}M^2$. *Thus, we have*

$$(H_3O^+)(OH^-) = 1.15 \times 10^{-15}M^2$$

Imposing the condition of electrical neutrality, $(H_3O^+) = (OH^-)$, *we find that*

$$(H_3O^+)(OH^-) = 1.15 \times 10^{-15}M^2$$

Solving,

$$(H_3O^+) = 3.39 \times 10^{-8}M$$

The acid–base condition that exists in pure water is said to be *neutral.* Solutions having a concentration of hydrated protons greater than $10^{-7}M$, which necessarily have a pH less than 7, are termed *acidic* and have a characteristic sour taste. Those with a lesser concentration, which necessarily have a pH greater than 7, are termed *basic* and have a characteristic bitter taste. The criteria for acidic and basic aqueous solutions are collected in Table 11-1.

Several related problems can be handled in exactly the same way as we found the pH of water. Attention is now directed to such cases: those involving strong acids or strong bases.

Solutions of strong acids and strong bases

Acids and bases can be characterized with respect to their strength, that is, with respect to their abilities to donate protons or accept protons. Later we shall examine a quantitative basis for describing the strengths of acids and bases. For the moment, a qualitative description will serve. **Strong acids and bases** dissociate essentially completely in aqueous solutions. **Weak acids and bases** do not. Note that our distinction between weak and strong may depend on the conditions for which the distinction is made. Thus, some acids that are almost completely dissociated in dilute solutions may be appreciably associated in concentrated ones. Some typical strong acids and strong bases are collected in Table 11-2. These species are completely dissociated in solutions as concentrated as 1M.

We can look at the above statements in a way that we can later generalize to a quantitative basis for acid and base strengths. The idea of complete dissociation of an acid or base is equivalent to the statement that reactions of the following type have very large equilibrium constants; that is, they proceed very nearly to completion.

$$HB + H_2O \rightleftharpoons B^- + H_3O^+ \qquad (B = Cl, Br, ClO_4, NO_3, etc.)$$
$$XOH \rightleftharpoons X^+ + OH^- \qquad (X = Na, Li, K, etc.)$$

Weak acids and weak bases, then, are those for which the above reactions do not have large equilibrium constants. In fact, known acids and bases show a graduation in strengths so that the associated equilibrium constants have values that vary from 10^{10} (very strong) to 10^{-20} or less (very weak). More about these matters follows later.

How can we calculate the pH of a solution of, for example, 0.1*M* hydrochloric acid? Qualitatively, it is easy to arrive at a very nearly exact solution. Hydrated protons can be derived from just two sources: dissociation of the HCl and dissociation of water. Since HCl is a strong acid (Table 11-2), we know that 0.1*M* hydronium ion comes from that source. This is much larger than the amount of this ion derived from the dissociation of water, which is, at most, $10^{-7}M$. In fact, the addition of a strong acid to water will tend to repress the ionization of water since, according to LeChatelier's principle, the equilibrium for water ionization will shift in such a way as to minimize the effect of the added acid. Thus, the total concentration of hydronium ion in 0.1*M* HCl must be almost exactly 0.1*M*. To prove this point and to illustrate methods employed to solve problems of this type, we can formulate the situation in terms of electrical neutrality and the ionization of water. Since the solution contains only hydronium ion as positive ions and chloride and hydroxide ion as negative ions, we have

$$(H_3O^+) = (Cl^-) + (OH^-)$$
$$(H_3O^+)(OH^-) = K_w \qquad (11\text{-}8)$$

Substituting the quantity for (OH^-) from the second equation into the first, we obtain

$$(H_3O^+) = (Cl^-) + K_w/(H_3O^+) \qquad (11\text{-}9)$$

Since the concentration of chloride ion is exactly 0.1*M*, the above equation can be solved directly for the concentration of hydronium ion from our knowledge that HCl is a strong acid (completely dissociated). Multiplying through by (H_3O^+) and rearranging yields

$$(H_3O^+)^2 - (H_3O^+)(Cl^-) - K_w = 0 \qquad (11\text{-}10)$$

This equation can be solved in the usual way that one treats quadratic equations to yield a final expression:

$$(H_3O^+) = \frac{(Cl^-) \pm \sqrt{(Cl^-)^2 + 4K_w}}{2} \qquad (11\text{-}11)$$

If we now substitute the numerical values for (Cl^-) and K_w into this expression, we obtain

$$(H_3O^+) = \frac{0.1 \pm \sqrt{0.01 + 4 \times 10^{-14}}}{2} \cong 0.1M \tag{11-12}$$

This result confirms our expectations. The amount of hydronium ion contributed by the ionization of water is trivial in comparison to the amount derived from the dissociation of HCl. Provided that the solution of HCl is not too dilute, the concentration of hydronium ion may always be approximated by the concentration of HCl added to the solution. This conclusion is, of course, true for other strong acids as well. Even if the solution of a strong acid is as dilute as $10^{-6}M$, an error of less than 2 percent is made in neglecting the contribution of hydronium ions from the solvent. Hence, we have the rather general relationship for solutions of strong acids:

$$(H_3O^+) \cong (HX) \qquad pH = -\log(HX) \tag{11-13}$$

Exactly the same considerations apply to solutions of strong bases:

$$(OH^-) \cong (YOH) \qquad pOH = -\log(YOH) \qquad pH = 14 - pOH \tag{11-14}$$

It is true, of course, that if the solution of a strong acid or strong base is sufficiently dilute, then the contribution of hydronium ion or hydroxide ion from the ionization of water cannot be neglected. For the case of strong acids, we then rely on the general equation developed above. For example, if the concentration of HCl is only $10^{-7}M$, we have

$$(H_3O^+) = \frac{10^{-7} \pm \sqrt{10^{-14} + 4 \times 10^{-14}}}{2} = 1.62 \times 10^{-7}M \qquad pH = 6.79 \tag{11-15}$$

If we separate the total hydronium ion concentration into the individual contributions from HCl and from H_2O, we note that $1 \times 10^{-7}M$ comes from the former and that $0.62 \times 10^{-7}M$ from the latter. This confirms our earlier conclusion that the addition of strong acid to water represses the ionization of water; were the HCl not present, ionization of water would have contributed $1.0 \times 10^{-7}M$ total hydronium ion.

Example *Calculate the concentration of hydrated protons in a solution of water containing $3 \times 10^{-7}M$ HCl.*

Solution *In this case the amount of hydrated protons contributed by the HCl, $3 \times 10^{-7}M$, is not very much greater than the maximum that could be contributed by ionization of the water itself, $1 \times 10^{-7}M$. Thus, an accurate calculation requires that we resort to the exact expression in Equation (11-11) rather than relying on the approximate one in Equation (11-13). Inserting the values for (Cl^-) and K_w and solving, we obtain*

$$(H_3O^+) = 3.3 \times 10^{-7}M$$

Note that this value is 10 percent larger than that predicted by using the approximate expression of Equation (11-13).

For very dilute solutions of strong bases, we can derive an equation analogous to that for strong acids. The equation of electrical neutrality now reads, for NaOH for example,

$$(Na^+) + (H_3O^+) = (OH^-) \tag{11-16}$$

Proceeding just as before for the case of HCl, we arrive at the following exact expression:

$$(H_3O^+) = \frac{-(Na^+) \pm \sqrt{(Na^+)^2 + 4K_w}}{2} \tag{11-17}$$

Generalizing to the case of any strong acid, HX, and any strong base, YOH, we have the twin exact expressions for the hydronium ion concentration:

$$\mathbf{(H_3O^+) = \frac{(X^-) \pm \sqrt{(X^-)^2 + 4K_w}}{2}} \tag{11-18}$$

$$\mathbf{(H_3O^+) = \frac{-(Y^+) \pm \sqrt{(Y^+)^2 + 4K_w}}{2}} \tag{11-19}$$

Whether or not it is necessary to use the exact expressions or the approximate ones derived earlier depends on the concentration of the acid or base and the desired degree of accuracy of the calculation. In practice, it seldom proves necessary to employ the exact expressions. Nevertheless, in those cases in which doubt exists, the exact expressions should be used. Their use may require some unnecessary arithmetic, but it will avoid errors in questionable cases.

Example *Suppose that you must do a series of calculations of the concentration of hydrated protons in aqueous solutions of a strong acid and that the results must be accurate to within 5 percent. To avoid the use of the exact expression in those cases in which the approximate one would do, how would you proceed to determine the minimum strong acid concentration for which the approximate expression would yield results within the required margin of error?*

Solution *We need to compare the results of calculating the exact and approximate concentrations of hydrated protons according to our expressions [see Equations (11-13) and (11-18)]:*

$$(H_3O^+)_{exact} = \frac{(X^-) \pm \sqrt{(X^-)^2 + 4K_w}}{2}$$

$$(H_3O^+)_{approx.} = (HX)_{total} = (X^-)$$

The exact calculation will always be larger than the approximate one, since the former includes the contribution of water to the total hydrated proton concentration and the latter does not. Therefore, our error limit can be expressed in the form

$$\frac{(H_3O^+)_{exact}}{(H_3O^+)_{approx.}} = 1.05$$

Using the above equations, this expression takes the form

$$\frac{(H_3O^+)_{exact}}{(H_3O^+)_{approx.}} = 1.05 = \frac{(X^-) \pm \sqrt{(X^-)^2 + 4K_w}}{2(X^-)}$$

Solving for (X^-):

$$\begin{aligned} 2.10(X^-) &= (X^-) \pm \sqrt{(X^-)^2 + 4K_w} \\ 1.10(X^-) &= \pm \sqrt{(X^-)^2 + 4K_w} \\ 1.21(X^-)^2 &= (X^-)^2 + 4K_w \\ 0.21(X^-)^2 &= 4K_w \end{aligned}$$

Inserting $K_w = 10^{-14}M^2$ *and solving, we find that*

$$(X^-) = 4.4 \times 10^{-7}M$$

It follows that for any solution containing this concentration or a greater concentration of the strong acid, the approximate equation will suffice. For more dilute solutions, the exact expression must be used.

The condition of electrical neutrality

Since electrical neutrality of solutions is important for solution of problems of acid-base equilibria, let us look at some examples of the use of this condition. In each case we simply wish to write an equation in which the *total number of positive charges is equated to the total number of negative charges*. Provided that we have only singly charged ions in solution, this is accomplished by equating the concentrations of all positively charged species to the sum of the concentrations of all negatively charged species. For example,

The solution contains:	The equation of electrical neutrality:
HNO_3	$(H_3O^+) = (NO_3^-) + (OH^-)$
LiOH	$(H_3O^+) + (Li^+) = (OH^-)$
$HClO_4$ and NaOH	$(H_3O^+) + (Na^+) = (ClO_4^-) + (OH^-)$

The situation is slightly more complicated if multiply charged ions are present. Thus, for solutions of sulfuric acid, both the HSO_4^- and SO_4^{2-} ions will be present. Since the latter contributes two negative charges per ion, we must account for this in writing the equation of electrical neutrality:

$$(H_3O^+) = (HSO_4^-) + 2(SO_4^{2-}) + (OH^-) \quad (11\text{-}20)$$

Table 11-3 Values of pK_a and K_a for some weak acids

Acid	Conjugate base	K_a (M)	pK_a
CH_3COOH	CH_3COO^-	1.76×10^{-5}	4.76
HCN	CN^-	4.0×10^{-10}	9.40
NH_4^+	NH_3	5.75×10^{-10}	9.24
H_3PO_4	$H_2PO_4^-$	7.6×10^{-3}	2.12
$H_2PO_4^-$	HPO_4^{2-}	6.16×10^{-8}	7.21
HPO_4^{2-}	PO_4^{3-}	4.8×10^{-13}	12.32
H_2CO_3	HCO_3^-	4.27×10^{-4}	3.77
HCO_3^-	CO_3^{2-}	4.68×10^{-11}	10.33

The coefficient 2 is employed as indicated to account for the fact that the number of negative charges contributed by the sulfate ion is twice as great as its molar concentration indicates. For the case of phosphoric acid, H_3PO_4, the equation is more complex:

$$(H_3O^+) = (H_2PO_4^-) + 2(HPO_4^{2-}) + 3(PO_4^{3-}) + (OH^-) \tag{11-21}$$

Example *Formulate the condition of electrical neutrality for aqueous solutions of the following salts:*

a. Na_2HPO_4
b. $Ca_3(PO_4)_2$

Solution *It is essential to include all the species that can be derived from these salts in the condition of electrical neutrality. Note that both of these salts can generate other ions through acid-base reactions [see Equation (11-21)]. The conditions of electrical neutrality are*

a. $(Na^+) + (H^+) = (H_2PO_4^-) + 2(HPO_4^{2-}) + 3(PO_4^{3-}) + (OH^-)$
b. $2(Ca^{2+}) + (H^+) = (H_2PO_4^-) + 2(HPO_4^{2-}) + 3(PO_4^{3-}) + (OH^-)$

SOLUTIONS OF WEAK ACIDS AND WEAK BASES

Our considerations above concerning calculation of the pH of aqueous solutions of strong acids and strong bases, which are completely dissociated in solution, can be readily generalized to treat the cases of weak acids and weak bases, which are not completely dissociated in solution. Let us begin by examining the behavior of a hypothetical weak acid, HB:

$$HB + H_2O \rightleftharpoons H_3O^+ + B^- \tag{11-22}$$

We have chosen HB to be electrically neutral so that the conjugate base, B^-, necessarily has a negative charge. Of course, not all weak acids are uncharged. The calculation indicated below for the case of an uncharged acid would be changed only in the way that the condition of electrical neutrality is formulated were the acid of interest anionic or cationic.

For the equilibrium of Equation (11-22), we can associate an equilibrium constant. Such constants are generally called *dissociation constants* and are denoted by K_a:

$$K_a = \frac{(H_3O^+)(B^-)}{(HB)} \tag{11-23}$$

The value of K_a is a measure of the strength of the acid: The greater the value of K_a, the greater the acid strength. In the case of strong acids, K_a is necessarily very large; in the case of weak acids, it is by definition not large. It is also customary to indicate acid strength in terms of the negative logarithm of K_a; such values are denoted, as usual, by **pK_a**. Note that the smaller the value of pK_a, the greater the strength of the acid. Values of pK_a for some common weak acids are collected in Table 11-3.

A rigorous solution to the problem of the dissociation of weak acids in aqueous solution requires that four equations be satisfied. These equations account for the following: (1) the self-dissociation of water, (2) the dissociation of the weak acid, (3) the electrical neutrality of the solution, and (4) the conservation of the added acid. Quantitatively, these factors take the form of Equations (11-24) through (11-27), respectively:

$$K_w = (H_3O^+)(OH^-) \tag{11-24}$$

$$K_a = \frac{(H_3O^+)(B^-)}{(HB)} \tag{11-25}$$

$$(H_3O^+) = (B^-) + (OH^-) \tag{11-26}$$

$$C_0 = (HB) + (B^-) \tag{11-27}$$

In Equation (11-27), C_0 is just the total concentration of HB initially added to the solution; clearly, this concentration must equal the sum of the concentrations of (HB) and (B^-) following dissociation.

The four equations above are adequate to solve our problem completely, since we have exactly four unknowns: (H_3O^+), (OH^-), (HB), and (B^-). However, the solution to four simultaneous equations frequently leads to complicated algebraic expressions which are difficult to work with. In fact, the exact solution to the above set of equations for the concentration of hydronium ion in terms of K_a is given by

$$K_a = \frac{(H_3O^+)\left[(H_3O^+) - \dfrac{K_w}{(H_3O^+)}\right]}{C_0 - (H_3O^+) + \dfrac{K_w}{(H_3O^+)}} \tag{11-28}$$

This equation is cubic in (H_3O^+); that is, if rearranged it takes the form

$$a_1(H_3O^+)^3 + a_2(H_3O^+)^2 + a_3(H_3O^+) + a_4 = 0$$

Equations of this form have exceedingly complicated solutions. Fortunately, it is generally possible to make some reasonable assumptions that greatly simplify the necessary calculations. It is very seldom necessary to resort to exact calculations that involve the solutions of equations such as that in (11-28). As we shall see, an exact solution is required only for very dilute solutions of very weak acids; happily, such cases are rare in real life, and we shall not deal further with them here.

The two useful assumptions that we can make are the following: (1) The amount of hydronium ion derived from the dissociation of HB is much greater than the amount derived from the self-dissociation of water; and (2) the fraction of HB that dissociates is small.

The first assumption is generally quite an adequate one. We already know that self-dissociation of water yields only $10^{-7}M$ hydronium ion and that the addition of acids to water represses such self-dissociation. Hence, $10^{-7}M$ is the upper limit to the concentration of hydronium ion that water can contribute to a solution of weak acid in water, and the actual contribution will generally be substantially less. As we shall see, this is usually very much less than that contributed by the dissociation of HB, so the assumption seldom fails. If it does fail, then recourse to the complete exact solution is necessary. Quantitatively, the assumption may be formulated as

$$(H_3O^+) = (B^-) \tag{11-29}$$

This expression follows from the stoichiometry of the dissociation process [Equation (11-22)] and the assumption. Note that it is equivalent to neglecting the contribution of hydroxide ion to the condition of electrical neutrality [Equation (11-26)]. This is a natural consequence, since if the dissociation of HB to generate H_3O^+ is more important than the dissociation of water to generate H_3O^+, then it must be true that the dissociation of HB to generate B^- is more important than the dissociation of water to generate hydroxide ion.

The second assumption is not quite as reliable as the first. It follows from the fact that we are dealing with *weak* acids, that is, acids that do not dissociate to a great extent. Therefore, it may frequently be true that only an insignificant fraction of the acid will dissociate under a given set of conditions. Quantitatively, we write

$$C_0 = (HB) \tag{11-30}$$

This expression is equivalent to neglecting the contribution of B^- to the equation of conservation of added acid [Equation (11-27)] in accordance with our assumption.

Let us now reformulate the expression for the dissociation constant, K_a, in terms of our assumptions [Equations (11-29) and (11-30)]:

$$K_a = \frac{(H_3O^+)(B^-)}{(HB)} = \frac{(H_3O^+)^2}{(HB)} = \frac{(H_3O^+)^2}{C_0} \tag{11-31}$$

Clearly, this expression may be solved directly for the concentration of hydronium ion:

$$(H_3O^+) = (C_0K_a)^{1/2} \qquad (11\text{-}32)$$

It is frequently convenient to use Equation (11-32) in a logarithmic form. If we take logarithms of this expression, we obtain

$$\log(H_3O^+) = \frac{\log K_a + \log C_0}{2}$$

Recognizing that $\log(H_3O^+) = -pH$ and that $\log K_a = -pK_a$, we obtain

$$pH = \frac{pK_a - \log C_0}{2} \qquad (11\text{-}33)$$

Let us employ this equation to calculate the pH of a solution of acetic acid, $pK_a = 4.76$, at a total concentration of 1*M*. Putting these values into Equation (11-33), we obtain

$$pH = \frac{4.76 - 0.00}{2} = 2.38 \qquad (H_3O^+) = 4.3 \times 10^{-3}M \qquad (11\text{-}34)$$

At this point, uncertainty remains as to the accuracy of the calculated value as a result of the assumptions that were employed in arriving at Equation (11-33). How can we know if the assumptions were justified or not? The simplest procedure to answer this question is to use the calculated results as a test of the assumptions themselves. That is, are Equations (11-29) and (11-30) in fact good approximations to reality?

Looking first at Equation (11-29), we recognize that this requires that the (H_3O^+) be greater than $10^{-7}M$, the maximum amount of hydronium ion contributed by the water itself. Clearly, this condition is satisfied; the calculated concentration of hydronium ion is several orders of magnitude greater than that contributed by the solvent. Therefore, Equation (11-29) is a good approximation to reality.

Turning our attention to Equation (11-30), we can calculate the actual concentration of HB from Equation (11-27) and compare that value with C_0, which was actually used in the calculation:

$$(HB) = C_0 - (B^-) = C_0 - (H_3O^+) = 1.0 - 4.3 \times 10^{-3} \cong 1.0$$

Thus, the degree of dissociation of HB is quite small, and the approximation of Equation (11-30) is justified. The error involved in our calculation is not greater than 1 percent. Generally, in calculations of this kind, errors of 5 percent or less are satisfactory.

Let us repeat our calculation for the same acid at a total concentration of 0.001*M*. Now Equation (11-33) tells us that

$$\text{pH} = \frac{4.76 + 3}{2} = 3.88 \qquad (H_3O^+) = 1.36 \times 10^{-4}M \tag{11-35}$$

The calculated concentration of hydronium ion is still much greater than that contributed by the solvent, so our first assumption is true. The actual concentration of HB is given by

$$(HB) = C_0 - (B^-) = C_0 - (H_3O^+) = 0.001 - 0.000136 = 0.000864M$$

Now, the calculated value of (HB) is about 14 percent less than the value employed in arriving at our answer; that is, the acid is about 14 percent dissociated, and our calculation assumed that it was not dissociated at all. An error of this magnitude is greater than we can tolerate. Hence, our calculation is not an adequate one. What do we do now?

Since the second assumption has proved inadequate, we can improve the calculation by employing the exact expression of Equation (11-27) instead of the approximate one of Equation (11-30). We can now reformulate the dissociation constant as follows:

$$K_a = \frac{(H_3O^+)(B^-)}{(HB)} = \frac{(H_3O^+)(B^-)}{C_0 - (B)} = \frac{(H_3O^+)^2}{C_0 - (H_3O^+)} \tag{11-36}$$

This expression can be solved for (H_3O^+) by making use of the general solution for quadratic equations. We obtain

$$(H_3O^+) = \frac{-K_a \pm \sqrt{K_a^2 + 4K_aC_0}}{2} \tag{11-37}$$

Note that this expression reduces to Equation (11-32) if $C_0 \gg K_a$. Hence, it is clear that expression (11-32) will be adequate if the concentration of the acid is not too dilute; it is inadequate for such dilute solutions, and recourse must be made to Equation (11-37). Repeating our calculation of $0.001M$ acetic acid, we obtain a value of $(H_3O^+) = 1.27 \times 10^{-4}M$. This value is about 7 percent less than that previously calculated.

Thus far, we have found that for sufficiently concentrated solutions of weak acids in water, we can calculate the pH (or the concentration of hydronium ion) by the use of Equation (11-32) or (11-33), and for more dilute solutions we can use Equation (11-37) for the same purpose. No firm statement can be made about the concentration at which it becomes necessary to use the more complex expression; this depends on the strength of the acid as well. *It is a good idea always to check results to be sure that they accord with the assumptions employed to generate them, just as we did in the cases above.*

Once the concentration of hydronium ion in solution has been established as described above, recourse to Equations (11-24), (11-26), and (11-27) suffices to determine the concentrations of hydroxide ion, B^-, and HB. Hence, a complete solution to the problem is in hand.

Example *Hydrogen cyanide (HCN) is an exceedingly poisonous gas. When dissolved in water it acts as a weak acid, generating the toxic cyanide ion:*

$$HCN \rightleftharpoons CN^- + H^+$$

The dissociation constant for HCN is $4 \times 10^{-10}M$. *For a solution containing an initial concentration of 0.1M HCN, calculate the equilibrium concentrations of all dissolved species: HCN,* CN^-, H_3O^+, *and* OH^-.

Solution: *The easiest way to get started is to use one of the approximate equations, either (11-32) or (11-33), to calculate the concentration of* H_3O^+. *Choosing the former, we find that*

$$(H_3O^+) = (C_0K_a)^{1/2} = (0.1 \times 4 \times 10^{-10})^{1/2} = (40 \times 10^{-12})^{1/2} = 6.3 \times 10^{-6}M$$

Before proceeding any further with the calculation, it is essential to check to see if use of the approximate equation is valid. First, we note that the calculated concentration of hydronium ion is small but considerably greater than the maximum produced by the dissociation of water. Hence, the assumption that the amount of acid produced from the dissociation of the weak base is much greater than that derived from water is justified. Second, our assumption that HCN dissociates to only a small extent is also well justified, since

$$(HCN) = C_0 - (CN^-) \cong C_0 - (H_3O^+) \cong 0.1 - 6.3 \times 10^{-6} \cong 0.1M$$

Having justified our assumptions, we can now proceed to complete the calculation. (You may wish to repeat this part of the calculation beginning with a solution of 0.001M HCN; it then turns out that the first assumption is not justified, although the second one is). We have already noted in checking the assumption that $(HCN) \cong 0.1M$ *and that* $(CN^-) \cong (H_3O^+) \cong 6.3 \times 10^{-6}M$. *It remains to calculate the concentration of hydroxide ion:*

$$K_w = (H_3O^+)(OH^-)$$

$$(OH^-) = \frac{K_w}{(H_3O^+)} = \frac{10^{-14}}{6.3 \times 10^{-6}} = 1.6 \times 10^{-9}M$$

Example *Dichloroacetic acid,* $Cl_2HC—COOH$, *is a weak acid having a dissociation constant of* $5 \times 10^{-2}M$. *For a solution containing an initial concentration of 0.1M dichloroacetic acid (0.1M DCA), calculate the equilibrium concentration of all dissolved species.*

Solution *As above, we can begin the calculation by employing the approximate Equation (11-32). The concentration of the hydrated proton is calculated to be*

$$(H_3O^+) = (C_0K_a)^{1/2} = (0.1 \times 5 \times 10^{-2})^{1/2} = (50 \times 10^{-4})^{1/2} = 7.05 \times 10^{-2}M = 0.0705M$$

Is this result consistent with the assumption made in deriving Equation (11-32)? The calculated concentration of hydronium ion is certainly much greater than that which could be contributed by the dissociation of water, so our neglect of the latter source of hydronium ion is well justified. On the other hand, it is clearly not true that dichloroacetic acid dissociates only to a small extent:

$$(\mathrm{DCA}) = C_0 - (\mathrm{DCA}^-) \cong C_0 - (\mathrm{H_3O^+}) \cong 0.1 - 0.07 \cong 0.03$$

That is, the calculation reveals that about 70 percent of the acid initially added is dissociated at equilibrium. Hence, the assumption that $C_0 \cong$ (DCA) *fails in this case. Hence, recourse must be made to the more precise expression of Equation (11-37):*

$$(\mathrm{H_3O^+}) = \frac{-K_a \pm \sqrt{K_a^2 + 4K_aC_0}}{2} = \frac{-0.05 \pm \sqrt{(0.05)^2 + 4 \times 0.05 \times 0.1}}{2} = 0.05M$$

The concentration of dichloroacetate ion is given by the expression

$$(\text{dichloroacetate}) \cong (\mathrm{H_3O^+}) = 0.05M$$

The concentration of undissolved acid is just (DCA) $= C_0 -$ *(dichloroacetate⁻)* $= 0.05M$. *Finally,* $(\mathrm{OH^-}) = K_w/(\mathrm{H_3O^+}) = 10^{-14}/0.05 = 2.0 \times 10^{-13}M$.

We can make exactly equivalent arguments for the case of aqueous solutions of weak bases. For an uncharged weak base, B, we formulate the equilibrium in the following way:

$$\mathrm{B + H_2O \rightleftharpoons HB^+ + OH^-} \tag{11-38}$$

The associated equilibrium constant, K_b, is given by

$$K_b = \frac{(\mathrm{HB^+})(\mathrm{OH^-})}{(\mathrm{B})} \tag{11-39}$$

In accordance with the procedures developed above, let us make the two simplifying assumptions

$$(\mathrm{HB^+}) = (\mathrm{OH^-}) \tag{11-40}$$

$$(\mathrm{B}) = C_0 \tag{11-41}$$

Putting these expressions into Equation (11-39), it follows directly that

$$(\mathrm{OH^-}) = (K_bC_0)^{1/2} \tag{11-42}$$

This expression is just analogous to that of Equation (11-32). In logarithmic form, Equation (11-42) becomes

$$\mathrm{pOH} = \frac{\mathrm{p}K_b - \log C_0}{2} \tag{11-43}$$

This expression, of course, is closely related to that of Equation (11-33).

Example $\mathrm{p}K_b$ *for ammonia is 4.75. What is pOH for a solution initially containing 0.2M* $\mathrm{NH_3}$?

Solution *Employing Equation (11-43),*

$$pOH = \frac{pK_b - \log C_0}{2} = \frac{4.75 + 0.7}{2} = 2.725$$

As always in problems of this type, it is now necessary to check that the assumptions employed in obtaining Equation (11-43) are justified. First, we need to convert a pOH of 2.725 into the concentration of hydroxide ion: $(OH^-) = 1.88 \times 10^{-3}M$. This value is clearly much larger than the concentration of hydroxide ion supplied by the dissociation of water, so Equation (11-40) is justified. It follows that $(NH_4^+) = 1.88 \times 10^{-3}M$. Using this value, we can now check the second assumption [Equation (11-41)]. $(NH_3) = C_0 - (NH_4^+) = 0.2 - 0.00188 = 0.19812$. This assumption is true within an error of less than 2 percent, which is satisfactory for most purposes.

If we are concerned with sufficiently dilute solutions of weak bases, then the assumption embodied in Equation (11-41) will fail as a good approximation to reality, and we must use the exact expression $(B) = C_0 - (HB^+)$. Our solution then becomes

$$(OH^-) = \frac{-K_b \pm \sqrt{K_b^2 + 4K_bC_0}}{2} \tag{11-44}$$

Example *The K_b for ammonia is $1.8 \times 10^{-5}M$. Calculate the concentration of hydroxide in a solution containing an initial concentration of ammonia of $1 \times 10^{-4}M$ employing both Equations (11-42) and (11-44).*

Solution *Using Equation (11-42) first, we obtain*

$$(OH^-) = (K_bC_0)^{1/2} = (1.8 \times 10^{-5} \times 10^{-4})^{1/2} = (18 \times 10^{-10})^{1/2} = 4.25 \times 10^{-5}M$$

Now, using Equation (11-42), we find

$$(OH^-) = \frac{-K_b \pm \sqrt{K_b^2 + 4K_bC_0}}{2}$$

$$= \frac{-1.8 \times 10^{-5} \pm \sqrt{(1.8 \times 10^{-5})^2 + (4 \times 1.8 \times 10^{-5} \times 10^{-4})}}{2}$$

$$= 3.42 \times 10^{-5}M$$

Note that the two results differ significantly. Thus, for a solution of ammonia of $1 \times 10^{-5}M$, the assumption embodied in Equation (11-41) fails, and use of the simple Equation (11-42) yields an erroneous result. You may wish to prove for yourself that Equation (11-41) is not a reasonable approximation to reality in this case.

The similarity of the expressions employed to calculate concentrations of hydronium ion and hydroxide ion for solutions of weak acids and bases implies that the two types of calculations are intimately related. We shall learn more about this relationship in the next section, but it is convenient to find out the relationship between K_a and K_b for a conjugate acid–base pair at this point. Returning to Equations (11-25) and (11-39), which define K_a and K_b, we note that

$$K_a K_b = K_w \tag{11-45}$$

In logarithmic form, this expression becomes

$$\mathbf{p}K_a + \mathbf{p}K_b = \mathbf{p}K_w = \mathbf{14} \tag{11-46}$$

Since the sum of the two quantities is equal to a constant, it follows that when the strength of an acid increases, the strength of its conjugate base necessarily decreases.

TITRATION OF A WEAK ACID WITH A STRONG BASE

In the last section we learned how to calculate values of pH for solutions of weak acids and weak bases. A slightly more complicated situation arises when we have mixture of a weak acid and strong base or a mixture of a weak base and strong acid. For the sake of convenience, let us consider the former of these two cases, although precisely the same considerations apply to both. We wish, then, to arrive at a method for the calculation of values of pH for a solution of a weak acid into which a known amount of strong base has been added. The reaction that occurs upon addition of a strong base to a weak acid is shown in Equation (11-47):

$$\text{HB} + \text{YOH} \longrightarrow \text{B}^- + \text{H}_2\text{O} + \text{Y}^+ \tag{11-47}$$

We want to note immediately that the amount of conjugate base, B^-, produced is equal to the amount of strong base added. This fact follows immediately from our definitions of weak acids and strong bases.

This situation can be treated in basically the same way as developed above, except that we need to reformulate the equation of electrical neutrality to include the concentration of the cation of the strong base:

$$(\text{Y}^+) + (\text{H}_3\text{O}^+) = (\text{B}^-) + (\text{OH}^-) \tag{11-48}$$

This equation of electrical neutrality, together with the dissociation constant for water, the dissociation constant for the weak acid, and the usual conservation of species equations, are sufficient to describe this system completely. However, as in the case above, an exact treatment of the situation in solution is cumbersome and generally not necessary. Let us, therefore, begin by making the same assumptions as we have found useful in the treatment of dilute solutions of weak acids and weak bases alone. To begin with, it will nearly always be true that the concentration of the conjugate base of the weak acid will be very much greater than that of hydroxide ion. This follows since, in weakly acidic solutions, the concentration of hydroxide ion is necessarily very, very low and since the amount of B^- will be quantitatively equal to the amount of added strong base. This last quantity is generally quite substantial. Making this assumption, the equation of electrical neutrality then reduces itself to the form:

$$(Y^+) + (H_3O^+) = (B^-) \tag{11-49}$$

If we define C_0 as the total concentration of weak acid present initially and X as the amount of strong base added, which is quantitatively equal to the concentration of Y^+, then we can formulate the dissociation constant for the weak acid as follows:

$$K_a = \frac{(H^+)(B^-)}{(HB)} = \frac{(H^+)(X + H^+)}{C_0 - (X + H^+)} \tag{11-50}$$

Equation (11-50) can be solved directly for the concentration of hydronium ion. Clearly, such a solution will be quadratic in the concentration of hydronium ion and, while it is sometimes necessary to use such equations as we have observed above, they are frequently unnecessarily complicated. Equation (11-50) can be simplified substantially by noting that ordinarily C_0 and $X \gg H^+$. If this relationship is obeyed, then we obtain the following expression:

$$K_a = \frac{(H^+)(X)}{C_0 - X} \tag{11-51}$$

This expression is now linear in the concentration of hydronium ion and is useful quite generally for deriving that quantity from the experimental information. As we have found in the past, it is frequently useful to employ such equations in the logarithmic form. If we take logarithms of both sides of Equation (11-51), we obtain

$$pH = pK_a + \log \frac{X}{C_0 - X} \tag{11-52}$$

In this expression we note that $X = (B^-)$ and $(HB) = C_0 - X$. Substituting these relationships in the above equation yields a more generally useful form:

$$\mathbf{pH = p}\boldsymbol{K_a}\mathbf{ + \log \frac{(B^-)}{(HB)} = p}\boldsymbol{K_a}\mathbf{ + \log \frac{(base)}{(acid)}} \tag{11-53}$$

This equation is called the **Henderson-Hasselbalch equation** and finds very wide use for the purpose of calculating values of pH for solutions of weak acids and strong bases or, generally, for solutions in which appreciable concentrations of both a weak acid and its conjugate base exist.

We can understand something about the conditions under which this equation is valid by recognizing that we could have derived it directly from the expression for the dissociation constant of the weak acid simply by taking logarithms of that expression. Hence, the assumptions that we have

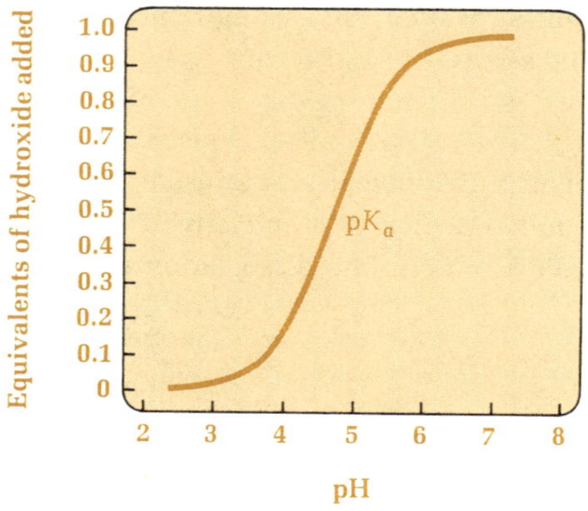

Figure 11-1 The titration of acetic acid.

made in arriving at the Henderson-Hasselbalch equation are equivalent to saying that the interconversions of the acid and its conjugate base through reaction with solvent are not very important. Thus, the dissociation of the acid to yield the conjugate base,

$$HB + H_2O \rightleftharpoons B^- + H_3O^+$$

and the association of the conjugate base with water,

$$B^- + H_2O \rightleftharpoons HB + OH^-$$

are not considered to be important reactions in the solution. Put a slightly different way, if we were to add to a solution a given molarity of a weak acid and a given molarity of a strong base or, what is equivalent, of the conjugate base of that weak acid, these molarities are considered to be the ones present at equilibrium.

From the Henderson-Hasselbalch equation (11-53), we note that when the concentration of a weak acid and its conjugate base are exactly the same, the value of pH is exactly equal to the pK_a for that weak acid. This, then, provides an experimental method for the determination of values of pK_a's. To see precisely how this is done, let us consider the way in which the pH of a dilute solution of a weak acid changes as we add increasing quantities of a strong base. For the sake of illustration, let us consider the addition of sodium hydroxide to acetic acid. The results of such an experiment are shown graphically in Figure 11-1.

We notice that as we begin to add sodium hydroxide, the pH first changes very rapidly but then begins to change more slowly, and near the midpoint of the curve changes quite slowly with increasing additions of hydroxide ion. Then, toward the end of the titration, the pH of the solution begins to change very rapidly as the amount of hydroxide ion added approaches the amount of acetic acid initially present. When the equivalents of hydroxide are exactly one-half of those of acetic acid, the midpoint of the curve is reached and the concentration of acetic acid must equal the concentration of acetate ion. This pH, then, must be the pK_a of acetic acid. Similar procedures employing other weak acids will suffice to establish the values of pK_a for those acids as well.

Let us return for a moment to our observation that values of pH in the course of the titration of acetic acid with sodium hydroxide change most slowly for a given change in hydroxide ion concentration near the pK_a of acetic acid. This fact gives rise to the concept of **buffering.** The term buffering refers to the capacity of a solution containing a weak acid and its conjugate base to accept appreciable amounts of hydronium ion or hydroxide ion with very small changes in the value of the pH of the solution. Buffering is of utmost importance for the maintenance of living systems. For example, a change in the pH of the blood by as little as 0.5 pH units can prove fatal. Consequently, elaborate systems for maintaining the pH of the blood constant have been developed. We shall return to these later. The concept of buffering is also important environmentally, since the maintenance of constant relative populations of different species in a body of water depends crucially on the value of pH.

Quantitatively, we can describe the ability of a given weak acid and its conjugate base to act as a buffer in terms of a buffering index, β. For our purposes it will be convenient to define β in the following way:

$$\beta \cong \alpha(1 - \alpha) \tag{11-54}$$

In this expression α is the fraction of total buffer species that exists in the basic form. That is,

$$\alpha = \frac{(B^-)}{(B^-) + (HB)} \tag{11-55}$$

We note that value of β is a maximum when α is equal to 0.5, that is, when the pH of the solution is equal to the pK_a of the weak acid or, what is equivalent, when the concentration of the conjugate acid and the conjugate base are equal. Under these conditions β has a value of 0.25. If the value of α is greater than 0.5, corresponding to solutions more acidic than this value, the value of β is necessarily less. This corresponds to a lowering of the buffering capacity of the solution. Hence, the buffering capacity of any acid and its conjugate base is maximal at a pH corresponding to the associated pK_a and becomes less for values of pH either more acidic or more basic. Generally, conjugate acid-base pairs prove to be useful buffers from about one pH unit below the pK_a to about one pH unit above this value. We shall return to specific examples of the utilization of buffer systems for the maintenance of constant pH in important systems later, but for the moment let us turn to the use of the Henderson-Hasselbalch equation for calculations of characteristic types.

To begin with, let us calculate the pH of a solution that initially contained 0.1*M* acetic acid to which has been added sufficient hydroxide to make that solution 0.05*M*. It follows from these conditions that the addition of the hydroxide ion will convert exactly half of the acetic acid initially present to sodium acetate. Consequently, both of these species will have concentrations of 0.05*M* at equilibrium. Use of the Henderson-Hasselbalch equation (11-53) then tells us that the pH of the solution is just the pK_a of acetic acid, 4.76.

It is easy to see that the result just obtained is the same as adding sufficient acetic acid and sodium acetate to pure water so that the solution will be 0.05*M* in each. As before, the Henderson-Hasselbalch equation requires that the pH of the solution be numerically equal to the pK_a of the conjugate acid.

Finally, we can note that the pH of a buffer solution is independent of the concentration of the buffer components and depends only on the ratio of the two. Thus, a solution that is 0.2*M* in acetic acid and 0.2*M* in sodium acetate has exactly the same value of pH as one that is 0.05*M* in each of these components. The buffering capacity of the former, more concentrated, solution is, of course, four times as great as the latter.

Example *Calculate the pH of a solution containing 0.05M acetic acid and 0.10M sodium acetate. The* pK_a *of acetic acid is 4.76.*

Solution *The concentrations of acetic acid and sodium acetate are sufficiently high to give confidence that the Henderson-Hasselbalch equation suffices for the calculation:*

$$\mathrm{pH} = \mathrm{p}K_a + \log\frac{\text{base}}{\text{acid}} = 4.76 + \log\frac{0.10}{0.05} = 4.76 + \log 2 = 5.06$$

Example *Dichloroacetic acid has a* pK_a *value of 1.29. Would you employ the Henderson-Hasselbalch equation to determine values of pH for solutions containing dichloroacetic acid and dichloroacetate ion?*

Solution *Use of this equation would be very risky in this case, since dichloroacetic acid is sufficiently strong to react appreciably with water:*

$$HA + H_2O \longrightarrow A^- + H_3O^+$$

When such reactions are important, the assumptions used in obtaining the Henderson-Hasselbalch equation are not valid.

CARBON DIOXIDE AND CARBONIC ACID

Carbon dioxide plays several important roles in nature. The process of respiration in living systems involves the conversion of complicated organic compounds to carbon dioxide and water (see Chapter 21).

respiration: organic compounds $\longrightarrow CO_2 + H_2O$

The energy derived from reactions of this type is employed for the maintenance of living systems as developed earlier. The opposite process to respiration is called photosynthesis (Chapter 19).

photosynthesis: $CO_2 + H_2O \xrightarrow{h\nu}$ organic compounds

Photosynthesis is thus the counterpart to respiration and helps to maintain the balance of carbon dioxide in nature relatively constant. Finally, we can note that carbon dioxide and the species derived from it are very important as buffers in natural systems. That is, these species function as devices for maintaining constant the pH of such naturally important solutions as the oceans and blood.

There are four equilibria in which carbon dioxide and carbonic acid participate in which we are interested. The first of these is shown in Equation (11-56), which refers simply to the way carbon dioxide distributes itself between the air and the oceans:

$$CO_2 \text{ (air)} \rightleftharpoons CO_2 \text{ (water)} \qquad (11\text{-}56)$$

The equilibrium constant for reactions of this sort is generally called A, the **absorption coefficient** for the gas in question, and is defined as shown in Equation (11-57):

$$A = \frac{\text{moles per liter in solution}}{\text{moles per liter in gas phase}} \qquad (11\text{-}57)$$

Thus, values of A simply give the relative molar concentrations of the appropriate gas in solution and in the gas phase. Carbon dioxide has a very unusual absorption coefficient. Although the values are dependent on temperature, they are generally in the vicinity of unity. Thus, at 20°C, A has a value of 0.94, at 25°C, a value of 0.84, and at 30°C, a value of 0.75. Most gases, including such simple ones as oxygen and nitrogen, have very small absorption coefficients. These are in the vicinity of 0.02 to 0.04. Consequently, the concentrations of these gases in air are very much greater than in water. Carbon dioxide, in contrast, has a nearly uniform partitioning between these two phases and it is, consequently, impossible to flush all of the carbon dioxide out of the oceans or, in contrast, to dissolve all of the carbon dioxide from the air into the oceans. This is certainly a favorable state of affairs since, as we have noted, carbon dioxide plays important roles in both phases in the maintenance of pH in the oceans and in the processes of respiration and photosynthesis in the air.

Example *Why is champagne served cold?*

Solution *The bubbly character in champagne is due to dissolved carbon dioxide. The temperature dependence of the absorption coefficient for this*

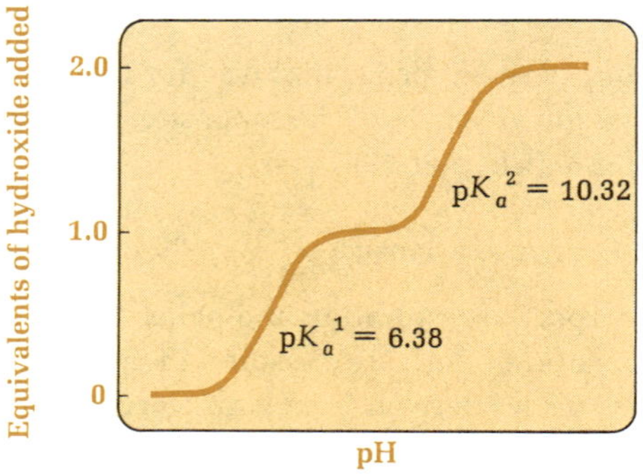

Figure 11-2 The titration of H_2CO_3.

gas decreases with increasing temperature. Hence, the higher the temperature of the champagne, the less the solubility of carbon dioxide at equilibrium. Champagne, then, is iced to help retain the maximum amount of dissolved CO_2*; that is, to keep it as bubbly as possible.*

The second equilibrium with which we need to deal is the hydration of carbon dioxide in aqueous solution. This is formulated as follows:

$$CO_2 + H_2O \rightleftharpoons H_2CO_3 \qquad K_{hyd} = \frac{(H_2CO_3)}{(CO_2)} = 0.00258 \tag{11-58}$$

As noted here, the equilibrium constant for the hydration of carbon dioxide is quite small. Consequently, the principal form in solution is carbon dioxide itself, not carbonic acid.

The final two equilibria of interest to us in this connection are those that describe the successive ionizations of carbonic acid. The first of these may be formulated as follows:

$$H_2CO_3 \rightleftharpoons HCO_3^- + H_3O^+ \qquad K_{H_2CO_3} = \frac{(H_3O^+)(HCO_3^-)}{(H_2CO_3)} = 1.7 \times 10^{-4} M \tag{11-59}$$

In this process carbonic acid loses one of its two hydrogens to generate the bicarbonate anion. Bicarbonate is itself a weak acid and can dissociate to yield the carbonate dianion:

$$HCO_3^- \rightleftharpoons CO_3^{2-} + H_3O^+ \qquad K_{HCO_3^-} = \frac{(H_3O^+)(CO_3^{2-})}{(HCO_3^-)} = 4.8 \times 10^{-11} M \tag{11-60}$$

These equilibria then, define the behavior of carbon dioxide and carbonic acid in the gaseous and aqueous phases.

Generally, we shall find it more convenient to deal with an overall first ionization constant for carbonic acid rather than the specific ionization constant as defined in Equation (11-59). The difference between the two arises from the fact that carbonic acid is in equilibrium with carbon dioxide in aqueous solution. We can define the overall first ionization constant as shown in Equation (11-61). In this equation we replace the concentration of carbonic acid in the denominator of the equilibrium constant by the total concentration of carbon dioxide and carbonic acid that are in equilibrium with each other. We can derive an explicit expression for the overall first ionization constant by replacing the concentration of carbon dioxide in the denominator of Equation (11-61) employing Equation (11-58):

$$K_{ov} = \frac{(H_3O^+)(HCO_3^-)}{(CO_2) + (H_2CO_3)} \tag{11-61}$$

$$K_{ov} = \frac{(H_3O^+)(HCO_3^-)}{\dfrac{(H_2CO_3)}{K_{hyd}} + (H_2CO_3)} \tag{11-62}$$

Rearrangement of this expression yields

$$K_{ov} = \frac{(H_3O^+)(HCO_3^-)}{(H_2CO_3)\left[1 + \dfrac{1}{K_{hyd}}\right]} = \frac{K_{H_2CO_3}K_{hyd}}{1 + K_{hyd}} \tag{11-63}$$

Realizing that K_{hyd} is very much smaller than unity, it follows that values of K_{ov} are actually very nearly equal to the product of the dissociation constant for carbonic acid itself and for the hydration of carbon dioxide:

$$K_{ov} = K_{H_2CO_3}K_{hyd} = 4.2 \times 10^{-7} \tag{11-64}$$

Since most of the total carbon dioxide that exists in aqueous solution does exist as carbon dioxide and not as carbonic acid, it follows that the titration curve for carbonic acid will reflect the overall first ionization constant rather than the specific one for carbonic acid itself. Consequently, the titration curve for carbonic acid has the appearance of that shown in Figure 11-2. Note in this case that there are two breaks in the profile; one corresponding to the overall first ionization constant and the second to the second ionization constant of carbonic acid. Hence, solutions of carbon dioxide will have two regions of substantial buffering power; one centering about the first pK_a, which has a value near neutrality, and another centering about a pH of 10.3, that is, under quite basic conditions.

Solutions of sodium bicarbonate

There are, of course, a variety of acid–base equilibrium problems that arise in solutions of dibasic acids, but we shall consider just one of these as a representative example of the kind of calculations involved. That one centers on solutions of sodium bicarbonate.

Sodium bicarbonate can function as both a weak acid and as a weak base. Consequently, the two pertinent equilibria may be formulated as

$$\begin{aligned} HCO_3^- + H_2O &\rightleftharpoons H_3O^+ + CO_3^{2-} \\ HCO_3^- + H_2O &\rightleftharpoons OH^- + H_2CO_3 \end{aligned} \tag{11-65}$$

If we simply sum these two equilibria, we obtain the overall expression

$$2HCO_3^- \rightleftharpoons CO_3^{2-} + H_2CO_3 \tag{11-66}$$

The equilibrium constant associated with Equation (11-66) is simply the ratio of the two dissociation constants for carbonic acid:

$$K = \frac{(H_2CO_3)(CO_3^{2-})}{(HCO_3^-)^2} = \frac{(H_3O^+)(CO_3^{2-})}{(HCO_3^-)} \frac{(H_2CO_3)}{(HCO_3^-)(H_3O^+)} = \frac{K_{HCO_3^-}}{K_{ov}} \tag{11-67}$$

To derive a value for the pH of these solutions, we can once again rely on the equations for charge balance (11-68) and material balance (11-69):

$$(Na^+) + (H_3O^+) = (HCO_3^-) + 2(CO_3^{2-}) + (OH^-) \tag{11-68}$$

$$(Na^+) = (H_2CO_3) + (HCO_3^-) + (CO_3^{2-}) = [NaHCO_3]^0 \tag{11-69}$$

Note that the latter equation follows from the fact that all of the species derived from the bicarbonate moiety of the original salt must be equivalent to the amount of sodium ion derived from that salt. If we simply subtract Equation (11-69) from Equation (11-68), we obtain

$$(H_3O^+) + (H_2CO_3) = (CO_3^{2-}) + (OH^-) \tag{11-70}$$

To put this equation into the most useful form, we wish to replace the concentrations of H_2CO_3 and CO_3^{2-} by the concentration of bicarbonate. A procedure in which the concentrations of acidic and basic species are given in terms of the major species in solution is, as noted above, generally profitable. We can accomplish this by employing the dissociation constants defined in Equations (11-60) and (11-61). We obtain

$$(H_3O^+) + \frac{(H_3O^+)(HCO_3^-)}{K_{ov}} = \frac{(HCO_3^-)K_{HCO_3^-}}{(H_3O^+)} + \frac{K_w}{(H_3O^+)} \tag{11-71}$$

Multiplying through by the concentration of hydronium ion yields

$$(H_3O^+)^2\left[1 + \frac{(HCO_3^-)}{K_{ov}}\right] = K_{HCO_3^-}(HCO_3^-) + K_w \tag{11-72}$$

Making the reasonable assumptions that (1) $(HCO_3^-) \gg K_{ov}$ and (2) $K_{HCO_3^-}(HCO_3^-) \gg K_w$, this expression simplifies to

$$(H_3O^+) = \sqrt{K_{ov}K_{HCO_3^-}} \tag{11-73}$$

Thus, equations of this form are generally valid for equimolar mixtures of weak acids and weak bases, even if, as in the present case, the acid and base are the same.

With this background in acid-base equilibria, we are in a position to consider the regulation of pH in blood. We turn now to a consideration of the chemistry of this complicated biological fluid and shall need to describe several basic facts before returning to acid-base aspects.

SOME ASPECTS OF THE CHEMISTRY OF BLOOD

Complicated organisms, such as man, face certain problems in sensing and communicating with their environment that are not shared by simple unicellular ones. Thus, for example, a bacterium takes in nutrients from its aqueous environment directly across the cell wall and excretes the products of its metabolism into the environment, also directly across that cell wall. Consequently, this organism is able to communicate with its environment without the necessity of having complicated systems for doing so. Obviously, more complicated organisms cannot survive in a similar fashion, and we can clearly recognize the development of a variety of complicated systems that permit higher organisms to interact with their environment in a number of ways. For example, man's complicated nervous system, together with his external organs of sense, permit him to communicate in a sensory way with the environment. Similarly, the blood and lymph play important functions in terms of the material interaction of the organism with its environment. At this point, we wish to focus attention specifically on the chemistry of human blood to understand something about how this complicated biological fluid is able to serve the living organism as a communication device.

Blood serves, of course, a substantial number of functions in man. To understand something about the relationship of these functions to the chemical components of the blood, let us divide this fluid into two parts. This can be accomplished by centrifuging whole human blood at a fairly high rate of speed. If we do so, the cellular material of blood will be sedimented, leaving above it a clear, slightly yellow liquid. This liquid contains all of the noncellular components of blood and is a complicated mixture of proteins, carbohydrates, inorganic ions, and other species. It is called *blood plasma*. The inorganic constituents of blood plasma are presented in Table 11-4.

Comparison of relative concentrations of a variety of these constituents with the same constituents of sea water reveals a marked similarity between the compositions of these two important solutions. In fact, some people have been led to suggest that blood had its origin in sea water. However, along with the similarities, a number of distinct differences exist as well, and it is currently believed that sea water and blood did have separate origins.

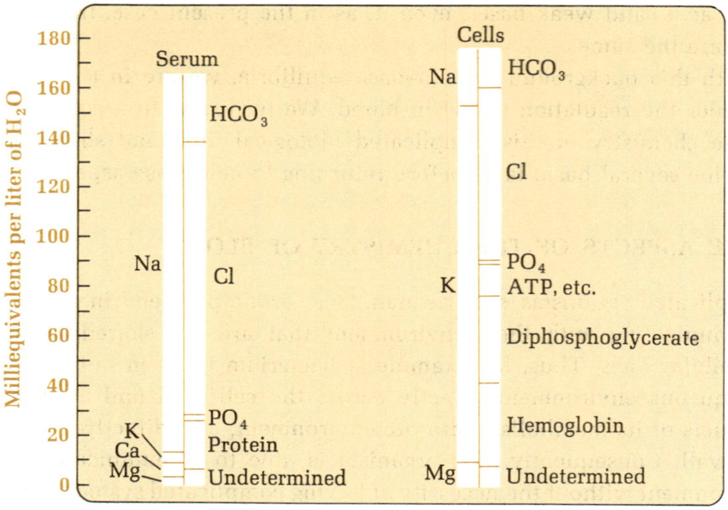

Figure 11-3 The electrolytic structure of oxygenated human erythrocytes. [From G. M. Guest, *Am. J. Diseases Children,* **64,** 401 (1942). Copyright 1942, American Medical Association.]

Table 11-4 Approximate ranges of values for the principal inorganic constituents of the blood plasma of man

Anions	Concentration (meq./liter[a])	Cations	Concentration (meq./liter[a])
Total	142–150	Total	142–158
Bicarbonate	24–30	Calcium	4.5–5.6
Chloride	100–110	Magnesium	1.6–2.2
Phosphate	1.6–2.7	Potassium	3.8–5.4
Sulfate	0.7–1.5	Sodium	132–150
Iodine (total)	8–15[b]	Iron	50–180[b]
Iodine (protein-bound)	6–8[b]	Copper	8–16[b]

[a] meq./liter, milligram equivalents per liter, or milliequivalents per liter; 1 meq./liter of an ion or a substance = (mg/liter × valence)/(atomic or formula weight).
[b] These concentrations are in terms of micrograms per 100 ml.

Source: A. White, P. Handler, and E. Smith, *Principles of Biochemistry,* 4th ed., McGraw-Hill, New York, 1968. Copyright © 1968, McGraw-Hill Book Company. Used with permission of McGraw-Hill Book Company.

In addition to inorganic constituents, blood plasma contains a variety of proteins. Let us focus our attention on four of these. First, the quantitatively most important constituent of blood plasma is *serum albumin*. Serum albumin is a relatively small protein, as plasma proteins go, with a molecular weight of 69,000. It appears to have two principal functions: It is important in the osmotic regulation of the blood, and it also assists transport of certain species that are not very soluble in aqueous solution. Serum albumin is able to function in the latter capacity since these species, including the fatty acids, are able to associate physically with serum albumin; hence, they can be transported by that molecule to the site of their utilization within the body. Second, a very important group of plasma proteins are the *immunoglobulins*. These are the proteins that are elaborated by the body in response to invasion by outside toxins. They are frequently referred to as *antibodies* and form one of the organism's primary lines of defense against poisoning and disease. Third, the plasma contains a number of proteins that are involved in the transport of metal within the body. One of these is the protein *transferrin*, with a molecular weight of 81,000, which is involved in the transport of iron. Two atoms of iron are tightly bound to each mole of transferrin. A second important protein is *ceruloplasmin*, which is involved in the transport of copper. Eight atoms of copper may be bound per molecule of ceruloplasmin, which has a molecular weight of 150,000. The ceruloplasmin–copper complex has a characteristic blue color. Fourth, and finally, the blood contains a substantial amount of a protein called *fibrinogen*. This protein is involved in the clotting of blood and may be converted, following injury, to the insoluble protein *fibrin*, from which the clot itself is formed.

If we turn attention away from the blood plasma and to the cellular material, which may be centrifuged out of the whole blood, we find generally two kinds of cells. Principal among these are the red blood cells or *erythrocytes*. These have the shape of a hockey puck that is slightly indented on both faces. That is, it is substantially thinner in the center than it is at the edge. Normal blood contains some 3 or 4 million erythrocytes per cubic centimeter. The erythrocytes are relatively simple cells and contain predominantly one protein. That protein is *hemoglobin*, which is, of course, concerned with the transport of oxygen from the lungs to the tissues, where it is employed as a terminal electron acceptor for a number of metabolic processes that are oxidative in nature. In the erythrocytes, hemoglobin is found in a concentration of about 34 g/100 ml. In terms of whole blood, this corresponds to a concentration of about 15 g of hemoglobin per 100 ml of blood.

The erythrocytes also contain substantial quantities of simple inorganic electrolytes. These differ, in a quantitative sense, quite markedly from the corresponding constituents of blood plasma. In Figure 11-3 the compositions of blood plasma and the erythrocytes are compared directly.

Since we shall be directly concerned with the function of hemoglobin as an oxygen-carrying species in blood, let us briefly consider its structure. You may wish to refer back to Chapter 7 to refresh yourself concerning certain aspects of the structure of proteins. Hemoglobin is a protein of molecular weight 68,000, which is characterized by quaternary structure. That is, hemoglobin consists not of one polypeptide chain organized in space, but of four independent polypeptide chains, each of which is coiled in its own specific way, and then these are joined together to form the intact hemoglobin molecule. Each of the subunits of hemoglobin has a molecular weight near 17,000 and resembles very markedly the myoglobin molecule. Myoglobin is the oxygen-storage pigment of mammalian muscle. Each of the hemoglobin subunits contains one heme prosthetic group. This prosthetic group consists of a tetrapyrrole nucleus with an atom of iron in the middle:

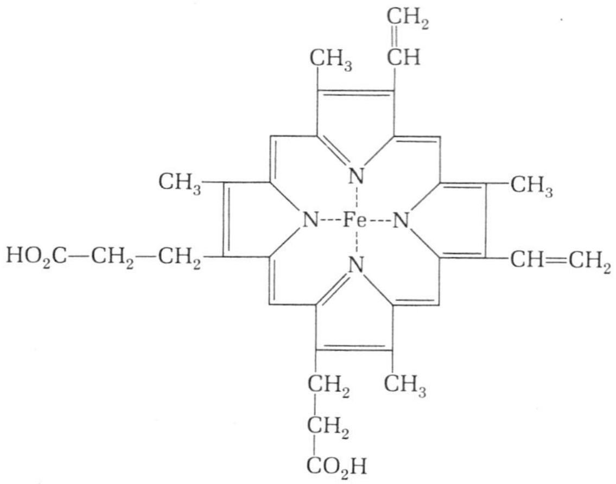

In hemoglobin, the iron is in the ferrous state. Ordinarily, ferrous ion will make bonds to six ligands. In the case of hemoglobin, these are formed from the four nitrogens of the tetrapyrrole structure as shown in the above illustration, a ligand donated from the protein and, finally, the molecule of oxygen, which is transported by the hemoglobin molecule. Thus, the oxygen interacts directly with ferrous iron atoms. In fact, it is a remarkable property of hemoglobin that the reduced ferrous ion is able to combine directly with molecular oxygen without undergoing an oxidation reaction. In contrast to hemoglobin, myoglobin contains just one heme group to which an oxygen atom may be bound.

Myoglobin, with just one heme group, interacts with just one molecule of oxygen per molecule of protein. Hemoglobin, on the other hand, can interact with four molecules of oxygen since it has four heme groups, one associated with each of the polypeptide chains. The pertinent equilibria are as shown in Equation (11-74), where Mb is myoglobin and Hb is hemoglobin:

$$Mb + O_2 \rightleftharpoons MbO_2$$
$$Hb + 4O_2 \rightleftharpoons Hb(O_2)_4 \qquad (11\text{-}74)$$

These two oxygen-carrying proteins differ in a most interesting way in terms of the response of the proteins to increasing oxygen pressure. Myoglobin exhibits the simpler behavior of the two. A plot of the percentage of myoglobin saturated with oxygen against the pressure of oxygen in the environment is shown in Figure 11-4. This curve is just a rectangular hyperbola and is exactly the type of behavior expected in terms of the simple formalism of Equation (11-74). The corresponding curve for the saturation of hemoglobin with oxygen is also shown in Figure 11-4. Note that in this case the curve is sigmoid in shape rather than hyperbolic. A curve of this shape is not expected in terms of a simple saturation process and indicates that the chemistry of oxygen binding by hemoglobin must be a complicated process. In fact, it is now known that this curve may be accounted for in terms of the interactions of the hemoglobin subunits with each other and how these change upon oxygenation. That is, the complexing of hemoglobin with the first molecule of oxygen changes the affinity of the remaining heme groups for oxygen. It does this in such a way that each successive molecule of oxygen is easier to put on than the first. Thus, the saturation of hemoglobin with oxygen is said to be a *cooperative process*. The first molecules of oxygen assist in the binding of subsequent ones.

The second striking aspect of the behavior of hemoglobin in its oxygenation reactions has to do with the effect of carbon dioxide concentration on the affinity of hemoglobin for oxygen. In Figure 11-5 the percent saturation of hemoglobin by oxygen is plotted as a function of the pressure of oxygen for three pressures of carbon dioxide. As we can readily see, as the pressure of carbon dioxide increases, the degree of saturation of hemoglobin with oxygen for a given pressure of oxygen decreases. As we already know, increasing the concentration of carbon dioxide in solution will lower the pH of that solution. This is, of course, a consequence of the fact that carbon dioxide can be hydrated to generate carbonic acid, which is a weak acid. In fact, the influence of changing the concentration of carbon dioxide on the affinity of hemoglobin for oxygen may be ascribed entirely to changes in the pH of the solution. Consequently, we may conclude that the acidity of the solution and the equilibration of hemoglobin with molecular oxygen are linked processes. It is important to recognize that not only does the acidity of the solution affect the oxygenation of hemoglobin, but the oxygenation of hemoglobin affects the acidity of the solution. In fact, one implies the other. It is now known that the oxygenation of hemoglobin changes the value of pK_a of some group on hemoglobin from a value of 7.71 to 6.17. While the exact nature of this group is not known, it is clear that oxyhemoglobin is a stronger acid than hemoglobin. We may express

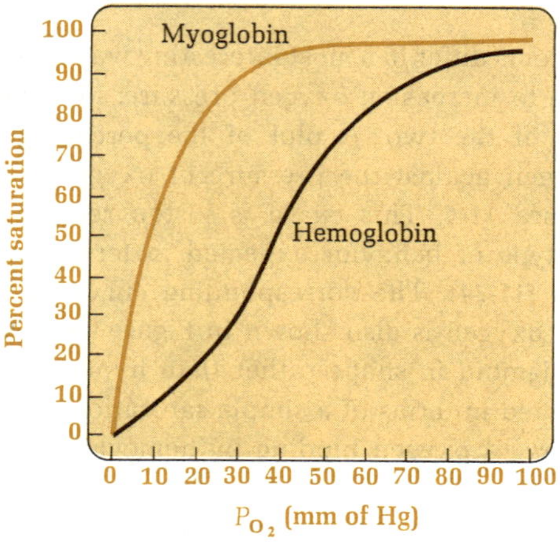

Figure 11-4 The percent saturation of hemoglobin and myoglobin by oxygen plotted as a function of the partial pressure of oxygen. Note the simple rectangular hyperbola for myoglobin and the more complex sigmoid curve for hemoglobin.

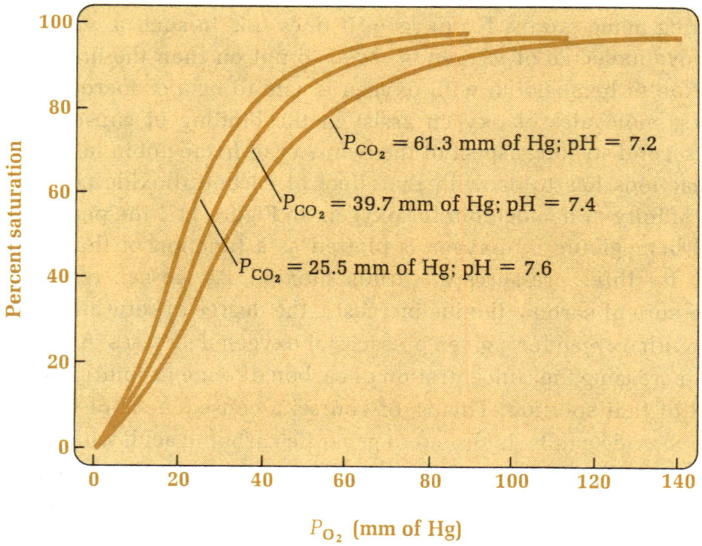

Figure 11-5 Effect of CO_2 tension on the dissociation of oxyhemoglobin in human blood. (From A. White, P. Handler, and E. Smith, *Principles of Biochemistry,* 4th Ed., McGraw-Hill, New York, 1968. Copyright © 1968, McGraw-Hill Book Company. Used with permission of McGraw-Hill Book Company.)

these ideas in the following way:

$$HbO_2 + H^+ \rightleftharpoons HbH^+ + O_2 \quad (11\text{-}75)$$

We should note that Equation (11-75) as formulated is not quite stoichiometric. In fact, only about 0.65 mole of oxygen is released per mole of hydronium ion bound.

As we shall see shortly, we are now in a position to understand the principal features of the functioning of hemoglobin as an oxygen-carrying pigment and device for the regulation of the acidity of blood. However, let us examine first some of the salient facts concerning the transport of oxygen and the transport of carbon dioxide in the blood. In this connection it is important to realize that the hemoglobin absorbs oxygen from the lungs and transports it to the tissues. In the tissues there are a number of oxidative reactions that are employed to generate energy to sustain the life process. These oxidative reactions involve the conversion of complicated organic structures to carbon dioxide and water. Consequently, we shall be concerned with how the blood is able to transport the carbon dioxide so formed back to the lungs where it can be respired. Thus, we must imagine a device by which hemoglobin can accept oxygen at the lungs, yet give it up at the tissues where carbon dioxide is released, and that this can in turn be expelled in the lungs subsequently.

Oxygen transport

Following oxygenation in the lungs, the blood is transported to the tissues via the arterial system. In the arterial blood the partial pressure of oxygen in the erythrocytes is about 100 mm of Hg, while the partial pressure of carbon dioxide in the plasma is about 40 mm of Hg. Referring to Figure 11-5, it is clear that under these conditions the hemoglobin is about 96 percent saturated with oxygen.

In the tissues the partial pressure of oxygen is near 35 mm of Hg, and that of carbon dioxide is near 50 mm of Hg. Consequently, both oxygen and carbon dioxide will diffuse down their concentration gradients. Oxygen from the erythrocytes diffuses through the plasma and into the tissues, and carbon dioxide diffuses out of the tissues and into the plasma, where a portion of it enters the erythrocytes although hemoglobin has no specific affinity for carbon dioxide. As a result of the interchange of oxygen and carbon dioxide between the blood and the tissues, the blood returning to the lungs via the venous system has a partial pressure of oxygen near 40 mm of Hg and a partial pressure of carbon dioxide of 45 mm of Hg. Reference once again to Figure 11-5 reveals that under these conditions hemoglobin is about 65 percent saturated by oxygen. Thus, about 31 percent of the oxygen-binding capacity of hemoglobin represents that amount of oxygen transferred from the erythrocytes to the tissues in the passage of the blood.

Carbon dioxide transport

The transport of carbon dioxide from the tissues to the lungs generates two quantitative questions for our consideration. As we noted above, the partial pressure of carbon dioxide differs between the arterial and venous blood by only about 5 mm of Hg. This is a very small quantity of carbon dioxide compared to the values obtained by direct quantitative analysis of arterial and venous blood. That this is true will be demonstrated by a simple calculation immediately below.

For the moment we can focus attention on the question of how the CO_2 is in fact transported so that the actual difference in concentration shows up in only a minor way in terms of a change in the partial pressure of carbon dioxide. The second problem that confronts us is that of the pH of the blood. The influx of substantial quantities of carbon dioxide from the tissues ought to have the effect of dramatically lowering the pH of the blood. Yet direct measurement reveals that the values of pH of arterial and venous blood are, in fact, very similar. By what mechanism, then, is the blood able to compensate for the influx of acidic species from the tissues without undergoing a dramatic change in pH?

How can we determine the total amount of carbon dioxide and its derivatives in the blood? By "total carbon dioxide" we mean the sum of the amounts of CO_2 itself, carbonic acid, bicarbonate, and carbonate:

$$\text{total carbon dioxide} = CO_2 + H_2CO_3 + HCO_3^- + CO_3^{2-} \tag{11-76}$$

In practice, the simplest procedure is to convert all of these forms to CO_2 through acidification with mineral acid. Clearly, increasing the amount of hydronium ion in the blood will convert both bicarbonate and carbonate to carbonic acid. As we already know from Equation (11-58), the preponderance of the newly formed carbonic acid will dehydrate to yield CO_2. Thus, through the processes of neutralization and dehydration, acidification will result in the existence of "total carbon dioxide" as CO_2 alone. By reducing the pressure over the blood, the CO_2 can be pumped off and measured. Utilization of such a procedure reveals that arterial blood has a concentration of "total carbon dioxide" near 50 ml/100 ml of blood, while venous blood has a concentration in the range of 55–60 ml/100 ml of blood. Such values are frequently referred to as *volumes percent*. Thus, we can say that the concentration of total carbon dioxide in arterial blood is 50 volumes percent.

The difference in total carbon dioxide concentration between arterial and venous blood is between 5 and 10 volumes percent. Can we account for this difference directly in terms of the difference in partial pressure of CO_2 between arterial and venous blood? This question can be approached using the equilibrium constant, A, which describes how a gas distributes itself between the gas and liquid phases [Equation (11-57)]. For the absorption of CO_2 by blood plasma, the value of A is 0.51. This value gives the number of millilters of CO_2 per milliliter of blood when the pressure of CO_2 in the gas phase is 760 mm of Hg:

$$
\begin{array}{ll}
\text{gas:} & P_{CO_2} = 760 \text{ mm of Hg} \\
 & \downarrow \\
\text{plasma:} & 0.51 \text{ ml of } CO_2/\text{ml of plasma}
\end{array}
$$

The equilibrium state for the absorption of CO_2 by plasma

As noted above, the difference in the partial pressure of CO_2 in the arterial and venous blood is about 5 mm of Hg. Hence, we can make the following calculation:

$$
\begin{aligned}
(CO_2)_{\text{plasma}} &= A(CO_2)_{\text{gas phase}} \\
\Delta(CO_2)_{\text{plasma}} &= A\,\Delta(CO_2)_{\text{gas phase}} = 0.51 \times \frac{5}{760} \text{ ml } CO_2/\text{ml plasma}
\end{aligned}
\tag{11-77}
$$

This corresponds to about 0.35 ml of CO_2/100 ml of plasma difference in CO_2 content between the arterial and venous blood. This is only a small part of the total concentration difference, and it follows that the majority of the total CO_2 transported from the tissues to the lungs through the blood is not transported as dissolved CO_2.

The major fraction of the transported carbon dioxide occurs in two forms: as bicarbonate and covalently bound to proteins. Let us look at these forms one at a time. To calculate the ratio of bicarbonate to carbon dioxide (the sum of concentrations of H_2CO_3 and CO_2) we can use the Henderson-Hasselbalch equation (11-53) and the overall dissociation constant for carbonic acid (11-64). Thus, we have for normal blood at pH 7.4:

$$
7.4 = 6.3 + \log \frac{(HCO_3^-)}{(H_2CO_3 + CO_2)}
\tag{11-78}
$$

A simple calculation reveals that the ratio of bicarbonate to total carbonic acid is near 13. Thus, by far the major fraction of total CO_2 in blood exists as bicarbonate, not as CO_2. Hence, a rather small change in the partial pressure of CO_2 between the arterial and venous blood reflects a substantial change (about 13-fold greater) in the total concentration difference.

Use of the Henderson-Hasselbalch equation in this way requires, of course, that carbon dioxide be in equilibrium with carbonic acid and bicarbonate. The equilibrium of carbonic acid and bicarbonate, which involves only the transfer of a single proton, is very rapid indeed. The hydration of carbon dioxide, in contrast, is not a rapid reaction in comparison with the rate of events involved in its diffusion from the tissues into the blood and transfer to the lungs. In fact, the diffusion of carbon dioxide from the tissues to the plasma and thence into the erythrocytes is fast enough to occur without significant hydration. In the erythrocyte the CO_2 is equilibrated with its hydrate, carbonic acid, through the intervention of a potent enzymatic catalyst, *carbonic anhydrase*. This enzyme has a molecular weight near 30,000 and contains one atom of zinc per molecule

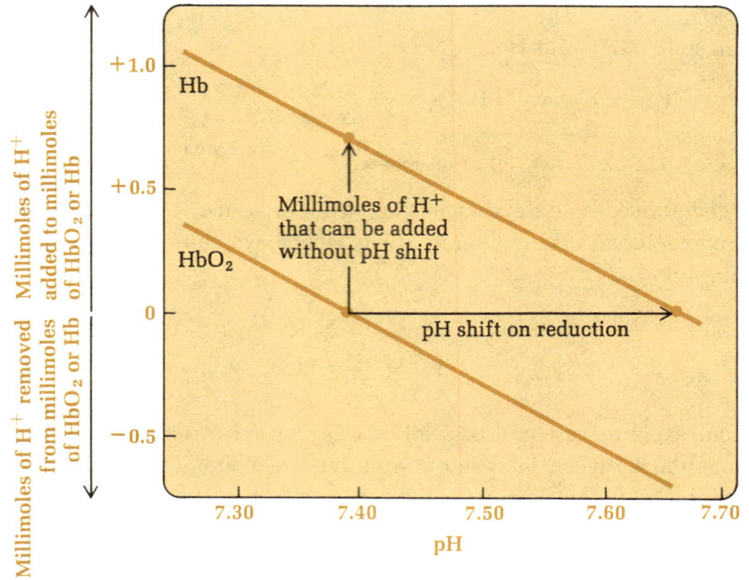

Figure 11-6 Titration curves of hemoglobin and oxyhemoglobin. (From H. W. Davenport, *The ABC of Acid-Base Chemistry,* 3rd ed., University of Chicago Press, Chicago, 1950.

of protein. One molecule of carbonic anhydrase is capable of catalyzing the hydration of about 1 million molecules of carbon dioxide per minute. It is this extraordinary catalytic capacity that accounts for the equilibration of CO_2 and H_2CO_3 and, hence, for the carbon dioxide transporting capacity of the blood.

The last important means for transporting carbon dioxide in the blood is through its combination with exposed amino groups of the plasma proteins:

$$\text{protein—NH}_2 + CO_2 \rightleftharpoons \text{protein—NH—COO}^- + H^+ \tag{11-79}$$

Compounds with the structure of the reaction product are called *carbamates.* These reactions are rapid and reversible and proceed in either direction depending on the pressure of carbon dioxide in the system. Thus, formation is favored in the venous blood and dissociation is favored at the lungs.

The final matter for our consideration is that of the maintenance of a constant pH in the blood. Specifically, why doesn't the infusion of substantial amounts of carbon dioxide into the blood dramatically lower the pH of the blood? On the basis of the amount of carbon dioxide transported as bicarbonate, we might anticipate that the pH of the venous blood would be about 0.3 pH units lower than that of the arterial blood. In fact, the pH changes in a barely detectable manner.

We must recall that the hemoglobin of the erythrocytes is itself an important buffer and that its acid–base properties change with oxygenation. Specifically, oxyhemoglobin is a stronger acid than hemoglobin itself. This point is made quantitatively in Figure 11-6. Here the pH of a 1 millimolar (m*M*) solution of oxygenated and unoxygenated hemoglobin is plotted as a function of the equivalents of acid or base added to the solution. We note that, at pH 7.40, conversion of 1 millimole (mmole) of oxyhemoglobin to hemoglobin would result in an increase in the pH to 7.67. To reduce the pH to the original value, we would have to add about 0.7 mmole of acid. But this is exactly what happens in the tissues: Oxyhemoglobin gives up its oxygen to the tissues, which would tend to cause the pH to rise, but carbon dioxide enters the plasma from the tissues, which would tend to lower the pH. Our calculation shows that oxygen transport can be balanced, in terms of no net pH change, by the influx of sufficient CO_2 to generate 0.7 equivalents of hydronium ion. Hence, the acid–base behavior of hemoglobin is perfectly suited to its biological function.

Myoglobin

Myoglobin serves as a storage site for molecular oxygen in the muscles of vertebrates and invertebrates. Myoglobin has a higher affinity for oxygen than does hemoglobin. Thus, myoglobin can accept oxygen from oxyhemoglobin and store it until its use becomes necessary.

In man, the highest concentration of myoglobin is found in the cardiac muscle. This is a sensible observation, since the heart may be called upon to form prodigious muscular feats in times of stress and must be able to respond to such calls when they occur. In diving mammals such as the whale, the myoglobin content of the muscles is much higher than in man. This probably accounts for the ability of the whale to remain submerged for extended periods of time.

Glossary

Absorption coefficient an equilibrium constant that measures the distribution of a gas between the gas phase and a liquid phase with which it is in contact; a measure of the solubility of a gas in a liquid.

Acid according to the Lowry-Brønsted view, any substance capable of donating a proton to an acceptor; according to the Lewis concept, any substance capable of accepting an unshared pair of electrons from a donor.

Base according to the Lowry-Brønsted view, any substance capable of

accepting a proton from a donor; according to the Lewis concept, any substance capable of donating an unshared pair of electrons to an acceptor.
Buffering the action of weak acids or bases in aqueous solution to minimize pH changes resulting from addition of acid or base to the solution.
Conjugate acid–base pair an acid and the corresponding base derived by the loss of a proton from that acid.
Henderson-Hasselbalch equation an equation for conveniently calculating the pH of a solution containing known amounts of an acid and its conjugate base; specifically:

$$pH = pK_a + \log \frac{(\text{base})}{(\text{acid})}$$

Lowry-Brønsted theory the concept that an acid is any substance capable of donating a proton to an acceptor and that a base is any substance capable of accepting a proton from a donor.
pH the negative logarithm of the concentration (more accurately, activity) of the hydrated proton.
pK_a the negative logarithm of dissociation constants of acids.
Principle of electrical neutrality the requirement that all solutions be electrically neutral overall; that is, the sum of the number of positive charges in the solution must be equal to the sum of the number of negative charges.
Strong acids and bases those acids and bases that dissociate completely upon solution in water.
Weak acids and bases those acids and bases that dissociate only partially upon solution in water. These species are characterized by small values for K_a and K_b.

Study problems

1. What is the pH of 0.150*M* NaOH solution?
2. What is the pH of $3.86 \times 10^{-3}M$ $Ba(OH)_2$ solution?
3. What is the pOH of $1 \times 10^{-3}M$ $HClO_4$ solution?
4. The pH of an HCl solution is 0.629. Calculate the concentration of HCl.
5. The pH of a KOH solution is 12.48. Calculate the concentration of KOH.
6. The dissociation constant for HNO_2 is 4.5×10^{-4}. Calculate the molar concentration of NO_2^- in 0.02*M* HNO_2.
7. The dissociation constant for HCN is 4×10^{-10}. Calculate the molar concentration of CN^- in 0.03*M* HCN.
8. The value of K_b for aqueous ammonia is 1.8×10^{-5}. Calculate the molar concentration of NH_4^+ in 0.02*M* NH_3.
9. A solution of 0.012*M* HCNO shows an H_3O^+ concentration that is $1.2 \times 10^{-3}M$. Calculate the dissociation constant of HCNO.
10. A solution of 0.20*M* hydrazine (N_2H_4) shows an OH^- concentration that is $4 \times 10^{-4}M$. Calculate K_b for hydrazine.
11. A solution of 0.01*M* $HAsO_2$ shows an H_3O^+ concentration that is $2.5 \times 10^{-6}M$. Calculate the dissociation constant of $HAsO_2$.

12. The value of K_b for aqueous ammonia is 1.8×10^{-5}. Calculate the pH of 0.55*M* NH_3.
13. The pH of 0.3*M* formic acid (HCOOH) is 2.13. Calculate the numerical value of the dissociation constant of formic acid.
14. The pOH of 0.5*M* pyridine (C_5H_5N) is 4.58. Calculate the numerical value of K_b for pyridine.
15. The dissociation constant for HCNO is 1.2×10^{-4}. To prepare a solution of HCNO that will have a pH of 2.65, what molarity of solute is required?
16. The value of K_b for ammonia is 1.8×10^{-5}. To prepare a solution of NH_3 that will have a pH of 12, what molarity of solute is required?
17. Calculate the pH of the solution that results when 500 ml of 1*M* CH_3COOH is titrated with 300 ml of 1*M* NaOH. The dissociation constant for CH_3COOH is 1.8×10^{-5}.
18. Calculate the pH of a CH_3COOH solution containing equimolar concentrations of CH_3COOH and CH_3COO^-.
19. Calculate the pH of the solution that results when 5×10^{-4} mole of NaOH is added to 25 ml of 0.1*M* CH_3COOH.
20. Calculate the pH of a solution that is 0.1*M* in ammonia and 0.01*M* in ammonium chloride. The value of K_a for NH_4^+ is 5.75×10^{-10}.
21. Calculate the pH of a solution that is 0.2*M* in CH_3COOH and 0.1*M* in CH_3COONa.
22. Calculate the pH of a solution made by adding 0.01 mole of NaOH to 100 ml of 0.50*M* CH_3COOH and 0.50*M* sodium acetate.
23. Calculate the pH of a 0.1*M* KNO_3 solution.
24. Calculate the pH of a 1.0*M* CH_3COONa solution.
25. Calculate the pH of a 0.001*M* NaF solution. The dissociation constant of HF is 6.7×10^{-4}.

Suggestions for further reading

S. W. Benson *Chemical Calculations*, 3rd. ed., Wiley, New York, 1962, Chap. 12.

J. N. Butler *Solubility and pH Calculations*, Addison-Wesley, Reading, Mass., 1964.

A. E. Grosser and I. S. Butler *Relevance in Chemical Science*, Benjamin, Menlo Park, Calif., 1971, Chap. 12.

M. J. Sienko *Stoichiometry and Structure*, Benjamin, Menlo Park, Calif., 1964, Chap. 14.

12

Solubility in water

No mercy, no power but its own controls it. Panting and snorting like a mad battle steed that has lost its rider, the masterless ocean overruns the globe.

Herman Melville, *Moby Dick*

Among the important properties of water is its ability to dissolve many substances. While not a "universal" solvent, water is a versatile one. We have already noted some of the consequences of dissolved materials for the structure of liquid water itself (Chapter 10).

One of the most remarkable consequences of the solvent properties of water is the composition of the oceans—detailed in the next chapter. The important fact is that the oceans are an immense soup containing a great many dissolved substances: The oceans serve as a giant reservoir for many of these. The oceans even contain, in the form of living organisms, mechanisms for the concentration of some of the dissolved substances present in low concentrations. There are, for example, organisms capable of accumulating high concentrations of vanadium within themselves even though very low concentrations of vanadium are dissolved in sea water. While no direct use has been made of these organisms up to the present, their existence and the enormous size of the oceans, which therefore contain much vanadium despite its low concentration, suggest that someday we may mine the oceans with their help. Our increasing rate of depletion of terrestial reservoirs of metals may hasten that day.

The extent to which various substances will dissolve in water varies greatly depending on their chemical makeup. The chemist makes use of this fact in his efforts to separate mixtures of substances for the purposes of analysis and purification. The understanding of the properties of water as a solubilizing medium is important for other purposes as well. One such example—the desalination of sea water—is described in some detail in the following chapter.

SOLUBILITY AS AN EQUILIBRIUM PHENOMENON

If we take a glass of water and add common table sugar, sucrose, to it little by little, the sugar will dissolve until we have added a specific amount. Further addition of sugar will just cause the sucrose to pile up as a solid on the bottom of the glass. The water is **saturated** with sucrose. The concentration of sugar dissolved in the water at saturation defines the **solubility** of sugar in water at the particular temperature of the experiment. If two substances dissolve completely in one another whatever proportions are used, they are said to be **miscible.** Chemists have thought up a lot of ways to describe solubility—the simplest and most generally useful way is in

Table 12-1 Some conventions for expressing solubility

Designation	Meaning
Molarity	Number of moles of a substance that dissolves in 1 liter of solution
Molality	Number of moles of a substance that dissolves in 1000 g of solvent
Normality	Number of equivalents of a substance that dissolves in 1 liter of solution

Table 12-2 Solubilities of some simple gases in water at 298°K

Monoatomic gases		Diatomic gases	
Gas	Solubility (moles/liter atm)	Gas	Solubility (moles/liter atm)
He	4.25×10^{-4}	H_2	8.34×10^{-4}
Ne	4.25×10^{-4}	O_2	1.38×10^{-3}
Ar	1.38×10^{-3}	N_2	7.23×10^{-4}
Kr	2.30×10^{-3}		
Xe	4.51×10^{-3}		

Table 12-3 The solubility of several alcohols in water at 298°K

Alcohol	Solubility (*M*)
Methanol, CH_3OH	Miscible
Ethanol, CH_3CH_2OH	Miscible
Propanol, $CH_3CH_2CH_2OH$	Miscible
Butanol, $CH_3(CH_2)_3OH$	1.2
Pentanol, $CH_3(CH_2)_4OH$	0.3
Propanediol, $HOCH_2CH_2CH_2OH$	Miscible
Butanediol, $HO(CH_2)_4OH$	Miscible
Hexanediol, $HO(CH_2)_6OH$	Slightly soluble
Glycerol, $HOCH_2—CH(OH)—CH_2OH$	Miscible

terms of **molarity,** the number of moles of a solute that dissolves in 1 liter of *solution* (not solvent). Other ways include **molality** and **normality.** These alternative ways to express solubility are collected in Table 12-1.

A solution of a substance such as sucrose in which the solute retains its identity in solution is said to be a **molecular solution.** There are a great many examples of molecular solutions, including almost all solutions of gases and of compounds based on carbon. We can formulate an equation to describe the solution process very simply:

$$\text{sucrose (solid)} \rightleftharpoons \text{sucrose (solution)} \tag{12-1}$$

Just as in the case of any other chemical reaction, we can write an equilibrium constant associated with it:

$$K_s = \frac{(\text{sucrose})_{\text{solution}}}{(\text{sucrose})_{\text{solid}}} \tag{12-2}$$

This expression can be simplified by noting that the activity of all solids is taken by convention to be 1.00. (Activity is defined in more detail below; for the moment we simply can consider it to be equivalent to a concentration in an equilibrium expression.) Stated differently, the chemist refers his equilibrium constants to a *standard state* in which the activity of the solid form is unity. In reality, any other choice of a standard state would have served as well, but this one gives maximum convenience. Equation (12-2) may then be written

$$K_s = (\text{sucrose})_{\text{solution}} \tag{12-3}$$

That is, the numerical value of the equilibrium constant at a given temperature is equal to the solubility at that temperature. If we wish, we may calculate a free-energy change for the solubilization reaction using the familiar equation $\Delta G^0 = -RT \ln K$. These considerations are, of course, quite general and apply with equal strength to solutions other than those formed with water as solvent.

Solution of gases in water can be treated in much the same way as that just developed. We have, in fact, encountered one such case earlier in our consideration of the solubility of oxygen and carbon dioxide in blood (Chapter 11).

The solubility of molecules in water depends on the balance of the interactions between the molecules as they exist in the solid or gaseous state and the interactions between the molecules and water in the dissolved state. The simplest case to consider is that of the solubilities of gases, since the intermolecular interactions in the gaseous state are rather weak; therefore, the relative solubilities depend principally on how well each gas interacts with water. The solubilities of various gases in water at 25°C are collected in Table 12-2. Note that the solubility of the monoatomic rare gases increases with increasing size (although none of the gases is very soluble). This behavior probably reflects the fact that the larger atoms disrupt the normal structure of water the least. The behavior of diatomic molecules is more complicated, and factors other than size must clearly be of im-

The solubility properties of minerals result in the formation of some of the most striking structures in nature, as testified to by this photograph of stalagmites and stalactites. (Courtesy of U.S. Department of the Interior.)

portance. Again, none of these gases is very soluble in water. Generally, solubilities can be qualitatively understood on the "like likes like" basis. That is, molecules that themselves have some of the structural properties of water usually prove to be soluble in it, while those that share no similarities with water are likely to be quite insoluble in it. Thus, sugars, which possess several hydroxyl groups, are very soluble in water. But hydrocarbons, which lack the polarity of water and the capacity to form hydrogen bonds with it, are relatively insoluble. The importance of the two kinds of properties can be appreciated from the variation in the solubility of several alcohols (compounds containing at least one hydroxyl group) as a function of the bulk of the hydrocarbon part of the molecule. The pertinent data are collected in Table 12-3. Note how solubility decreases as the hydrocarbon part of the molecule becomes larger and also how an increase in the number of hydroxyl groups raises solubility.

The solubility of a substance that dissociates to form ions in water can be understood on the same basis as that just developed. For example, when sodium chloride dissolves in water, ionization occurs, and the dissolved species are sodium ion and chloride ion, not sodium chloride. Such solutions, in which the essential character of the solute is lost, are called **ionic solutions.** Ionic solutions and molecular solutions can be distinguished in many ways. Perhaps the simplest is on the basis of electrical conductivity: The production of ions in solution greatly increases the conductivity of water, since these ions act as mobile charge carriers. Consequently, a solution of sucrose in water has about the same conductivity as water itself; but a solution of sodium chloride in water has a conductivity several orders of magnitude greater.

Let us consider the solution of silver chloride, AgCl, in water. This salt is only sparingly soluble, and the process is described by the equation

$$\mathrm{AgCl(s)} \rightleftharpoons \mathrm{Ag^+(aq)} + \mathrm{Cl^-(aq)} \tag{12-4}$$

We can formulate an equilibrium constant for this reaction in the usual way:

$$K_{eq} = \frac{(\mathrm{Ag^+})(\mathrm{Cl^-})}{(\mathrm{AgCl})} \tag{12-5}$$

If, as in Equations (12-2) and (12-3), the activity of the solid AgCl is taken as unity, the equilibrium expression reduces to a simpler one known as the **solubility product:**

$$K_{sp} = (\mathrm{Ag^+})(\mathrm{Cl^-}) \tag{12-6}$$

Knowledge of the solubility product for silver chloride under a specified set of conditions will permit us to calculate the solubility of this salt for those conditions.

Table 12-4 Solubility products of salts

Expression for K_{sp}	Value (286°K)
Fluorides	
$(Ba^{2+})(F^-)^2$	$= 1.7 \times 10^{-6}$
$(Ca^{2+})(F^-)^2$	$= 4.0 \times 10^{-11}$
$(Mg^{2+})(F^-)^2$	$= 6.5 \times 10^{-9}$
$(Pb^{2+})(F^-)^2$	$= 3.7 \times 10^{-8}$
Chlorides	
$(Ag^+)(Cl^-)$	$= 1.8 \times 10^{-10}$
$(Cu^+)(Cl^-)$	$= 1.2 \times 10^{-6}$
$(Hg_2^{2+})(Cl^-)^2$	$= 1.3 \times 10^{-18}$
$(Pb^{2+})(Cl^-)^2$	$= 1.6 \times 10^{-5}$
$(Tl^+)(Cl^-)$	$= 1.7 \times 10^{-4}$
Iodides	
$(Ag^+)(I^-)$	$= 8.3 \times 10^{-17}$
$(Cu^+)(I^-)$	$= 5.1 \times 10^{-12}$
$(Hg_2^{2+})(I^-)^2$	$= 4.5 \times 10^{-29}$
$(Pb^{2+})(I^-)^2$	$= 7.1 \times 10^{-9}$
Chromates	
$(Ba^{2+})(CrO_4^{2-})$	$= 2.4 \times 10^{-10}$
$(Pb^{2+})(CrO_4^{2-})$	$= 1.8 \times 10^{-14}$
$(Ag^+)^2(CrO_4^{2-})$	$= 2.4 \times 10^{-12}$
Sulfates	
$(Ba^{2+})(SO_4^{2-})$	$= 1.1 \times 10^{-10}$
$(Sr^{2+})(SO_4^{2-})$	$= 3.8 \times 10^{-7}$
$(Ca^{2+})(SO_4^{2-})$	$= 1.9 \times 10^{-4}$
$(Pb^{2+})(SO_4^{2-})$	$= 1.6 \times 10^{-8}$
$(Ag^+)^2(SO_4^{2-})$	$= 1.6 \times 10^{-5}$

Expression for K_{sp}	Value (286°K)
Hydroxides	
$(Al^{3+})(OH^-)^3$	$= 2 \times 10^{-2}$
$(Mg^{2+})(OH^-)^2$	$= 1.9 \times 10^{-13}$
$(Fe^{3+})(OH^-)^3$	$= 4.0 \times 10^{-38}$
$(Fe^{2+})(OH^-)^2$	$= 8.0 \times 10^{-16}$
$(Mn^{2+})(OH^-)^2$	$= 4 \times 10^{-14}$
$(Ni^{2+})(OH^-)^2$	$= 6.5 \times 10^{-18}$
$(Cu^{2+})(OH^-)^2$	$= 1 \times 10^{-19}$
$(Zn^{2+})(OH^-)^2$	$= 1.2 \times 10^{-17}$
$(Ag^+)(OH^-)$	$= 1 \times 10^{-8}$
$(Be^{2+})(OH^-)^2$	$= 7 \times 10^{-22}$
$(Cd^{2+})(OH^-)^2$	$= 5.9 \times 10^{-15}$
$(Ca^{2+})(OH^-)^2$	$= 5.5 \times 10^{-6}$
Sulfides	
$(Zn^{2+})(S^{2-})$	$= 1.0 \times 10^{-21}$
$(Cd^{2+})(S^{2-})$	$= 9.8 \times 10^{-27}$
$(Hg^{2+})(S^{2-})$	$= 4.0 \times 10^{-53}$
$(Cu^{2+})(S^{2-})$	$= 9.0 \times 10^{-36}$
$(Ni^{2+})(S^{2-})$	$= 3.0 \times 10^{-19}$
$(Co^{2+})(S^{2-})$	$= 8 \times 10^{-22}$
$(Fe^{2+})(S^{2-})$	$= 3.7 \times 10^{-18}$
$(Mn^{2+})S^{2-})$	$= 2.5 \times 10^{-10}$
$(Ag^+)^2(S^{2-})$	$= 2.0 \times 10^{-49}$
$(Pb^{2+})(S^{2-})$	$= 8 \times 10^{-28}$
$(Sn^{2+})(S^{2-})$	$= 1 \times 10^{-25}$
Carbonates	
$(Ba^{2+})(CO_3^{2-})$	$= 8.1 \times 10^{-9}$
$(Ca^{2+})(CO_3^{2-})$	$= 8.7 \times 10^{-9}$
$(Mg^{2+})(CO_3^{2-})$	$= 1.0 \times 10^{-5}$
$(Pb^{2+})(CO_3^{2-})$	$= 3.3 \times 10^{-14}$
$(Fe^{2+})(CO_3^{2-})$	$= 3.5 \times 10^{-11}$
$(Sr^{2+})(CO_3^{2-})$	$= 1.1 \times 10^{-10}$

Example *At 15°C, the solubility product for silver chloride is $1.8 \times 10^{-10} M^2$. What is the solubility of this salt at this temperature?*

Solution *Each time one molecule of silver chloride dissolves, one Ag^+ and one Cl^- is produced. Consequently, it must be true that $(Ag^+) = (Cl^-)$. Using this information and the solubility product expression, we can calculate the solubility:*

$$K_{sp} = (Ag^+)(Cl^-) = (Ag^+)^2 = 1.8 \times 10^{-10} M^2$$

or

$(Ag^+) = 1.3 \times 10^{-5}M$

This must also be the molar solubility of AgCl.

Solubility products can be used in another way as well: to calculate the solubility of salt in water containing one of the derived ions.

Example *Calculate the solubility of AgCl in a solution that contains 0.1M chloride ions.*

Solution *It follows directly from the conditions stated in the problem and the solubility product expression that*

$K_{sp} = (Ag^+)(Cl^-) = (Ag^+) \times 0.1 = 1.8 \times 10^{-10}$

or

$(Ag^+) = 1.8 \times 10^{-9}M$

Note that we have made an assumption in this calculation. Specifically, we have neglected the contribution of dissolved AgCl to the total concentration of dissolved chloride ion. But it is easy to show that the neglect of this chloride ion does not affect our result appreciably. The calculation shows that only about 10^{-9}M AgCl dissolves in a solution containing 0.1M chloride ion. Therefore, the solution process can only contribute 10^{-9}M chloride ion to the total, a quantity completely insignificant compared to the 0.1M Cl^- initially present. This kind of assumption is frequently made; it will fail only if the initial concentration of the common ion is very low or if the salt in question is fairly soluble. If you are in doubt as to the validity of the assumption, you can always check it just as we did above.

A number of solubility products are collected together with the expressions for K_{sp} in Table 12-4. Before continuing with this chapter, we should examine this table carefully to be certain that we understand precisely how to formulate the solubility product expressions and that we understand the two examples just worked out.

In the cases of some salts that dissolve in water to form ionic solutions, our considerations are slightly more complicated than those already developed. For example, in the next chapter we shall be concerned with the solubility of magnesium hydroxide, $Mg(OH)_2$. Each time that one molecule of this species dissolves, *one* Mg^{2+} ion and *two* OH^- ions will be produced. How, then, do we express the concentration of the solution that results? The usual procedure is to express the concentration in terms of the amount of salt that was added to the water in the first place. Hence, if 10^{-4} mole of magnesium hydroxide is added to 1 liter of water, the concentration is said to be 10^{-4}M in magnesium hydroxide; this is how we express the fact that the solution actually contains 10^{-4}M Mg^{2+} and 2×10^{-4}M OH^- but no $Mg(OH)_2$ at all. We can, of course, make the same type of calculation as in the simpler case of silver chloride.

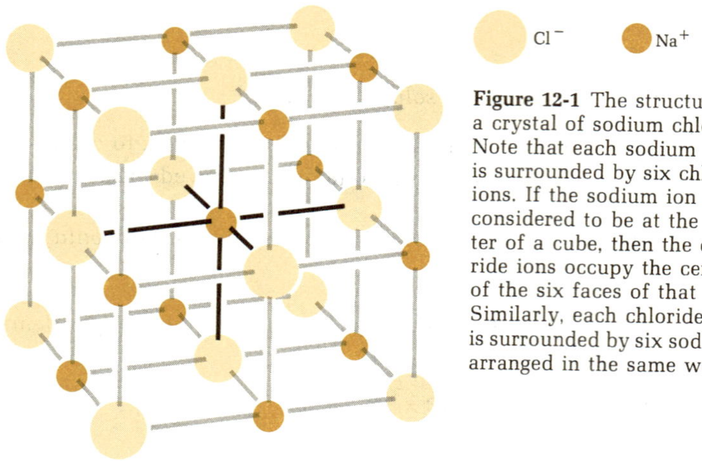

Figure 12-1 The structure of a crystal of sodium chloride. Note that each sodium ion is surrounded by six chloride ions. If the sodium ion is considered to be at the center of a cube, then the chloride ions occupy the centers of the six faces of that cube. Similarly, each chloride ion is surrounded by six sodiums arranged in the same way.

Example *At 25°C, the solubility product of magnesium hydroxide is $1 \times 10^{-11}M^3$. What is the solubility of this base at this temperature? What is the pH of the resulting solution?*

Solution *To begin with, we must recognize that the composition of the salt requires that* $(OH^-) = 2(Mg^{2+})$. *Substituting this information into the solubility product expression, we obtain*

$$K_{sp} = (Mg^{2+})(OH^-)^2 = (Mg^{2+})(2Mg^{2+})^2 = 4(Mg^{2+})^3 = 1 \times 10^{-11}M^3$$

Solving, we obtain

$$\mathbf{(Mg^{2+}) = 1.38 \times 10^{-4}M}$$

which is equal to the solubility of the base. From this result, it follows that the concentration of hydroxide ion is $2.76 \times 10^{-4}M$; this corresponds to

$$\mathbf{pH = 10.44}$$

Up to this point, we have made a basic assumption that is, in fact, not quite true. Equilibrium constants are correctly expressed in terms of **activities,** not *concentrations*. The two are linked by a simple equation:

$$\textbf{activity} = \gamma \times \textbf{concentration} \tag{12-7}$$

where γ is called the **activity coefficient.** For sufficiently dilute solutions the activity coefficient is approximately unity, and concentrations and activities are nearly equal. For more concentrated solutions this is no longer true, and it is necessary to correct for the difference between activities and concentrations. The principal problem that we wish to discuss in terms of solubility is the avoidance of scale formation in the desalination of sea water (see Chapter 13). Since this is a relatively complex example, let us begin our discussion in terms of simpler examples.

ACTIVITIES, ACTIVITY COEFFICIENTS, AND SOLUBILITY

In the foregoing discussion we have assumed that the solubility of a salt could be determined from the solubility product and the concentrations, if any, of dissolved common ions. In reality, the situation is not quite so simple: Solubilities of salts are *a function of the concentrations of all ions in solution,* whether they are common to those of the salt in question or not. Even this statement is too limited. All equilibria involving ions, the dissociation of acetic acid for example, are affected by the concentration of dissolved salts. To understand this important point, let us examine a little more closely the nature of solutions of salts.

We know that a salt such as sodium chloride exists in the crystalline state with sodium ions surrounded by chloride ions and vice versa (Figure 12-1). When salt is introduced into water, interaction with water molecules overcomes the forces holding the ions together in rigid array and allows the ions to move freely into solution. Each ion tends to be solvated by water molecules (Chapter 10), but it is overly simplistic to view the ions as totally independent of each other. The electrical attractions that brought the ions to an orderly array in the crystalline solid must still be at work in solution. Some balance is achieved between these forces and the kinetic motion of the molecules in solution that tends to prevent orderly arrangement. If solutions are sufficiently dilute, the average distance between ions can become quite large. Since electrical forces between particles fall off as the square of the distance separating them, ions present in sufficiently dilute solutions tend to behave as independent particles, each unaware of the existence of the other. In more concentrated solutions the distances between ions is less, and they "feel" the presence of each other more strongly.

With these considerations in mind, let us ask what the influence of the concentration of dissolved ions on other equilibria may be. One example is provided by the influence of sodium chloride concentration on the solubility of calcium sulfate, a reaction of some interest in connection with sea water desalination as developed in Chapter 13. In dilute solutions of calcium sulfate the dissolved calcium and sulfate ions would behave independently of each other. As the concentration of calcium and sulfate ions is increased, the interactions between them increase, leading to the eventual precipitation of calcium sulfate from the solution as the solubility of this salt is exceeded. How does this situation change if we now add appreciable concentrations of sodium chloride to the dilute solution of calcium sulfate?

Just as we have noted above that sodium and chloride ions interact with each other electrostatically, so they will interact with other dissolved ions. Thus, we may expect that each dissolved calcium ion will be a gathering place for one or more chloride ions and that each sulfate ion will be a gathering place for one or more sodium ions. This is shown diagrammati-

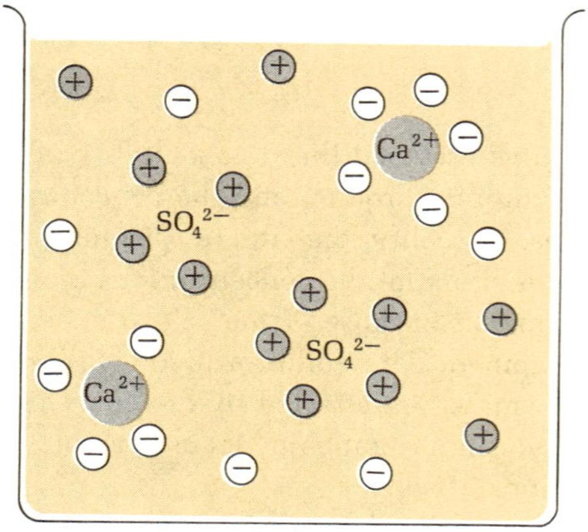

Figure 12-2 A visualization of the interactions between ions in aqueous solution. Note the mutual attraction of those ions of opposite charge.

cally in Figure 12-2. As a result of the presence of these *counterions* in the vicinity of the calcium and sulfate ions, the properties of the latter are changed appreciably. For example, the precipitation of calcium sulfate from solution will require that individual calcium ions and sulfate ions be stripped of their respective counterions. Only then can we form a crystalline lattice containing only calcium sulfate. We might write the equation

$$Ca^{2+}\ (+\ Cl^-\ \text{ions}) \longrightarrow Ca^{2+}(\text{free}) + Cl^-\ \text{ions} \qquad (12\text{-}8)$$

showing such a step. Formation of the insoluble solid requires driving this reaction to the right, and thus a given concentration of calcium appears to be less effective than we might expect. It is therefore convenient to consider this effective concentration to be related to the actual concentration by some appropriate factor which we call the *activity coefficient*. The solubility product law we have derived above actually applies to the individual ion *activities* rather than to their concentrations. For many relatively insoluble salts, the equilibrium concentrations are so low that we need not concern ourselves with this complication. For example, the molar solubility of silver chloride in water is only about $10^{-5}M$ for a solution prepared by dissolving silver chloride to its limit in water, and the activities of both silver and chloride ions are very nearly identical with their concentrations. However, for higher salt concentrations, activities of all ions will be affected by the others.

The relationship of activity coefficient to solution composition has been studied extensively. Since the interaction between ions and counterions is electrostatic in nature, we might expect stronger interaction between ions of higher charge. For example, the solubility of calcium sulfate might be expected to be more sensitive to solutions of magnesium chloride than those of sodium chloride owing to the greater interaction between magnesium and sulfate ions than between sodium and sulfate ions. To correlate effects of this sort, it is convenient to define a function known as **ionic strength, *I*,** which simultaneously takes into account both charge and concentration of ions present. The ionic strength of a solution may be calculated by adding terms for the concentration of an ion multiplied by the square of its charge for each ion in solution and dividing the resulting sum by two. Mathematically, we write

$$I = \tfrac{1}{2}\Sigma_i C_i Z_i^2 \tag{12-9}$$

where the summation is carried out over all ions present. In this equation C_i is the concentration of the ith ionic species and Z_i is the charge of that species. Note that the ionic strength is linearly proportional to the concentration of ions but depends on the square of the ionic charge.

Example *Calculate the ionic strength of a solution containing 0.1M* $MgCl_2$.

Solution *To solve this problem, we must put in the appropriate concentrations and charges into Equation (12-9). Note that the concentration of dissolved magnesium ion is 0.1M and that of dissolved chloride ion is 0.2M. Then, we have*

$$I = \tfrac{1}{2}(0.1 \times 2^2 + 0.2 \times 1^2) = \tfrac{1}{2}(0.4 + 0.2) = 0.3$$

Example *Calculate the ionic strength of a solution containing 0.2M* $CaBr_2$ *and 0.5M* $MgSO_4$.

Solution *As before, we must use Equation (12-9); note that the dissolved concentration of bromide ion is 0.4M. The result is*

$$\begin{aligned} I &= \tfrac{1}{2}(0.2 \times 2^2 + 0.4 \times 1^2 + 0.5 \times 2^2 + 0.5 \times 2^2) \\ &= \tfrac{1}{2}(0.8 + 0.4 + 2.0 + 2.0) = 2.6 \end{aligned}$$

Peter Debye and Erich Hückel studied the relationship between activity coefficients of ions and the ionic strength of a solution and derived a remarkably useful correlation. Their theory is based on the assumption that salts are completely dissociated into ions in solvents such as water. They assumed that electrical interactions occur between ions and thereby obtained an equation for the activity of coefficient of an ion in aqueous solution at 25°C:

$$\log \gamma_i = -0.509 Z_i^2 (I)^{1/2} \tag{12-10}$$

If solutions are sufficiently dilute, the **Debye-Hückel limiting law** is remarkably successful. By studying the solubility of slightly soluble salts in solutions containing added inert salts such as sodium chloride, barium chloride, or potassium nitrate, activity coefficients can be obtained as a

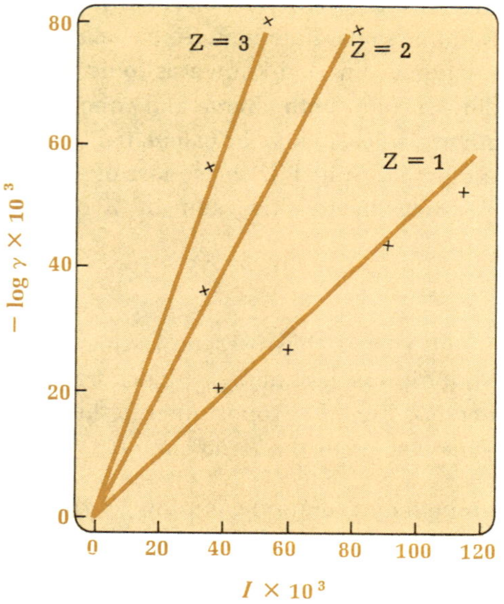

Figure 12-3 Activity coefficients for several insoluble salts as a function of ionic strength. Lines are given for the Debye-Hückel limiting law at different Z values.

function of total ionic strength. Figure 12-3 compares the results of measured and calculated activity coefficients for three types of salts at low ionic strengths. The solid straight lines are those given by the Debye-Hückel limiting law, and the points are experimental. The agreement between experiment and theory is good.

Example *Calculate the activity coefficient of the sodium ion in a solution of total ionic strength 0.01. Repeat the calculation for a solution of total ionic strength 0.03.*

Solution *The Debye-Hückel limiting law expressed as Equation (12-10) can be employed directly. Noting that the charge on the sodium ion, Z, is unity, we write*

$$\log \gamma = -0.509 \times 1^2 \times (0.01)^{1/2}$$
$$= -0.509 \times 0.1 = -0.509 = \bar{1}.9491$$
$$\gamma = 0.89$$

For the case of ionic strength 0.03, we have

$$\log \gamma = 0.509 \times 1^2 \times (0.03)^{1/2}$$
$$= -0.509 \times 0.173 = -0.088 = \bar{1}.912$$
$$\gamma = 0.816$$

At higher concentrations of salts the theory of interaction between ions becomes more complicated, and the simple limiting law is no longer well obeyed. Great progress has been made in recent years in understanding more concentrated solutions, but we shall not study this in more detail at the present time.

The laws governing behavior of solutions quite generally depend on the *activities* of the substances present rather than their *concentrations*. This point is so important that we should illustrate it with examples. Let us return to the description of the ionization of a weak acid in water solution (p. 348). In our earlier considerations we have implicitly assumed that the activity coefficients for each of the species present in the equilibrium

$$HA \rightleftharpoons H^+ + A^- \qquad (12\text{-}11)$$

is unity and that the concentrations are therefore actually activities. In the Debye-Hückel limiting law described by Equation (12-10), the log of the activity coefficient is related to the square of the charge on the particle. Since the undissociated acid carries no charge, the logarithm of its activity coefficient is zero or the activity coefficient is one. Consequently, the assumption that its concentration is equal to its activity is accurate. However, neither the activity of H^+ nor that of A^- will be quite unity. For dilute solutions of weak acids, concentrations are still nearly identical to activities in the absence of added salts. Thus, for acetic acid in 0.1*M* solution,

$$\frac{(H^+)(CH_3COO^-)}{(CH_3COOH)} = 1.8 \times 10^{-5}$$

$$(H^+)(CH_3COO^-) = (1.8 \times 10^{-5})(0.1) \qquad (12\text{-}12)$$

$$(H^+) = (CH_3COO^-) = 1.34 \times 10^{-3}M$$

The hydrogen ion and acetate ion concentrations are 0.00134*M*. Since both ions are singly charged, we find that the ionic strength of the solution is also 0.00134*M*. According to the Debye-Hückel limiting law for this solution,

$$\log \gamma = -0.509(1)^2(1.34 \times 10^{-3})^{1/2}$$

$$= -0.0186 = \bar{1}.9814 \qquad (12\text{-}13)$$

$$\gamma = 0.958$$

Again, since both particles are univalent, the values of γ of both the H^+ and the CH_3COO^- are 0.958. The difference between the calculation that considers the activity coefficients and the one that ignores them is significant although not large. If added salt is present, however, activity coefficients may be strikingly different from unity, and the errors made in calculation of concentrations of the various ionic species will be correspondingly large. This will typically be the case for buffer solutions containing added salts, but for simple illustration it will perhaps be sufficient to consider a solu-

Table 12-5 Solubility of silver bromate in aqueous lithium perchlorate solutions at 25°C

$(LiClO_4)$ (*M*)	$(AgBrO_3)$ (*M*)	K_{sp}	I	γ_+ or γ_-	$\gamma_+\gamma_-$	K_s
0.000	8.09×10^{-3}	6.54×10^{-5}	8.09×10^{-3}	0.999	0.826	5.42×10^{-5}
0.025	8.75×10^{-3}	7.62×10^{-5}	3.37×10^{-2}	0.839	0.79	5.36×10^{-5}
0.050	9.09×10^{-3}	8.26×10^{-5}	5.91×10^{-2}	0.806	0.650	5.37×10^{-5}
0.075	9.25×10^{-3}	8.74×10^{-5}	8.44×10^{-2}	0.783	0.613	5.36×10^{-5}
0.100	9.52×10^{-3}	9.06×10^{-5}	10.95×10^{-2}	0.773	0.598	5.42×10^{-5}

Average K_s^0 $(5.39 \pm 0.03) \times 10^{-5}$

Source: Adapted from data in R. W. Ramette and E. A. Dratz, *J. Phys. Chem.*, **67**, 940 (1963).

tion containing an inert salt, say 0.10*M* NaCl. For this solution the ionic strength is

$$I = \tfrac{1}{2}(0.10 \times 1^2 + 0.10 \times 1^2) = 0.10 \qquad (12\text{-}14)$$

We can ignore the contribution to the total ionic strength from the proton and the acetate ions, since this is insignificant compared to that derived from the added sodium chloride. Applying the Debye-Hückel limiting law, we see that the activity coefficients of both species are

$$\begin{aligned} \log \gamma &= -0.509(1)^2(0.10)^{1/2} \\ &= -0.163 = \bar{1}.837 \\ \gamma &= 0.687 \end{aligned} \qquad (12\text{-}15)$$

Under these conditions, the activities of the ions are significantly less than the actual concentrations. We may continue the calculation:

$$\begin{aligned} \frac{(\gamma_{H^+})(\gamma_{C_2H_3O_2^-})(H^+)(CH_3COO^-)}{(CH_3COOH)} &= 1.8 \times 10^{-5} \\ \frac{(0.687)(0.687)(H^+)(CH_3CO_2^-)}{(CH_3COOH)} &= 1.8 \times 10^{-5} \\ (H^+)(CH_3COO^-) &= \frac{1.8 \times 10^{-5} \times (0.1)}{(0.687)(0.687)} \\ (H^+)(CH_3COO^-) &= 3.81 \times 10^{-6} \\ (H^+) &= 1.95 \times 10^{-3} \end{aligned} \qquad (12\text{-}16)$$

This calculation shows that in our earlier solution, where activity coefficients were implicitly assumed to be unity, we miscalculated the actual concentrations of the ions by nearly 50 percent (see p. 391).

Let us explore the effect that activity coefficients have in an entirely different area, the calculation of solubilities of slightly soluble salts. Consider the solubility of silver bromate in solutions containing added lithium perchlorate. The expression for the solubility products should now read

$$K_{sp}^0 = (a_{Ag^+})(a_{BrO_3^-}) = (Ag^+)(BrO_3^-)\gamma_+\gamma_- \qquad (12\text{-}17)$$

We can experimentally determine the solubility of silver bromate in moles per liter in solutions of various concentrations in lithium perchlorate and for such solutions calculate the solubility product constant using concentrations alone. Alternatively, we may calculate the total ionic strength of the solution and compute the solubility product constant including the activity coefficient correction. Such a calculation is shown in Table 12-5. Examination of the two columns will show that the one in which activity coefficient corrections have been made is indeed nearly a constant, with only slight variation as the salt concentration is increased. In contrast, the solubility product calculated ignoring this effect changes by nearly 50 percent over the same concentration region.

Example *Calculate the relative solubility of calcium sulfate in pure water and in 0.1M sodium chloride solution.*

Solution *We may formulate the solubility product in terms of activities:* $K_{sp} = A_{Ca^{2+}}A_{SO_4^{2-}} = \gamma_+\gamma_-(Ca^{2+})(SO_4^{2-})$. *In pure water the activity coefficients will be assumed to be unity. Therefore, we can write*

$$K_{sp} = (Ca^{2+})(SO_4^{2-}) = (Ca^{2+})^2$$

or

$$(Ca^{2+})_{H_2O} = (K_{sp})^{0.5}$$

In 0.1M NaCl, *the activity coefficients can be calculated from Equation (12-10):*

$$\begin{aligned}\log \gamma_+ = \log \gamma_- &= -0.509 \times (2)^2 \times (0.1)^{1/2} \\ &= -0.509 \times 4 \times 0.316 = -0.643 = \bar{1}.357 \\ \gamma &= 0.227\end{aligned}$$

Then we have

$$K_{sp} = \gamma_+\gamma_-(Ca^{2+})(SO_4^{2-}) = 0.227 \times 0.227 \times (Ca^{2+})^2$$

or

$$(Ca^{2+})_{0.1M\ NaCl} = \frac{(K_{sp})^{0.5}}{0.227}$$

Therefore, the relative solubility in H_2O *and 0.1M* NaCl *is*

$$\frac{(Ca^{2+})_{0.1M\,NaCl}}{(Ca^{2+})_{H_2O}} = \frac{(K_{sp})^{0.5}/0.227}{(K_{sp})^{0.5}} = \frac{1}{0.227} = 4.4$$

Table 12-6 Residence times of several species in sea water

Species	Residence time (millions of years)
Cl	100
Na	70
SO_4	10
Mg	10
Ca	1
K	7
HCO_3	0.1
SiO_2	0.02

Source: K. E. Chave, *J. Chem. Educ.*, **48,** 148 (1971). Copyright © 1971 by the American Chemical Society. Reprinted by permission of the copyright owner.

SOME COMMENTS ABOUT THE HISTORY OF SEA WATER

In the following chapter there is ample opportunity to put quantitative calculations concerning solubilities to good use. For the moment, let us conclude this chapter with a qualitative discussion concerning the history of the oceans. Since the events of interest occurred over geological time periods, the evidence on which our arguments are based is necessarily indirect.

Life almost certainly originated in the early oceans. This simple fact leads us to an important conclusion concerning the early history of the hydrosphere: As long as 3 billion years ago, the chemical and physical properties of the oceans must have been similar to the current ones. This follows directly from our observations of the critical dependence of living organisms on their chemical and physical environment. For example, a number of higher marine organisms are very sensitive to the pH of the oceans: A variation of as little as 0.5 pH units can spell the difference between survival and extinction. Although simpler organisms are characteristically less sensitive, a number of them cannot tolerate a change of more than one pH unit in their aqueous environment. It is certainly true that there is some danger in arguing about the properties of early life forms in terms of those of current organisms, but it seems unlikely that really profound changes have occurred. It is, of course, both possible and probable that slow evolutionary changes have taken place in terms of what constitutes a fit environment for living organisms. The current record indicates that organisms are dependent not only on the acidity of the oceans but on the salinity as well. Moreover, many organisms have specific requirements, on the one hand, and specific tolerances, on the other, for many ions, including sodium, potassium, and calcium. The ancestors of these organisms may well have had equally stringent requirements, or perhaps even more stringent ones.

All of the above demands that there exist means of controlling the composition of the oceans. Prior to examining these directly, let us recognize that the early history of the oceans and the atmosphere must have been intimately related. The most important means of communication between atmosphere and hydrosphere is carbon dioxide (see the discussion of CO_2 solubility in Chapter 11). Thus, variations in atmospheric CO_2 would be reflected in changes in the CO_2 content, and hence the pH, of the oceans. Other evidence leads us to the conclusion that carbon dioxide, and other volatiles, were added to the atmosphere little by little through outgassing of the earth's interior (see Chapter 17). With these ideas in mind, let us re-create, very briefly, the stages in the development of the oceans.

Initially, the atmospheric concentration of CO_2 was much higher than it is at the present. Consequently, the pH of the oceans must have been quite low, perhaps as low as 0.3. This acid solution would have rapidly attacked rocks with which it came into contact, solubilizing basic elements. This would lead to a rapid increase in pH until the solubility product of calcium carbonate is reached. At this early stage, the pH of the oceans would have been near 5.7, a value quite a bit more acidic than that for the current ocean, pH 8.2, but perhaps not so different as to be inconsistent with living organisms. Further additions of carbon dioxide from the atmosphere or from the weathering of rocks would then result in the deposition of calcite and dolomite (limestone), with only slow and modest changes in the chemical composition of the seas. Thus, the precipitation of calcium carbonate must have been an important stabilizing factor in progress toward the current hydrosphere composition. Just how important can be appreciated from the fact that there is about *600 times* more carbonate present in sedimentary rocks than is present in the atmosphere (as CO_2) and dissolved in the oceans combined.

Despite the apparent overriding importance of equilibria involving carbonates in the early history of the seas, there is considerable evidence to suggest that the current ocean composition is maintained largely by other equilibria. While the atmospheric level of carbon dioxide is essentially constant at the present time, many elements are being continuously added to the sea water in the form of suspended matter, derived from the weathering of rocks, carried by the rivers of the world. The composition of the river water is very different from that of sea water—the former is basically a silica-rich calcium bicarbonate solution, whereas the latter is essentially a sodium chloride solution. The control of sea water composition can be appreciated from the data collected in Table 12-6. In this table we have collected *residence times* for a variety of chemical species. Residence time is defined as the ratio of the total amount of the species in the sea to the amount added yearly to the sea by rivers. The residence times of many of these species are quite long; but that for all of them is short in comparison with the geological age of the oceans—3 billion years at least. Consequently, all of these elements must be continuously removed from the oceans as they enter. This cannot be due entirely to calcium carbonate deposition, since this reaction provides no means to control levels of sodium, potassium, and silica, for example. Moreover, the precipitation of calcium and magnesium carbonates does not seem to be at equilibrium in the oceans, possibly because of inhibition of carbonate nucleation by dissolved organic matter. But another possibility does exist—reactions between dissolved ions and aluminosilicates brought to the oceans in the form of mud by the rivers. These are very complicated reactions; two are provided in the following equations:

$$3.5Al_2Si_{2.4}O_{5.8}(OH)_4 + Na^+ + 2.6SiO_2 + HCO_3^- \longrightarrow 3Na_{0.33}Al_{2.33}Si_{3.67}O_{10}(OH)_2 + CO_2 + 4.5H_2O \qquad (12\text{-}18)$$

$$2.5Al_2Si_{2.4}O_{5.8}(OH)_4 + K^+ + SiO_2 + HCO_3^- \longrightarrow 2K_{0.5}Al_{2.5}Si_{3.5}O_{10}(OH)_2 + CO_2 + 3.5H_2O \qquad (12\text{-}19)$$

The product of the former reaction is a type of monomorillonite, and the latter is illite. The nonstoichiometry is simply a consequence of the complicated crystal structures of these minerals. These reactions, and others like them, apparently serve to regulate the concentrations of sodium, potassium, magnesium, and silica in the oceans. The regulation of calcium levels appears to involve yet other reactions, including those of living organisms. While a quantitative grasp of these regulatory processes remains out of reach, the importance of precipitation reactions in this process is evident.

Glossary

Activity the *effective* concentration of a species in solution. This may differ from the *actual* concentration by an amount depending on the exact composition of the total solution.

Activity coefficient the proportionality constant relating activities to concentrations. Specifically, activity = $\gamma \times$ concentration, where γ is the activity coefficient.

Debye-Hückel limiting law a theoretical statement linking the activity coefficient of an ionic species to the ionic strength of the solution:

$$\log \gamma = -0.059 \times Z^2 \times (I)^{1/2}$$

Ionic solution a solution in which the essential character of the solute is changed in the solution process through dissociation into ions.

Ionic strength a property of a solution depending on the nature and concentration of dissolved salts:

$$I = \tfrac{1}{2} \Sigma_i \, C_i \, Z_i^2$$

Miscibility the property of having two substances mutually soluble in all proportions.
Molality the concentration of a solution given as the number of moles of a substance dissolved in 1000 g of solvent.
Molarity the concentration of a solution given as the number of moles of a substance dissolved in 1 liter of solution.
Molecular solution a solution in which the essential character of the solute is maintained following the solution process.
Normality the concentration of a solution given as the number of equivalents of a substance dissolved in 1 liter of solution.
Saturation the state that exists when the maximum amount of a solute is dissolved in a solvent at a particular temperature, pressure, and so forth.
Solubility the maximum amount of a substance that can be dissolved in a given solvent under a specified set of conditions.
Solubility product the equilibrium constant for the solution of a substance that forms an ionic solution:

$$AB \rightleftharpoons A^+ + B^- \qquad K_{sp} = (A^+)(B^-) \text{ at saturation}$$

Study problems

1. Using the values of solubility collected in Table 12-4, calculate the solubilities of the following salts:
   a. $CuCl$ e. Hg_2Cl_2 i. $Fe(OH)_3$
   b. AgI f. Ag_2CrO_4 j. $Al(OH)_3$
   c. CdS g. PbI_2 k. Ag_2S
   d. $BaSO_4$ h. $Zn(OH)_2$
2. At 25°C, the experimentally determined solubility of AgN_3 is $5.4 \times 10^{-5}M$. Calculate the solubility product for this salt.
3. At 25°C, the experimentally determined solubility of $SrSO_4$ is $8.7 \times 10^{-4}M$. Calculate the solubility product for this salt.
4. At 25°C, the experimentally determined solubility of CaF_2 is $2.15 \times 10^{-4}M$. Calculate the solubility product for this salt.
5. At 25°C, the experimentally determined solubility of $Pb(IO_3)_2$ is $8.66 \times 10^{-5}M$. Calculate the solubility product for this salt.
6. Using the data collected in Table 12-4, calculate the solubilities of the following salts in solutions containing 0.01M iodide ion.
   a. AgI c. HgI_2
   b. CuI d. PbI_2
7. Using the solubility products collected in Table 12-4, calculate the solubilities of the following bases in solutions at the indicated values of pH.
   a. $AgOH$ at pH 8 c. $Zn(OH)_2$ at pH 9
   b. $Ni(OH)_2$ at pH 8 d. $Al(OH)_3$ at pH 10
8. At what pH would a solution containing 0.01M cupric ion just begin to form a $Cu(OH)_2$ precipitate? (See solubility product in Table 12-4.)

9. Calculate the ionic strength of the following solutions.

a. 0.2*M* KBr	d. 1.0*M* $MgSO_4$
b. 0.1*M* $MgCl_2$	e. 0.01*M* K_2CO_3
c. 0.2*M* Na_3PO_4	f. 0.2*M* KCl + 0.1*M* K_3PO_4

10. Calculate the activity coefficient for Na^+, Mg^{2+}, and PO_4^{3-} in each of the solutions of Problem 9.
11. Using the data in Table 12-4, calculate the solubility of NiS in 0.2*M* $MgCl_2$.
12. How will the activity coefficient change with increasing ionic strength for
    a. potassium ion
    b. chloride ion
    c. sucrose
13. Xenon is more soluble in water than neon. Which of the following qualities of these atoms is most directly responsible for this difference?
    a. size
    b. electronegativity
    c. polarizability
    d. hydrogen bonding ability
    e. dipole moment

    Justify your answer. Would you expect the entropy change of solution of these two gases in water to be different? Why?
14. How will the acid strength of (a) acetic acid and (b) ammonium ion change with increasing ionic strength?

Suggestions for further reading

S. W. Benson *Chemical Calculations,* 3rd ed., Wiley, New York, 1962, Chaps. 8, 12.

J. N. Butler *Solubility and pH Calculations,* Addison-Wesley, Reading, Mass., 1964.

K. E. Chave "Chemical Reactions and the Composition of Sea Water," *J. Chem. Educ.,* **48,** 148 (1971).

A. B. Grosser and I. S. Butler *Relevance in Chemical Science,* Benjamin, Menlo Park, Calif., 1971, Chaps 9, 10.

M. J. Sienko *Stoichiometry and Structure,* Benjamin, Menlo Park, Calif., Chap. 14.

13

The desalination of water

No water resources program is of greater long-range importance—for relief not only of our shortages, but for arid nations the world over—than our efforts to find an effective and economical way to convert water from the world's greatest, cheapest natural resources—our oceans—into water fit for consumption in the home and by industry. Such a break-through would end bitter struggles between neighbors, states, and nations—and bring new hope for millions who live out their lives in dire shortage of usable water and all its physical and economical blessings, though living on the edge of a great body of water throughout that parched life-time.

John F. Kennedy

In the opening pages of this section attention was directed to the crucial need for ever-greater supplies of fresh water for human uses. The sea provides such an immense source of water that it is natural for us to turn to it for a solution to the problem. Furthermore, about two-thirds of our population lives in the 24 states bordering on the oceans. Of course, these areas of dense population are not necessarily the ones where large agricultural water uses occur, and it would at first glance appear that their problem cannot be solved by water from the oceans. However, at present, water must be transported from remote sources to supply urban uses. For example, much of the water supply for the city of Los Angeles comes from the Colorado River (and, we might add, is a source of conflict between the states of Arizona and California). Additional water supply is currently being arranged from the Feather River in northern California through an extensive aqueduct system to southern California. If the needs of the city of Los Angeles could be met from readily available ocean water, these water supplies could be diverted to the central California valley and to the arid regions of the inland West. Furthermore, within the central part of the United States there are large, but inadequately surveyed, quantities of brackish water containing lower concentrations of salt. Water desalination here would also be highly valuable.

The conversion of salt water into fresh water is by no means a dream for the future. Throughout the world, commercial installations are currently producing more than 25 million gallons/day of fresh water. Costs of operation of such installations vary remarkably, depending on whether sea water or brackish water of lower salt content is used and on the scale of the operation itself. From smaller installations operating with sea water, the initial cost of pure water may be $1 or more per 1000 gallons. Larger or more efficient plants may reduce this cost 70 cents per thousand gallons or less, and in favorable cases brackish water may be desalted for as little as 30 cents/1000 gallons. Prior to considering methods for desalination of sea water, let us review the total resources of the oceans to understand better both the problems to be faced and the magnitude of the total resources available.

THE OCEANS: PHYSICAL RESOURCES

The physical resources of the oceans are staggering. The oceans occupy a total volume of some 350 million cubic miles. One cubic mile of water weighs 4.7 billion tons. Dissolved solids in the oceans total about 35,000 parts per million (ppm). Translated into weight units, this corresponds to 165 million tons of dissolved solids per cubic mile. The major constituents of the sea water are summarized in Table 13-1. Clearly, the bulk of dissolved solids consists of sodium and chloride ions; however, a total of 77 of the elements have been detected in sea water, and it seems likely that all of them are present. The quantities of some of these elements per cubic mile of sea water are collected in Table 13-2. To understand the formidable technological problems in conversion of sea water into potable fresh water, compare the quantities of the most important elements in sea water listed in Table 13-1 with those of some common water supplies in the United States listed in Table 13-3. Clearly, the quantities of sodium and chloride ions that occur in typical fresh water sources are less than 0.0001 of those that occur in sea water.

Despite the enormous quantities of minerals present per cubic mile of sea water, as outlined in Table 13-2, only three minerals are recovered from the ocean in commercially important amounts. These are sodium chloride, magnesium, and bromine. Sodium chloride, or common table salt, recovered from the oceans has a commercial value of about $175 million/year. Magnesium recovered has a value of about $70 million/year. These quantities dwarf the commercial value of recovered bromine, which is only about $5 million/year, but that does represent some 70 percent of the total world supply of bromine. This element is used particularly to produce a common gasoline additive. In addition to these elements extracted directly from the sea itself, it is appropriate to mention that some $6\frac{1}{2}$ million barrels of oil/day are recovered from offshore drilling operations and that this quantity represents about 17 percent of the total oil produced in the noncommunist countries. So important has recovery of offshore oil become, that $6\frac{1}{2}$ million acres of the continental shelf of the United States have been leased to various oil companies. The resulting income to the United States government has thus far amounted to $3.4 billion. Of less financial impact, but important nevertheless, is the recovery of sulfur from the cap rock of salt domes. This amounts to some 2 million tons of sulfur/year.

Finally, we may note that the food harvested from the ocean currently

Table 13-1 Major constituents of sea water (ppm)

Sodium (Na^+)	10,561
Magnesium (Mg^{2+})	1,272
Calcium (Ca^{2+})	400
Potassium (K^+)	380
Chloride (Cl^-)	18,980
Sulfate (SO_4^{2-})	2,649
Bicarbonate (HCO_3^-)	142
Bromide (Br^-)	65
Other solids	34
Total dissolved solids	34,483
Water (balance)	965,517
Specific gravity (20°C)	1.0243

Source: K. S. Spiegler, *Salt Water Purification,* Wiley, New York, 1962.

Table 13-2 Concentrations of 57 elements in sea water

Element	Tons per cubic mile	Element	Tons per cubic mile
Chlorine	89,500,000	Nickel	9
Sodium	49,500,000	Vanadium	9
Magnesium	6,400,000	Manganese	9
Sulfur	4,200,000	Titanium	5
Calcium	1,900,000	Antimony	2
Potassium	1,800,000	Cobalt	2
Bromine	306,000	Cesium	2
Carbon	132,000	Cerium	2
Strontium	38,000	Yttrium	1
Boron	23,000	Silver	1
Silicon	14,000	Lanthanum	1
Fluorine	6,100	Krypton	1
Argon	2,800	Neon	0.5
Nitrogen	2,400	Cadmium	0.5
Lithium	800	Tungsten	0.5
Rubidium	570	Xenon	0.5
Phosphorus	330	Germanium	0.3
Iodine	280	Chromium	0.2
Barium	140	Thorium	0.2
Indium	94	Scandium	0.2
Zinc	47	Lead	0.1
Iron	47	Mercury	0.1
Aluminum	47	Gallium	0.1
Molybdenum	47	Bismuth	0.1
Selenium	19	Niobium	0.05
Tin	14	Thallium	0.05
Copper	14	Helium	0.03
Arsenic	14	Gold	0.02
Uranium	14		

Table 13-3 Chemical composition of two typical fresh water supplies in the United States (ppm)

	Na^+	Mg^{2+}	Ca^{2+}	Cl^-	SO_4^{2-}	HCO_3^-
Lake Erie at Buffalo, N.Y.	7	7	31	9	13	114
Shenandoah River at Mileville, W.Va.	7	8	32	3	6	132

Source: K. S. Spiegler, *Salt Water Purification,* Wiley, New York, 1962.

amounts to some 55 million tons/year, 50 percent of which is consumed directly and 50 percent of which is converted to fish meal for other nutritional purposes. The total commercial value of food taken from the ocean is about $8 billion/year. We certainly expect that the ocean will become an increasingly important food source as the population of the countries, and hence the crowding of people on the land mass of the earth, increases.

Prior to turning our attention to some aspects of the chemistry of water, we may note the composition of certain other bodies of water on earth. The concentrations of various salts in three of these, the Dead Sea, Great Salt Lake, and the Caspian Sea, are collected in Table 13-4. Clearly, these bodies of water contain even greater quantities, and in some cases much greater quantities, of dissolved solids than do the oceans. Consequently, they may serve as important commercial sources of certain of these minerals, but at the same time generation of fresh water from them will prove to be a much more difficult and expensive task.

With these considerations in mind, we turn attention to the methods that we can utilize to gain fresh water from sea water.

FUNDAMENTAL PRINCIPLES OF WATER DESALINATION

In its simplest form, the fundamental problem of water desalination can be stated in the following way: We desire to take a sample of raw sea water weighing 1000 lb and separate this into a mixture of dry salts weighing 35 lb and 965 lb of pure water. Stated in this manner, it should be apparent that we are introducing a degree of organization that did not previously exist and have thereby affected a decrease in entropy. To accomplish this, the second law of thermodynamics requires that we use at least some absolute minimum amount of energy regardless of what process is used to effect the separation. Since the minimum energy requirement does not depend on the process used (it is a state function), let us consider an

"It still tastes salty to me." (Drawing by Carl Rose; © 1970, *The New Yorker Magazine,* Inc.)

Table 13-4 Concentration of salts in different salt waters (ppm)

	Dead Sea	Great Salt Lake, Utah	Caspian Sea
Sodium	32,000	67,300	3,200
Magnesium	35,700	5,600	773
Calcium	12,700	300	297
Potassium	6,400	3,400	70
Chloride	178,600	112,900	5,500
Sulfate	400	13,600	2,970
Carbonate	Trace	200	48
Bromide	5,200	Trace	—

Source: K. S. Spiegler, *Salt Water Purification,* Wiley, New York, 1962.

extremely simple scheme for purification and calculate the energy required. In this scheme we shall transfer water vapor from a salt-water container to one containing pure water. To consider this problem in detail, we must first examine another property of solutions.

In the late nineteenth century François Raoult carried out an extensive experimental study of the pressure of vapor in equilibrium with pure liquids and with solutions. As a result of this experimental study, he advanced **Raoult's law,**

$$P_a = X_a \times P_a^0 \qquad (13\text{-}1)$$

where P_a is the observed pressure of vapor in equilibrium with the solution, X_a is the **mole fraction** of component A, and P_a^0 is the vapor pressure of the pure liquid A at the same temperature. If two or more liquids are present in the same container, then P_a is referred to as the **partial pressure** of A, and the total pressure of gas in the container is the sum of partial pressures,

$$P_a + P_b + \cdots = P_{\text{total}} \qquad (13\text{-}2)$$

for as many components as are present. To calculate the mole fraction, X, of a particular component, we simply calculate the number of moles of each substance present, add them to obtain the total number of moles of all components, and then calculate the fraction for the substance of interest. To illustrate, a solution composed of 1000 g of water and 100 g of sugar, $C_{12}H_{22}O_{11}$, contains 55.555 moles of water and 0.292 moles of sugar or a total of 55.847 moles in all. We may readily compute that the mole fraction of water is 55.555/55.847 = 0.9948 and that of sugar is 0.0052.

Example *A dry vodka martini contains very little besides water and ethanol (CH_3CH_2OH). Such a martini (properly prepared according to one author's taste) contains, by weight, about 65 percent water and 35 percent ethanol. What is the mole fraction of ethanol in such a martini?*

Solution *First we must calculate the number of moles of water and ethanol for any given quantity of martini. For example, in 100 g of martini we have 65 g of water and 35 g of ethanol. The molecular weight of water is 18 and that for ethanol is 46. Consequently, the number of moles of water is 65 g/18 g/mole = 3.6 moles. For ethanol, we have 35 g/46 g/mole = 0.76 mole. It follows that the mole fraction of ethanol is*

$$X_{\text{ethanol}} = \frac{\text{moles of ethanol}}{\text{moles of water} + \text{moles of ethanol}}$$

$$= \frac{0.76}{3.6 + 0.76} = \frac{0.76}{4.36} = 0.174$$

From our previous discussion of the composition of sea water, it is apparent that to obtain the mole fraction of water we must add the number

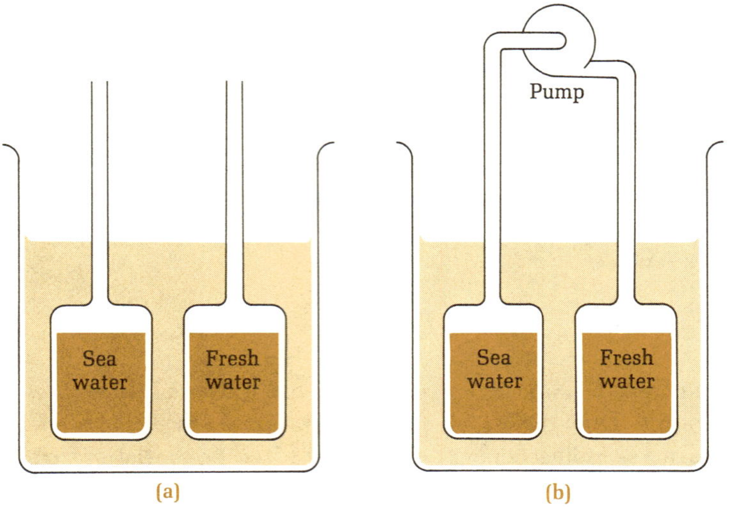

Figure 13-1 Diagram of devices to study water transport between salt water and fresh water. The compressor pump in (b) must increase the pressure from the sea water and deliver at a higher pressure to the fresh water container. (From K. S. Spiegler, *Salt Water Purification,* Wiley, New York, 1962.)

of moles for a number of different components. Furthermore, for a substance such as sodium chloride, which is dissociated in water solutions, each ion must be considered separately. (Complications arise at the salt concentration present in sea water, but we shall deal with these later since they do not interfere with the basic flow of our argument at present.) Rather than to calculate the pressure of water at equilibrium with sea water for such a complicated mixture, it is simpler to measure it directly; it has been found that at 25°C the vapor pressure of sea water is 1.84 percent lower than that of pure water.

Example *What is the mole fraction of water in sea water?*

Solution *The vapor pressure of sea water is 1.84 percent less than the vapor pressure of pure water at the same temperature. Using Raoult's law,*

$$P_a = X_a \times P_a^0$$

where $P_a/P_a^0 = 0.9814$, *we find*

$$X_a = \frac{P_a}{P_a^0} = 0.9814$$

Suppose, then, that we consider a simple system in which a container of sea water and one of fresh water are placed in a closed container as illustrated in Figure 13-1(a). Evaporation of water would take place from the surface of each sample to build up the natural vapor pressure. When the water vapor pressure in the container reached that natural for sea water, however, it would still be nearly 2 percent less than the vapor pressure of fresh water and evaporation from the fresh water surface would continue to take place. As evaporation continues, however, the pressure exceeds that maintained over the sea water. Under these circumstances, water vapor condenses into the sea water container. Eventually, all of the fresh water would evaporate and condense into the sea water container, and finally the pressure would fall to that characteristic of the new solution (now only half as concentrated in salt). What we observe in this experiment is simply another example of the tendency of nature to seek a disordered system and is precisely the reverse of what we wish to have take place.

If the sea water is to evaporate and condense in the fresh water container, we must find some means of increasing the water pressure. Figure 13-1(b) schematically shows a compressor introduced to raise the pressure of the vapor evaporated from the salt water container at least to, or slightly above, the pressure in the fresh water container. The process of water transfer under these conditions will occur in the desired direction. The calculation of the minimum energy requirement is straightforward. In Chapter 8 we have observed that work of this sort can be calculated from the formula

$$w = \Delta PV \tag{13-3}$$

At 25°C the vapor pressure of pure water is 0.0312 atm, and the volume occupied by 1 g of water vapor at this temperature and pressure is 43.4 liters. We may now calculate that for 1 g of water it will be necessary to expend

$$w = (0.0312 - 0.9816 \times 0.0312) \times 43.4 = 0.0249 \text{ liter atm/g} \tag{13-4}$$

of energy. Converting this to units more convenient for economic discussion, we obtain a minimum energy expenditure of 2.65 kilowatt hours (kW hr) per 1000 gallons.

The above calculation represents an absolute minimum of work that can possibly be used to produce fresh water from the sea. Any real process will require more work for several reasons. The calculation above implicitly assumes that the compression of the water vapor has been carried out in a reversible fashion, and this implies a careful and slow process. To work at reasonable rates, real compressors do more work than the minimum required. Furthermore, the driving energies supplied to a compressor are not all usefully applied, and some are dissipated as heat energy to the surroundings. Most practical installations operate at between 10 and 50 percent efficiency, so that as much as 26.5 kW hr/1000 gallons may be

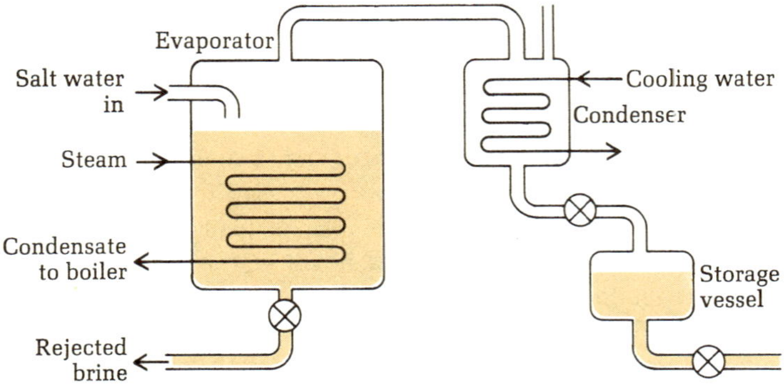

Figure 13-2 A simplified distillation apparatus. (From K. S. Spiegler, *Salt Water Purification,* Wiley, New York, 1962.)

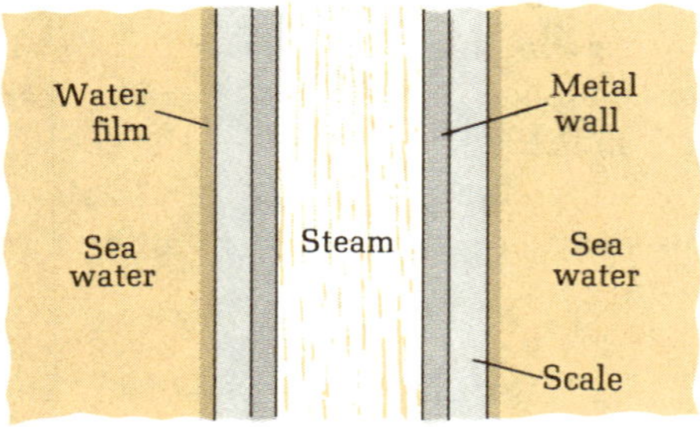

Figure 13-3 Section of heating element for sea water distillation.

needed for a workable process. Finally, as water is removed from the sea water, the mole fraction of salt increases. When half of the water has been removed, the mole fraction of water remaining is reduced and, according to Raoult's law, the pressure of water vapor over the remaining solution will be correspondingly lower; still more energy will be required. The problem for the engineer becomes therefore simply this: Find a process as efficient as possible and approach the minimum required work as nearly as possible. We shall next examine several methods of operation that have actually been used with varying degrees of efficiency.

DIRECT DISTILLATION

One of the simplest techniques that can be used for desalination of water is direct distillation. Water for shipboard use has been distilled from the sea for well over a century, and a few communities in the world have for several years been totally supplied with fresh water from large-scale stills. The basic elements of a distillation device are shown in Figure 13-2. The unit operates at two temperatures. Salt water is brought into a container and heated with steam. Water evaporates from the sea water container, and eventually a concentrated brine is discharged. The vapor is transported to the cooler condensing unit, where it is converted once more to the liquid phase and admitted to a suitable storage container. Although it closely resembles the device used in most chemistry laboratories to produce distilled water from natural waters of much lower salt content, this simple device is relatively inefficient for several reasons. For example, the cooling water that flows through the condensing unit removes the heat that has been supplied from the steam coil. If this heat is not recovered, an obvious inefficiency is introduced. One way of recovering at least a portion of this heat is to use the sea water itself to cool the condenser. When it enters the evaporation unit it is therefore at a higher temperature, and less steam will be required for evaporation. Even at its best, such a simple apparatus wastes a considerable portion of the heat supply.

A second important problem lies in the inefficiency of transfer of heat from the steam line to the brine. In Figure 13-3 we show a greatly enlarged cross section of the steam pipe to illustrate the problem that arises. Steam condenses on the metal tube and the heat thereby liberated flows through to the outside of the tube, which is immersed in salt water. The sea water immediately in contact with the tube is raised in temperature more than the bulk water farther away, and under favorable conditions a thermally insulating scale is deposited which reduces the efficiency of heat transfer. The problem of scale formation is one of the most serious that arises in the distillation method and deserves a more careful study.

SCALE

As water evaporates from a sea water sample, concentrations of remaining salt increase and eventually precipitation occurs. Scale formation is primarily the result of precipitation of four insoluble salts, and we shall examine each separately.

Sea water is an essentially saturated solution of calcium carbonate. Moreover, the solubility of this compound is strikingly dependent on the amount of carbon dioxide present. As the concentration of carbon dioxide increases, carbonic acid is generated according to the equation

$$CO_2 + H_2O \rightleftharpoons H_2CO_3 \qquad (13\text{-}5)$$

Although carbonic acid is a relatively weak acid, an increase in its concentration tends to shift the equilibrium

$$H_2CO_3 + CO_3^{2-} \rightleftharpoons 2HCO_3^- \qquad (13\text{-}6)$$

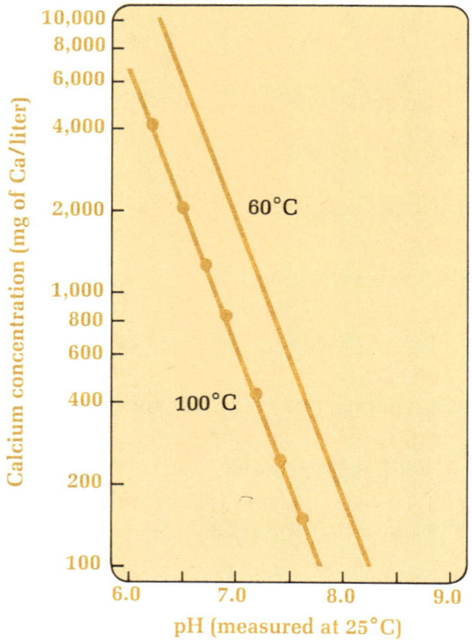

Figure 13-4 Curves for precipitation of $CaCO_3$ from sea water at 60°C and 100°C. (From K. S. Spiegler, *Salt Water Purification*, Wiley, New York, 1962.)

thus reducing the concentration of the carbonate ion. Whereas calcium carbonate is only very slightly soluble in pure water, the bicarbonate salt is much more soluble. Precipitation of calcium carbonate when sea water is heated and evaporated is the result of two factors. Simple heating of sea water decreases the solubility of carbon dioxide and shifts the equilibrium

$$HCO_3^- \rightleftharpoons CO_2 + OH^- \tag{13-7}$$

to the right as carbon dioxide is lost. The hydroxide ion liberated by this reaction increases the pH of the solution. This shifts the equilibrium

$$HCO_3^- + OH^- \rightleftharpoons H_2O + CO_3^{2-} \tag{13-8}$$

and produces more carbonate ion.

To understand the way in which solubility of calcium carbonate depends on the concentration of each of the components, we examine the equilibrium

$$CaCO_3(s) \rightleftharpoons Ca(aq)^{2+} + CO_3(aq)^{-2} \tag{13-9}$$

In this equation the notation (aq) specifically notes that the ions are present in dissolved state in which some degree of interaction occurs between solvent water and the ion. The exact nature of this interaction is dependent on the concentration of the ions or of the presence of other salts (see Chapter 12). For the moment we ignore this dependence and simplify our expressions by dropping this notation in the remainder of this discussion. It will be implicit in each equation we write that some degree of interaction occurs between the ions and the solvent. Furthermore, once solid calcium carbonate is present, its concentration is unchanging and we can simplify the overall equilibrium and formulate the solubility product constant, K_{sp}, in the usual way:

$$K_{sp} = (Ca^{2+})(CO_3^{2-}) \tag{13-10}$$

For calcium carbonate at 25°C, $K_{sp} = 9 \times 10^{-9}$. A typical calcium concentration in sea water is 0.4 g/liter, corresponding to a molar concentration of 0.01. We may use this information together with the solubility product constant to determine the maximum amount of carbonate ion that may be present without formation of solid. Substituting into the expression given above,

$$\begin{aligned} 9 \times 10^{-9} &= (0.01)(CO_3^{2-}) \\ (CO_3^{-2}) &= 9 \times 10^{-7}M \end{aligned} \tag{13-11}$$

According to this calculation, if the carbonate ion concentration in sea water equals or exceeds approximately $10^{-6}M$, the prediction of thermodynamics is that calcium carbonate precipitation from solution is possible. Of course, as with all thermodynamic calculations, we know nothing of the speed with which deposition of the solid might occur. In actual fact, the carbonate ion concentration in average sea water somewhat exceeds this amount and $CaCO_3$ should precipitate, a statement which should not be surprising. Bear in mind that coral and the shells of many other sea animals such as oysters are basically calcium carbonate!

Since several factors will determine whether calcium carbonate can precipitate as scale, it is easiest to see their interaction in graphical form. Figure 13-4 shows the dependence of scale formation on pH at two temperatures. Note that $CaCO_3$ solubility is sharply dependent on the acidity of the solution; a change of two pH units increases permissable Ca^{2+} ion concentration by 10,000-fold.

CALCIUM SULFATE

The acidification of sea water prior to distillation can prevent the formation of calcium carbonate, but other calcium salts are sufficiently insoluble to present a scale problem. When a solution containing calcium ions and sulfate ions is evaporated, either of two insoluble products may eventually precipitate. These products are $CaSO_4$ (known as anhydrite) and

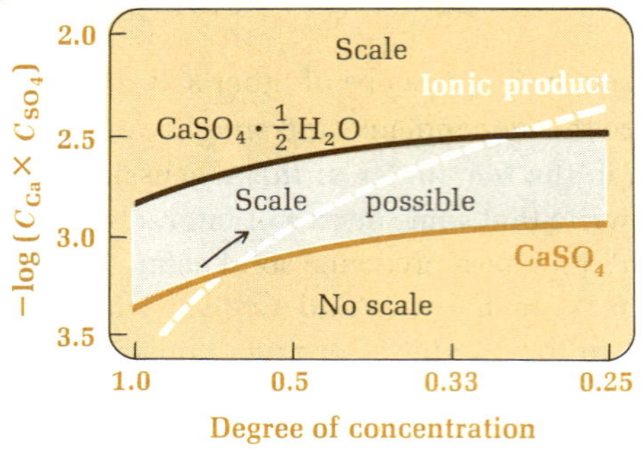

Figure 13-5 Solubility products of $CaSO_4$ and $CaSO_4 \cdot \frac{1}{2}H_2O$ in sea water at 100°C. Degree of concentration shows fraction of total volume remaining. (From K. S. Spiegler, *Salt Water Purification*, Wiley, New York, 1962.)

Table 13-5 Calculation of ionic strength for sea water of average composition

Ion	Parts per million	Atomic weight	C_i	$C_i Z_i^2$
Na^+	10,561	22.997	0.4592	0.4592
Cl^-	18,980	35.46	0.5353	0.5353
Mg^{2+}	1,272	24.31	0.0523	0.2092
Ca^{2+}	400	40.08	0.0100	0.0400
SO_4^{2-}	2,649	96.066	0.0276	0.1104
K^+	380	39.102	0.0097	0.0097
HCO_3^-	142	61.01	0.0023	0.0023
Br^-	65	79.909	0.0008	0.0008
				1.3669

$$I = \tfrac{1}{2}(1.3669) = 0.683$$

$CaSO_4 \cdot \frac{1}{2}H_2O$ (the hemihydrate more familiarly known as plaster of paris). Figure 13-5 shows the relationship between scale formation and degree of concentration defined as the fraction of the original volume remaining after evaporation. The dashed line shows the product of calcium ion concentration and sulfate ion concentration that would result upon the removal of water. When the ion product reaches the solubility product lines, precipitation may occur. As with all equilibrium calculations, this diagram shows only what is permitted by thermodynamics. In fact, it is not uncommon to observe that deposition of calcium sulfate scale does not occur at an appreciable rate even after the solubility product limit has been crossed. Such a solution is *supersaturated* in anhydrite. Continued evaporation eventually makes the ion product cross the hemihydrate line, and for this salt supersaturated solutions are much less common. In actual distillation practice, therefore, it is generally possible to remove about two-thirds of the water from a sea water sample before the probability of formation of scale becomes large.

Note carefully that the product $(Ca^{2+})(SO_4^{2-})$ in Figure 13-5 is dependent on the degree of concentration of the sea water. Specifically, this product increases with increasing concentration. This is precisely what we would have predicted on the basis of our discussion of the effect of ionic strength on activities of dissolved ions in Chapter 12. The oceans contain more than sufficient ionic strength to make the activity coefficients of ionic species present quite different from unity. An approximate calculation of the ionic strength of the oceans is presented in Table 13-5. Note that the ionic strength is so high that the Debye-Hückel limiting law is not really useful for making quantitative calculations of activity coefficients. But the qualitative prediction of decreasing activity coefficient with increasing ionic strength is still accurate, as is borne out by the data in Figure 13-5. You may wish to return briefly to Chapter 12 at this time to review the quantitative aspects of the solubility behavior of calcium sulfate.

MAGNESIUM HYDROXIDE

The last slightly soluble salt that must be considered in the formation of scale is magnesium hydroxide. For average sea water, the concentration of magnesium is near 0.054*M*. The solubility product of magnesium hydroxide is 1×10^{-11}, and we may calculate the concentration of hydroxide ion in a saturated solution:

$$(Mg^{2+})(OH^-)^2 = 1 \times 10^{-11}$$

$$(OH^-)^2 = \frac{1 \times 10^{-11}}{0.054} = 1.85 \times 10^{-10} \qquad (13\text{-}12)$$

$$(OH^-) = 1.26 \times 10^{-5}M$$

For average sea water of pH about 8, the hydroxide concentration is $10^{-6}M$ and the solution is unsaturated with respect to magnesium hydroxide. However, when sea water is distilled at elevated temperatures, two factors change the situation. First, the dissociation of water becomes significantly larger as the temperature is raised; above about 70°C, the hydroxide ion concentration becomes sufficient to commence precipitation of magnesium hydroxide. Furthermore, as carbon dioxide is lost from the warm water according to the equation

$$HCO_3^- \longrightarrow CO_2 + OH^- \qquad (13\text{-}13)$$

additional hydroxide is formed, enabling further precipitation of magnesium hydroxide to occur. We have observed above that the deposition of calcium carbonate can be avoided by adding acid to the water prior to distillation. This addition simultaneously suppresses the precipitation of magnesium hydroxide.

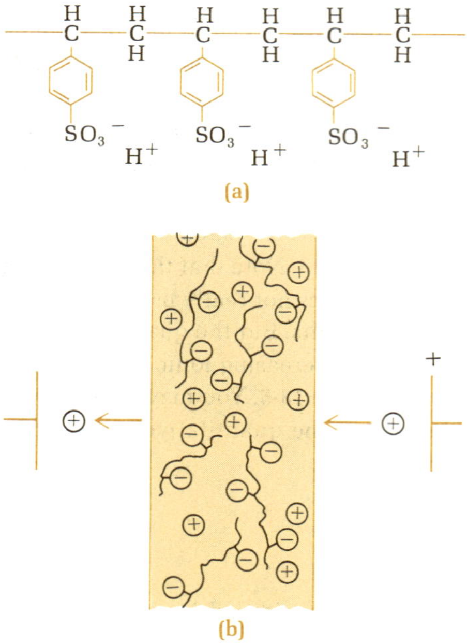

Figure 13-6 A possible type of ion-selective membrane. In (a) a substituted polystyrene molecule is shown (see Chapter 7). The negative sulfonate group is bound to the polymer molecule and cannot move through the solution. The dissociated protons from the acid groups are free to move or be replaced by other positive ions such as sodium. In (b) a schematic membrane is shown. As positive ions approach the membrane from the right under the influence of the electrical field, positive ions on the left are attracted from the membrane. Ions within the membrane migrate to replace them and are in turn replaced by positive ions from the solution on the right, keeping the total charge within the membrane constant.

In summary, if sea water is concentrated without addition of acid, the deposition of scale depends on the conditions of the distillation. If the sea water is distilled at relatively low temperatures (by working at reduced pressures), the scale consists mainly of calcium carbonate. If the still operates at a temperature between about 60°C and 100°C, a mixed calcium carbonate–magnesium hydroxide scale is formed. Finally, if the still is operated above 100°C, and particularly for maximum recovery of water, deposition of calcium sulfate commences and becomes more important the higher the operating temperature and degree of concentration.

MEMBRANES

In contrast to the desalination of sea water by distillation, which depends on the use of heat to separate physically the water from the dissolved solids, other desalination methods depend on the properties of membranes. As we shall see, these may be employed in a variety of ways depending on the particular kind of membrane employed.

Basically, a membrane is a device that acts as a physical barrier between two fluids while permitting a degree of communication between them. Thus, some of the components of the two fluids will pass through the membrane, while others will be prevented from doing so. On this basis, we can classify membranes into several categories. First, there are **osmotic membranes.** These permit diffusion of the solvent only between two compartments that are separated by the membrane. Osmotic membranes are impermeable to all solute species, which therefore cannot cross. Second, there are *non-osmotic membranes,* which have just the opposite properties. The solute species are permitted to pass the membrane, while the solvent is prevented from doing so. Third, there are membranes that make distinctions between molecules based on their size. These are called **size-selective membranes.** The idea here is a simple one: The membrane has pores through which molecules may pass, and these are of a definite size so that molecules smaller than the pore size may pass through the membrane while those that are larger may not. It is precisely this type of size discrimination on which most osmotic membranes depend for their ability to exclude solute molecules. Finally, there are **ion-selective membranes,** which sort out species on the basis of their charge. Ion-selective membranes are themselves formed from polymers to which a number of fixed charges are attached. Thus, if the membrane possesses a large number of cationic groups, anions can be absorbed to the membrane, desorbed from it, and in so doing, pass through the membrane. Cations, on the other hand, will be electrostatically repelled by the membrane, and the membrane will be impermeable to them. In contrast, if the membrane is constructed from a polymer that is itself anionic, then cations will be the mobile species and have the capacity to pass through the membrane, while anions will be electrostatically repelled and will be unable to cross the membrane. An example of how this can be done is shown in Figure 13-6. We shall see how both osmotic membranes and ion-selective membranes may be used to desalinate sea water.

OSMOTIC PRESSURE

Let us consider a vessel divided into two compartments by an osmotic membrane. On one side of the membrane we shall have pure solvent and on the other side of the membrane solvent in which some solute is present.

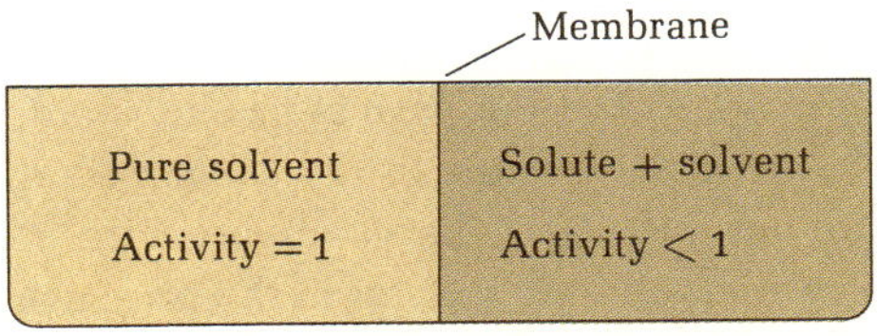

In a system of this kind, the equilibrium situation is reached when the activity of those species to which the membrane is permeable is the same

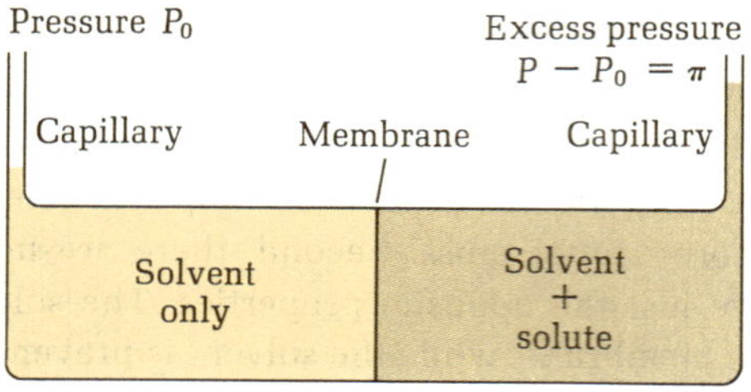

Figure 13-7 Schematic representation of an osmotic pressure apparatus. (From C. Tanford, *Physical Chemistry of Macromolecules,* Wiley, New York, 1961.)

on both sides of the membrane. Since we are dealing here with an osmotic membrane, only the solvent is permeable. The equilibrium situation, then, is the state in which the activity of solvent on both sides of the membrane is equal. Clearly, the situation diagrammed above cannot be an equilibrium situation. In the pure solvent we chose the activity to be unity. In the compartment containing the solvent and solute, the activity of the solvent must be less than unity, since its mole fraction is less than unity. We can, if we wish, employ the partial pressures of the solvent above the two compartments as measures of the relative activities. We know from Raoult's law that the pressure of the solvent above the second compartment will be less than that of the pure solvent by an amount that depends on the mole fraction of dissolved solute. As a result of these considerations, solvent will tend to flow across the membrane from the side containing pure solvent to the side containing the solute dissolved in the solvent. Since equilibrium can be reached only when the activity of the solvent on both sides is the same, this process will tend to occur until the solute has been infinitely diluted out. The tendency of the solvent to move from one side of the membrane to the other to equilibrate the activities of solvent on both sides is called the **osmotic pressure.**

The most basic relationship between osmotic pressure and the composition of the two basic fluids is

$$\Pi = \frac{RT}{V_0} \ln \frac{P_0}{P} \tag{13-14}$$

In this equation Π is the osmotic pressure, R is the perfect gas constant, T is the absolute temperature, V_0 is the **molal volume** of the solvent (that volume occupied by 1 mole of pure solvent), P_0 is the vapor pressure of the pure solvent, and P is the vapor pressure of the solvent in the presence of solute. As we know from Raoult's law, the ratio of the pressures of solvent in the two compartments is a measure of the mole fraction of the solvent in those compartments. Provided that the solute is reasonably dilute, the above equation can be written in somewhat more useful forms. It can be formulated either in terms of the solute concentration in moles per liter,

$$\Pi = RTC \tag{13-15}$$

or in terms of the mole fraction of the solute, N_2,

$$\Pi = \frac{RT}{V_0} N_2 \tag{13-16}$$

These latter two equations are identical, since the concentration of the solute is in fact given by the ratio of the mole fraction of the solute to the molal volume of the solvent. The latter two equations fail only if the solution is not dilute. Consequently, if we know the concentration of the solute in the solvent, we can use these equations to calculate the osmotic pressure. Alternatively, we can measure the partial pressures above the two solutions and use these as a means of calculating the osmotic pressure. It is also quite simple to measure the osmotic pressure experimentally. This is accomplished by measuring the hydrostatic pressure above the fluid containing the solute that is just necessary to balance the osmotic pressure. A device for performing this kind of measurement is shown in Figure 13-7. Note that as the solvent moves across the membrane, it will tend to force the solution in the other compartment up the capillary tube on the right. This, then, will develop an opposing hydrostatic pressure. When the osmotic pressure and the hydrostatic pressures are equal, no additional movement of solvent across the membrane will occur. A situation of equilibrium will have been reached. By measuring the hydrostatic pressure, we obtain a measure of the osmotic pressure.

Example *Calculate the osmotic pressure of a solution that contains 4 g of sucrose, molecular weight 362, per 100 ml of solution at 298°C.*

Solution *Provided that we calculate the concentration of sucrose in this solution, we can then apply Equation (13-15) to obtain the osmotic pressure. The concentration of sucrose is given by*

$$C\ (\text{moles/liter}) = \frac{4\ \text{g}}{362\ \text{g/mole}} \times \frac{1}{0.1\ \text{liter}}$$

$$= 0.11\ \text{mole/liter}$$

Then

$$\Pi = RTC = 0.083\ \frac{\text{liter atm}}{\text{mole deg}} \times 298\ \text{deg} \times 0.11\ \frac{\text{mole}}{\text{liter}}$$

$$= 2.72\ \text{atm}$$

Note that this is a substantial pressure.

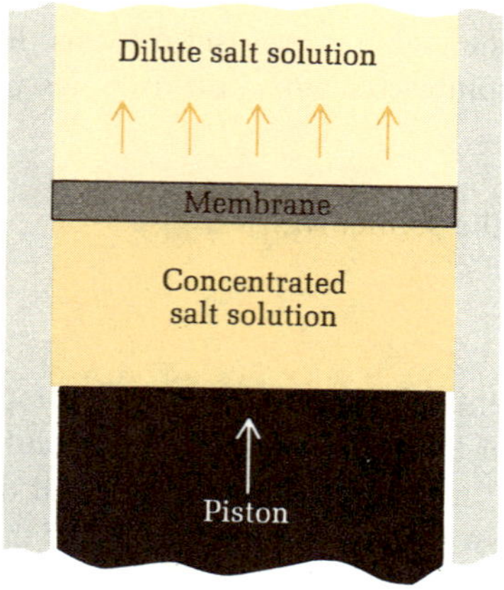

Figure 13-8 Salt filtration through a membrane. (From K. S. Spiegler, *Salt Water Purification,* Wiley, New York, 1962.)

REVERSE OSMOSIS

We can employ the considerations developed so far as a means of generating an additional technique of sea water desalination. Note that the process of osmosis in which the solvent flows across the membrane to dilute the solvent containing solute is precisely the opposite of the desalination procedure. What we should like to do is to force the solvent from the solute solution across the membrane into the pure solvent side. The measurement of osmotic pressure suggests how we might do that. Suppose, for example, that the hydrostatic pressure exerted on that compartment containing the solute is greater than the osmotic pressure. This will have the effect of reversing the entire process of osmosis. That is, the movement of solvent will now be in the opposite direction: It will flow from the side containing the solute into the pure solvent side. Hence, this desalination process is referred to as **reverse osmosis** or **ultrafiltration.** A device for carrying out a reverse osmosis process is shown in Figure 13-8.

Our earlier considerations about the minimum work required for sea water desalination permit us to calculate the osmotic pressure of sea water and to understand the pressures that must be employed to effect desalination. We decided earlier that about 0.7 kW hr is required to desalinate 1 m^3 of sea water. Put in more convenient units for our purposes, we note that 1 kW hr is equal to 3.55×10^4 liter atm or, what is equivalent, 35.5 m^3 atm. It follows, then, that $35.5 \times 0.7 = 24.8$ atm is the osmotic pressure of sea water. That is, the minimum pressure that will be required to prevent solvent from moving from a pure solvent side into the sea water is 24.8 atm. If we exert more pressure on sea water than that, the reverse process will occur and pure water can be forced across the membrane from the sea water.

In practice, of course, it requires rather more pressure than 24.8 atm to effect this process. In the first place, most of the membrane materials that are sufficiently nearly osmotic so that they are permeable to little besides water do exert a substantial resistance to water flow. Consequently, pressures in the neighborhood of 50 or 100 atm are ordinarily required to assure that reverse osmosis will occur at a useful speed. Moreover, as the reverse osmosis process continues, the sea water becomes continually more concentrated; as a consequence, the osmotic pressure rises and additional force must be exerted to overcome it.

The most commonly employed materials for membranes useful in the reverse osmosis process come from cellophane, cellulose acetate, and polyvinyl alcohol. These membranes have rather good combinations of permeability to water, selectivity for water, and strength. In addition, complicated membranes called *composition membranes* are also used. One of these is illustrated below:

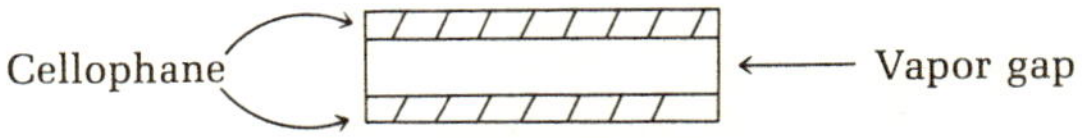

A composition membrane

In this membrane two pieces of cellophane are separated by a thin vapor gap. The permeation process involves the penetration of water through one layer of cellophane, its evaporation into water vapor between the two, and recondensation and permeation on the opposite surface. Since the evaporation–condensation processes are impossible for ions, membranes of this type have very high selectivities for water.

The various technical and economic considerations involved have so far restricted reverse osmosis applications of desalination principally to brackish waters rather than sea water. Brackish waters have rather too much salt to permit their use for human consumption or for irrigation purposes. At the same time they have as much as ten times less salt than typical sea water samples. Large-scale applications of reverse osmosis to sea water desalination must await further technical developments.

As an example of the utility of reverse osmosis in purifying brackish water, let us consider the example of the small town of Plains, Texas. This village of 1250 inhabitants was served by wells that yielded water containing about 2000 parts per million (ppm) of dissolved solids. This is far in excess of the maximum content recommended by the U.S. Public Health Service, which is 500 ppm. The result was corroded plumbing, scale-clogged pipes, discolored teeth, latherless soaps, and foul-tasting water. Late in 1970, a reverse osmosis water purifying plant was completed. This plant now

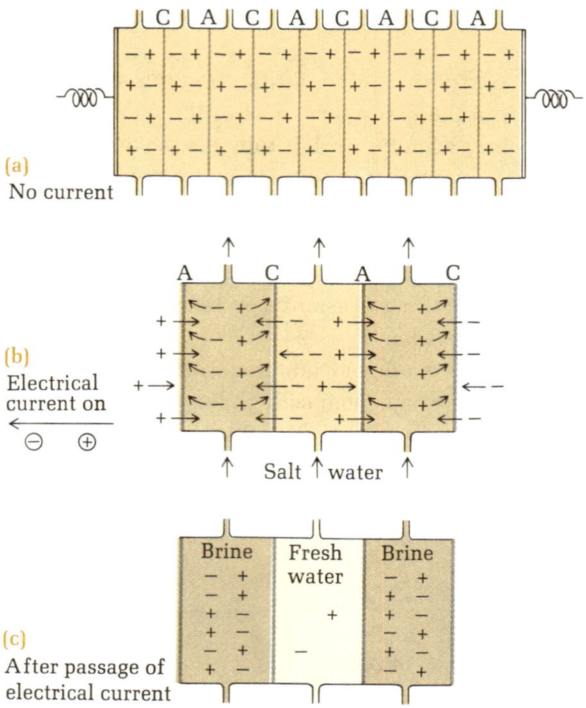

Figure 13-9 Principle of electrodialysis. Positive ion (+) (e.g., sodium); negative ion (−) (e.g., chloride). C and A are cation- and anion-permeable membranes, respectively. Ion migration under the action of electrical current causes salt depletion in alternate compartments and salt enrichment in adjacent ones. (From K. S. Spiegler, *Salt Water Purification,* Wiley, New York, 1962.)

provides 100,000 gallons of water a day for the town with a content of dissolved solids of 300 ppm, well within tolerable limits. The price of the purified water is about 75 cents/1000 gallons.

To produce such large quantities of water, it is essential to maximize the water–membrane surface contact. This was accomplished by synthesis of membranes in the form of tiny hollow fibers. Millions of these are incorporated into a single purification unit. When brackish water is forced into a metal pipe containing these fibers, water, containing much less dissolved solids, passes into the interior of the fibers and finally to a collection point. Each such unit can supply about 10,000 gallons of purified water per day.

The impact of this plant on the lives of the citizens of Plains is substantial. Because of it, this town may grow and prosper, rather than slowly die. We can certainly anticipate many other similar installations in the near future.

In addition to water purification, reverse osmosis should provide useful treatment of municipal and industrial waste effluents, for the concentration of juices and other food products, and for the concentration of medicinally useful solutions.

ELECTRODIALYSIS

The last method of sea water desalination that we shall consider also makes use of membranes, but of quite a different sort than those employed to effect reverse osmosis or ultrafiltration. In **electrodialysis** membranes are used that are ion-selective; that is, they are permeable to cations but impermeable to anions or permeable to anions and impermeable to cations. Moreover, the movement of ions across the membranes is driven in this case by an electrical field. A typical electrodialysis setup is shown in Figure 13-9. The important features of this setup are the following. First, an alternating series of cation-permeable and anion-permeable membranes are used to separate the entire apparatus into a series of compartments and second, electrodes are placed at the extreme ends of the apparatus. Now we need to ask ourselves what will happen to the ions in this system when the electricity is turned on, that is, when a current flows across the various compartments from one electrode to the other. If we look at the situation in a compartment that is bounded on the left side by a cation-permeable membrane and on the right side by an anion-permeable one, it is clear that the current will cause the cations to migrate across the boundary to the left toward the negative electrode and the anions to migrate across the boundary to the right toward the positive electrode. Consequently, this compartment will be depleted of ions of both types. Just the opposite situation is true for those compartments bounded on the left by the anion-permeable membrane and on the right by a cation-permeable one. While the anions and cations will tend to move in the same directions as in the previous case, they are prevented from moving from one compartment to the other by the permeability characteristics of the membranes that bound the compartment. As a result of these considerations, it is easy to see that in half of the compartments we shall tend to deplete the concentration of ions, and hence generate pure or nearly pure water, and in the other compartments we shall tend to concentrate ions into a brine saltier than normal sea water.

As in the case of desalination by reverse osmosis, the process of electrodialysis is crucially dependent on the quality and characteristics of the membranes involved. Membrane materials presently available have permitted the construction of several large-volume desalination plants, and these are currently functioning to provide as much as several hundred thousand gallons of water per day. As in the other cases of desalination processes, scale formation is again a problem in electrodialysis. Clearly, as the concentration of the ions in the water increases by electrodialysis in the appropriate compartments, we have the possibility of precipitation of calcium carbonate, calcium sulfate, or magnesium hydroxide. The deposition of these solids on the membranes, of course, renders the electrodialysis process ineffective or impossible.

Glossary

Electrodialysis a method of water purification in which ionic solutes are driven across ion-selective membranes by an electrical field.
Ion-selective membranes membranes that distinguish between solutes on the basis of electrical charge. Thus, cation-selective membranes are permeable to cations but impermeable to anions; anion-selective membranes have just the opposite properties.
Mole fraction the mole fraction of any substance is obtained by dividing the number of moles of that substance present in a given system by the total number of moles of all substances present in the same system:

$$X_a = \frac{N_a}{(N_a + N_b + N_c + \cdots)}$$

Osmotic membranes membranes that are permeable to solvent but impermeable to all solutes.
Osmotic pressure the tendency of solvent to move from one side of an osmotic membrane to the other to equalize its activity on both sides of the membrane. Quantitatively, the osmotic pressure is given by $\Pi = RTC$, where C is expressed in moles per liter for reasonably dilute solutions.
Partial pressure in a multicomponent system, the pressure that would be due to individual components were they present individually.
Raoult's law an expression linking partial pressure to mole fractions in multicomponent systems:

$$P_a = X_a \times P_a^0$$

Size-selective membranes membranes that distinguish between solutes on the basis of size. Such membranes usually are permeable to solutes up to a certain size but are impermeable to solutes that are larger.
Ultrafiltration (reverse osmosis) a method of water purification in which solutes are removed by forcing the water sample through a size-selective membrane.

Study problems

1. Commercial isopropyl alcohol contains about 70 percent by weight of the alcohol, $(CH_3)_2$—CHOH, and about 30 percent by weight water. What is the mole fraction of water in commercial isopropyl alcohol?
2. Suppose that 1 kg of a ZnI_2 solution contains 420 g of this salt. What is the mole fraction of water in this solution? Calculate the vapor pressure of this solution relative to the vapor pressure of pure water at the same temperature.

3. Given that the vapor pressure of water from the Dead Sea is 8 percent less than that of pure water at the same temperature, calculate the mole fraction of water in Dead Sea water. Will it require more or less work to desalinate Dead Sea water or ordinary ocean water? Defend your answer.
4. The vapor pressure of water from the Great Salt Lake is 0.956 of that of pure water at the same temperature. Calculate the minimum energy required to desalinate 100 g of Great Salt Lake water at 25°C. Remember that the vapor pressure of pure water at this temperature is 0.0312 atm and that 1 g of water at this temperature and pressure occupies 43.4 liters.
5. Calculate the ionic strength for water from the Dead Sea. Will the activity coefficients for ions dissolved in this water be greatly different from zero? Is it possible to make a reliable quantitative calculation of activity coefficients of ions in solutions at this ionic strength?
6. In water from Lake Erie, the concentration of calcium is near 0.031 g/liter. What is the molar concentration of calcium in this water? What is the maximum equilibrium concentration of carbonate ion, CO_3^{2-}, that could be present in Lake Erie water without the precipitation of solid $CaCO_3$?
7. In the Caspian Sea, the concentration of magnesium ion is near 0.8 g/liter. At what value of pH would magnesium hydroxide just begin to precipitate from this water? If the pH were maintained constant at 8, how many fold could we concentrate Caspian Sea water before magnesium hydroxide began to precipitate?
8. Calculate the osmotic pressure of a solution that contains 10 mg/ml of myoglobin, molecular weight 16,800. Would it be easier to measure the osmotic pressure of a solution of myoglobin or sucrose? Why? What difficulties might be encountered in each case?
9. Given that the osmotic pressure of a solution containing 5 mg/ml of hemoglobin is 0.002 atm calculate the molecular weight of hemoglobin.
10. Below are listed several pairs of substances. For each pair, indicate the type of membrane that you would select to effect separation of the two.
    a. sucrose and myoglobin
    b. sodium chloride and water
    c. sodium ion and chloride ion

Suggestions for further reading

K. E. Chave "Chemical Reactions and the Composition of Sea Water," *J. Chem. Educ.*, **48**, 148 (1971).

R. A. Horne *Marine Chemistry*, Wiley, New York, 1969.

M. C. Porter and A. S. Michaels "Membrane Ultrafiltration," *Chem. Tech.*, January, 1971, pp. 56–63.

J. P. Riley and G. Skirow *Chemical Oceanography*, Academic, New York, 1965.

Part three

LIGHT AND ENERGY

In the six chapters that follow, comprising the third stage of our investigation into chemistry, we shall encounter a great variety of new things and ideas. We begin with a discussion of light and its properties, continue with inquiries into how light interacts with matter, how it is generated in the sun and other stars, the ways in which light and other energy sources influenced the development of the earliest living organisms, and conclude with a brief examination of how sunlight serves the needs of man today. In tracing out this pathway we shall encounter such diverse topics as nuclear fusion, solar stills, and photosynthesis. But two common threads run through the entire discussion: *light* and *energy*. These are not independent matters, since light is one of the forms that energy takes. Prior to entering into the chemical aspects of these topics, let us pause just long enough to develop some insight into a matter of great general interest: the needs of man for energy and how he supplies them.

It is possible to trace out a history of man in terms of the various sources of energy that have been harnessed to serve his needs: an adequate supply of food, protection from the environment, provision for basic physical comforts, means of transportation and communication, and so forth. The generation and utilization of energy have been of fundamental importance at every stage of development in the quality of life beyond the rudimentary subsistence level.

Yet it is common knowledge that the rapid increase in power consumption in the United States and elsewhere has generated very significant problems. These include (1) pollution and (2) exhaustion of the natural resources of the earth. The problem of environmental pollution associated with energy production is discussed in Chapters 16 and 18 and need not be examined in detail here. Suffice it to say that each of the important means of energy generation currently in use does have its own associated environmental problems. A crucial factor in our search for new and better energy sources is that they can be "cleaner" in the sense of contributing less damage to the ecosphere.

What alternatives are available to us? One possibility has been suggested by the "counterculture": turn back the clock to an earlier era in which man lived more simply, in a rural setting, and consumed much less energy on a per person basis. It is difficult to imagine this as a realistic solution when the great majority of people on earth are striving to attain a style of life that consumes more energy per capita, not less. The alternative is to develop new energy sources. Among these, there are three that are promising but almost completely untapped at present.

Perhaps the most hopeful of these is nuclear fusion, discussed in Chapter 16. The basic fuel for fusion reactions that might provide useful energy on earth is the isotope of hydrogen known as deuterium. Extraction of just *1 part per million* of the deuterium present in the oceans would provide a fuel reserve about 100-fold greater than the total of all stored fossil fuels on earth! Moreover, energy generation by nuclear fusion reactions is "clean"; that is, it yields no pollutants. The problem is solving the engineering problems in converting fusion reactions into a useful energy source.

A second possible source of energy, and one employed to a modest extent now, as described in Chapter 19, is the radiant energy from the sun. The main problem in using radiant energy from the sun is the low *density* of energy: Enormously large reflectors would be required to harvest significant quantities of energy. But since this energy is free and noncontaminating, a great deal of thought and work has been put into ways in which more of it might be employed.

The last untapped energy source is the heat of the earth itself. Well below the surface of the earth are molten magmas, and it may be possible to extract heat from them by drilling down to them at spots where they reach fairly close to the surface. The heat of the magma could then be used to convert water to steam, which could be employed to turn turbines.

The September, 1971, issue of *Scientific American* is devoted entirely to the matter of energy. If you wish to read more about this problem, this source is heartily recommended.

The energy crisis is of profound importance. The better informed we are about possible alternatives, the better we will be able to deal with it.

14

The nature and properties of light and energy

Shedding some light on the good old days:
Things aren't as good as they used to be—and they never were.

Joe Sutton

Light and the interaction of light with matter are topics of fundamental importance. On earth, the great bulk of our light comes to us directly from the sun. This light provides the external energy source on which all life depends. We recall that the second law of thermodynamics requires that all real chemical and physical processes occur with a degree of inefficiency (Chapter 8). Thus, were it not for an external energy source, our local energy reservoirs would eventually be consumed and life would run out.

The process of *photosynthesis* provides the organic compounds that serve as foods for all living things (the *biosphere*). Photosynthesis is a process carried on by green plants and certain microorganisms by which the energy from the sun's light is employed to reduce carbon dioxide in the atmosphere to organic compounds. This topic is developed further in Chapter 19.

The development of photosynthetic organisms is partially responsible for the formation of our atmosphere as we know it, as discussed in Chapter 17. Thus, light both provides the environment necessary for the development of living things and yields the energy required to maintain them.

Light serves many more purposes than those explicitly mentioned so far. Let us consider some of the range of ways light affects us. For example, starlight gives us a means of navigation and the basis of our system for the measurement of time. The impact of light on the retina of the eye provides the foundation for *vision* (Chapter 15). Many bioluminescent organisms, such as the firefly, generate their own light, which they use for navigation, protection, or as a sex attractant. Light even acts to affect strongly the psychological state of the human. Both the behavior of individuals and societies is strongly influenced by the quality and quantity of light they receive. Suicide rates increase during winter months when the sun is seldom seen. The tendency of people to follow the sun is well known. In a city such as Lima, Peru, for example, which is hidden from the sun a large portion of the year due to an overcast sky resulting from the cold Humboldt current, much of the population makes a weekly pilgrimage outside the city in search of the sun. Sun gods have been regularly worshiped during the history of man.

Humans are not the only living beings to seek out light; certain microorganisms and plants do so as well. The tendency of microorganisms to swim toward light is called *phototaxis;* that of plants to grow toward light is called *phototropism*. Let us try to define light from a fundamental point of view, since we shall encounter many examples of its importance in chemistry.

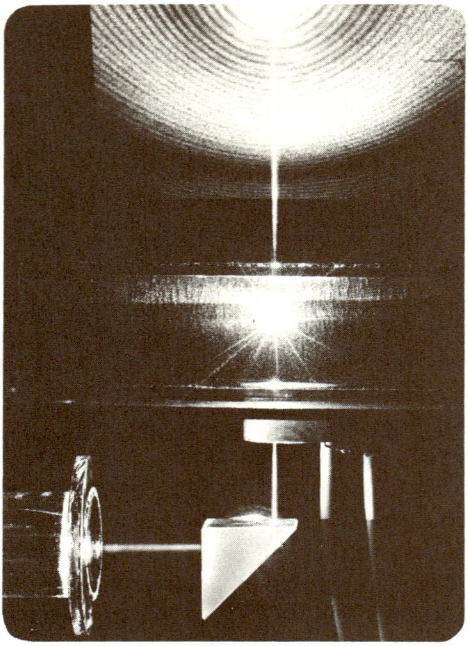

Optical levitation of a very small particle by means of a vertically directed beam of green laser light is shown in this photograph. The particle, a transparent glass sphere 20 microns in diameter, is shown levitated in air about 1 centimeter above a glass plate by the 250-milliwatt beam. Although the particle is barely visible to the unaided eye, it scatters enough laser light to sparkle brilliantly in this experiment. This is a very straightforward illustration that light has an associated pressure and that this pressure reflects the momentum of individual photons. [Ben Rose, *Sci. Am.*, **226** (February, 1972).]

THE NATURE OF LIGHT

The physical nature of light is an elusive quantity to pin down. This is true because most people try to understand the properties of a new object in terms of the properties of objects with which they are familiar. Hence, we might describe a pebble as small, gray, hard, and shiny. Since we frequently have contact, through our senses, with objects that have these properties, we can form a good mental image of the pebble in question from this description. But what happens when we attempt to arrive at a mental picture of something that has properties, or a combination of properties, not found in things to which we are accustomed? For example, from the discussion presented in Chapter 4, in what terms do you view an electron? What adjectives do you use to describe it? While it is possible to describe an electron and its behavior in a purely mathematical way, such descriptions alone are not satisfactory to most of us: We strive to understand the nature of such objects in terms that are more familiar to us. This is perfectly acceptable, provided that we bear in mind the cautionary words of Albert Einstein and Leopold Infeld:

> Science is not just a collection of laws, a catalogue of unrelated facts. It is a creation of the human mind, with its freely invented ideas and concepts. Physical theories

try to form a picture of reality and to establish its connection with the wide world of sense impressions. Thus, the only justification for our mental structures is whether and in what way our theories form such a link. . . .

In an effort to understand the nature of light in terms of a theory that links predictions about the properties of light to our sense impressions, we shall arrive at the conclusion that light has properties usually associated with both waves and particles. That is, light behaves as if it were at once a smooth and continuous substance yet particulate or grainy. For the moment, let us think of light in terms of a wave.

The wave nature of light

What we have been calling light is frequently called **electromagnetic radiation;** or, what is equivalent, light can be said to be composed of *electromagnetic waves.* We commonly use a number of terms to describe electromagnetic radiation: **gamma rays** and **X-rays** are electromagnetic radiation of very high energy; **ultraviolet light, visible light,** and **infrared light** are electromagnetic radiation of progressively decreasing energy. Finally, **microwaves** and **radio waves** are electromagnetic radiation of quite low energy.

How can we describe electromagnetic radiation in terms of the properties of waves? Suppose we throw a pebble into a lake and shortly thereafter take a "snapshot" across the surface of the lake. We would observe, in one dimension, something similar to the diagram shown in Figure 14-1. That is, at any moment in time we find alternate hills and valleys in the water. If we take a second snapshot a moment later, we observe that the hills and valleys have moved a little, perhaps to the positions shown by the light yellow lines in Figure 14-1. We can say that the wave is propagating in the indicated direction. We should mention that the water itself does not move in the direction of the wave. A cork floating on the lake bobs up and down but does not move along the surface. There are two basic quantities that are employed to describe our wave quantitatively: the **amplitude,** a measure of the height of the hills or the depths of the valleys (for simple kinds of waves, these are equal); and the **wavelength,** λ, the distance between corresponding parts of two waves. That is, the wavelength is the distance from the top of one hill to the top of the next one, from the bottom of one valley to the bottom of the next, and so on. We can also measure a **velocity** for the wave, the rate at which it is propagated from the site of its origin.

Electromagnetic radiation can be thought of in precisely the same terms, except that the amplitude of the wave does not measure the height or depth of the water at a particular point but the strength of the *electrical field,* E, of the wave. The laws of physics require that such a wave have an

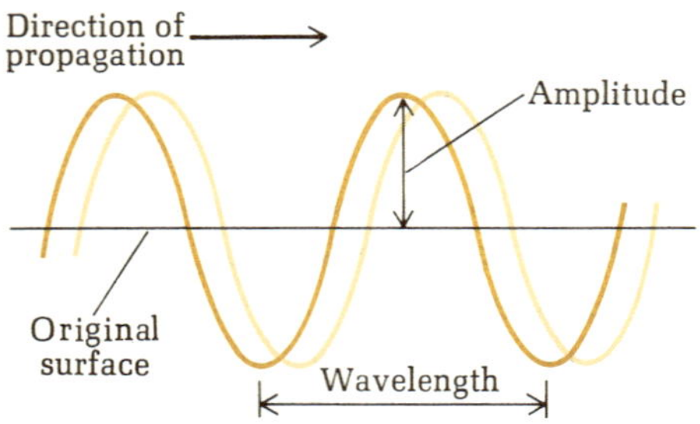

Figure 14-1 A diagrammatic view of a wave at one instant in time. The light yellow line indicates the position of the wave at some later instant. Note that the wave is described by the amplitude and the wavelength.

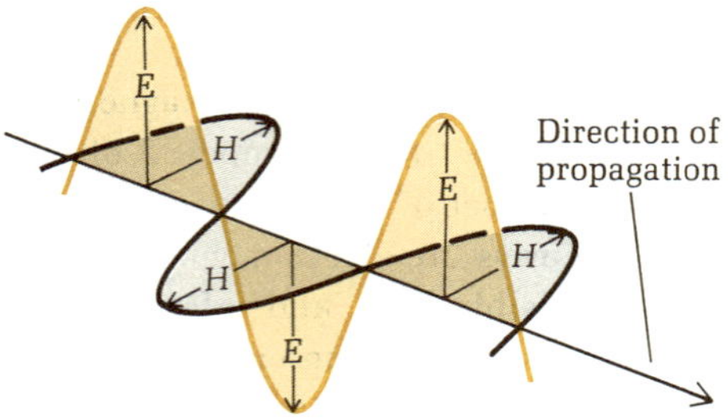

Figure 14-2 The electrical and magnetic components of an electromagnetic wave.

associated *magnetic field, H,* as well and that this be perpendicular to the electrical one (Figure 14-2). Hence, the designation *electromagnetic.* All electromagnetic radiation is propagated (in a vacuum) with the same velocity, $c = 2.997 \times 10^{10}$ cm/sec. While this velocity is certainly enormous, light does in fact require appreciable time to travel astronomical distances. Thus, the sun rises about 9 min before the first light reaches us; it requires about 1.3 sec for light to travel from the earth to the moon. The latter fact accounts for the slight delay that we noticed in listening to conversation between moon-bound astronauts and the Apollo Control Center in Houston.

In terms of the wave velocity and wavelength, we can calculate the wave **frequency,** ν, which has units of sec^{-1} and measures how fast the wave oscillates:

$$\text{frequency} = \nu = \frac{\text{velocity}}{\text{wavelength}} = \frac{c}{\lambda} \tag{14-1}$$

Finally, one additional term, the **wave number,** $\bar{\nu}$, which is the reciprocal of the wavelength and has units of inverse centimeters, is useful:

$$\bar{\nu} = \text{wave number} = \frac{1}{\lambda} \tag{14-2}$$

The particle nature of light

Our description of light as a wave will suffice for many of our purposes; yet the description cannot be regarded as complete or fully adequate. About the turn of the century, a series of experiments dealing with the **photoelectric effect** turned out to have most unusual results. The basis of the photoelectric effect is the observation that shining light on a metal surface may result in ejection of electrons from that surface (Figure 14-3). The curious observation made was that there was a lower limit to the frequency of light that is effective in causing the emission of these *photoelectrons*. For zinc, for example, it has been found that ultraviolet light with a wavelength shorter than 3500 Å can produce photoelectrons. This maximum wavelength (3500 Å) below which photoelectrons appear is called the **photoelectric threshold.** For sodium metal, the photoelectric threshold is about 6500 Å, and light of a much longer wavelength can cause emission of an electron from sodium than it can from zinc (Figure 14-4). How can we explain this finding? In 1905, Albert Einstein provided an explanation in terms of a *quantum theory* for electromagnetic radiation, first suggested by Max Planck in 1900. The explanation rests on the theory that electromagnetic radiation comes in packages or **quanta.** The energy of one quantum of electromagnetic radiation is proportional to the frequency:

$$E = h\nu \qquad (14\text{-}3)$$

In Equation (14-3), the proportionality constant relating energy to frequency is called **Planck's constant** and has a value of 6.625×10^{-27} erg sec (1 erg is 1 g cm^2/sec^2). Note carefully that Planck's theory treats electromagnetic radiation as particles or packages of energy, not as a wave. This generates what has been called the **wave-particle duality** of light and gives rise to the conceptual difficulties noted above. More about these matters follows shortly.

The quantum theory of light accounts for the observations concerning the photoelectric effect in a very nice manner. Let us remember that in Chapter 4 (Figure 4-9 and Table 4-2) we defined a quantity, the ionization energy, that measures the amount of energy needed to pull an electron completely away from the nucleus of an atom; this is related to what must be done to generate the photoelectric effect. It follows, then, that one quantum of light may strike one atom and, if it has sufficient energy, may kick out an electron. But if the light has a low frequency, and hence a low energy per quantum (red light), and if this energy is less than the energy required to kick out one electron, the photoelectric effect cannot occur regardless of the number of the quanta that strike the metal. There is, thus, a *threshold* of energy for the photoelectric effect: Quanta having less energy than this threshold cannot cause emission of electrons, quanta having an

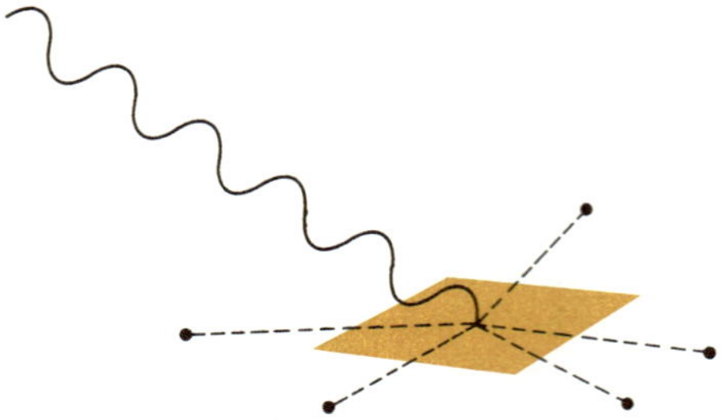

Figure 14-3 The light-induced emission of electrons from a metallic surface, the photoelectric effect.

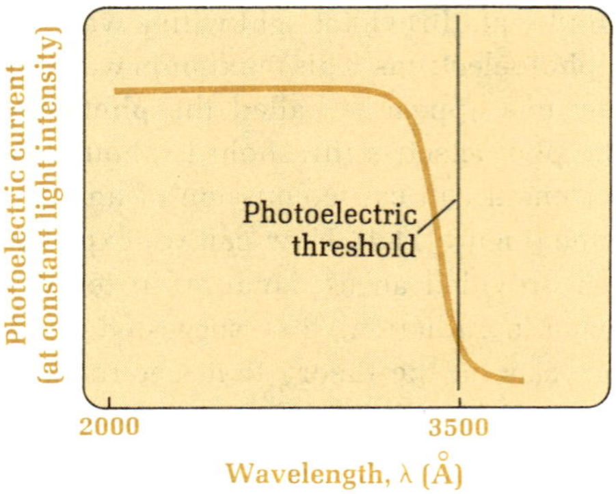

Figure 14-4 Plot of photoelectric current versus wavelength of incident light for zinc.

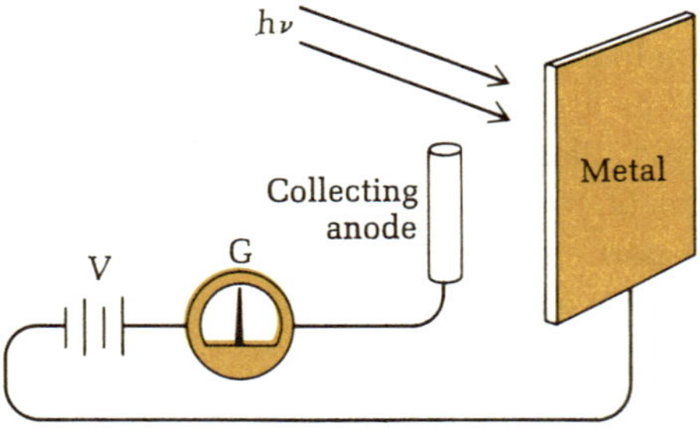

Figure 14-5 Physical arrangement for measurement of kinetic energy of photoelectrons. The retarding voltage, V, is adjusted so as to just keep the photoelectrons from reaching the collecting anode. When the galvanometer, G, indicates zero current, we know that V has been set correctly.

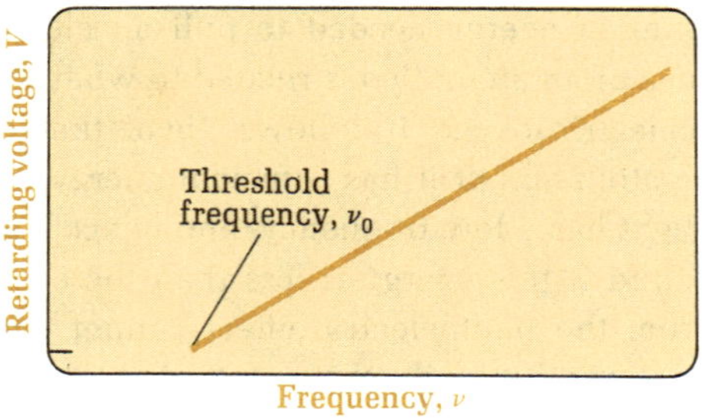

Figure 14-6 Plot of retarding voltage, V, versus the frequency of the incident light. From the value of the threshold frequency, we can evaluate the ionization energy, E_i. It is then possible to compare this experimental value of E_i with the theoretically calculated one.

energy equal to this threshold cause emission of electrons having zero kinetic energy, quanta having increasingly greater energies cause emission of electrons having correspondingly greater kinetic energies. The mathematical relation that describes these observations is

$$h\nu = E_i + \tfrac{1}{2}mv^2 \qquad (14\text{-}4)$$

where E_i is the (ionization) energy required to overcome the attraction between an electron and the nucleus of the atom from which it is ejected and $\frac{1}{2}mv^2$ is the extra kinetic energy of the photoelectron. The photoelectron can possess extra kinetic energy only if the energy of the original photon is greater than the ionization energy E_i.

To test the validity of the Einstein photoelectric equation, it is necessary to know the value of the ionization energy (E_i) and the value of the velocity, and hence the kinetic energy, of the photoelectron. We can calculate the value of E_i for any specified metal. However, it is very difficult to measure the velocity of an electron directly, so what is done is to determine the voltage necessary to prevent the photoelectrons from reaching a collecting anode placed near the metal sample (Figure 14-5).

When the voltage is adjusted so that no photoelectrons strike the collecting anode, the photoelectric current will be zero. Under these conditions, the kinetic energy of a photoelectron is given by the energy of the retarding voltage:

$$\tfrac{1}{2}mv^2 = eV \qquad (14\text{-}5)$$

where V is the retarding voltage and e is the charge of an electron.

If we introduce this equality into the earlier Einstein photoelectric equation, we have

$$h\nu = E_i + eV \qquad (14\text{-}6)$$

Rearranging,

$$eV = h\nu - E_i \qquad (14\text{-}7)$$

Solving for V,

$$V = \frac{h\nu}{e} - \frac{E_i}{e} \qquad (14\text{-}8)$$

A more accurate method of determining the threshold voltage is to plot V versus the frequency of the incident light as shown in Figure 14-6. The phenomenon of the photoelectric effect provided one of the strongest pieces of evidence in favor of the particle nature of light.

It is of practical interest to note that the *photoelectric cell* makes use of the principles we have been discussing. A photoelectric cell is fabricated by depositing a thin layer of an alkali metal, such as sodium, rubidium,

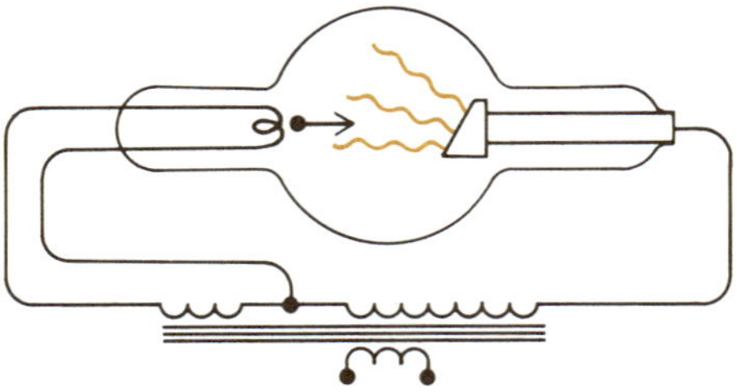

Figure 14-7 A diagram of an X-ray tube. Electrons are emitted from the heated filament and accelerated to high energies by electrical forces. When the high-energy electrons strike the positively charged target, they kick out *inner* nuclear electrons from the target material. When electrons more distant from the nucleus fall into the gap thus created, the energy is released in the form of very high energy photons—X-rays.

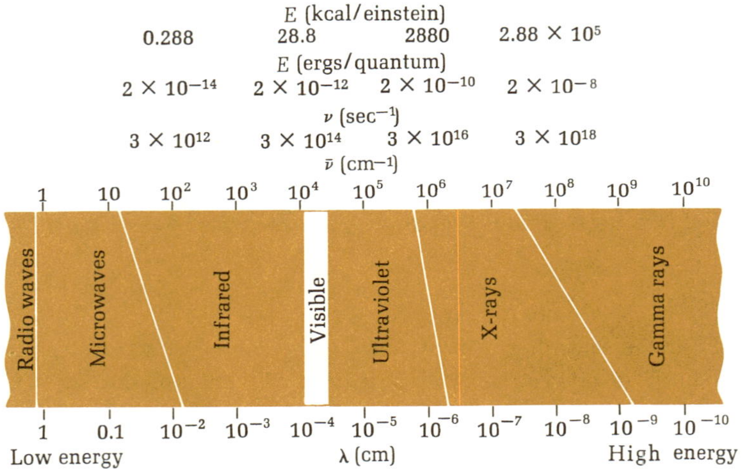

Figure 14-8 The electromagnetic spectrum.

or cesium, on the inner surface of a small vacuum tube. The collecting anode is positively charged, so the photoelectrons are attracted to it. Illumination of the metal surface by any radiation with a wavelength shorter than the photoelectric threshold causes emission of photoelectrons and a resulting flow of electrical current; this current is proportional to the intensity of the incident light. Needless to say, the flow of current (or perhaps its interruption when something or someone comes between the light source and the photoelectric cell) can be utilized to turn switches or alarm systems on or off as desired.

An inverse photoelectric effect, namely the production of a photon by slowing down a fast-moving electron, is involved in the production of X-rays (Figure 14-7). In an X-ray tube, electrons from the cathode are accelerated to a high velocity by a large potential difference, V, between the cathode and the anode. The kinetic energy of the electrons is equal to eV. When such a fast-moving electron strikes the anode, it is quickly brought down to very low (and perhaps zero) velocity, and all of its energy, eV, is converted into light (X-radiation) having an energy $h\nu$. (The ionization energy term, E_i, can be neglected here, because it is so small compared to the other energy terms.)

The quantum nature of light shows up in many other ways. It is common knowledge that ordinary white light can be broken up into its individual components (red, yellow, orange, etc.) as it is when a rainbow is formed or when light is reflected from the surface of a soap bubble. Using modern scientific instruments (spectroscopes), it is possible to refine this concept enormously; some aspects are developed in Chapter 15.

Returning now to Equation (14-3), we can calculate the energy associated with one light quantum for several wavelengths. Green light, for example, has a wavelength of about 5×10^{-5} cm. Thus, its frequency must be

$$\nu = \frac{c}{\lambda} = \frac{3 \times 10^{10} \text{ cm/sec}}{5 \times 10^{-5} \text{ cm}} = 6 \times 10^{14} \text{ sec}^{-1}$$

It follows that the energy of one quantum of this frequency is

$$\begin{aligned} E = h\nu &= 6.62 \times 10^{-27} \text{ erg sec} \times 6 \times 10^{14} \text{ sec}^{-1} \\ &= 39.72 \times 10^{-13} \text{ erg} \end{aligned}$$

This is a very small amount of energy indeed. It is somewhat more common to calculate the amount of energy associated with 1 mole of light quanta, called an *einstein*. There are Avogadro's number, 6.023×10^{23}, of quanta in 1 einstein. It follows that the energy of 1 einstein of green light is $39.72 \times 10^{-13} \times 6.02 \times 10^{23} = 2.4 \times 10^{12}$ ergs. Remembering that 1 erg is equal to 2.39×10^{-11} kcal, it follows that 1 einstein of green light contains $2.4 \times 10^{12} \times 2.39 \times 10^{-11} = 57.6$ kcal of energy. This quantity of energy is somewhat smaller than typical bond strengths (Table 8-2), so we would not expect that green light would elicit many bond-changing chemical reactions. However, it is clear that light with slightly shorter wavelengths (ultraviolet light) will have sufficient energy per quantum to rupture most chemical bonds. We reserve detailed discussion of the consequences for Chapter 15. The spectrum of electromagnetic radiation with associated energies is shown in Figure 14-8.

Summarizing briefly, we find that, on the one hand, light is described as a kind of wave, called electromagnetic radiation; on the other hand, light proves to have particulate properties, and we also refer to it in terms of

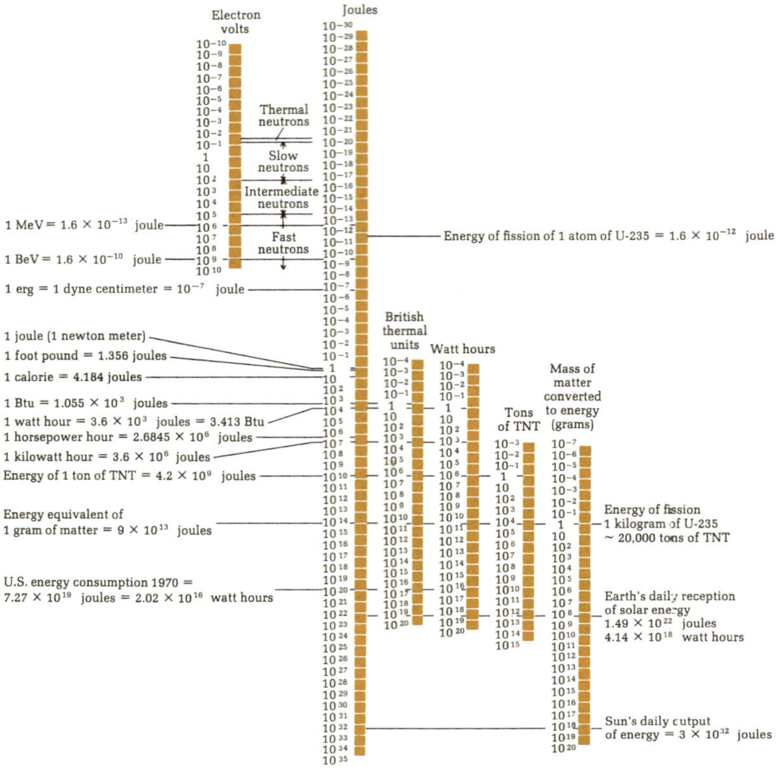

A summary of useful energy units and conversion factors. Physicists find it convenient to use electron volts, ergs, and joules. Biologists and nutritionists think in calories. Engineers deal in British thermal units and watt hours. Since Hiroshima, energy release is commonly expressed in tons of TNT. It is less often observed that a ton of ordinary coal contains three times as much energy as a ton of TNT. The illustration is based on one that appears in *The New College Physics: A Spiral Approach* by Albert V. Baez. [Chauncey Starr, "Energy and Power," *Sci. Am.*, **225**, 37 (September, 1971).]

photons or quanta. Light, then, is said to exhibit *wave-particle duality*.

But if we review carefully what has previously been developed in this text, we shall discover a counterpart to such behavior. In Chapter 7, the diffraction of X-rays, high-energy electromagnetic radiation, was discussed as a method for elucidating the structures of crystalline solids, including very complicated ones such as proteins. *Diffraction*, the breaking up of a beam into its components resulting from the interference of one wave with another, is certainly a property associated with waves, not particles. So far, so good. In Chapter 2 there is an *electron diffraction* pattern of a simple organic molecule. If diffraction is a property of waves, how can we speak of electron diffraction? An electron, after all, is a particle. Or is it? If light can behave as a particle, is it also possible that particulate matter can behave as a wave? Are both energy and matter part of the same basic currency? Is there any exchange rate between them? It is to these questions that we now address ourselves.

THE UNITY OF MATTER AND ENERGY

Some of the simplest experiments are also the most revealing. Let us consider an experiment that is conceptually simple although somewhat difficult to carry out in the laboratory. The electron is a familiar particle: It has a characteristic mass and carries one unit of negative charge. Somewhat less familiar is the *positron,* a particle that is identical to the electron in all respects except that it bears one unit of positive charge. Thus, it is appropriate to consider a positron as an antielectron. What happens if an electron and a positron collide? The result of this simple encounter between two fundamental particles is exceedingly important: A process called **annihilation** occurs in which *the positron and electron disappear and two gamma rays are formed:*

$$e^- + e^+ \longrightarrow 2h\nu \tag{14-9}$$

Two particles with well-defined masses have been converted into two energy quanta, species without mass: Matter has been converted into energy. It follows that *matter and energy are two forms of the same basic cosmic fabric.* Moreover, they are interconvertable in processes such as that shown in Equation (14-9) and in a great many others as well. Other examples are developed in Chapter 16. The fundamental unity of matter and energy was first clearly recognized by Einstein in 1905. He proposed an equation through which one is related to the other:

$$E = mc^2 \tag{14-10}$$

This simple equation, in which the square of the speed of light appears as a proportionality constant linking matter and energy, is the most important single statement in all of science. Einstein's equation was initially tested and corroborated by J. D. Cockcroft and E. T. S. Walton in studies dealing with nuclear chemistry (Chapter 16).

To gain some understanding of the equivalence of matter and energy, let us calculate the wavelength of the two photons emitted in the annihilation process in Equation (14-9).

Example *Given that the mass of an electron or positron is 0.911* $\times$ *10*$^{-27}$ *g, what is the wavelength of the two photons produced when an electron and positron undergo annihilation?*

Solution *The total mass of the annihilated particles is* $2 \times 0.911 \times 10^{-27} = 1.822 \times 10^{-27}$ *g. The corresponding energy is given by Einstein's relation,* $E = mc^2$.

$$E = 1.822 \times 10^{-27} \text{ g} \times (3 \times 10^{10} \text{ cm/sec})^2 = 16.4 \times 10^{-7} \text{ g cm}^2/\text{sec}^2$$
$$= 16.4 \times 10^{-7} \text{ erg} = 8.2 \times 10^{-7} \text{ erg/photon emitted}$$

Having calculated the total energy involved, we can now calculate the wavelength of the photons from Planck's equation, $E = hc/\lambda$ *or* $\lambda = hc/E$.

$$\lambda = \frac{hc}{E} = \frac{6.62 \times 10^{-27} \text{ erg sec} \times 3 \times 10^{10} \text{ cm/sec}}{8.2 \times 10^{-7} \text{ erg}}$$
$$= 2.4 \times 10^{-10} \text{ cm} \quad \text{or} \quad 0.024 \text{ Å}$$

The result of this calculation is striking: The wavelength of this light is exceptionally short, well into the gamma-ray region (Figure 14-8). To derive such high-energy radiation from such light particles requires that enormous amounts of energy be stored in the form of matter. The energy of 1 einstein of light quanta of this wavelength is 12 million kcals! To understand this matter more clearly, consider the following problem.

Example *To how much energy does 1 atomic mass unit (amu) correspond? Extend the calculation to 1 g of matter.*

Solution *1 amu weighs* 1.66×10^{-24} *g. Using Einstein's formula, it follows that*

$$E = 1.66 \times 10^{-24} \text{ g} \times (3 \times 10^{10} \text{ cm/sec})^2 = 1.5 \times 10^{-3} \text{ erg}$$

By multiplying this number by Avogadro's number, we can calculate the amount of energy produced from conversion of 1 g of matter:

$$E = 9 \times 10^{20} \text{ ergs} = 2.16 \times 10^{10} \text{ kcal}$$

Translated into more common terms, this energy amounts to more than 2 million kW hr, or enough energy to run 2 million 100-W lamps for 10 hr, all from just 1 g of matter. Witnesses to nuclear explosions will testify to the devastating effect of release of such quantities of energy over short time periods.

Philosophically, we have resolved our dilemma posed earlier: Matter and energy are one. Particles are expected to have wave properties and waves are expected to have particulate properties. There are a great many consequences that follow from this realization. Perhaps the most important is Heisenberg's uncertainty principle.

THE UNCERTAINTY PRINCIPLE

One of the distinguishing features of particles and waves, as we ordinarily think of them, is that particles can be localized precisely and waves cannot. A particle, a baseball for example, can be said to occupy a certain spot in space. But a wave cannot be so characterized. What restrictions are placed on our ability to know exactly where an electron, or any other object, is at a given moment? It follows that some limitation ought to exist since we must now think of an electron, at least partially, as a wave. Heisenberg provided an answer to this question in 1927: His **uncertainty principle** states that the uncertainty in the position of an object Δx, multiplied by the uncertainty in its momentum, $\Delta(mv)$, is equal to or greater than Planck's constant divided by 4π:

$$(\Delta x)[\Delta(mv)] \geq \frac{h}{4\pi} \tag{14-11}$$

Example *Suppose that an electron is moving with a velocity of 10^8 cm/sec and that we measure its position with an accuracy of 0.01 Å. Calculate the uncertainty in the momentum of this electron. Compare this uncertainty with the momentum itself.*

Solution *From Heisenberg's uncertainty principle, Equation (14-11), the uncertainty in the momentum is given by*

$$\Delta(mv) = \Delta p = \frac{h}{4\pi\ \Delta x} = \frac{6.62 \times 10^{-27}\ \text{g cm}^2/\text{sec}}{4 \times 3.1416 \times 0.01 \times 10^{-8}\ \text{cm}}$$

$$= 5 \times 10^{-18}\ \text{g cm/sec}$$

The momentum of the electron itself is given by the product of its mass and velocity:

$$p = mv = 0.91 \times 10^{-27}\ \text{g} \times 10^8\ \text{cm/sec} = 0.91 \times 10^{-19}\ \text{g cm/sec}$$

We obtain the surprising result that the uncertainty in the momentum is more than 50 times as large as the momentum itself. Having determined the position of the electron so precisely, we deprive ourselves of any useful information at all concerning its momentum.

To understand more about the basis of the uncertainty principle, let us consider how we go about determining the position of an electron; that is, how do we make a measurement on very small systems? Basically, a measurement requires that there be some interaction between the object of interest and some entity involved in the measuring process, some entity that can be viewed by an observer. We might, for example, try to determine the position of an electron by bouncing a photon of light off it. But the impact of the light quantum on the electron will perturb it to some degree; that is, it will render the momentum uncertain by changing that momentum in the measuring process. At the same time, our knowledge of its position will be uncertain to some extent, since the resolving power of the system is the wavelength of the light used. To see how this works in practice, let us consider the interaction between light of wavelength 5×10^{-5} cm (green light) and the electron of the previous example. Remembering that the

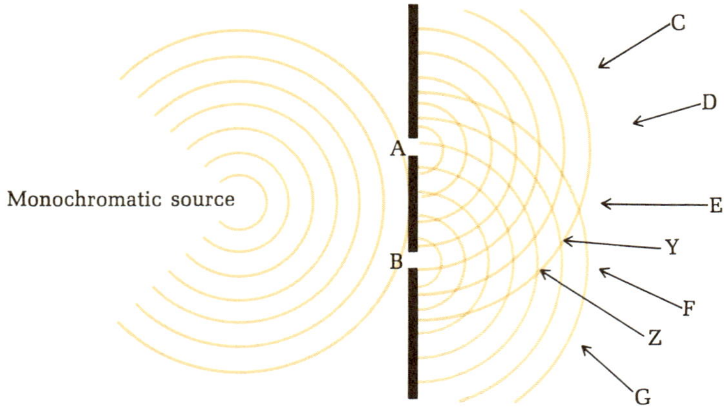

Figure 14-9 Interference pattern set up by monochromatic light passing through two pinholes, A and B. Note that observers at points C, D, E, F, and G would see a reinforcement of the light intensity; between these points there would be complete cancellation of the light waves to produce darkness. Note how reinforcement occurs; at point Y the observer is a distance of seven wavelengths from A and six wavelengths from B, whereas at point Z the observer is seven wavelengths from A but only five wavelengths from B. Reinforcement (constructive interference) occurs at each point that is an integral number of wavelengths from each pinhole.

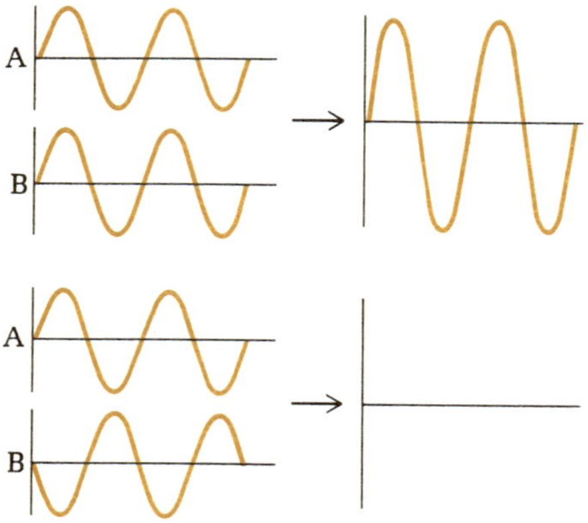

Figure 14-10 Diagrams showing constructive (top) and destructive (bottom) interference. Notice that the peak amplitude of the composite wave in the top diagram is twice that of the two original waves, but that the wavelength of the composite is the same as the individual waves from A and B. In the bottom diagram, destructive interference arises because of the point-by-point cancellation of the amplitude of wave A by wave B. These phenomena are what give rise to the regions of lightness and darkness in Figure 14-11.

energy of the electron is given by $E = \frac{1}{2}mv^2$, we can calculate its energy to be 4.55×10^{-12} erg. From Planck's relationship, the energy of the photons of this light is about 4×10^{-12} erg, nearly as large. Hence, when one photon strikes one electron, that electron will be badly disturbed in the process. It is more or less like trying to find out something about the momentum of a Cadillac by bouncing a Ford LTD off it. But even if the energy matter were not so serious, light of this wavelength is completely inadequate to locate the electron properly. The wavelength of the light is about 5000 times the diameter of a typical atom. If we try to do a better job of locating the electron by using light of shorter wavelength, the energy problem gets worse. If light of lower energy is used, the location problem becomes hopeless. In this matter, you just can't win.

LIGHT AND ELECTRONS AS WAVES

Earlier in this chapter we described light in terms of an electromagnetic wave, but we have not defended this description on the basis of experiment. In fact, the wave character of light has been revealed in a large number of ways.

One of the earliest pieces of experimental evidence in support of the wave nature of light arose from work by Thomas Young, a London physician, who in 1803 examined the behavior of a monochromatic beam of light as it passed through two closely spaced pinholes. This monochromatic light, or light of a single wavelength or frequency, set up an interference pattern that resembled the interference patterns encountered with sound or water waves. The requirement that the light be monochromatic is essential. If a source of polychromatic light were used, a whole range of wavelengths would emerge from each pinhole, and reinforcement of light would be viewed from virtually every position behind the pinholes.

Somewhat later, Maxwell proposed his electromagnetic theory of light, which we noted earlier. According to the theoretical picture, light could be described as a rapid variation (oscillation) of the electrical and magnetic fields surrounding an oscillating charged particle. These oscillating electrical and magnetic disturbances were presumed to be propagated through space. If another electrical charge were placed in the path of this electromagnetic radiation, the second charged particle could be shown to oscillate, which suggests the wavelike quality of light. By taking a side view of an electromagnetic wave as shown in Figure 14-2, we can perceive how constructive and destructive interference occurs in the experiment described in connection with Figures 14-9 and 14-10.

The wave character of electrons can be demonstrated in exactly the same way as for light. That is, electrons also yield interference patterns when monochromatic beams are passed through two slits or pinholes. Interference

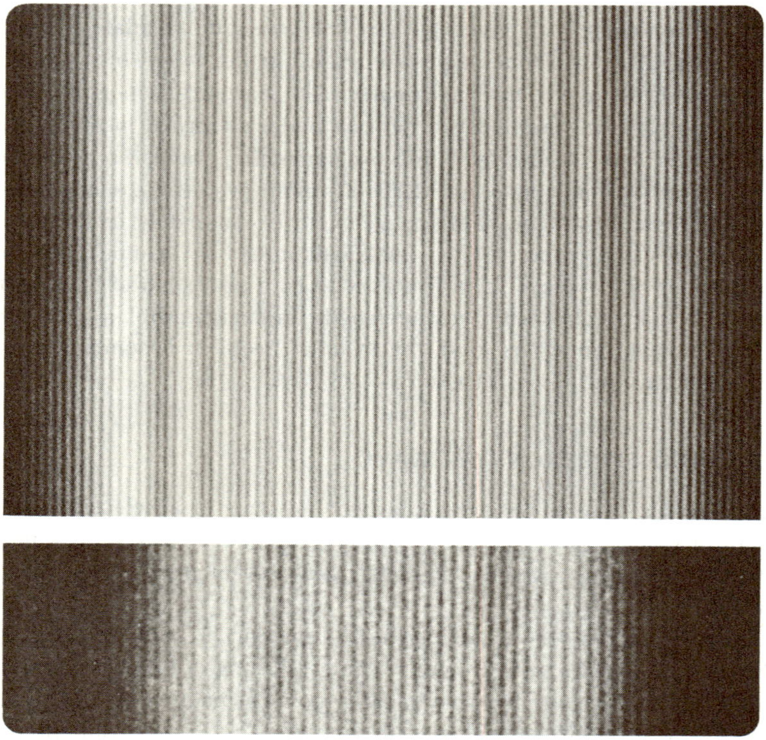

Figure 14-11 Comparison of interference patterns generated from the double-slit arrangement shown in Figure 14-9. The pattern above was generated by light, that below by electrons. (Top: From J. Valasek, *Introduction to Theoretical and Experimental Optics,* Wiley, New York, 1949. Bottom: Photograph by Prof. G. Mollenstedt, University of Tubingen, Germany.)

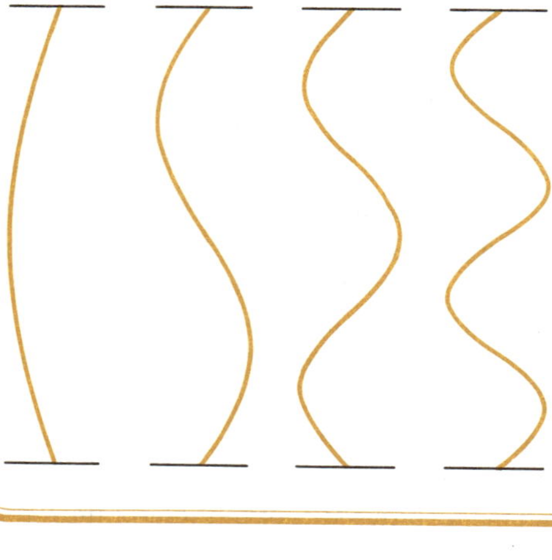

Figure 14-12 Standing waves of certain fixed frequency and wavelength generated by the boundary conditions that the two ends of the rope be stationary.

patterns generated by light and electrons are compared in Figure 14-11. As we can readily judge from these photographs, the character of the interference patterns is similar. Thus, the wave character of electrons is clearly demonstrated. In fact, the concept of an electron as wave was deduced before such experiments were done. Let us turn to the seminal considerations of Louis de Broglie.

In the 1920s, the atom was thought of in terms of the considerations of Niels Bohr, in which electrons were viewed, correctly, as having certain definite energies. If we consider the possibility of wave character for electrons, what is the relationship between this wave character and the quantization of electronic energies? To approach this matter, let us briefly consider the properties of **standing waves.** A standing wave is one in which the wave adopts a time-independent certain fixed pattern; that is, the positions of maxima, minima, and nodes do not change as a function of time. One example is provided by the waves that a man can generate by swinging a rope attached to a fixed support. Photographs of four standing waves are provided in Figure 14-12. Note carefully that only certain frequencies or wavelengths produce fixed patterns. Specifically, the generation of a standing wave requires that the total distance from one end of the rope to the other be an integral number of half-wavelengths. This condition is imposed by a *boundary condition:* The fixed ends of the rope cannot move. Satisfaction of this condition restricts the wavelengths possible to certain ones. That is, we have introduced the phenomenon of quantization by imposing boundary conditions on our system.

To see the point a little more clearly, compare the acceptable and unacceptable waves of a rope shown in Figure 14-13. Those waves in part (a) are acceptable standing waves, since the boundary condition of no motion of the two ends is met. But those waves in part (b) do not meet this condition and cannot be standing waves. Note that only those waves with $n\lambda = 2a$, where a is the length of the rope and n is any integer, are standing waves. If the waves were those of electrons or light quanta, it follows that only certain discrete energies are permitted.

Precisely the same considerations apply to waves that trace out a circular orbit, such as those occupied by electrons in the Bohr atom. Our boundary condition is now stated somewhat differently: The circumference of the orbit must be equal to an integral number of wavelengths, $2\pi r = n\lambda$. This point is shown graphically in Figure 14-14. If the boundary condition is not met, a mismatch occurs as the wave moves around the orbit and a standing wave cannot be achieved. These are precisely the considerations that led de Broglie, in 1923, to interpret the quantization of electron energies in terms of the wave character of these particles.

The angular momentum of an electron in an orbit about a nucleus is given by the expression

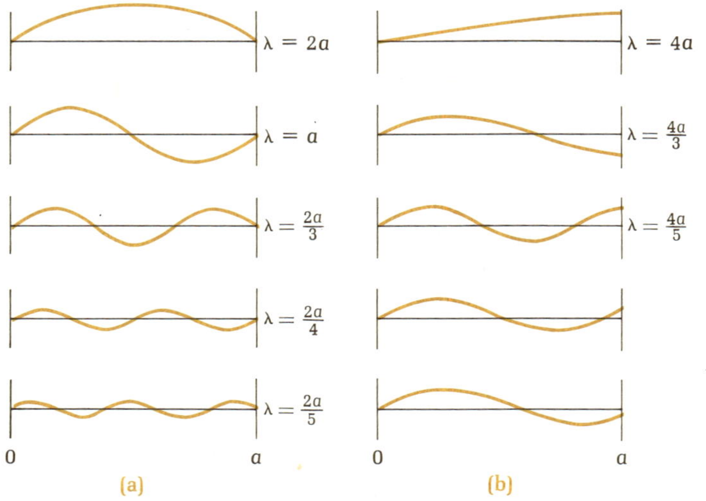

Figure 14-13 (a) Acceptable and (b) unacceptable standing waves according to the boundary condition that the ends of the rope be motionless. This condition requires that the wavelength of the waves be $\lambda = 2a/n$, where a is the length of the rope and n is any integer. (From R. E. Dickerson, H. B. Gray, and G. P. Haight, Jr., *Chemical Principles,* Benjamin, Menlo Park, Calif., 1970. Copyright © 1970, W. A. Benjamin, Inc.)

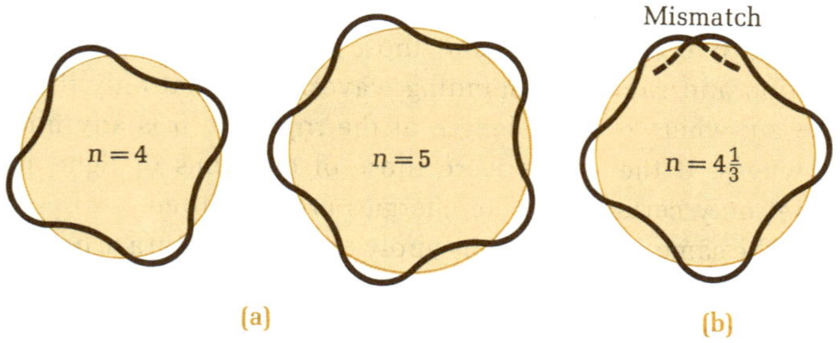

Figure 14-14 (a) Acceptable and (b) unacceptable waves in a circular orbit. To achieve a standing wave in a circular orbit, the boundary conditions require that the circumference of the orbit be equal to an integral number of wavelengths, $2\pi r = n\lambda$. (From R. E. Dickerson, H. B. Gray, and G. P. Haight, Jr., *Chemical Principles,* Benjamin, Menlo Park, Calif., 1970. Copyright, © 1970, W. A. Benjamin, Inc.)

$$\text{angular momentum} = mvr = \frac{nh}{2\pi} \tag{14-12}$$

or

$$2\pi r = n\frac{h}{mv} \tag{14-13}$$

where m is the mass of the electron and v is its velocity. Now if only those atomic orbits are possible for which a standing wave can be achieved,

$$2\pi r = n\lambda \tag{14-14}$$

it follows that

$$\lambda = \frac{h}{mv} \tag{14-15}$$

Equation (14-15) applies not only to electrons but is true in general. That is, each particle has a wavelength associated with it given by Equation (14-15), whether that particle is a proton or a baseball.

Example *Calculate the wavelength of a baseball weighing 100 g thrown with a velocity of 100 m/sec.*

Solution *From Equation (14-15),*

$$\lambda = \frac{h}{mv} = \frac{6.62 \times 10^{-27}\text{ g cm}^2/\text{sec}}{10^4\text{ cm/sec} \times 100\text{ g}} = 6.62 \times 10^{-33}\text{ cm}$$

THE BOHR ATOM: A WAVE MODEL

It is precisely those considerations just developed regarding the properties of standing waves that underlie a fruitful, though incorrect, model of the atom. This model was proposed by the Danish physicist Neils Bohr in 1913 and was the first such model to incorporate any of the ideas of modern quantum theory. Let us think a little bit about the properties of standing waves in circular orbits to see how these relate to the model of Bohr. In the first place, we have seen that only orbits having certain definite states of motion, and certain definite states of energy, are allowed by the properties of standing waves [Equation (14-13)]. This is the first of four postulates that define the Bohr model. Second, the model suggests that when an electron associated with an atom changes orbit from one of higher to one of lower energy, the atom will emit a photon whose energy, $h\nu$, is equal to the difference in energy of the two orbits. This is an exceedingly important observation, since it accounts for the well-defined lines of fixed frequency seen in the emission spectra of atoms (see Chapter 15). Third, Bohr assumed that the orbits of nuclear electrons were circular. Finally, he postulated that the states of allowed electronic motion are those in which the angular momentum of the electron is an integral multiple of $h/2\pi$ [Equation (14-12)]. Central to the whole Bohr concept is the idea of quantization of states that the electrons occupy, a matter that follows directly from the properties of standing waves.

How much of the Bohr concept can we retain in light of current quantum theory? If we compare the Bohr atom with our model developed in Chapter 4, we find that postulates 1 and 2 are completely correct. But the third one is completely incorrect; we no longer speak of a precisely defined orbit but of a wave function the square of which yields information concerning the probability of finding the electron at a given point of space. The fourth and final postulate is partially correct: The angular momentum of electrons is quantized, but not exactly as Bohr thought.

Within 12 years of the formulation of the Bohr model of the atom, it had to be abandoned in the light of new findings. Nevertheless, the novel character of this model and the inclusion of important results from quantum theory made it an important step forward in our understanding of atomic structure. Like many revolutionary ideas, it contained the seeds of its own destruction, yet succeeded in advancing concepts of the nature of the physical world.

In this chapter we have discussed the most fundamental properties of light and matter and their interaction. We are now in a position to consider *how* light and matter interact. We shall discover this in the next chapter.

Glossary

Amplitude the characteristic of a wave that is measured by the difference in its average and extreme ranges.

Annihilation the mutual destruction of a particle and its corresponding antiparticle (an electron and positron, for example) with the generation of electromagnetic radiation.

Diffraction the breaking up of a wave into its component parts caused by deflection of the wave at the edge of an opaque object.

Electromagnetic radiation radiation consisting of electrical and magnetic fields oscillating perpendicularly to each other and to the direction of propagation. Such radiation includes X-rays, gamma rays, ultraviolet, visible, and infrared radiation as well as microwaves and radio waves.

Emission spectrum the totality of the light emitted by a sample of atoms or molecules in electronically excited states.

Frequency a property of waves measured by the number of complete waves that pass a given point per unit time; given mathematically by the ratio of the speed of propagation of the wave and the wavelength.

Gamma ray a high-energy form of electromagnetic radiation emitted from some radioactive materials.

Infrared radiation electromagnetic radiation having wavelengths slightly longer than those that can be detected by the human eye.

Interference the mutual interaction of two waves leading to their reinforcement or cancellation.
Microwaves electromagnetic radiation of very low energy, having wavelengths longer than 0.01 cm.
Photoelectric effect the ejection of electrons from a metallic surface elicited by the impingement of photons having more than a certain critical energy content.
Photoelectric threshold the minimum energy of photons having the capacity to elicit ejection of electrons from a metallic surface.
Quantum a discrete, indivisible package of energy.
Radio waves electromagnetic radiation of exceedingly low energy; characterized by wavelengths greater than 1 cm.
Standing wave a wave for which the positions of the maxima, minima, and nodes do not change as a function of time.
Ultraviolet radiation electromagnetic radiation having wavelengths slightly shorter than those that can be detected by the human eye.
Uncertainty principle a law of physics indicating that it is impossible to know precisely both the position and momentum of a particle at the same time.
Visible radiation electromagnetic radiation that can be detected by the human eye; the wavelength of this radiation roughly spans the range 4000–7500 Å.
Wavelength λ, the distance between corresponding parts of two successive wave forms.
Wave number $\bar{\nu}$, the reciprocal of the wavelength.
Wave-particle duality the idea that light, usually thought of as a wave, also has particulate properties and that electrons, for example, usually thought of as particles, have wave properties. More generally, wave-particle duality indicates that it is really improper to think of anything solely in terms of a wave or of a particulate model.
X-rays electromagnetic radiation of exceedingly high energy, characterized by wavelengths less than 10^{-6} cm.

Study problems

1. If the position of an electron has been ascertained to an accuracy of 0.1 Å, calculate the uncertainty in its momentum.
2. Suppose that the uncertainty in the momentum of an electron moving with a velocity of 10^8 cm/sec is 5×10^{-19} g cm/sec. Calculate the ratio between the momentum of the electron and the uncertainty in its momentum.
3. Calculate the velocity of a neutron, mass 1.67×10^{-24} g, having a momentum of 2×10^{-16} g cm/sec.
4. Calculate the momentum of a proton, mass 1.67×10^{-24} g, moving with a velocity of 1.4×10^8 cm/sec.

5. Given a rope 10 ft long that is secured at both ends, calculate the four longest allowed wavelengths that can be formed by the rope.
6. For standing waves in a circular orbit of radius 5×10^{-9} Å, calculate the three longest allowed wavelengths.
7. If a standing wave in a circular orbit with a radius of 3×10^{-9} cm has a wavelength of 4.7×10^{-9} cm, calculate the number of waves in one orbit.
8. What is the longest wavelength of a standing wave in a circular orbit of circumference 1 Å?
9. Suppose an electron is moving with a velocity so that its de Broglie wavelength is 2 Å. What is the radius of the smallest orbit that this electron might take around a nucleus?
10. Calculate the wavelength of a golf ball weighing 100 g moving at a velocity of 1000 m/sec.
11. Calculate the de Broglie wavelength of an electron, mass 9.11×10^{-28} g, moving at half the speed of light.
12. Calculate the de Broglie wavelength of an electron moving around a proton at a distance of 0.524 Å.
13. Calculate the velocity of an electron in the lowest orbital of the hydrogen atom, radius 0.524 Å.
14. Calculate the momentum of an electron having a wavelength of 1 Å.
15. Calculate the momentum of a photon having a wavelength of 6000 Å.
16. If potassium metal, $E_i = 2.2$ eV, is irradiated with ultraviolet light of wavelength 2000 Å, what will be the maximum kinetic energy of emitted photoelectrons?
17. If the threshold wavelength for emission of photoelectrons from a zinc surface is 3500 Å, what is the ionization energy for zinc?
18. The ionization energy of cesium metal is 1.81 eV. What is the longest wavelength of light that can produce photoelectrons from cesium?
19. It has recently been observed that shining a sufficiently intense beam of laser light on small particles will keep the particles suspended above a surface; that is, the light overcomes the gravitational forces tending to pull the particles down. In terms of the nature of light, how do you explain this observation?

Suggestions for further reading

A. Ashkin "The Pressure of Laser Light," *Sci. Am.*, **226,** 62 (February, 1972).

S. Hecht *Explaining the Atom,* Viking, New York, 1964, pp. 53–71.

Lasers and Light, Readings from Scientific American, Freeman, San Francisco, 1969.

F. Murphy and D. E. Yount "Photons as Hadrons," *Sci. Am.*, **225,** 94 (July, 1971).

Sci. Am., **219** (March, 1968). An entire issue devoted to the subject of light.

15

Some aspects of photochemistry

Concerning photochemistry and everything else:
Nothing helps.

Alfred Diamant

In Chapter 14 we developed some understanding of the nature and variety of the properties of light. At this point we wish to lay a foundation for learning about how light interacts with the components of the normal and contaminated atmosphere considered in detail in Chapter 18. To accomplish this goal, we must begin with some simpler examples of the interaction of light with matter. It is to this objective that the present chapter is devoted.

THE ABSORPTION SPECTRUM OF ATOMIC HYDROGEN

Let us begin by examining the interaction of light with the hydrogen atom, the simplest atom known, with only one proton in the nucleus and a single electron in the 1s orbital. To find out just what does happen when we shine light on a collection of hydrogen atoms, we utilize a **spectrometer.** Basically, a spectrometer consists of a light source, a prism or grating to isolate light of a particular wavelength together with the capacity to select the wavelength isolated (a *monochromator*), a sample holder into which the hydrogen atoms (as a gas) may be placed and through which the light from the monochromator is passed, and, finally, a photocell, which detects the intensity of the light passing through the sample (see Figure 15-1).

The purpose of this experimental setup is to measure the fraction of incident light that is absorbed by the sample. Generally, the experiment is carried out by slowly changing the monochromator and measuring the percent of light absorbed as a function of wavelength. When we perform such an experiment, it is found that light of a few selected wavelengths is absorbed by the hydrogen atoms, while light of other wavelengths passes through the sample unchanged in intensity. Graphically, the results of the experiment are shown in Figure 15-2.

How can we interpret the observation that light of certain wavelengths is absorbed by the atomic hydrogen? Reference to Figure 15-2 reveals that there are several absorption lines at lower energies than that needed to cause ionization of the hydrogen atoms. Hence, it follows that the energy of the photons absorbed must be stored in the undissociated atoms. In what form might this energy be stored? Let us recall that there are any number of atomic orbitals available to the single electron of the hydrogen atom (see Figure 4-8). Ordinarily, the electron will occupy the 1s orbital, which is the orbital of lowest energy. But were we to supply the necessary energy, it would be possible to "kick the electron upstairs," that is, to excite it to an

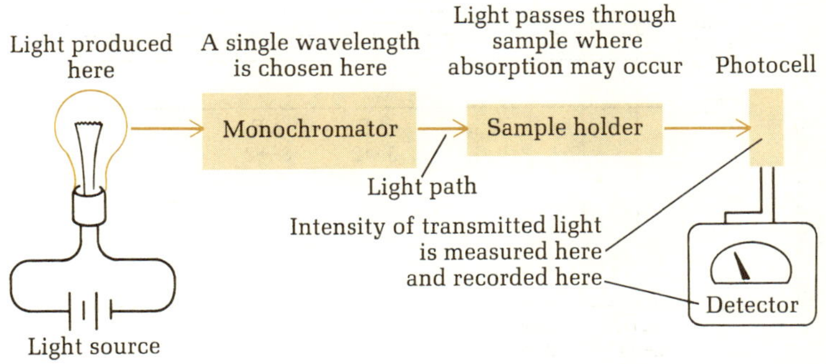

Figure 15-1 A schematic view of a spectrometer, a device for quantitative measurement of the extent of absorption of light by atoms and molecules.

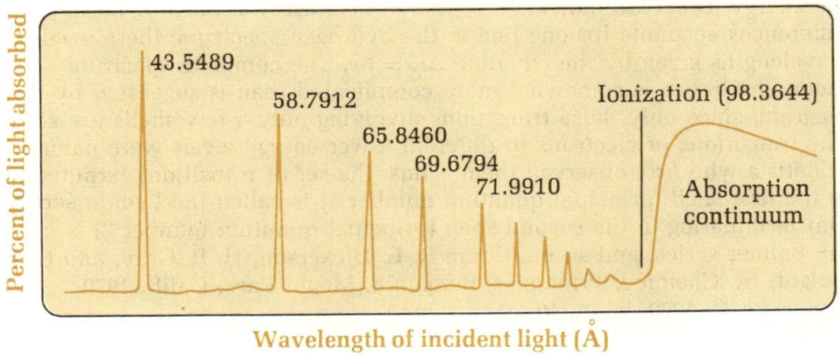

Figure 15-2 The absorption spectrum of atomic hydrogen. (From J. Arthur Campbell, *Chemical Systems, Energetics, Dynamics, and Structure,* Freeman, San Francisco, 1970. Copyright © 1970.)

orbital of higher energy. One way to supply the necessary energy is in the form of electromagnetic radiation, and this energy might be employed to excite an electron from a 1s orbital to a 2s orbital or any of higher energy. Moreover, if the hydrogen atom were in an excited state, we could imagine exciting an electron from a 2s orbital to a 3s, 3*p*, 3*d*, 4s, and so on, orbital. The condition for such promotion is that the *energy of the photon just equals the energy difference between the orbitals involved.* Thus, if we wish to excite an electron from an orbital of energy E_1 to one of energy E_2, the following condition must be satisfied:

$$E_2 - E_1 = h\nu \qquad (15\text{-}1)$$

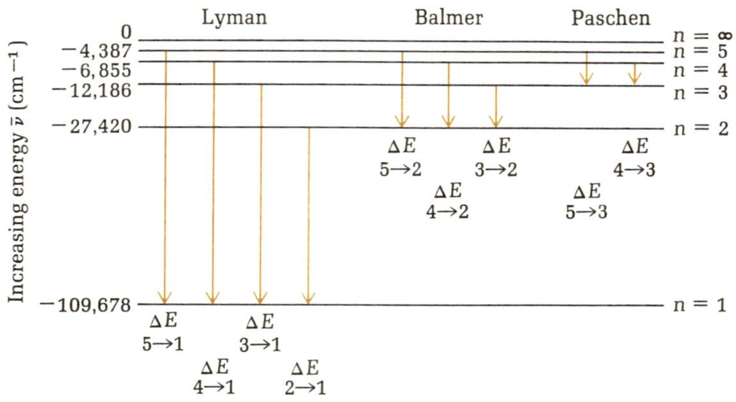

Figure 15-3 An energy-level diagram accounting for the emission (or absorption) spectrum of atomic hydrogen. On the left, the energies of orbitals in the first few shells are given in wave numbers. Possible transitions between the energy levels are indicated by the vertical arrows. Each of these energy differences accounts for one line in the hydrogen spectrum; these occur at wavelengths given by the equation $\Delta E = h\nu$. The complete spectrum of atomic hydrogen is somewhat more complicated than is suggested by this diagram, since only those transitions involving only a few shells are shown. The transitions of electrons to different lower energy levels were named for scientists who first observed them. Thus, the set of transitions terminating in the first shell (principal quantum number 1) is called the *Lyman series*, that terminating in the second shell (principal quantum number 2) is called the Balmer series, and so on. (From R. E. Dickerson, H. B. Gray, and G. P. Haight, Jr., *Chemical Principles*, Benjamin, Menlo Park, Calif., 1970. Copyright © 1970, W. A. Benjamin, Inc.)

In practice, the experiment described above is difficult to perform for hydrogen atoms for two reasons. First, almost all hydrogen atoms will be combined to form hydrogen molecules at ordinary laboratory conditions. Further, very few atoms will not have the electron in the 1s orbital, jumps between higher orbitals will be very rare, and negligible light absorption will occur. Fortunately, a much more productive but equivalent experiment can be carried out in this particular case. If a large amount of energy is suddenly added to a collection of hydrogen molecules, perhaps by an electrical spark, many molecules will dissociate into atoms, and in fact in many cases the electron will be completely ripped from the atom or at least boosted to a much higher orbital than the 1s. The energy relationships for an electron in a number of different orbitals of the hydrogen atom are shown in Figure 15-3. When a large number of atoms are produced, there will be some in which the $n = 5$ state occurs, in others $n = 4$, and so on. When the electrons return to a lower state, energy is lost, and from Equation (15-1) we can calculate the wavelengths corresponding to the discrete **emission spectrum** that results. The energy differences between various types of orbitals in one shell (between 3s, 3*p*, and 3*d* orbitals, for example) are very much less than those between orbitals in different shells. Consequently, we have ignored the energy differences between orbitals in a given shell in Figure 15-3. These differences, however, are real and lend fine structure to spectra observed in very careful measurements.

Note that absorption spectra, such as that in Figure 15-2, and emission spectra are, for atomic cases, precisely the same. They reflect two different ways of measuring the same set of quantities: the energy differences between the atomic orbitals that the electrons may occupy. The only difference is the experimental method of obtaining the spectra. Absorption spectra are measured by exciting the electron to higher energy levels using photons of the appropriate specific wavelength corresponding to the energy differences. Emission spectra are obtained by exciting the electron to higher energy levels by heating or other means and then measuring the wavelength of the photons emitted when the electron falls to lower levels. The wavelength of a photon absorbed in a specific excitation process must be exactly the same as that for a photon emitted when the electron falls back to the original energy level:

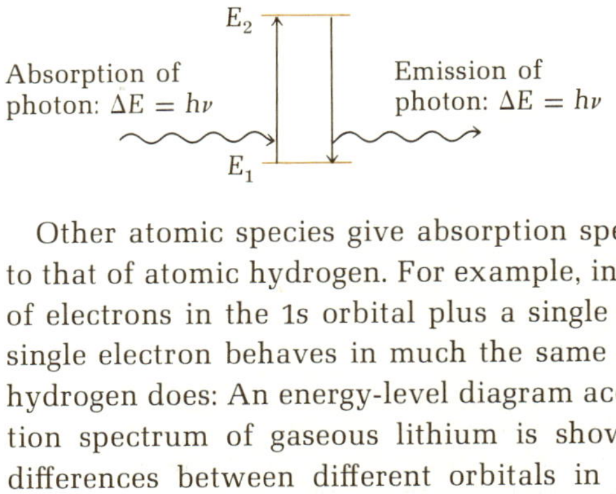

Other atomic species give absorption spectra quite similar in character to that of atomic hydrogen. For example, in atomic lithium there are a pair of electrons in the 1s orbital plus a single electron in the 2s orbital. This single electron behaves in much the same way that the single electron in hydrogen does: An energy-level diagram accounting for part of the absorption spectrum of gaseous lithium is shown in Figure 15-4. The energy differences between different orbitals in a single shell have now been included. In Figure 15-4(a) the actual spectrum of atomic lithium is shown. We can see that the energies at which absorption lines occur correspond to those indicated in the diagram in part (b); not all of the possible transitions are observed in the spectrum. For reasons provided by quantum mechanical arguments, some of the transitions are forbidden.

Empirically, it has been found that the energy of any line in a hydrogen-like spectrum (Figures 15-2 and 15-4) can be calculated with great accuracy

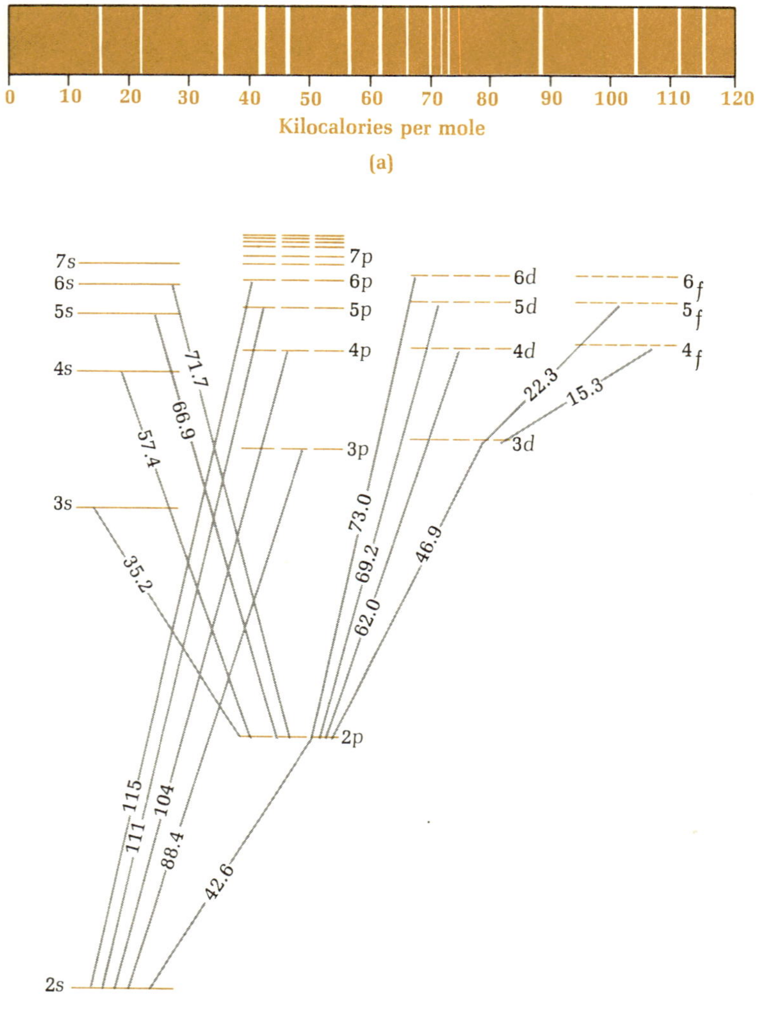

Figure 15-4 (a) Part of the spectrum of gaseous lithium. (b) An energy-level diagram corresponding to the observed spectrum. Note that energies have been expressed in kilocalories per mole in both parts of this figure. Note that not all the absorption lines suggested by the energy-level diagram are actually observed in the experimental spectrum. The others are forbidden. Can you assign each of the lines in the spectrum to a particular transition in the energy-level diagram? Can you see the correlation between those transitions that are allowed and those that are forbidden? (From J. A. Campbell, *Chemical Systems, Energetics, Dynamics, and Structures*, Freeman, San Francisco, 1970. Copyright © 1970.)

from the **Rydberg equation:**

$$\Delta E = KZ^2\left(\frac{1}{n_1^2} - \frac{1}{n_2^2}\right) \tag{15-2}$$

In this equation K is the Rydberg constant, 313.64 kcal/mole or 109,678 cm^{-1}, Z is the atomic number, and n_1 and n_2 are principal quantum numbers of the shell involved in the transition of interest, $n_2 > n_1$. Derivation of this equation and of the numerical value of the Rydberg constant by Bohr marked the real beginning of our understanding of the electronic structure of atoms (see Chapter 14).

Example *Calculate the wave number of the second line in the Lyman series for hydrogen.*

Solution *From Figure 15-3, we see that the second line in the Lyman series involves a transition from $n_1 = 1$ to $n_2 = 3$. From the Rydberg equation (15-2), it follows that*

$$\Delta E = 109{,}678\ \text{cm}^{-1}\left(\frac{1}{1^2} - \frac{1}{3^2}\right) = 97{,}492\ \text{cm}^{-1}$$

Example *Calculate the ionization energy of atomic hydrogen.*

Solution *For ionization to occur, the electron must move from its ground state, $n_1 = 1$, to infinity, $n_2 = \infty$. Since the reciprocal of infinity is zero, it follows that the ionization energy of atomic hydrogen must be equal to the Rydberg constant:*

$$\Delta E = 109{,}678\ \text{cm}^{-1}\left(\frac{1}{1^2} - 0\right) = 109{,}678\ \text{cm}^{-1} \quad \text{or} \quad 313.64\ \text{kcal/mole}$$

BASIC CONCEPTS OF PHOTOCHEMISTRY

Absorption spectra

The **absorption spectra** for most complex molecules are essentially no more complicated than those for atomic species just discussed. The interpretation of such spectra, on the other hand, is more complicated, since we deal with simple transitions of an electron from one atomic orbital to another in the latter case, but from one molecular orbital to another in the former one. In favorable cases for simple molecules, we may be able to define these molecular orbitals rather well; in more complicated cases, our information may be poorly defined. At any event, molecular orbitals are more complicated to handle than atomic ones, and we shall be concerned only with the qualitative aspects of light absorption by complex molecules.

In one important way, absorption spectra for complex molecules appear to be different from that for atomic hydrogen or for other atomic species:

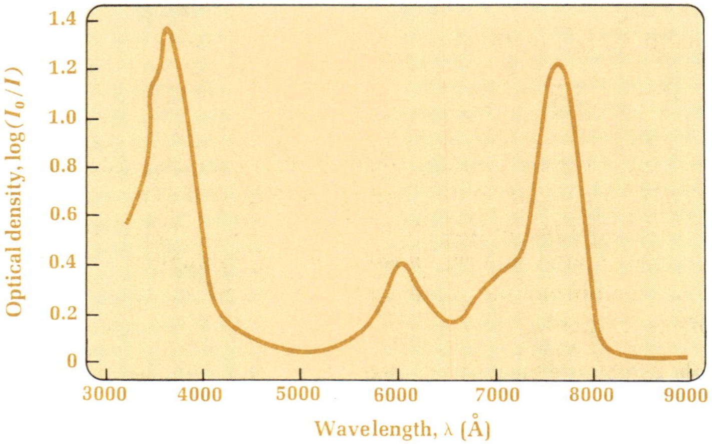

Figure 15-5 An absorption spectrum of bacteriochlorophyll isolated from Rhodosprillum rubrum dissolved in methanol.

Rather than sharp absorption lines, we obtain broad absorption curves. A typical example is provided in Figure 15-5, in which the absorption spectrum for the photosynthetic pigment bacteriochlorophyll (a type of chlorophyll found in bacteria capable of carrying out photosynthesis, see p. 580) is shown. The reason for the lack of sharpness of these absorption bands is well known but is not of direct concern to us.

Notice that the absorption spectrum of bacteriochlorophyll shown in Figure 15-5 is reasonably complex. It would seem unlikely that very many other molecules would exhibit precisely the same absorption spectrum. A great deal of experience indicates that this feeling is quite true. Hence, it is possible to use absorption spectra as "fingerprints" for molecules. Spectrophotometry proves to be an invaluable tool for the identification of molecules. Large collections of spectra of complex molecules have been prepared, and reference may be made to these when we wish to ascertain the nature of a compound of unknown structure, perhaps a compound isolated from nature, one brought back from the moon on an Apollo mission, or an unusual molecule found in the blood or urine of a patient suffering from some disease.

In addition to providing information concerning the nature of a particular material, absorption spectrophotometry is also useful for estimating the quantity of a material present in a solution or in the gas phase. Commercial spectrophotometers, which are functionally similar to the instrument shown in Figure 15-1, can measure a quantity called the **optical density,** which is defined as follows:

$$\text{optical density} = \log \frac{I_0}{I_t} \tag{15-3}$$

In this expression I_0 is the intensity of the incident light falling on the sample and I_t is the intensity of the transmitted light that emerges from the sample. The difference between the two values is, of course, a measure of the fraction of the light absorbed by the sample. We may relate the optical density of a solution of a light-absorbing material to the concentration of the solution by the **Beer-Lambert law:**

$$\log \frac{I_0}{I_t} = \varepsilon \times c \times l \tag{15-4}$$

In Equation (15-4), c is the concentration of the dissolved substance in moles per liter, l is the distance in centimeters that the light travels through the absorbing medium, and ε is a proportionality factor called the **extinction coefficient,** which measures how intensely the material in question absorbs light of the wavelength being used:

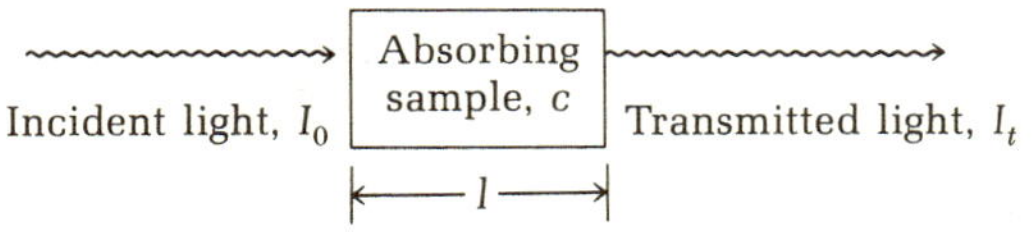

Bacteriochlorophyll, for example, has an extinction coefficient at 7730 Å of about 100,000 liters/cm mole. That is, a 1*M* solution of chlorophyll in a sample vessel of path length 1 cm would exhibit an optical density of 100,000. Using Equation (15-4), can you calculate what fraction of the light incident on this sample would be transmitted through it?

Example *What concentration of bacteriochlorophyll in a sample container of 1 cm path length would absorb one-half of 7730 Å light shining on it?*

Solution *Since half of the light is absorbed, it follows that the other half of the light must be transmitted through the sample. Hence, I_0/I_t is just 2 (the intensity of the transmitted light is one-half that of the incident light). The optical density then is log 2 = 0.3. Hence, $0.3 = 100{,}000 \times c \times 1$ [from Equation (15-4)]; or $c = 3 \times 10^{-6}M$, a very dilute solution indeed. Obviously, bacteriochlorophyll absorbs light of this wavelength quite strongly.*

Example *A solution of myoglobin in water contained in a sample holder of 1 cm path length has an optical density at 4090 Å of 0.8. Given that a 1 mg/ml solution of this protein has an optical density of 1.0 at this wavelength, calculate the molarity of the protein solution, remembering that the protein has a molecular weight of 17,000.*

Solution *Since the Beer-Lambert law tells us that optical density is linearly related to concentration, the data given require that the protein solution contain 0.8 mg of protein/ml or 0.8 g of protein/liter. Hence, the molarity is given by*

$$M = \frac{0.8 \text{ g/liter}}{17{,}000 \text{ g/mole}} = 4.7 \times 10^{-5} \text{ mole/liter}$$

Since the measurement of absorption spectra for most molecules can be

The popular new glasses that darken when exposed to sunlight provide an interesting and common example of photochemical behavior. A simple way to produce a photochromic glass is to precipitate microcrystals of a silver halide in the glass as it is made. On exposure to light the halide is decomposed to silver and free halogen:

$$AgCl \xrightarrow{h\nu} Ag + Cl\cdot$$

The silver particles thus deposited cut down sharply on the amount of light transmitted through the glass. Since the products of photodissociation are held in a rigid matrix, recombination takes place to reform the silver halide. The net result is that an equilibrium amount of silver is formed which depends on the light intensity. When more photons strike the glass, it darkens, and when the light intensity falls, the glasses "bleach" again. (Courtesy of Corning, New York.)

easily accomplished in a short time, this method provides a most useful quantitative analytical method.

Photochemical laws

As is developed below, absorption of a quantum of light by a molecule can lead to a number of subsequent events, including chemical reaction. Since the activation energy for these reactions is supplied by the light, such reactions are said to be *photochemical*. Photochemical reactions have been found empirically to obey a set of just four rules. The first two of these are readily understandable in terms of our discussion up to this point. The third and fourth rules will require additional explanation, which is provided in a later subsection of this chapter. The four laws of photochemistry are as follows.

1. Only those light quanta absorbed by a system are effective in producing chemical change.
2. Each proton (quantum) absorbed activates just one molecule in the excitation step of the whole photochemical process.
3. Each photon absorbed by a molecule has a certain probability of populating either the lowest excited singlet state S_1, or the lowest triplet state, T_1.
4. The lowest excited singlet and triplet states are the starting points (in solution, although not necessarily in the gas phase) for most photochemical processes.

The process of absorption of light quanta by molecules

The absorption process involved in exciting an electron of a molecule (or atom) from its stable ground state to an orbital of higher energy is controlled by just a few factors. The most basic of these, the requirement that the energy of the exciting radiation be equal to the energy difference of the electron in the two orbitals, has already been developed in our discussion of atomic hydrogen.

How rapid is the excitation process? How much time is required for the molecule to go from its ground state to an excited state in the process of the absorption of a quantum of light? It turns out that the excitation process is exceedingly rapid, on the order of 10^{-15} sec. This length of time is very short compared to that typical of nuclear motions (as in the vibration of a chemical bond, for example), which occur on a time scale of 10^{-12} sec. This fact has consequences that are expressed in the **Franck-Condon principle:** *Electronic transitions are so fast in comparison to nuclear motion that immediately after the transition, the nuclei have nearly the same relative positions and velocities as they did just before the transition.* While this principle does not require (and indeed it is not true) that the ultimate structure of the excited state be exactly the same as the structure of the ground state, it does have the consequence that the most likely transitions between different electronic states are those in which the nuclear positions do not change very much. That is, the transitions that occur most frequently yield excited states in which a minimum amount of molecular reorganization has occurred.

In view of the Franck-Condon principle, what are the most usual types of excitation process? Thinking back to Chapter 5, we recall that electrons in molecules occupy one of three types of orbital: (1) sigma orbitals, which are cylindrically symmetrical about the axis joining two nuclei; (2) pi orbitals, which have a nodal plane bisecting the axis connecting two nuclei; and (3) nonbonding, or *n*, orbitals, which contain those electrons not involved in the formation of chemical bonds. In water, for example, there are two pairs of unshared electrons that occupy nonbonding orbitals. For each bonding molecular orbital, there exists a corresponding antibonding orbital at higher energies: hence, we have σ^*, π^*, and n^* orbitals into which electrons may be placed as a result of the excitation process. Of the various types of excitation processes possible, just four are usually encountered. These are

Transition	Excited state designation
$n \longrightarrow \pi^*$	n, π^*
$\pi \longrightarrow \pi^*$	π, π^*
$n \longrightarrow \sigma^*$	n, σ^*
$\pi \longrightarrow \sigma^*$	π, σ^*

This set is derived from the observation that electrons in sigma orbitals are seldom excited in photochemical reactions and that those electrons that are excited seldom are placed into n^* orbitals as a result.

The nature of excited states

As developed thus far, excited states of molecules are those reached by promoting an electron in the ground-state structure from its most stable orbital to another at higher energy. What is the consequence of having done so in terms of the properties of the molecule? Providing answers to a question of this type is not so easy. In the case of ground-state, stable structures, many opportunities exist to determine the parameters that describe the molecule. These might include bond lengths and angles, bond strengths, hybridization about certain atoms, dipole moment, acid strength, and the like. But excited states, by virtue of the excess energy pumped into them by capture of a photon or by other means, are reactive and fleeting species indeed. A typical excited state may endure for no more than 10^{-10} sec, hardly sufficient time to perform the usual kind of experiments. Nevertheless, methods for measuring certain properties of some excited states have been developed, and a fair amount of information is currently available. Although the Franck-Condon principle suggests minimal structural reorganization of the nuclei resulting from the excitation process, the properties of the molecule may be drastically altered by the change in electronic structure. Indeed, it has been possible to detect structural changes in the geometry of molecules following excitation: bond angles and bond distances may change appreciably, reflecting (or causing) changes in hybridization of some atoms. But really striking changes are also observed in properties of molecules that depend basically on the electronic structure. For example, dipole moments of molecules may change, either increase or decrease, severalfold on excitation. Acidity, as measured by the equilibrium,

$$\text{R—COOH} \rightleftharpoons \text{R—COO}^- + \text{H}^+$$

is even more sensitive to change on excitation. Typically, a carboxylic acid is 1 million-fold less dissociated in an excited state than in the ground state.

A very important difference in structure that *may* result from electronic excitation is measured by the **spin multiplicity.** This quantity reflects the number of unpaired electrons in the molecule and is defined by the equation

$$\text{spin multiplicity} = S + 1 \tag{15-5}$$

where S is the number of unpaired electrons. Hence, if there are no unpaired electrons, the spin multiplicity is 1 and the state is called a *singlet*. If there is one electron whose spin is not paired, the spin multiplicity is 2 and the state is termed a *doublet;* two electrons with unpaired spins yield a *triplet* state, and so on. Now let us inquire what the consequence of these considerations is for photochemistry. Most common molecules have no unpaired electrons in the ground state; that is, the ground state is a singlet. What are the possibilities upon excitation of such a molecule? Suppose that we focus our attention on a single filled orbital. The *Pauli exclusion principle* requires that the spins be paired; that is, if one electron has a spin of $+\frac{1}{2}$, then the other must have a spin of $-\frac{1}{2}$. Upon excitation, the two electrons become orbitally unpaired, so that Pauli's principle no longer requires the spins to be paired. There are, then, two possibilities. Excitation may occur *without* inversion of the spin of the excited electron, in which case the electron spins remain paired even though the electrons now occupy different obitals. The excited state is, therefore, a singlet. In contrast, excitation may occur *with* inversion of spin of the excited electron so that there are now *two* electrons with unpaired spins. In this (rare) case, the excited state is a triplet. Diagrammatically, the two possibilities are

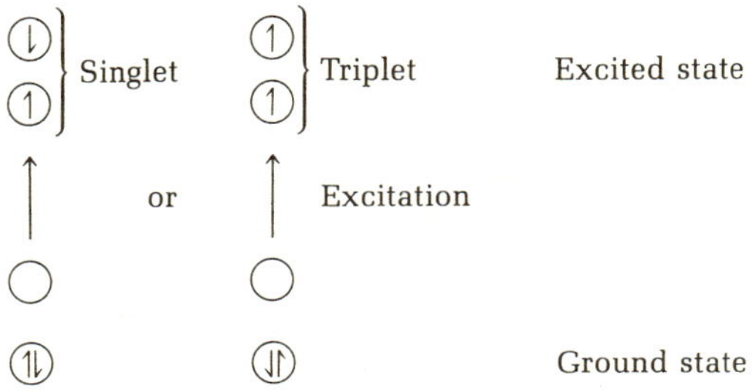

Whether the excited state is a triplet or not makes a great deal of difference to its subsequent fate. Processes in which electron spins are inverted are frequently quite slow. Hence, the lifetime of a triplet excited state is usually several orders of magnitude greater than that for a singlet excited state. This point is developed further in the following subsection.

The fates of electronically excited states

Let us suppose that the absorption of a photon by a molecule with a singlet ground state yields (as it almost always does) a singlet electronically excited

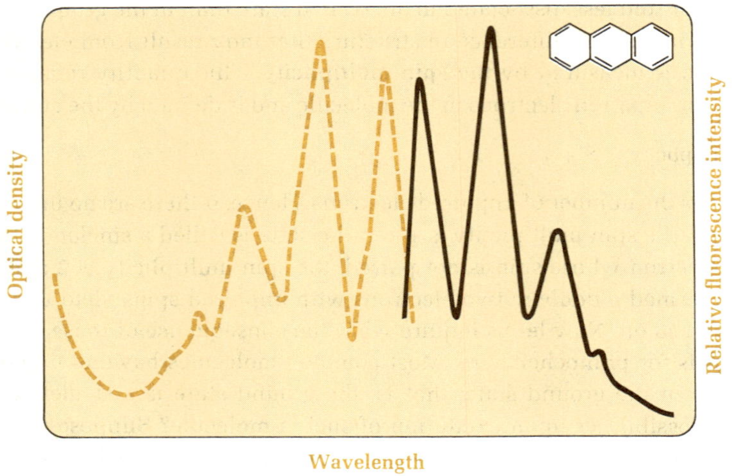

Figure 15-6 A comparison of the absorption spectrum (dashed line) and fluorescence spectrum (line) for anthracene. Note that the two are nearly mirror images of one another.

state. What, then, are the possible fates of this excited state? They are just four in number. As mentioned earlier, this excited state is not stable and will search for ways to return to its most stable electronic structure. In fact, three of the fates of the excited singlet state result in reformation of the ground-state structure. The fourth possibility is occurrence of a chemical reaction. This last possibility is deferred for the moment, and we turn our attention to the three routes for return to the ground state.

First, the absorbed light may simply be emitted as a photon with return to the original electronic structure. The emission of light from a singlet excited state is called **fluorescence.** Just as the absorption spectrum of a molecule provides a distinctive fingerprint for that molecule, so does the fluorescence spectrum. Since both types reflect basically the same considerations, the absorption and fluorescence spectra for a molecule are closely related, as can be seen in Figure 15-6. This case is similar to the relation between absorption and emission spectra of atomic species (see above). The existence of fluorescence has some interesting industrial uses. For example, most brands of household detergents contain small amounts of optical brighteners. These are complex molecules that are adsorbed to the surface of cloth during the washing process. These molecules absorb light (sunlight, for example) in the ultraviolet region of the spectrum. This light has a wavelength slightly too short to be detected by the human eye. A portion of the absorbed light is emitted as fluorescence, but the emitted light has a longer wavelength than the absorbed light and is very pale blue in color. This additional light has the effect of causing yellowed cloth to appear more nearly white. Many will have observed the striking glow of an ordinary white shirt when exposed to a strong source of "black light," particularly when observing psychedelic posters, whose glow is also an example of fluorescence.

Second, the singlet excited state may return to the ground state by a *radiationless transition*, frequently the consequence of a collision with another molecule, then called *collisional deactivation*:

$$A^* + M \longrightarrow A + M \quad (15\text{-}6)$$

All processes that result in return from a singlet excited state to a singlet ground state without emission of radiation are called **internal conversions.** Both the phenomena of fluorescence and internal conversion involve the conversion of a singlet excited state to a singlet ground state. Since no change in spin multiplicity is involved, these processes are frequently very fast, with rate constants in the vicinity of 10^{10} sec^{-1}. Put another way, they usually occur in a period of about 10^{-10} sec. The same considerations apply to the case of return of the triplet excited state to a triplet ground state.

Third, the excited *singlet* state may undergo a process called **intersystem crossing,** which yields the excited *triplet* state. Excited triplet states are usually formed in this way rather than by a direct excitation process. Usually, the excited triplet state lies at a slightly lower energy than the excited singlet. Now we must worry about the possible fates of the excited triplet state. These are exactly analogous to those for the excited singlet, except that return to the singlet excited state is unlikely because it is energetically uphill. Thus, the excited triplet state may return to the ground state by emission of radiation, by intersystem crossing (radiationless transition), or it may undergo chemical reaction. But it is important to recognize that return to the ground state now involves a change in spin multiplicity; consequently, those processes for return to the ground state are now much slower than before. Return of a triplet state to a singlet ground state with emission of radiation is called **phosphorescence.** If the triplet excited state is sufficiently long lived, it may be possible to detect the phosphorescence visually after the exciting light has been turned off. Most of us have, at one time or another, witnessed this phenomenon.

A summary of all these considerations is provided in Figure 15-7, sometimes called a **Jablonski diagram.**

Of the several possible events following excitation, what event (or events) will actually take place in a given situation? Clearly, the answer to this question depends on the relative rates of the various possible processes. In Figure 15-8, the Jablonski diagram for benzophenone, which has the structure

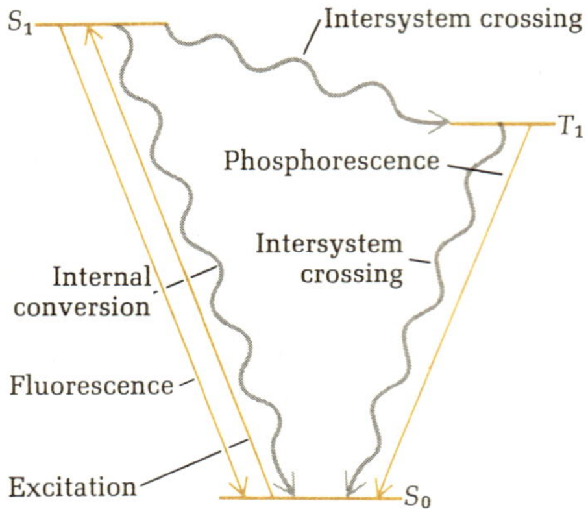

Figure 15-7 A typical energy diagram (Jablonski diagram) illustrating the possible events following an initial activation process. Transitions involving radiation are shown as straight lines, those not involving radiation are shown as wavy lines. The singlet ground state, singlet excited state, and triplet excited state are denoted S_0, S_1, and T_1, respectively.

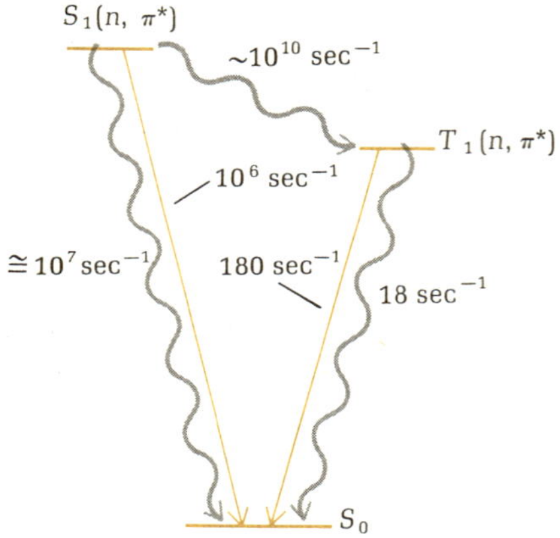

Figure 15-8 A Jablonski diagram for benzophenone at 77°K. The rate constant for each process is indicated on the appropriate arrow. The symbolism employed is the same as in Figure 15-7.

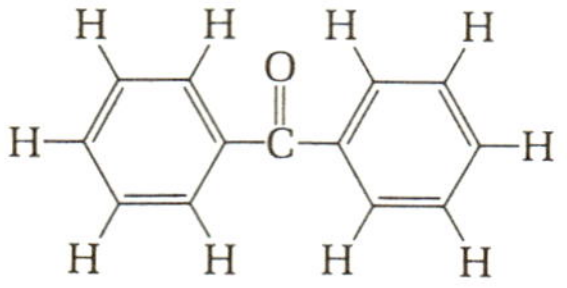

is shown. What can we conclude from it? Note that the rate constant for intersystem crossing from singlet to triplet is very much greater than those for either internal conversion or fluorescence. Hence, it must be true that essentially all of the excited singlet states formed will decay by conversion to the excited triplet state. What will happen to the excited triplet state so formed? We see that the rate constant for phosphorescence is ten times greater than that for intersystem crossing. It follows that about 90 percent of the time the triplet will decay by the route of phosphorescence and 10 percent of the time by intersystem crossing.

PHOTOCHEMICAL REACTIONS

Since most of Chapter 18 deals directly with photochemical reactions that occur in the atmosphere, our account here will be brief and confined largely to the principles involved.

All photochemical reactions are initiated by the absorption of a quantum yield of light leading to the formation of an electronically excited state (see the first law). Each excited state thus formed does not necessarily result in a chemical reaction. Indeed, we have just outlined the possible fates of excited states, and only one of these leads to overall chemical change. To have a measure of the efficiency of a photochemical reaction, it is convenient to define a **quantum yield** for it:

$$\text{quantum yield} = \Phi = \frac{\text{number of molecules that undergo reaction}}{\text{number of quanta absorbed by the system}} \tag{15-7}$$

If every quantum absorbed by the system leads to a particular chemical reaction, then the quantum yield for that process is unity. If none of them do, then the quantum yield is zero. The great bulk of photochemical reactions studied have quantum yields in the range 0.1–0.8.

cis-trans isomerism

A great many kinds of photochemical reactions are known. The simplest of these involve small rearrangements in molecular geometry. The most complex may involve rather complete structural reorganization of a molecule or the generation of a set of entirely new structures. In the former class of reaction, one of the characteristic reactions is the interconversion of *cis* and *trans* isomers. A typical case is that of the isomerization of the stilbenes:

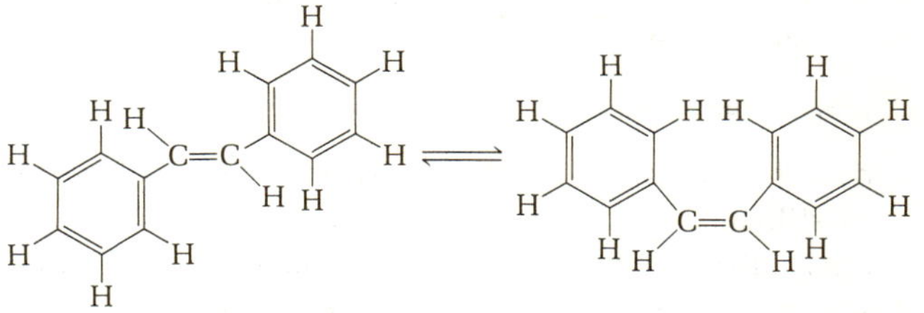

(15-8)

trans *cis*

From our discussion in Chapter 2, you will recognize that the *cis* isomer is sterically very crowded: The two large phenyl rings will bump into each other rather badly. Hence, the *trans* isomer, in which the phenyl groups are as far apart as possible, will be much the more stable of the two. Nevertheless, photochemically, it is possible to generate an equilibrium mixture in which the *cis* isomer predominates. To see how this can be so, let us consider the mechanism of the isomerization.

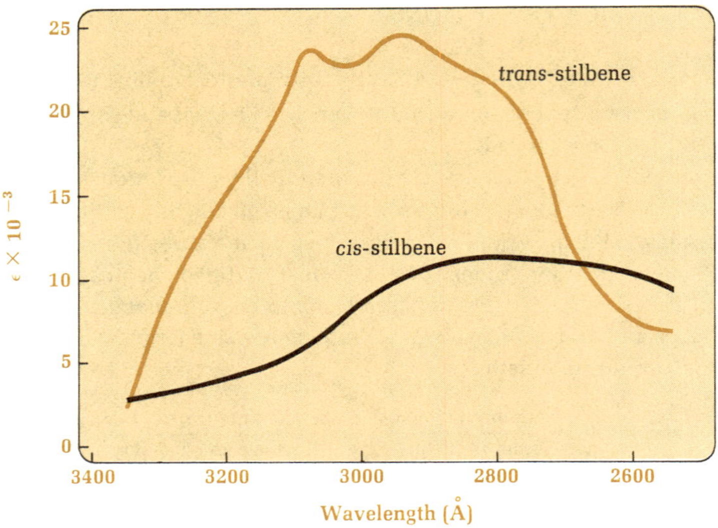

Figure 15-9 Absorption spectra for *cis*- and *trans*-stilbene.

In the ground state of stilbene, *cis* or *trans*, the electronic configuration of the double bond poses a strong barrier to rotation. But in the excited state (probably a triplet), the electronic structure is altered so that most of the double-bond character is lost. Consequently, only a very small barrier to rotation exists. Hence, we may formulate the reaction pathway in terms of the two stilbenes, excited states from them, and the thermal interconversion of these excited states:

$$\textit{cis}\text{-stilbene} \rightleftharpoons \mathrm{C}^* \rightleftharpoons \mathrm{T}^* \rightleftharpoons \textit{trans}\text{-stilbene} \tag{15-9}$$

To continue with the argument, consider the absorption spectra of the *cis* and *trans* isomers of stilbene, which are reproduced in Figure 15-9. Note that these differ rather substantially. Specifically, at most shorter wavelengths, the *trans* isomer absorbs much more strongly than the *cis* isomer. Hence, if we use light of short wavelength to drive the photochemical reaction, we must end up with principally the *cis* isomer. This follows since, to a first approximation, only the *trans* isomer absorbs light in this region. To be somewhat more quantitative, let us suppose that the quantum yield for conversion of the *cis* isomer to the *trans* isomer is the same as for the same reaction in the opposite direction. This is the same as assuming that the probability that the excited *cis* isomer will yield the *trans* isomer as product and the probability that the excited *trans* isomer will yield the *cis* isomer as product are the same. In this case the product composition when the *photostationary state* is reached will directly reflect the extinction coefficient of the two isomers at the wavelength of light used. Thus, if the *trans* isomer absorbs light ten times more strongly than the *cis* isomer at a given wavelength (has an extinction coefficient ten times greater at this wavelength), then the photostationary state will be composed of approximately 91 percent *cis* isomer and 9 percent *trans* isomer. In order for us to see this point more clearly, let us consider the following problem.

Example *Suppose that for a photochemical isomerization,* A $\rightleftharpoons$ B, *the quantum yield in the forward direction,* Φ_{AB}, *is 0.6 and that in the reverse direction,* Φ_{BA}, *is 0.2. At the wavelength of light employed, the extinction coefficient of* B *is six times greater than that of A. What is the composition of the mixture in the photostationary state?*

Solution *To achieve the photostationary state, it is necessary that the rates of the reaction in the two directions be equal; that is, the rate of conversion of* A *to* B *must be the same as that for* B *to* A. *The rate is proportional to the product of the extinction coefficient, the quantum yield, and the concentration of absorbing species: Rate* $\propto$ *extinction coefficient* $\times$ *quantum yield* $\times$ *concentration.*

$$\varepsilon_A \times \Phi_{AB} \times (A) = \varepsilon_B \times \Phi_{BA} \times (B)$$

or

$$\frac{(A)}{(B)} = \frac{\varepsilon_B \Phi_{BA}}{\varepsilon_A \Phi_{AB}} = \frac{6}{1} \times \frac{0.2}{0.6} = \frac{1.2}{0.6} = 2$$

Hence, at the photostationary state we shall have 67 percent A *and 33 percent* B. *Note directly that the photostationary state, which is the equilibrium state reached under the influence of the light used to promote the reaction, may be very different indeed from the thermodynamic equilibrium state determined by the overall free-energy change for the reaction. Thus, the light energy may be employed to maintain a system in a steady state far from thermodynamic equilibrium.*

The visual process

Photochemical *cis–trans* isomerism lies at the very heart of the visual process. Basically, vision involves a process of reception of a quantum of light by certain *visual pigments*, which are localized in the rod or cone cells of the eye. The energy of these photons is converted into electrical energy, and an electrical impulse, which contains the information concerning what the eye "saw," is conducted along the optic nerve to the brain, where the information is decoded, that is, where a mental image is formed. Generation of this electrical impulse is caused by a photochemical reaction involving the pigments that absorb the light. A great many people have contributed to our understanding of the chemistry of vision. Perhaps foremost among these is George Wald of Harvard University, who was awarded a Nobel Prize for his work in this area.

The basic chemistry involved in vision is strikingly similar in a great many organisms. Differences that do occur are only subtle variations on a constant basic theme. Let us then focus our attention on the series of chemical reactions that occur in the rod cells of cattle, the best studied case. The visual pigments are derived from vitamin A, now more commonly called *retinol:*

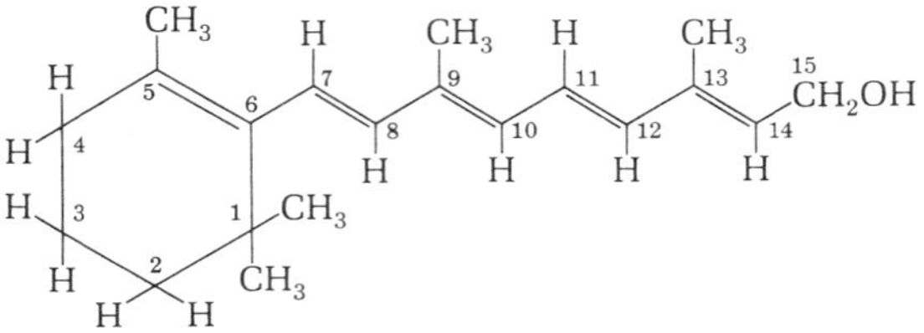

retinol (vitamin A)

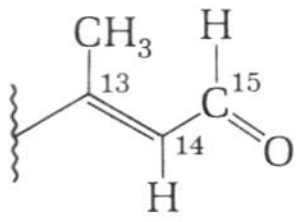

retinal

In the cells of the retina, retinol can be oxidized, under the influence of an appropriate enzyme, to retinal:

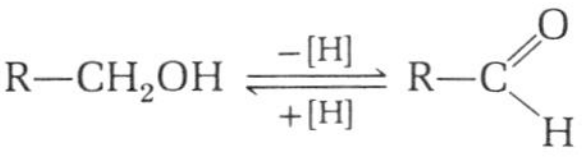

(15-10)

retinol retinal

Under physiological conditions the equilibrium for reaction (15-10) lies far to the left side; that is, at equilibrium there is much more retinol than retinal. However, there is a protein in these same cells, called *opsin,* that reacts specifically with the retinal to form an adduct with it. This adduct involves condensation of an amino group of a lysine residue (see Table 7-2) with the terminal carbon-oxygen double bond of retinal to form a *Schiff base* linkage:

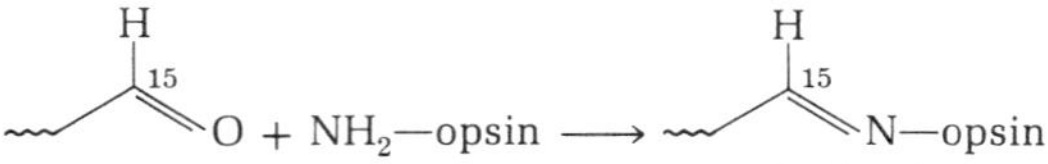

(15-11)

rhodopsin

The product is called *rhodopsin,* and this species is the actual pigment that absorbs the light quanta that strike the retina. Rhodopsin has a molecular weight near 40,000 and absorbs light maximally at 4980 Å.

Thinking back to our discussion of *cis–trans* isomerism on p. 62, Chapter

2, we can recognize that there are quite a few possible isomers of retinal: The one depicted above is the all-*trans* one. In fact, the form of retinal bound to opsin is the 11-*cis* isomer:

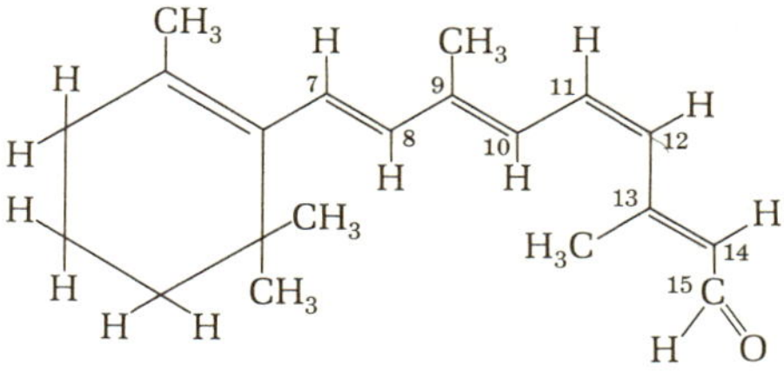

(Note that the hydrogens are on the same side of the bond between carbon atoms 11 and 12, thus giving rise to the *cis* designation.) The 11-*cis* isomer is not as stable as the all-*trans* one; this point can be readily appreciated by noting the steric hindrance that exists between the methyl group on carbon 13 and the proton on carbon 10 in the 11-*cis* isomer. At any event, it is this isomer that forms part of the total visual pigment.

When rhodopsin is exposed to visible light, its maximum absorption changes from 4980 Å to about 3870 Å; thus, rhodopsin is said to be "bleached" by light. This spectral change is accompanied by liberation of the retinal from the opsin. It is especially interesting that the liberated retinal is the all-*trans* isomer. Consequently, one chemical consequence of the absorption of light is a *cis–trans* isomerization in which 11-*cis* retinal is converted to all-*trans* retinal. Prior to reattachment of this pigment to opsin, which is necessary to prepare the system to respond to the next light stimulus, it must be reconverted by a specific enzyme to the 11-*cis* form. Thus, the overall cycle involved can be formulated as

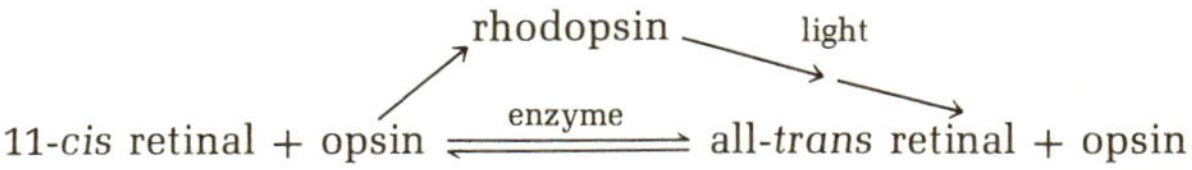

(15-12)

The transformation of rhodopsin into all-*trans* retinal plus opsin is a complicated chemical process involving several intermediates; the details of the reaction scheme remain uncertain. The important intermediates, together with their light absorption maxima, are shown in Figure 15-10. It is possible, though not certain, that the initial reaction, converting rhodopsin into prelumirhodopsin, involves the crucial *cis-trans* isomerization. Subsequent steps would be viewed as changes in the detailed structure of the opsin or retinal. Clearly, a binding site on a protein into which fits 11-*cis* retinal would be distorted by conversion of that species into the all-*trans*

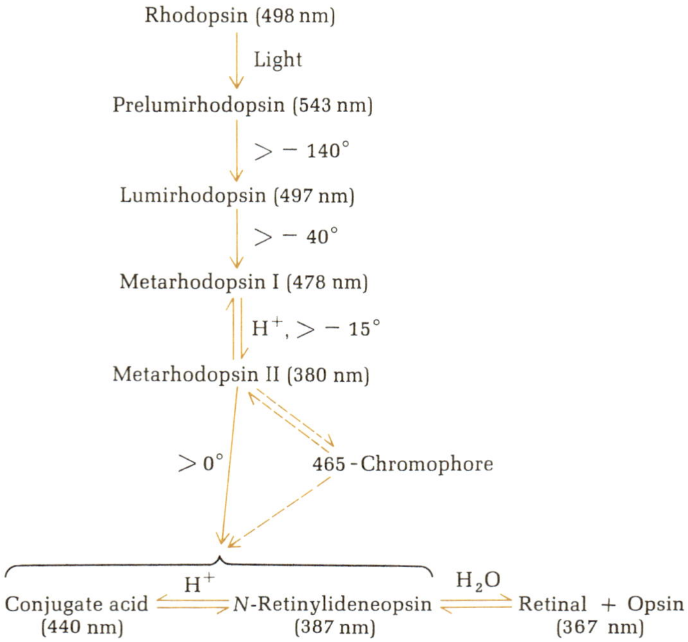

Figure 15-10 The sequence of intermediates in the bleaching of rhodopsin.

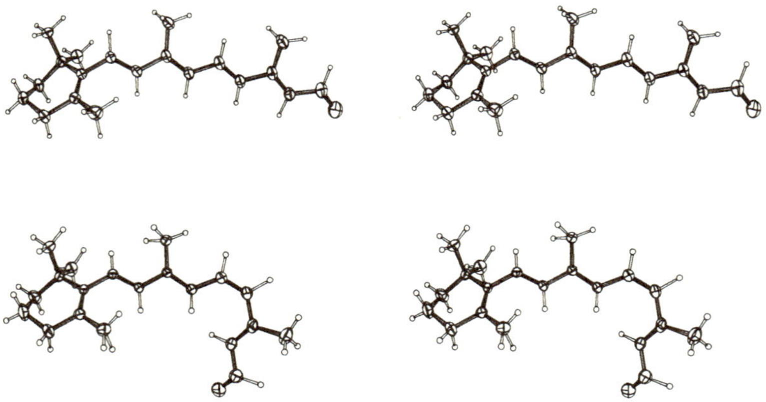

Figure 15-11 Stereodiagrams of all-*trans* retinal (top) and 11-*cis* retinal (bottom). Note the very different shape of these closely related isomers. (Courtesy of J. Karle, Laboratory for the Structure of Matter, U. S. Naval Research Laboratory.)

isomer. The structural differences between these two isomers may best be appreciated by studying the stereodiagrams in Figure 15-11.

How does the chemistry resulting from the photochemical *cis–trans* isomerization of rhodopsin generate the phenomenon of vision? This is a question to which we do not yet have a firm answer. In fact, most of the problems relating to matters of sensory reception pose fascinating questions that have not yielded to scientific investigation. But it is possible to construct reasonable models to account for the visual process in terms of what we now know. The crucial matter is how the electrical stimulus is generated as a result of the photochemistry described above. One possibility is that the *cis–trans* isomerization reaction followed by structural changes in the protein opens up an enzymatic site on the opsin moiety. Thus, the photochemical reaction would have the effect of converting rhodopsin from an inactive to an active form. In consequence, rhodopsin might then catalyze a chemical reaction leading to the development of an electrical potential and providing for an electrical stimulus. Such chemical reactions are said to be *electrogenic*. A crude visualization of this theory is provided in Figure 15-12. It is important to remember that this is just a theory, not an established fact. Further experimental and theoretical work is required prior to development of a really full understanding of the chemical basis for vision and other sensory reception processes.

Chemiluminescence

There are a number of chemical reactions known that spontaneously emit light in the course of the reaction; such reactions are called **chemiluminescent.** We are in a good position to formulate a hypothesis concerning the origin of this light. It has been established above that light may be emitted by a molecule in its excited singlet or triplet states. Consequently, if a chemical reaction happens to yield a product molecule in an excited state, that molecule might return to the ground state by emission of a photon; that is, the reaction would exhibit chemiluminescence:

$$A + B + \cdots \longrightarrow P^* + R + \cdots \qquad (15\text{-}13)$$

$$P^* \longrightarrow P + h\nu \qquad (15\text{-}14)$$

One of the simplest examples of chemiluminescence is provided by the reaction between hypochlorite ion (derived in basic solution from hypochlorous acid, HClO, the active reagent in chlorine bleaches such as Clorox) and hydrogen peroxide. Oxygen is produced in an excited state and decays to ground-state oxygen with emission of a faint red light. If the reaction is carried out in a dark room, the emitted light is quite obvious.

Toadstools of genus *Mycena* glow by self-emitted light. (Courtesy of National Science Foundation.)

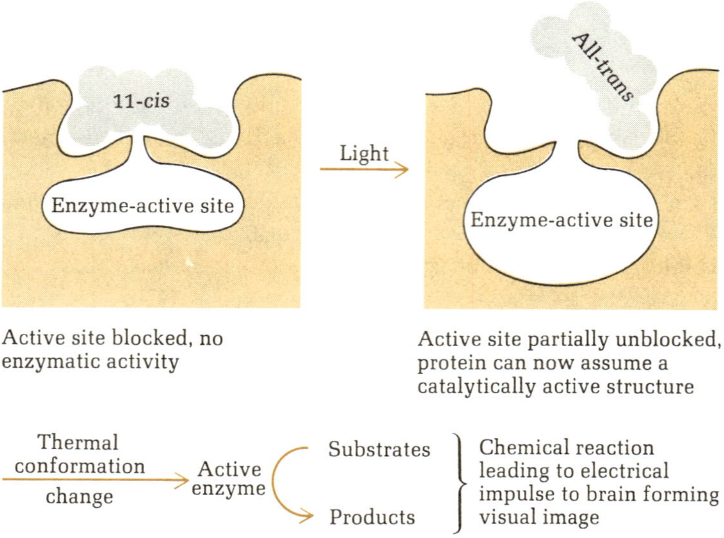

Figure 15-12 A possible scheme of processes underlying the generation of an electrical stimulus following a photochemical reaction in the visual process.

A classic example of a chemiluminescent reaction is the oxidation of luminol:

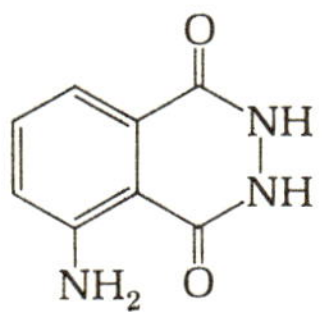

When this complex molecule is treated with potassium hydroxide, hydrogen peroxide, and sodium hypochlorite (other combinations of reagents will also work), luminol is oxidized, nitrogen gas is evolved, and a brilliant blue light is emitted. As you may have guessed, this is a complicated reaction, the details of which still have not been completely unraveled. Perhaps the most usual and interesting chemiluminescent reactions are those that occur in living organisms.

Bioluminescence

The emission of light by living organisms—**bioluminescence**—is quite a common phenomenon in nature. Almost everyone is familiar with fireflies and the greenish light that they emit. Many people have also seen one of the oceans brilliantly lit up by bioluminescent microorganisms swimming near the surface. In fact, a great many species are known that are bioluminescent, and many of these have received appreciable experimental study. These include jellyfish, the sea firefly (*Cypridinia*), several fish, clams, limpets, worms, dinoflagellates, bacteria, fungi, fireflies, and the sea pansy.

Bioluminescence has long been an object of fascination and study for man. Pliny the Elder, in the first century A.D., made the earliest recorded observations concerning this phenomenon. His commentary concerned a luminous jellyfish. Studies continued off and on for a number of centuries; bioluminescence receives a good deal of study in the present day. It is interesting to note that Robert Boyle demonstrated in 1667, prior to the discovery of oxygen, that air is necessary for bioluminescence.

Research work has indicated that the reactions leading to bioluminescence differ quite markedly from species to species. In the firefly, the best studied case, it has been established that there is a molecule, a **luciferin,** that reacts with adenosine triphosphate (ATP, see Chapter 21) and oxygen under the influence of an enzyme, a **luciferase,** to generate an oxidized product in an excited state which then emits light. The overall reaction may be formulated as follows (LH_2. luciferin; AMP, adenosine monophosphate; E, luciferase):

$$\mathrm{E + LH + ATP} \rightleftharpoons [\mathrm{E \cdot LH \cdot AMP}] \longrightarrow \left[\begin{array}{c} \mathrm{OOH} \\ | \\ \mathrm{E{-}L{-}AMP} \\ | \\ \mathrm{H} \end{array}\right] \longrightarrow$$

$$\left[\begin{array}{c} \mathrm{O} \\ \| \\ \mathrm{E{-}L{-}AMP} \end{array}\right]^{*} \longrightarrow \begin{array}{c} \mathrm{O} \\ \| \\ \mathrm{E{-}L{-}AMP} \end{array} + h\nu \qquad (15\text{-}15)$$

The structure of the firefly luciferin is

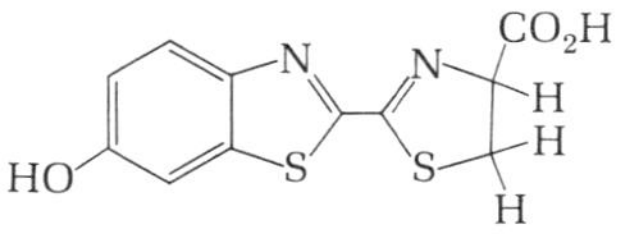

The structure of the product that emits the light has not been established.

Glossary

Absorption spectrum a plot of optical density against wavelength of incident light; obtained experimentally by measuring the intensity of light transmitted through a sample as a function of wavelength.

Beer-Lambert law the quantitative expression linking optical density to extinction coefficient, concentration of absorbing species, and optical path length:

$$\text{optical density} = \varepsilon \times c \times l$$

Bioluminescence emission of light from living organisms; the consequence of formation of the product of an enzymatic reaction in an electronically excited state.

Chemiluminescence emission of light from a chemical reaction; the consequence of the formation of a reaction product in an electronically excited state.

Electronic excitation the promotion of an electron from one atomic or molecular orbital to another at higher energy; may be accomplished through irradiation of the substance with light of an appropriate wavelength or through heating of the sample.

Emission spectrum a plot of intensity of light emitted from an electronically excited sample as a function of wavelength.

Extinction coefficient a measure of the extent to which a given substance absorbs light of a specific wavelength; quantitatively, it is given by the optical density of a 1*M* solution of the substance of interest contained in a vessel having a 1 cm optical path length.

Fluorescence emission of light from an electronically excited state of an atom or molecule that accompanies an electronic transition *not* involving a change in spin multiplicity.

Franck-Condon principle the assertion that electronic transitions are so fast in comparison to nuclear motions that immediately after the transition, the nuclei have nearly the same relative positions and velocities as they did just before the transition.

Internal conversion a radiationless electronic transition not involving a change in spin multiplicity.

Intersystem crossing a radiationless electronic transition that does involve a change in spin multiplicity.

Jablonski diagram a pictorial representation of the process of electronic excitation and the various possible fates of the electronically excited state.

Luciferase a general term for the enzyme catalyzing the transformation of a luciferin to an electronically excited product, responsible for bioluminescence.

Luciferin the substrate that, when acted on by a luciferase, is converted into that product that emits light in the process of bioluminescence.

Optical density the logarithm of the ratio of the intensity of incident to transmitted light:

$$\text{optical density} = \log(I_0/I_t)$$

Phosphorescence emission of light from an electronically excited state of an atom or molecule that accompanies an electronic transition involving a change in spin multiplicity.

Quantum yield a measure of the efficiency of a photochemical reaction given by the ratio of the number of molecules that react to the number of quanta absorbed by those molecules.

Rydberg equation an equation that relates energy of absorption or emission lines in hydrogenlike spectra to quantum numbers of the orbitals involved in the electronic transition. Specifically, the equation reads

$$E = KZ^2\left(\frac{1}{n_1^2} - \frac{1}{n_2^2}\right)$$

where K is the Rydberg constant and Z is the atomic number.

Spectrometer an instrument employed for the experimental determination of absorption or emission spectra.

Spin multiplicity a measure of the number of unpaired electrons in an atom or molecule; spin multiplicity = $S + 1$, where S is the number of unpaired electrons.

Study problems

Problems concerning the quantitative aspects of light absorption

1. Calculate the optical density for each of the following cases.
   a. ten percent of the light incident on the sample is transmitted through it
   b. eighty percent of the light incident on a sample is transmitted through it

c. sixty percent of the light incident on a sample is absorbed by it
2. Light absorption is sometimes measured in terms of the *transmittance*, T, the fraction of the light incident on a sample that passes through it: $T = I_t/I_0$. Derive a quantitative relationship between T and optical density and construct a plot of one as a function of the other.
3. For chlorophyll *a*, a pigment important for photosynthesis, the extinction coefficient at 600 nanometers (nm) is $10^5 M^{-1}\ cm^{-1}$. Calculate the optical density for a $10^{-6}M$ solution of chlorophyll *a* in a vessel having a 5 cm light path at this wavelength.
4. Using the data from the above problem, calculate the molarity of a solution of chlorophyll *a* in a vessel having a 1 cm light path, given that the optical density at 600 nm is 0.850.
5. An important biological electron carrier is cytochrome c, a small iron-containing protein. Cytochrome c exists in both oxidized and reduced forms:

$$\text{cyt c(Fe}^{III}) + e^- \rightleftharpoons \text{cyt c(Fe}^{II})$$

At 550 nm the extinction coefficient for reduced cytochrome c is $25 \times 10^3 M^{-1}\ cm^{-1}$, and for oxidized cytochrome c at this wavelength it is $8 \times 10^3 M^{-1}\ cm^{-1}$. For a solution of $5 \times 10^{-5}M$ cytochrome c(Fe^{II}) in a vessel having a 1 cm path length, calculate the optical change that would occur on complete oxidation of the sample.
6. Chlorophylls *a* and *b*, two types of chlorophyll, have the following extinction coefficients for the pigments dissolved in acetone:

	ε (cm²/μg) at:	
Type	645 nm	662 nm
Chl *a*	0.017	0.082
Chl *b*	0.045	0.010

A particular sample, held in a 1 cm light path, has an optical density of 0.50 at 645 nm and 1.00 at 662 nm. What are the concentrations of Chl *a* and Chl *b* in the sample? (From R. K. Clayton, *Light and Living Matter, Vol. 1: The Physical Part*, McGraw-Hill, New York, 1970.)
7. Two substances, A and B, are dissolved together at concentrations (A) and (B). The path length is 1 cm. The extinction coefficients are ε_1^A and ε_1^B at one wavelength, λ_1, and ε_2^A and ε_2^B at a second wavelength, λ_2. The optical densities measured at λ_1 and λ_2 are OD_1 and OD_2. Derive formulas for (A) and (B) in terms of the εs and ODs. The correct formulas should give the right answers for problem 6. (From R. K. Clayton, *Light and Living Matter, Vol 1: The Physical Part*, McGraw-Hill, New York, 1970.)

Problems concerning electronic transitions

8. Calculate the ionization energy for He^+ and for Li^{2+}.
9. Calculate the wavelength of light emitted for each of the following electronic transitions:

a. the first line in the Balmer series for hydrogen (i.e., the transition from $n = 3$ to $n = 2$)
b. the fourth line in the Lyman series for hydrogen
c. the first line in the Lyman series for Li^{2+}

10. Interpret the following Jablonski diagram as completely as you can. Identify the nature of all the transitions indicated.

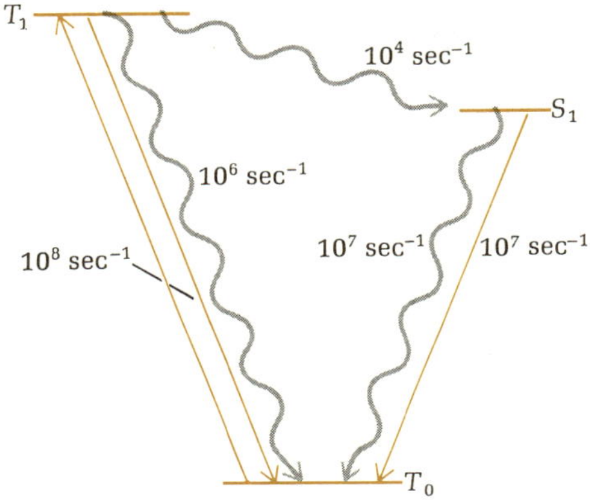

Some problems concerning photochemical reactions

11. Calculate the photostationary state for each of the following cases:
a. $A \rightleftharpoons B$ $\Phi_{AB} = 0.1$; $\Phi_{BA} = 0.5$; $\varepsilon_A = 10^3 \times \varepsilon_B$
b. $A \rightleftharpoons B$ $\Phi_{AB} = 0.6$; $\Phi_{BA} = 0.02$; $\varepsilon_A = 0.1 \times \varepsilon_B$
c. $A \rightleftharpoons B \rightleftharpoons C$ $\Phi_{AB} = \Phi_{BC} = \Phi_{BA} = \Phi_{CB}$; $\varepsilon_A = 10 \times \varepsilon_B$; $\varepsilon_C = 2 \times \varepsilon_B$
12. Explain clearly why the photostationary state for a given reaction is a function of the wavelength of light used to obtain it.
13. For a given wavelength of light and a particular reaction, $A \rightleftharpoons B$, will the same photostationary state be reached regardless of whether the reaction is started with A, B, or with a mixture of A and B? Defend your answer.
14. Suppose that for a given isomerization reaction, $A \rightleftharpoons B$, you wish to measure the rate of interconversion of A and B under the influence of light when these reactants are present at the photostationary state. Can you suggest an experiment that would enable you to make this measurement?
15. How does the rate of attainment of the photostationary state depend on the intensity of the light employed to attain it?

Suggestions for further reading

R. K. Clayton *Light and Living Matter*, McGraw-Hill, New York, 1970, Vols. I, II.

J. R. Edisbury *Practical Hints on Absorption Spectrometry,* Hilger and Watts, London, 1966.
E. G. Harvey *Bioluminescence,* Academic, New York, 1952.
W. D. McElroy and B. Hass, Eds. *Light and Life,* Johns Hopkins Press, Baltimore, 1961.
M. H. L. Pirenne *Vision and the Eye,* Chapman & Hall, London, 1967.
H. H. Seliger and W. D. McElroy *Light, Physical and Biological Action,* Academic, New York, 1965.

16

Nuclear reactions: Source of light, energy, and elements

Two hundred years without much progress:
The rapid progress true Science now makes occasions my regretting sometimes that I was born too soon. It is impossible to imagine the heights to which may be carried, in a thousand years, the power of man over matter. O that moral Science were in as fair a way of improvement, that men would cease to be wolves to one another, and that human beings would at length learn what they now improperly call *humanity.*

Benjamin Franklin
in a letter to Joseph Preistley,
February 8, 1780

From the second law of thermodynamics, we know that nature continuously strives for a maximum amount of disorder: The entropy of the world constantly increases. To avoid eventually running out of energy, it must be continually supplied from outside the earth. Our most important source of energy comes to us in the form of sunlight (Chapters 17 and 19). To be sure, we make considerable use of fossil fuels—coal and oil—for supplying our energy needs, but even these were (and are) derived from formerly living things, which depended on sunlight for survival.

In this chapter attention is focused on the question: How is sunlight generated? Later, we shall be concerned with the role of sunlight in the formation of our atmosphere, in the process of chemical evolution that resulted in life, in the generation of photochemical smog, and in how we utilize the radiant energy of sunlight.

As we search for an answer to our question about how sunlight is generated, we shall encounter several related matters of much interest, including what may be the most fundamental question of all for chemistry: How are the chemical elements themselves synthesized in the universe? Understanding the chemistry of the sun will require that we develop some sophistication in *nuclear chemistry,* those reactions in which atomic nuclei actually undergo change. Let us begin, then, with a description of a few of the basics of nuclear chemistry prior to turning our attention to solar chemistry and, later, to the chemistry of hotter stars.

SOME BASICS OF NUCLEAR CHEMISTRY

The atomic nucleus is an exceedingly small structure, with a radius less than 10^{-12} cm, about four orders of magnitude less than the radius of a typical atom. How was such a small structure, even in atomic terms, discovered in the first place? As we might well imagine, development of ideas about the atomic nucleus did not come from a single experiment but from a series of investigations, great insight on the part of a few scientists, and a fair measure of good luck. For the discovery of nuclei, a typical nuclear reaction was employed as the essential tool.

Perhaps the first crucial observation was that of the French scientist Henri Becquerel in the latter part of the nineteenth century, who discovered **radioactivity**, the spontaneous emission of charged particles from certain elements. Early workers were able to detect two types of particles emitted during **radioactive decay, alpha particles**, which are nothing more than helium nuclei, and **beta particles**, which are simply electrons moving at very high speed. Using this information, Ernest Rutherford, in 1909, performed a series of elegant experiments which greatly advanced our understanding of atomic structure.

Rutherford's experiments consisted of examining the fate of alpha particles that were shot, like small bullets, at thin films of gold and other elements. His experimental setup is shown in Figure 16-1. He and his coworkers were able to observe the alpha particles with the aid of a zinc sulfide screen which emitted a small flash of light, visible through a microscope, when struck by one of these particles. The crucial observation made by Rutherford was that a certain small fraction of the alpha particles actually left the foil *from the same side as they entered* [Figure 16-1(b)]; that is, the particles bounced back toward their source. This result would never have been expected on the basis of the uniform-density atom of Dalton and Thomson (Chapter 4); in this case only low-angle scattering is anticipated [Figure 16-1(a)]. Rutherford concluded:

> It was almost as incredible as if you had fired a 15-inch shell at a piece of tissue paper and it came back and hit you . . . when I made calculations, I saw that it was impossible to get anything of that order of magnitude unless . . . the greater part of the mass of the atom was concentrated in a minute nucleus. It was then that I had the idea of an atom with a minute massive center carrying a charge.

This central observation gave rise to the concept of the planetary atom, described in Chapter 4.

Subsequent experimentation has led to a great deal of more sophisticated information, some of which we shall have occasion to refer to in our discussion below. In the simplest terms, we now believe that the nucleus includes both **protons**, positively charged particles, and **neutrons**, which carry no charge. Chemists have adopted a shorthand notation for indicating the composition of the nucleus: $^{m}_{n}X$. In this designation X is the symbol for the chemical element involved, the subscript n indicates the **atomic number**, the number of protons in the nucleus, and the superscript m indicates the **atomic mass**, the total of the number of protons and neutrons. It follows that the number of neutrons in the nucleus is just $m - n$. Atomic masses are referred to a scale in which the mass of the common isotope of carbon,

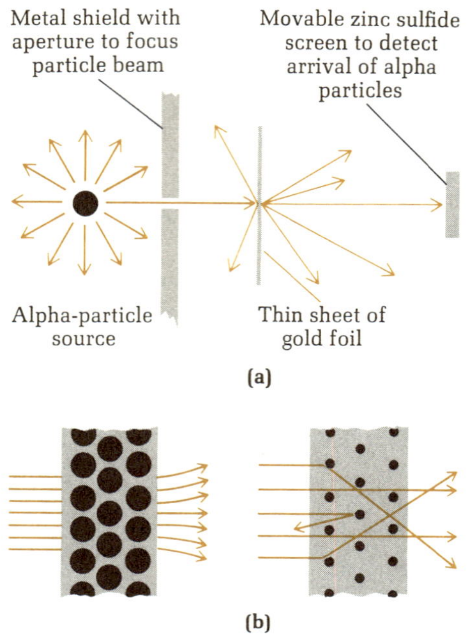

Figure 16-1 A diagrammatic view of the Rutherford experiment, which established the basic structural organization of the atom. In part (a) the experimental setup is shown. A beam of alpha particles is focused on the foil target. Scattering of the alpha particles is studied using the zinc sulfide detector. In part (b), the two possible results of the experiment are shown. Were atoms bodies of uniform density, only low-angle scattering of the alpha particles would be expected; this possibility is shown on the left. With the nuclear model, shown on the right, we expect most alpha particles to pass through the film unimpeded, but those that happen to strike the nucleus may be scattered through very large angles. This is the experimental result that in fact was obtained by Rutherford. (From G. C. Pimental and R. D. Spratley, *Understanding Chemistry,* Holden-Day, San Francisco, 1971.)

$^{12}_{6}C$, is taken as 12.000000 atomic mass units (amu). One amu is equal to 1.66024×10^{-24} g. The masses of the fundamental particles of atoms are collected in Table 16-1 along with several conversion factors that we shall find useful later in this chapter. A complete list of the proper designations and masses of many common isotopes is provided in Table 16-2.

To begin to understand something about the *changes* that occur in nuclear structure, let us consider the modes of radioactive decay in slightly more detail.

Table 16-1 Some fundamental quantities and conversion factors

Energy
1 eV = 1.6021×10^{-12} erg = 3.814×10^{-23} kcal
1 erg = 6.2420×10^{11} eV = 2.39×10^{-11} kcal = 1 g cm^2/sec^2
1 kcal = 4.19×10^{10} ergs = 2.622×10^{22} eV
1 eV/molecule = 23.069 kcal/mole = 8066 cm^{-1}
1 kcal/mole = 350 cm^{-1} = 0.0434 eV/molecule

Masses
1 amu = 1.66024×10^{-24} g
1 electron rest mass = 0.000549 amu = 0.91091×10^{-27} g
1 neutron rest mass = 1.008665 amu = 1.67482×10^{-24} g
1 proton rest mass = 1.007277 amu = 1.67252×10^{-24} g
1 (proton + electron) rest mass = 1.007826 amu

Mass-energy conversion
1 amu of mass = 931.4×10^6 eV = 931.4 MeV

Alpha decay

Many of the heavier isotopes decay spontaneously with the emission of an alpha particle. In the most general terms we can write the equation for **alpha decay** as

$$ {}^{m}_{n}X \longrightarrow {}^{m-4}_{n-2}Y + {}^{4}_{2}He \tag{16-1}$$

Note that the atomic mass of the new element is four less and the atomic number of the new element is two less than that for the original. Overall, Equation (16-1) is balanced both with respect to total atomic mass units (although not, as we shall see, in exact mass) and with respect to charge. Typical examples of alpha decay include

$$ {}^{226}_{88}Ra \longrightarrow {}^{222}_{86}Rn + {}^{4}_{2}He \tag{16-2}$$
$$ {}^{238}_{92}U \longrightarrow {}^{234}_{90}Th + {}^{4}_{2}He \tag{16-3}$$

The energy with which the alpha particle is emitted, which is reflected in the velocity of the particle, varies from isotope to isotope. Generally, however, the alpha particle is produced with a very high velocity; that produced via Equation (16-2), for example, has a velocity of 1.5×10^9 cm/sec, less than tenfold lower than the speed of light. This corresponds to an energy of 4.777 million electron volts (MeV, see Table 16-1). Implicit in this statement is a fact of great importance: Alpha particles emitted during radioactive decay have one, or a few, well-defined energies. That is, all alpha particles emitted during the decay of ^{226}Ra have exactly the same energy. Those emitted during the decay of ^{238}U have one of two energies—77 percent

Table 16-2 Table of selected isotopic weights

Isotope	Atomic mass	Isotope	Atomic mass
$^{1}_{1}H$	1.007825	$^{46}_{22}Ti$	45.95263
$^{2}_{1}H$	2.01410	$^{47}_{22}Ti$	46.9518
$^{3}_{2}He$	3.01603	$^{48}_{22}Ti$	47.94795
$^{4}_{2}He$	4.00260	$^{49}_{22}Ti$	48.94787
$^{6}_{3}Li$	6.01512	$^{50}_{22}Ti$	49.9448
$^{7}_{3}Li$	7.01600	$^{50}_{23}V$	49.9472
$^{8}_{4}Be$	8.00531	$^{51}_{23}V$	50.9440
$^{9}_{4}Be$	9.01218	$^{52}_{24}Cr$	51.9405
$^{9}_{5}B$	9.01333	$^{53}_{24}Cr$	52.9407
$^{10}_{5}B$	10.01294	$^{54}_{24}Cr$	53.9389
$^{11}_{5}B$	11.00931	$^{55}_{25}Mn$	54.9381
$^{12}_{6}C$	12.00000	$^{54}_{26}Fe$	53.9396
$^{13}_{6}C$	13.00335	$^{56}_{26}Fe$	55.9349
$^{14}_{7}N$	14.00307	$^{57}_{26}Fe$	56.9354
$^{15}_{7}N$	15.00011	$^{58}_{26}Fe$	57.9333
$^{16}_{8}O$	15.99491	$^{59}_{27}Co$	58.9332
$^{17}_{8}O$	16.99914	$^{58}_{28}Ni$	57.9353
$^{18}_{8}O$	17.99916	$^{60}_{28}Ni$	59.9332
$^{16}_{9}F$	16.01171	$^{61}_{28}Ni$	60.9310
$^{19}_{9}F$	18.99840	$^{62}_{28}Ni$	61.9283
$^{20}_{10}Ne$	19.99244	$^{64}_{28}Ni$	63.9280
$^{21}_{10}Ne$	20.99395	$^{63}_{29}Cu$	62.9298
$^{22}_{10}Ne$	21.99138	$^{65}_{29}Cu$	64.9278
$^{23}_{11}Na$	22.98977	$^{64}_{30}Zn$	63.9291
$^{24}_{12}Mg$	23.98504	$^{66}_{30}Zn$	65.9260
$^{25}_{12}Mg$	24.98584	$^{67}_{30}Zn$	66.9271
$^{26}_{12}Mg$	25.98259	$^{68}_{30}Zn$	67.9249
$^{27}_{13}Al$	26.98153	$^{69}_{31}Ga$	68.9257
$^{28}_{14}Si$	27.97693	$^{71}_{31}Ga$	70.9249
$^{29}_{14}Si$	28.97649	$^{70}_{32}Ge$	69.9243
$^{30}_{14}Si$	29.97376	$^{72}_{32}Ge$	71.9217
$^{31}_{15}P$	30.99376	$^{73}_{32}Ge$	72.9234
$^{32}_{16}S$	31.97207	$^{74}_{32}Ge$	73.9219
$^{33}_{16}S$	32.97146	$^{76}_{32}Ge$	75.9214
$^{34}_{16}S$	33.96786	$^{75}_{33}As$	74.9216
$^{36}_{16}S$	35.96709	$^{74}_{34}Se$	73.9225
$^{35}_{17}Cl$	34.96885	$^{76}_{34}Se$	75.9192
$^{37}_{17}Cl$	36.96590	$^{77}_{34}Se$	76.9199
$^{36}_{18}Ar$	35.96755	$^{78}_{34}Se$	77.9173
$^{38}_{18}Ar$	37.96272	$^{80}_{34}Se$	79.9165
$^{40}_{18}Ar$	39.96238	$^{82}_{34}Se$	81.9167
$^{39}_{19}K$	38.96371	$^{79}_{35}Br$	78.9183
$^{41}_{19}K$	40.96184	$^{81}_{35}Br$	80.9163
$^{40}_{20}Ca$	39.96259	$^{78}_{36}Kr$	77.9204
$^{42}_{20}Ca$	41.95863	$^{80}_{36}Kr$	79.9164
$^{43}_{20}Ca$	42.95878	$^{82}_{36}Kr$	81.9135
$^{44}_{20}Ca$	43.95549	$^{83}_{36}Kr$	82.9141

Table 16-2 *(Continued)*

Isotope	Atomic mass
$^{84}_{36}Kr$	83.9115
$^{86}_{36}Kr$	85.9106
$^{209}_{83}Bi$	208.9804
$^{211}_{83}Bi$	210.9873
$^{206}_{84}Po$	205.9805
$^{207}_{84}Po$	206.9816
$^{208}_{84}Po$	207.9813
$^{209}_{84}Po$	208.9825
$^{210}_{84}Po$	209.9829
$^{211}_{84}Po$	210.9866
$^{212}_{84}Po$	211.9889
$^{213}_{84}Po$	212.9928
$^{214}_{84}Po$	213.9952
$^{215}_{84}Po$	214.9995
$^{216}_{84}Po$	216.0019
$^{218}_{84}Po$	218.0089
$^{211}_{85}At$	210.9875
$^{213}_{85}At$	212.9963
$^{214}_{85}At$	213.9963
$^{215}_{85}At$	214.9987
$^{216}_{85}At$	216.0024
$^{217}_{85}At$	217.0046
$^{218}_{85}At$	218.0086
$^{219}_{85}At$	219.0114
$^{210}_{85}Rn$	209.9897
$^{211}_{86}Rn$	210.9906
$^{212}_{86}Rn$	211.9907
$^{215}_{86}Rn$	214.9987
$^{216}_{86}Rn$	216.0002
$^{217}_{86}Rn$	217.0039
$^{218}_{86}Rn$	218.0056
$^{219}_{86}Rn$	219.0095
$^{220}_{86}Rn$	220.0114
$^{212}_{87}Fr$	211.996
$^{217}_{87}Fr$	217.0048
$^{218}_{87}Fr$	218.0095
$^{219}_{87}Fr$	219.0092
$^{220}_{87}Fr$	220.0123
$^{221}_{87}Fr$	221.0142
$^{219}_{88}Ra$	219.0100
$^{220}_{88}Ra$	220.0110
$^{221}_{88}Ra$	221.0139
$^{222}_{86}Ra$	222.0154
$^{223}_{88}Ra$	223.0186
$^{224}_{88}Ra$	224.0202
$^{226}_{88}Ra$	226.0254
$^{221}_{89}Ac$	221.0157
$^{222}_{89}Ac$	222.0178
$^{223}_{89}Ac$	223.0191
$^{224}_{89}Ac$	224.0217
$^{225}_{89}Ac$	225.0231
$^{227}_{89}Ac$	227.0278
$^{223}_{90}Th$	223.0209
$^{224}_{90}Th$	224.0214
$^{225}_{90}Th$	225.0237
$^{226}_{90}Th$	226.0249
$^{227}_{90}Th$	227.0278
$^{228}_{90}Th$	228.0287
$^{229}_{90}Th$	229.0316
$^{230}_{90}Th$	230.0331
$^{232}_{90}Th$	232.0382
$^{226}_{91}Pa$	226.0278
$^{227}_{91}Pa$	227.0289
$^{228}_{91}Pa$	228.0310
$^{231}_{91}Pa$	231.0359
$^{234}_{91}Pa$	234.043
$^{227}_{92}U$	227.0309
$^{228}_{92}U$	228.0313
$^{229}_{92}U$	229.0332
$^{230}_{92}U$	230.0339
$^{231}_{92}U$	
$^{232}_{92}U$	232.0372
$^{233}_{92}U$	233.0395
$^{234}_{92}U$	234.0409
$^{235}_{92}U$	235.0439
$^{236}_{92}U$	236.0457
$^{238}_{92}U$	238.0508
$^{231}_{93}Np$	231.0383
$^{233}_{93}Np$	233.0406
$^{236}_{93}Np$	236.0466
$^{237}_{93}Np$	237.0480
$^{232}_{94}Pu$	232.0411
$^{234}_{94}Pu$	234.0433
$^{236}_{94}Pu$	236.0461
$^{237}_{94}Pu$	237.0483
$^{238}_{94}Pu$	238.0495
$^{239}_{94}Pu$	239.0522
$^{240}_{94}Pu$	240.0540
$^{242}_{94}Pu$	242.0587

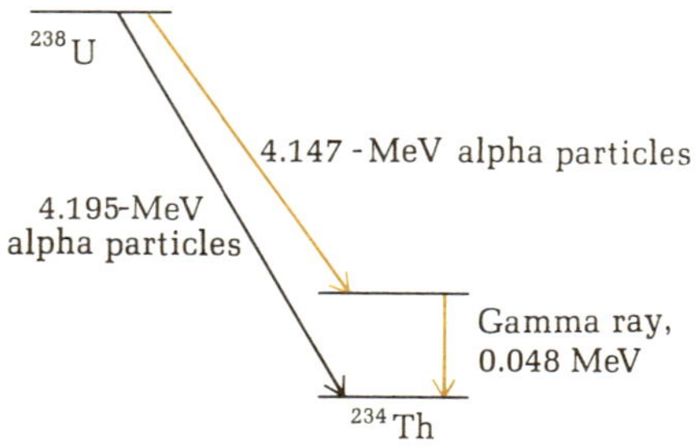

Figure 16-2 Alpha-particle decay from ^{238}U. Seventy-seven percent of the particles have an energy of 4.195 MeV and lead directly to the ^{234}Th nuclear ground state. But 23 percent of the particles have an energy of only 4.147 MeV, and their emission results in formation of the ^{234}Th excited state; decay of this to the nuclear ground state is accompanied by emission of a gamma ray having an energy of 0.048 MeV.

have an energy of 4.195 MeV and 23 percent have an energy of 4.147 MeV. It is easier to account for the production of alpha particles having a single energy than for those involving two or more energies. In the former case we simply conclude that the parent nucleus had one energy content, the product nucleus another, and that the energy of the alpha particle is equal to the difference between these. How can we generalize this idea to accord with the production of alpha particles from the same nucleus having different energies? In fact, we encountered an idea in Chapter 15, electronically excited states, that we can apply to explain this case. We learned that it is possible to obtain an electronically excited state for a molecule or atom through energy input in the form of a photon. In some cases the decay of that excited state back to the ground state occurs with the emission of a photon (fluorescence or phosphorescence). Just as there exist electronically excited states, so there also exist nuclear excited states. Thus, if radioactive decay were to yield a product nucleus in an excited state, the energy of the alpha particle emitted would be less than the maximal amount by the energy difference between the nuclear excited and ground states. Diagrammatically, this situation is shown in Figure 16-2 for the case of ^{238}U decay. Just as the decay of electronically excited states can occur with emission of a photon, so can the decay of nuclear excited states; photons emitted in this process are called **gamma rays.** Gamma rays are not essentially different from any photon but have very short wavelengths indeed (see Figure 14-8).

Example *Calculate the wavelength of a gamma ray emitted during the decay of the nuclear excited state of* ^{234}Th *(see Figure 16-2).*

Solution *The difference in energy between the nuclear excited and ground states is* $4.195 - 4.147$ MeV $= 0.048$ MeV*. This, then, is the energy of the* *gamma ray emitted. We can convert this to the more usual energy unit of ergs by employing the conversion factor in Table 16-1:*

$$0.048 \text{ MeV} \times 1.602 \times 10^{-6} \text{ erg/MeV} = 7.68 \times 10^{-8} \text{ erg}$$

We can now apply the Planck equation (14-3) to obtain the wavelength:

$$E = h\nu = \frac{hc}{\lambda}$$

$$\lambda = \frac{hc}{E} = \frac{6.62 \times 10^{-27} \text{ erg sec} \times 3 \times 10^{10} \text{ cm/sec}}{7.68 \times 10^{-8} \text{ erg}}$$

$$\boldsymbol{\lambda = 2.59 \times 10^{-9}} \textbf{ cm}$$

Beta decay

Another common mode of radioactive decay occurs with the emission of an electron, or beta particle. The general equation for **beta decay** is

$${}^{m}_{n}\text{X} \longrightarrow {}^{m}_{n+1}\text{Y} + {}^{0}_{-1}\text{e} \qquad (16\text{-}4)$$

Since the mass of the electron is small compared to that of the proton or neutron, beta decay does not result in an appreciable change in atomic mass; so the atomic mass number is unchanged (a small change in mass does occur in these reactions and will be discussed later). The atomic number actually *increases* by one in this process. Note that this is necessary to balance Equation (16-4) with respect to charge: The emission of an electron requires that the number of protons increase by one. Thus, the basic nuclear process that has occurred is

$${}^{1}_{0}\text{n} \longrightarrow {}^{1}_{1}\text{p} + {}^{0}_{-1}\text{e} \qquad (16\text{-}5)$$

A nuclear neutron is converted into a nuclear proton and the electron which escapes. A common example of an isotope that decays by emission of beta particles is provided by ^{14}C:

$${}^{14}_{6}\text{C} \longrightarrow {}^{14}_{7}\text{N} + {}^{0}_{-1}\text{e} \qquad (16\text{-}6)$$

In some respects the energetics of beta decay and alpha decay are related and in others they are distinct. Like alpha decay, beta decay can yield one or more nuclear excited states and so be accompanied by gamma radiation. Like alpha decay, beta decay occurs with emission of a particle of variable energy depending on the nature of the emitting nucleus. But there is a central distinction: Beta decay of a single type of nucleus occurs with emission of particles having a broad, continuous spectrum of energies, quite unlike alpha decay. Now, this observation just does not fit into our picture of the energetics of radioactive decay (Figure 16-2) and appears to violate the conservation of energy principle. In order to avoid violation of this law

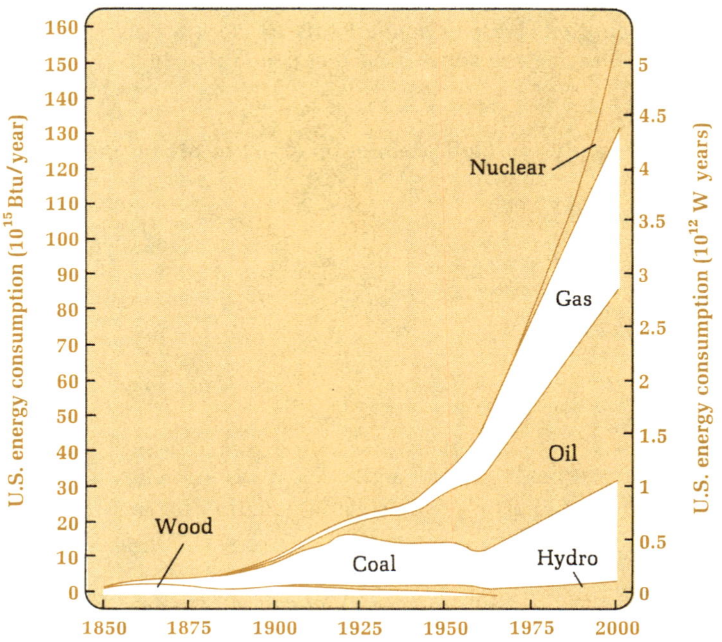

This figure traces out some features of the history of energy utilization in the United States. Energy consumption has multiplied some 30 times since 1850, when wood supplied more than 90 percent of all the energy units. By 1900, coal had become the dominant fuel, accounting for more than 70 percent of the total. Fifty years later, coal's share had dropped to 36.5 percent, and the contribution from oil and natural gas had climbed to 55.5 percent. In 1970, coal accounted for 20.1 percent of all energy consumed, oil and gas 75.8 percent, hydropower 3.8 percent, and nuclear energy 0.3 percent. Energy consumption figures are from the U.S. Bureau of Mines. [Chauncey Starr, "Energy and Power," **225,** 37 (September, 1971).]

of physics, it was necessary to postulate the existence of a new particle, the **neutrino,** considered to be emitted simultaneously with the beta particle and to have the necessary energy to account for energy conservation. The neutrino was a difficult particle to identify, since it was considered to have zero charge and zero mass; nevertheless, it has been identified.

Positron emission

Positron emission is closely related to beta decay. The **positron** is a particle identical to the electron in all respects except that it bears a positive unit charge rather than a negative one. Many isotopes decay by the emission of a positron according to the general equation

$$^{m}_{n}\text{X} \longrightarrow {}^{m}_{n-1}\text{Y} + {}^{0}_{+1}\text{e} \tag{16-7}$$

As in the case of beta decay, the overall atomic mass remains almost unchanged. But in this process the atomic number of the product atom is *decreased* by one from that of the reactant. This again reflects the nature of the actual nuclear event and conserves charge:

$$^{1}_{1}p \longrightarrow {}^{1}_{0}n + {}^{0}_{+1}\text{e} \tag{16-8}$$

In this case it is a nuclear proton that decays, generating a neutron and the emitted positron.

Orbital electron capture

For the sake of completeness, we note that the final mode of radioactive decay is the **capture of an orbital electron** by the nucleus. The orbital electron combines with a nuclear proton to form a neutron. Thus, the general equation is

$$^{m}_{n}\text{X} \longrightarrow {}^{m}_{n-1}\text{Y} \tag{16-9}$$

Note that the atomic number decreases by one, just as in the case of positron emission.

Example *Complete each of the following reactions:*

a. $^{15}_{8}\text{O} \xrightarrow{\text{orbital electron capture}} \text{X}$
b. $\text{X} \longrightarrow {}^{203}_{82}\text{Pb} + {}^{4}_{2}\text{He}$
c. $^{207}_{84}\text{Po} \longrightarrow \text{X} + {}^{0}_{+1}\text{e}$

Solution *In reaction (a) orbital electron capture occurs with conversion of a nuclear proton to a neutron. Hence, the atomic mass is unchanged, but the atomic number decreases by one. Thus, the product must be* $^{15}_{7}\text{N}$. *In reaction (b) an alpha particle is emitted. Therefore, the parent atom must have had an atomic mass four units greater and an atomic number two units greater than the product. Hence, the parent must have been* $^{207}_{84}\text{Pa}$. *In reaction (c)* positron emission conserves atomic mass but decreases *atomic number by one. It follows that* X *is* $^{207}_{83}\text{Bi}$.

Now we have acquired some of the basic language of nuclear reactions, and we can move on to a consideration of solar chemistry.

NUCLEAR CHEMISTRY IN THE SUN

We shall find that in several respects the chemistry of the sun is quite distinct from the chemistry with which we are accustomed to dealing in the laboratory. To begin with, the sun does chemistry on a monumental

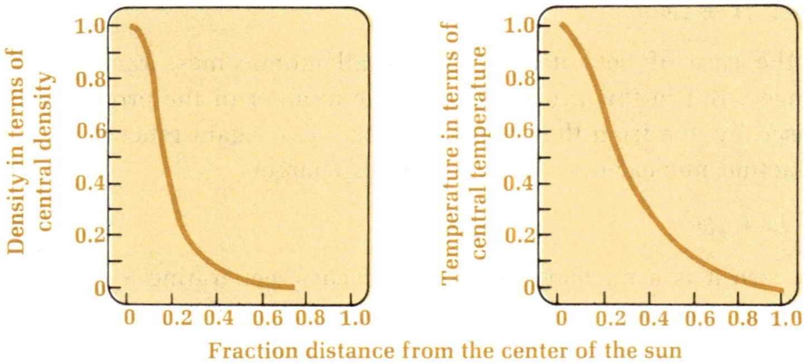

Figure 16-3 Density and temperature inside the sun as a function of distance from its center.

scale. Calculations of astronomers reveal that the sun weighs about 2×10^{33} g and is approximately 1.39×10^{11} cm in diameter. A simple calculation shows that the volume of the sun is about 1.4×10^{33} cm^3 (we recall that $V = \frac{4}{3}\pi r^3$). To give these numbers more meaning, we may note that the volume of the sun is about 300,000 times that of the earth. Knowing the volume of the sun and its mass, we can calculate its density:

$$\text{density} = \frac{\text{mass}}{\text{volume}} = \frac{2 \times 10^{33}\ \text{g}}{1.4 \times 10^{33}\ \text{cm}^3} = 1.4\ \text{g/cm}^3 \qquad (16\text{-}10)$$

Thus, the sun is about 40 percent more dense, on the average, than water, which has a density of 1 g/cm^3. This rather high density, when combined with knowledge of the composition of the sun, leaves us with a problem. Specifically, about half of the atoms of the sun are hydrogen, and a good deal of the remainder are helium. But under normal conditions, the density of hydrogen gas is only about one-millionth that of water. At the surface of the sun, a density similar to this must prevail; It follows that the interior of the sun must be much denser than the surface. In fact, the density of the sun near its center is about 50 times that of water. Such a density is greater than that for our heaviest metals; the density of lead, for example, is only 19 times that of water. How is it possible for hydrogen to be denser than lead?

To understand the density of the sun, we must understand something about its temperature. At the surface of the sun the temperature is a modest 6000°C, but this value rises to near 10 million °C at the sun's center. In Figure 16-3, both the density and temperature of the sun as a function of the distance out from the center are shown. Note that the two quantities parallel each other, suggesting perhaps that one accounts for the other.

Specifically, we speculate that the increase in temperature accounts for the increase in density. At temperatures characteristic of the interior of the sun, the thermal energies are so great that virtually all of the electrons are stripped from their nuclei. Hence, much of the sun resembles a sea of bare protons and rapidly moving electrons. This has an important consequence when we recall that almost all of the volume of an atom is occupied by the electron cloud, not by the nucleus (Chapter 15). A typical atomic radius is 10^{-8} Å, but a typical nuclear radius is about 10^{-12} Å. Hence, by stripping off the electrons from the nucleus, it is possible to pack matter at an exceedingly high density. There are, for example, some stars, known as *white dwarfs,* that have densities on the order of 1 million g/cm^3; 1 in.^{3} of such a star weighs about 18 tons. (There are other stars, neutron stars, which are still more dense.) We certainly do not require such high densities to explain the density of the sun; consequently, we may think of the sun as a "nuclear gas" rather than, for example, a "nuclear solid." Another simple calculation will require that the chemistry involved in energy production in this nuclear gas be something out of the ordinary. We turn now to that matter.

The energy emitted by the sun is about 4×10^{33} ergs/sec. This corresponds to about 2 ergs g^{-1} sec^{-1}. If the sun produced energy at this rate by a normal combustion process, it would soon burn itself out. It is known that the sun has existed for at least 5 billion years; the energy produced by the sun in this period exceeds by a factor of more than 1 million any known *chemical* process of energy production. Energy generation in the sun actually arises from **nuclear reactions.** Let us consider some of these in more detail.

Under the conditions that exist in the sun's interior, nuclear fusion reactions such as

$$2\,{}^{1}_{1}\text{H} + 2\,{}^{1}_{0}n \longrightarrow {}^{4}_{2}\text{He} \qquad (16\text{-}11)$$

can occur with the liberation of large amounts of energy. [Equation (16-11) should be taken as representative of an overall process and not a precise description of a single reaction. We shall examine it in more detail later.] Let us calculate the mass change for this reaction. The mass of the proton and its electron [taken together, since they are usually tabulated that way (see Table 16-2) even though they are dissociated in the sun's interior] is 1.007826 amu, that of the neutron is 1.008665 amu; the product helium atom has a mass of 4.002599 amu. We can then calculate for production of 1 mole of helium:

2H	$2 \times 1.007826 =$	2.015652
2n	$2 \times 1.008665 =$	2.017330
	total mass $=$	4.032982
	less one He atom $=$	4.002599
		0.030383 amu

Thus, we find that this reaction occurs with the loss of a small amount of mass. The energy released by a reaction in which mass is lost may be calculated from the Einstein relationship

$$E = mc^2 \tag{16-12}$$

where m is the mass loss and c is the speed of light. For the conversion represented in Equation (16-11), we calculate

$$\begin{aligned} E &= 0.030383 \text{ amu} \times (1.66024 \times 10^{-24} \text{ g/amu}) \times (3 \times 10^{10} \text{ cm/sec})^2 \\ &= 4.48 \times 10^{-5} \text{ g cm}^2/\text{sec}^2 = 4.48 \times 10^{-5} \text{ erg} \end{aligned}$$

It will prove convenient to express this energy change in other units as well. Using the conversion factors in Table 16-1, we find

$$\begin{aligned} E &= 4.48 \times 10^{-5} \text{ erg} \times 6.24 \times 10^{11} \text{ eV/erg} \\ &= 28.29 \times 10^{6} \text{ eV} = 28.29 \text{ MeV} \\ &= 6.52 \times 10^{8} \text{ kcal/mole} \end{aligned}$$

The enormity of this energy may be appreciated by recalling that the oxidation of 2 moles of hydrogen to water will release only 116 kcal of energy. The fusion reaction releases more than 5 million times as much energy. Correspondingly, for equal rates of energy release, if the sun's mass were consumed by chemical reaction in a few thousand years, nuclear energy release can readily be sustained for some tens of billions of years.

Example *If the sun has been producing 4×10^{33} ergs/sec for 3×10^{9} years, how much mass has been consumed during this time?*

Solution *The total energy produced is 4×10^{33} ergs/sec $\times\ 3 \times 10^{9}$ years $\times\ 3.2 \times 10^{7}$ sec/year $= 3.84 \times 10^{50}$ ergs. Using Einstein's equation (16-12),*

$$m = \frac{3.84 \times 10^{50} \text{ ergs}}{9 \times 10^{20} \text{ cm}^2/\text{sec}^2} = 0.43 \times 10^{30} \text{ g}$$

Granted that this is an enormous mass, it still represents only about one part in 5000 of the total mass of the sun. We may confidently expect that the sun will continue to shine for quite some time yet.

Even under extreme conditions, such as those that exist in the interior of the sun, it is very unlikely that a reaction involving four particles, such as that of Equation (16-11), will occur in a single step. Reactions involving many particles almost always occur by a mechanism involving several successive reactions, each of which involves not more than two particles (the same thing is true even for simple chemical reactions of the usual kinds). Two proposals have been advanced to provide pathways for the synthesis of helium by the fusion of hydrogen nuclei, a reaction nearly identical with that of Equation (16-11). The simpler of the two was first proposed by Critchfield and is known as the **H–H reaction.** In this process the first step is the fusion of two protons to form a deuterium nucleus and a positron:

$$^{1}_{1}H + ^{1}_{1}H \longrightarrow ^{2}_{1}D + ^{0}_{+1}e + \text{neutrino} \qquad (16\text{-}13)$$

which is followed in turn by two reactions leading to formation of the helium 4 nucleus:

$$^{2}_{1}D + ^{1}_{1}H \longrightarrow ^{3}_{2}He + \text{gamma ray}$$

$$^{3}_{2}He + ^{3}_{2}He \longrightarrow ^{4}_{2}He + 2^{1}_{1}H \qquad (16\text{-}14)$$

$$\text{sum:} \quad 4^{1}_{1}H \longrightarrow ^{4}_{2}He + 2^{0}_{+1}e \qquad (16\text{-}15)$$

Overall, then, four protons are converted into one helium with the release of two positrons.

The sum of the reaction in Equations (16-13) through (16-15) is called a **nuclear fusion** reaction, since several nuclei have been fused into one. This sequence of reactions is the only important contributor to energy generation in our own sun. However, this is not the only important reaction sequence (or even a very important one) in many hotter stars. In these, a more complicated series of reactions proposed by Hans Bethe and Carl von Weizsacker is the important generator of energy. These are collectively known as the **carbon cycle** and are presented in the following equations:

$$^{12}_{6}C + ^{1}_{1}H \longrightarrow ^{13}_{7}N + \text{gamma ray} \qquad (16\text{-}16)$$

$$^{13}_{7}N \longrightarrow ^{13}_{6}C + ^{0}_{+1}e \qquad (16\text{-}17)$$

$$^{13}_{6}C + ^{1}_{1}H \longrightarrow ^{14}_{7}N + \text{gamma ray} \qquad (16\text{-}18)$$

$$^{14}_{7}N + ^{1}_{1}H \longrightarrow ^{15}_{8}O + \text{gamma ray} \qquad (16\text{-}19)$$

$$^{15}_{8}O \longrightarrow ^{15}_{7}N + ^{0}_{+1}e \qquad (16\text{-}20)$$

$$^{15}_{7}N + ^{1}_{1}H \longrightarrow ^{12}_{6}C + ^{4}_{2}He \qquad (16\text{-}21)$$

$$\text{sum:} \quad \mathbf{4^{1}_{1}H \longrightarrow ^{4}_{2}He + 2^{0}_{+1}e}$$

Note that the overall reaction sum is precisely the same in this case as in the simpler H–H reaction: In the carbon cycle the heavier elements act as catalysts to facilitate the nuclear fusion reaction.

Let us analyze the reactions of the carbon cycle in more detail. Four of the reactions, (16-16), (16-18), (16-19), and (16-21), are nuclear fusion reactions in which protons fuse with heavier nuclei. In the first three of these reactions, nuclear fusion is accompanied by emission of a gamma ray, the result of the production of the new nucleus in an excited state as we have already noted. The other two reactions in the series, (16-17) and (16-20), involve radioactive decay by positron emission from isotopes that are not common in nature. A common feature of all radioactive decay processes is that they obey a decay law similar to that for first-order kinetics (Chapter 9). For example, for the process indicated in Equation (16-17), the kinetic law is

$$\text{rate of disappearance of } ^{13}N = \lambda(^{13}N) \qquad (16\text{-}22)$$

where λ is the radioactive decay constant and (^{13}N) indicates the number of ^{13}N atoms present. One characteristic of all first-order processes is that the time required for one-half of the material present at any instant to undergo reaction is simply 0.693/rate constant. [For readers with a background of calculus, the proof of this relation is easy: Integration of the rate equation (16-22) leads to the relation $^{13}N/^{13}N_0 = e^{-\lambda t}$ where $^{13}N_0$ is the number of ^{13}N atoms present at the beginning and ^{13}N is the number remaining at any time *t*. If we wait until one-half of the atoms have decayed, we get $\frac{1}{2} = e^{-\lambda t}$. Taking the logarithm and rearranging this becomes $\text{time}_{1/2} = \ln 2/\lambda = 0.693/\lambda$.]

The **half-lives** of both the nitrogen and oxygen isotopes are short, 10.1 and 2.1 min, respectively. Thus, for a given number of original ^{15}O atoms, after 2.1 min only half that number remain. After another 2.1 min, only half of the remaining atoms are still present, and so on. After an integral number of half-lives, *n*, the fraction of the original atoms remaining is $(\frac{1}{2})^n$. In the case of ^{15}O, after only 21 min only $(\frac{1}{2})^{10} = 1/1024$ of the original remains. Of course, in stars both the nitrogen and oxygen isotopes are continually produced by the carbon cycle at the same time that the ones produced earlier decay. Eventually, a *steady state* is reached where the number of atoms formed just equals the number that decay and no net change in number occurs.

Example *Suppose that for a particular isotope, 12.5 percent of an original sample remains after a period of 6 min. What is the half-life for this isotope?*

Solution *From the above discussion, we know that*

$$\text{fraction of original sample remaining} = (\tfrac{1}{2})^n$$

In our case,

$$\tfrac{1}{8} = (\tfrac{1}{2})^n \quad \text{or} \quad n = 3$$

If 6 min is three half-lives, it follows that 2 min is one half-life.

It is important to be able to make simple calculations of energy changes accompanying nuclear fusion (or fission, for that matter) reactions correctly. For the most part, these are quite straightforward. The following set of rules will help you to understand the calculations and to avoid making mistakes.

CALCULATION OF MASS CHANGES IN NUCLEAR REACTIONS

1. The mass quantities usually employed in these calculations are (a) that of the *neutron* and (b) that of the *proton plus electron*. Addition of these masses gives values that may be compared directly with the masses of atoms, which are experimentally determined and tabulated values. These are collected in Table 16-2. Even in the case of reactions involving nuclei from which the electrons have been stripped, it is convenient and customary to include the electronic mass. This leads to no real difficulties, as developed below.

2. In making these calculations it is important to compare carefully the actual products and reactants with the masses employed. For example, for the spontaneous decay of carbon 14:

   $$^{14}_{6}C \longrightarrow {}^{14}_{7}N^+ + {}^{0}_{-1}e$$

   we can calculate the mass change directly from the mass of a $^{14}_{6}C$ *atom* and that for a $^{14}_{7}N$ *atom*. To understand this point, note carefully that the reaction products include a nitrogen *ion*, which contains only *six* orbital electrons rather than the seven required to attain electrical neutrality. The electron emitted in the reaction, then, provides the total mass of a $^{14}_{7}N$ atom. With the exception of those processes involving positron emission, all similar calculations can be made simply on the basis of *atom* masses.
3. For reactions in which a positron is emitted, we must be careful to examine the reaction products in particular detail. For example, Equation (16-13) is of this type:

   $$^{1}_{1}H + {}^{1}_{1}H \longrightarrow {}^{2}_{1}D^- + {}^{0}_{+1}e + \text{neutrino} \qquad (16\text{-}13)$$

   On the reactant side, we have two hydrogen atoms represented (which really react as hydrogen nuclei, see point 4 below) and on the product side a deuterium *ion*, which contains *one extra orbital electron* (two rather than one) plus a positron. Hence, the products contain two electronic masses in addition to the mass of a deuterium atom. These two electronic masses must be accounted for explicitly in making the mass-change calculation. Since each hydrogen has a single orbital electron, two are carried to the deuterium atom. The positron has its source in the reaction within a nucleus:

   $$\text{proton}^+ \longrightarrow \text{neutron} + \text{positron}^+$$

4. Finally, let us note that calculating the mass change of a reaction such as (16-13) may be done perfectly well on the basis of *atomic* weights (proton plus electron) even though it is the *isolated nuclei* that, in fact, react. If we look at the reaction exactly as it occurs, then two protons react to form a deuterium nucleus (proton plus neutron) plus a positron. We can certainly do the mass-change calculation on the basis of these particles. If we choose to use atomic weights, this simply involves adding two electrons to both sides of the equation: When the difference is taken, the mass of these electrons cancels out exactly. Hence, we shall get precisely the same quantitative result. If we are careful to see that *charges are balanced* in our equations, no problems will occur in energy calculations.

Example *Calculate the energy released, in million electron volts, for the reaction*

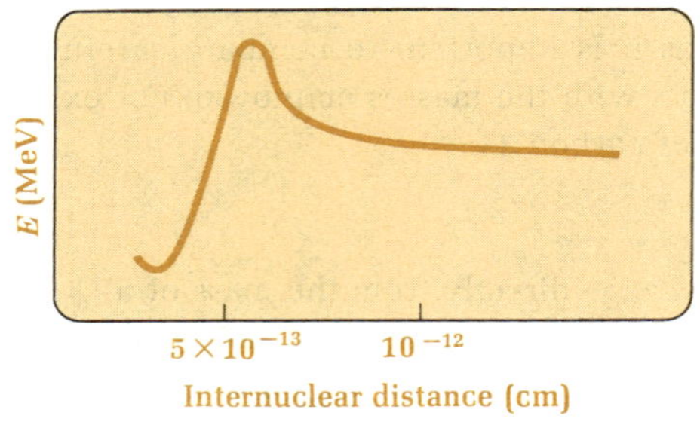

Figure 16-4 Electrostatic repulsion energy between two nuclei as a function of the distance between them.

$^{15}_{7}N + ^{1}_{1}H \longrightarrow ^{12}_{6}C + ^{4}_{2}He$

Solution *We need, first of all, to calculate the masses of the reactants and products. Referring to Table 16-2, we have*

Reactants		Products	
$^{15}_{7}N$ mass =	15.00011	$^{12}_{6}C$ mass =	12.00000
$^{1}_{1}H$ mass =	1.007325	$^{4}_{2}He$ mass =	4.00260
sum:	16.007435	sum:	16.00260

We now calculate the difference between the two to obtain the decrease in mass:

$$\begin{array}{r} 16.007435 \\ \underline{16.00260} \\ 0.004835 \text{ amu} \end{array}$$

Finally, we can convert this change in mass to energy by employing Einstein's equation, $E = mc^2$. This is most simply done by using the conversion factor in Table 16-1:

$$1 \text{ amu} = 931.4 \text{ MeV}$$

It follows that

$$\begin{aligned} \text{energy released} &= (0.004835 \times 931.4) \text{ MeV} \\ &= 4.5 \text{ Mev} \end{aligned}$$

THERMONUCLEAR REACTIONS ON EARTH

The reactions that take place in the stars are highly desirable sources of energy for man on earth. The reactions described, fusion of nuclei to form heavier elements, have been imitated by man principally in the development of hydrogen bombs. So far his efforts to control these reactions to produce useful forms of energy have not been strikingly successful, although progress is being made and the practical problems involved might be solved by the end of the 1970s. Let us examine the problem somewhat more carefully.

In general, the fusion of two nuclei to form a third is a process that must bring the two charged nuclei close enough so that eventually they fuse to form a single product nucleus. The laws of classical physics show that the Coulombic repulsion to be expected is given by the equation

$$\frac{Z_1 Z_2 e^2}{R^2} = \text{energy} \tag{16-23}$$

where Z_1 and Z_2 are the number of positive charges on the two nuclei, R is the distance between them, and e is the electronic charge. If we postulate that the two nuclei must come together until they are touching before fusion occurs, R is the nuclear diameter and has a value of about 5×10^{-13} cm. The charge on the electron is 4.8×10^{-10} esu; placed into the proper units, we can calculate that the Coulombic energy at the touching distance is

$$E(\text{Coulombic}) = 0.28 Z_1 Z_2 \text{ MeV} \tag{16-24}$$

A plot of the Coulombic repulsion energy as a function of distance between the two nuclei is shown in Figure 16-4. Once the two nuclei are fused into the new element, the mass difference between reactants and products is converted into energy as the system drops to the stable energy level of the new nucleus.

From Equation (16-24) we can see why the carbon cycle is important only in hot stars. Compare only the first steps in each sequence. Reaction (16-13) involves fusion of two hydrogen nuclei with unit charges on each; the Coulombic energy at the touching distance is 0.28 MeV. In contrast, for the carbon cycle (16-16), the first step involves one nucleus of unit charge and a second with six positive charges. Therefore, the energy required to initiate fusion is six times as great for the carbon cycle as for the H–H reaction and becomes even greater in subsequent steps. The energy to overcome the Coulombic barrier to nuclear fusion must be supplied principally by the kinetic energy of the individual nuclei. Only if the temperature of a star is greater than that of our sun will a large number of the atoms have sufficient energy to cross the barrier to effect the carbon cycle reactions.

Reactions between nuclei are actually somewhat more complicated than has been implied in our discussion so far. For example, it is not true that the kinetic energies of the nuclei are the only important factors in determining the probability of a successful fusion reaction. An additional factor that affects this probability is called the reaction **cross section.** This measures the probability of having a successful (leading to fusion) collision between two nuclei. In part, this depends on the size of the particles. After all, it is more likely that two basketballs thrown into the air will collide than it is that two baseballs will collide, simply because the former sweep out larger pathways in space. In addition, cross sections for nuclear reactions depend on particle energies, a matter that we return to below when nuclear fission is discussed.

Direct fusion of hydrogen nuclei is unsuitable for the production of useful power on earth even though it is energy releasing. Thus, reactions of the type

$$^{1}_{1}H + ^{1}_{1}H \longrightarrow ^{2}_{1}D \qquad (16\text{-}25)$$
$$^{1}_{1}H + ^{2}_{1}D \longrightarrow ^{3}_{2}He \qquad (16\text{-}26)$$
$$^{1}_{1}H + ^{3}_{1}T \longrightarrow ^{4}_{2}He \qquad (16\text{-}27)$$

are energy releasing, but the cross section for reaction is too small for the reaction to be self-sustaining at temperatures that can be obtained in the laboratory. On the other hand, reactions of deuterium or tritium atoms such as

$$^{2}_{1}D + ^{2}_{1}D \longrightarrow ^{3}_{2}He + ^{1}_{0}n + 3.27\text{ MeV} \qquad (16\text{-}28)$$
$$^{2}_{1}D + ^{2}_{1}D \longrightarrow ^{3}_{1}T + ^{1}_{1}H + 4.03\text{ MeV} \qquad (16\text{-}29)$$
$$^{2}_{1}D + ^{3}_{1}T \longrightarrow ^{4}_{2}He + ^{1}_{0}n + 17.6\text{ MeV} \qquad (16\text{-}30)$$
$$^{2}_{1}D + ^{3}_{2}He \longrightarrow ^{4}_{2}He + ^{1}_{1}H + 18.3\text{ MeV} \qquad (16\text{-}31)$$

are self-sustaining at much lower temperatures, but still near 50 million °C, and in principle offer a useful source of energy. The typical problems that presently face development of fusion for energy purposes are simply these: The nuclear reactions are not self-sustaining at low temperatures, and no suitable way has yet been found for containing the materials at high temperatures long enough to take energy from the system. The energy required to hold the particles together exceeds the energy that is recovered from the reaction. There are, in fact, solutions to this problem, and it is similar to the problem of combustion in ordinary internal combustion engines. The temperature of the ignited flame in an automobile is somewhere in the vicinity of 2000°K, but the outside of the motor is maintained at temperatures near the boiling point of water by adequate cooling means. Useful energy is generated, but the motor does not become overly hot. The temperatures of **thermonuclear reactions** are so much greater that the cooling problems are correspondingly magnified, but the problems to be solved are similar in principle.

The importance of finding a means of usefully extracting energy from thermonuclear reactions on earth can scarcely be overemphasized. The demands of man for more and more energy have escalated dramatically in recent years and will, in the United States, probably double during the next decade. Production of the amounts of energy that we currently consume has created grave problems of environmental pollution and natural resources depletion. Not only do our most important typical fuels, coal and oil, generate atmospheric pollutants when they undergo combustion, but their supply on earth is being rapidly consumed. Clearly, we must not depend on them forever. Falling water can be used to generate electricity and frequently is. However, this source too has its limitations, particularly since the construction of additional hydroelectric plants threatens some of America's most beautiful unspoiled country. Finally, as is developed below, nuclear fission reactions, already rather widely used, create problems of thermal pollution and of disposal of radioactive wastes. Thermonuclear reactions alone, among possible energy supplies on the horizon, appear to have the promise of supplying sufficient energy to meet our demands without, at the same time, rendering the earth, in one way or another, less fit for man to live in.

SYNTHESIS OF THE ELEMENTS

Since the whole field of chemistry is concerned with the one-hundred-plus elements, we might pause at this point to see what can be said concerning their origin. It is estimated that approximately 93 percent of all atoms in the universe are hydrogen atoms and most of the remaining 7 percent are helium. All heavier elements combined account for much less than 1 percent of the total. Since helium is heavier than hydrogen, on a weight basis it accounts for about 23 percent of the universe and hydrogen for about 76 percent. Most of the helium of the universe was formed very early in its history (see below). In addition, we have just seen in the sections above that in the stars hydrogen is continuously converted to helium. Such processes also account for some part of the total helium content of the universe.

Theories concerning the synthesis of the elements of necessity must consider the problem of the creation of the universe. This fascinating problem has received the attention of some of the best minds of man. (Not only "modern" man is interested in this problem, of course; primitive societies generally have some concept of "the beginning.") Relatively recently it has been suggested that perhaps there was no moment of creation at all and that, in fact, matter is being continuously created even now. Older theories start with the assumption of a single creation. In either event, hydrogen is considered to be the primitive element from which other elements have been derived.

Questions concerning the creation of the universe and its subsequent history fall into the science of **cosmology.** By its very nature, cosmology is unique among the sciences, and there are those who dispute that it is a science at all. After all, do not sciences rest their claims to truth on their ability to repeat and reproduce experimental evidence? There are very few important measurements in physics, chemistry, or biology, among other sciences, that have not been repeatedly checked and rechecked to ensure that theory coincides, as nearly as we can judge, with reality. But the cosmologist cannot repeat the past, and he cannot check his results against

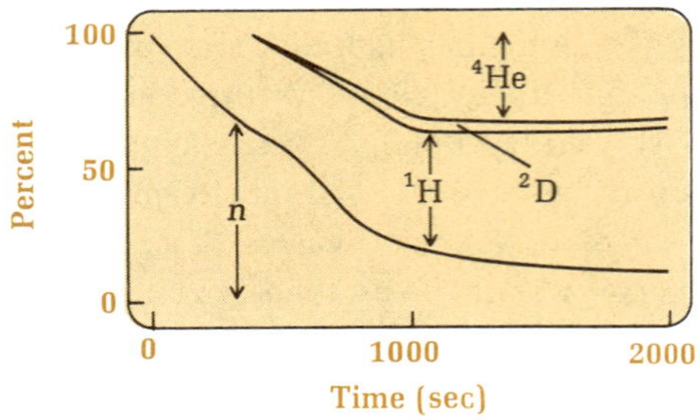

Figure 16-5 The concentrations of neutrons, protons, deuterons, and helium nuclei during the first half-hour of the expanding universe.

those obtained in other universes. He deals with a unique object, the universe, and with a series of unique events. Perhaps the wonder of this area of study is that any plausible theories can be suggested at all that are subject to some experimental check. Although we cannot repeat the past, the past has left its traces; with the aid of these, some imagination, and some theoretical calculations, cosmologists have been able to develop reasonable and coherent theories of the creation and development of the universe. As noted above, not all cosmologists are of one mind in their thinking; further study will undoubtedly bring additional possibilities to mind.

According to the "primeval fireball" or "big bang" theory, the starting material for the universe consisted of a soup of elementary particles and intense radiation in thermal equilibrium at exceedingly high temperatures, $> 10^{10}$°K. Following some initial event (the "big bang"), the universe began to expand and cool down. It is still expanding and cooling down, some 10 billion or so years later. As the cooling process continues, the types of nuclear processes that can occur change very rapidly. The results of some calculations concerning the evolution of elementary particles and light elements during the first half-hour following the big bang are presented in Figure 16-5. Note that after about 30 min, the universe became more or less stable at the distribution of elements that still exists. That is, almost everything was hydrogen and helium, and these are calculated to be present in approximately their correct abundances. From here on, whatever creation theory we may use must account for the same set of facts concerning the synthesis of the heavier elements.

Creation of the heavier elements is now thought to have occurred in a succession of reactions and, actually, in a succession of stars. When first brought together from dispersed material, a star is relatively cool and derives most of its energy from the H–H reaction. Gradually, the hydrogen is consumed, and the core of the star becomes more and more concentrated in helium. The H–H reaction occurs in the shell surrounding the core. As the reaction proceeds, the star contracts more and more, and the temperature of the core is raised until a reaction called **helium burning** can begin. In this process a series of nuclear fusion reactions occurs, accounting for the synthesis of light elements perhaps up to ^{24}Mg.

$$2\,^{4}_{2}He \longrightarrow \,^{8}_{4}Be \tag{16-32}$$

$$^{4}_{2}He + \,^{8}_{4}Be \longrightarrow \,^{12}_{6}C \tag{16-33}$$

$$^{4}_{2}He + \,^{12}_{6}C \longrightarrow \,^{16}_{8}O \tag{16-34}$$

$$^{4}_{2}He + \,^{16}_{8}O \longrightarrow \,^{20}_{10}Ne \tag{16-35}$$

$$^{4}_{2}He + \,^{20}_{10}Ne \longrightarrow \,^{24}_{12}Mg \tag{16-36}$$

This sequence of reactions is very appealing from the chemist's point of view. Specifically, the isotopes produced in the sequence are all very common in nature, with the single exception of $^{8}_{4}Be$; that exception requires further comment.

The $^{8}_{4}Be$ isotope can be produced in the laboratory by a number of methods. Whenever it is made, it promptly splits into two alpha particles with the release of 94 keV of energy. Even though the decay of $^{8}_{4}Be$ is very rapid, in the very hot and dense core of a star it may be produced sufficiently rapidly so that moderate quantities of this isotope are present at all times. Fusion with $^{4}_{2}He$ then yields carbon atoms [Equation (16-33)], and the remainder of the chain is established. Some indirect evidence for the occurrence of Equation (16-33) is available from what might appear to be an unlikely source: the study of fast-moving deuterons in collision with ^{11}B atoms. The observed reactions are

$$^{2}_{1}D + \,^{11}_{5}B \longrightarrow \,^{12}_{5}B + \,^{1}_{1}H \tag{16-37}$$

$$^{12}_{5}B \longrightarrow \,^{12}_{6}C^{*} \qquad t_{1/2} = 0.02\text{ sec} \tag{16-38}$$

$$^{12}_{6}C^{*} \longrightarrow \,^{12}_{6}C + \text{energy} \tag{16-39}$$

$$^{12}_{6}C^{*} \longrightarrow 3\,^{4}_{2}He \tag{16-40}$$

where C* is an excited state of the carbon nucleus. Note that the excited carbon nucleus in small part loses energy to become an ordinary carbon atom [in reaction (16-39)], but the majority of such nuclei break down into alpha particles as in reaction (16-40). It is precisely on this reaction and the preceding one that we need to focus our attention: Reading Equations (16-40) and (16-39) backwards is equivalent to the synthesis of ^{12}C by the helium-burning sequence [Equations (16-32) and (16-33)]. This is very important when we realize that a fundamental scientific law states that reactions possible in one direction are also possible in the reverse direction. That is, the observation of decomposition of ^{12}C into helium atoms implies the existence of a route for the synthesis of ^{12}C from helium atoms. This provides supporting evidence for the synthesis of several light elements by the helium-burning reactions.

In any event, helium burning seems to be a suitable source for many of the light nuclei but is unlikely to form elements with atomic numbers much

The Crab Nebula, the remains of a supernova explosion, was first detected and recorded by Asian astronomers in 1054. It was not until the eighteenth century that this structure was observed in the Western world. (Courtesy of Hale Observatories.)

beyond that for Mg. With the synthesis of each succeeding element, the Coulombic repulsion given in Equation (16-24) becomes larger for the next step. Eventually, we reach the stage where even the temperature at the interior of the star is insufficient to supply the necessary kinetic energy to overcome the barrier.

When much of the helium has been converted to heavier elements, the star contracts still further and the temperature once more increases significantly. Many of the lightweight atoms synthesized in the helium-burning process begin to fuse in reactions yielding heavier nuclei releasing many neutrons, protons, and alpha particles. Typical reactions are

$$^{12}_{6}C + ^{12}_{6}C \longrightarrow ^{23}_{11}Na + ^{1}_{1}H \qquad (16\text{-}41)$$

and

$$^{16}_{8}O + ^{16}_{8}O \longrightarrow ^{31}_{15}P + ^{1}_{1}H \qquad (16\text{-}42)$$

At this stage in star development, reactions become very rapid indeed. The temperature rises rapidly, and synthesis of elements having atomic numbers up to that of iron occurs. The increasing density of the star causes its contraction due to gravitational forces to accelerate. Eventually, the density of the star approaches the density of nuclear matter, enormous repulsion forces build up, the star explodes, and its matter is dispersed far and wide, eventually to be incorporated, along with much new hydrogen, into a new system to begin almost the same process once more but with significant differences. We shall turn to the differences in a moment.

The explosion of a star creates a *supernova,* an intensely bright cloud of luminous gas and dust. The first supernova to be observed, called the Crab Nebula, was reported by Asian astronomers in 1054. Such a supernova gave rise to our solar system something more than 5 billion years ago.

First let us see why elements heavier than iron are not important products of the synthesis process at this stage and in fact cannot be explained by the relatively straightforward process we have described. Earlier, we calculated the mass loss for the simple reaction written in Equation (16-11) and showed that the mass loss corresponded to a very large energy release. A continual theme in our studies has been that nature likes energy-releasing processes and dislikes energy-consuming ones. The same criteria apply for nuclear reactions, and if we can decide which reactions are most energy releasing, we can readily tell which will proceed. For each nucleus we may consider a hypothetical synthesis from the simplest building blocks, the proton (plus electron) and the neutron. Comparison of the mass of the building blocks with the mass of the resulting nucleus then gives a measure of the energy released as the reaction proceeds. As two examples, let us compare the energetics of the conversion of protons and neutrons to $^{11}_{5}B$, a typical light element, and $^{56}_{26}Fe$.

$^{11}_{5}B$		$^{56}_{26}Fe$	
$5p^+ + 5e^- = 5 \times 1.007826 =$	5.039130	$26p^+ + 26e^- = 26 \times 1.007826 =$	26.2034
$6n = 6 \times 1.008665 =$	6.051990	$30n = 30 \times 1.008665 =$	30.2599
sum =	11.091120	sum =	56.4633
$-\,^{11}B$ mass =	11.009305	$-\,^{56}Fe$ mass =	55.9349
mass loss =	0.081815	mass loss =	0.5284
energy (MeV) =	76.18	energy (MeV) =	492.03
energy/nucleon =	6.93	energy/nucleon =	8.78

The comparison clearly shows that if protons and neutrons are used to make iron in preference to boron, 1.85 MeV more energy is released *per nucleon*

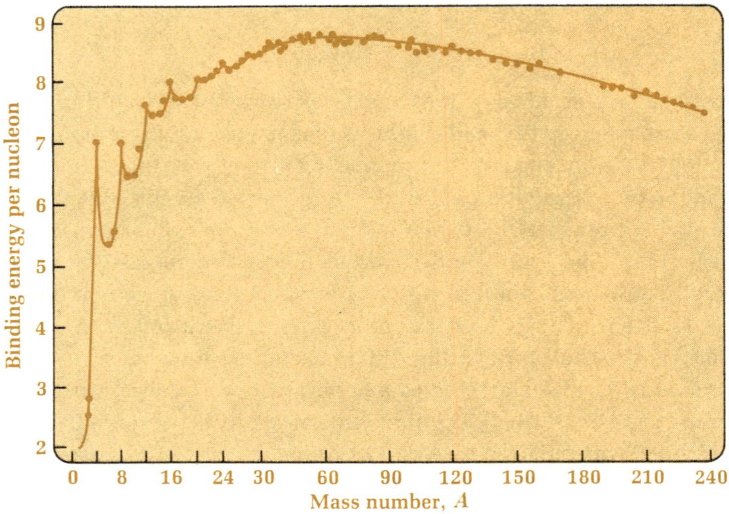

Figure 16-6 Binding energy, in million electron volts, per nucleon as a function of mass number, A.

used. This energy is called the **binding energy** of the nucleus and is shown as a function of mass for the stable isotopes in Figure 16-6. It is clear that maximum energy release per nucleon occurs for elements with masses near that of iron. Energy release per nucleon is slightly less for isotopes of either higher or lower mass and, consequently, synthesis in the stars tends to stop with a preponderance of elements near iron.

Part of the shape of the curve in Figure 16-6 is not difficult to understand. The binding energy of a nucleus depends on the number of mutual interactions that the nucleons experience. Clearly, in the hydrogen nucleus there can be no binding energy, since there is only one nucleon. In the helium nucleus there are four nucleons, so that, on the average, each nucleon can interact with three others. As we continue to build up the size of the nucleus, the number of internucleon interactions will continue to increase and, hence, the binding energy per nucleon will continue to increase. Eventually, we shall arrive at a nuclear size in which each of the nucleons has a maximum number of nearest neighbors. At this point the binding energy per nucleon no longer increases with increasing mass number. The actual decrease observed with further increase in atomic mass has a complex basis that need not concern us here. The important point to remember is that the nuclear binding energy per nucleon depends in large measure on the number of nucleonic interactions.

If the star undergoes an explosive expansion at this stage in its development, debris scattered through the heavens will contain a relative excess of the elements near iron in the periodic table. In fact, such a nonuniform distribution of elements seems to exist in the universe. When new stars are formed, they will include some of the elements made in earlier stars, these may be involved in element synthesis in second-generation stars. In particular, carbon so introduced may lead to the carbon cycle when the temperature becomes high enough.

If the star is massive enough to withstand explosion for a time, totally new processes may begin. These new processes involve the capture of neutrons by nuclei, followed by beta decay (loss of an electron). Capture of the neutron increases the atomic mass by one unit but leaves the atomic number unchanged. When a neutron subsequently decays, the element of next higher atomic number is formed. As we shall see below, a particularly important example of nuclear synthesis by this route is provided by the conversion of molybdenum to technetium:

$$^{98}_{42}\text{Mo} + ^{1}_{0}n \longrightarrow ^{99}_{42}\text{Mo} \quad (16\text{-}43)$$

$$^{99}_{42}\text{Mo} \longrightarrow ^{99}_{43}\text{Tc} + ^{0}_{-1}\text{e} \quad (16\text{-}44)$$

Neutron-capture beta-decay routes can be subdivided into two categories. The first, called the ***s* process,** involves very slow neutron capture. Consecutive captures of a neutron by the same nucleus occur with a frequency of 100 years to 10^5 years. This long time scale yields plenty of time to ensure that beta decay will intervene between essentially every successive neutron capture. The s process is nicely represented by Equations (16-43) and (16-44).

On the other hand, the ***r* process** involves successive neutron captures on a very short time scale, 0.01–10 sec. Thus, neutron capture is rapid compared to the rate of beta decay. Consequently, a nucleus may capture several neutrons prior to emission of a beta particle with formation of the element with the next highest atomic number. The s process is considered to be responsible for the formation of a number of isotopes in the range $23 <$ atomic mass < 46 and in the range $63 <$ atomic mass < 209. The r process is responsible for a large number of elements with atomic masses between 70 and 209, and for the synthesis of uranium and thorium.

How is it possible to have any real confidence in the conclusions that we have just stated? We are speaking of events that occur at a great distance from observers on earth under conditions that we cannot duplicate experimentally very readily. Can we be sure, for example, that element synthesis is really an ongoing stellar process? Is it possible that all of the elements were actually formed in some single cataclysmic event? Surprisingly perhaps, it turns out that some of our conclusions rest on reasonably firm ground. For example, let us return to the synthesis of the element technetium as shown in Equations (16-43) and (16-44). Historically, technetium occupies

Table 16-3 Summary of processes involved in element synthesis

Synthesis of	Process
1. Helium from hydrogen	H-H reaction Carbon cycle
2. ^{8}Be, ^{12}C, ^{16}O, ^{20}Ne, ^{24}Mg from helium	Helium burning
3. Many elements up to atomic number near that of iron	Fusion reactions yielding heavier isotopes plus neutrons, protons, and alpha particles
4. Most elements heavier than iron plus some lighter ones	Neutron capture followed by beta decay (includes both the *s* process and the *r* process)

an interesting niche in the periodic table. A gap existed at atomic number 43 from the time of Mendeleev, and the element was long searched for. No evidence for its existence was found until the nuclear bomb program of the 1940s, when technetium was synthesized by man in a way similar to that shown in these equations. As it turns out, technetium has no stable isotopes: All of them are radioactive and decay spontaneously. The longest-lived isotope has a half-life of only 216,000 years. While this may seem like a long time, it is much short of the age of any natural materials on earth. Consequently, any technetium originally on the earth would have decayed to other elements long before man began his study of chemistry.

One of the more significant discoveries for cosmology that has been made in recent years has been the discovery that technetium is present in some stars. This provides a strong line of evidence that element synthesis is a stellar phenomenon that takes place steadily and continually. Were this not the case, no technetium could exist in current stars. A summary of the various processes involved in element synthesis is provided in Table 16-3.

NUCLEAR FISSION

Although man has not yet learned to control in the laboratory or power plant the reactions that occur in the stars, he has had success in controlling energy released by **nuclear fission** processes. The use of **nuclear reactors** for the production of electrical energy has now become commonplace in Britain and Western Europe. In August 1969, 2.3×10^6 MW hr of power was generated in Britain alone, and 1.25×10^6 MW hr in the balance of Europe. Slightly less than 900,000 MW hr was generated in the United States during the corresponding period. The lower figure in the United States compared to Britain reflects principally the fact that other fuel sources for the production of power are still cheaper in this country. Consequently, most of the nuclear generating plants built in this country are at least in part experimental.

In contrast to the nuclear fusion process described above, fission of nuclei in general involves neutron bombardment of a heavy metal nucleus (although spontaneous fission of some transuranium metals occurs). Since the neutron carries no positive nuclear charge, the same Coulombic repulsion problem that is so important in fusion does not arise in fission. One model for the fission process suggests combination of two particles to form a compound nucleus of one greater mass number, followed by splitting into two roughly, but not identically, equal particles. For example, the $^{235}_{92}U$ atom undergoes the reaction $^{235}_{92}U + ^{1}_{0}n \longrightarrow ^{236}_{92}U$. The $^{236}_{92}U$ isotope is a known isotope that does not undergo spontaneous fission. It is instructive to consider the masses of reactants and products in this reaction to see when fission does accompany neutron irradiation. The ^{235}U isotope has a mass of 235.1170 amu, and the neutron has a mass of 1.008665 amu. The known ^{236}U isotope has a mass of 236.1191 amu. Clearly, the sum of masses of the reactants is 0.0069 mass units greater than the known mass of the resulting isotope. This mass corresponds to an energy of 6.4 MeV, and the nucleus first formed by combination of the two must be excited by this amount of excess energy relative to the known ground state of the isotope. In fact, with this much excess energy the nucleus promptly splits into two smaller particles, and the excess energy is carried as kinetic energy of the fragments:

$$^{235}_{92}U + ^{1}_{0}n \longrightarrow [^{236}_{92}U] \longrightarrow \text{fission particles} \qquad (16\text{-}45)$$

A similar calculation for the reaction of a neutron with the ^{238}U isotope shows that the mass difference corresponds to an energy of only 4.8 MeV. Neutron radiation of this isotope with low-energy neutrons in an atomic reactor produces ^{239}U, which does not undergo fission. Apparently, more than 4.8 but somewhat less than 6.4 MeV is required to break apart the compound uranium nucleus. If energy is added to this reaction by using neutrons with excess kinetic energy, a test of this conclusion can be made. In fact, if the neutron enters the reaction with 1 MeV of excess energy, the compound nucleus formed contains 5.8 MeV of excess energy and fission occurs.

An immense amount of energy is released in the fission process. In contrast to the nuclear reactions of fusion mentioned above, energy from fission is an order of magnitude greater and is in the neighborhood of 200 MeV. Furthermore, 85 percent of this energy is available as kinetic energy of the fission particles and is thus available in the form of heat energy. The remaining energy is largely in the form of beta- and gamma-ray production not directly available for useful conversion. Since fission of the ^{236}U nucleus emits an average of 2.5 neutrons [Equation (16-45)], more neutrons arise

This is a reproduction of an original painting by Gary Sheahan, *Chicago Tribune* staff artist, that shows the scene when atomic scientists observed the world's first nuclear reactor as it became self-sustaining. The time—3:22 p.m., December 2, 1942. The place—a squash court under the West Stands of Staff Field, The University of Chicago. (Courtesy of The Chicago Historical Society.)

from the reaction than are required to initiate it. The reaction is therefore self-sustaining if these neutrons can be captured by other ^{235}U nuclei. One complication is introduced by the fact that the probability of fission of a ^{235}U nucleus is 500 times greater with neutrons of low energy than with the high-energy neutrons formed by the fission process. That is, the ^{235}U nucleus appears to be a much smaller target when the neutron is moving very rapidly than when it is moving more slowly. Arrangements must therefore be made to remove the excess energy. This is most effectively accomplished by allowing neutrons to collide with many lightweight particles, imparting some of their kinetic energy to these and, of course, raising the temperature of these atoms in the process. Maximum efficiency of energy transfer takes place if the particles are of approximately equal mass. One of the principal drawbacks to the use of nuclear fission processes for energy generation is the necessity of neutron energy moderation. This is typically accomplished by using water, either directly as a moderator or as a coolant for the moderator, which is diverted either from a river or from coastal waters. The former situation creates greater problems. A large nuclear power plant produces sufficient heat to raise the temperature of a whole river by several degrees. In one extreme case, a river in this country reached the temperature of 140°F downstream from a nuclear power plant! Increasing the water temperature lowers the solubility of oxygen and disturbs the balance of living organisms in the waterway. While not as menacing as other types perhaps, thermal pollution does pose a real threat to the life of a number of rivers in this country and elsewhere.

Thermal pollution caused by nuclear power plants can be fought in a number of ways. Instead of heating large quantities of inland bodies of water, the heat may be used to evaporate much smaller quantities of water (recall the high heat of vaporization of water, Chapter 10). This is accomplished through the use of a "wet" cooling tower. The disadvantage of this method is that large amounts of water vapor are discharged into the atmosphere, which can result in sizable increases in local humidity. This might result in production of heavy fog, particularly in valley areas. The proper choice of a site for a power plant would, of course, minimize these difficulties. It is also possible to use "dry" cooling towers to dissipate unwanted heat. These work by heating air through the use of heat exchangers, much in the same way that an automobile radiator works. Finally, the heat produced in power plants can be employed in useful ways. For example, the heat generated in one nuclear reactor in the Soviet Union is used for the desalination of sea water; the fresh water produced satisfies the needs of a city of 65,000 people.

ISOTOPE SEPARATION

As it is found in nature, uranium contains 0.006 percent ^{234}U, 0.71 percent ^{235}U, and 99.28 percent ^{238}U. Unfortunately, ^{238}U is not an efficient source of energy by fission under conditions normally found in power-generating nuclear reactors. In fact, the ^{238}U isotope interferes with the basic reaction required for power generation, since it absorbs neutrons without releasing them by fission. If the ^{238}U isotope is present in overwhelming quantity, the chain reaction is terminated. Consequently, it is necessary to separate the ^{235}U in relatively pure form. Since both isotopes are of the same element, their chemical reactions and the physical properties of their compounds are almost identical. Isotope separation is particularly difficult for heavy atoms with small mass differences. The most successful separation at present is based on the diffusion of UF_6 as a gas; those molecules containing the lighter isotope diffuse more rapidly. A complex and expensive plant is required for the separation. Figure 16-7 shows a schematic representation of the separation process.

Separation of isotopes by gaseous diffusion depends wholly on the slight differences in velocity of molecules of different mass. The kinetic energy of a molecule is dependent only on the absolute temperature, and thus for two isotopic molecules at the same temperature,

$$\tfrac{1}{2}m_h v_h^2 = \tfrac{1}{2}m_l v_l^2 \qquad (16\text{-}46)$$

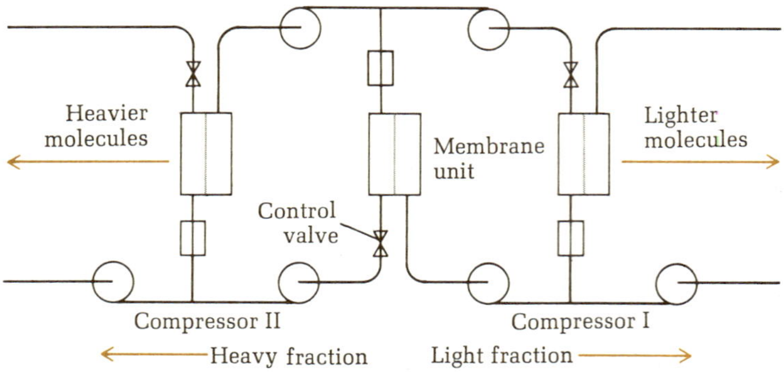

Figure 16-7 Schematic representation of a process for separation of uranium isotopes by diffusion of UF_6.

where the h and l subscripts refer to the heavy and light molecules. Rearranging, we see that the relative velocities of the heavy and light molecules are related to the square root of their mass ratios,

$$\frac{v_l}{v_h} = \left(\frac{m_h}{m_l}\right)^{1/2} \tag{16-47}$$

For the case of diffusion of hexafluorides made from ^{235}U and ^{238}U, the masses are 349 and 352, respectively. Substituting in Equation (16-47), we determine that the ^{235}U-containing gas moves 1.0043 times as fast as the heavier one. The small ratio between the velocities simply means that an operating plant must repetitively make partial separation. Each of the separation chambers shown in Figure 16-7 is called a *stage*, and a very large number of stages is required to separate the two isotopically different species completely.

Alternatively, it may be easier to use a fissionable material that need not be separated from an unwanted isotope. The reaction of ^{238}U with neutrons proceeds through the sequence

$$^{238}U + {}^1_0n \longrightarrow {}^{239}_{92}U \xrightarrow{23.5\ \text{min}} {}^{239}_{93}Np + {}_{-1}^{0}e$$

$$^{239}_{93}Np \xrightarrow{2.33\ \text{days}} {}^{239}_{94}Pu + {}_{-1}^{0}e$$

$$^{239}_{94}Pu \xrightarrow{24{,}000\ \text{years}} {}^4_2He + {}^{235}_{92}U$$

By this route the ^{239}Pu isotope is produced and, with a long half-life of 24,000 years, is readily processed. This isotope also undergoes fission with thermal neutrons, and under suitable circumstances can be prepared in large quantities by deliberately introducing ^{238}U into a reactor. Since plutonium is a distinct chemical element and different in its properties from uranium, it is possible to take advantage of these chemical differences and separate plutonium in a fashion much simpler than the isotope-separation procedure. Plutonium has consequently become the preferred nuclear fuel.

The use of plutonium as a nuclear fuel brings us to one last consideration—the use of **breeder reactors.** When Pu undergoes fission, excess neutrons are produced which can be employed to convert more ^{238}U to Pu. Overall, more Pu is produced than is consumed. Consequently, it is in principle practicable to build a breeder reactor that will use ^{238}U as a fuel directly, provided some Pu is supplied to initiate the fission process. As energy is generated, sufficient Pu will be produced to supply other nuclear reactors. The development of successful breeder reactors will markedly reduce the cost of power production through nuclear fission means.

There are, in fact, several types of breeder reactors under development. In addition to that just mentioned, it is possible to use excess neutrons from the fission of ^{235}U to convert ^{238}U into ^{239}Pu or to convert ^{232}Th, which is nonfissionable, into ^{233}U, which is.

Considerable modern-day research is taking place on preparation and study of elements with atomic numbers slightly over 100. Recent theoretical studies suggests that those elements with substantially greater atomic numbers, in the mid 120s, may be significantly longer lived than the lighter transuranium elements.

Glossary

Alpha decay that mode of radioactive decay involving the expulsion of an alpha particle from the nucleus.

Alpha particles helium nuclei emitted from the nucleus in the process of alpha decay.

Atomic mass number the total number of nuclear protons plus neutrons characterizing a particular isotope.

Atomic number the number of nuclear protons characteristic of a particular element.

Beta decay that mode of radioactive decay involving the expulsion of a beta particle from the nucleus.

Beta particle a high-energy electron emitted from the nucleus in the process of beta decay.

Binding energy the energy resulting from mutual interactions among nuclear particles, which holds the nucleus together.

Breeder reactor a type of nuclear reactor in which the power-generating process also provides additional nuclear fuel.

Carbon cycle a route for helium synthesis through fusion of hydrogen nuclei that is catalyzed by ^{12}C and involves the transient formation of unstable isotopes of oxygen and nitrogen.

Cosmology the science that deals with the creation and history of the universe.
Cross section a parameter measuring the probability that a collision between two particles will be "successful," that is, will lead to a chemical or nuclear reaction.
Gamma rays very high energy light quanta which may accompany beta-particle emission.
Half-life that length of time, characteristic for each unstable isotope, during which half of an initial sample will decay spontaneously to some other isotope.
Helium burning nuclear fusion reactions in which helium nuclei undergo successive fusion reactions leading to increasingly heavy isotopes.
H–H reaction helium synthesis through a series of fusion reactions involving only isotopes of hydrogen and helium.
Neutrino an elementary particle having zero charge and zero rest mass.
Neutron the uncharged elementary particle characteristic of atomic nuclei.
Nuclear fission a nuclear reaction in which a nucleus is split into smaller ones, frequently through bombardment with neutrons.
Nuclear fusion a nuclear reaction in which smaller nuclei are joined together to form a larger one.
Nuclear reactions those reactions, fission and fusion, involving the transformation of one or more nuclei.
Nuclear reactor a device for carrying out nuclear reactions (fission) at controlled rates, usually for the generation of useful power.
Orbital electron capture that mode of radioactive decay in which the nucleus captures an orbital electron.
Positron an elementary particle identical to the electron in all respects except that it is positively rather than negatively charged.
Positron emission that mode of radioactive decay accompanied by emission of a positron.
Proton the elementary positively charged particle of unit mass found in the atomic nucleus.
***r* process** the process of element synthesis in which a nucleus captures several neutrons prior to undergoing a beta-decay process.
Radioactive decay the spontaneous decay of unstable nuclei, accompanied by one of several sorts of radiation.
Radioactivity radiation accompanying radioactive decay.
***s* process** the process of element synthesis in which capture of one neutron by a nucleus is followed by a beta-decay process.
Thermonuclear reaction a synonym for nuclear fusion reaction.
White drawf a star having an exceedingly high density; essentially a "nuclear solid."

Study problems

1. Calculate the energy released, in million electron volts, for each of the following reactions:
   a. $^{1}_{1}\text{H} + ^{1}_{1}\text{H} \longrightarrow ^{2}_{1}\text{D}^{-} + ^{0}_{+1}\text{e}$

b. $^{3}_{2}He + ^{3}_{2}He \longrightarrow ^{4}_{2}He + 2^{1}_{1}H$
c. $^{12}_{6}C + ^{12}_{6}C \longrightarrow ^{23}_{11}Na + ^{1}_{1}H$
d. $^{17}_{8}O + ^{4}_{2}He \longrightarrow ^{20}_{10}Ne + ^{1}_{0}n$

2. What is the total mass, in atomic mass units, of the particles from which each of the following nuclei is composed?
   a. $^{8}_{4}Be$ b. $^{12}_{5}B$ c. $^{23}_{11}Na$ d. $^{31}_{15}P$
3. Calculate the binding energy, in million electron volts, for each of the following nuclei (use your results from Problem 2 to help answer these questions):
   a. $^{8}_{4}Be$ b. $^{12}_{5}B$ c. $^{23}_{11}Na$ d. $^{31}_{15}P$
4. In the fusion of tritium and hydrogen to yield helium, 0.021 amu of mass is lost. Calculate the energy yield of this reaction in ergs.
5. Write each of the following reactions:
   a. positron emission from $^{120}_{51}Sb$
   b. electron emission from $^{31}_{16}S$
   c. alpha-particle emission from $^{226}_{88}Ra$
6. Complete each of the following reactions:
   a. $^{238}_{92}U + X \longrightarrow ^{239}_{92}U$
   b. $^{17}_{8}O + X \longrightarrow ^{20}_{10}Ne + ^{1}_{0}n$
   c. $^{99}_{42}Mo \longrightarrow ^{99}_{43}Tc + X$
   d. $^{15}_{8}O \longrightarrow ^{15}_{7}N + X$
7. Explain how the energy barrier to nuclear fusion varies as a function of atomic number.
8. Which of the following nuclear fusion reactions requires the least thermal energy of the reacting particles?
   a. carbon–carbon
   b. oxygen–oxygen
   c. carbon–hydrogen
   d. oxygen–helium
   e. helium–helium
9. The half-life of $^{13}_{7}N$ is about 10 min. After a period of 30 min, what fraction of an original sample of this isotope would remain?
10. Following a period of 40 min, 25 percent of an original sample of a particular isotope remains. What is the half-like of this isotope?
11. The density of the sun is 1.6 times greater than that of water. How do you explain this fact in view of the composition, mostly hydrogen and helium, of the sun?
12. Explain clearly but briefly why plutonium is preferable to uranium as a fuel for nuclear fission reactions.

Suggestions for further reading

L. Aller *The Abundance of the Elements,* Wiley (Interscience), New York, 1961. Much too detailed for general reading, but Chapter 10 is a good treatment of the origin of the elements.

G. Choppin *Nuclei and Radioactivity,* Benjamin, Menlo Park, Calif., 1964. A good paperback.

"Energy and Power," *Sci. Am.*, **225** (September, 1971). The entire issue.

G. Gamow *The Creation of the Universe,* Viking, New York, 1952. Somewhat out of date but entertaining reading.

G. Gamow *Matter, Earth and Sky,* Prentice-Hall, Englewood Cliffs, N.J., 1958. Especially Chapters 19–21. Slightly out of date but in general good.

W. C. Gough and B. J. Eastlund "Prospects for Fusion Power," *Sci. Am.*, **224,** 50 (February, 1971).

B. G. Harvey *Nuclear Chemistry,* Prentice-Hall, Englewood Cliffs, N.J., 1965. A good paperback.

J. Holdren and P. Herrera, *Energy,* Sierra Club, San Francisco, 1971.

M. J. Lubin and A. P. Fraas, "Fusion by Laser," *Sci. Am.*, **224,** 21 (June, 1971).

Resources and Man, Freeman, San Francisco, 1969. A study of energy resources of the earth by the Committee on Resources and Man of the Division of Earth Sciences, National Academy of Sciences-National Research Council.

G. T. Seaborg and J. L. Bloom "Fast Breeder Reactors," *Sci. Am.*, **223,** 13 (November, 1970).

17

History of the atmosphere and the origin of life

Of course I'm a nut about the environment, I don't have anywhere else to live.

The Fox

The planet earth has a thin skin. From its center to its crust there is a distance of some 8000 miles; but the distance from the crust to the edge of the **atmosphere** is just a few miles. The great bulk of the world's population live within 10,000 ft (less than 2 miles) of sea level. At an altitude of 16,000 ft (three short miles), the atmosphere is already so dilute that it will not maintain human life for extended periods. Trees and other plant growth are even less tolerant in terms of the altitudes at which they survive. Thus, humans, both directly and in terms of the survival of other animals and plants on which we depend, are critically dependent on the environment provided by the thin skin of the earth: We are born, live, work, love, and die in this fragile atmosphere.

The state of the atmosphere is of utmost importance to man. On the average, each of us breathes about 22,000 times each day and takes in about 35 lb of air—a weight greater than the total food and drink we consume in the same time period. Although some natural atmospheric phenomena are violent and destructive (witness the hurricane that caused loss of perhaps half a million lives in East Pakistan in the winter of 1970), for the most part the atmosphere and nature interact in processes that are supportive of human life. From the atmosphere we derive the rain from which all plants and animals obtain their nourishment. An ozone layer in the outer atmosphere shields us from biologically harmful intense rays of the sun (see Chapter 18). Plants and the atmosphere combine in the benevolent conversion of atmospheric carbon dioxide into organic foods and, at the same time, replenish the supply of oxygen in the air. In all these processes, nature and the atmosphere conspire to make the earth a pleasant place for man.

An observer from another planet might well conclude that man is engaged in a deliberate attack on the very atmosphere that supports and nurtures him. The observer would, of course, be only partly correct: Although the reality of the attack is all too apparent to most people, it is not so much deliberate as mindless. Whatever the underlying causes, almost uncountable tons of pollutants are pumped into the atmosphere yearly, and the quantity increases steadily.

Urban dwellers in the United States are necessarily aware of the contamination of the atmosphere. For them, the sun is more often a hazy dull glow than a clearly etched ball of fire. The air in which they live is yellow or brown, not colorless, murky, not clear, foul smelling, not fresh, abrasive rather than refreshing. Eye irritation and a mild discomfort in breathing are part of the daily routine. The air in which these people live is a constant threat to health and life. Diseases of the respiratory system are much more frequent among city dwellers than among the population generally, and the incidence of these diseases is increasing. At times, the reality of the condition of our air is brought home with striking clarity, not merely revealed by our discomforts and statistics concerning life and death. Thus, there have been several tragic episodes resulting from air pollution that have attracted worldwide attention. In 1930, 60 people died in the Meuse River Valley in Belgium as a direct result of air contamination. In 1948, in Donora, Pennsylvania, the situation became so bad that the entire town was evacuated. Serious episodes occurred in London in 1952 and again in 1962. The former incident claimed 4000 lives! The point to be understood is simple: In many of our cities, *the condition of the air is already a disaster*. And the situation is worsening.

Air pollution is by no means confined to cities. Air travelers will testify to the yellow haze that clings to the ground even in the great plains and desert regions. Visible contamination of the atmosphere is common in Maine and Nebraska, in Wyoming and Montana. The thin skin of the earth is being turned into a menace to man.

Atmospheric pollution is by no means a modern problem. Over 700 years ago, the Queen of England moved from London to escape the insufferable smoke. Some 400 years ago, the brewers of Westminister offered to use wood in place of coal because of Queen Elizabeth's allergy to coal smoke. Legislative action—probably ineffective—was instituted some time later to prohibit the use of coal in London, but only while Parliament was sitting! Despite all the investigative commissions and legislative acts that have followed the problem, atmospheric pollution persists.

Nor is atmospheric pollution a problem peculiar to the United States. Perhaps the worst conditions exist in highly industrial, modern Japan. Children may be seen wearing gas masks in the choking cities. Their exercise at school is sometimes restricted due to the condition of the air. Coin machines on the street dispense a few breaths of pure oxygen. In Italy, Greece, and other European areas, priceless works of architecture and sculpture are being destroyed by corrosive acids and other chemicals carried in the air.

The pollution of the atmosphere is obviously a matter for study by chemists. The processes that lead to direct pollution, largely the combustion of various sorts of fuels, are of interest to the chemist. The photochemical processes that follow and worsen the initial direct contamination also fall within the province of chemistry. Knowledge concerning the chemistry of air pollution will not guarantee a solution to the problem. However, an informed and concerned citizenry is an essential prerequisite to such a solution. In this chapter and the following one, we deal with the atmosphere:

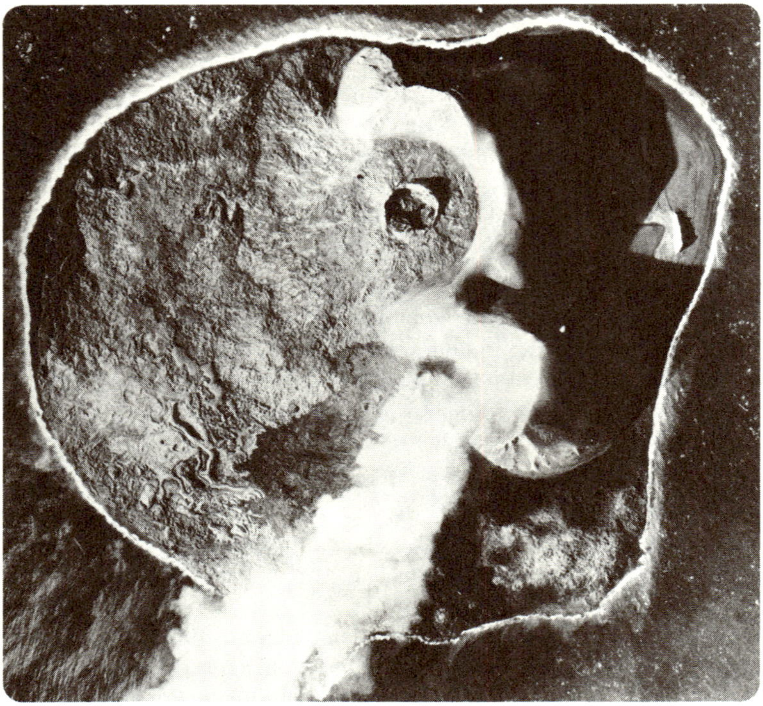

This photograph shows Surtsey Island shortly after it was formed in the North Sea by an undersea volcanic eruption. Such eruptions not only have been responsible for formation of new land areas but also may have been crucially involved in provision of energy for the development of living forms in the very early history of the earth. (Courtesy of U.S. Department of the Interior.)

What it is, where it came from, and where it is going. We shall have many occasions to refer to the photochemistry covered in Chapter 15 and its relation to the problem. Following a brief discussion of the atmosphere itself and its relationship to the origin of life in this chapter, the chemistry of photochemical smog generation is developed in some detail in the following one.

THE COMPOSITION OF THE ATMOSPHERE

Before we can understand the processes that contribute to atmospheric pollution, we must understand the composition of the normal atmosphere. This is not a simple task, since atmospheric pollution has long preceded

Table 17-1 Average composition of the atmosphere

Gas	Composition by weight (ppm)	Composition by volume (ppm)	Total mass $\times 10^{20}$ g
N_2	755,100	780,900	38.648
O_2	231,500	209,500	11.841
Ar	12,800	9,300	0.6555
CO_2	460	300	0.0233
Ne	12.5	18.2	0.000636
He	0.72	5.24	0.000037
CH_4	1.2	2.2	0.000062
Kr	2.9	1.14	0.000146
N_2O	1.5	1	0.000077
H_2	0.03	0.5	0.000002
Xe	0.36	0.087	0.000018

Source: P. E. Cloud, Jr., "Atmospheric and Hydrospheric Evolution of the Primitive Earth," *Science,* **160,** 729 (1968). Copyright © 1968 by the American Association for the Advancement of Science.

our earliest analyses of the atmospheric composition. Nevertheless, we can be reasonably sure about some aspects of the normal atmosphere. Among the components of the atmosphere are several whose concentration is reasonably independent of time. These components, together with their concentrations in the atmosphere, are collected in Table 17-1. Notice that the great bulk of atmospheric gases consists of nitrogen and oxygen. Together, these simple gases contribute more than 99 percent of the atmosphere. The third most abundant gas is argon. Other noble gases are present but in exceedingly low concentrations: Neon is present at a level about 1 percent that of argon, and the others are present in even smaller amounts. Among the minor atmospheric constituents, much importance attaches to carbon dioxide. As noted in Chapter 19, this gas is the building block for complex molecules synthesized in the photosynthetic process. As we shall see below and have already noted in Chapter 13, the equilibrium between atmospheric carbon dioxide and carbon dioxide dissolved in the seas is of crucial importance.

In addition to these constant components, some variable ones have considerable importance. The amount of water in the atmosphere, for example, varies greatly from place to place on earth and from time to time in a given location. The quantities of hydrocarbons, nitrogen oxides, and sulfur oxides are a function of the degree of pollution due to automobiles and industrial sources (see Chapter 18). Lightning causes variations in the ozone concentration. The reproduction process of plant life causes varia-

Table 17-2 Variable constituents of natural origin present in the earth's atmosphere (ppm by volume)

Water
Methane, 1.5–2.
Carbon monoxide, 0.06–1
Meteoritic dust
Sodium chloride (ocean spray)
Airborne soil
Nitrogen dioxide (electrical discharge), 5×10^{-4}–2×10^{-2}
Nitrous oxide, 0.5
Sulfur dioxide, 0–1
Hydrogen chloride } volcanic origin
Hydrogen fluoride }
Hydrogen sulfide (sulfide bacteria, volcanic origin)
Ozone (formed photochemically and by electrical discharge), 0.01–0.1
Biological material (pollen, bacteria, spores, etc.)

Table 17-3 Relative abundances of elements in the earth and the cosmos (atoms/10,000 atoms of Si)

	Atomic number	Whole earth (*a*)	Cosmos (*b*)	Deficiency factor log(*b*/*a*)
H	1	84	3.5×10^{8}	6.6
He	2	3.5×10^{-7}	3.5×10^{7}	14
C	6	71	80,000	4.0
N	7	0.21	160,000	5.9
O	8	35,000	220,000	0.8
F	9	2.7	90	1.5
Ne	10	1.2×10^{-6}	50,000	10.6
Na	11	460	462	0
Mg	12	8,900	8,870	0
Al	13	940	882	0
Si	14	10,000	10,000	0
P	15	100	130	0.1
S	16	1,000	3,500	0.5
Cl	17	32	170	0.7
Ar	18	5.9×10^{-4}	1,200	6.3
Kr	36	6×10^{-8}	0.87	7.2
Xe	54	5×10^{-9}	0.015	6.5

tions in the amount of airborne pollen, and so the list goes. In Table 17-2 we have collected some of the variable components of the atmosphere together, where possible, with estimates of their concentration.

It will be helpful in our understanding of the chemistry of the atmosphere to examine first its origin and the chemical factors important in producing the present atmosphere. Indeed, the origin of the earth itself and its primeval atmosphere are intriguing problems in science. An understanding of the conditions existing at the time of the earth's formation is fundamental in geology, in the intriguing problem of the origin of life, and important in judging the process of evolution through the ages. Obviously, these processes are deeply buried in time, but the laws of the physical universe were as true billions of years ago as they are today. Thus, much can be deduced from indirect evidence.

THE PRIMEVAL ATMOSPHERE

The composition of the present atmosphere probably does not resemble that of the original one very closely. As a starting point, we may compare the relative abundances of elements on earth with those of the cosmos. These data are collected in Table 17-3. As noted in Chapter 16, hydrogen and helium are the most abundant elements in the universe but are rather rare elements on earth. Note that the deficiency of hydrogen on earth compared to the universe is about 7 million-fold; that of helium is 14 powers of ten! If these volatile elements were present in cosmic proportions on earth at its formation, they have long since escaped into interstellar space. In contrast, a number of the heavier, relatively involatile elements are present on earth in about the same relative concentrations as they are in the universe. This suggests to us that the original material of the earth may have resembled that of the universe as a whole rather closely. Some of the volatile elements escaped, other elements have changed on earth through synthesis and loss through radioactive decay, and still others maintain their original concentrations rather closely.

The nature of the primeval atmosphere is unknown, chiefly because no rocks survive from the earliest stages of earth history. At the moment, there are two principal schools of thought. These theories are interesting in themselves and provide us with some insight concerning the nature of the evidence that reflects early stages in the history of the earth.

The atmosphere of the earth cannot be older than the earth itself. Recent estimates are that the universe is at least 12 billion, our galaxy 7.4 billion, and our solar system 4.6 billion years old. The best estimate that we can make concerning the age of the earth is 4.55 billion years. The oldest rocks on earth that have been found have an age of about 3.5 billion years. Thus, for about the first billion years of the history of the earth, we have as yet discovered no clear historical records. The older theory regarding the state of the atmosphere during this early period reasons from the composition of the universe as a whole (Table 17-3). Since hydrogen is the most abundant element, it has been argued that all of the reactive atmospheric elements must have been in their reduced states, that is, in combination with hydrogen. This suggests that the early atmosphere was composed principally of

Table 17-4 Equilibrium constants involving reduction of various species by hydrogen at 25°C

Reaction	K
$C + 2H_2 \longrightarrow CH_4$	8×10^8
$N_2 + 3H_2 \longrightarrow 2NH_3$	7×10^5
$H_2 + \frac{1}{2}O_2 \longrightarrow H_2O$	4×10^{41}
$CO_2 + 4H_2 \longrightarrow CH_4 + 2H_2O$	7×10^{21}
$S + H_2 \longrightarrow H_2S$	6×10^{15}

CH_4, NH_3, H_2O, and H_2S, plus hydrogen and some of the noble gases. Some equilibrium constants that illustrate this point are given in Table 17-4. Others have challenged the idea that the original atmosphere was in thermodynamic equilibrium with hydrogen and argue that the hydrogen originally present had escaped very rapidly. These workers support the theory that a secondary atmosphere must have been formed from the **outgassing** of volcanos and hot springs, bringing **juvenile gases** from the interior of the earth to the surface. Reasoning from the current composition of volcanic gases, this theory suggests that the principal gases in this atmosphere were H_2O, CO_2, CO, N_2, SO_2, HCl, and a few others of lesser importance. Note that this composition, having formed in equilibrium with rocks, not H_2, is much more oxidized than that suggested above. At this time, evidence from the composition of the oldest rocks suggests that the volcanic outgassing theory has the better of the argument, at least as of 3.4 billion years ago. However, it seems certain that the last word in this controversy has yet to be said.

One important point on which both theories of the primeval atmosphere agree: There was little or no molecular oxygen present. The geochemical record yields good information concerning the date of appearance of oxygen in the atmosphere. For example, in a section of South Africa there are appreciable amounts of the compound UO_2 in gold-uranium deposits. The age of these deposits has been estimated by studying their isotopic composition, and they appear to be between 1.8 and 2.5 billion years old. In deposits of more recent origin, UO_2 is not present. The equilibrium between the various oxides of uranium and the elements can be studied in the laboratory, and it can be shown that UO_2 is converted to UO_3 in the presence of only trace amounts of oxygen. Therefore, at the time of formation of these deposits, the UO_2 could not have been exposed to molecular oxygen. Since these deposits were in all probability exposed to the gases of the atmosphere at the time of their formation, we tentatively conclude that little or no oxygen was present in the atmosphere 2 billion years ago.

Additional evidence pointing to the same conclusion comes from the fact that these same deposits contain sulfides of the heavy metals. These sulfides are thermodynamically unstable and are oxidized to sulfates if molecular oxygen is present. For example, oxidation of lead sulfide occurs at room temperature according to the equation

$$PbS + 2O_2 \rightleftharpoons PbSO_4 \tag{17-1}$$

Studies of shales from deposits on three continents show that sufficiently old samples contain lead sulfides mixed with other minerals. In newer sediments, no lead sulfides are found. Thus, the ancient weathering process that led to the laying down of the old shales must have occurred in an atmosphere containing little or no oxygen. Just how little oxygen must have been present can be understood from the fact that the partial pressure of oxygen in equilibrium with $PbSO_4$ is only 10^{-63} atm! That is, exceedingly little oxygen in contact with PbS is thermodynamically sufficient to cause conversion of this compound of the sulfate.

Such observations are suggestive but not conclusive. Better estimates of the time of appearance of oxygen in the atmosphere will no doubt be made in the coming years, but for our purposes we may assume that significant amounts of the element were present in the atmosphere for most of the last 2 billion years.

The importance of molecular oxygen, of course, stems from the dependence of many of our current living species on it. Moreover, it appears likely that some of the earlier organisms may have been partially responsible for formation of the oxygen-containing atmosphere. Let us, therefore, trace the probable history of molecular oxygen in the atmosphere and its relation with living organisms.

THE HISTORY OF MOLECULAR OXYGEN IN THE ATMOSPHERE

The formation of oxygen in the primitive atmosphere, as well as the formation of more complex molecules, depends on the energy sources available. We cannot be certain about the relative importance of various free-energy sources on the primitive earth, but there is no reason to believe that these were grossly different from those of the contemporary earth; these are collected in Table 17-5. Note that by far the most important energy source is the sun (see Chapter 16). Whereas the other sources noted may have been more important on the primitive earth than they are currently, it seems unlikely that they would have been so much more important as to have competed with solar radiation as the prime driving force for atmospheric evolution.

During the early evolution of the atmosphere, the sun provided a driving force for the formation of oxygen by the photolysis of water:

$$H_2O \xrightarrow{h\nu} O_2 + \text{other products} \tag{17-2}$$

Table 17-5 Free-energy sources on the contemporary earth

Source	Energy (cal/cm^2/year)
Total radiation from sun	260,000
Ultraviolet light	
$\lambda < 2500$ Å	570
$\lambda < 2000$ Å	85
$\lambda < 1500$ Å	3.5
Electrical discharges	4
Cosmic rays	0.0015
Radioactivity (to 1.0 km depth)	0.8
Volcanos	0.13

The photolysis of water requires short-wavelength light; that is, light of high energy. Since there was no barrier to the penetration of high-energy radiation from the sun, substantial photolysis of water could have occurred. This is, however, a self-limiting process. As molecular oxygen is formed, part of it will be converted, again by the action of sunlight, to **ozone,** O_3. Ozone has the property of absorbing the short-wavelength radiation from the sun. Hence, as more oxygen is formed, more ozone is formed, preventing the further photolysis of water. This mechanism is a more powerful limiting force than might appear at first glance. Oxygen and ozone are volatile molecules that have the capacity to distribute themselves throughout the atmosphere. Water, on the other hand, cannot rise above an altitude of 10 km, since the temperature at this altitude condenses the water into ice crystals. Thus, oxygen and ozone can rise above the level at which water exists and effectively blanket the water from the sunlight capable of dissociating it to form additional oxygen. This control mechanism is thought to have limited the atmospheric oxygen content to about 0.001 of the present atmospheric level.

The intense sunlight that first caused the photolysis of water is thought also to have initiated the process of **chemical evolution,** in which simple organic molecules were converted into more complicated ones characteristic of living organisms (see the discussion at the end of this chapter). As a result, the early seas were converted into an organic soup from which the earliest living forms developed. The first organisms probably lacked many of the usual biochemical capacities. In fact, it is difficult to believe that the first organisms had more than the minimum capacities to sustain life. It is probable that most species formed in the early stages of earth's development did not survive the development of an oxygen atmosphere once the ozone shield developed. Only those capable of using the longer wavelengths of light probably persisted.

These early organisms probably evolved through a long time period, acquiring new biochemical abilities as they went. Eventually, organisms capable of **photosynthesis** were formed (Chapter 19). This begins the second chapter in the development of atmospheric oxygen. Photosynthetic organisms evolve molecular oxygen in the process of conversion of carbon dioxide to complex organic molecules. Thus, development of these organisms provides a second source of oxygen. Early biological O_2 production was principally consumed in the seas through reaction with ferrous iron, resulting in the deposition of "red beds" of ferric oxides. Later, with the development of more sophisticated oxygen-producing organisms, molecular oxygen began to escape from the seas into the atmosphere. However, this oxygen increase, too, is self-limiting. The more oxygen available, the stronger the evolutionary pressure is to develop organisms that live by aerobic respiration, that is, that consume oxygen in energy-generating reactions. The faster oxygen is made, the faster it is consumed by the development of new organisms that use it. This mechanism is considered to have limited the atmospheric concentration of oxygen to about 0.01 of its present level. Eventually, this limit too was broken through, perhaps by the development of photosynthetic plants on land, and atmospheric oxygen attained its present level. In passing, we should observe that oxygen is in many respects a gas not compatible with life as we know it. All living things (at least in our world) are based on the chemistry of hydrocarbon derivatives. Each of these is thermodynamically unstable with respect to carbon dioxide and water in the presence of oxygen. All aerobic organisms have developed means to circumvent the toxic qualities of oxygen. A summary of these events as a function of time is presented in Figure 17-1.

THE CHEMISTRY OF THE ORIGIN OF LIFE

In our discussion of the origin of the atmosphere, we have pointed out the crucial role of living organisms in terms of their contribution of oxygen. But we have said nothing concerning how life developed in the first place. This is really quite an old topic of discussion, although it does not seem that we are very near a satisfactory solution.

In the eighteenth and nineteenth centuries, there were a number of experiments that purported to show that life could be generated from nonliving matter, that is, that **spontaneous generation** occurred. All of these experiments were dealt a mortal blow by Louis Pasteur in the 1860s. He performed a series of careful experiments that established that "spontaneous generation" was, in fact, the result of bacterial contamination. Rather little has been said concerning this type of experiment in the last century. We must recognize that Pasteur's experiments showed that early observa-

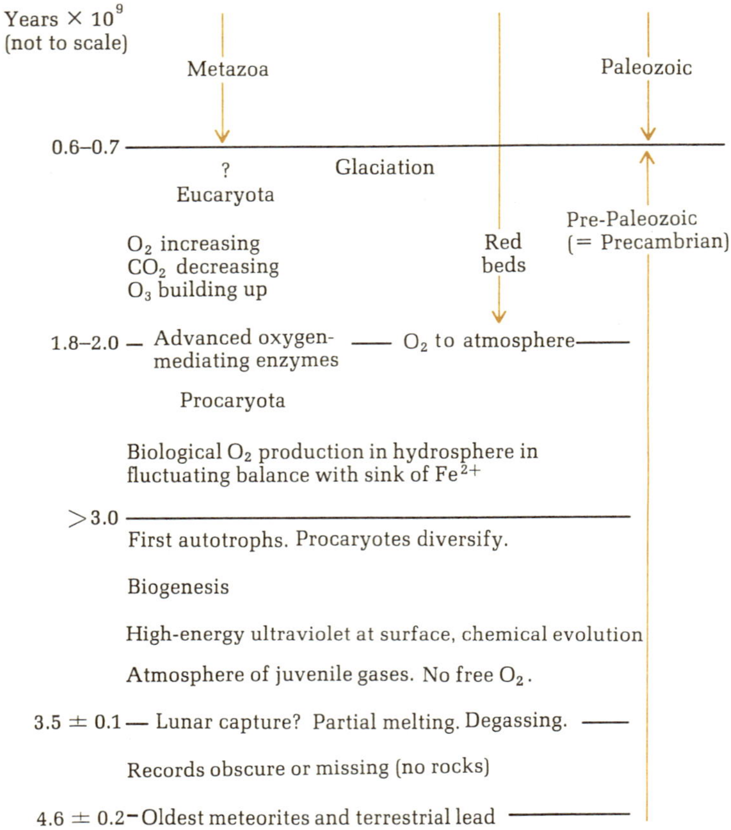

Figure 17-1 Some of the postulated principal features of the evolution of the atmosphere and biosphere on the primitive earth. The left-hand column is a time scale, in billions of years, counting back from the present. Note the absence of records of the evolutionary process prior to 3.5 billion years ago. Then, with formation of the atmosphere, perhaps by volcanic outgassing, the series of chemical reactions leading to the first living forms about 3 billion years ago began. These early organisms were almost certainly procaryotes; that is, they had no clearly defined nucleus or organelles. Certain of these organisms are considered to have developed into photosynthetic species, which resulted in the release of molecular oxygen into the oceans (hydrosphere) where it reacted with ferrous ion and formed iron deposits characteristic of this period. Later, molecular oxygen was able to escape into the atmosphere in appreciable quantities. About 1.5 billion years ago, life forms with nucleated cells, eucaryotes, developed and, finally, more advanced forms having many cells, the metazoans, came into being.

tions suggesting spontaneous generation of intact organisms were in error, but this does not prove that spontaneous generation at the molecular level is an impossibility. Indeed, although the earth is about 4.5 billion years old, evidence for life goes back only about 3.4 billion years. Thus, more than 3.4 billion years ago, an act of spontaneous generation must have occurred. How?

Most people believe that we are not very close to having the answer to this question. However, a number of scientists have been doing experiments designed to indicate some of the features of the chemistry of the origin of life. How can we approach experimentally in the chemistry laboratory a process that occurred 3 billion years ago? What experiments might we do to try to identify individual steps in what must have been a most complex process? As you can imagine, it is not easy to design meaningful experiments. Nevertheless, some very suggestive results have been obtained.

The idea is to re-create the primitive scenario as nearly as possible. A simplified picture of what we have to deal with, summarizing the arguments previously developed in this chapter, is shown in Figure 17-2. The crucial features include (1) a collection of atmospheric gases to serve as the basic materials from which other more complex molecules will be synthesized; (2) sources of energy to promote chemical transformation of the atmospheric gases (see Table 17-5); and (3) oceans to trap the complicated molecules and provide the medium for further chemical transformations. It follows that we wish to study the chemistry of mixtures of gases that might have been present in the primitive atmosphere in the presence of suitable sources of energy.

The most important of the early experiments of this type was carried out by Stanley Miller in 1953. Miller attempted to re-create chemical events in the primitive atmosphere. Using an electrical arc as an energy source, he examined the products of reactions possible in a mixture of methane, ammonia, hydrogen, and water. The experimental setup for carrying out this experiment is shown in Figure 17-3. After circulating the gases past the arc for several days and trapping the products that were formed, chemical analysis showed the presence of a number of the common amino acids found in proteins as well as some simple organic acids (which are themselves of biochemical interest). The results are summarized in Table 17-6. The observation of the formation of the amino acids is crucial. We have already emphasized in Chapter 7 that the two types of molecules that appear to be indispensable to life as we know it are the proteins and nucleic acids. Therefore, primary attention in studies of the chemistry of the origin of life must center about a search for the monomeric units of these polymers. Amino acids are such monomers.

Miller's original experiment might be criticized on the basis that he chose a highly reduced atmosphere as the model, a view due to his mentor, Harold

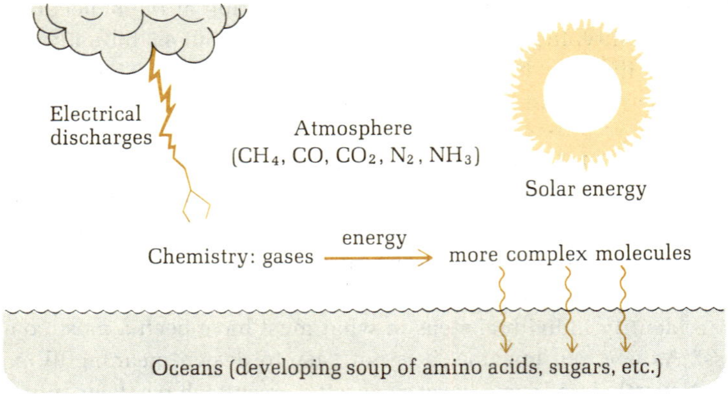

Figure 17-2 A schematic view of the primitive scenario in which the early steps leading to the first living forms may have occurred.

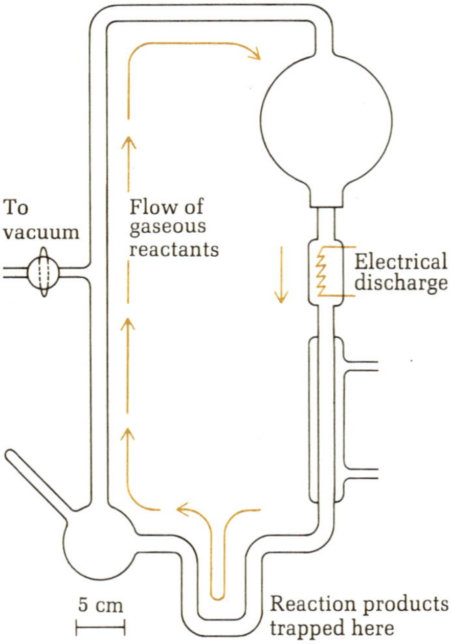

Figure 17-3 Miller's experimental apparatus, designed to examine the chemistry of what he believed to represent the primitive atmosphere. The important point is the circulation of gases past an electrical discharge, trapping the product molecules as they are formed.

Table 17-6 Organic compounds obtained by sparking a mixture of methane, ammonia, hydrogen, and water for 1 week

Amino acids	Structure	Organic acids	Structure
Glycine[a]	$H_2N—CH_2—COOH$	Formic acid	$HCOOH$
Alanine[a]	CH_3 $H_2N—CH—COOH$	Acetic acid	CH_3COOH
Aspartic acid[a]	$COOH$ CH_2 $H_2N—CH—COOH$	Propionic acid	CH_3CH_2COOH
Glutamic acid[a]	$COOH$ CH_2 CH_2 $H_2N—CH—COOH$	Glycolic acid	$HO—CH_2—COOH$
Sarcosine	$CH_3—NH—CH_2—COOH$	Lactic acid	$CH_3—CHOH—COOH$
β-Alanine	$H_2N—CH_2—CH_2—COOH$		
α-Aminobutyric acid	CH_3 CH_2 $H_2N—CH—COOH$		
N-Methylalanine	CH_3 $CH_3—NH—CH—COOH$		

[a]These amino acids occur in proteins.

Source: D. Kenyon and G. D. Steinman, *Biochemical Predestination*, McGraw-Hill, New York, 1969. Copyright © 1969, McGraw-Hill Book Company. Used with permission of McGraw-Hill Book Company.

Urey. As noted earlier in this chapter, the best evidence now favors a more oxidized atmosphere on the primitive earth (although not one containing any oxygen). But the importance of the seminal observations made by Miller, Urey, Oparin, and others must not be underestimated: The observation of the formation of amino acids generated great interest in this field and has led to most subsequent experimentation. Before briefly considering more recent results, let us think a little about the chemistry involved in the synthesis of amino acids in Miller's experiment. An experiment was carried out in which samples from the reaction mixture were withdrawn from time to time and subjected to chemical analysis. The results are shown in Figure

(Drawing by Charles Addams; © 1963, *The New Yorker Magazine*, Inc.)

17-4. The important observation is that the appearance of amino acids is associated with the disappearance of ammonia, hydrogen cyanide, and aldehydes. Thus, it appears that (1) the original gases were partially transformed to HCN and aldehydes and (2) further reaction of these immediate products and the remaining gases generated the amino acids. One reasonable route for these reactions is

$$CH_4,\ NH_3,\ H_2,\ H_2O \xrightarrow{\text{energy}} HCN + RCHO \tag{17-3}$$

$$RCHO + NH_3 \longrightarrow R{-}CH{=}NH + H_2O \tag{17-4}$$

$$RCH{=}NH + HCN \longrightarrow R{-}CH\begin{matrix}\diagup NH_2 \\ \\ \diagdown C{\equiv}N\end{matrix} \tag{17-5}$$

$$R{-}CH\begin{matrix}\diagup NH_2 \\ \\ \diagdown C{\equiv}N\end{matrix} + 2H_2O \longrightarrow \underset{\alpha\text{-amino acid}}{R{-}CH\begin{matrix}\diagup NH_2 \\ \\ \diagdown COOH\end{matrix}} + NH_3 \tag{17-6}$$

This synthetic route is actually closely related to that developed by chemists a number of years ago for the synthesis of the α-amino acids.

The exact nature of the atmosphere of gases subjected to sparking is not crucial in terms of the synthesis of amino acids. Thus, many of these compounds have been obtained with mixtures of carbon monoxide, ammonia, and water and with carbon dioxide, nitrogen, hydrogen, and water, among others. Hence, the importance of the experimental observation does not rest on a knowledge of the primitive atmosphere so detailed that it may be difficult to pin down. In addition, closely related results have been obtained when mixtures of simple gases are exposed to ultraviolet radiation, probably a better model for the energy sources of the primitive earth than an electrical discharge. Finally, use of heat also leads to the synthesis of amino acids from simple gases. The results of one experiment of this type are collected in Table 17-7. Note that more than half of the amino acids that normally occur in proteins result from heating methane, ammonia, and water in the presence of a suitable catalyst.

Experiments involving the reaction of simple gases under the influence of a variety of energy sources have been followed up by studies of the chemistry of reactive intermediates formed from these gases. Thus, heating of ammonium cyanide (NH_4CN, derived from ammonia and HCN) leads to synthesis of aspartic acid, threonine, serine, glycine, alanine, isoleucine, and leucine. In addition, adenine, one of the heterocyclic bases of the nucleic acids (p. 218), is produced in this same reaction. Heating formaldehyde in water solution leads to the formation of a number of sugars, including those that occur in nucleic acids. Without citing additional examples and without worrying about the details of the chemistry involved, we can appreciate

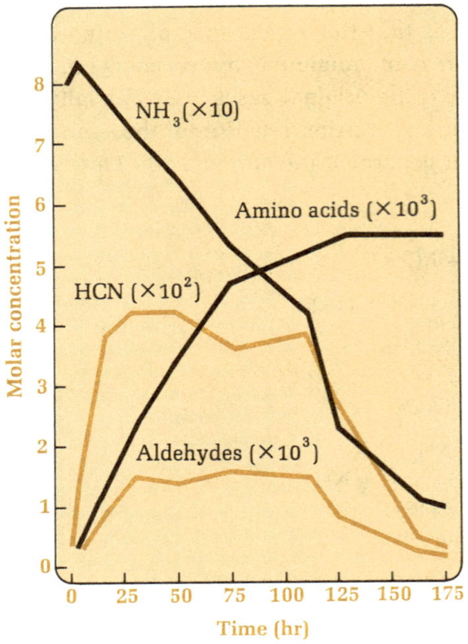

Figure 17-4 Time course for the appearance and disappearance of various compounds during the sparking of a mixture of methane, ammonia, hydrogen, and water. (From D. Kenyon and G. D. Steinman, *Biochemical Predestination*, McGraw-Hill, New York, 1969. Copyright © 1969, McGraw-Hill Book Company. Used with permission of McGraw-Hill Book Company.)

Table 17-7 Yields of amino acids resulting from heating methane, ammonia, and water at 950°C in the presence of a quartz sand catalyst

Amino acid	Percent yield
Aspartic acid	3.4
Threonine	0.9
Serine	2.0
Glutamic acid	4.8
Proline	2.3
Glycine	60.3
Alanine	18.0
Valine	2.3
Alloisoleucine	0.3
Isoleucine	1.1
Leucine	2.4
Tyrosine	0.8
Phenylalanine	0.8
α-NH_2-Butyric acid	0.6
β-Alanine	?
Sarcosine	
N-Methylalanine	

Source: D. Kenyon and G. D. Steinman, *Biochemical Predestination*, McGraw-Hill. New York, 1969. Copyright © 1969, McGraw-Hill Book Company. Used with permission of McGraw-Hill Book Company.

the significance of these observations. Simply, *efforts to reproduce the primitive scenario lead specifically to those compounds on which the polymers crucial to life are based* (Chapter 7).

As interesting and important as these observations are, we must be careful to understand that formation of the monomeric units of proteins and nucleic acids is not equivalent to accounting for the formation of the polymers themselves. In fact, formidable problems remain. Consider the matter of assembling protein molecules in the oceans. The equation is

$$\mathrm{R{-}COOH} + \mathrm{RNH_2} \rightleftharpoons \mathrm{R{-}\overset{\overset{\displaystyle O}{\|}}{C}{-}NH{-}R} + \mathrm{H_2O} \tag{17-7}$$

Note that formation of the peptide bond requires that we liberate a molecule of water. From the point of view of thermodynamics, this is likely to be an uphill process, since the high concentration of water in the oceans will tend to drive the reaction toward hydrolysis, not peptide bond formation.

Example *The standard free energy of hydrolysis of the peptide bond in neutral aqueous solution is near −500 cal/mole. What is the equilibrium constant for the synthesis of peptides at 300°K as shown in Equation 17-7?*

Solution *The synthesis reaction is just the reverse of hydrolysis; hence, the free-energy change for synthesis is just +500 cal/mole. We can now calculate the associated equilibrium constant in the usual way:*

$$\Delta G^\circ = -RT \ln K = -2.3\ RT \log K$$

$$500 \text{ cal/mole} = -2.3 \times 1.98 \text{ cal/deg mole} \times 300^\circ\text{K} \log K$$

$$\log K = \frac{500}{-1360} = -0.368 = \bar{1}.632$$

$$K = 0.46$$

Assembly of the complete monomers of the nucleic acids and association of these to form polynucleotide chains involves the same thermodynamic problem. Thus, we must find some way to provide an essentially water-free environment within the oceans in order to drive the formation of peptide bonds and phosphate diester formation to completion. It has been suggested that this occurs through the formation of very small aggregates of organic molecules, called *coacervates*, from which water is almost completely excluded.

To trace the possible steps from formation of polymers characteristic of living systems through to the development of living systems is beyond the scope of our interests. A time course for the various stages that might have been involved is provided in Figure 17-5. Note that there are many complicated questions for which answers are not yet available that we have completely neglected. For instance, we may wish to consider how it came about that proteins contain only L-amino acids, or how it happened that the distribution of amino acids along the polypeptide backbone of proteins

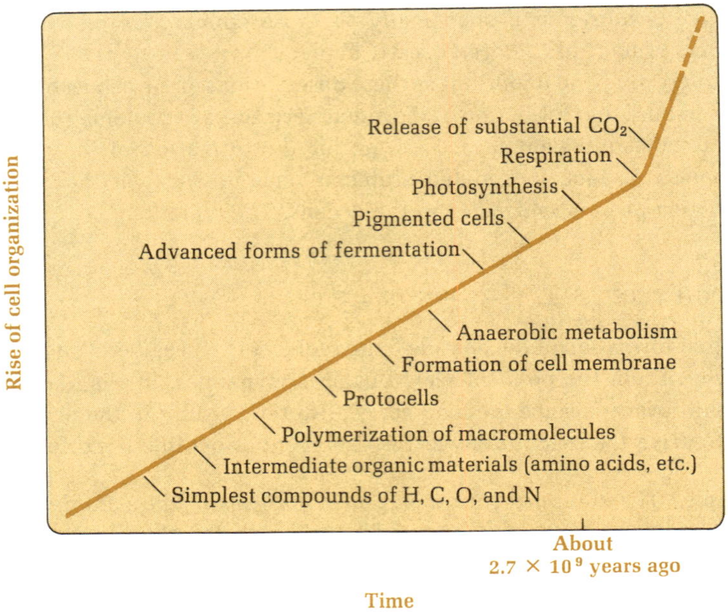

Figure 17-5 A simplified representation of the various stages involved in the formation of complex living organisms. We cannot be certain about either the precise timing or the various stages, but this diagram does meet the criteria of chemical and biological reasonableness. (From D. Kenyon and G. D. Steinman, *Biochemical Predestination*, McGraw-Hill, New York, 1969. Copyright © 1969, McGraw-Hill Book Company used with permission of McGraw-Hill Book Company.)

is nonrandom. Other questions of the same type may occur to us as we think about all the aspects of the chemistry of those molecules that are important in maintaining living systems.

Glossary

Atmosphere the layer of gases, mostly oxygen and nitrogen, that immediately surrounds the earth and provides the environment in which living things can survive.

Chemical evolution the sequence of chemical reactions leading from the simple molecules present in the primeval atmosphere to the complex macromolecules typical of living systems and, ultimately, to living systems themselves.

Juvenile gases those gases that have not previously been present on the surface of the earth; usually produced by volcanic outgassing.

Ozone O_3, a gas present in vanishly low concentrations in the atmosphere at the surface of the earth but present in sufficient quantities at higher elevations to screen out the harmful short-wavelength radiation of the sun.

Photosynthesis the light-dependent conversion of atmospheric carbon dioxide into sugars and related molecules; usually accompanied by the liberation of molecular oxygen and considered to be a prime source of atmospheric O_2.

Spontaneous generation the formation of living matter from nonliving matter; originally a theory discredited by the work of Pasteur in the last century, but now sometimes spoken of as the end product of the process of chemical evolution.

Volcanic outgassing the release of gases from the interior of the earth into the atmosphere through volcanos.

Study problems

1. Summarize the three principal stages of formation of atmospheric oxygen, indicating the source and the factors that limited its accumulation at each stage.
2. Explain clearly why the generation of atmospheric oxygen through the photolysis of water is a self-limiting process.
3. Summarize briefly the role of solar energy in the development of our current atmosphere from the primeval one.
4. Explain the reasoning that suggests that oxygen first appeared in appreciable quantities in the atmosphere about 2 billion years ago.
5. Assuming that the partial pressure of hydrogen gas in the primeval atmosphere was, at some time, 1 atm, calculate the relative concentrations of O_2 and H_2O in this atmosphere at equilibrium. Use the data in Table 17-4.
6. In Miller's synthesis of amino acids under "primeval conditions," yields are doubled if methane is replaced by methanol, CH_3OH. What can you conclude from the observation? Can you construct the chemistry involved in your conclusion?
7. In the example on p. 535 we found that the equilibrium constant for synthesis of a single peptide bond is near 0.46. What would be the equilibrium constant, under the same conditions, for synthesis of a protein containing 101 amino acids in a single linear chain?

Suggestions for further reading

P. Cloud *Adventures in Earth History,* Freeman, San Francisco, 1970. An anthology including many chapters on events on the primitive earth.

V. R. Eshleman "The Atmospheres of Mars and Venus," *Sci. Am.,* **220,** 78 (March, 1969).

S. W. Fox et al. "Chemical Origins of Cells," *Chem. Eng. News,* June 22, 1970, p. 80.

D. Kenyon and G. D. Steinman *Biochemical Predestination,* McGraw-Hill, New York, 1969.

B. Mason "The Lunar Rocks," *Sci. Am.,* **225,** 48 (October, 1971).

S. L. Miller *Science,* **117,** 528 (1953). Miller's first paper on the synthesis of complex molecules from simple substances present at the time of the earth's formation.

C. Ponnamperuma and N. Gabel "Current Status of Chemical Studies on the Origin of Life," *Space Life Sciences,* **1,** 64–96 (1968). Includes a review of the earliest stages of earth.

H. C. Urey *The Planets: Origin and Development,* Yale University Press, New Haven, Conn., 1952. One of the early scientific looks at the overall problem and a classic.

L. White, Jr. "Medieval Uses of Air," *Sci. Am.,* **223,** 92 (August, 1970).

18

Chemistry of the atmosphere

Love thy neighbor as thyself:
We are all joint tenants of the thin layer of oxygen which envelops our planet. The cool air mass which New Yorkers hope will bring relief from summer's heat will have been enjoyed yesterday by our Canadian neighbors ... and tomorrow will have been passed along to our neighbors on the continent of Europe ... to be returned to us three weeks hence.

Glenn T. Seaborg

At the outset of the last chapter, the crucial importance of the atmosphere in the life of man was developed and the threat to it, which ironically comes from man, was briefly explored. That man places exceedingly large quantities of a variety of unnatural substances into the atmosphere each year is common knowledge. It is also common knowledge that doing so presents hazards to health and life. Regrettably, the details of the chemistry of air pollution are much less well known than they might be. This chemistry asks at least two questions. First, what are the chemical changes in the atmosphere that result from contamination? Second, what are the chemical changes that result from the interaction of a contaminated atmosphere with both living and nonliving matter? If we had in hand detailed answers to these questions, it would be possible to pinpoint those contaminants that pose the most direct threats and to focus attention on minimizing their release into the atmosphere. Important as this matter is, only the most superficial pictures of the chemistry of atmospheric contamination are available. We cannot begin to explore the second of these two questions; the necessary research to provide answers in this area simply has yet to be done. But we can usefully explore the first question, basing our discussion on the general treatment of photochemical reactions provided in Chapter 15. What we wish to accomplish, then, is to develop as complete a picture as possible in the light of current information about the photochemistry of smog generation.

We can make one of two basic observations. To begin with, atmospheric contaminants can be divided into two classes: those substances that do not undergo chemical change following their release into the atmosphere and those that do. Contaminants that fall into the latter class can generate new contaminants through their chemical transformation. Thus, many of the more noxious atmospheric pollutants are not added directly by man but are products of chemical reactions of those that are. As we shall see, many of the chemical reactions involved in the generation of these "secondary" contaminants are photochemical in nature. This brings us to a second basic observation. We need to ask if solar radiation provides enough light quanta

Table 18-1 National air pollutant emissions (millions of tons/year)

	Carbon monoxide	Sulfur oxides	Hydrocarbons	Nitrogen oxides	Dustfall	Total
Industry	2	9	4	2	6	23
Power plants	1	12	—	3	3	19
Cars and trucks	66	1	12	6	1	86
Heating	2	3	1	1	1	8
Trash Burning	1	—	1	—	1	3
Total	72	25	18	12	12	139

Source: U.S. Department of Health, Education and Welfare, *The Sources of Air Pollution*, Public Health Service Publication No. 1548, Washington, D.C., revised, September, 1968.

of sufficient energy to promote extensive atmospheric photochemical activity.

The first law of photochemistry (p. 461) states that only light quanta absorbed by a system are effective in producing chemical change. Let us make a simple calculation to assess the importance of photochemically induced reactions in nature. Over most of the visible spectrum, the mean number of photons arriving at the atmosphere from the sun is 5×10^{15} each second for each square centimeter. We may then calculate that for 1 mole, 6×10^{23}, of photons to arrive at each square centimeter we must wait for 10^8 sec, about 3 years. Our first reaction to this information might be that no significant amount of chemical reaction would be photochemically produced in a reasonable time. However, we must remember that each square centimeter simultaneously accepts the same number of photons per second and that there are a great many square centimeters in a large land area. It may be more useful to calculate that the area of the earth's surface receiving 1 mole of photons each second is less than 3 acres, about half of a city block. If all of the light were absorbed and used to produce photochemical reactions with perfect efficiency, some 1200 moles of pollutant would be produced over each acre each hour. Since many pollutants are noxious or toxic at concentrations below 1 ppm, it is clear that the amount of sunlight normally impinging on the atmosphere is sufficient to produce overwhelming atmospheric contamination.

SOME ATMOSPHERIC POLLUTANTS AND SMOG

A great deal of detective work has gone into unraveling the complex causes of air pollution, and extensive studies are still underway. Table 18-1 lists

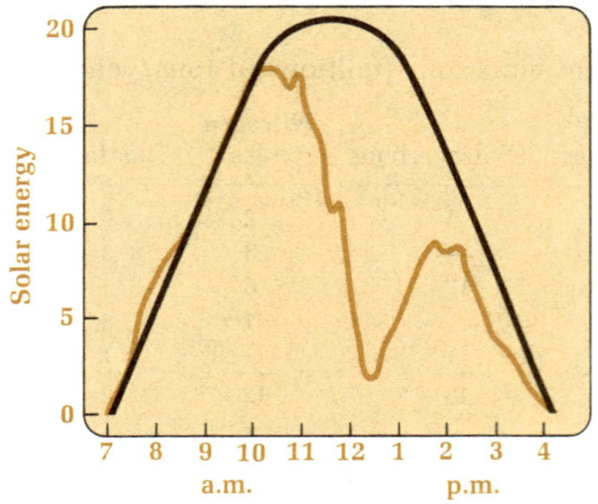

Figure 18-1 Solar energy arriving at the surface of the earth in Pasadena, California, on May 24, 1954, as a function of the time of day. The brown line is a calculated one based on no atmospheric contamination; the yellow line is the experimental one. The degree of difference between the two lines is a measure of the extent of smog formation. (From P. A. Leighton, *Photochemistry of Air Pollution,* Academic, New York, 1961.)

Table 18-2 Summary of maximum hourly oxidant levels for eight U.S. cities, 1964–1965, from continuous air monitoring stations

			Days having maximum hourly oxidant equal to or greater than indicated concentration					
			0.15 ppm		0.10 ppm		0.05 ppm	
	Year	Days of valid data	No.	%	No.	%	No.	%
Chicago	1964	254	0	0	15	6	149	59
	1965	275	0	0	9	3	120	44
Cincinnati	1964	303	5	1.5	36	12	137	45
	1965	310	5	1.5	19	6	182	59
Denver	1965	285	14	5	51	18	226	79
Los Angeles	1964	349	83	24	149	42	221	63
	1965	365	121	33	185	51	275	75
Philadelphia	1964	289	9	3	37	13	124	43
	1965	266	4	1.5	23	9	109	41
St. Louis	1964	253	6	2.5	26	10	156	62
	1965	329	8	2.5	33	10	206	63
San Francisco	1964	298	1	0.5	13	4	72	24
Washington	1964	293	4	1	40	14	163	56
	1965	284	3	1	25	9	150	53

Source: Air Quality Data from the National Air Sampling Networks and Contributing State and Local Networks, 1964–1965, U.S. Department of Health, Education and Welfare, Washington, D.C., 1966.

some of the principal compounds involved in air pollution together with their important sources.

Carbon monoxide, stemming significantly from automobiles, is the quantitatively largest single air pollutant, but oxides of sulfur, oxides of nitrogen, and a variety of hydrocarbons are significant pollutants as well. Furthermore, chemical studies have shown that the presence of strong oxidizing agents, ozone and nitrogen oxides, in the air is related to air pollution. Table 18-2 surveys the oxidant level in eight U.S. cities for 1964–1965. It is apparent that unusually high levels of oxidants are common in many cities.

What data suggest that formation of smog may be in part photochemical? A photochemical origin of smog requires that there be a direct relationship between the intensity of solar radiation impinging on a contaminated atmosphere and smog levels. One of the easiest ways to measure the extent of smog formation, though by no means the only one, is suggested by our observation that smog dims the light from the sun and obscures visibility. It follows that the more smog in the atmosphere, the less sunlight will reach the earth. Thus, we can measure the extent of smog formation by measuring the amount of solar radiation that arrives at the earth's surface. Figure 18-1 plots such intensity measurements for a day, some years ago, in Pasadena, California, when the smog was particularly bad. Normally, after sunrise the amount of light reaching the surface of the earth steadily increases until the sun is overhead at noon and then decreases once more until it reaches zero at sunset. The brown curve in Figure 18-1 shows the ideal situation for a clear day in the fall. The yellow curve on the same figure displays the energy actually observed and shows a striking departure from the ideal. The extent of smog formation is measured by the degree of difference between the brown and yellow lines. In fact, the day was exceptionally clear at dawn, but about 10:30 a.m. smog began to form in the Los Angeles Basin and the amount increased rapidly until about 12:30 p.m. At that time, only about 10 percent of the light energy reaching the limits of the atmosphere above Pasadena actually penetrated and reached the surface of the earth. Ninety percent of the light was absorbed or scattered. The curve shows that the pollution decreased significantly in the early hours of the afternoon, but even as late as 3:30 p.m. only half of the light of the sun actually reached the earth's surface. Note that the general features of this curve are just what we would expect on the basis of photochemical smog formation. There is a good parallel between solar light intensity impinging on the upper atmosphere and the extent of smog formation. This data, of course, does not prove that smog formation is related to light, and we must turn to a more fundamental examination of photochemical reactions involving components of the atmosphere.

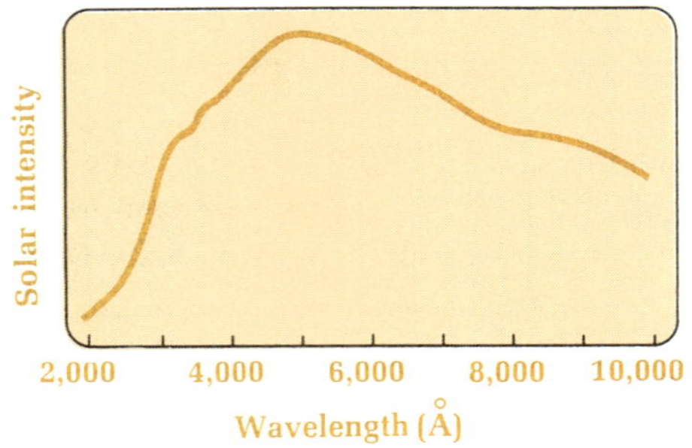

Figure 18-2 Mean solar intensity (smoothed) as a function of wavelength.

Table 18-3 Absorbers and nonabsorbers in the visible region of the spectrum

Absorbers	Nonabsorbers
Oxygen	Nitrogen
Ozone	Water
Nitrogen dioxide	Carbon monoxide
Sulfur dioxide	Carbon dioxide
Nitric acid, alkyl nitrites, and nitro compounds	Nitric oxide
	Sulfur trioxide and sulfuric acid
Aldehydes and ketones	Hydrocarbons
Acyl nitrites, pernitrites, nitrates, and peroxides	Alcohols
	Organic acids

Source: P. A. Leighton, *Photochemistry of Air Pollution,* Academic, New York, 1961.

PHOTOCHEMISTRY OF THE NORMAL ATMOSPHERE: BASIC POINTS

For each of the substances present in the atmosphere, light-induced reactions may occur with other components under the proper circumstances. For each substance, we must consider its concentration in the atmosphere, its absorption spectrum, the path length of light through the atmosphere, and finally the intensity of light of the proper wavelength that arrives at the outermost edges of the atmosphere. These quantities correspond to the terms c, $\varepsilon(\lambda)$,l, and I_0 of the Beer-Lambert law discussed earlier (p. 459).

Figure 18-2 shows the distribution of light energy incident at the outer edges of the atmosphere for wavelengths between 2200 and 10000 Å. The curve shown has been smoothed somewhat to remove some of the minor variations in intensity arising from many sharp absorption lines superimposed across the spectrum. Despite this smoothing, it is clear that incident light intensity varies significantly over the spectrum.

In terms of numbers of photons per second, the maximum in solar intensity occurs near 5000–6000 Å and varies relatively little in the visible region. From the Planck-Einstein relation (p. 433), we calculate that the energy associated with 1 mole of photons of wavelength 6000 Å is about 45 kcal/mole. In Chapter 8 we have seen that most chemical bonds are significantly stronger than this (Table 8-2). We may therefore conclude that many of the most important photochemical reactions will have insufficient energy for dissociation to be a major factor; most reactions of interest will be of molecules in excited electronic states. Only when molecules strongly absorb at short wavelengths in the ultraviolet will primary photochemical dissociation be an important process.

Table 18-3 divides the common components of the atmosphere into those that do and those that do not absorb light in the photochemically active region.

Note that the most abundant gas in the atmosphere, nitrogen, does not absorb light in the visible region. Neither does water vapor, although we noted in Chapter 17 that at much shorter wavelengths water does absorb and dissociates into hydrogen and oxygen.

Oxygen, on the other hand, does absorb light in the red end of the spectrum, at about 7620 and 12,690 Å. The absorption bands are very weak indeed. Only about 30 percent of the incident radiation is absorbed by the 7620 Å band when sunlight passes through the entire atmosphere. Nevertheless, molecular oxygen in an excited state may be produced by photon absorption in sufficient quantity to be of some importance in atmospheric chemistry. Furthermore, indirect processes also lead to the formation of excited molecular oxygen, thereby, perhaps, allowing it to assume a more important role than would be possible if only that formed directly were considered.

EXCITED MOLECULAR OXYGEN

Let us first seek to understand why the oxygen absorption bands are so weak. In Chapter 5 we discussed the electronic structure of the oxygen molecule on the basis of molecular orbital theory. It will be recalled that the highest filled molecular orbital is the antibonding pi orbital formed from the p_x and p_y atomic orbitals and that this is occupied by only two electrons. Furthermore, in the ground state of the molecule the two electrons are not spin-paired and the molecule is expected to be (and is) paramagnetic. That is, the ground state of oxygen is a *triplet*. Light absorption in the red region of the spectrum provides sufficient energy to rearrange the electrons within

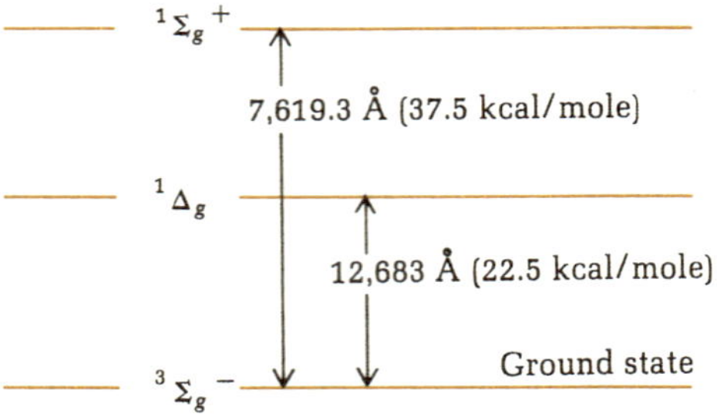

Figure 18-3 Lower energy levels of the O_2 molecule.

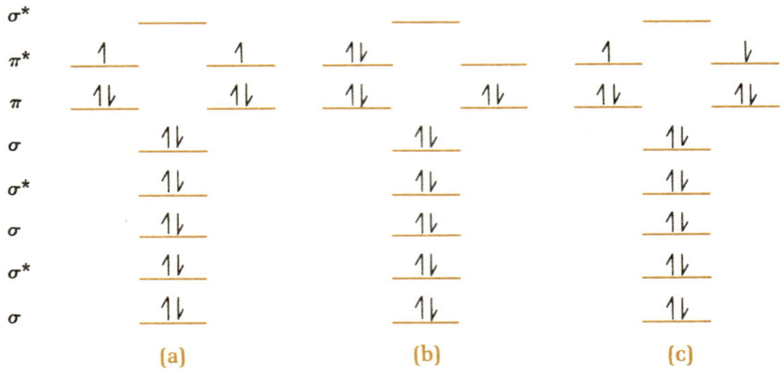

Figure 18-4 Electronic structures for (a) the ground state, $^3\Sigma_g{}^-$, (b) the first excited state, $^1\Delta_g$, and (c) the second excited state, $^1\Sigma_g{}^+$, of molecular oxygen.

the occupied orbitals but is insufficient to promote electrons to unoccupied molecular orbitals. With the energies available, the electron rearrangements that occur involve only pairing or unpairing of electron spins. As previously noted (see Chapter 15), transitions in which spins are inverted are generally improbable. This accounts for the weakness of light absorption by oxygen at these wavelengths.

Figure 18-3 shows an energy-level diagram for the low-energy transitions of the oxygen molecule. Each electronic state is given its proper name, although we shall not be concerned with detailed interpretation of the names. The ground state is $^3\Sigma_g{}^-$ (which is read "triplet sigma g minus"; the left superscript gives the spin multiplicity). Both low excited states are singlets and can therefore be reached only by spin inversion of one electron.

The exact description of electron distribution in these singlet states is beyond our present interest, but approximate descriptions may be useful. The pi molecular orbitals of interest to us are derived by overlapping of the p_x and p_y orbitals of the individual atoms. If this view is preserved, the higher excited state, $^1\Sigma_g^+$, allows one electron to occupy the pi orbital formed by overlap of the two atomic p_x orbitals and the other to occupy, with opposite spin, the orbital formed by overlap of the two atomic p_y orbitals. In the lower excited state, $^1\Delta_g$, both electrons are to be found (spin-paired of course) in one or the other of the two pi orbitals and the two are therefore degenerate (p. 157). As pointed out in Chapter 5, the actual molecular orbital has the full symmetry of a cylinder, and thus we cannot in fact distinguish the individual components of the degenerate set as we have done above. A schematic description of the electronic structure of the ground and first two excited states of molecular oxygen is in Figure 18-4.

Adsorption of light in the near infrared results in formation of the $^1\Sigma_g^+$ excited state and is of low intensity because a change in spin multiplicity is required, that is, is *spin-forbidden*. Once the singlet excited state is formed, internal conversion to the excited $^1\Delta_g$ state can occur. Since this process is *spin-allowed*, it occurs relatively rapidly, and the principal excited state that may be involved in reaction with other components of the atmosphere is $^1\Delta_g$. Once formed, this excited state has a relatively long life, since its conversion to the ground state is a spin-forbidden process. Measurements at very low pressures tend to limit as much as possible the collisions that occur with other molecules and show that only half of the excited molecules decay to the ground state in 45 min. During this period, the excited oxygen molecule will have an immense number of collisions with other molecules in the upper atmosphere. Ample opportunity therefore exists for excited oxygen to react with other molecules in the atmosphere. Furthermore, the molecule now carries an extra 22.6 kcal/mole of energy and may enter into reactions that are not thermodynamically favorable for ordinary oxygen in its ground state. That is, the energy of excitation that the excited oxygen molecule carries with it may be employed as a thermodynamic driving force for those reactions in which this molecule participates. It follows that excited oxygen is more reactive than ground-state oxygen.

Example *We have calculated in Chapter 8 (p. 271) that the formation of* H_2O_2 *from* H_2O *and* O_2 *is endergonic by 29.94 kcal/mole. Calculate the standard free-energy change for this reaction assuming that the* O_2 *molecule involved is (a) in the* $^1\Delta_g$ *excited state and (b) in the* $^1\Sigma_g^+$ *excited state.*

Solution *Note that the product of the reaction is unchanged as a result of the excitation of the reactant* O_2 *molecule. Consequently, the only energetic change is destabilization of the reactants relative to the products*

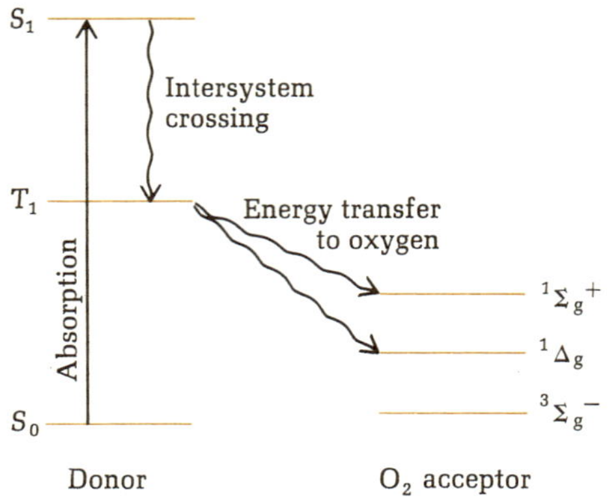

Figure 18-5 An idealized representation of indirect energy transfer to oxygen. Photons are absorbed by a suitable organic molecule (benzene, benzaldehyde, etc.) and, eventually, the energy is used to excite oxygen to the singlet state.

Table 18-4 Rates of production of $O_2(^1\Delta_g)$ from various sources

Source	Production rate (pphm/hr)[a]
$O_2 + h\nu \longrightarrow O_2^*$	0.01
$O_3 + h\nu \longrightarrow O_2^* + O$	0.1
Donor + $O_2 \longrightarrow O_2^*$	80
$NO + O_3 \longrightarrow O_2^* + NO_2$	1100

[a] pphm/hr = parts per hundred million per hour.

by the excitation energy involved. Consequently, for the case of the $^1\Delta_g$ excited state, the standard free-energy change is $29.94 - 22.6 = 7.34$ kcal/mole. For the $^1\Sigma_g^+$ excited state, we have $29.94 - 37.7 = -7.76$ kcal/mole. In the latter case, formation of H_2O_2 is actually an exergonic reaction.

The direct photochemical excitation of molecular oxygen is not the only mechanism for the formation of the excited molecule. In fact, there are at least two other means by which excited O_2 can be formed. First, there is a route in which some other molecule absorbs a quantum of light and is thereby promoted to an excited state and subsequently collides with an oxygen molecule, transferring its energy to it. The product is an excited O_2. This phenomenon is called *sensitized excitation*. Second, any chemical reaction that produces O_2 and that is exothermic by at least 37.7 or 22.6 kcal/mole, the energy needed to attain the $^1\Sigma_g^+$ or $^1\Delta_g$ state, is capable of producing O_2 in an excited state. Remember that production of molecules in excited states leads to chemiluminescence and bioluminescence (pp. 473, 475). Let us consider the two possibilities for O_2 excitation individually.

In air contaminated with common products from automotive pollution, indirect formation of excited-state oxygen may become important. Both unburned aromatic hydrocarbons and partial oxidation products such as aldehydes can effectively serve as energy carriers, resulting in formation of singlet oxygen. Figure 18-5 shows how this can occur. Benzaldehyde, benzene, and similar compounds generally are strong absorbers of light in the near ultraviolet (2000–4000 Å). The electronic transitions that occur in these molecules are between singlet ground states and singlet excited states, are therefore spin-allowed and much more likely than the spin-forbidden transitions of oxygen itself. Rapid intersystem crossing to triplet states of the organic molecules can occur, and further decay to the ground state will be slow (spin forbidden) (see p. 463). As a result, energy may remain stored in the lower triplet state of these molecules until collision occurs with an oxygen molecule and transfers energy to it.

Ozone is an important minor component of the atmosphere and one we shall discuss in some detail later. For the moment, we shall limit our comments to its role as a source of excited molecular oxygen in at least two separate ways.

Gaseous ozone absorbs strongly in the ultraviolet below 3100 Å with a maximum at 2550 Å with an extinction coefficient of 3250 liters/mole cm. This coefficient is about 10^8 greater than that of oxygen at 7620 Å. The principal reaction that occurs when a photon is absorbed is the cleavage of the ozone molecule into an oxygen atom and an excited oxygen molecule:

$$O_3 + h\nu \longrightarrow O_2(^1\Delta_g) + O \qquad (18\text{-}1)$$

Even in the absence of photodecomposition, ozone may provide excited oxygen through a variety of chemical reactions. Consider, as a typical example,

$$NO + O_3 \longrightarrow NO_2 + O_2 \qquad \Delta H = -47.77 \text{ kcal/mole} \qquad (18\text{-}2)$$

Nitric oxide is a significant component of polluted atmospheres. Furthermore, both NO and NO_2 contain an odd number of electrons and must necessarily have one unpaired electron (they therefore have *doublet* ground states). Since all electrons are paired in the ozone molecule, it is tempting to guess that the oxygen produced will also have all electrons paired, thereby avoiding unpairing of electrons during the course of the reaction. This reaction of ozone, as well as similar reactions, is currently under study in a number of laboratories in an attempt to confirm its importance in the air pollution field.

Excited oxygen is produced by all of the above sources. Table 18-4 gives estimates for its rate of formation from the individual reactions. Note that

PARIS INN
TELEVISION
TWA
SHELL

both sensitized excitation and chemical reactions are potentially more important contributors to the store of singlet oxygen than is direct photochemical excitation. Consequently, the quantities of this species produced are principally a reflection of the amounts of ozone and atmospheric contaminants present: High quantities of these will result in high levels of singlet oxygen.

The precise fate of excited oxygen molecules generated in the atmosphere is at present unknown. Laboratory studies have been carried out comparing the rate of reaction of ground-state and excited-state oxygen for many systems. It is known that excited oxygen molecules react rapidly with many organic systems that appear to be totally unreactive towards ordinary oxygen. Furthermore, compounds so produced are identical to substances known to be present in photochemically produced smog. For example, olefins (compounds with carbon-carbon double bonds) produced in automobile exhaust are known to react to produce hydroperoxides, which in turn readily decompose to form aldehydes:

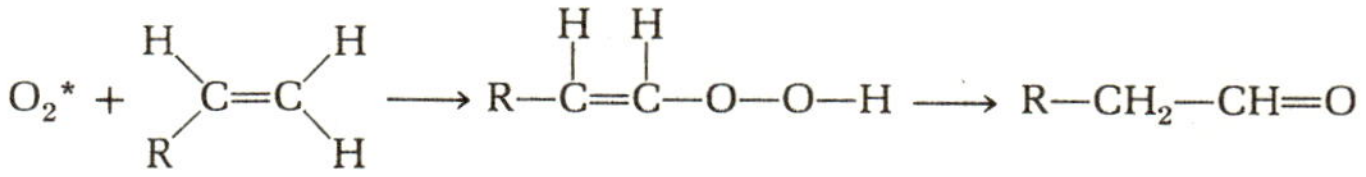
(18-3)

However, as we shall see later, these compounds may also be produced by other reactions. Further work is required to define the relative importance of excited oxygen and of other oxidizing reactions in generation of photochemical smog. At present we can only be sure that the excited oxygen molecules are indeed present in normal and polluted atmospheres and are capable of reacting to form undesirable pollutants. Beyond this, we must await definitive studies.

THE ROLE OF NITROGEN OXIDES

The oxides of nitrogen are thought to be particularly important photochemically activated reactants in polluted atmospheres. Nitrogen oxides themselves, although simple compounds, undergo relatively complex reactions, and their ordinary chemistry should be examined before we turn to their interactions with light.

The principal binary compounds of nitrogen and oxygen are N_2O, NO, N_2O_3, NO_2, N_2O_4, and N_2O_5 (see p. 92). All of these compounds are of sufficient stability so that their properties can be studied in the laboratory. In addition, there is some evidence for the existence of unstable nitrogen

Los Angeles before, during, and after the generation of photochemical smog. (Courtesy of Los Angeles County Air Pollution Control District.)

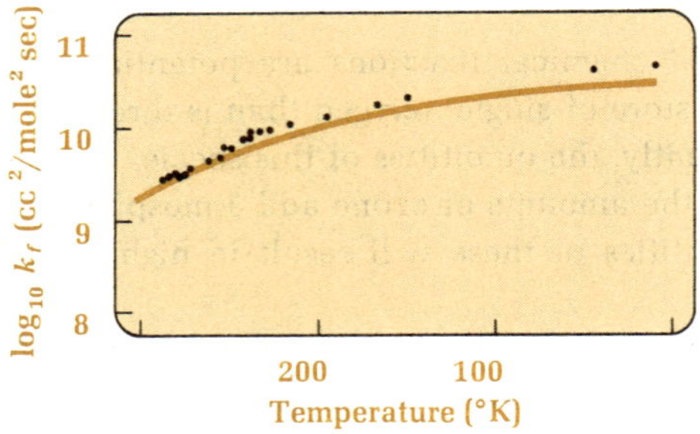

Figure 18-6 Rate constants for oxidation of nitric oxide plotted as a function of temperature. (From D. Yost and H. Russell, Sr. *Systematic Inorganic Chemistry*, Prentice-Hall, Englewood Cliffs, N.J., 1946.)

Table 18-5 Rate constants for oxidation of nitric oxide

T(°K)	$k_f \times 10^{-9}$ (cm^6/mole^2 sec)
273.1	7.88
333.2	5.58
470.0	3.34
563.6	2.82
661.9	2.54

Source: D. Yost and H. Russell, Sr., *Systematic Inorganic Chemistry*, Prentice-Hall, Englewood Cliffs, N.J., 1946.

oxides as reaction intermediates. For our present purposes, we shall restrict our discussion almost completely to the chemistry of NO, NO_2, and its dimer, N_2O_4.

Nitric oxide, NO, is a compound of technical importance. It is formed by the reaction

$$N_2 + O_2 \longrightarrow 2NO \qquad \Delta H = 43.2 \text{ kcal/mole} \tag{18-4}$$

provided sufficient energy is available to drive the reaction to completion. In nature, this reaction is associated with lightning strokes; it can also be driven in the laboratory by electrical discharges. When excess oxygen is present, nitric oxide is rapidly oxidized to **nitrogen dioxide,** NO_2. Nitrogen dioxide dissolves in water with formation of nitric acid and regeneration of nitric oxide. Thus, the sequence of reactions

$$\begin{aligned} N_2 + O_2 &\longrightarrow 2NO \\ 2NO + O_2 &\longrightarrow 2NO_2 \\ 3NO_2 + H_2O &\longrightarrow 2HNO_3 + NO \end{aligned} \tag{18-5}$$

is a useful preparative sequence and was, in the past, a significant industrial source of nitric acid.

Recent actions of man have increased the amounts of nitrogen oxides introduced into the atmosphere. Electrical discharges from machinery simulate the effects of lightning, and internal combustion engines add still greater amounts (see Table 18-1). Electrical storms alone, however, are estimated to fix some 4×10^7 *tons* of nitrogen each year. The majority is oxidized quickly to nitric acid and swept from the atmosphere by rain. The nitrates thus delivered to the soil are useful nutrients for plant life. Inevitably, small quantities remain in the atmosphere and are involved in the photochemical reactions discussed below.

The crucial reaction in the scheme described above is the oxidation of nitric oxide to nitrogen dioxide. We shall see later that this is of considerable importance in the atmosphere, and we wish now to examine its course in more detail. The oxidation of nitric oxide is unusual in that it is one of the very few known reactions that occurs in the gas phase with third-order kinetics (see Chapter 9). The rate law governing oxidation is

$$\frac{d(\mathrm{NO_2})}{dt} = k(\mathrm{NO})^2(\mathrm{O_2}) \qquad (18\text{-}6)$$

Rate constants for this reaction at various temperatures are collected in Table 18-5 and are plotted as a function of temperature in Figure 18-6. Observe that, in contrast to most reactions, oxidation of nitric oxide takes place more slowly at elevated temperatures. Direct application of the ideas presented in Chapter 9 would lead to the apparent conclusion that a negative activation energy exists for this reaction, an idea not at all appealing to the chemist.

We can rationalize the apparent negative activation energy for the oxidation of nitric acid with the interpretation of the third-order rate law. To begin with, we need to make one basic observation: The simultaneous collision of two nitric oxide molecules and one oxygen molecule to initiate the reaction between them is so infrequent that we cannot explain the rate of the reaction on this basis. As noted in Chapter 16, reactions involving three species almost always occur by successive reactions between two reactants at a time. Thus, we must search for more complicated mechanisms that will account for the third-order kinetics and also provide an explanation for the decrease in reaction rate with increasing temperatures. There are two attractive possibilities:

$$\begin{array}{ll} 2\mathrm{NO} \rightleftharpoons \mathrm{N_2O_2} & \text{(fast)} \\ \mathrm{N_2O_2} + \mathrm{O_2} \longrightarrow 2\mathrm{NO_2} & \text{(rate limiting)} \\ \text{mechanism 1} & \end{array}$$

$$\begin{array}{ll} \mathrm{NO} + \mathrm{O_2} \rightleftharpoons \mathrm{NO_3} & \text{(fast)} \\ \mathrm{NO_3} + \mathrm{NO} \longrightarrow 2\mathrm{NO_2} & \text{(rate limiting)} \\ \text{mechanism 2} & \end{array}$$

In mechanism 1, the reaction is viewed as occurring with rate-determining attack of molecular oxygen on a molecule formed in a rapid reaction between two nitric oxide molecules. Each nitric oxide molecule has one unpaired electron, so formation of a covalent bond between them is not unreasonable. This equilibrium has been observed to occur in the gas phase at high pressure and also to take place when nitric oxide is solidified. In mechanism 2, the rate-determining step is viewed as attack of nitric oxide on NO_3, formed in a rapid preequilibrium reaction between NO and O_2. Formation of NO_3 from NO and O_2 in the gas phase has yet to be demonstrated in the laboratory. Both of these mechanisms appear chemically reasonable and both account for third-order kinetics. Distinguishing between them is a difficult chemical problem and one not yet resolved. The bulk of the evidence favors mechanism 1, although it is important to keep an open mind in this case.

Example *Derive rate laws for the oxidation of nitric oxide by mechanisms 1 and 2.*

Solution *Let us begin with mechanism 1. The rate of formation of* NO_2 *is given by*

$$\frac{d(NO_2)}{dt} = k_1(N_2O_2)(O_2)$$

This equation can be put in a more useful form by substituting for (N_2O_2) *using the equilibrium expression for dimerization of nitric oxide:*

$$2NO \rightleftharpoons N_2O_2 \qquad K_{eq_1} = \frac{(N_2O_2)}{(NO)^2}$$

or

$$(N_2O_2) = K_{eq_1}(NO)^2$$

Substituting into the rate law:

$$\frac{d(NO_2)}{dt} = k_1K_{eq_1}(O_2)(NO)^2$$

For mechanism 2, we proceed in the same way:

$$\frac{d(NO_2)}{dt} = k_2(NO_3)(NO)$$

$$K_{eq_2} = \frac{(NO_3)}{(NO)(O_2)}$$

Substituting:

$$\frac{d(NO_2)}{dt} = k_2K_{eq_2}(O_2)(NO)^2$$

Either mechanism can account for the rate decrease as a function of temperature increase. Basically, the reaction rate is determined by the product of the concentration of a reaction intermediate and the rate constant for the conversion of the intermediate to products. This rate constant certainly increases with increasing temperature, as do all rate constants. But the equilibrium constant that determines the concentration of the intermediate may well decrease with increasing temperature (corresponding to a positive ΔH for formation of the intermediate). Provided that the decrease in the equilibrium constant for intermediate formation is greater than the increase in rate constant for its conversion to product, the overall rate of the reaction will decrease with increasing temperature. In cases of this type it is important to remember that the effects of temperature on rate constants and equilibrium constants may be opposite, and that the behavior of the complete system depends on how these opposing tendencies balance out.

There is a puzzle associated with nitric oxide oxidation that we shall uncover now and discuss in a later section. First, let us calculate the total rate of conversion of nitric oxide to nitrogen dioxide. Using 0.1 ppm as a reasonable NO concentration in an atmosphere with some air pollution and 23.15 percent as the fraction of oxygen, we calculate their concentrations to be (remembering that 1 mole of a perfect gas at 25°C occupies 22,400 cm^3 = 22.4 liters):

$$(\text{NO}) = \frac{1 \times 10^{-7}}{22400} = 4.46 \times 10^{-12} \text{ mole/cm}^3$$

$$(\text{O}_2) = \frac{1}{22400} \times 0.2315 = 1.03 \times 10^{-5} \text{ mole/cm}^3$$

Taking the rate constant for 0°C from Table 18-5 and inserting these figures in the rate expression gives

$$\frac{d(\text{NO}_2)}{dt} = 7.88 \times 10^9(4.46 \times 10^{-12})^2(1.03 \times 10^{-5}) = 1.6 \times 10^{-18} \text{ mole/cm}^3 \text{ sec}$$

Note that only about one-millionth of the NO is converted to NO_2 each second and that the rate of the reaction is seen to be far too low to account for its rapid disappearance. We shall return to this problem later.

Nitrogen dioxide also exhibits some unexpected behavior that complicates our discussions of its role in the atmosphere. The nitrogen dioxide molecule has 23 electrons and consequently must have at least one unpaired electron in the ground state. A simple Lewis structure for the molecule would be

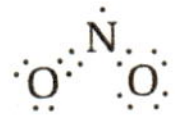

with a single unpaired electron residing on the nitrogen atom. Note that we have shown a structure with a double bond to one of the oxygens and only a single bond to the second. The structure of NO_2 has been determined, and it is known that the two N—O bonds are exactly equal in length and are joined in a nonlinear fashion with an O—N—O angle of 132°. In valence bond terms we then prefer to describe the electronic structure as a re-

sonance hybrid with partial double bonding between the nitrogen and each of the oxygens:

In molecular orbital terms the ground state of the NO_2 molecule (ignoring the electrons in the 1s shells of each atom) is

$$(2s_a)^2(2s_b)^2(\sigma_z)^2(\pi_y)^2(2p_{x_a})^2(2p_{y_b})^2(\pi_y)^2(\sigma_x)^1$$

When we recall that nature prefers systems with electron pairs, it may seem reasonable that nitrogen dioxide dimerizes readily to dinitrogen tetroxide,

$$2NO_2 \rightleftharpoons N_2O_4 \tag{18-7}$$

and the relative concentrations of the two species will be governed by an equilibrium constant of the form

$$K_{eq} = \frac{(N_2O_4)}{(NO_2)^2} \tag{18-8}$$

In our earlier discussion in Chapter 8 we did not deal with equilibria in the gas phase of the type represented here, in which the total number of molecules present changes as the reaction proceeds. To be able to deal with such reactions, we must briefly consider the general problem.

For reactions involving gases, we can express concentrations in two convenient ways: in moles per liter as usual or in terms of the partial pressures of the gases (Chapter 13). Thus, for a reaction of the type

$$aA + bB \longrightarrow cC + dD \tag{18-9}$$

we can formulate the equilibrium constant in terms of concentrations or pressures:

$$K_c = \frac{(C)^c(D)^d}{(A)^a(B)^b} \quad \text{or} \quad K_p = \frac{(P_C)^c(P_D)^d}{(P_A)^a(P_B)^b}$$

Are K_c and K_p the same? To provide an answer to this question, let us see if it is possible to relate them directly to one another. To do this, we must find a way to express pressures as concentrations.

At low pressures for most gases, pressures and concentrations can be related through the use of the **ideal gas law.** This approximate relationship, derived on the two assumptions that the gas molecules are infinitesimally small and do not interact with one another, has the form

$$PV = nRT \tag{18-10}$$

where P is the pressure in atmospheres, V is the volume in liters, n is the number of moles of gas, R is the gas constant, 0.082 liter atm/mole deg, and T is the absolute temperature. In a system containing many gases, we can write the ideal gas law for each of them. The total pressure of the system is just given by the sum of the individual partial pressures:

$$P_t = P_A + P_B + P_C + \cdots \tag{18-11}$$

Now, let us solve Equation (18-10) for P and substitute this expression into that for K_p:

$$P_iV_t = n_iRT \qquad i = A, B, C, \ldots$$

or

$$P_i = \frac{n_iRT}{V_t}$$

and

$$K_p = \frac{\left(\frac{n_C}{V_t}\right)^c (RT)^c \left(\frac{n_D}{V_t}\right)^d (RT)^d}{\left(\frac{n_A}{V_t}\right)^a (RT)^a \left(\frac{n_B}{V_t}\right)^b (RT)^b}$$

$$= \frac{\left(\frac{n_C}{V_t}\right)^c \left(\frac{n_D}{V_t}\right)^d}{\left(\frac{n_A}{V_t}\right)^a \left(\frac{n_B}{V_t}\right)^b} (RT)^{c+d-a-b} \tag{18-12}$$

Finally, we need to recognize that, for example, n_A/V_t is just (A). So we can simplify Equation (18-12):

$$K_p = \frac{(C)^c(D)^d}{(A)^a(B)^b} (RT)^{c+d-a-b} \tag{18-13}$$

or

$$K_p = K_c(RT)^{c+d-a-b} \tag{18-14}$$

Equation (18-14) is the explicit relationship between K_p and K_c that we started out to obtain. We see that if the number of moles of gaseous products is equal to the number of moles of gaseous reactants, $c + d = a + b$, then $K_c = K_p$. If this is not true, then these two equilibrium constants will not be numerically equal. At all ordinary temperatures, RT is greater than unity. Hence, if $c + d > a + b$, K_p will be greater than K_c; if $c + d < a + b$, then K_c will be greater than K_p.

Example *For each of the following reactions, indicate the qualitative relationship between K_p and K_c:*

a. $2NO + O_2 \longrightarrow 2NO_2$
b. $NO_2 + O_3 \longrightarrow NO_3 + O_2$
c. $C(s) + O_2 \longrightarrow CO_2$

Solution *In reaction (a) there are two product gas molecules and three reactant gas molecules. It follows that* $K_c > K_p$. *In reaction (b) the number of gas molecules on the two sides of the equation is equal; consequently,* $K_p = K_c$. *This is also true for reaction (c); note that one of the reactants is a solid, not a gas.*

Dinitrogen tetroxide is a curious substance indeed. The simple structure of the monomer with a single electron occupying an orbital of the nitrogen atom suggests that dimerization will occur with formation of an N—N bond:

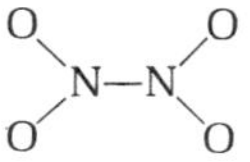

X-Ray diffraction of single crystals in the solid state has confirmed this structure and provided the additional information that the molecule is planar, with an N—N bond length of 1.75 Å. In comparison, the N—N bond distance in hydrazine (Chapter 3) is 1.47 Å. When we remember that for similar bonds the longer one is weaker, we might expect the N_2O_4 molecule to undergo bond cleavage readily, and in fact the enthalpy associated with dimerization is only about 14 kcal/mole at room temperature. Although there is no doubt about the fact that dinitrogen tetroxide exists in the structure described above when crystals are carefully grown, there is also strong evidence that other forms of the dimer exist under suitable conditions. For example, when monomeric NO_2 is rapidly condensed onto a very cold surface in the presence of an inert gas such as argon, in addition to the N—N bonded planar structure there are also forms with the structures

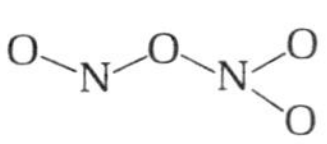

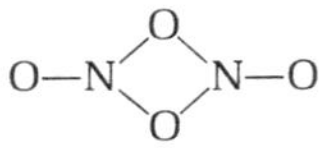

The first of these structures is of special significance, since many reactions of N_2O_4 are most easily interpreted in terms of a reaction where cleavage of the molecule occurs to produce ionic fragments:

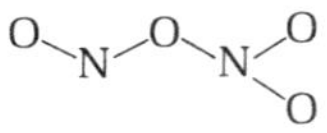 $\longrightarrow ON^+ + NO_3^-$ (18-15)

A complete understanding of the structure and reactions of nitrogen dioxide remains a challenging problem for the chemist.

At the very low concentrations of nitrogen dioxide usually present in the atmosphere, the material is almost exclusively present as NO_2. Nitrogen dioxide monomer is brown in appearance in the laboratory owing to the absorption of light in the 3000–6000 Å region. The N—O bond energy is slightly over 70 kcal/mole, and this corresponds in energy content to light of wavelength near 4000 Å. Thus, energy supplied by visible light is sufficient to cause the reaction

$$h\nu + NO_2 \longrightarrow NO + O \qquad (18\text{-}16)$$

The oxygen atoms produced by the reaction are very important reactants in the atmosphere, and we shall turn to them shortly. Before doing so, let us observe that if a suitable pathway does exist for reoxidation of NO, a cyclic, photochemical process exists that can supply a continuous pool of oxygen atoms to the atmosphere. The importance of this cycle makes it particularly important to resolve the rate question discussed above for NO oxidation.

THE ROLE OF OXYGEN ATOMS

Atomic oxygen is highly reactive, and it is important to examine its chemistry in the atmosphere. Oxygen atoms undergo a particularly important reaction with oxygen molecules:

$$O_2 + O \longrightarrow O_3 + 24.6 \text{ kcal/mole} \qquad (18\text{-}17)$$

The free energy of formation of ozone is +39.06 kcal/mole; thus, we do not expect molecular oxygen to be converted spontaneously to ozone in appreciable quantities. However, oxygen atoms carry with them nearly 60 kcal/mole of excess energy (half the bond energy of the O_2 molecule), and this is sufficient to overcome the thermodynamic barrier and make ozone formation exothermic. The kinetics of this reaction are interesting. As in the case of NO oxidation, the rate of ozone formation decreases as the temperature increases. A three-step mechanism offers an explanation for this behavior. The primary step is combination of an oxygen atom with an oxygen molecule to form an electronically excited ozone molecule:

$$O + O_2 \longrightarrow O_3^* \qquad (18\text{-}18)$$

The excited ozone molecule may now undergo either of two reactions:

$$O_3^* \longrightarrow O + O_2 \qquad (18\text{-}19)$$

$$O_3^* + M \longrightarrow O_3 + M \qquad (18\text{-}20)$$

If the excited O_3^* molecule collides with another molecule (perhaps N_2), energy can be transferred to the second molecule and the ozone molecule returns to the ground state. Both formation of O_3^* and its collisional deactivation are expected to occur more rapidly at higher temperatures. If, on the other hand, the rate of dissociation of O_3^*, Equation (18-19), increases more rapidly with increasing temperature than does the rate for Equation (18-20), it follows that dissociation will become relatively more important than formation of ground-state O_3. The net result is that the reaction to form ozone becomes slower as the temperature increases.

Oxygen atoms are so reactive that they react with most of the components of normal or polluted air. The fate of a particular atom depends on the

Table 18-6 Some important fates of oxygen atoms in the atmosphere

Reaction	Concentration of reactant other than O (pphm)[a]	Estimated relative rate at 25°C
$O + O_2 \longrightarrow O_3$	2×10^7	100
$O + O_3 \longrightarrow 2O_2$	50	10^{-4}
$O + O \longrightarrow O_2$	—	6×10^{-10}
$O + N_2 \longrightarrow N_2O$	8×10^7	7×10^{-4}
$O + NO \longrightarrow NO_2$	10	0.01
$O + NO_2 \longrightarrow NO + O_2$	10	0.02
$O + NO_2 \longrightarrow NO_3$	19	0.07
$O + CH_4$	250	0.001
O + higher paraffins (estimated average collision yield $\sim 10^{-4}$)	25	0.001
$O + C_2H_4$	50	0.014
O + higher olefins	25	0.1

[a]pphm = parts per hundred million.

Source: P. A. Leighton, *Photochemistry of Air Pollution,* Academic, New York, 1961.

Table 18-7 Summary of rate constants of ozone reactions

Reactant other than ozone	k (liters/mole sec at 25°C)	Representative rate in polluted air (pphm/hr)[a]
NO	1×10^7–3×10^7	~200
NO_2	2×10^4–4×10^4	3–6
Olefin hydrocarbons	10^3–10^5	2–20
Paraffin hydrocarbons	Too small to measure	—
Acetylene	~46	$\sim 10^{-2}$
Aromatic hydrocarbons	Not known; probably small	Probably $<10^{-1}$
Formaldehyde	Possibly ~10	Probably $<10^{-2}$
Aliphatic aldehydes and ketones	Not known; possibly between 10^1 and 10^3	?

[a]pphm/hr = parts per hundred million per hour.

Source: P. A. Leighton, *Photochemistry of Air Pollution,* Academic, New York, 1961.

relative probability of collision with the various reactants (controlled by their relative concentrations) and on the rate constants for each reactant. Thus, an air collision with an N_2 molecule is quite common, since N_2 accounts for three out of every four molecules; however, the rate constant for reaction of oxygen atoms with N_2 is very small, and the result is that few oxygen atoms react with N_2. On the other hand, the rate constant for reaction of O atoms with NO molecules is very large, but so few molecules of NO are present that the probability of collision is small. The net result is that the rate of reaction of O atoms with NO is small.

The last column in Table 18-6 gives the estimated relative rates of the various reactions of atmospheric oxygen atoms. By far the most important reaction is combination with the oxygen molecule. A major factor in the importance of this reaction is that oxygen is so much more abundant than any other molecule with the exception of nitrogen. Oxygen atoms formed by photochemical decomposition of NO_2 are almost exclusively converted into ozone by the oxygen reaction.

Ozone plays a most important role in atmospheric chemistry, and we discuss it below. For the moment, it is important to us, however, not only because it is produced from oxygen atoms but also because it produces them [Equation (18-1)]. Ozone absorbs strongly in the near ultraviolet, with a maximum at about 2500 Å. Light at this wavelength supplies sufficient energy to *photolyze* O_3 to molecular oxygen plus oxygen atoms. This reaction supplies a common alternative source for oxygen atoms. Since these atoms commonly simply react with another oxygen molecule to form ozone, the photolytic reaction has little net effect on either oxygen atom or ozone concentration. However, the absorption coefficient for ozone is so large (3250 liters/mole cm) that even the small amounts of ozone present at equilibrium serve to screen out quite effectively most of the ultraviolet light that reaches the upper atmosphere. Since many life forms, including man, are very sensitive to high intensities of ultraviolet light, this screening is very fortunate indeed. If all the ozone in the atmosphere were present as a single layer of the pure gas surrounding the earth as a shell, it would be only about 2.2 mm thick. This may seem like remarkably little ozone to act as a shield against ultraviolet radiation from the sun; in fact, it works very well indeed. For example, at a wavelength of 2500 Å, only about one part in 10^{10} of the light impinging on the upper atmosphere finds its way to the surface of the earth.

Finally, let us observe that reactions of oxygen atoms with hydrocarbons are important. Particularly, reactions with ethylene and other olefins leads to (among other products) the formation of oxidized compounds (aldehydes) that are eye irritants. Similar products are produced through oxidation with ozone.

PROPERTIES OF OZONE

Ozone is involved in a number of reactions in the atmosphere that are of interest to us. A few rate constants for these are collected in Table 18-7. Observe that the most common fate of ozone is the reaction

$$NO + O_3 \longrightarrow NO_2 + O_2 \qquad (18\text{-}21)$$

However, the subsequent photolysis of NO_2 and reaction of the oxygen atom with an oxygen molecule eventually replaces the ozone and NO. This is a cycle that serves to stabilize atmospheric concentrations of the various reactants.

Reaction also occurs between ozone and nitrogen dioxide, although at a much lower rate. Here a sequence of reactions,

$$NO_2 + O_3 \longrightarrow NO_3 + O_2 \qquad (18\text{-}22)$$
$$NO_3 + NO_2 \longrightarrow N_2O_5 \qquad (18\text{-}23)$$
$$N_2O_5 + H_2O \longrightarrow 2HNO_3 \qquad (18\text{-}24)$$

leads to eventual removal of both the ozone and the nitrogen oxide from the atmosphere.

Observe from Table 18-7 that the rate of attack of ozone on olefins is substantial. Again, irritating oxidized species are produced particularly in areas where high hydrocarbon emission from automobiles is common. In a typical study it was found that 1-pentene,

$$CH_2{=}CH{-}CH_2{-}CH_2{-}CH_3$$

reacts with ozone to produce $\frac{1}{2}$ mole of formaldehyde, CH_2O, and $\frac{1}{2}$ mole of butyraldehyde, C_4H_8O, per mole of ozone consumed:

$$CH_2{=}CH{-}CH_2{-}CH_2{-}CH_3 + O_3 \longrightarrow CH_2{=}O + O{=}CH{-}CH_2{-}CH_2{-}CH_3 + \text{other products} \qquad (18\text{-}25)$$

The annoying properties of formaldehyde are familiar to those who have had a laboratory course in elementary biology.

Ozone also displays a number of complex reactions that are responsible for several of the detrimental properties of smog. For example, in serious smog attacks in Los Angeles, there is significant destruction of rubber. This is noticed particularly in automobile tires, in rubber tubing, and other rubber products under tension. In fact, observation of the quantitative rate of cracking of rubber bands has been used as a precise measure of smog concentration.

Let us examine the cause of cracking a bit more closely. Rubber is a product isolated from nature which can be closely imitated by synthesis in the laboratory. When natural rubber samples are heated, they decompose to form a hydrocarbon called *isoprene,* which has the molecular formula C_5H_8 and the structure

$$CH_2{=}C(CH_3)CH{=}CH_2$$

Synthetic rubber can be made by polymerization (Chapter 7) of isoprene to form a polymer of basic structure

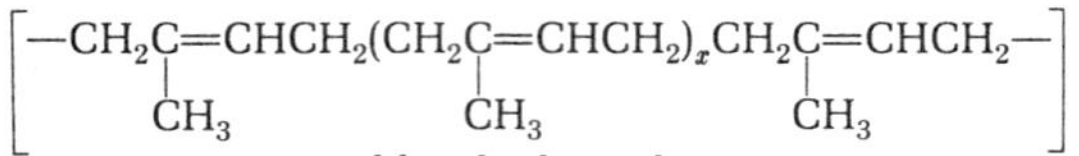

rubber hydrocarbon

The structure of natural rubber is probably quite similar to that of the artificial polymer. Either natural rubber or its laboratory-produced imitation tends to be relatively soft and break down under gentle heating. Furthermore, when cooled to low temperatures, it becomes hard and brittle. Both of these properties can be substantially eliminated by the process of *vulcanization,* discovered by Charles Goodyear in 1839. The process he developed was one of heating natural rubber with elemental sulfur, producing cross-linking between chains.

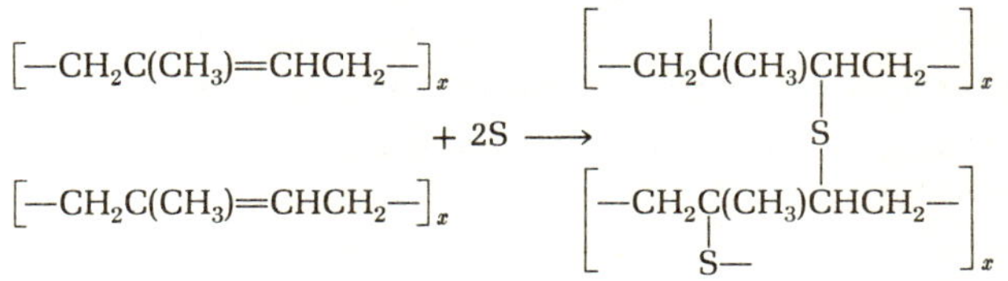

(18-26)

Only a few cross-links are necessary to increase substantially the molecular weight of the polymer and to change its character markedly (see Chapter 7). If only 1 percent of the double bonds in crude rubber are cross-linked, a suitable product for many common uses is produced. The process to form hard rubber introduces sulfur into only about one-third of the bonds. Ozone is known to be fairly reactive towards double-bonded systems, and in polluted atmospheres rubber is attacked at the remaining double bonds by reactions such as

$$\text{rubber} \xrightarrow{O_3} O{=}CHCH_2CH_2C(CH_3){=}O \qquad (18\text{-}27)$$

levulinic aldehyde

Although cross-linked bonds introduced by sulfur addition remain intact in this process, the molecules are cleaved at a variety of sites. A smaller polymer with lowered tensile strength results, and cracking eventually occurs.

Ozone also has been shown to attack green plants directly. Plant damage in areas of high smog concentration is frequently observed.

POSTULATED MECHANISM OF SMOG FORMATION

Many studies have been made in an effort to understand completely the chemistry of smog formation. Although much remains to be learned, a general picture has emerged, which is summarized in the sequence of reactions in Table 18-8. It is apparent that this scheme incorporates some reactions other than those we have discussed above. Many of the details remain to be worked out, but it should be apparent why emission of nitrogen

Table 18-8 A simplified reaction scheme for photochemical smog formation[a]

$NO_2 + h\nu \longrightarrow NO + O\cdot$	Formation of atomic oxygen
$O\cdot + O_2 \longrightarrow O_3$	Ozone production, principal fate of O·
$O\cdot$ + hydrocarbon(Hc) $\longrightarrow HcO\cdot$	Oxidation of hydrocarbons (first step)
$O_3 + NO \longrightarrow NO_2 + O_2$	Production of NO_2 leading to HNO_3
$HcO\cdot + O_2 \longrightarrow HcO_3\cdot$	Second stage in hydrocarbon oxidation
$HcO_3\cdot + Hc \longrightarrow$ aldehydes, ketones	A terminal step in hydrocarbon oxidation
$HcO_3\cdot + NO \longrightarrow HcO_2\cdot + NO_2$	Intermediate step in hydrocarbon oxidative chemistry
$HcO_3\cdot + O_2 \longrightarrow O_3 + HcO_2\cdot$	Second route for production of $HcO_2\cdot$
$HcO_x\cdot + NO_2 \longrightarrow$ peroxyacylnitrates	Peroxyacylnitrate production

[a] $HcO\cdot$, $HcO_2\cdot$, $HcO_3\cdot$, and $HcO_x\cdot$ are free radicals (species with unpaired electrons) derived from hydrocarbons and oxygen.

oxides and hydrocarbon residues from automobiles and other sources are a problem of primary concern to pollution control boards. The discussion above should focus our attention on the important role played by light. Furthermore, even in the absence of man, many of the reactions listed have been going on for millions of years. Our search for solutions to some of our air pollution problems will be difficult, but the basic pathways are now understood.

FINAL COMMENT ON THE NITRIC OXIDE OXIDATION PROBLEM

It should be clear that one major gap still exists in our knowledge. The nitrogen dioxide–nitric oxide cycle for oxygen atom generation, and thus a controlling step in rate of ozone formation, requires further study. Direct oxidation appears to be too slow, and the ozone oxidation of NO is unproductive: It consumes the important product. One suggestion that has been advanced is that excited state oxygen molecules may play an important role, but further studies will be required to confirm this suggestion.

Glossary

Atomic oxygen a highly reactive species important in photochemical smog generation. Its most important reactions are (1) with O_2 to form ozone; (2) with NO to form NO_2; and (3) with a variety of hydrocarbons to form several oxidized products, most of which are toxic or irritants or both.

Ideal gas law an equation relating the pressure, volume, and temperature of a sample of an idealized gas; specifically, $PV = nRT$, where n is the number of moles of gas in the sample. Most real gases obey this equation rather accurately at low pressures.

Nitric oxide NO, another important atmospheric contaminant; converted to nitrogen dioxide and excited molecular oxygen through oxidation by ozone; a precursor of atmospheric nitric acid.

Nitrogen dioxide NO_2, an important atmospheric contaminant; it is quite toxic itself and is converted to nitric acid through oxidation and to nitric oxide and atomic oxygen through photolysis.

Oxygen O_2, the second most abundant gas, after N_2, in the atmosphere. O_2 has a triplet ground state, $^3\Sigma_g^-$, but the two lowest excited states are both singlets, $^1\Delta_g$ and $^1\Sigma_g^+$. Photochemical and sensitized excitation yield these excited states. Excited O_2 may play an important part in photochemical smog formation. O_2 is also an important trap for oxygen atoms.

Ozone O_3, important in atmospheric photochemistry; (1) photolysis yields atomic oxygen; (2) reaction with NO yields NO_2 and excited molecular oxygen; and (3) reaction with hydrocarbons yields a variety of oxidized products.

Photolysis the photochemical cleavage of a molecule.

Study problems

1. Summarize briefly the principal sources of excited oxygen in the atmosphere, emphasizing why direct photochemical excitation is not the most important among these. Qualitatively, how would you expect the importance of these sources to vary as a function of the degree of contamination of the atmosphere?
2. Describe how the photochemistry of the atmosphere would change were those electronic transitions involving a change in spin multiplicity not forbidden.
3. Draw Lewis electron dot structures for O_3, NO, and NO_2. What is the spin multiplicity of the ground state of each of these species?
4. In areas having high levels of atmospheric contamination, rain is often quite acidic. How can you account for this fact? What is the sequence of reactions leading to the formation of an acidic product in the atmosphere?
5. Suppose that it were possible to exclude completely the oxides of nitrogen from the atmosphere. What changes in atmospheric photochemistry would you expect?
6. Describe as clearly as you can the nature of the changes involved in the photochemical formation of the two lowest excited states of the oxygen molecule. Is promotion of an electron to a higher energy molecular orbital involved?
7. The extinction coefficient of ozone at 2550 Å is 3250 liters/mole cm. Calculate the fraction of light of this wavelength that emerges from a

1-mm layer of pure ozone. Assume that ozone behaves as a perfect gas and that the measurement is performed at a pressure of ozone of 1 atm and at 298°K.

8. The terminal reaction in the atmosphere leading to the formation of nitric acid is $3NO_2 + H_2O \longrightarrow 2HNO_3 + NO$. From this stoichiometry, what can you conclude about the rate law for this reaction? What can you conclude about the mechanism of this reaction?
9. Both NO and NO_2 react with ozone in second-order reactions having rate constants $k_{NO} = 2 \times 10^7 M^{-1}\ min^{-1}$ and $k_{NO_2} = 3 \times 10^4 M^{-1}\ min^{-1}$. If both of these nitrogen oxides are present together with ozone in a single reaction mixture and if $(NO_2) = 50(NO)$, will ozone react more rapidly with NO or with NO_2?
10. The reaction of oxygen atoms with nitric oxide is chemiluminescent:

    $O + NO \longrightarrow NO_2$

    From this information, what can you conclude about (a) the thermodynamics of the reaction; (b) the rate of the reaction; (c) the electronic structure of the product; (d) the change in spin multiplicity during the reaction?
11. Qualitatively, how do K_p and K_c compare for each of the following reactions?

    a. $2NO \longrightarrow N_2 + O_2$
    b. $2CO \longrightarrow CO_2 + C(s)$
    c. $2SO_3 \longrightarrow 2SO_2 + O_2$
    d. $2H_2 + S_2(g) \longrightarrow 2H_2S$
    e. $2NH_3 \longrightarrow 3H_2 + N_2$
12. Provide two possible explanations for an observation that the rate of a particular reaction decreases with increasing temperature.

Suggestions for further reading

P. A. Leighton *Photochemistry of Air Pollution,* Academic, New York, 1961.

R. E. Newell "The Global Circulation of Atmospheric Pollutants," *Sci. Am.,* **224,** 32 (January, 1971).

H. S. Stoker and S. E. Seager *Environmental Chemistry: Air and Water Pollution,* Scott, Foresman, Glenview, Ill., 1972.

A. M. Trozzolo, Ed. "Singlet Molecular Oxygen and Its Role in Environmental Sciences," *Ann. N.Y. Acad. Sci.,* **171,** 1 (1970).

B. Weinstock and H. Niki "Carbon Monoxide Balance in Nature," *Science,* **176,** 290 (1972).

19

Utilization of solar light energy

An early example of the utilization of solar energy:
Your serpent of Egypt is bred of your mud by the operation of your sun: so is your crocodile.

William Shakespeare
Antony and Cleopatra, act II, scene vii

In the earlier chapters of this section we have described some of the properties of light and its interaction with matter as well as the chemistry of light generation in the sun. Now we turn our attention to the ways in which man utilizes the radiant energy that arrives at the earth from the sun. In part, our uses are indirect. For example, sunlight drives the photosynthetic process, described later in the chapter, which generates our foodstuffs. The sunlight-dependent growth of plants in the geological past indirectly has supplied us with fossil fuels—oil, gas, and coal—which are very important as energy sources. Finally, we may note that light energy is responsible for *ocean recycle*. In this process, heat from sunlight causes evaporation of water from the oceans, which is carried, in the form of clouds, over land. There, the water returns to earth as rain, is collected by streams and rivers, and finally returns to the ocean. This overall cycle permits us to generate power hydroelectrically, taking advantage of energy liberated when water falls from the highlands back to sea level.

In addition to indirect uses of radiant energy from the sun, limited *direct* use is made of this energy source in the form of **solar cells, solar stills,** and **solar furnaces.** While these are not quantitatively important energy uses, they do promise to receive additional attention in the years to come and are worthy of our attention.

SOLAR CELLS

In a solar cell, light is employed to accomplish photochemical reactions in a way that can generate electrical energy. To understand the operation of such a device, one example of which is the **photovoltaic cell,** we must first explore the properties of **semiconductors,** materials that exhibit electrical conductivity intermediate between that of metals and insulators. The unique properties of semiconductors depend in many cases on crystal imperfections—foreign atoms incorporated into the crystalline lattice, small deviations from stoichiometry, or lattice defects. Among the most familiar semiconductor materials are germanium, silicon, indium antimonide, zinc oxide, and cadmium sulfide.

A schematic two-dimensional projection of the covalently bonded carbon atoms in a perfect diamond lattice looks like

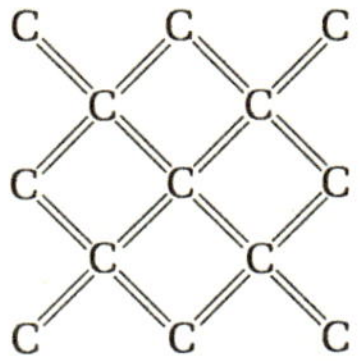

Each carbon atom contributes four valence electrons and is tetrahedrally bonded to four neighboring carbons, so all available electrons are utilized in forming the covalent bonds.

In many semiconductors it is important to recognize two kinds of carriers of electrical current—*electrons* and *holes*. If an extra electron is added to this structure, no empty low-lying orbitals are available, and the electron is more or less free to wander through the solid.

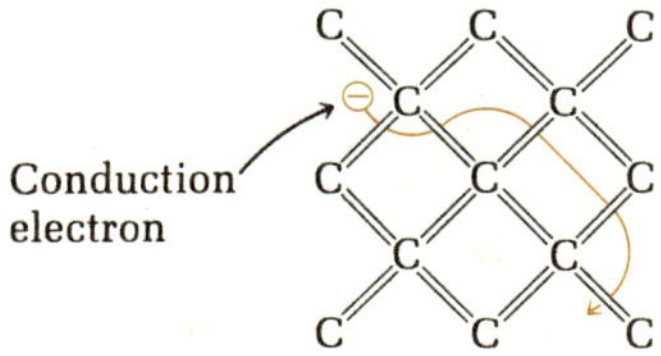

If an electrical field is applied, the electron will move in the direction opposite to the field, and can contribute to the electrical conductivity. Such an electron is called a **conduction electron;** if enough conduction electrons are present, the carbon sample will no longer be an insulator.

A second way in which the total number of electrons fails to match the available number of electron bonding sites is that there may be too *few* electrons. There is then only one electron in some of the bonds,

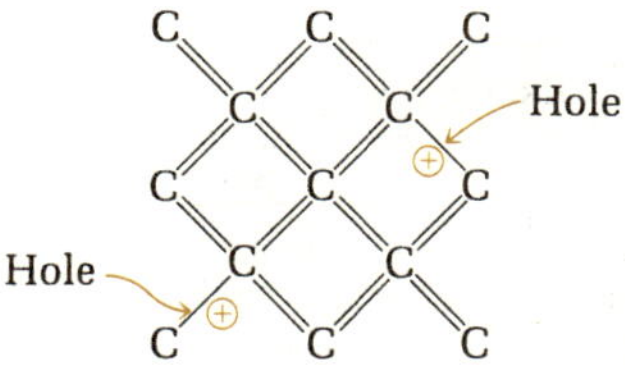

and the position where an electron is missing is called a **hole.** A hole (which is simply a half-occupied orbital) moves through a crystal as an electron from an adjacent bond jumps into the vacant position. This movement is in the opposite direction from that of a conduction electron:

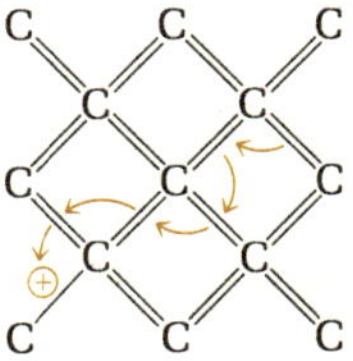

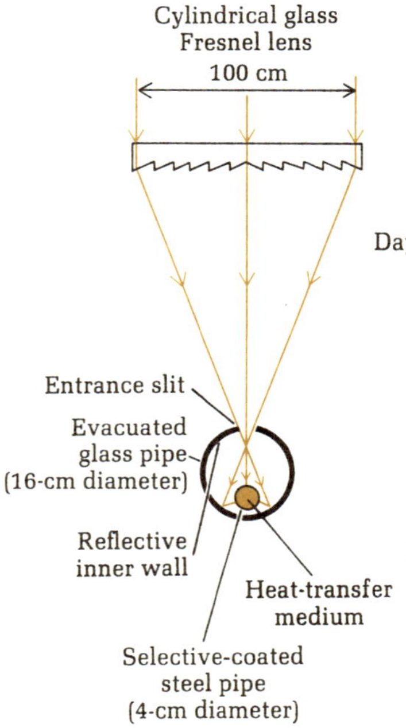
Cylindrical glass
Fresnel lens
100 cm
Entrance slit
Evacuated
glass pipe
(16-cm diameter)
Reflective
inner wall
Heat-transfer
medium
Selective-coated
steel pipe
(4-cm diameter)

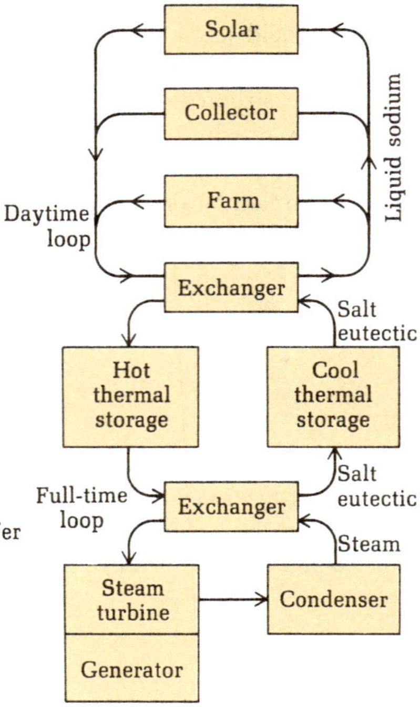
Solar
Collector
Farm
Liquid sodium
Daytime
loop
Exchanger
Salt
eutectic
Hot
thermal
storage
Cool
thermal
storage
Salt
eutectic
Full-time
loop
Exchanger
Steam
Steam
turbine
Condenser
Generator

A hole can contribute to electrical conductivity just as effectively as a conduction electron; the total conductivity results from motion of both kinds of carriers in the presence of an applied electrical field:

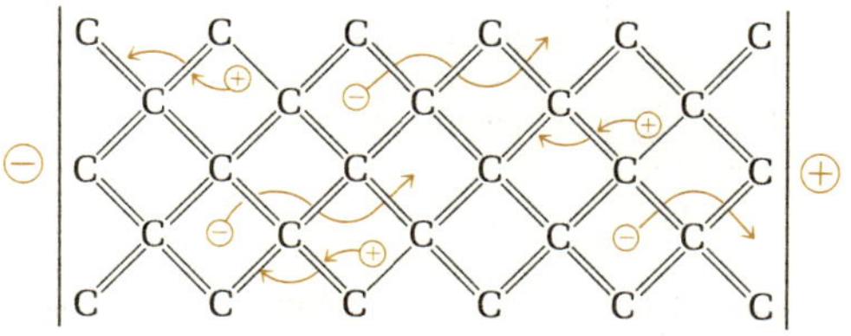

We shall turn next to a consideration of the **band theory of solids.** For simplicity, let us examine the molecular orbital energy-level diagram for a pair of lithium atoms and then for a collection of many, say *N*, lithium atoms:

One of the most innovative and significant suggestions for utilization of solar energy is the "solar power ranch" illustrated here and developed by A. B. Meinel and M. P. Meinel of the University of Arizona. At the top is an artist's conception of a solar power ranch showing the extensive areas required for a 1000-megawatt plant. The long ramadas are spaced by sun-angle requirements and allow a dual use of land. The collector arrays use cylindrical Fresnel lenses to concentrate sunlight on the absorbing surface of a steel pipe enclosed in a vacuum envelope. Some details of how this works are shown at the bottom on the left. A heat-transfer fluid is pumped through the steel pipe to carry the collected energy to a central location in the background where the energy is stored until it is used by the thermodynamic cycle of the power plant. The schematic design of the solar energy ranch is shown in the diagram on the right. Note that thermal storage to provide energy for the power plant overnight and on cloudy days is part of the ranch design.

The high temperatures needed by the thermodynamic cycle are accomplished by use of spectrally selective optical coatings on the steel pipe. Such a "selective coating" absorbs sunlight but suppresses thermal reemission.

The heat-transfer fluid and thermal storage medium can be different materials or a common material. Candidate materials are compressed gases, organic liquids, molten salt eutectics, or liquid metals. Thermal energy storage can utilize the specific heat of the material as well as changes of phase of materials. Such a system as is represented by solar power ranches involves many trade-offs between technological and environmental factors relating to the choice of materials.

Thermodynamic factors are also important. For example, even though such a system utilizes energy that would reach the earth in any event (and therefore not significantly alter the thermal balance of the earth), micro-environmental questions need to be addressed since any thermodynamic cycle producing work out of heat has "waste" heat to be rejected into the environment. Hence, the slight cooling in the shadow of each ramada is offset by the heating at the condenser towers of the power plant. (Courtesy of the University of Arizona.)

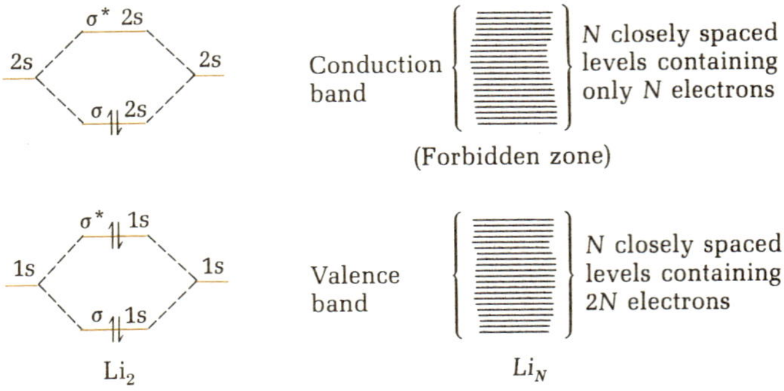

The band theory of solids is somewhat related to molecular orbital theory for molecules (see Chapter 5). If we imagine a crystal formed just from the nuclei that occupy their fixed positions, we can ask what happens if we begin to pour in electrons. If the nuclei were isolated, the electrons would simply associate with individual nuclei. But in a solid the nuclei are not isolated, and the orbital extends over more than a single nucleus. Consequently, what matters are not the atomic orbitals of the individual nuclei but orbitals, or energy levels, that extend over the whole crystal. Each atomic orbital contributes one energy level (orbital) to the crystal. Since there are a great many atomic orbitals, corresponding to the enormous number of atoms in the crystal, a very large number of closely spaced energy levels will be present. These are called a **band.**

For a more concrete picture, let us consider a simple case of atoms spaced equally along a line at separation r and gradually reduce the distance r to improve overlap between individual atomic orbitals.

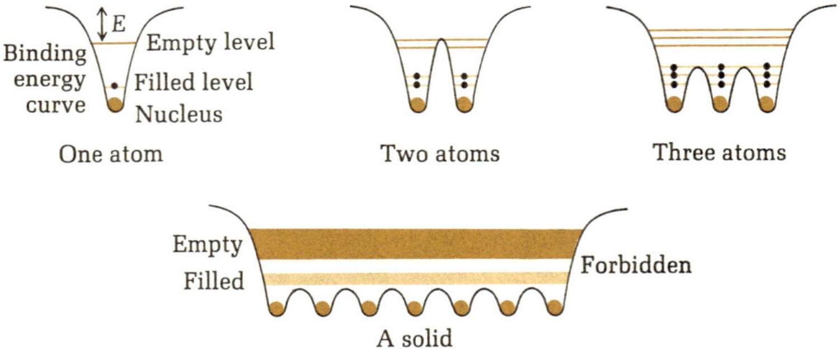

If we are considering an array of sodium atoms, for example, there are just sufficient electrons to fill completely the 1s, 2s, and 2p orbitals, with a single electron left over to occupy the 3s atomic orbital. In the diagram above (top center), when the atomic separation is large the electrons just fill the lower lying orbitals (with one left over in the 3s orbital) and are in quite well-defined energy states. As the atoms move closer together (top right of diagram above), the possible energy spread becomes larger for each state.

Here an electron in the 3s level has enough energy to "pour" over the barrier to the next atom and thus conduct electricity. Note particularly that with sufficient atoms in the array, the widths of the 3s and the 3p bands become large enough so that there is actual overlap and an electron that is in the 3s state in an isolated atom can move through the broad 3p band of the solid. Thus, the electrons in a solid such as magnesium (with a filled 3s shell) are not restricted to the atomic locations but are also free to move and make magnesium a good conductor.

The band theory of solids provides a convenient way to discuss a few of the characteristic properties of metals. First, the large number of mobile electrons renders a metal highly conductive. Second, the electrical conductivity of metals *decreases* as temperature increases, because increased vibrations of atoms in a metal lattice interfere with movement of conduction electrons. Third, the mobile electrons in a conduction band absorb and transport thermal energy, which accounts for the high thermal conductivity of metals. Fourth, the ductility and malleability of metals are due to the facile making and breaking of metal–metal bonds, since the bonding electrons are not localized with respect to any particular atoms.

Using the band theory, we can distinguish easily metals, insulators, and semiconductors. As noted in the picture above, orbitals in a solid are organized into a *valence band,* in which the electrons have a minimum freedom to move about, and a *conduction band,* in which the electrons are mobile. Between these two, there is an energy gap, the *forbidden zone,* which is unavailable for occupancy by electrons. These considerations are shown schematically in Figure 19-1.

There are two types of semiconductors, those with no electrons in the conduction band and those with either holes in the valence band or a small number of electrons in the conduction band. *Intrinsic* semiconductors are those solids with a sufficiently narrow forbidden zone so that simply heating the solid may suffice to move some electrons to the conduction band. For example, gray tin has a forbidden zone only 2.3 kcal wide, and electrons can relatively easily jump the forbidden zone to the conduction band. In principle, any solid would become a semiconductor at sufficiently high temperature, but in practice it may melt or vaporize first. In any event, in contrast to metals, the conductivity of semiconductors *increases* with increasing temperature as more electrons are fed into the conduction band.

Impurity (or *extrinsic*) semiconductors, on the other hand, depend on deliberate contamination of a pure solid to introduce holes into the valence band or a few electrons into the conduction band. Pure germanium, for example, has a conduction band that is separated from the valence band by 16.6 kcal, and at normal temperatures the metal is not at all a good electrical conductor. Deliberate introduction of impurities can drastically

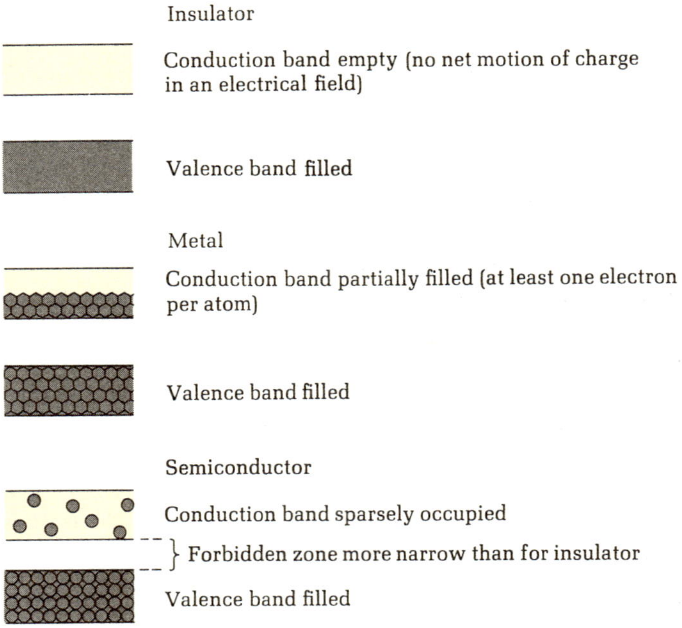

Figure 19-1 Distinction between an insulator, a metal, and a semiconductor in terms of the band theory of solids.

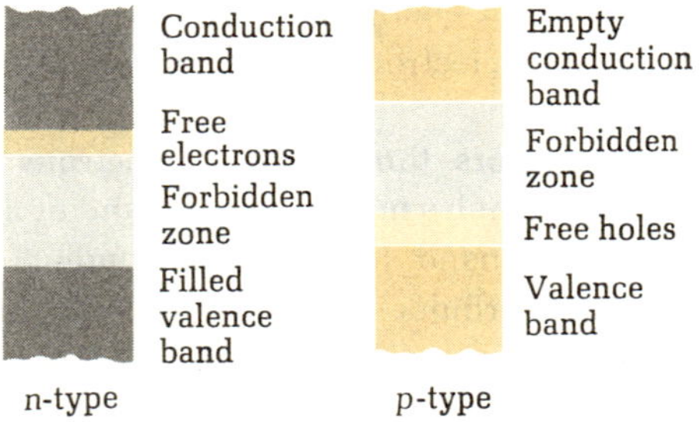

Figure 19-2 Energy-band models for *n*-type and *p*-type semiconductors.

alter the situation, however. For example, if an atom very similar to the germanium atom but with a different number of electrons can be introduced into the solid structure, either extra electrons or holes will result. Indium and arsenic atoms can successfully replace individual germanium atoms in the solid lattice. If pure germanium is deliberately *doped* with small amounts of arsenic, the resulting solid has an occasional extra electron contributed by the impurity and the solid becomes a semiconductor. On the other hand, if the crystal is doped with indium, each indium atom carries with it one too few electrons to fill the valence band and a hole results. Again the metal becomes a more useful semiconductor.

A diagrammatic view of impurity semiconductors of these two types is shown in Figure 19-2. Some properties of a variety of semiconductors are collected in Table 19-1.

The photovoltaic cell

The photovoltaic cell takes advantage of the properties of semiconductors just described. Two types of semiconductors are employed: a *p* layer, which is electron-deficient so that positive holes may move within it when a sufficient exciting force is applied; and an *n* layer, which is electron-rich so that negative electrons may be excited to move within it. An arsenic-doped germanium crystal is an *n*-type semiconductor. In contrast, an indium-doped germanium crystal is a *p*-type semiconductor.

Example *If a germanium crystal is doped with (a) phosphorus and (b) aluminum, what type(s) of semiconductors would result?*

Solution *Inspection of the periodic table (Figure 3-1) reveals that phosphorus possesses one more valence electron than germanium. This phosphorus, like arsenic, will form an n-type semiconductor when added to germanium. Aluminum, in contrast, contains one less valence electron than germanium and will, therefore, form a p-type semiconductor.*

What happens if we now bring these two crystals together so as to form a ***p*-*n* junction** between them as shown in Figure 19-2? At first, electrons will tend to flow from the *n* layer to the *p* layer and positive holes will flow in the opposite direction, since electrons and holes have an affinity for each other. As this process continues, a net charge will develop on each layer, positive on the *n* layer and negative on the *p* layer. The result is the formation of a potential difference across the *p*–*n* junction, and this will tend to impede further migration of electrons and holes across this junction. At equilibrium, the potential difference across the junction will just balance the affinity of electrons for holes, and migration will cease. Now, if light of sufficient energy to promote electrons into the conduction band impinges on the *p*–*n* junction, the potential difference present will immediately pull the electrons toward the positive *n* layer and the holes toward the negative *p* layer. If there are no external connections linking the two sides of the junction, this will result in a current flow across the junction, restoring the equilibrium condition. However, if such connections are present, a portion of this current will flow through the external circuit, generating useful electrical energy. These considerations are summarized in Figure 19-3.

The way that a solar cell of this type operates places some conditions on how it is constructed. First, either the *p* layer or the *n* layer must be quite thin to permit solar light quanta to penetrate to the *p*–*n* junction. Light absorbed far from this junction forms electrons and holes, but there is no

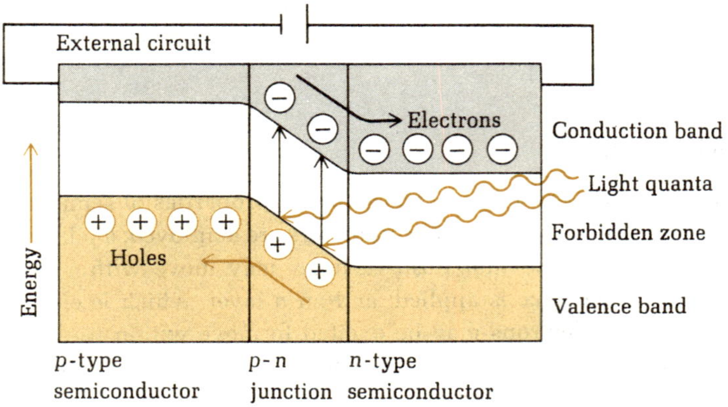

Figure 19-3 A schematic view of photovoltaic cell. The conduction band is shown as the shaded area above, the valence band as the shaded area below, and the forbidden zone as the clear area between. The light energy falling on the *p-n* junction causes electrons to be swept to the *n*-type layer and positive holes to be swept to the *p*-type layer as shown. This will cause a current to flow in the external circuit, generating useful energy.

Table 19-1 Some typical semiconductors and their associated forbidden energy gaps

Name	Chemical symbol	Forbidden energy gap (eV)
Germanium	Ge	0.72
Silicon	Si	1.1
Diamond	C	5.3
Indium antimonide	InSb	0.18
Gallium arsenide	GaAs	1.34
Gallium phosphide	GaP	2.25
Cadmium sulfide	CdS	2.45
Cadmium selenide	CdSe	1.74
Cadmium telluride	CdTe	1.45
Lead sulfide	PbS	0.37
Lead selenide	PbSe	0.27
Lead telluride	PbTe	0.33
Zinc oxide	ZnO	3.3
Magnesium oxide	MgO	7.3

driving force for their separation and they simply fall back together. Second, the amount of energy generated depends on the total radiant energy captured. Hence, cells are usually constructed with large *p–n* junction areas.

Example *What is the principal consideration that would dictate the nature of a semiconductor employed in a solar cell?*

Solution *The crucial consideration is the magnitude of the energy gap between the valence and conduction bands. The bulk of the light that reaches the surface of the earth from the sun has an energy near 2.5 eV (as you may verify by a simple calculation, see Chapters 14 and 18). Hence, any semiconductor having an energy gap greater than this value would be useless for the construction of solar cells. In fact, silicon, with a gap of 1.1 eV (Table 19-1) is the most frequently employed semiconductor for solar cells.*

There are several advantages to the use of solar cells for generation of electricity. First, the ultimate energy source costs nothing. Second, solar cells are simple to construct and are quite long lived. Third, solar cells operate without pollution of the environment. On the other hand, solar cells sufficiently large to generate the massive amounts of electrical energies that urban populations demand will be difficult to construct. Hence, at the moment, they are generally used to provide modest quantities of energy. Every satellite or space ship, for example, depends heavily on solar cells for its energy needs.

SOLAR STILLS

One of man's growing needs is a supply of fresh water. As our population has grown, water requirements for industry, agriculture, and to satisfy needs of individuals have raced ahead. Moreover, the increasingly severe pollution of our lakes, rivers, and oceans threatens to make some current supplies of fresh water unusable for many purposes. This whole matter was dealt with in some detail in the introduction to Part Two and Chapter 13.

One means to obtain pure drinkable water is to distill sea water or brackish water, as was developed in detail in Chapter 13. In the distillation process, water is heated so that evaporation occurs at some reasonable rate; the water that evaporates is quite pure, since salts and other contaminants are not volatile and are therefore left behind. A conceptually simple way of accomplishing this purpose is use of a *solar still*. The unique feature of the solar still is that sunlight is used as the energy source for hastening the evaporation of water. The basic setup is shown in Figure 19-4. A glass or other transparent material covers the reservoir of water so that the hot rays of the sun fall directly on this reservoir. As the water evaporates, it

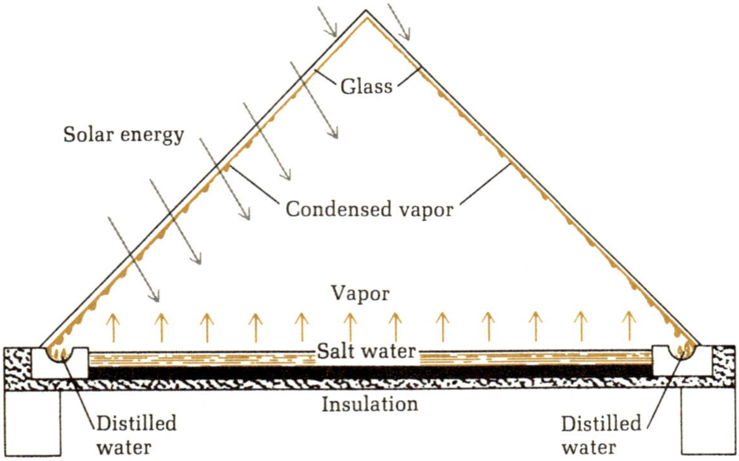

Figure 19-4 A typical solar still. (From K. S. Spiegler, *Salt Water Purification*, Wiley, New York, 1962.)

condenses on the glass cover and runs down it to be collected at the bottom. This pure water may then be drained off and used directly.

There are several advantages to the use of solar stills: There are no power costs, there is no machinery to malfunction, and no pollutants are generated. At the same time, several limitations to the use of the method are also present. One of the most severe is the slowness of the process due to the modest amount of energy available from the sunlight. For example, to generate sufficient water by this means to use for irrigation of farm lands, the solar still itself would have to be so large that it would occupy an area comparable to the area farmed. Clearly, solar stills are not suitable for this purpose. They do serve admirably, however, to satisfy the more modest needs of small groups of people living in areas where an adequate supply of fresh water is lacking. In fact, small solar stills can be purchased that may be transported behind a vehicle and used by those who investigate or explore arid areas.

To understand the limited ability of a solar still more clearly, let us make an approximate calculation concerning the amount of water that we might produce per day per square foot of space occupied by a solar still. The energy of solar radiation is typically

$$\text{energy of solar radiation} = 260{,}000 \text{ cal/cm}^2 \text{ year}$$

To calculate the energy of solar radiation per square foot per day, we simply have to change units, remembering that there are about 900 cm^2 per square foot:

$$\text{energy of solar radiation} = 2.6 \times 10^5 \text{ cal/cm}^2 \text{ year} \times \frac{1}{365 \text{ days/year}} \times \frac{900 \text{ cm}^2}{1 \text{ ft}^2}$$

$$= 6.4 \times 10^5 \text{ cal/day ft}^2$$

Now that we have the amount of energy available in the proper terms, we must ask how much energy is required to vaporize 1 g of water. From Chapter 10, we know that the heat of vaporization is near 540 cal/g and is about the same regardless of the temperature of the water. We can now calculate the total amount of water that we can produce in our still:

$$\text{amount of water} = \frac{6.4 \times 10^5 \text{ cal/day ft}^2}{540 \text{ cal/g}} = 1185 \text{ g/day ft}^2 = 2.6 \text{ lb/day ft}^2$$

We must remember that this is the theoretical upper limit to the amount that can be produced, assuming that all the solar energy is captured and used with 100 percent efficiency. In fact, solar stills use radiant energy at an efficiency near 40–60 percent, so that a typical *actual* yield of water is nearer to 1 lb/day per square foot of still area. Bear in mind that water weighs about 8.3 lb/gallon, and you can see that quite large solar stills are required to produce useful amounts of water.

Example *Suppose that, at a particular site, the solar energy density is 200,000 cal/cm² year. If it is required to produce 100 kg of water per day using a solar still that operates at 50 percent efficiency, what area must the still occupy?*

Solution *First, we can calculate the solar energy density per day:*

$$\frac{200{,}000 \text{ cal/cm}^2 \text{ year}}{365 \text{ days/year}} = 550 \text{ cal/cm}^2 \text{ day}$$

We can now convert this figure into the amount of water produced per square centimeter by the still:

$$\frac{550 \text{ cal/cm}^2 \text{ day}}{540 \text{ cal/g}} \cong 1.0 \text{ g/cm}^2 \text{ day} \times 0.5 \text{ (efficiency)}$$

$$= 0.5 \text{ g/cm}^2 \text{ day}$$

Since we require 100 kg or 10^5 g of water per day, it follows that the area of the still must be

$$\frac{10^5 \text{ g/day}}{0.5 \text{ g/cm}^2 \text{ day}} = 2 \times 10^5 \text{ cm}^2 = 20 \text{ m}^2$$

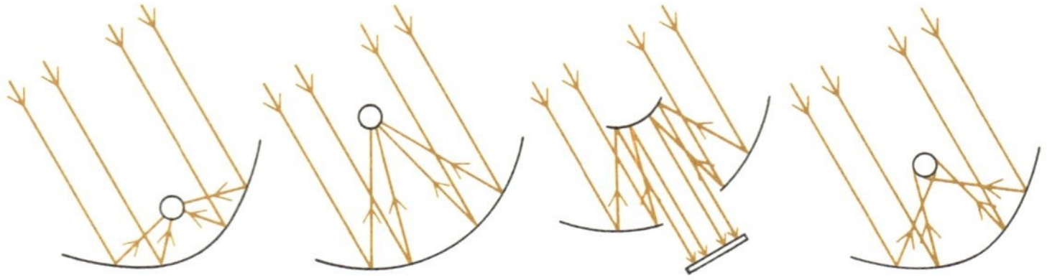

Figure 19-5 Cross sections of circular solar collectors.

SOLAR FURNACES

It is possible, and sometimes convenient, to use the direct rays of the sun to generate local regions of very high temperature. Devices to accomplish this task are called *solar furnaces*. Since the energy per unit area that falls on the earth is not sufficient to generate high temperatures, it is necessary to collect solar radiation over a wide area and then to focus it onto a very small area with lenses or mirrors. Probably, at some time, we have all used a magnifying glass to focus sunlight onto a piece of paper and caused it to ignite. Some other ways in which this can be accomplished by using circular solar collectors are shown in Figure 19-5. In this way, exceedingly high local temperatures can be achieved—in the neighborhood of 3000°C.

Solar furnaces have limited utility but do prove useful for certain specialized purposes. These include high-temperature studies of a variety of materials such as alloys used in rocket propulsion systems or as heat shields during the reentry of space capsules into the atmosphere. In addition, a solar furnace can provide an intense light source for photochemical studies (Chapter 15), although this light source is not monochromatic.

PHOTOSYNTHESIS

The preceding parts of this chapter have outlined several man-made devices for capture and utilization of the radiant energy of the sun. In living systems, other kinds of devices for the same general purposes have evolved; the most important of these are the complex structures involved in the process of **photosynthesis.**

In its most basic aspects, photosynthesis reverses the chemical process, *respiration,* through which organisms such as man derive the energy necessary to maintain the living state. In man, for example, life is supported from the energy released in the combustion of sugars according to the following stoichiometry:

$$C_6H_{12}O_6 + 6O_2 \longrightarrow 6CO_2 + 6H_2O \qquad \Delta H = -673 \text{ kcal/mole} \tag{19-1}$$

An elaborate mechanism is employed to capture a large part of the total energy released in this oxidative process in the form of the biochemical energy currency, ATP (adenosine triphosphate) (see Chapter 21). In the process, organic molecules are converted to carbon dioxide and water with the consumption of atmospheric oxygen. In photosynthesis, the radiant energy of the sun is employed as an energy source to drive reaction (19-1) in the reverse direction:

$$6CO_2 + 6H_2O + 48h\nu \longrightarrow C_6H_{12}O_6 + 6O_2 \tag{19-2}$$

Thus, carbon dioxide is removed from the atmosphere, organic compounds are synthesized, and the store of atmospheric oxygen is replenished. Clearly, photosynthesis is of overriding importance in nature. Not only do we depend on photosynthesis to provide necessary foodstuffs and to maintain a balance between carbon dioxide and oxygen in the atmosphere, but early photosynthetic organisms appear to have played an important role in the development of higher life forms in the first place (see Chapter 17).

Photosynthetic organisms

Those organisms capable of carrying out the photosynthetic process include quite a few species. Among them are, of course, plants—whether higher plants, ferns, mosses, diatoms, or algae (green, blue-green, red, or brown). In addition, certain microorganisms are photosynthetic, including some protozoa and a number of different bacteria (purple sulfur bacteria, purple nonsulfur bacteria, and green sulfur bacteria). With the exception of the bacteria, all of these organisms carry out the reaction of Equation (19-2). Bacterial photosynthesis employs substances other than water as oxidizable hydrogen donors and does not produce oxygen. Formally, bacterial photosynthesis may be written

$$CO_2 + 2H_2D + nh\nu \longrightarrow \{CH_2O\} + H_2O + 2D \tag{19-3}$$

Typical hydrogen donors, H_2D, include H_2S, hydrogen gas, thiosulfate, and simple organic substances. In what follows, we shall focus our attention on the reactions that occur in the photosynthetic process typical of green plants. While differences in the chemistry do occur from species to species, the basic considerations remain remarkably the same. Most of what we have to say below applies to bacterial photosynthesis as well as to that for green plants.

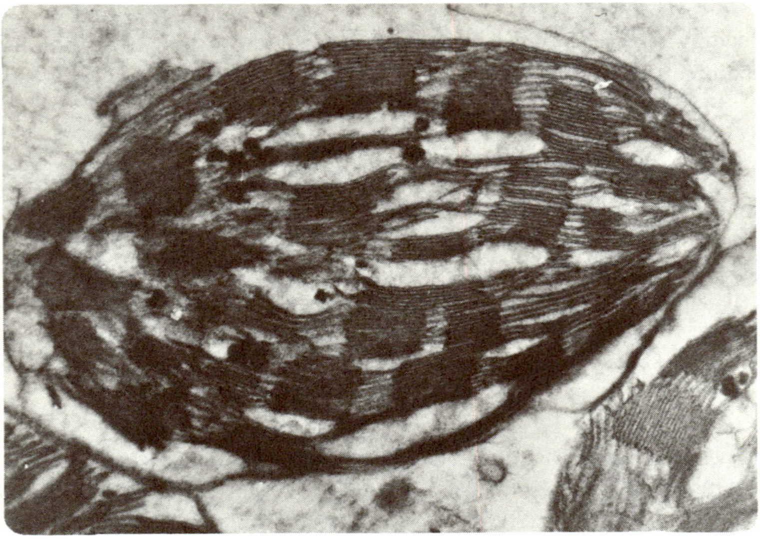

(a)

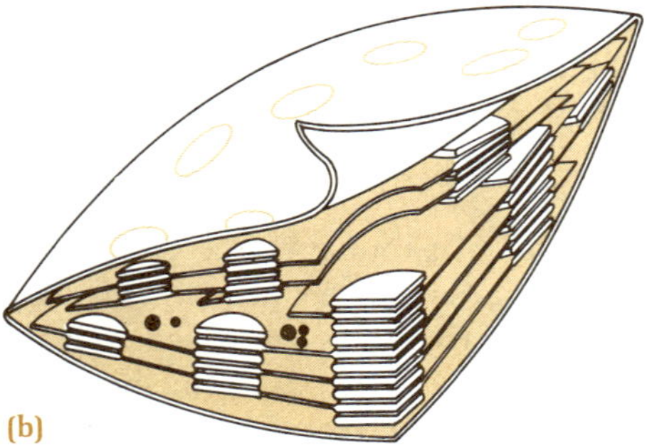

(b)

Figure 19-6 (a) An electron micrograph of a typical chloroplast at a magnification of 24,000 times. Note the lamellar structures, grana, which occupy much of the interior of the chloroplast. (b) A schematic three-dimensional view of a chloroplast of a higher plant. (Courtesy of K. Mühlethaler, Zurich, and D. von Wettstein, Copenhagen.)

The machinery of photosynthesis

In green plants, photosynthesis takes place in subcellular structures called **chloroplasts.** An electron micrograph of a typical chloroplast is shown in Figure 19-6. Note especially the small, dark membranous bodies that are pervaded by a distinct lamellar structure. These are called **grana.** These lamellar membranes contain all of the photosynthetic pigments, which absorb the light quanta and are capable of carrying out several of the characteristic reactions of photosynthesis. They do, however, require several enzymes from the rest of the chloroplast to be capable of performing the entire operation.

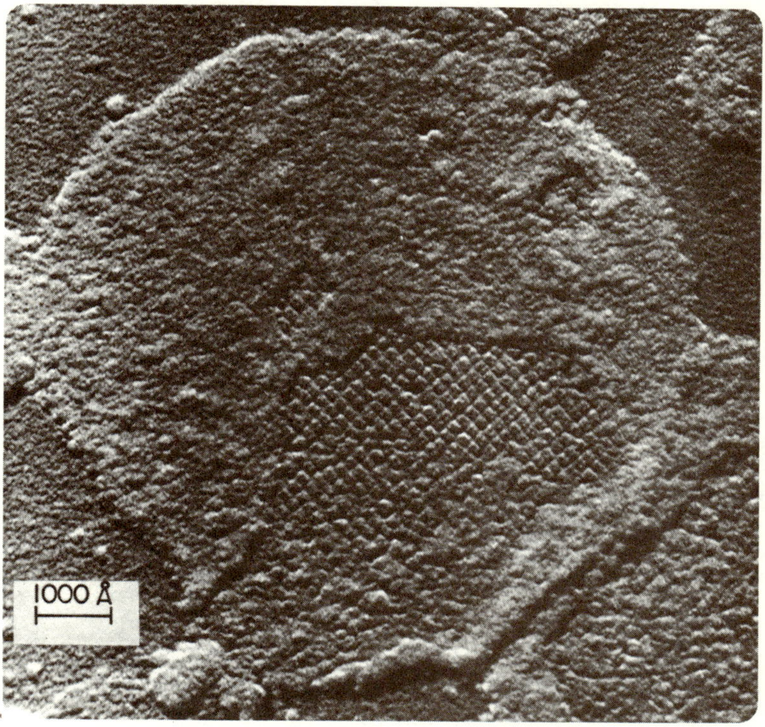

Figure 19-7 Lamellae of spinach chloroplasts showing arrays of quantasomes. This picture is that of a chromium-shadowed chloroplast lamella at a magnification of 71,000. Note the crystalline array of structural units (quantasomes with dimensions 185 × 155 × 100 Å). [From R. B. Park and J. Biggins, *Science,* 144, 1009 (1964). Copyright © 1964 by the American Association for the Advancement of Science.]

The lamellar membranes of chloroplasts are, in part, composed of repeating subunits as shown in Figure 19-7. Each of these subunits, called **quantasomes,** is believed to be the basic physiological unit of photosynthesis. They are about 150 Å in diameter and 100 Å thick. A molecular weight of the photosynthetic unit is 2×10^6, and each unit contains about 230 chlorophyll molecules, 48 carotenoids (accessory pigments), and many other complex organic molecules. These account for about half of the quantasomal weight; the other half is contributed by proteins.

In passing, it is worth noting the several levels of structural organization

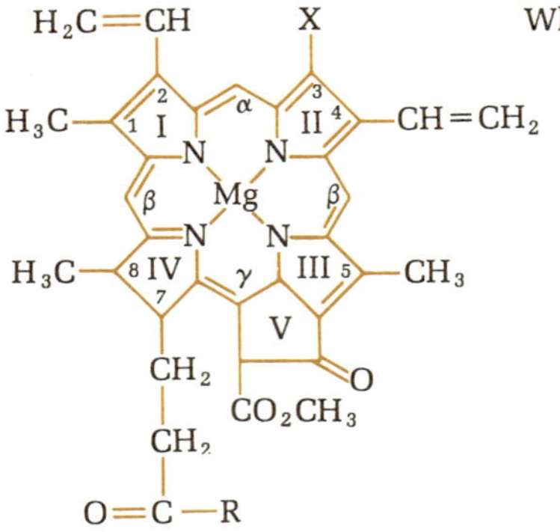

When X = —CH_3 : chlorophyll *a* (Chl *a*)
X = —CHO : chlorophyll *b* (Chl *b*)

R = phytyl ($C_{20}H_{39}O$—)

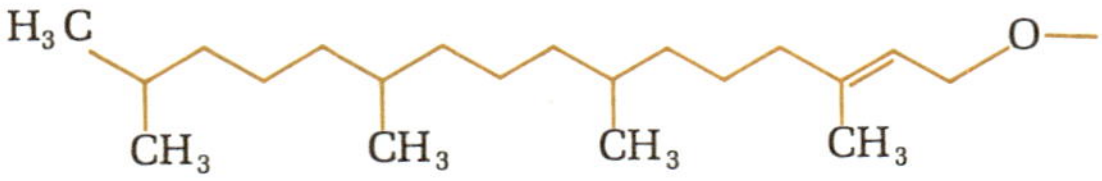

Note the chlorin (reduced porphin in ring IV) tetrapyrrole ring system IV, plus the fused cyclopentanone ring V. The complete parent ring system IV is called phaeoporphyrin. In bacteriochlorphyll, ring II is reduced as well; that is, positions 3 and 4 are —CH.

(a) The chlorophylls

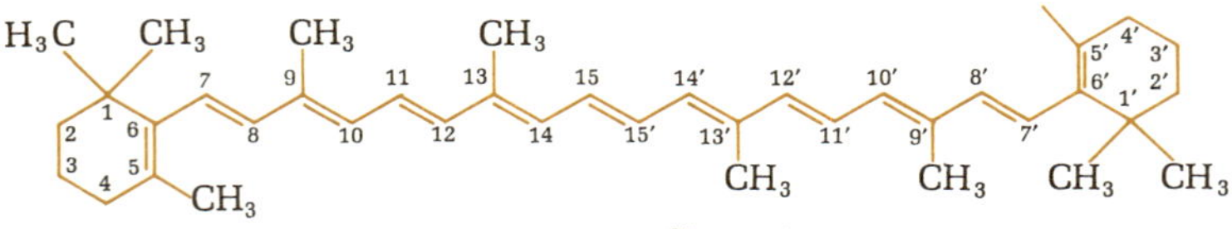

β-Carotene (all *trans*),
a *carotenoid*

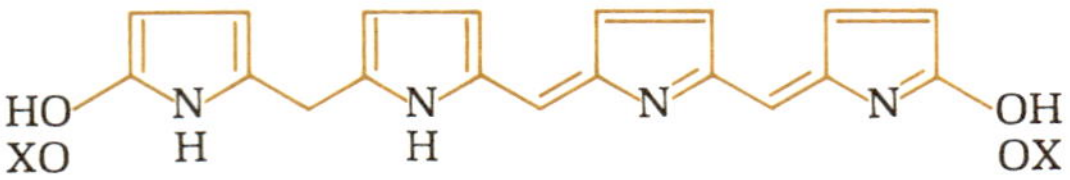

Phycoerythrobilin,
a *phycobilin*

This open-chain tetrapyrrole is covalently linked, probably through the —OH group(s) to a protein (X). The resulting complex is called *phycoerythrin*.

(b) Examples of accessory pigments

Figure 19-8 Structures for the most important of the photosynthetic pigments.

that occur in the complete photosynthetic machinery:

basic molecules (proteins plus pigments, etc.)
↓
quantasomes
↓
grana
↓
chloroplasts

The chloroplasts, in turn, become part of the organization of cells, which themselves are organized into tissues from which, finally, the complete organism is formed. Such a multitude of organization levels is typical of living systems and provides for control and regulation of each of the biochemical processes required for life.

The photosynthetic pigments

The photosynthetic process requires three things: (1) pigments to harvest the quanta from the sunlight that falls on the plant; (2) a system for transferring the energy of the absorbed quanta to a reaction site; and (3) some way to use the energy of light to cause chemical change. Let us look first at the former two points.

A crucial pigment for photosynthesis is **chlorophyll,** really a family of closely related molecules that differ slightly in structure from species to species. The structures of several members of the family are collected in Figure 19-8. Note the structural similarity between chlorophyll and heme, the prosthetic group of myoglobin and hemoglobin (Chapter 11).

One of the curious properties of chlorophyll is that its light absorption behavior seems to be just the opposite of what is wanted in a photosynthetic pigment. This fact becomes evident from the comparison of the absorption spectrum of chlorophyll with the intensity of sunlight reaching the surface of the earth, both of which are depicted in Figure 19-9. Surprisingly, chlorophyll does not absorb light maximally at those wavelengths where solar radiation is most intense.

Nevertheless, chlorophyll possesses several properties that override its disadvantageous spectrum. First, its ring system consists of a conjugated system of single and double carbon–carbon bonds; this lends chlorophyll its color, and the pi electrons associated with the conjugated system make absorption of light a highly probable event. Second, once the chlorophyll absorbs light, it remains in an excited electronic state, a triplet, for an unusually long time (*ca.* 0.0005 sec), and one such excited chlorophyll molecule is able to transfer this energy to other chlorophyll molecules; this accounts for the remarkably high efficiency of photosynthesis. The transfer

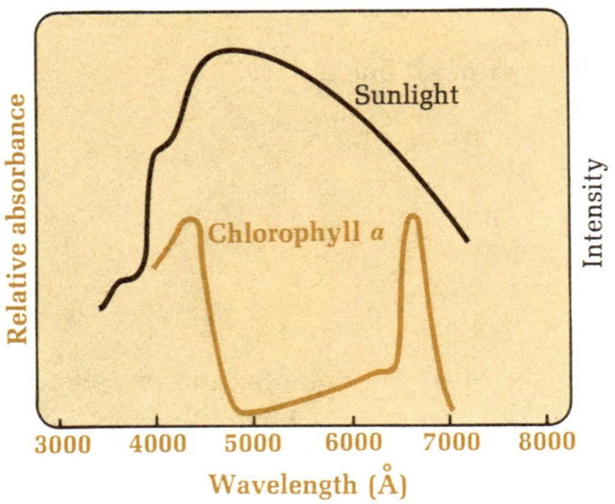

Figure 19-9 A comparison of the absorption spectrum of chlorophyll and the wavelength dependence of radiant energy from the sun.

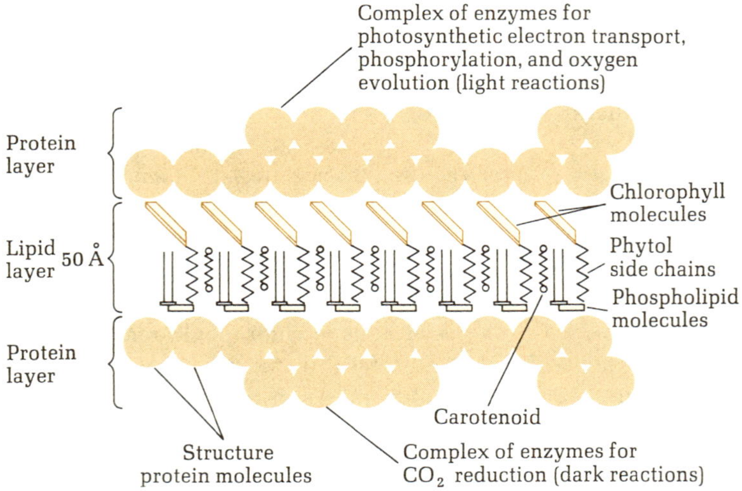

Figure 19-10 A schematic representation of the molecular organization of chlorophyll-lipid lamellae in the grana of chloroplasts. (From A. L. Lehninger, *Bioenergetics*, Benjamin, Menlo Park, Calif., 1965. Copyright © 1965, W. A. Benjamin, Inc.)

of energy is accomplished by a fluorescence process, corresponding to light quanta of energy content near 40 kcal/mole.

Efficient energy transfer from one chlorophyll molecule to another requires that these molecules form an organized array. It is the whole array that is active in absorbing light quanta, which then may be transferred to the *reaction site* itself, where the energy of the quantum will be employed to cause chemical change. Thus, a quantum absorbed anywhere in the array, which acts as an umbrella for capturing photons, is equally effective in promoting photosynthesis. A schematic view of such an organized array of chlorophyll (and other pigments) is provided in Figure 19-10.

In addition to chlorophyll, there are several accessory pigments that are also crucial to the photosynthetic process. The structures of some of these are included in Figure 19-8. Note that the **carotenoids** are related structurally to retinal and vitamin A (Chapter 15).

In green plants, there appear to be two distinct centers for the absorption of light quanta; in one, chlorophyll is the principal absorbing pigment and in the other, which absorbs quanta at shorter wavelengths than does chlorophyll itself, the accessory pigments are the primary absorbers. Just how these two systems interact is discussed a bit later in this chapter.

Chemical requirements for photosynthesis

To what uses must the radiant energy of sunlight be put to effect photosynthesis? What chemical changes must result from the absorption of the light quanta? To understand these matters, we need only focus our attention on the two basic features of photosynthesis: It is *reductive* and *energy-requiring*. Hence, we must plan to synthesize a reducing agent and an energy-supplying molecule. The nature of both of these is known. The immediate reducing agent that converts carbon dioxide into complex organic molecules is called *reduced nicotinamide adenine dinucleotide phosphate* or *NADPH*. Its structure is

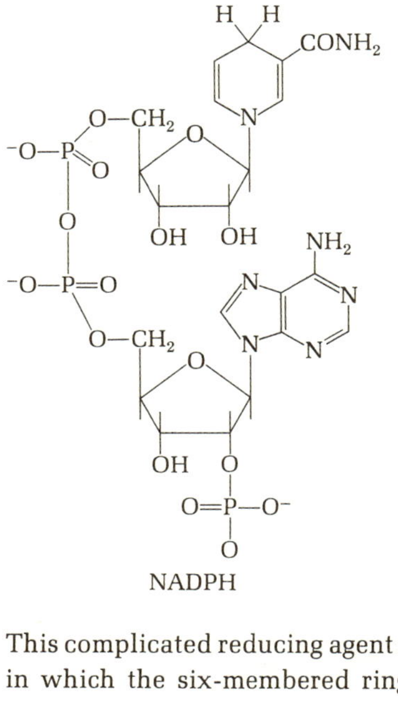

NADPH

This complicated reducing agent is derived from the corresponding molecule in which the six-membered ring is oxidized; this form is called $NADP^+$.

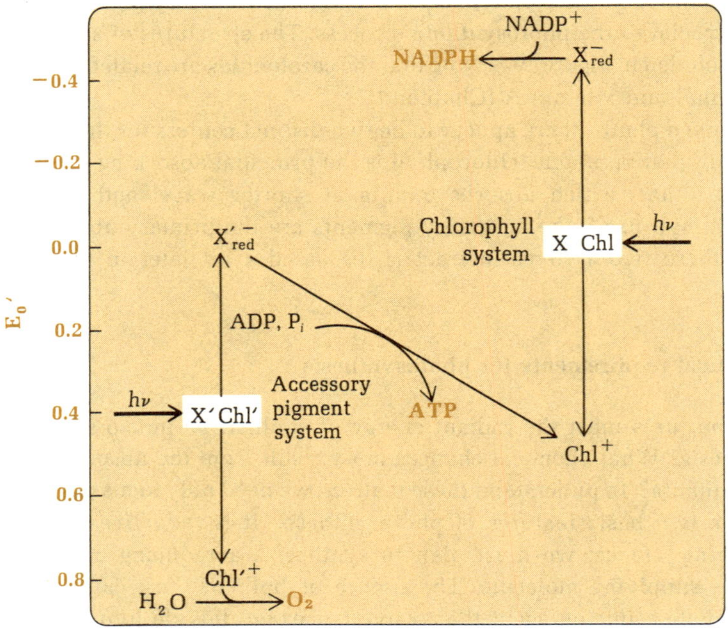

Figure 19-11 A formulation for the photochemical generation of NADPH and ATP. Absorption of light quanta by the two photochemical systems generate oxidants, Chl^+ and Chl'^+, and reductants, X^-_{red} and X'^-_{red}. Further reactions of these result in formation of the necessary cofactors plus generation of molecular oxygen. (Adapted from H. R. Mahler and E. H. Cordes, *Biological Chemistry*, 2nd ed., Harper & Row, New York, 1971.)

$NADP^+$ is converted to NADPH through addition of a hydride ion, H^-, or the equivalent:

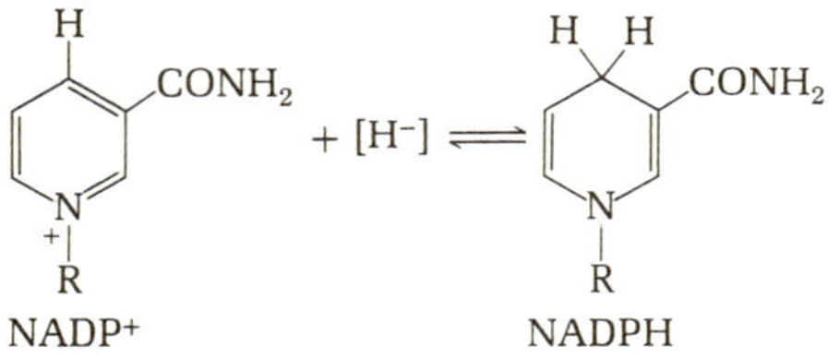

(19-4)

The energy-supplying molecule is ATP:

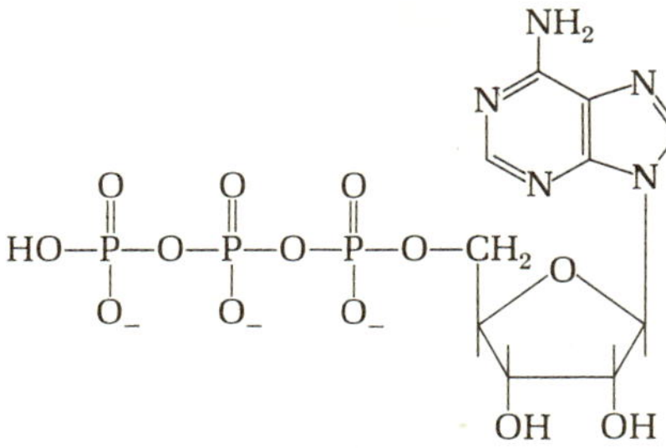

ATP

ATP is formed from ADP (the diphosphate) plus inorganic phosphate (usually abbreviated P_i):

$$ADP^{3-} + H_2PO_4^{-} \longrightarrow ATP^{4-} \qquad (19\text{-}5)$$

Let us turn our attention, then, to the matter of how the absorption of light quanta by the two pigment systems results in formation of ATP, NADPH, and in the process generates molecular oxygen. The scheme of necessary events is shown in Figure 19-11.

When a quantum of light strikes the chlorophyll system, an electron is kicked out of a certain chlorophyll molecule, forming a reductant, X^-_{red}, and an oxidant, Chl^+. In Figure 19-11, these are shown at different values of standard electrode potential, E_0'. As developed later (Chapter 20), the value of E_0' is a measure of the tendency of a system to donate electrons, negative values of E_0', or accept electrons, positive values of E_0'. Note that the value of E_0' for X^-_{red} is large and negative, indicating that it is a good electron donor. In fact, it is involved in the reduction of $NADP^+$ to yield NADPH. The chemical nature of X^-_{red} is not known.

The absorption of light quanta by the accessory pigment system also results in the generation of a reductant, X'^-_{red}, and an oxidant, Chl'^+. Note that Chl'^+ has a large positive value of E_0' (Figure 19-11) and is therefore a strong oxidizing agent. It serves to oxidize water, generating molecular oxygen. There remain two species with which we have not dealt: X'^-_{red} and Chl^+. These interact with one another (through the intermediacy of several electron carriers) in a process in which electrons from X'^-_{red} are transferred to Chl^+. This is an energy-yielding chemical reaction, and it is coupled to the chemical synthesis of ATP. Overall, the initial photochemical reactions serve to generate intermediates whose return to their initial state results in generation of NADPH, ATP, and molecular O_2.

The dark reactions of photosynthesis

The synthesis of the species just enumerated is the sole function of the light reactions of photosynthesis. The actual reduction of carbon dioxide and water to form carbohydrates occurs perfectly well in the dark. The pathway for carbohydrate formation involves a great many individual reactions of no particular interest to us here. Suffice it to say that the overall stoichiometry for the synthesis of one molecule of a hexose (a sugar containing six carbon atoms) is

$$6CO_2 + 18ATP + 12NADPH + 12H^+ \longrightarrow \text{hexose} + 18P_i + 18ADP + 12NADP^+ \qquad (19\text{-}6)$$

Note that this process consumes the NADPH and ATP produced in the light reaction and generates the precursors of these species, $NADP^+$, ADP, and P_i.

The energetics of photosynthesis

To understand a little about the energetics of photosynthesis, we need to recall that the energy of a photon is given by

$$E = h\nu \qquad (19\text{-}7)$$

or, in terms of its wavelength,

$$E = \frac{hc}{\lambda} \qquad (19\text{-}8)$$

Now, light that is useful in driving the photosynthetic reactions has a wavelength typically near 7500 Å $= 7.5 \times 10^{-5}$ cm. The energy associated with one photon of such light is then given by

$$E = \frac{6.62 \times 10^{-27} \text{ erg sec} \times 3 \times 10^{10} \text{ cm/sec}}{7.5 \times 10^{-5} \text{ cm}} = 2.65 \times 10^{-12} \text{ erg}$$

For 1 mole of such quanta, the energy is just the energy of a single quantum multiplied by Avogadro's number, $N = 6.023 \times 10^{23}$:

$$E(\text{per mole}) = 2.65 \times 10^{-12} \times 6.023 \times 10^{23} = 1.59 \times 10^{12} \text{ ergs} = 38 \text{ kcal}$$

Returning to Equation (19-2), we note that it requires 8 moles of light quanta to fix 1 mole of carbon dioxide. The overall energy input of 8 moles of light quanta for 7500 Å light is $38 \times 8 = 304$ kcal. The overall enthalpy change for fixation of 1 mole of CO_2 according to reactions (19-1) and (19-2) is positive by about 112 kcal/mole. Thus, overall, photosynthesis is about $112/304 = 37$ percent efficient.

Glossary

Band a large number of very closely spaced energy levels.

Band theory of solids a theory of solid state structure in which the solid is viewed as having a valence band, in which electrons are tightly bound to nuclei, and a conduction band, in which electrons are free to move throughout the solid.

Carotenoids (accessory pigments) complex organic molecules that function as photosynthetic pigments.

Chlorophyll a complex molecule involved in harvesting solar radiation for the photosynthetic process.

Chloroplast the complete subcellular organelle in which photosynthesis occurs.

Conduction electron an electron in a conduction band and therefore capable of carrying electrical current in a solid.

Grana membranous bodies contained within chloroplasts, which are capable of carrying out several of the reactions of photosynthesis.

Hole a positively charged site formed by the loss of an electron from that site.

Ocean recycle the succession of events comprised by (1) evaporation of water from the seas, (2) migration of clouds inland, (3) rainfall, and (4) return of the runoff to the seas by way of rivers and streams.

***p–n* junction** a union formed between an electron-rich, *n*-type semiconductor and an electron-poor, *p*-type semiconductor.

Photosynthesis the light-dependent conversion of atmospheric carbon dioxide into complex organic molecules.

Photovoltaic cell a device for the conversion of solar energy into electrical energy through the irradiation of a *p–n* junction.

Quantasome the ultimate photosynthetic unit.

Semiconductor a substance having conductivity properties intermediate between those of an insulator and those of a metal. In terms of the band theory of solids, semiconductors are characterized by a small but significant energy gap between the valence and conduction bands.

Solar cell any device for the conversion of solar energy into electrical energy.

Solar furnace a device for collecting solar radiation over a large area and focusing it on a small area, providing high temperatures at that spot.

Solar still a device employing solar radiation for the purification of water through distillation.

Study problems

1. Provide a clear but brief explanation accounting for the fact that the conductivity of semiconductors increases with increasing temperature.
2. How would you go about choosing impurities to add to a particular substance so as to prepare *n*-type and *p*-type semiconductors?

3. Briefly summarize the band theory of metals. In terms of this theory, how would you distinguish between metals, semiconductors, and insulators?
4. Solar cells and solar stills share several of the same advantages and disadvantages. What are these?
5. Calculate the maximum amount of water that could be produced per day by a solar still occupying 10 ft^2 operating with a solar energy intensity of 10^6 cal/ft^2 day.
6. Suppose that a solar still that is 40 percent efficient and occupies 1 ft^2 produces 3 kg of pure water per day. Calculate the solar energy intensity at the site of this solar still.
7. Calculate the efficiency of a solar still operating at a solar energy intensity of 5×10^5 cal/ft^2 day and occupying 1 ft^2 producing 500 g of water per day.
8. What are the principal roles of photosynthesis in nature?
9. Discuss the role of the photosynthetic pigments in the process of photosynthesis.
10. Suppose you wished to improve on chlorophyll as a photosynthetic pigment. What properties would you look for in your search for a better molecule?
11. Suppose that a green plant were irradiated with monochromatic light of a wavelength that is strongly absorbed by chlorophyll but only weakly absorbed by carotenoids (the accessory pigment system). In comparison to irradiation of the same plant with sunlight, what would happen to the rate of fixation of carbon dioxide? to the rate of evolution of oxygen? to the rate of synthesis of organic molecules?
12. Suppose that photosynthesis in a green plant were driven by two beams of monochromatic light, one of which is strongly absorbed by chlorophyll and the other strongly absorbed by the accessory pigment system. How would the rate of photosynthesis in this system compare with the maximal rate elicited by sunlight?
13. Explain how light drives the formation of NADPH and ATP. How are these compounds employed in the fixation of carbon dioxide?
14. Is light required for the specific reactions in which carbon dioxide is fixed? Explain your answer.
15. Suppose that photosynthesis were carried out in water containing excess oxygen 18, $H_2{}^{18}O$. In what product(s) of the photosynthetic process would the oxygen 18 appear?
16. The herbicide DCMU [3-(3, 4-dichlorophenyl)-1-1-dimethylurea)] prevents oxygen evolution in photosynthesizing systems. What is the most likely site of DCMU action?
17. Mutant photosynthetic algae have been isolated that lack the accessory pigment system. Nevertheless, if these algae are supplied with a reducing agent (electron donor), they fix carbon dioxide. Account for this fact. How would this photosynthetic process differ from that carried out by normal algae?

Suggestions for further reading

J. A. Bassham "The Path of Carbon in Photosynthesis," *Sci. Am.*, **206,** 88 (June, 1962).

F. Daniels *Direct Use of the Sun's Energy,* Yale University Press, New Haven, Conn., 1964.

P. E. Glaser "Power from the Sun," *Science,* **162,** 857 (1968).

R. P. Levine "The Mechanism of Photosynthesis," *Sci. Am.*, **221,** 58 (December, 1969).

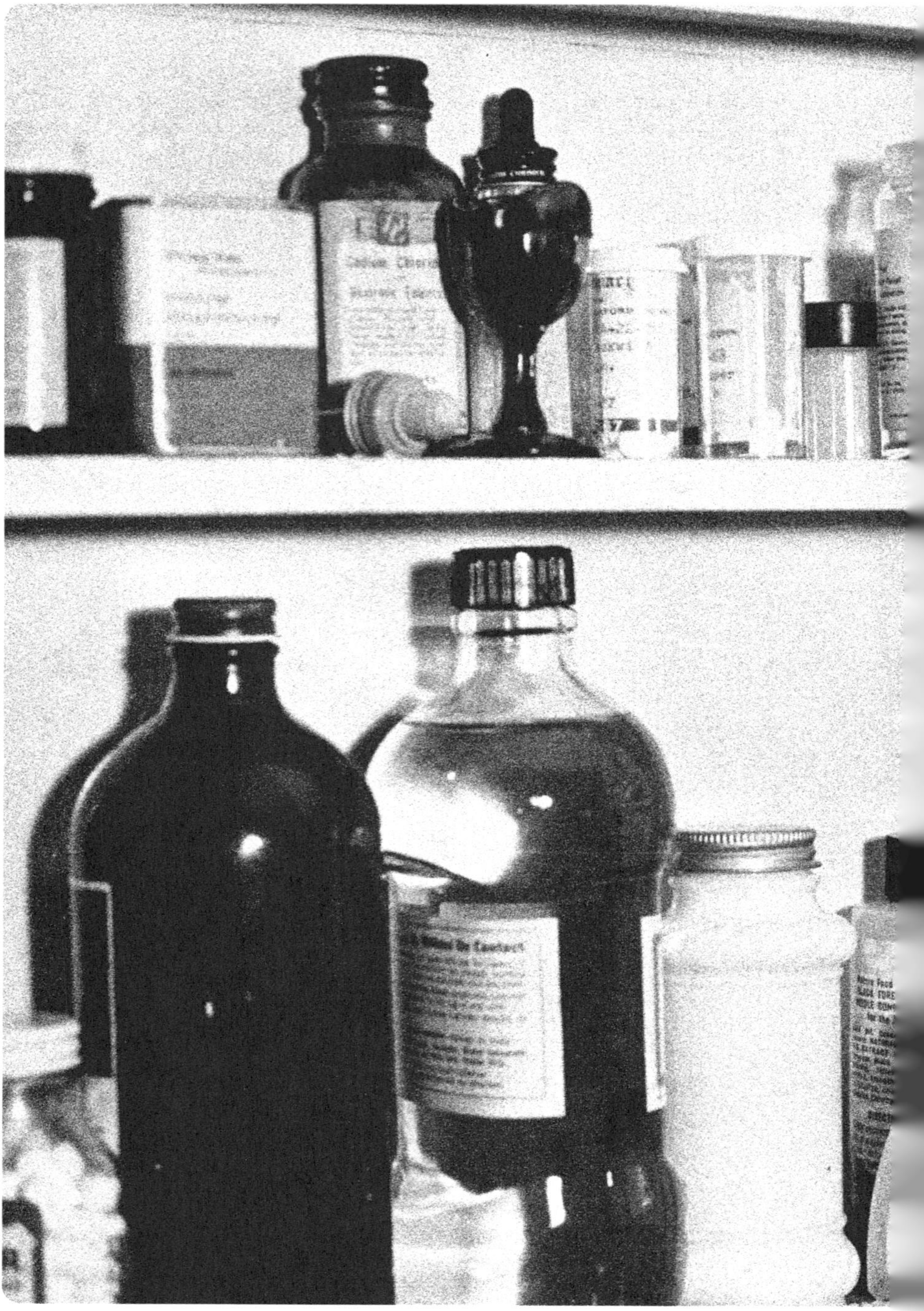

Part four

ELECTRON TRANSFER REACTIONS

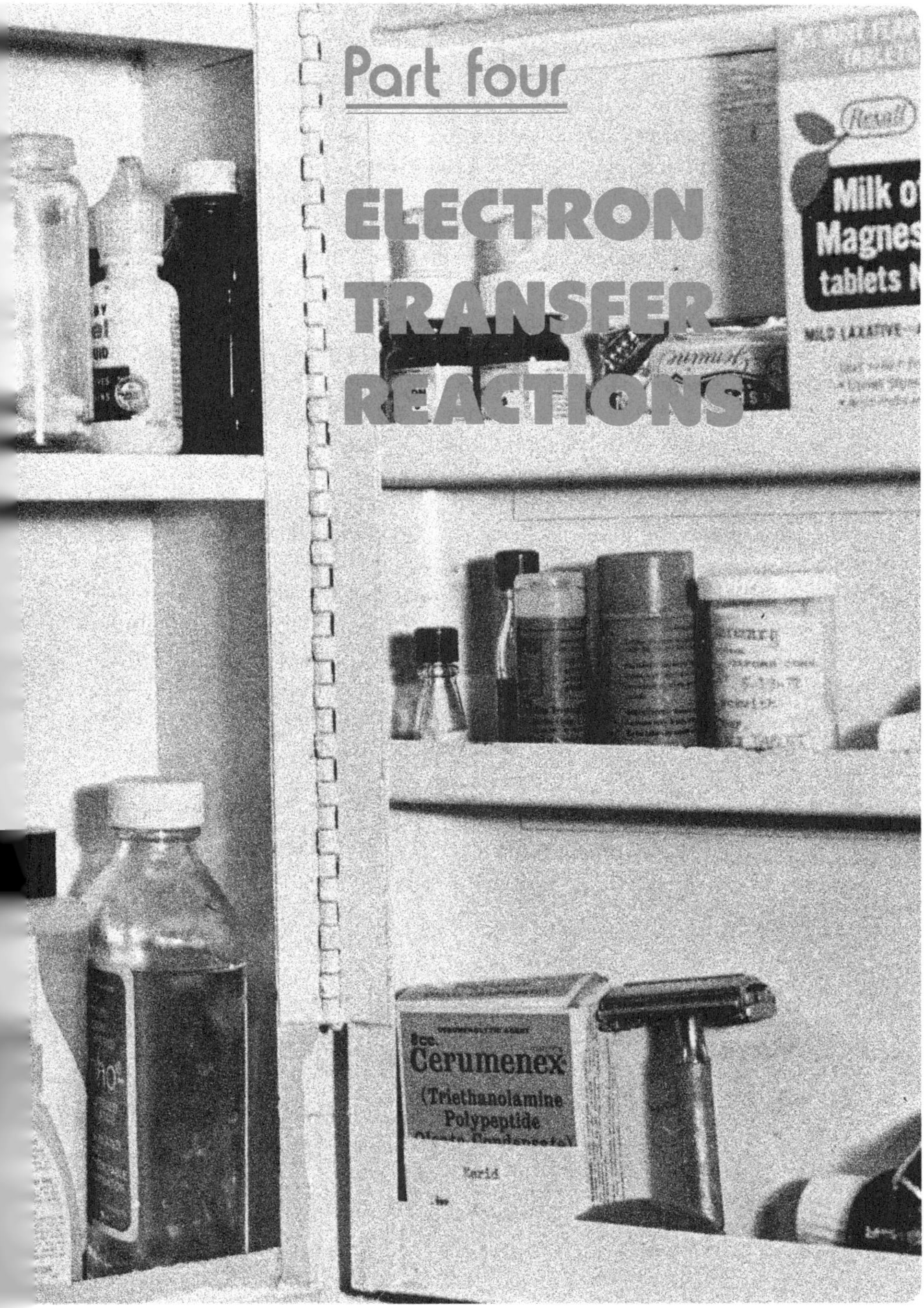

The final section of our exploration of chemistry is largely given to important groups of chemical reactions in living systems. In the first chapter of this section, attention is devoted to the basic aspects of reactions in which electrons are transferred from one atom, ion, or molecule to another. These are called *oxidation–reduction* reactions. Relatively simple systems are used to illustrate the principles involved. The principal thrust of this discussion deals with the thermodynamics of oxidation–reduction reactions or, more specifically, the free-energy changes associated with them. This information complements and extends that developed in Chapter 8 for reactions not involving the interchange of electrons. In the following chapter, this approach is used to explore the free-energy changes of several oxidation–reduction reactions of biochemical interest. Some ideas are developed about how the useful free energy resulting from these reactions is captured and used for the purposes of living organisms.

The basic problem faced by living organisms is how to extract energy from the environment to provide for their needs in the synthesis of highly specialized and specific molecules and the detailed organization of these into systems capable of carrying out very complex systems of chemical reactions. We shall find that living organisms proceed by degrading complex foodstuff molecules found in the environment, frequently through oxidation to carbon dioxide and water, and capturing the liberated energy in the form of molecules of adenosine triphosphate (ATP), a universal energy currency in nature. Although we indicate some general features of how all of this is done, some of the basic features of energy generation in living systems remain unknown, even after extensive efforts to unravel the story in the laboratories of many of the finest biochemists of this century. But nature yields her secrets grudgingly or, as one cynical scientist puts it: "Nature sure is a tricky old bitch."

Even though we do not understand in detail how ATP is generated in living systems, a good deal is known about how it is utilized for several purposes. We examine two of these: synthesis of polysaccharides and proteins. In the latter case, attention is especially directed to understanding of information flow in biological systems.

The last chapter of the book is in some ways a natural continuation of its forerunners in this section, but in other ways tends to stand apart. It deals not with oxidation–reduction reactions as such but with some of the properties of molecules that are synthesized in living organisms as a result of energy generated in such reactions. In part it is analogous to Chapter 9, in which reaction mechanisms are discussed, but this time the focus is on reactions of particular importance to living systems, and very large and complex molecules are involved. Perhaps it is appropriate that this final chapter brings together a significant fraction of everything that has gone before: A proper appreciation of the material presented depends on understanding something about the structure of both small molecules and polymers, how these interact with each other to form complexes, and how such complexes lead to chemical reaction or physiological response.

20

Oxidation-reduction reactions

It is worth knowing something about everything:
Even the dogs may eat the crumbs which fall from the rich man's table; and in these days, when the rich in knowledge eat such specialized food at such separate tables, only the dogs have a chance of a balanced diet.

Sir Geoffrey Vickers
The Art of Judgment

Throughout our study of chemistry, the importance of structure has been a continuing theme both from an atomic and an electronic point of view. At the level of isolated atoms, we have seen that electrons seek to occupy the lowest energy levels available but, if a suitable energy source is supplied, they may be promoted to excited states. For example, in Chapter 15 we learned that electronic excitation of hydrogen can be achieved by an electrical discharge through a tube filled with this gas. Later in the same chapter we saw that light of a suitable wavelength also serves to excite electrons of atoms or molecules to higher energy levels. In this chapter we shall examine chemical changes in which electrons are transferred from one molecule or ion to another rather than simply rearranged in a single structure. Such changes are called *oxidation–reduction reactions*. We shall here focus our attention on simple inorganic systems for which complete transfer of one or more electrons from one atom to another commonly occurs. In Chapter 21 we shall examine the more complex electron transfers of vital importance in the chemistry of living systems.

Our discussion will be limited to reactions of inorganic substances that occur in aqueous solutions. Earlier we have studied in some detail the nature of simple ionic substances dissolved in water and, in particular, the fact that one or more water molecules are always intimately associated with the individual ions (Chapter 10). For the moment we shall ignore this complication when we write equations for simple reactions. Reactions of $CrCl_2$, for example, will be written in the ionic form

$$Cr^{2+} + 2Cl^-$$

There is strong evidence that the Cr^{2+} ion in aqueous solution is strongly associated with six water molecules arranged in an octahedral array about the central atom. The Cr^{2+} ion in water is always associated with this water of hydration, and therefore any measurements we make refer to the *aquated* ion. This fact can be taken explicitly into account in an equation by writing

$$Cr^{2+}(aq) + Fe^{3+}(aq) \longrightarrow Cr^{3+}(aq) + Fe^{2+}(aq) \qquad (20\text{-}1)$$

The symbol (aq) is frequently omitted if it is clear that the reaction takes place in water solution.

The reaction between Cr^{2+} and Fe^{3+} (20-1) proceeds very nearly to completion under ordinary conditions. We shall return to examine the thermo-

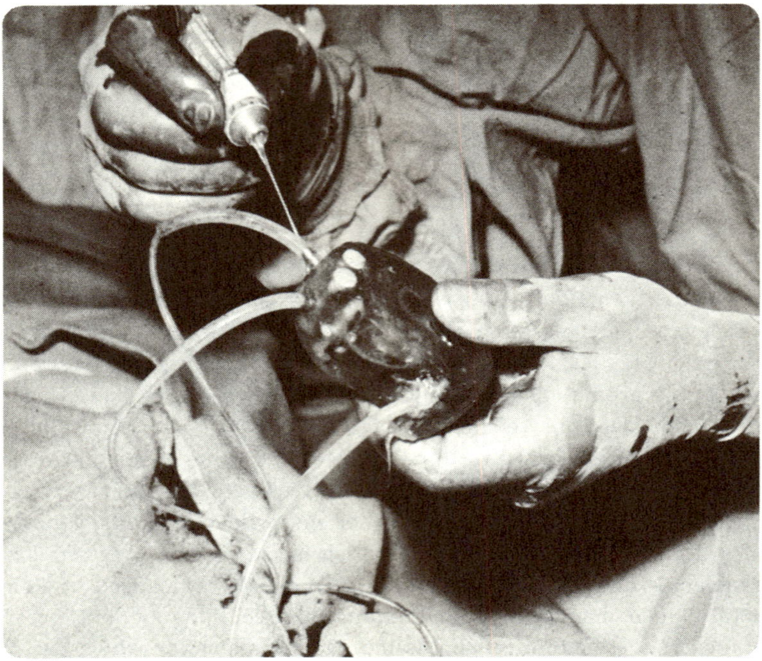

The energy generated in oxidation-reduction reactions can be utilized in many ways; one of the more unusual applications is shown in this figure. In the photograph at the top, one stage in the surgical implantation of a type of biological battery, a biogalvanic system, in a dog is shown. This battery operates on materials normally found in tissues of living organisms and uses body fluids as the electrolyte. It will supply energy to run a cardiac pacemaker that regulates the beating of the dog's heart. In the lower part of the figure, the voltage and energy output of a device of this type implanted in a dog over a period of four months is shown. Note that the device is working quite well, producing essentially constant outputs over time. Further developments along this line should lead to biological batteries that can be placed in humans requiring pacemakers; this would make the periodic replacement of the primary energy source for the pacemaker unnecessary. Thus, the chemistry of the body itself will provide a permanent energy source for the operation of the pacemaker necessary to sustain life. (Courtesy of ESB Research, Yardley, Pa.)

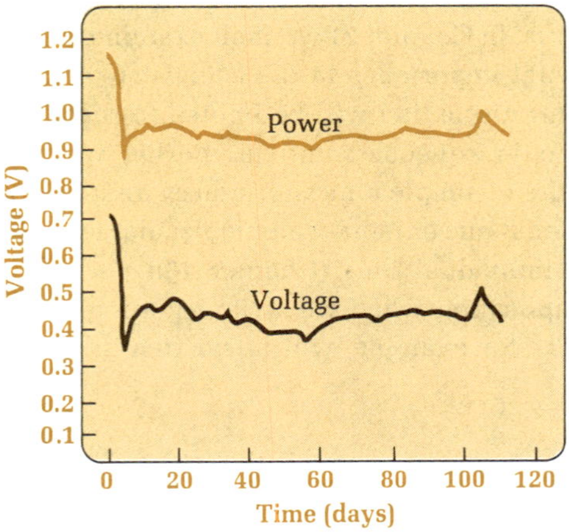

dynamics of this reaction later, but for the moment observe that the charge of the iron ion has decreased from +3 to +2 and that of the chromium ion has increased from +2 to +3. This change in charge is the result of transfer of a single electron from chromium to iron. Loss of an electron by the Cr^{2+} ion converts it to the Cr^{3+} ion and is a process called **oxidation.** Addition of the electron to the Fe^{3+} ion converts it to the Fe^{2+} ion and is an example of the process called **reduction.** Electrons are real objects that can be neither created nor destroyed in the course of a chemical reaction; they can only be traded between atoms. The conservation of electrons in the course of a reaction is of considerable use in balancing many equations. Before turning to this topic, it will be necessary to introduce a new concept that will enable us to keep track of electrons by a simple bookkeeping procedure.

OXIDATION STATE

Assignment of an *oxidation state* to an element in a chemical compound is a method of comparing its electronic environment to that of an identical atom in a sample of the pure element. If we consider a pure element, we assign to it an *oxidation number* of 0 to record that fact. Since each electron lost by an element leaves an unbalanced positive charge on the atom and each electron gained adds a negative charge, it is convenient to record these changes in oxidation number by giving the sign and magnitude of charge that results. Thus, in reaction (20-1) the reactants are chromium in the oxidation state +2 and iron in the oxidation state +3. The products, in contrast, are chromium with an oxidation number of +3 and iron with an oxidation number of +2. (Note the interchangeable way of using the terms "oxidation state" and "oxidation number.") In the reaction

$$2Br^- + Cl_2 \longrightarrow Br_2 + 2Cl^- \qquad (20\text{-}2)$$

the oxidation state of bromine changes from −1 to 0 as the oxidation state of chlorine changes from 0 to −1. The fact that Br_2 and Cl_2 are diatomic molecules is of no significance in calculating their oxidation numbers. We are interested in each atom in the molecule and the comparison of its total number of electrons to that in the elementary state. Since both atoms in each molecule are in fact in their elementary state and have neither gained nor lost electrons as yet, they have an oxidation number of 0. It follows from what we have said above that, for any simple monoatomic ion, its oxidation state and its actual electrical charge are the same. In the rare cases in which an ion contains two or more identical atoms, the charge is split equally between them; Hg_2^{2+} and the I_3^- ions are examples. In the former case, the Hg atoms are each in the +1 oxidation state and, in the latter, each iodine atom has an oxidation number of $-\frac{1}{3}$. Occurrence of a

Table 20-1 Oxidation states for some common compounds

$\overset{+1}{\mathrm{Na}}\overset{-1}{\mathrm{Cl}}$	$\overset{+1}{\mathrm{Li}}\overset{+5}{\mathrm{Cl}}\overset{-2}{\mathrm{O}}_3$	$\overset{+1}{\mathrm{K}}_2\overset{+5}{\mathrm{S}}_2\overset{-2}{\mathrm{O}}_6$
$\overset{+2}{\mathrm{Fe}}\overset{-1}{\mathrm{Br}}_2$	$\overset{+1}{\mathrm{Na}}\overset{+5}{\mathrm{N}}\overset{-2}{\mathrm{O}}_3$	$\overset{+2}{\mathrm{Mg}}\,\overset{+2.5}{\mathrm{S}}_4\,\overset{-2}{\mathrm{O}}_6$
$\overset{+2}{\mathrm{Ca}}\overset{-2}{\mathrm{O}}$	$\overset{+2}{\mathrm{Ca}}_3\overset{-3}{\mathrm{P}}_2$	$\overset{+4}{\mathrm{Sn}}\overset{-1}{\mathrm{Cl}}_4$
$\overset{+3}{\mathrm{Al}}_2\overset{-2}{\mathrm{S}}_3$	$\left(\overset{+6}{\mathrm{S}}\overset{-2}{\mathrm{O}}_4\right)^{2-}$	$\left(\overset{+3}{\mathrm{Cr}}\overset{-1}{\mathrm{Cl}}_6\right)^{3-}$

fractional oxidation number is properly interpreted to mean that one extra electron is distributed over all three atoms. It certainly does not imply existence of a fractional portion of an electron.

For those cases in which atoms are joined by covalent, electron-sharing bonds, some agreement must be reached on the manner of assigning oxidation states to each atom. The exact way in which this assignment is made is of no great importance so long as the method chosen is followed consistently. By analogy, each large corporation has its own unique accounting scheme and so long as it is followed interested company officers and stockholders can assess the financial health of the company as time passes. If IBM were suddenly to change its accounting methods to those used by Du Pont, however, vast confusion would almost certainly follow. In chemistry we adopt a straightforward convention: *Assign both electrons in a covalent bond to the atom with the greater electronegativity.* Let us apply this principle to HCl and SO_2. In the former case the electronegativity of chlorine is 3.0 and that of hydrogen is 2.1 (see Table 4-5). Both of the bonding electrons are therefore assigned to chlorine and none to hydrogen; therefore, chlorine is assigned an oxidation number of -1 and hydrogen is assigned $+1$. The case of SO_2 is somewhat more complicated, and to assist us we may first write a Lewis electronic structure for the molecule (Chapter 5, p. 171); electronegativities of sulfur and oxygen are 2.5 and 3.5, respectively, and therefore we assign all bonding electrons to the oxygen atoms for bookkeeping purposes. Only two electrons remain with the sulfur, and therefore it has a *formal charge* (a name intended to show the bookkeeping origin of the charges) of $+4$. Each oxygen has two extra electrons and therefore an oxidation state of -2. Note that the *sum* of all oxidation states in the molecule is zero, since the SO_2 molecule itself has no electrical charge. When oxidation states are calculated for ions, the sum of the states must just equal the actual ionic charge. Table 20-1 lists a number of common substances with oxidation states indicated for many elements. With a little practice you should be able to assign states to each element rather quickly. It is of considerable help in making assignments to learn certain of the common and relatively constant oxidation states: +1 for hydrogen except in hydrides, −2 for oxygen except in peroxides, +1 for alkali metals (Li, Na, K, Rb, Cs), and +2 for alkaline earth metals (Be, Mg, Ca, Sr, Ba).

BALANCING REACTIONS INVOLVING OXIDATION AND REDUCTION

Two common schemes are used as devices to aid in balancing oxidation-reduction equations. In the first method a record is kept in the chemical reaction itself of the electron transfers that occur. This record is used to complete the balancing process. In the second method the total reaction is first separated into its oxidation and reduction components, electrons are accounted for, and finally the two parts are recombined to give the balanced equation. We shall here illustrate the first method, but in subsequent sections we shall develop and work principally with the latter, **half-reaction,** method.

Consider the reaction that occurs when metallic copper is dissolved in dilute nitric acid:

$$Cu + HNO_3 \longrightarrow Cu(NO_3)_2 + NO \qquad (20\text{-}3)$$

We first calculate oxidation numbers for each element and fix our attention on those elements for which a change in oxidation state has occurred during the reaction:

$$\overset{0}{Cu} + \overset{+1}{H}\overset{+5}{N}\overset{-2}{O_3} \longrightarrow \overset{+2}{Cu}(\overset{+5}{N}\overset{-2}{O_3})_2 + \overset{+2}{N}\overset{-2}{O} \qquad (20\text{-}4)$$

We next connect with lines the elements undergoing change and show the number of electrons transferred to or from each atom:

loss of two electrons (Cu → Cu)

$$\overset{0}{Cu} + H\overset{+5}{N}O_3 \longrightarrow \overset{+2}{Cu}(NO_3)_2 + \overset{+2}{N}O \qquad (20\text{-}5)$$

gain of three electrons (N → N)

We now observe that copper loses electrons only in units of two per atom that reacts, whereas nitrogen gains three for each atom converted; since electrons can neither be created nor destroyed, it will be necessary for reaction to involve unequal numbers of copper atoms and nitrogen atoms. Reaction of three copper atoms will supply six electrons, which will be just sufficient to convert two nitrogens from nitrate ions into the form of NO. We can then write

$$3Cu + 2HNO_3 \longrightarrow 3Cu(NO_3)_2 + 2NO \qquad (20\text{-}6)$$

Note that the equation is not balanced as it is now written. No hydrogen is indicated on the product side although it is present on the reactant side. Furthermore, eight nitrogens appear on the right, and only two are present on the left. This latter deficiency comes from the fact that the original nitric acid suffers two fates, with some being converted to NO and a portion remaining in the nitrate group in $Cu(NO_3)_2$ without change in the oxidation state of the nitrogen. We have thus far only assured that the electron change is balanced, and we must now continue to balance the remainder of the equation.

In any succeeding steps it is essential that we do nothing to disturb the ratio of Cu consumed to NO formed, since this ratio is fixed by the electron-conservation requirement. We can observe, however, that the three $Cu(NO_3)_2$ molecules must have required six additional HNO_3 molecules to supply the necessary nitrate groups without change in oxidation state of the nitrogen. To show this in explicit fashion, let us rewrite the equation as

$$3Cu + 2HNO_3 + 6HNO_3 \longrightarrow 3Cu(NO_3)_2 + 2NO \qquad (20\text{-}7)$$

We now observe that the reactants contain eight hydrogens that have not appeared among the products, and that only 20 oxygens are accounted for out of the 24 that appear on the reactant side. The very natural solution to this dilemma is that four molecules of water are also produced. Introducing these and also combining the separate listings of HNO_3 molecules, we arrive at the balanced equation:

$$3Cu + 8HNO_3 \longrightarrow 3Cu(NO_3)_2 + 2NO + 4H_2O \qquad (20\text{-}8)$$

Example *Balance the reaction below by the electron transfer method.*

$$K_2Cr_2O_7 + HCl \longrightarrow KCl + CrCl_3 + Cl_2$$

Solution *Assign oxidation numbers.*

$$\overset{+1\ +6\ -2}{K_2Cr_2O_7} + \overset{+1-1}{HCl} \longrightarrow \overset{+1-1}{KCl} + \overset{+3\ -1}{CrCl_3} + \overset{0}{Cl_2} + \overset{+1-2}{H_2O}$$

Find the elements undergoing oxidation-reduction.

$$K_2\overset{+6}{Cr_2}O_7 + H\overset{-1}{Cl} \rightarrow KCl + \overset{+3}{Cr}Cl_3 + \overset{0}{Cl_2} + H_2O$$

Calculate the number of electrons lost (or gained) in each oxidation-reduction step and adjust the coefficients of each reactant and product to conserve electrons (note that chromiums must be taken two at a time in $K_2Cr_2O_7$ *and that chlorides must be oxidized in pairs to give a* Cl_2 *molecule).*

$2 \times 3e^- = 6e^-$ gained

$$K_2\overset{+6}{Cr_2}O_7 + 6H\overset{-1}{Cl} \rightarrow KCl + 2\overset{+3}{Cr}Cl_3 + 3\overset{0}{Cl_2} + H_2O$$

$6 \times 1e^- = 6e^-$ lost

Account for K^+ *and* O^{2-} *contained in* $K_2Cr_2O_7$.

$$K_2Cr_2O_7 + 6HCl \longrightarrow 2KCl + 2CrCl_3 + 3Cl_2 + 7H_2O$$

Adjust the coefficient of the HCl *to provide the additional* H^+ *required by the water formed.*

$$K_2Cr_2O_7 + 6HCl + 8HCl \longrightarrow 2KCl + 2CrCl_3 + 3Cl_2 + 7H_2O$$

Finally, check to see that the equation is balanced with respect to the chlorine atoms.

$$K_2Cr_2O_7 + 14HCl \longrightarrow 2KCl + 2CrCl_3 + 3Cl_2$$
$$14Cl \longrightarrow 2Cl + 6Cl + 6Cl = 14Cl$$

HALF-REACTIONS

Much of chemistry can be simplified by observing that the oxidation-reduction reactions that actually occur for any specific element usually are confined to a small number of possibilities. For example, in most of its reactions, iron appears either in the +2 or the +3 oxidation state, related by the reaction

$$Fe^{3+} + e^- \longrightarrow Fe^{2+} \qquad (20\text{-}9)$$

Since the reaction shown in (20-9) cannot occur in the absence of some source of electrons, it is clearly only half of a reaction. Two additional half-reactions serve to describe almost completely the **redox** behavior of iron,

$$Fe^{2+} + 2e^- \longrightarrow Fe \qquad (20\text{-}10)$$
$$Fe^{4+} + e^- \longrightarrow Fe^{3+} \qquad (20\text{-}11)$$

and of these two reactions, (20-11) is of little importance.

As we have written the above equations, each corresponds to a reduction of the reactant. If we were dealing with species in the gas phase, the thermodynamics of the electron captures indicated would be measured by the electron affinities of the reactants or, alternatively, the reverse reactions would have energy requirements precisely measured by the ionization potentials (see Chapter 4). In aqueous solution the situation is more complicated, and the energy changes are somewhat different from those in the gas phase. This largely reflects the details of the interaction of the metal ion(s) with the solvent water. This interaction is highly dependent on the charge on the atom and makes a substantial contribution to the overall energy change for each reaction. Our present knowledge of these interactions is not precise enough to enable us to take a very fundamental

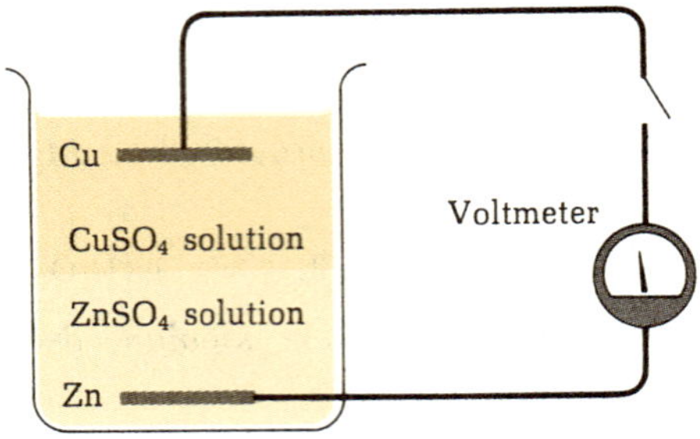

Figure 20-1 An electrochemical cell with direct solution contact.

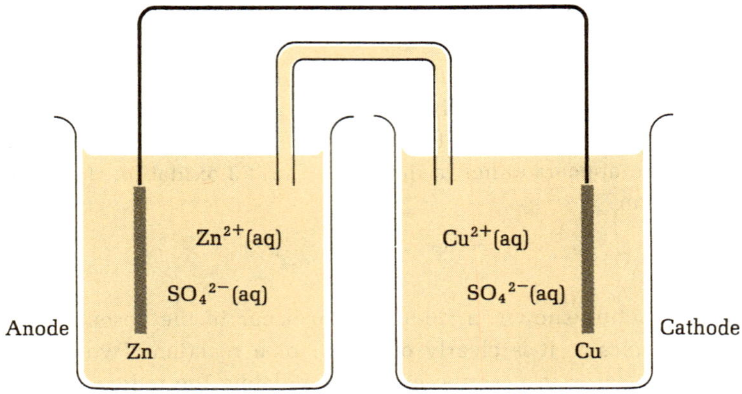

Figure 20-2 An electrochemical cell with a salt bridge.

approach to describing the energetics of a particular half-reaction. Consequently, we take an indirect approach. If we can successfully assign an energetic factor to each half-reaction that is proportional to the thermodynamic tendency for the reaction to occur as written, then combination of half-reactions may give very useful information concerning the driving force for an overall reaction. We shall, therefore, turn our attention to the problem of assigning some suitable thermodynamic quantity to specific half-reactions before dealing directly with the method by which half-reactions can be combined. The success of this approximation rests on the fact that, although the exact hydration energies are often unknown, they are constant for each redox system.

ELECTROCHEMICAL CELLS

In the case of a very large number of redox reactions that occur in solution, it is possible to produce an experimental setup in which the electrons are transferred from one reagent to another through an external conductor. Figure 20-1 shows an **electrochemical cell,** a particularly simple way of accomplishing this for the reaction

$$Cu^{2+} + Zn \longrightarrow Cu + Zn^{2+} \qquad (20\text{-}12)$$

and Figure 20-2 shows a slightly more complicated version of the same system. Direct reaction between the metallic zinc and the copper ions is prevented in the first instance by surrounding the zinc electrode with a solution of $ZnSO_4$. It is true that contact occurs between the $ZnSO_4$ and the $CuSO_4$ solutions, but no reaction can occur at that juncture since no electron source is present in this zone. In Figure 20-2, contact between the two solutions is made still more indirect by connecting them through a tube filled with a salt solution. We shall see later why contact between the two solutions is necessary. In the two separate portions of the solution, different half-reactions are possible. At the zinc electrode the reaction is

$$Zn^{2+} + 2e^- \longrightarrow Zn \qquad (20\text{-}13)$$

and at the copper electrode the reaction is

$$Cu^{2+} + 2e^- \longrightarrow Cu \qquad (20\text{-}14)$$

As written, both reactions appear to consume electrons. However, a better way of stating the situation is this: At each electrode an equilibrium exists such that if electrons are available, the reaction will proceed towards the right, but if electrons are removed, one of the reactions goes to the left instead. When the external switch is closed connecting the two electrodes, both compete for the electrons in the external circuit. If electrons are removed at one end by reaction (20-14), the deficiency of electrons in the circuit will be felt immediately by the zinc electrode at the other end. If some of the metallic zinc dissolves to form Zn^{2+} ion and leave its electrons behind, these can flow to the copper electrode and discharge more copper ions. This is precisely what happens in the present case, and electrons flow through the external circuit from the zinc to the copper. We should rewrite Equation (20-13) to accommodate the facts of the present case as

$$Zn \longrightarrow Zn^{2+} + 2e^- \qquad (20\text{-}15)$$

If we now add the two half-reactions of Equations (20-14) and (20-15), we get an overall reaction

$$Cu^{2+} + 2e^- + Zn \longrightarrow Cu + Zn^{2+} + 2e^- \qquad (20\text{-}16)$$

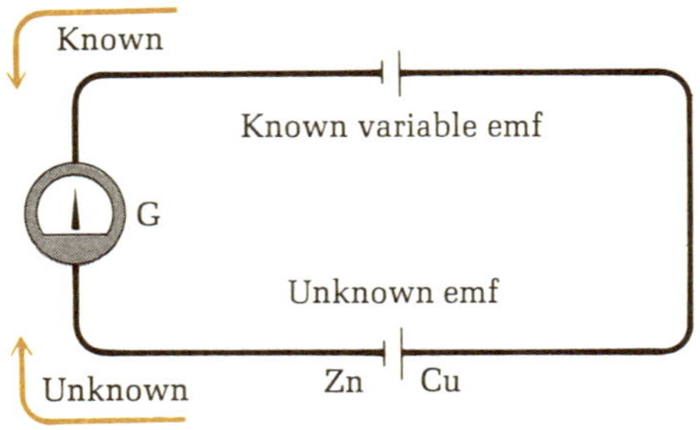

Figure 20-3 Diagram to illustrate the principles of the potentiometric measurement of electromotive force.

Since the electrons appear in identical number as both reactants and products, there is no purpose served by retaining them in the descriptive equation (except to remind us that they are indeed transferred in equal numbers), and we rewrite the net reaction as

$$Cu^{2+} + Zn \longrightarrow Cu + Zn^{2+} \tag{20-17}$$

Our choice of reversing reaction (20-13) rather than (20-14) was made so that (20-17) expresses the reaction as it proceeds in nature.

Example *Balance the reaction below using the half-reaction method.*

$$Hg_2Cl_2 + Ag \longrightarrow AgCl + Hg$$

Solution *Separate the equation into reactions of the elements involved in oxidation-reduction and write half reactions for each.*

a. $Ag + Cl^- \longrightarrow AgCl + 1e^-$
b. $Hg_2Cl_2 + 2e^- \longrightarrow 2Hg + 2Cl^-$

Note that reaction (b) requires two electrons per mole of reactant and therefore 2 moles of electrons must be produced by reaction (a).

c. $2Ag + 2Cl^- \longrightarrow 2AgCl + 2e^-$

Add reactions (b) and (c) and cancel elements or electrons that appear as both reactants and products.

a. $Hg_2Cl_2 + 2e^- \longrightarrow 2Hg + 2Cl^-$
b. $2Ag + 2Cl^- \longrightarrow 2AgCl + 2e^-$

$$Hg_2Cl_2 + 2Ag + 2Cl^- + 2e^- \longrightarrow 2Hg + 2AgCl + 2Cl^- + 2e^-$$

Net balanced equation:

$$Hg_2Cl_2 + 2Ag \longrightarrow 2Hg + 2AgCl$$

Let us now examine why the solutions must be in contact either directly, as in Figure 20-1, or indirectly through a salt bridge, as in Figure 20-2. Transfer of electrons from zinc metal is accompanied by release of the positive ions into solution. The aqueous solution cannot tolerate significant charge buildup, and the release of more zinc ions would be severely retarded. In the cells shown, negative ions diffuse into this region to balance the positive charge. These may be either SO_4^{2-} ions in the case of the direct contact between solutions in Figure 20-1 or Cl^- ions from the salt bridge shown in Figure 20-2. Although we have concentrated our attention at the zinc electrode, a precisely similar situation is set up at the copper electrode. There, as the Cu^{2+} ions receive electrons to become neutral atoms on the electrode surface, they leave behind an excess of negative ions in solution. Diffusion of positive ions into this region occurs to keep it electrically neutral.

Thus far we have focused our attention on the electrodes and surrounding solution. It is also worthwhile to give close attention to the electron flow through the external circuit. For this purpose we construct a slightly more complicated apparatus, shown in Figure 20-3, in which a second source of electrical current is opposed to the flow of electrons from the Zn–Cu cell. A sensitive galvanometer (G) is also introduced to detect electron flow through the circuit. We adjust the second electromotive force (emf) to balance the Zn–Cu cell and completely stop the flow of electrons. The voltage necessary to accomplish this is closely related to the tendency of reaction (20-17) to occur as written. In the present case, if solutions of both $ZnSO_4$ and $CuSO_4$ are maintained at 1*M*, the voltage required to stop electron flow is just 1.10 V. We can think of the total voltage developed by the cell as a measure of the combined tendency of Zn to give up electrons and of Cu^{2+} to accept electrons. The total effect of the two tendencies is to force electrons through the circuit at a voltage of 1.10 V.

It would be very nice if we could devise a simple procedure whereby two separate measurements could be made. Ideally, we would like independent measures of the tendency of Zn to give up electrons and of Cu^{2+} to accept them. Unfortunately, neither half-reaction will occur without the other. The principal problem of concern to us is to find some way to partition total voltages observed for many combinations of half-reactions so that combination of any two individual potentials will lead to the actual voltage produced by a cell based on them. Since we cannot measure the individual potentials, we adopt a different approach entirely and settle for comparing voltages with some standard half-reaction for which we *define* a zero half-cell potential. By convention, the reaction

$$H^+ + e^- \longrightarrow \tfrac{1}{2}H_2 \qquad (20\text{-}18)$$

has been selected. We shall see later that solution concentrations are important in determining cell potentials, and so we must also select a standard H^+ concentration. For most work, 1*M* acid has been adopted as standard (we shall see in Chapter 21 that a different choice is made in biochemical work). A suitable electrochemical cell is constructed in which hydrogen gas at 1 atm pressure is maintained in contact with a specially treated metal

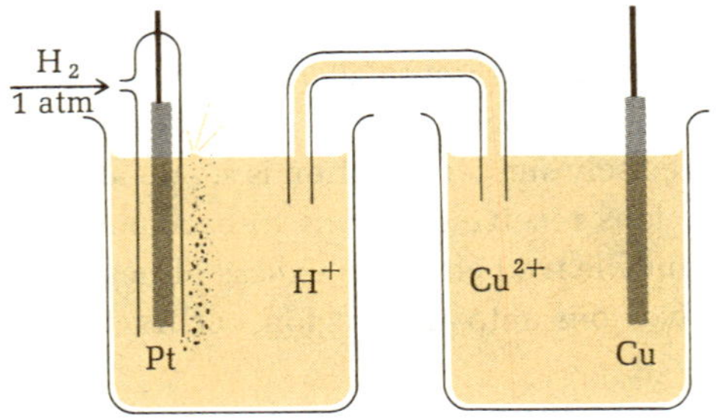

Figure 20-4 A cell based on the Cu^{2+}/Cu half-reaction.

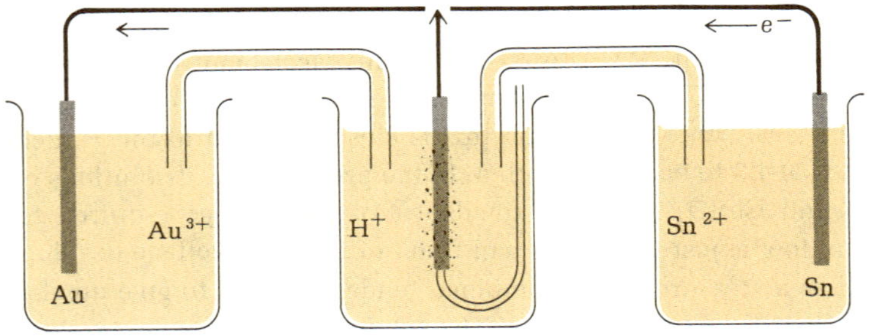

Figure 20-5 A system in which a single hydrogen electrode can be connected to either an Au/Au^{3+} or an Sn/Sn^{2+} electrode. Direction of electron flow is shown for either connection. If the tin and gold metals are connected directly together, the electrons still flow as indicated.

electrode in acid solution. This in turn is connected to a second cell through both a salt bridge and through an external measuring circuit similar to that of Figure 20-3.

A pair of cells as described above is shown in Figure 20-4, where measurement of the cell potential is taken against a half-cell for the Cu^{2+}/Cu reaction. At 1*M* Cu^{2+} concentration the measured potential is 0.34 V, and the electrons are observed to flow from the hydrogen cell to the copper cell. We assign a reduction potential of 0.34 V to the copper couple. Table 20-2 gives **standard reduction potentials** for a number of common half-reactions. Each of the cell potentials listed in Table 20-2 was measured in a solution with hydrogen ion concentration maintained at 1*M* by addition of a suitable acid and 1*M* concentrations of other soluble reactants. In a number of cases (those involving oxyions) the half-reaction either consumes water and releases hydroxide ion or consumes hydrogen ion with the formation of water. The potentials measured for these reactions are particularly sensitive to the concentration of acid in solution, and it is for this reaction that a standard 1*M* hydrogen concentration is adopted for measurement. Table 20-3 shows a similar list of reduction potentials measured in solution, where the OH^- ion concentration is maintained at 1*M*. Comparison of common couples will show just how sensitive the cell potentials for some reactions are to pH. The potential for zinc, for example, is 0.43 V more negative in basic than in acidic solution. Addition of four electrons to the O_2 molecule, on the other hand, is 0.83 V greater in acidic than in basic solution.

Such tables of reduction potentials are extremely useful in correlating a great deal of chemistry, just as tables of bond energies are useful in much the same way (Chapter 8). Combination of any half-reaction in the table with any other half-reaction leads to the equation for a complete chemical reaction. Furthermore, since the contribution that each half-reaction makes to the overall cell potential is listed as well, combination of the two half-potentials leads to this important experimental information. This use is so important that we shall illustrate it with two examples.

Consider the two half-reactions

$$Au^{3+} + 3e^- \longrightarrow Au \qquad E^0 = 1.50 \tag{20-19}$$

$$Sn^{2+} + 2e^- \longrightarrow Sn \qquad E^0 = -0.14 \tag{20-20}$$

In these two reactions the signs of the electrode potentials are opposite, and we should state the significance of this fact. The positive potential listed for the reaction with gold implies simply that the reaction will proceed as written if the half-reaction is combined with the standard hydrogen electrode. In contrast, the negative sign associated with the tin electrode implies that the reaction will proceed in the reverse direction if coupled with the hydrogen cell. In other words, if each cell were coupled with a hydrogen electrode, the Au^{3+} ion would accept electrons from the gaseous hydrogen (through the external circuit of course) but, in contrast, the metallic tin at the electrode would serve as a source of electrons to convert hydrogen ions to the element. If tin donates its electrons to hydrogen ions but gold, in contrast, accepts them from hydrogen molecules, we might expect that in direct reaction between them electron flow would be from the metallic tin to discharge Au^{3+} from solution. The situation is shown in Figure 20-5. It would now be more appropriate to rewrite Equation (20-20) to show the release of electrons as described above as an oxidation half-reaction:

$$Sn \longrightarrow Sn^{2+} + 2e^- \qquad E^0 = 0.14 \tag{20-21}$$

Observe that when the direction of the reaction is reversed, we change the sign of the electrode potential. To combine the two half-reactions to produce

Table 20-2 Standard reduction potentials[a] at 25°C

Reaction	Potential, E^0 (V)
$F_2 + 2e^- \longrightarrow 2F^-$	2.87
$MnO_4^{2-} + 4H^+ + 2e^- \longrightarrow MnO_2 + 2H_2O$	2.26
$Co^{3+} + e^- \longrightarrow Co^{2+}$	1.82
$H_2O_2 + 2H^+ + 2e^- \longrightarrow 2H_2O$	1.77
$MnO_4^- + 4H^+ + 3e^- \longrightarrow MnO_2 + 2H_2O$	1.70
$PbO_2 + 4H^+ + SO_4^{2-} + 2e^- \longrightarrow PbSO_4 + 2H_2O$	1.70
$Ce^{4+} + e^- \longrightarrow Ce^{3+}$	1.61
$MnO_4^- + 8H^+ + 5e^- \longrightarrow Mn^{2+} + 4H_2O$	1.51
$Au^{3+} + 3e^- \longrightarrow Au$	1.50
$Mn^{+3} + 1e^- \longrightarrow Mn^{2+}$	1.50
$Cl_2(g) + 2e^- \longrightarrow 2Cl^-$	1.36
$Cr_2O_7^{2-} + 14H^+ + 6e^- \longrightarrow 2Cr^{3+} + 7H_2O$	1.33
$MnO_2 + 4H^+ + 2e^- \longrightarrow Mn^{2+} + 2H_2O$	1.23
$O_2(g) + 4H^+ + 4e^- \longrightarrow 2H_2O$	1.23
$2IO_3^- + 12H^+ + 10e^- \longrightarrow I_2 + 6H_2O$	1.20
$Br_2 + 2e^- \longrightarrow 2Br^-$	1.06
$MnO_2 + 4H^+ + 1e^- \longrightarrow Mn^{3+} + 2H_2O$	1.0
$NO_3^- + 4H^+ + 3e^- \longrightarrow NO + 2H_2O$	0.96
$2Hg^{2+} + 2e^- \longrightarrow Hg_2^{2+}$	0.92
$Cu^{2+} + I^- + e^- \longrightarrow CuI$	0.85
$Ag^+ + e^- \longrightarrow Ag$	0.80
$Hg_2^{2+} + 2e^- \longrightarrow 2Hg$	0.79
$Fe^{3+} + e^- \longrightarrow Fe^{2+}$	0.771
$O_2 + 2H^+ + 2e^- \longrightarrow H_2O_2$	0.68
$Cu^{2+} + Cl^- + e^- \longrightarrow CuCl$	0.566
$MnO_4^- + 1e^- \longrightarrow MnO_4^{2-}$	0.56
$I_2 + 2e^- \longrightarrow 2I^-$	0.54
$Cu^{2+} + 2e^- \longrightarrow Cu$	0.34
$Hg_2Cl_2 + 2e^- \longrightarrow 2Hg + 2Cl^-$	0.270
$Hg_2Cl_2 + 2e^- \longrightarrow 2Hg + 2Cl^-$ (sat. KCl)	0.244
$AgCl + e^- \longrightarrow Ag + Cl^-$	0.222
$Cu^{2+} + e^- \longrightarrow Cu^+$	0.15
$Sn^{4+} + 2e^- \longrightarrow Sn^{2+}$	0.13
$2H^+ + 2e^- \longrightarrow H_2$	0.000
$Pb^{2+} + 2e^- \longrightarrow Pb$	−0.13
$Sn^{2+} + 2e^- \longrightarrow Sn$	−0.14
$CuI + e^- \longrightarrow Cu + I^-$	−0.17
$Ni^{2+} + 2e^- \longrightarrow Ni$	−0.25
$Co^{2+} + 2e^- \longrightarrow Co$	−0.28
$PbSO_4 + 2e^- \longrightarrow Pb + SO_4^{2-}$	−0.31
$Cd^{2+} + 2e^- \longrightarrow Cd$	−0.40
$Fe^{2+} + 2e^- \longrightarrow Fe$	−0.44
$Cr^{3+} + 3e^- \longrightarrow Cr$	−0.74
$Zn^{2+} + 2e^- \longrightarrow Zn$	−0.763
$Mn^{2+} + 2e^- \longrightarrow Mn$	−1.18
$Al^{3+} + 3e^- \longrightarrow Al$	−1.66
$H_2 + 2e^- \longrightarrow 2H^-$	−2.25

Table 20-2 *(Continued)*

Reaction	Potential, E^0 (V)
$Mg^{2+} + 2e^- \longrightarrow Mg$	−2.37
$Ce^{3+} + 3e^- \longrightarrow Ce$	−2.48
$Na^+ + e^- \longrightarrow Na$	−2.71
$Ca^{2+} + 2e^- \longrightarrow Ca$	−2.87
$Ba^{2+} + 2e^- \longrightarrow Ba$	−2.90
$Cs^+ + e^- \longrightarrow Cs$	−2.92
$K^+ + e^- \longrightarrow K$	−2.93
$Li^+ + e^- \longrightarrow Li$	−3.05

[a] $E^0 = E^0_{cell}$ when combined with H_2 (1 atm) $\longrightarrow$ $2H^+$ (1*M*) $+ 2e^-$

Table 20-3 Redox couples in basic solution

Couple	E^0(V)
$F_2 + 2e^- = 2F^-$	2.890
$BrO_3^- + 3H_2O + 6e^- = Br^- + 6OH^-$	0.584
$ClO_4^- + 8e^- + 4H_2O = Cl^- + 8OH^-$	0.560
$O_2 + 4e^- + 2H_2O = 4OH^-$	0.401
$PbO_2 + 2H_2O + 2e^- = Pb(OH)_3^- + OH^-$	0.214
$O_2 + H_2O + 2e^- = HO_2^- + OH^-$	0.065
$Pb(OH)_3^- + 2e^- = Pb + 3OH^-$	−0.538
$Fe(OH)_3 + e^- = Fe(OH)_2 + OH^-$	−0.547
$Cd(OH)_2 + 2e^- = Cd + 2OH^-$	−0.824
$2H_2O + 2e^- = H_2 + 2OH^-$	−0.828
$Zn(OH)_4^{2-} + 2e^- = Zn + 4OH^-$	−1.189
$Mn(OH)_2 + 2e^- = Mn + 2OH^-$	−1.533
$Mg(OH)_2 + 2e^- = Mg + 2OH^-$	−2.690
$Ca(OH)_2 + 2e^- = Ca + 2OH^-$	−3.017

a balanced equation, we now need only assure ourselves that electrons are neither created nor destroyed in the final reaction. If reaction (20-19) is written for reaction of 2 moles of Au^{3+} and reaction (20-20) is written for 3 moles of Sn, the half-reactions may be directly added with exact cancellation of the electrons from both sides:

$$\begin{array}{lll} 2Au^{3+} + 6e^- \longrightarrow 2Au & E^0 = 1.50 & \\ 3Sn \longrightarrow 3Sn^{2+} + 6e^- & E^0 = 0.14 & \\ \hline 2Au^{3+} + 3Sn + 6e^- \longrightarrow 2Au + 3Sn^{2+} + 6e^- & E^0 = 1.64 & (20\text{-}22) \\ 2Au^{3+} + 3Sn \longrightarrow 2Au + 3Sn^{2+} & & \end{array}$$

When we combined the two half-reactions to derive Equation (20-22), we multiplied the quantities of reactants and products by appropriate numbers so that the reactant electrons and the product electrons would exactly balance. We *did not*, however, multiply the individual cell potentials, and this may at first be difficult to understand. The important point is this: *The cell potential is a measure of the thermodynamic tendency of the reaction to proceed, regardless of whether it takes place or not.* Perhaps this can best be seen by returning to Figure 20-3, where the reaction has been set up to prevent any electron flow at all. So long as the circuits are complete, we must maintain exactly the same potential to prevent reaction from occurring. The total amounts of reagents are not important, only their nature and their concentrations. Precisely the same potential must be used to prevent electron flow if 2 liters or 1 of 1*M* solution are used. We shall return to this point later when we show that the cell potential is a measure of the free-energy release by the reaction. For the moment, we make only the final point that we have carefully selected the reactions above to ensure that the reaction proceeds from left to right as written and that the corresponding cell potential is positive.

As a second illustration of the use of half-reactions and corresponding potentials, let us consider the possibility that iodine might be made by reaction of an iodide with an iodate. From Table 20-2 we extract the pertinent half-reactions:

$$2IO_3^- + 12H^+ + 10e^- \longrightarrow I_2 + 6H_2O \qquad E^0 = 1.20 \tag{20-23}$$

$$I_2 + 2e^- \longrightarrow 2I^- \qquad E^0 = 0.54 \tag{20-24}$$

Reasoning as before, we reverse the direction of reaction (20-24) (and the sign of the electrode potential), make a suitable multiplication of it so that cancellation of electrons will occur, and add the reactions to obtain the overall equation:

$$\begin{array}{lll} 2IO_3^- + 12H^+ + 10e^- \longrightarrow I_2 + 6H_2O & E^0 = & 1.20 \\ 10I^- \longrightarrow 5I_2 + 10e^- & E_0 = & -0.54 \\ \hline 2IO_3^- + 10I^- + 12H^+ + \cancel{10e^-} \longrightarrow 6I_2 + 6H_2O + \cancel{10e^-} & E^0 = & 0.66 \\ 2IO_3^- + 10I^- + 12H^+ \longrightarrow 6I_2 + 6H_2O & & \end{array} \tag{20-25}$$

We see that a balanced equation results and that the cell potential is positive. We conclude that the reaction will take place as written, since we have seen in the preceding example that *a spontaneous chemical change is associated with a positive cell potential.*

Example *Will elemental lead react spontaneously with* Cr^{3+} *in acidic solution to give* Pb^{2+} *and elemental chromium according to the equation*

$$3Pb + 2Cr^{3+} \longrightarrow 3Pb^{2+} + 2Cr$$

Solution *Find the proper half-reactions from Table 20-2.*

$$Pb^{2+} + 2e^- \longrightarrow Pb \qquad E^0 = -0.13$$

$$Cr^{3+} + 3e^- \longrightarrow Cr \qquad E^0 = -0.74$$

Rearrange the first equation to start with elemental lead (as required in the final reaction) and adjust the coefficients so that equal numbers of electrons are required in each half-reaction.

$$3Pb \longrightarrow 3Pb^{2+} + 6e^- \qquad E^0 = 0.13$$
$$2Cr^{3+} + 6e^- \longrightarrow 2Cr \qquad E^0 = -0.74$$

Add the reactions and corresponding potentials with cancellation of electrons.

$$3Pb + 2Cr^{3+} \longrightarrow 3Pb^{2+} + 2Cr \qquad E^0 = -0.61\text{ V}$$

Since the potential of the total cell is negative, we conclude that the reaction will not take place to a very great extent as written. The reverse reaction,

$$3Pb^{2+} + 2Cr \longrightarrow 3Pb + 2Cr^{3+} \qquad E^0 = +0.61$$

will occur spontaneously and proceed well toward completion.

RELATION OF CELL POTENTIAL TO FREE ENERGY

By qualitative reasoning and comparison of individual half-reactions with that of a standard hydrogen electrode, we have deduced that a positive cell potential means that the reaction will occur from left to right as written. In this section we justify this conclusion in somewhat more fundamental terms and extend the interpretation of cell potentials. We shall first review some common facts about electrical cells.

Electrical storage devices as sources of electrical energy are very familiar to us. The automobile storage battery and the flashlight battery are two of the familiar objects of our daily life. An attempt to start a car with a 6-V battery is doomed to failure if the starter requires 12 V; experience tells us that the energy that can be delivered depends on the voltage. On the other hand, if a flashlight is set aside without turning it off, we are made uncomfortably aware of a fact of nature the next time we try to use it: The total amount of electrical energy that can be delivered from the battery—a simple chemical system—is limited by the quantity of chemicals available to react.

The above facts can be applied immediately to the electrochemical cells we have been considering. The energy that can be delivered by a reaction such as (20-25) depends both on the cell potential and the number of electrons transferred. This can be expressed by the simple relation

$$\Delta G^0 = -n\mathfrak{F}E^0 \tag{20-26}$$

where ΔG^0 is the standard free energy associated with the reaction, n is the number of electrons transferred in the simplest balanced equation for the reaction, $\mathfrak{F}$ is a constant of nature called the *Faraday constant* and E^0

is the cell potential. Since a spontaneous chemical reaction is always characterized by a decrease in standard free-energy change, a positive cell potential is characteristic of spontaneous reactions. The value of the Faraday constant convenient to use to compare standard free-energy changes with those calculated in Chapter 8 is 23.06 kcal/V. The standard free-energy change associated with reaction (20-25) is then

$$\Delta G^0 = -10 \times 23.06 \times 0.66 = -152.2 \text{ kcal/mole}$$

The reaction is highly exergonic and may be expected to proceed to completion if a suitable mechanism for reaction exists (as, in fact, it does).

Example *Calculate the standard free-energy change associated with reaction (20-22).*

Solution *The cell potential calculated is 1.64 V, and six electrons are transferred from Sn to* Au^{3+}.

$$\Delta G^0 = -6 \times 23.06 \times 1.64 = 226.9 \text{ kcal/mole}$$

The possibility that tables of half-reactions give us to calculate standard free-energy changes for many reactions is so great that we should consider one further example before turning to other applications. Suppose, for example, that we wish to know whether Cu^+ ion in solution is stable or whether it undergoes reaction with itself (*disproportionates*) to give Cu^{2+} and copper metal according to the equation

$$2Cu^+ \longrightarrow Cu^{2+} + Cu \qquad (20\text{-}27)$$

In Table 20-2 we find the following data:

$$Cu^{2+} + e^- \longrightarrow Cu^+ \qquad E^0 = 0.15$$
$$Cu^+ + e^- \longrightarrow Cu \qquad E^0 = 0.52$$

Subtracting the first from the second leads to the result:

$$2Cu^+ \longrightarrow Cu + Cu^{2+} \qquad E^0 = 0.37$$
$$\Delta G^0 = -1 \times 0.37 \times 23.06 = -8.53 \text{ kcal/mole} \qquad (20\text{-}28)$$

We conclude from this result that cuprous ion, Cu^+, will be unstable in aqueous solution, and that any attempt to find a reaction that would produce it would lead ultimately to the formation of elemental copper and cupric ion instead. As always, thermodynamics tells us only that disproportionation is energetically favored and provides us with no knowledge as to how rapidly disproportionation will occur. In the present case the disproportionation reaction is rapid, and cuprous ion has a short lifetime in water solutions.

Since quantitative information is available concerning the free energy associated with the disproportionation of cuprous ion, we are in a position to make quantitative statements concerning the extent of the reaction. Recall from Chapter 8 that the relationship between standard free-energy change and the equilibrium constant for a reaction is

$$\Delta G^0 = -RT \ln K_{eq} \quad \text{or} \quad \Delta G^0 = -2.3RT \log K_{eq} \tag{20-29}$$

At a temperature of 25°C, the free energy of reaction (20-28) leads to

$$\frac{-8510}{-2.3 \times 2.0 \times 298} = \log K_{eq} = 6.245 \qquad K_{eq} = 1.76 \times 10^6 = \frac{(Cu^{2+})(Cu)}{(Cu^+)^2}$$

If we adopt the metal as the standard state for Cu and therefore enter its activity as unity in the above expression, we calculate that for a 1*M* solution of Cu^{2+} ion, the equilibrium concentration of Cu^+ ion will be

$$(Cu^+) = \left[\frac{(Cu^{2+})}{1.76 \times 10^6}\right]^{1/2} = 7.5 \times 10^{-4}M \tag{20-30}$$

OXIDATION POTENTIAL DIAGRAMS AND GRAPHS

Table 20-2 is convenient for the summary of large amounts of data, but examination will show that the reactions of a particular element are found widely separated in the table. To discuss the chemistry of a single element, it is convenient to use a shorter form that collects only the relevant potentials. For historical reasons, these collections have generally been written for oxidation reactions rather than the reductions as listed in the table above. The only consequence of this change is that the equations and signs of the potentials must be reversed. Our discussions of the stability of copper would then be based on the following diagram

$$Cu \xrightarrow{-0.52} Cu^+ \xrightarrow{-0.15} Cu^{2+}$$

and we could have proceeded with our calculations in the same manner as above.

Once we become familiar with this method of presentation, certain reactions stand out on brief inspection as either possible or forbidden by thermodynamics.

To ascertain what chemical reactions are possible, we need only to determine the sign of the overall potential. One important question for all metals is whether they can be dissolved readily in water or in acid solution to give hydrogen and the metal ion. At a pH of zero, the contribution of the hydrogen couple to any cell potential is always zero. It must always be the case, therefore, that the overall electrode potential has the sign (and magnitude) of the half-reaction listed for the specific element. From the diagram for copper given above, we see that the sign of the reaction potential is negative and conclude that copper will not be attacked by dilute acid solutions—a fact that we certainly should expect when we remember that nearly pure copper is used extensively in plumbing fixtures. Of course, if we use an acid that contains an oxidizing anion, a quite different reaction (one not evolving hydrogen) might result. In Equation (20-8) we have written

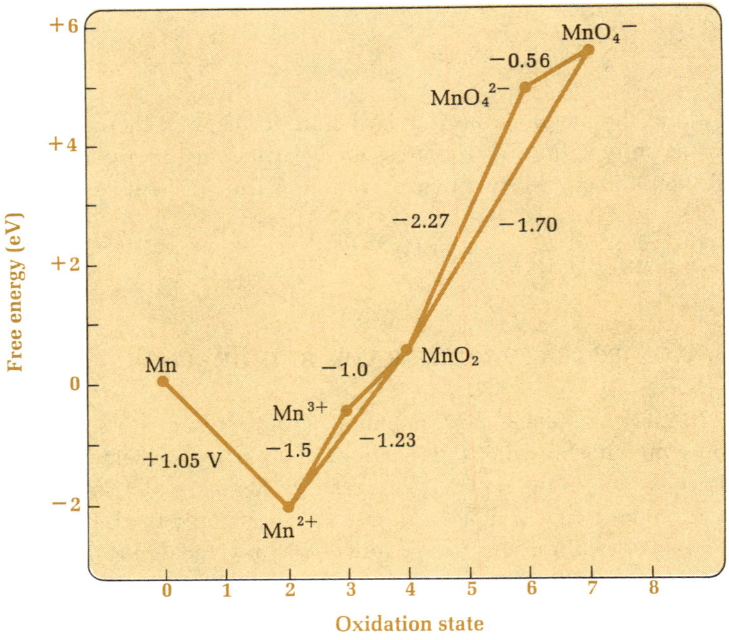

Figure 20-6 Free energy versus oxidation state diagram for the manganese system. [Adapted from A. Frost, *J. Amer. Chem. Soc.*, 73, 2680 (1951). Copyright 1951 by the American Chemical Society.]

just such a reaction between copper and nitric acid. Observe in Table 20-2 that the potential for reduction of nitrate ion is 0.96 V; it is this fact that is responsible for oxidation of copper by nitric acid.

We can also often easily tell when an element in a particular oxidation state will disproportionate to higher and lower states. If equal numbers of electrons are transferred as we move to the right or left from a particular state (as for Cu^+ above, for example), then a quick and simple algebraic sum can be used to determine stability towards disproportionation. For Cu^+ the change to the left is associated with +0.52 V (note the sign change when the reaction is reversed from its normal left-to-right reading). The sum of this potential with the oxidation potential to the next state, −0.15 V, is clearly +0.37 V, and we expect instability for the particular state, Cu^+.

To illustrate the general potential of this method, consider the more complicated diagram for manganese:

$$Mn \xrightarrow{1.18} Mn^{2+} \xrightarrow{-1.51} Mn^{3+} \xrightarrow{-0.95} MnO_2 \xrightarrow{-2.26} MnO_4^{2-} \xrightarrow{-0.564} MnO_4^-$$

The following chemistry is extracted with comparative ease:

1. The metal Mn will dissolve in 1*M* acid to give Mn^{2+} but will not be converted to Mn^{3+}.
2. If we generate Mn^{3+}, it will disproportionate to Mn^{2+} and MnO_2, since the cell potential is $(0.95) - (-1.51) = +0.56$ V.
3. MnO_2 is stable with respect to disproportionation into higher and lower oxidation states (all cell potentials are negative).
4. In 1*M* acid solution, MnO_4^{2-} will disproportionate into MnO_2 and MnO_4^- with a cell potential of $(0.564) - (-2.26) = +1.696$ V.

Thus, a great deal of the chemistry of a complex system is briefly summarized in the oxidation potential diagram.

An alternative form for displaying oxidation potential relationships of a single element is shown in Figure 20-6. In this figure we have taken advantage of relationship (20-26) between free energy and cell potential. The figure shows, for the element manganese, free energies relative to the elementary state versus oxidation number at first glance. The figure shows that a minimum free energy is associated with manganese in the +2 oxidation state. The simplest way to use the diagram is to think of the analogy of an object subject to a gravitational field. The steeper a slope, the harder it is to raise an object, and conversely, the more readily an object rolls down. In the physical situation, objects tend to seek the minimum, and so too in the chemical case. Thus, in acid solution we might expect Mn metal to dissolve to form Mn^{2+} but not to be oxidized further. Observe also that the instability of Mn^{3+} is also apparent, since it is found at an energy above the line connecting Mn^{2+} and MnO_2. More energy is generated when one Mn^{3+} drops to a lower oxidation state than is required to raise another ion to the higher state.

Finally, let us observe one other feature that we have not examined before: The cell potential for two successive changes in oxidation state does not appear to be simply related to the single potential linking the first to the third. Thus, the potential for the reaction

$$MnO_2 \longrightarrow MnO_4^{2-} + 2e^-$$

is -2.26 V, and that for

$$MnO_4^{2-} \longrightarrow MnO_4^- + e^-$$

is -0.56 V. The potential for the direct reaction

$$MnO_2 \longrightarrow MnO_4^- + 3e^-$$

is -1.70 V.

In the diagram it is immediately clear that the slope of the latter (and thus the cell potential) is intermediate between the former but not simply the average of the two. Calculation of the half-reaction potential is simple: The free-energy difference is taken between the initial and final states and divided by $n\mathfrak{F}$ (where n equals the number of electrons transferred and $\mathfrak{F}$ is, as usual, the Faraday constant).

$$E_{MnO_2:MnO_4^-} = \frac{-2 \times 23.06 \times (-2.27) - 1 \times 23.06 \times (-0.56)}{-3 \times 23.06}$$

$$= 1.70$$

Example *Calculate the potential for the oxidation*

$$Co \longrightarrow Co^{3+} + 3e^-$$

Solution *Table 20-2 lists two separate reduction potentials, which may be reversed to give the oxidation potentials.*

$$Co \longrightarrow Co^{2+} + 2e^- \qquad E = +0.28\ V$$
$$Co^{2+} \longrightarrow Co^{3+} + e^- \qquad E = -1.82\ V$$

The free energy for the oxidation from the element to the +3 oxidation state is independent of the pathway chosen (since it is a state function—see Chapter 8) and is therefore the sum of the free-energy changes for the two steps:

$$\Delta G^0_{overall} = \Delta G^0_{Co \to Co^{2+}} + \Delta G^0_{Co^{2+} \to Co^{3+}}$$
$$\Delta G^0_{overall} = -2(23.06)(+0.28) - 1(23.06)(-1.82)\ \text{kcal/mole}$$
$$E^0_{overall} = \frac{\Delta G^0}{-n\mathfrak{F}} = \frac{+29.06}{-3(23.06)} = -0.43\ V$$

EFFECTS OF CONCENTRATION ON CELL POTENTIALS

If a reaction system is at equilibrium, $\Delta G = 0$ and both Equations (20-26) and (20-29) must apply for this condition just as they do in any other situation. Since neither the Faraday constant nor the number of electrons transferred in the reaction can be equal to zero, we at once conclude from (20-26) that the cell potential must be equal to zero when equilibrium concentrations of reactants and products are present. Measured cell potentials are equal to those calculated from tables of reduction potentials only when all reactants are present at unit activity. (See Chapter 13 for a discussion of activity coefficients. We shall assume in our present discussion that this means that the concentration of soluble substances is 1*M* and that some of the solid phase is present for insoluble substances.) We have seen in (20-30) one case for which the equilibrium concentration is slightly over three powers of ten less than the standard 1*M*. For the case of the coupled copper reactions discussed above, the cell potential may vary from 0.37 V, when solutions are freshly prepared and at 1*M* concentrations, gradually down to zero as the equilibrium concentrations are achieved. The latter condition corresponds to one all too familiar to us—a "dead" battery.

To describe the cell potential as a function of concentration, we adopt the following procedures. The cell potential at standard conditions is given a special symbol, E^0, and is readily derived from tables such as 20-2. For any conditions other than standard, some corrections will be necessary. The free energy of a system not at equilibrium follows the relation

$$\Delta G = \Delta G^0 + RT \ln Q \tag{20-31}$$

where Q is an expression involving activities of reactants and products of the same form as the equilibrium constant K and differing only in the fact that the activities need not be those at equilibrium. For the cuprous disproportionation case, following this rule, Q takes the form

$$Q = \frac{(Cu^{2+})(Cu)}{(Cu^+)^2} \tag{20-32}$$

In the presence of copper metal the activity of Cu will always be unity, and the expression may be simplified to

$$Q = \frac{(Cu^{2+})}{(Cu^+)^2} \tag{20-33}$$

Substitution into (20-31) and entering the relation to cell potential leads to

$$\Delta G = \Delta G^0 + RT \ln \frac{(Cu^{2+})}{(Cu^+)^2} = -n\mathfrak{F}E_{cell} \tag{20-34}$$

Since ΔG^0 is related to E^0 by Equation (20-26), we may rewrite this as

$$\Delta G = -n\mathfrak{F}E^0 + RT \ln \frac{(Cu^{2+})}{(Cu^+)^2} = -n\mathfrak{F}E_{cell} \tag{20-35}$$

and finally derive the expression for the cell potential as a function of concentration:

$$\boldsymbol{E^0 - \frac{RT}{n\mathfrak{F}} \ln \frac{(Cu^{2+})}{(Cu^+)^2} = E_{cell}} \tag{20-36}$$

In the general case we write the **Nernst equation,**

$$\boldsymbol{E_{cell} = E^0 - \frac{RT}{n\mathfrak{F}} \ln Q} \tag{20-37}$$

We are frequently interested in cell potentials at room temperature. Let us assume a temperature of 25°C as common and convert this to 298°K. The gas constant R has the value 1.987 cal/mole deg, and the Faraday constant is 23.06 kcal/V. If we also introduce the relation between natural and base ten logarithms,

$$\ln N = 2.303 \log N$$

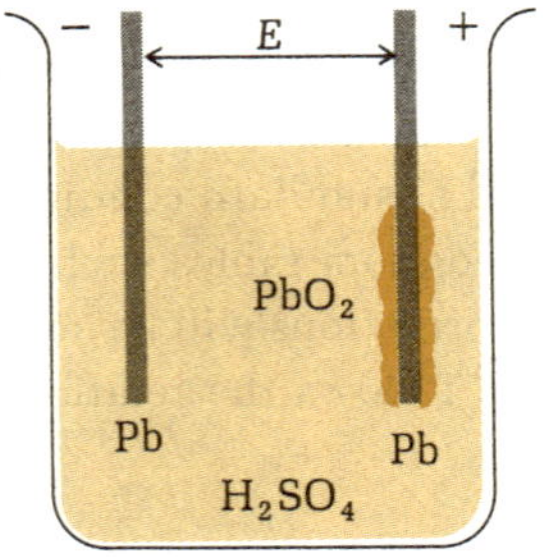

Figure 20-7 The lead storage battery. The key to obtaining electrical energy from a chemical reaction is to separate physically the two half-cell reactions, so that the electrons generated at the anode must pass through an external circuit to reach the oxidant.

we can substitute these values into Equation (20-37) and obtain

$$E = E^0 - \frac{RT}{n\mathcal{F}} \ln Q = E^0 - \frac{2.0 \times 298 \times 2.3}{23060 \times n} \log Q = E^0 - \frac{0.059}{n} \log Q \qquad (20\text{-}38)$$

Note that this simplified expression is strictly valid *only* at 25°C, but cell potentials for many reactions change only slightly with temperature.

USES OF CELL POTENTIALS

Let us consider two applications of cell potentials as illustrations of their general value. Figure 20-7 shows a schematic diagram of the common lead storage battery used in automobiles. The reaction that produces the electrical current is

$$Pb + PbO_2 + 2H_2SO_4 \longrightarrow 2PbSO_4 + 2H_2O \qquad (20\text{-}39)$$

The pertinent half-reactions from Table 20-2 are

$$\begin{array}{rlr} PbO_2 + 4H^+ + SO_4^{2-} + 2e^- \longrightarrow PbSO_4 + 2H_2O & E^0 = & 1.70 \\ PbSO_4 + 2e^- \longrightarrow Pb + SO_4^{2-} & E^0 = & -0.31 \\ \hline Pb + PbO_2 + 2H_2SO_4 \longrightarrow 2PbSO_4 + 2H_2O & E^0 = & 2.01 \end{array} \qquad (20\text{-}40)$$

In common application, six such cells are connected in series to produce the nominal 12-V battery. For each cell, the Nernst equation in form (20-38) becomes

$$E = 2.01 - \frac{0.0591}{2} \log \frac{(PbSO_4)^2(H_2O)^2}{(Pb)(PbO_2)(H_2SO_4)^2} \qquad (20\text{-}41)$$

However, considerable simplification is possible, since each of the lead compounds is insoluble and is therefore present at unit activity. The expression simplifies to

$$E = 2.01 - 0.0296 \log \frac{(H_2O)^2}{(H_2SO_4)^2} \qquad (20\text{-}42)$$

Most of us have observed the testing of a "weak" battery in a service station. Commonly, some of the battery fluid is drawn into a transparent tube containing a calibrated float. The density of the fluid is thus determined. The significance of this observation can be deduced from Equation (20-42). A large ratio of water to sulfuric acid would result in a significant decrease in the cell potential. Solutions of sulfuric acid become more dense as the concentration of the acid is increased. Thus, density of the fluid in each cell is directly related to its cell potential and thus to the "health" of the battery.

As a final example of the practical importance of cell potentials, let us see how they can be used to calculate solubility products. In Chapter 12 we have used solubility products extensively. Let us see how cell potentials can be used to determine K_{sp} for AgCl.

From Table 20-2 we find two suitable half-reactions and subtract to obtain the cell potential and associated free-energy change

$$\begin{array}{ll} AgCl(s) + e^- \longrightarrow Ag + Cl^- & E^0 = 0.222 \\ Ag^+ + e^- \longrightarrow Ag & E^0 = 0.80 \\ \hline AgCl(s) \longrightarrow Ag^+ + Cl^- & E^0 = -0.578 \end{array}$$

$$\Delta G^0 = -(1)(23.06)(-0.578) = 13.329 \text{ kcal}$$

Using Equation (20-29), we then write

$$\Delta G^0 = 13.329 = -RT(2.3) \log \frac{(Ag^+)(Cl^-)}{(AgCl)} \qquad (20\text{-}43)$$

At 25°C (observing that the solid AgCl is at unit activity) this becomes

$$\log(Ag^+)(Cl^-) = \frac{-13.329}{1.987(298)(2.3)} = -9.745$$

$$(Ag^+)(Cl^-) = 1.8 \times 10^{-10} \qquad (20\text{-}44)$$

This is just the value of K_{sp} listed for AgCl in Table 12-4.

USE OF ELECTROCHEMICAL CELLS TO PRODUCE CHEMICAL REACTION

In all of the cases we have so far examined we have used various chemical reactions to produce a cell potential. In each case we can use the electrical potential so caused for useful purposes. The lead storage battery we discussed above is in no way special as a portable energy source except that it is relatively easy to package the reaction in a convenient and portable container of suitable size and weight to be carried in a car. Before leaving the subject of electrochemical cells, it is important to note that in many

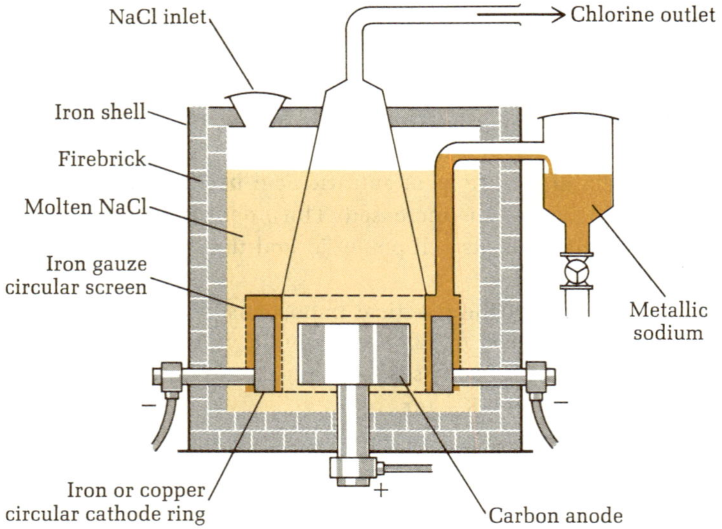

Figure 20-8 The Downs cell for producing sodium metal. (From H. H. Sisler, C. A. Van der Werf, and A. W. Davidson, *College Chemistry*, 3rd ed., Macmillan, New York, 1967. Copyright © 1967 by The Macmillan Company.)

cases it is desirable to use the cells in an opposite manner: It is convenient to force electrical energy through a cell in the reverse direction and thereby produce a desirable chemical reaction.

The lead storage battery provides a simple and common example of this process. When the battery has been discharged according to reaction (20-39), we simply attach it to a source of direct current that passes in the opposite direction and causes the reation

$$2PbSO_4 + 2H_2O \longrightarrow Pb + PbO_2 + 2H_2SO_4 \tag{20-45}$$

To cause electrical flow in the reverse direction, it is only necessary to attach a source of higher potential, as shown in Figure 20-3. In the most common case this potential is supplied by the generator or alternator carried in the car and driven by the motor. If we are careless and discharge the battery too far, we are forced to find an external charging source to recharge the battery sufficiently to start the motor.

Although the recharging of a storage battery is one of the most common events of daily life since it continually takes place in each automobile each time it is driven, the chemical industry provides many additional and important examples of electrochemically produced materials. Magnesium, aluminum, copper, sodium, chlorine, and many other elements are frequently prepared from their compounds by electrochemical means. We shall here consider only a single example of the very large number of important reactions.

If sodium chloride is melted and subjected to **electrolysis,** both metallic sodium and elemental chlorine are produced. A suitable cell for carrying out this reaction on a large scale is shown in Figure 20-8. Pure sodium chloride melts at slightly over 800°C, but the addition of sodium carbonate or calcium chloride can be used to lower the cell operating temperature to about 600°C.

$$2NaCl \longrightarrow 2Na + Cl_2 \qquad (20\text{-}46)$$

We should note that a rather elaborate cell is used to prevent mixing of the products once they have been formed.

There are two important operating conditions of the cell that we should examine closely. First, a sufficiently high potential must be applied to the cell to cause current to flow. In practice, only a few volts are required. Of more importance, one electron must be supplied for every single sodium atom converted to the element. The current flow, measured in amperes, must be very large to accomplish reduction of significant amounts of the metal in a reasonable time period. It should be clear that to produce 1 mole of sodium, 23 g, it would be necessary to supply 6.02×10^{23} electrons, since every atom requires just a single electron. Counting each electron passed in a system would be difficult, and fortunately is quite unnecessary. The amount of electrons flowing per unit time is simply measured by the number of amperes of current. Once we measure the amount of current required to produce 23 g of sodium, we know that an **equivalent weight** of any other substance could be produced by flow of an identical amount of current. The number of *coulombs* (C) of current is measured by the amperes (A) multiplied by the time the current flows measured in seconds.

$$\text{coulombs} = \text{amperes} \times \text{seconds}$$

In the early nineteenth century, the English scientist Michael Faraday, measured the number of coulombs per equivalent and found that 96,500 C was equivalent to the flow of a single mole of electrons. This unit is now commonly called a **faraday.**

To illustrate the use of this type of information, let us calculate the weight of sodium that could be produced per day from a cell such as shown in Figure 20-8 if a current of 100 A is maintained. First, the number of seconds per day is 24 hr × 60 min × 60 sec/min = 86400 sec/day. A current of 100 A during this time results in the passage of

$$100 \times 86{,}400 = 8.64 \times 10^6 \text{ C}$$

or

$$\frac{8.64 \times 10^6}{9.65 \times 10^4} = 8.95 \times 10^1 = 89.5 \text{ faradays}$$

Oxidation-reduction reactions find lots of common uses. One is to provide energy to power this experimental electric car. This particular model was developed by Electric Fuel Propulsion, Inc., Detroit, and is in use in Fairbanks, Alaska. In this cold clime, ordinary internal combustion engines contribute an unusual form of atmospheric pollution, ice fog, which results from crystallization of water vapor emitted in the exhaust when the temperature dips below $-30°F$. (Courtesy of Electric Fuel Propulsion Corp.)

Since each faraday reduces just 1 mole or 23 g of sodium, the total production is

$$23 \times 89.5 = 2060 \text{ g of Na}$$

Clearly, to produce the large quantities of sodium required by industry, a large number of cells must be operated and the currents used in each must also be large. It is for this reason that most electrochemical processes are carried out in areas where large amounts of relatively inexpensive hydroelectric power are available. For example, Niagara Falls, New York, has a very heavy concentration of electrochemically based industries.

Observe that to cause 89.5 faradays to flow through a cell and produce 2 kg of sodium, an equal number of electrons must be taken away from chloride ions at the other electrode. The reaction is

$$Cl^- \longrightarrow \frac{1}{2}Cl_2 + e^- \qquad (20\text{-}47)$$

where we have written the reaction with a single chloride ion to emphasize the relation to the total electron flow. Then 89.5 moles of chloride ion are discharged to form $89.5/2 = 44.75$ moles of Cl_2. Note the factor of two introduced when we calculate the number of moles of chlorine produced. Since 1 mole of chlorine is about 71 g, the total weight of chlorine produced per day is about 3170 g per cell.

Glossary

Electrochemical cell a device in which two half-reactions are coupled in such a fashion that electrical current flows through an external conductor.
Electrolysis the process in which chemical reaction is produced by use of an external electrical source.
Equivalent weight the molecular weight of a compound divided by the change of oxidation state during reaction. A single compound may have more than one equivalent weight depending on particular reactions it undergoes. Oxidation of Mn^{2+} to Mn^{3+} gives an equivalent weight of $54.93/1 = 54.93$, but oxidation of the same substance to MnO_2 gives an equivalent weight of $54.93/2 = 27.465$.
Faraday one equivalent of electrons, equal to 96,500 coulombs.
Half-reaction a reaction in which electrons are a reactant or product.
Nernst equation the equation that shows the concentration dependence of the electrical potential of a cell.
Oxidation the process in which one or more electrons are lost by an atom or molecule.
Redox a term used to describe a reaction in which both oxidation and reduction occur.
Reduction the process in which one or more electrons are added to an atom or molecule.
Standard reduction potential the potential assigned to an individual half-reaction when each reactant is present at unit activity and the reaction is written to use electrons as reactants.

Study problems

In all problems where temperature is not specified, assume 25°C.

1. Balance the following redox equations. In each equation identify the oxidizing and reducing agents.
   a. $HI + H_2SO_4 \longrightarrow I_2 + H_2S + H_2O$
   b. $HBr + H_2SO_4 \longrightarrow Br_2 + SO_2 + H_2O$
   c. $H_2S + FeCl_3 \longrightarrow S + FeCl_2$

d. $MnO_2 + KOH + O_2 \longrightarrow K_2MnO_4 + H_2O$
e. $PbS + PbSO_4 \longrightarrow Pb + SO_2$
f. $SnCl_2 + HgCl_2 \longrightarrow SnCl_4 + Hg_2Cl_2$
g. $KMnO_4 + Mn(OH)_2 \longrightarrow MnO_2 + KOH + H_2O$
h. $KClO_3 \longrightarrow KClO_4 + KCl$
i. $Br_2 + Na_2CO_3 \longrightarrow NaBrO_3 + NaBr + CO_2$
j. $NH_4NO_3 \longrightarrow N_2O + H_2O$
k. $Mg + HNO_3 \longrightarrow Mg(NO_3)_2 + NH_4NO_3 + H_2O$
l. $KI + KNO_2 + H_2SO_4 \longrightarrow K_2SO_4 + NO + I_2 + H_2O$
m. $HOCl + NaI \longrightarrow HIO_3 + NaCl + HCl$

2. For each of the redox couples written below, calculate the standard cell potential. Calculate the standard free-energy change and write the complete equation to show the actual reaction proceeding to the right. Use Table 20-2 to obtain standard reduction potentials.
   a. MnO_4^-/Mn^{2+} with Hg^{2+}/Hg_2^{2+}
   b. Co^{2+}/Co with I_2/I^-
   c. Br_2/Br^- with IO_3^-/I_2
3. Calculate the standard potential and equilibrium constant for the reaction

$$2Cu^{2+} + Sn^{2+} \longrightarrow 2Cu^+ + Sn^{4+}$$

4. For the reaction in Problem 3, calculate the cell potential if each reactant is present at an activity of 0.01.
5. Use potentials from Table 20-2 to determine whether Fe^{2+} can be reduced to elemental iron by metallic aluminum.
6. Pick two couples from Table 20-2 and write a balanced equation for the reaction expected. Calculate E^0 for the reaction.
7. Show oxidation states for all elements of
   a. $Ni(CN)_4^{2-}$
   b. WCl_6
   c. $Na_2S_2O_4$
   d. $Cr(NH_3)_6^{3+}$
   e. $H_2S_2O_7$
   f. IF_5
   g. XeF_2
8. Is Fe^{2+} stable with respect to disproportionation to Fe^{3+} and Fe?
9. Use information from Table 20-2 to construct an oxidation state–free energy diagram for the element chromium. Briefly describe the chemistry of chromium by use of this diagram.
10. From information in Table 20-2, compute the solubility product of CuI.

Suggestions for further reading

R. B. Fischer and D. G. Peters *Chemical Equilibrium,* Saunders, Philadelphia, 1970 (paperback), Chap. 5.

E. H. Lyons, Jr. *Introduction to Electrochemistry,* Heath, Lexington, Mass., 1967 (paperback), Chaps. 4, 9.

W. H. Nebergall, F. C. Schmidt, and H. F. Holtzclaw, Jr. *General Chemistry,* Heath, Lexington, Mass., 1972, Chap. 20.

21

The life processes: Thermodynamic aspects

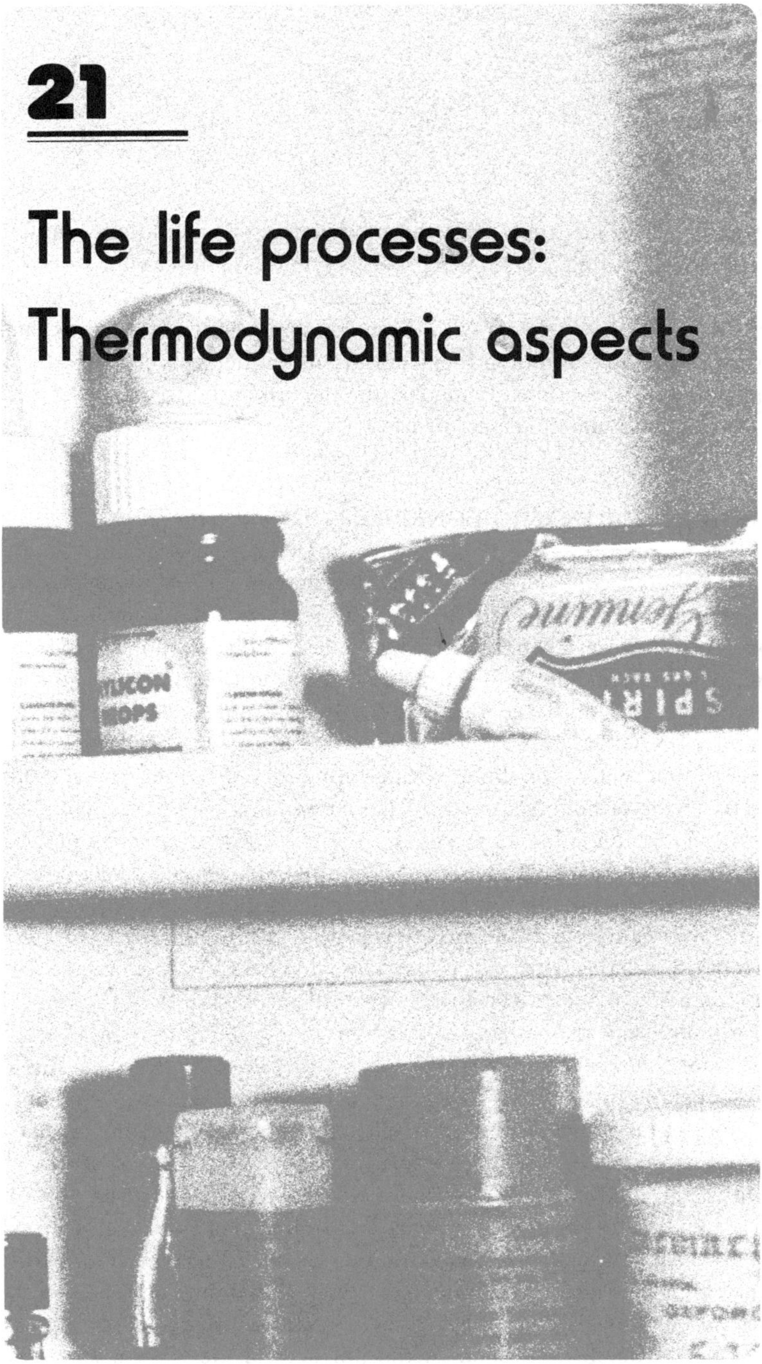

From here on, it's all downhill:
Shee is dead; and all which die
To their first elements resolve . . . .

John Donne

In Chapter 8 we have seen how thermodynamic considerations contribute to our understanding of chemical processes. In this chapter we wish to focus our attention on a specific class of chemical transformations: those involved in the life process. We begin with thermodynamic considerations of the life process as a whole and then focus our attention increasingly on individual chemical transformations as models for the whole body of reactions that constitute the dynamic aspect of life.

LIFE: THERMODYNAMIC CONSIDERATIONS

Let us begin our considerations of the thermodynamics of living organisms by reviewing briefly the structure of the molecules that account for the properties of these organisms and by considering some points that derive from these. We remember from the discussion in Chapter 7 that the basic information storehouse of the living cell is a molecule of nucleic acid, generally DNA. The information is stored as an ordered sequence of bases along the main chain of the molecule; this ordered sequence corresponds directly to the ordered sequence of the letters of the alphabet, which may constitute an intelligible message. The cellular information is employed, in the sequence of reactions depicted in Figure 7-18, for the direction of the synthesis of a second class of biomacromolecule, the proteins. The proteins retain the organizational information content of the nucleic acids: In the place of an ordered sequence of bases, we have now an ordered sequence of amino acids. Proteins serve a number of functions in living systems, but that of most importance to us at this time is the function of catalyzing the individual chemical reactions of the organism (Chapter 22). From the most basic considerations of life—the processes of digestion, respiration, muscular contraction, and the like—it is clear that living organisms must carry out a very large number of chemical reactions. As we shall see in more detail in the next chapter, specification of just what reactions will occur and at what rates they will occur is controlled by a subclass of proteins, the *enzymes*. The enzymes, then, directly control and account for the totality of the chemical reactions of a living system, its **metabolism.** The metabolic processes include, among other things, synthesis of yet other classes of ordered macromolecules such as the **polysaccharides** and certain lipids. These molecules serve functions that include architecture (cell walls, etc.)

and energy storage. Up to this point, we have identified several classes of macromolecules of definite, ordered structure performing a number of functions essential to the living organism. Note well that the ordering of the structure is essential for proper functioning.

What we need to recognize now is that the life process is dependent on a type of organization that is not directly expressed in the structures of the crucial macromolecules, such as the proteins. In other words, the structure of the DNA molecule confers on it the capacity to act as an information storage and transmission device, the structure of proteins permits them to act as efficient and specific catalysts for many reactions, the structure of polysaccharides renders them suitable to act as cell wall components and the like, but these capacities in themselves do not account for the properties of the living system. That is, superimposed on the functions themselves is *control* of these functions. At an elementary level, everyone is aware that a living system can change its chemistry to meet changing conditions: The human body responds to demands placed on it by vigorous exercise through an increase in rate and intensity of heartbeat, some microorganisms are able to modify their metabolism in such a way as to become immune to antibiotics, and so on. An orderly sequence of metabolic processes that permits the organism to develop, differentiate, respond, and recover requires that the metabolism as a whole be sensitive to an organized set of control mechanisms.

Perhaps the most dramatic example of such control in action is the developmental process. At inception, human life begins with a single cell containing a DNA molecule with information for the synthesis of proteins and of other nucleic acids. After a few cell divisions, all of the daughter cells still appear similar to the original parent one. Nevertheless, when we consider the adult organism, we are struck by the multiplicity of cell types. Certainly, the metabolism of a liver cell is distinct from that of a brain cell, a lung cell is distinct from an epithelial cell, and so on. Nevertheless, each of these cells contains all of the basic information content; that is, the DNA molecules of each of them are identical. It follows inescapably, then, that not all of the information present in a cell is actually expressed at one time and that information expressed in a cell varies as a function of time and environmental condition. Thus, the turning off and on of "informational switches" permits the orderly development of the organism through control of what enzymes are synthesized, which in turn control the remainder of the cellular chemical reactions. Now, the important point for our use at the moment is the necessity for *order* and *organization* in the control processes, leading to order and organization in the metabolic processes at any one time and to an orderly development and change in these processes as a function of time. In summary, life is utterly dependent on organization: that expressed in terms of the structure of the molecules of life and that expressed in terms of the sequences and interrelationships of those reactions involving these molecules.

With these points in mind, we can return to thermodynamics. From the standpoint of the physical scientist, what we have just said and its consequences has been elegantly summarized by one of the principal authors of quantum mechanics, Erwin Schrödinger, in his essay "What is Life?." Schrödinger points out the thermodynamic argument thus: ". . . the device by which an organism maintains itself stationary at a fairly high level of oderliness [= fairly low level of entropy] really consists in continually sucking orderliness from its environment." For the most part, this realization does describe the life process. The order that we have just described does require that maintenance of the life process involve a continual negative entropy change; this must be accomplished at the expense of order (or more generally, free energy) of the environment. Thus, ordering necessary for life is achieved at the expense of disordering the environment. In fact, much of the energy necessary is derived from the transformation of organized molecules of nature—proteins, polysaccharides—into much simpler molecules such as carbon dioxide, water, and ammonia. Thus, life is, thermodynamically, the derivation of order from order, the transformation of order in one set of molecules into that of another set.

Lest our view become too narrow, we should remember that entropy and enthalpy are both part of the same quantity—the free energy. Consequently, the negative entropy change involved in life maintenance can be overcome by the input of energy (enthalpy) as well as by the input derived from positive entropy changes in the environment. We shall see one example of this below.

To this point, our discussion of the thermodynamic changes that describe the life process has been quite general. Let us now focus our attention more specifically on some of the actual chemical changes involved and the energetics of these chemical changes.

GENERATION OF METABOLIC ENERGY

The energetics involved in the maintenance of life may be meaningfully discussed in three parts: the generation of metabolic energy, the transference of such energy, and the utilization of it. Taken together, the understanding of these points will suffice to understand the mechanisms by which the living organism extracts order from its environment as a means of sustaining itself. Let us begin with the first point: How is metabolic energy generated?

Perhaps the first real insight into metabolic energy generation was gained by Antoine Lavoisier following a series of calorimetric experiments in the late eighteenth century. By noting that the amount of heat produced by a guinea pig in the process of exhaling a given amount of carbon dioxide is similar to the amount of heat produced in burning an amount of carbon that will yield the same amount of carbon dioxide, Lavoisier concluded: "Respiration is a combustion, slow it is true, but otherwise perfectly similar to that of charcoal." The process of combustion, whether the item be charcoal, a candle, wood, or such, involves the transference of electrons from carbon-containing substances to molecular oxygen. Similarly, we know that in the process of **respiration**, inhaled oxygen is employed as an electron acceptor for the oxidation of carbon-containing (organic) substances, converting these into carbon dioxide. Thus, in this basic respect, respiration and combustion are, indeed, closely related processes. In general, living organisms do sustain themselves through the utilization of energy generated from the transfer of electrons (oxidation) from one substance to another. The transfer of electrons need not necessarily be from an organic substance to oxygen, as it is in mammals for example; many microorganisms utilize other electron transfer processes for energy generation. The point we wish to emphasize at the moment is that such electron transfer processes can be energy-yielding; later we shall investigate how such energy can be trapped and employed usefully. How much energy can be generated from such processes? One way of providing some information regarding this point is simply to measure the heat of combustion for the oxidation of some substance in a calorimeter. For example, the conversion of the sugar glucose to carbon dioxide and water,

$$C_6H_{12}O_6 + 6O_2 \longrightarrow 6CO_2 + 6H_2O \qquad (21\text{-}1)$$

liberates 673 kcal of heat per mole of glucose oxidized. Note that this is a measure of enthalpy, not a free-energy change. Alternatively, we might make calculations of standard free-energy or enthalpy changes using the tables of free energies or bond energies of Chapter 8. However, the most generally useful means of deriving free-energy changes for oxidation-reduction is the use of *reduction potentials* as developed in the last chapter. Since most biochemical reactions occur near pH 7, and since many standard reduction potentials do vary with pH, it is convenient to tabulate standard reduction potentials referred to a standard state of pH 7 rather than to one of 1M H^+, pH 0 as we did in Chapter 20. Such standard reduction potentials are referred to by the symbol $\boldsymbol{E_0'}$: a brief compilation of values of biochemical interest is provided in Table 21-1. As developed in Chapter 20, to calculate free-energy changes from electrode potentials, the necessary equation is simply

$$\Delta G^{0\prime} = -n\mathfrak{F}\,\Delta E_0' \qquad (21\text{-}2)$$

where n is the number of electrons transferred in the reaction process and $\mathfrak{F}$ is the faraday.

Example *An important biochemical oxidation-reduction reaction is conversion of β-hydroxybutyric acid into acetoacetic acid employing molecular*

Table 21-1 Standard reduction potentials for systems of biochemical importance

System	E_0', pH 7 (V)
Oxygen/water	0.815
Ferric/ferrous	0.77
Nitrate/nitrite	0.42
Ferricyanide/ferrocyanide	0.36
Oxygen/hydrogen peroxide	0.30
Cytochrome a; ferric/ferrous	0.29
Cytochrome c; ferric/ferrous	0.22
Crotonyl-SCoA/butyryl-SCoA	0.19
Methemoglobin/hemoglobin	0.17
Cytochrome b_2; ferric/ferrous	0.12
Ubiquinone; ox/red	0.10
Dehydroascorbate/ascorbate	0.08
Metmyoglobin/myoglobin	0.046
Fumarate/succinate	0.03
Methylene blue, ox/red	0.01
Yellow enzyme; FMN/$FMNH_2$	−0.122
Pyruvate + ammonium/alanine	−0.13
α-Ketoglutarate + ammonium/glutamate	−0.14
Oxaloacetate/malate	−0.17
Pyruvate/lactate	−0.19
Acetaldehyde/ethanol	−0.20
Riboflavin, ox/red	−0.21
Glutathione, ox/red	−0.23
Acetoacetate/β-hydroxybutyrate	−0.293
Lipoic acid, ox/red	−0.29
NAD^+/NADH	−0.32
Pyruvate/malate	−0.33
Uric acid/xanthine	−0.36
Carbon dioxide/formate	−0.42
H^+/H_2	−0.42
Acetate/acetaldehyde	−0.60
Succinate/α-ketoglutarate	−0.67
Acetate + carbon dioxide/pyruvate	−0.70

Source: H. R. Mahler and E. H. Cordes, *Biological Chemistry*, 2nd ed., Harper & Row, New York, 1971.

oxygen as electron acceptor (oxidizing agent):

$$CH_3-\underset{\displaystyle H}{\overset{\displaystyle OH}{\underset{|}{\overset{|}{C}}}}-CH_2-CO_2^- + \tfrac{1}{2}O_2 \longrightarrow CH_3-\overset{\displaystyle O}{\overset{\|}{C}}-CH_2-CO_2^- + H_2O$$

Calculate the standard free-energy change for this reaction at pH 7.

Solution *Reference to the data in Table 21-1 provides the necessary values for the reduction potentials at* pH 7.

$$\Delta G^{0\prime} = -n\mathfrak{F}\,\Delta E_0{}'$$
$$= -2 \times 23{,}060 \times [0.815 - (-0.293)]$$
$$= -46{,}000 \times (1.11) = -51{,}000 \text{ cal/mole}$$

With these considerations regarding free-energy changes in mind, let us return to our problem of the generation of metabolic energy. In living systems, a number of individual oxidation-reduction reactions are employed for the generation of energy; their exact nature does not concern us. What is important is that the oxidized substrate is usually a C—H bond and that the electrons are usually transferred to oxygen. In most common substrates employed for energy-generation purposes, these C—H bonds have standard reduction potentials near −0.2 V. Thus, transfer of a pair of electrons from such a substrate to oxygen will generate 2 × 23 × (0.8 + 0.2) = 46 kcal for use by the living system.

Returning to the analogy of combustion and respiration, we may note that the former process releases the energy from the oxidation reactions in the forms of light and heat and that these may be dissipated uselessly, as in the burning of a candle, or harnessed in part for the production of useful work, as in the functioning of an internal combustion engine. Thus, we have the matter of the harnessing of the liberated energy from the oxidative reactions in a form useful metabolically, that is, in a form useful for the accomplishment of the tasks inherent in the maintenance of life. While not all details of the harnessing of this energy are clear at this time, basically we do understand the essentials of the matter. The energy generated in the oxidation of foodstuffs is employed for the synthesis of chemical intermediates, which possess this energy in a form that may be employed to drive certain chemical reactions. The key chemical intermediate generated for this purpose is the complicated molecule **adenosine triphosphate, ATP,** the structure of which is

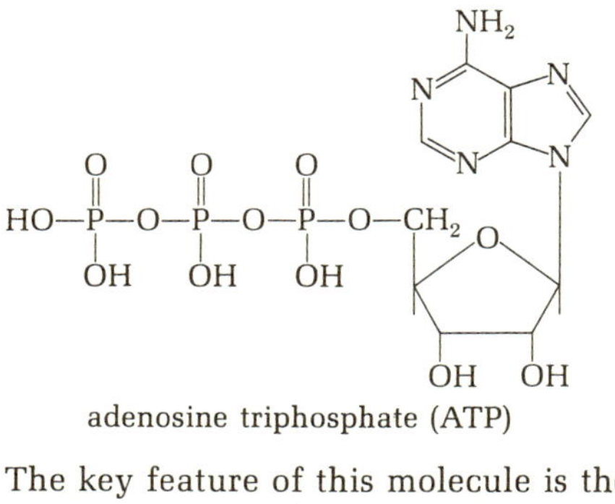

adenosine triphosphate (ATP)

The key feature of this molecule is the bonds that exist between the individual phosphate groups at the terminus of the molecule. These chemical bonds are frequently referred to as "high energy," since their cleavage proceeds with the release of much free energy. The simplest reaction of ATP is its hydrolysis: Cleavage of the terminal phosphate ester bond yields inorganic phosphate (P_i) and adenosine diphosphate (ADP):

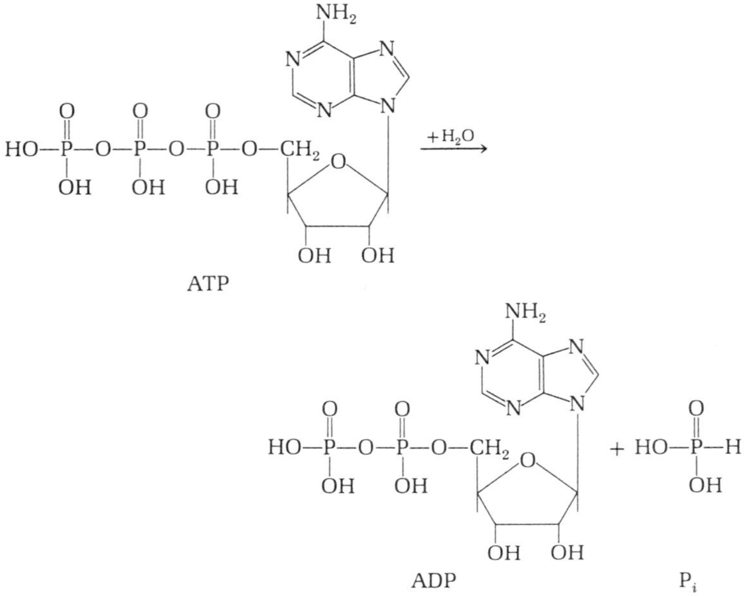

(21-3)

Hydrolysis of ATP can also occur to yield inorganic pyrophosphate (PP_i) and adenosine monophosphate (AMP):

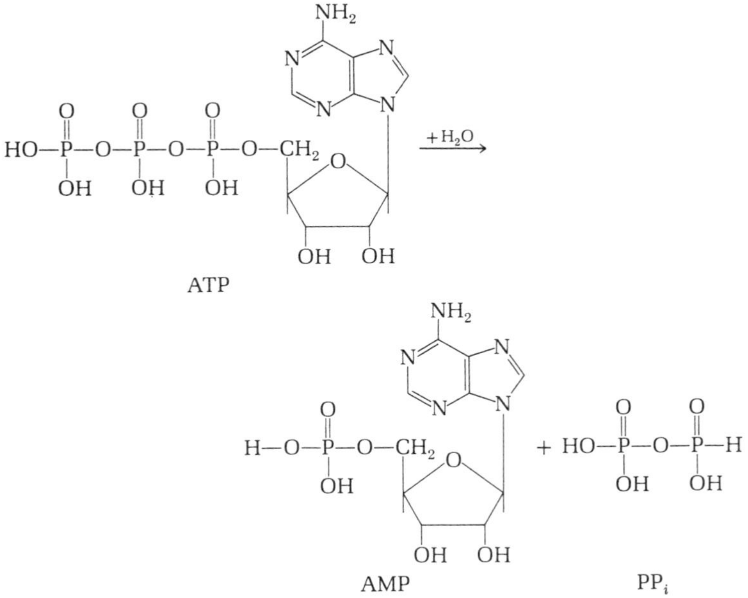

(21-4)

In both these cases the reaction is strongly exergonic (see the example below). This is of great importance, since many biochemical reactions that are themselves endergonic are driven through coupling with the hydrolysis of ATP to yield ADP and P_i or AMP and PP_i. Just how this coupling is achieved is considered later. For the moment let us focus on the mechanisms for the synthesis of ATP.

Example *Under physiological conditions and at 37°C, the equilibrium constant for the hydrolysis of* ATP *to* ADP *and* P_i *is*

$$K = \frac{(\text{ADP})(\text{P}_i)}{(\text{ATP})} = 1.78 \times 10^5 M$$

What is the standard free energy of hydrolysis of ATP *under these conditions?*

Solution *We can apply the familiar equation relating equilibrium constants and free-energy changes:*

$$\begin{aligned}\Delta G^0 &= -RT \ln K = -2.3\, RT \log K \\ &= -2.3 \times 1.98 \text{ cal/deg mole} \times 310 \text{ deg} (\log 1.78 \times 10^5) \\ &= -1410 \text{ cal} \times 5.25 = -7.46 \text{ kcal/mole}\end{aligned}$$

Thus, the standard free energy of hydrolysis of ATP *is slightly more negative than 7 kcal/mole. Since all of the species involved undergo acid-base reactions, this value varies with* pH. *As the solution becomes more basic than* pH *7, the actual free energy of hydrolysis becomes substantially more negative.*

ATP SYNTHESIS

The mechanism of transfer of electrons from a C—H bond to molecular oxygen is complicated by consisting not of a single electron transfer process but of a sequence of them. This sequence of oxidation-reduction reactions occurs along the **electron transport chain,** which, in the case of higher organisms, is located in subcellular particles called *mitochondria,* the powerhouse of the cell. The electron transport chain consists of a number of components. These include the following:

1. The **nicotinamide nucleotide coenzymes,** which undergo oxidation-reduction reactions of the sort

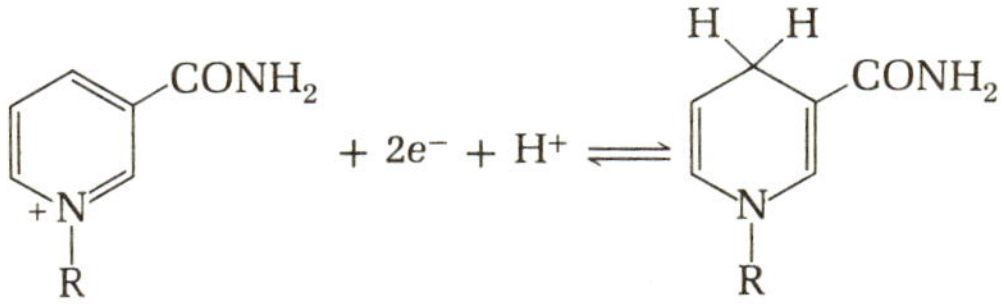

NAD+ NADH

(21-5)

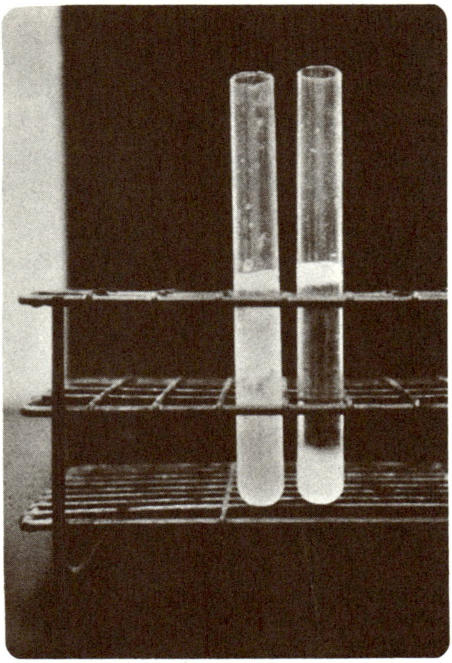

Adenosine triphosphate (ATP) is involved in many physiological phenomena, including the process of muscle contraction. One manifestation of this involvement is shown here. In the test tube on the left is a thick gel of actomyosin, a combination of proteins from muscle that is believed to change shape when muscle contracts. In the test tube on the right is the same preparation to which 0.002-molar ATP has been added. Note that structure of the gel is broken, reflecting a change in the shape of each molecule that formed it. This transformation is completely reversible; if the added ATP is now destroyed in some way, the actomyosin gel reforms. (From Robert M. Dowben, *General Physiology*, Harper & Row, New York, 1970.)

The complete structure of NAD has been provided in Chapter 19 in connection with our discussion of photosynthesis. Nicotinamide adenine dinucleotide, NAD, differs from NADP only in not having the phosphate group attached to the ribose linked to adenine. Otherwise the structures of these two species are identical. Note that all the bond-changing reactions occur at the nicotinamide ring. In the oxidized form, this ring exists as a positively charged structure as shown. Upon reduction, which can be viewed as the addition of a hydride ion, H^-, this positive charge is lost and a dihydronicotinamide is formed. Transfer of a hydride ion to some suitable acceptor would regenerate the oxidized form. **Nicotinamide** itself,

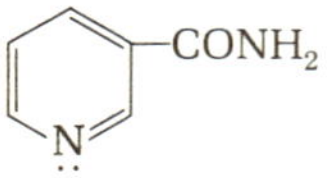

is required in the diet of man since we cannot synthesize this molecule, and it is commonly found in vitamin pills (niacin). It is used exclusively for the synthesis of NAD and NADP.

2. **Flavin nucleotides** have a structure partially identical to that of ATP:

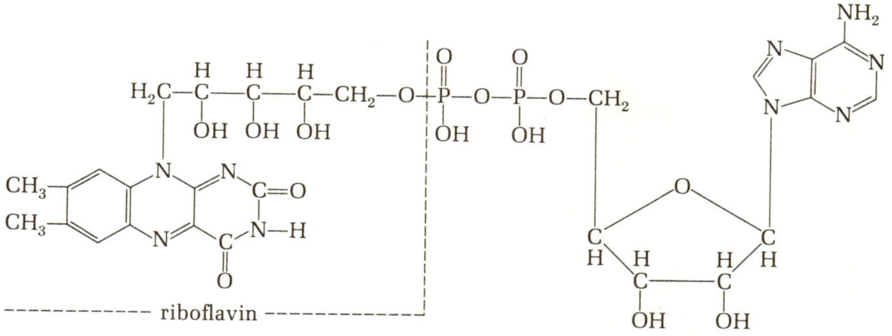

This complex molecule is called *flavin adenine dinucleotide* and is usually abbreviated *FAD*. Reduction of this molecule occurs at the complex isoalloxazine ring:

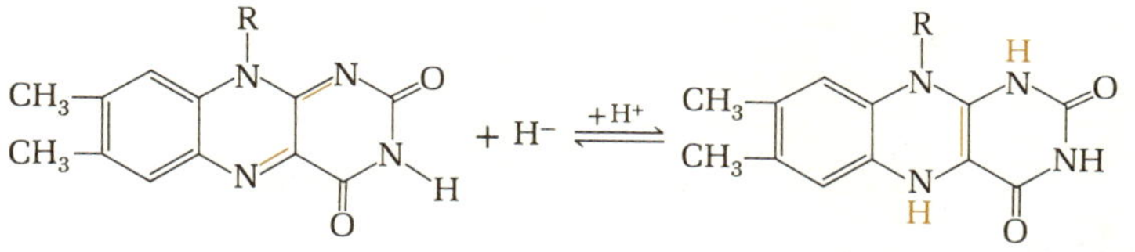

FAD FADH$_2$

As shown in the complete structure for this molecule, it is partially derived from **riboflavin,** accounting for the nutritional requirement of man for this vitamin. In the biochemical scheme of things, the nicotinamide nucleotide coenzymes are usually involved in the oxidation of organic substrates and become reduced in the process. They are reoxidized by the flavin nucleotides, which in turn become reduced. Thus, we can imagine a series of reactions involving these components:

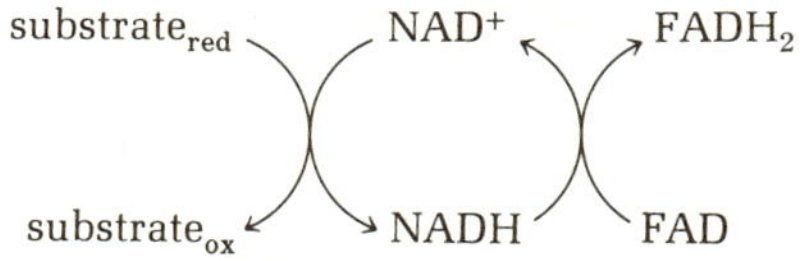

(21-6)

In this equation NAD acts as a carrier of electrons from the substrate to the FAD. The overall reaction is just

$$\text{substrate}_{red} + \text{FAD} \longrightarrow \text{substrate}_{ox} + \text{FADH}_2 \qquad (21\text{-}7)$$

3. A number of **cytochromes,** which are proteins similar in some respects to myoglobin, are components of the electron transport chain. They all contain one or more atoms of iron, which can be interconverted between the ferric and ferrous forms in one-electron transfer processes:

$$\text{cytochrome } (Fe^{3+}) \underset{}{\overset{+1e^-}{\rightleftharpoons}} \text{cytochrome } (Fe^{2+}) \qquad (21\text{-}8)$$

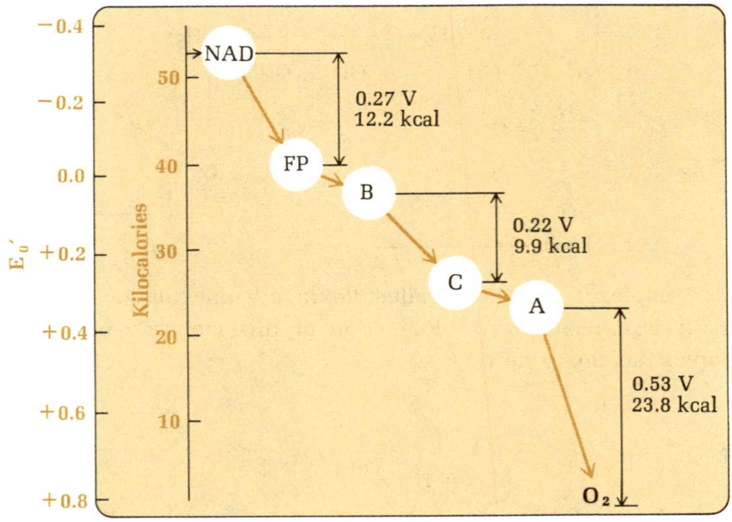

Figure 21-1 The flow of electrons and free-energy changes for the electron transport chain. The designations B, C, and A refer to cytochromes b, c, and a, respectively. The three reactions for which changes in reduction potential and free energy are explicitly noted are thought to be the sites of synthesis of ATP molecules. (From A. L. Lehninger, *Biochemistry*, Worth, New York, 1970.)

Just as FAD acts to reoxidize NADH [Equation (21-6)], so the cytochromes act to reoxidize $FADH_2$, and become themselves reduced in the process. Overall, we have, then, a sequence of these electron carriers, NAD, FAD, and the cytochromes, which catalyze the overall transfer of a pair of electrons from suitable organic substrates (glucose, ethanol, fats, and so forth) to molecular oxygen or other terminal electron acceptor.

The electron carriers are organized along the electron transport chain as shown in Figure 21-1, in which the standard reduction potentials of all carriers are noted as well as the standard free-energy change for the transfer of a pair of electrons at several stages of the process. As a result of the bit-by-bit release of energy along the electron transport chain, we have the opportunity to couple the synthesis of a number of molecules of ATP, three to be exact, to the overall reaction. These syntheses probably occur at the sites indicated in Figure 21-1. In summary, the negative free-energy change at three steps along the pathway of electron transfer is used to drive the endergonic synthesis of ATP from its constituent parts, ADP and P_i. Since we have calculated that the standard free energy of hydrolysis of ATP to ADP and P_i is about −7 kcal/mole, it follows that synthesis of ATP from ADP and P_i will have a standard free-energy change near +7 kcal/mole. Note that each of the reactions along the electron transport chain at which ATP synthesis occurs is exergonic by at least this amount. Thus, the exergonic oxidation–reduction reactions are employed to drive the synthesis of ATP, an endergonic reaction.

Example *As noted in Figure 21-1, the transfer of a pair of electrons from cytochrome b to cytochrome c is exergonic by 9.9 kcal/mole. If this reaction is coupled to the synthesis of one molecule of* ATP, *what is the overall free-energy change for the complete process?*

Solution *Free-energy changes for coupled reactions can be calculated by adding or subtracting, as appropriate, the free-energy changes for each individual reaction. In this case we have*

$$\begin{array}{rcll}
\text{cyt b(red)} + \text{cyt c(ox)} & \rightleftharpoons & \text{cyt b(ox)} + \text{cyt c(red)} & \Delta G^0 = -9.9 \text{ kcal/mole} \\
\text{ADP} + P_i & \rightleftharpoons & \text{ATP} & \Delta G^0 = +7.4 \text{ kcal/mole} \\
\hline
\text{cyt b(red)} + \text{cyt c(ox)} + \text{ADP} + P_i & \rightleftharpoons & \text{cyt b(ox)} + \text{cyt c(red)} + \text{ATP} & \Delta G^0_{\text{overall}} = -2.5 \text{ kcal/mole}
\end{array}$$

Just how is this coupling achieved? Despite the intensive studies of scientists for a number of years now, we still cannot give a definite answer to this question. To provide some insight into the matter, we can cite a possible model reaction, however, whose mode of generation of ATP is understood. This is the reaction catalyzed by the enzyme glyceraldehyde-3-phosphate dehydrogenase, which is the enzyme that promotes the dehydrogenation of glyceraldehyde-3-phosphate. The overall reaction is

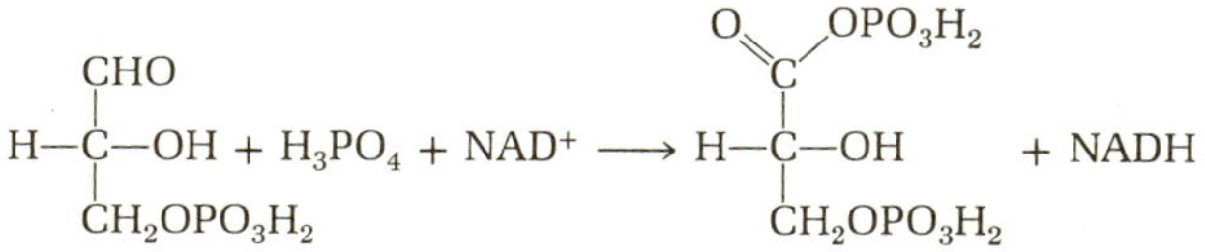

(21-9)

Thus, the process of oxidation of the substrate is coupled with the formation of a bond to phosphate. Now, this bond turns out to be a high-energy bond. Thus, the energy of the oxidation process is conserved through the transformation of a low-energy bond into a high-energy one. The removal of electrons from a chemical group can, and in this case does, markedly affect its properties, including its energy content. In the case just above, we can complete the process by transferring the molecule of phosphate from the product molecule to adenosine diphosphate, forming ATP:

$$\begin{array}{l} O \\ \| \\ C{-}OPO_3H_2 \\ | \\ CHOH \\ | \\ CH_2O{-}PO_3H_2 \end{array} + ADP \longrightarrow \begin{array}{l} O \\ \| \\ C{-}OH \\ | \\ CHOH \\ | \\ CH_2O{-}PO_3H_2 \end{array} + ATP \qquad (21\text{-}10)$$

Thus, the overall oxidative process has resulted in the formation of a molecule of ATP. Perhaps this sequence of reactions resembles those used along the electron transport chain, which is the major source of ATP for the living organism. Perhaps not, we simply do not know. We can be sure that some mechanism exists for the transformation of the free energy liberated in the oxidative processes into the free energy of the phosphate bonds of ATP.

TRANSFERENCE AND UTILIZATION OF CHEMICAL ENERGY

In the preceeding section we developed the means for the generation of chemical energy via the linkage of ATP synthesis to oxidative reactions. Now we need to turn our attention to the opposite side of the coin: How can we utilize this chemical energy for the variety of functions characteristic of living organisms?

In some cases it is useful to consider the transference and utilization of chemical energy as separate topics, and in others these prove to be indistinguishable. Thus, generally, we must transfer the chemical energy of ATP to those molecules actually involved in metabolic processes such as muscle contraction or nerve conduction; in other cases, such as the synthesis of some classes of macromolecules, the transference of chemical energy from ATP is equivalent to energy utilization for that synthesis. As a starting point, let us consider the synthesis of two classes of macromolecules—the polysaccharides and the proteins—for which we can ignore the difference between chemical energy transference and utilization.

Some simple polysaccharides are constructed of linear chains of sugar molecules: **Cellulose,** for example, has a structure in which glucose molecules are linked end to end, forming a very long chain. In nature, such molecules are assembled in the simplest possible way: The successive glucose molecules are added one at a time to the end of a growing chain. Thus, we can write the synthetic process in the following way:

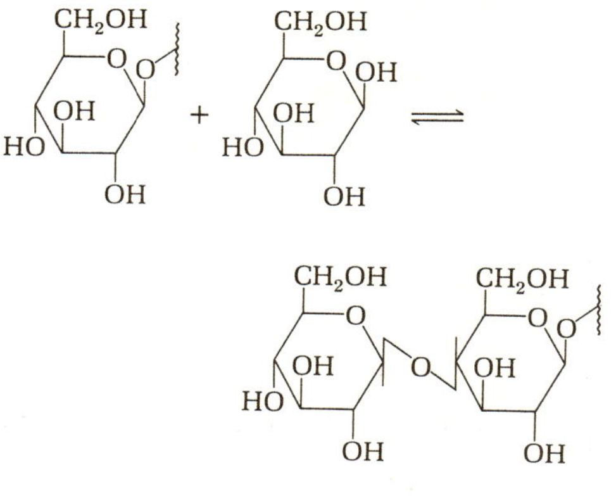

$\Delta G^0 = +3$ kcal/mole (21-11)

This reaction as written has a free-energy change of about +3000 cal/mole; that is, thermodynamics requires that the equilibrium lies in the direction of hydrolysis of the polysaccharide rather than in the direction of synthesis. Consequently, it follows that we shall be unable to generate long polysaccharide chains without the input of chemical energy. This energy is supplied by ATP. Although the exact details of just how ATP functions in supplying this energy differ a little from polysaccharide to polysaccharide, the following pathway is representative and illustrates how the hydrolysis of ATP is coupled to polysaccharide synthesis. First, an ATP molecule reacts with a glucose molecule to generate the corresponding *glucose-1-phosphate:*

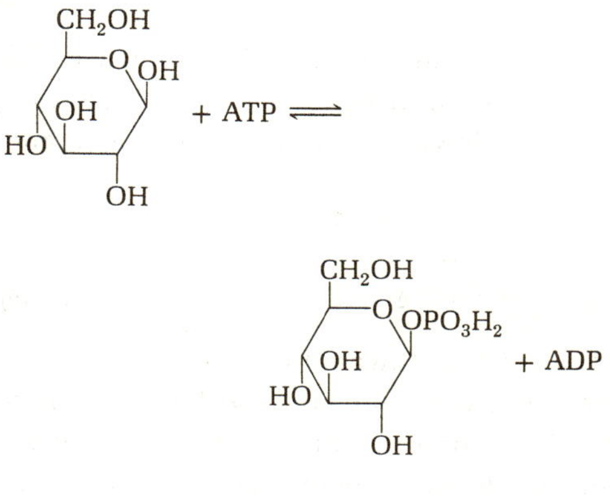

$\Delta G^0 = -1$ kcal/mole (21-12)

As written, this reaction has a slightly negative free-energy change, and hence tends to proceed to the right. This is equivalent to saying that the chemical energy in the pyrophosphate bond of ATP is slightly greater than that of the glucose-1-phosphate. The latter substance now undergoes another reaction with ATP (or similar molecule which can be generated from ATP without a significant free-energy change) in which the glucose is transferred to the terminal phosphate group of the adenosine diphosphate molecule:

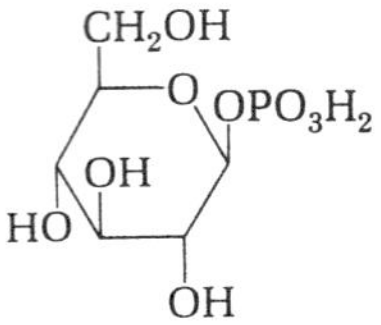

$+ \text{ATP} \rightleftharpoons$

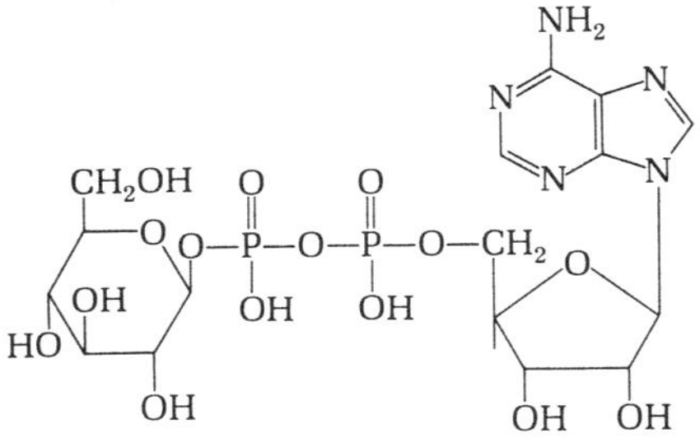

$+ H_4P_2O_7 \quad \Delta G^0 \cong 0$ kcal/mole (21-13)

adenosine diphosphoglucose (ADP-glucose)

This is essentially an isoenergetic reaction; that is, the overall free-energy change is about zero. It is the *ADP-glucose* that is the actual donor of the glucose group to the growing end of the polymer chain:

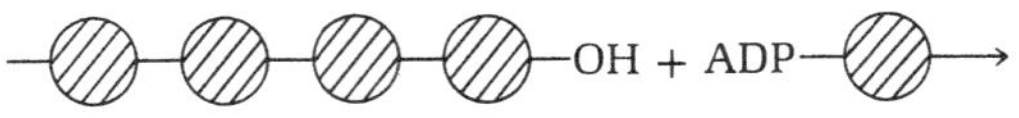

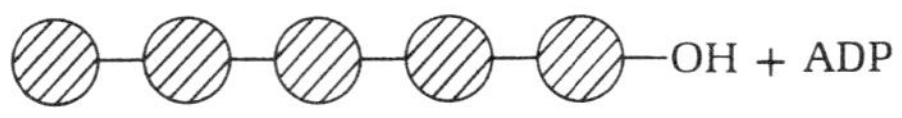

$\Delta G^0 = -3.5$ kcal/mole (21-14)

This reaction is energetically favorable by about 3.5 kcal/mole. If we add up these reactions involved in the addition of one glucose molecule to a growing cellulose chain, we get the following overall stoichiometry and energetics:

$$(\text{glucose})_n + \text{glucose} + 2\text{ATP} \rightleftharpoons$$

$$(\text{glucose})_{n+1} + 2\text{ADP} + \text{PP}_i \quad \Delta G^0 = -4.5 \text{ kcal/mole} \qquad (21\text{-}15)$$

Thus, by adopting the indicated reaction pathway, the free energy of hydrolysis of the ATP molecules has been used to convert a thermodynamically unfavorable reaction into a thermodynamically favorable one. Basically, this is accomplished by using the chemical energy of the ATP molecule to convert a "low-energy" molecule, glucose, into the "high-energy" intermediates glucose-1-phosphate and ADP-glucose.

Mechanistically, the assembly of protein molecules is a more complicated process than that for assembly of cellulose and similar polysaccharides. It is easy to see why this should be so, since in protein synthesis we must provide a mechanism for the orderly assembly of amino acids into a precise sequence along the polypeptide chain; this problem does not exist for cellulose, in which only one type of monomeric unit is present. Returning again to our discussion of the employment of nucleic acids as information storage and transfer devices, we remember that the information for specification of the exact sequence of amino acids is contained in the sequence of bases along the polynucleotide chain of the nucleic acid (Chapter 7). We now wish to develop a picture of just how the sequence of monomeric units along one type of polymer chain can be translated into the sequence of monomeric units along another type of such chain. At the same time, we ought to recognize that the thermodynamic problem here is very similar to that just encountered for the synthesis of cellulose. Thus, formation of a peptide bond between amino acids is an energetically unfavorable process (by about 1 kcal/mole) and one that therefore must be driven by the donation of energy from an outside source, the ATP molecule. Just as in the previous case, reaction of ATP with the monomeric unit converts it into a reactive intermediate, in this case an **aminoacyl adenosine phosphate (aminoacyl-AMP):**

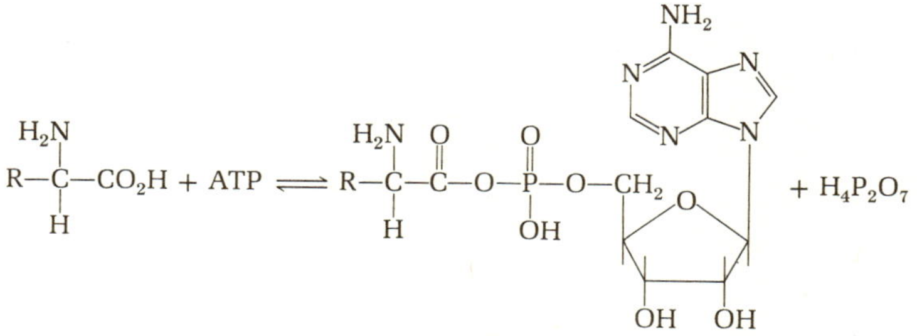

(21-16)

This reaction proceeds with little overall free-energy change. Consequently, the chemical energy of the ATP molecule has been transferred to the chemical energy of the aminoacyl-AMP intermediate; the unreactive amino acid has been converted into a reactive form. Thus, from the standpoint of the thermodynamics, we have accomplished the necessary task.

What we must worry about now is introducing specificity into the synthetic process. This is accomplished in several stages. First, the aminoacyl group is transferred from ATP to a terminal position in a small RNA molecule, called *transfer RNA* or *t*RNA:

$$\text{aminoacyl-AMP} + t\text{RNA} \xrightleftharpoons{\Delta G^0 \cong 0} \text{aminoacyl-}t\text{RNA} + \text{AMP} \qquad (21\text{-}17)$$

The important points here are the following: First, the chemical energy of the aminoacyl-AMP molecule is retained in the **aminoacyl-*t*RNA** molecule; second, the transfer is a specific one in the sense that one or more types of *t*RNA exist for each amino acid and that each type of *t*RNA can "recognize" its own type of amino acid. Such recognition is characteristic of reactions in biological systems, and we shall return to this matter below.

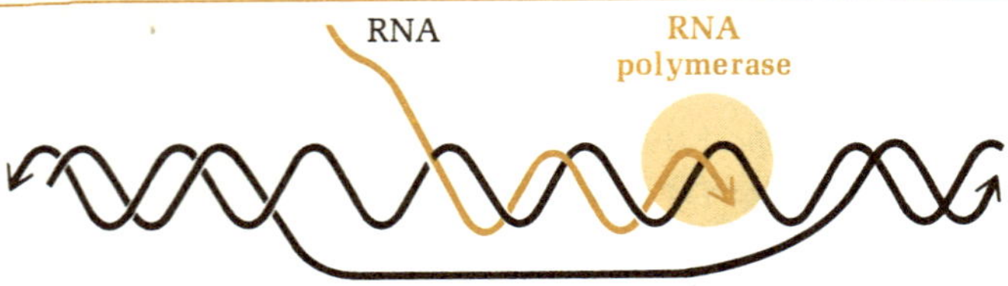

Figure 21-2 A visualization of the synthesis of an RNA molecule on a unique strand of its DNA template. The attachment of the enzyme RNA polymerase to a DNA molecule opens up a short section of the double helix, thereby allowing free bases on one of the DNA strands to base pair with the precursors of the RNA molecule being synthesized.

Table 21-2 Genetic dictionary[a]

First (X)↓	Second (Y): U	C	A	G	Third (Z)
U	Phe	*Ser*	Tyr	Cys	U
					C
	Leu	*Ser*	Terminate	Terminate	A
			Terminate	Trp	G
C	*Leu*	*Pro*	His	*Arg*	U
					C
	Leu	*Pro*	Gln	*Arg*	A
					G
A	Ile	*Thr*	Asn	Ser	U
					C
	Ile	*Thr*	Lys	Arg	A
	Met				G
G	*Val*	*Ala*	Asp	*Gly*	U
					C
	Val (internal)	*Ala*	Glu	*Gly*	A
	Met (initiate)				G

[a]This table provides the dictionary with which we may translate the three-letter words of *m*RNA into specific amino acids. Empty spaces in the table indicate identity of the codon with the amino acid above. Two or more codons exist for a particular amino acid; Phe is coded for by both UUU and UUC. But note carefully that the opposite situation does not occur: A single codon never serves for more than one amino acid. Those codons marked "terminate" do not correspond to any amino acid; hence, the occurrence of such a codon along the *m*RNA molecule causes interruption in the process of protein synthesis and cessation of further growth. Thus, these codons serve as indicators that the protein synthesis process is complete and that the complicated molecule may be released from the synthetic machinery.

Source: H. R. Mahler and E. H. Cordes, *Biological Chemistry*, 2nd ed., Harper & Row, New York, 1971.

Simply put, there is on each type of *t*RNA molecule a sequence of bases, or **recognition sites,** that permits it to react only with the appropriate type of aminoacyl-AMP. Third, there exists another type of RNA molecule that is important in the overall process of protein synthesis. This is called *messenger RNA* or *m*RNA. We know that the basic information for the structures of all cellular proteins is contained in the DNA of the nucleus, but that the actual process of protein synthesis occurs in the cellular cytoplasm. Consequently, we need a messenger to carry this information from the storehouse to the site of synthesis. This is the function of *m*RNA. It is synthesized in the nucleus by making a complementary copy of one of the two complementary chains of the DNA (its sequence is therefore identical to that of the other DNA chain). The specificity of this synthesis relies on the tendency of the bases involved to recognize each other through specific hydrogen bond formation just as they do in the DNA molecule itself. A schematic representation of the process of DNA-directed synthesis of *m*RNA is provided in Figure 21-2. Fourth, the *t*RNA molecule with its attached amino acid has the capacity to recognize specific sites on the *m*RNA molecule. We know from our discussion in Chapter 7 that DNA, and by virtue of the method of synthesis, *m*RNA, contains sequential groups of three bases, each of which is a word or codon corresponding to one amino acid. This **genetic code** is provided in Table 21-2. Now, if we have an anticodon on the *t*RNA molecule that has a base sequence that is complementary to that of a codon in the *m*RNA, these ought to be able to recognize each other by complementary base pairing of the type characteristic of the DNA molecule itself. A schematic visualization of such recognition is provided in Figure 21-3. Fifth, the whole process of recognition between the *m*RNA and the aminoacyl-*t*RNA takes place in a subcellular organelle called a *ribosome,* which interacts with both of these components. Finally, specific enzymes connect the specifically organized amino acids to form the peptide. Repetition of this process yields, eventually, a complete protein molecule whose structure is specified by the base sequence of the *m*RNA molecule whose structure, in turn, is determined by that of the nuclear DNA.

IMPLICATIONS CONCERNING PROTEIN SYNTHESIS

The above description serves to indicate the flow of information from DNA to RNA to protein. The net result is that the genetic information of the cell is expressed in terms of specific sequences of amino acids along the backbones of the proteins that constitute the cell. As developed in Chapter 7, these proteins then act directly to determine the biochemical specificity of the cell. Protein synthesis is involved in a great many important biological processes. Let us consider briefly a few of these.

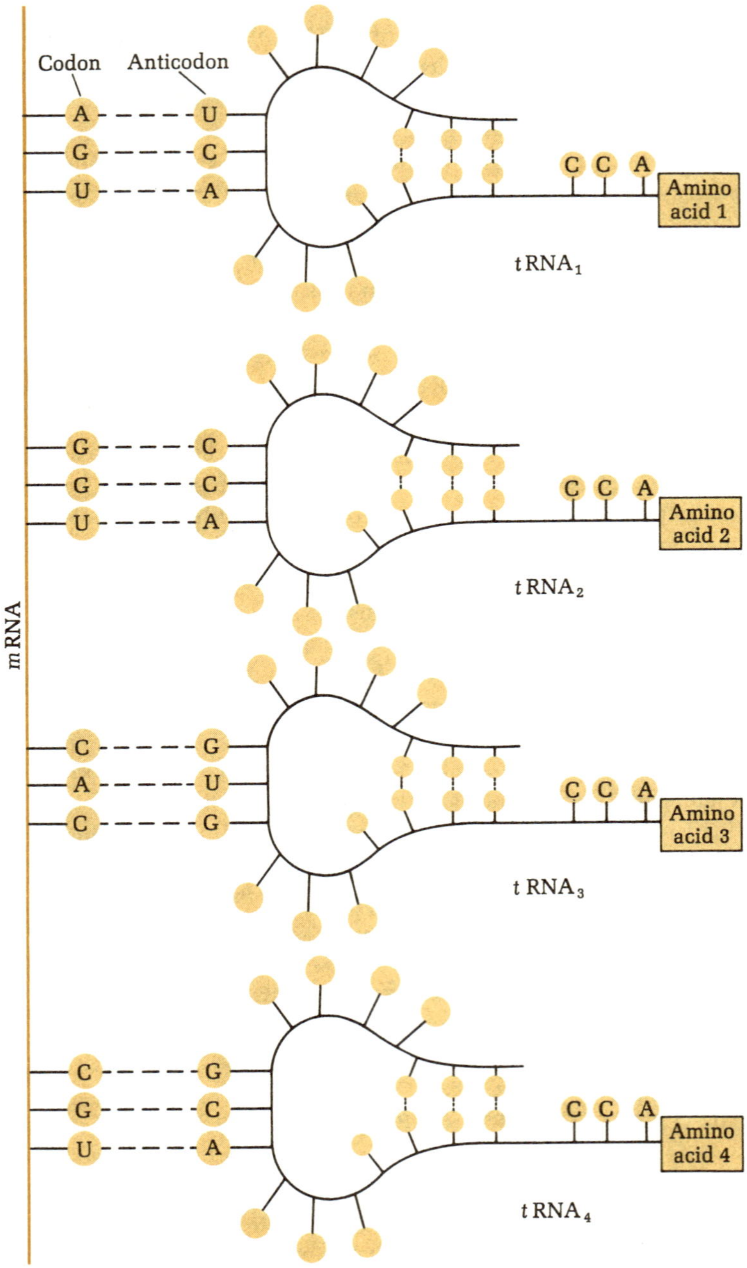

Figure 21-3 The recognition of aminoacyl-*t*RNA and sites on *m*RNA preparatory to peptide bond formation.

Inherited diseases

There are currently known a substantial number, perhaps 100, diseases of man that are **genetically inherited.** That is, the causative agent of the disease is not a bacterium, virus, or parasite, nor is it a nutritional deficiency. The causative agent is DNA defective in some respect, which is passed from parent to progeny. We have briefly mentioned one such disease in Chapter 7—sickle cell anemia. In this disease the hemoglobin in the red blood cells in afflicted persons is defective in that there is *one* mistake in the amino acid sequence. As a result of this mistake, the oxygenated form of this hemoglobin precipitates within the red blood cells, causing them to assume their characteristic "sickled" appearance; anemia results. The error in the hemoglobin structure must ultimately lie in the structure of the DNA from which it is derived. That is, the change in the nature of a single base pair causes the code word for one amino acid to be changed into that for another. The abnormal hemoglobin is the consequence.

Such errors in DNA can be hidden for many generations. Each person has two copies of the genetic information: One is donated by the father and the other by the mother. If one of these copies has a mistake in the message for hemoglobin, for example, but the other copy is accurate, much of the hemoglobin synthesized will be normal, and the person may suffer no clear effects from the abnormal copy. But if two people each with one defective copy of DNA for this protein are the parents of a child, that child will have the disease in one case out of four.

Many inherited diseases are exceedingly rare; others are common enough to be serious medical problems. In the latter category is *phenylketonuria,* a disease elicited by the lack of a single enzyme involved in the degradation of the amino acid phenylalanine. As a result of this lack, excessive concentrations of other metabolites build up, with very serious consequences for the patient. Profound mental retardation will result if the disease is not detected very early and treated by carefully controlling the dietary intake of phenylalanine.

Clearly, we wish to avoid diseases of this type. Many of them are serious or fatal and, for the majority, there is no treatment at all. In no case is there a cure. What causes the mistakes in the DNA in the first place? We do not have a complete answer to this question yet, although some damaging agents are known. Foremost of these is high-energy radiation. This fact led to the bitter and drawn-out debates in the early 1960s that terminated in the cessation of the testing of nuclear weapons in the atmosphere by those nations, including the United States and the Soviet Union, that signed the nuclear test ban treaty. The resulting fallout of radioactive material would have caused a terrible number of genetic errors, in the opinion of many scientists. Actually, such high-energy radiation is more common in our lives than we might suppose. One important source is the X-rays taken for medical and dental purposes. X-Rays are, of course, exceedingly useful and important diagnostic tools. Nevertheless, prudent physicians and dentists endeavor to minimize the number of X-rays for each patient. A more frivolous source of X-rays comes from color television sets. Particularly in some earlier models, unacceptable quantities of X-rays were received by persons seated close to the set. These sources of unwanted radiation are, of course, subject to minimization by proper legislation.

In addition to radiation, a number of chemicals are known that induce mistakes in DNA. An unresolved dispute centers about whether or not LSD, the hallucinogenic agent, is one of these.

Antibiotic action

There are many different reasons why antibiotics work in the ways that they do. Sulfa drugs interfere with the synthesis of folic acid (p. 667), and penicillin inhibits bacterial cell wall formation, for example. Other antibiotics act to prevent normal protein synthesis. Perhaps the most important of these is **streptomycin:**

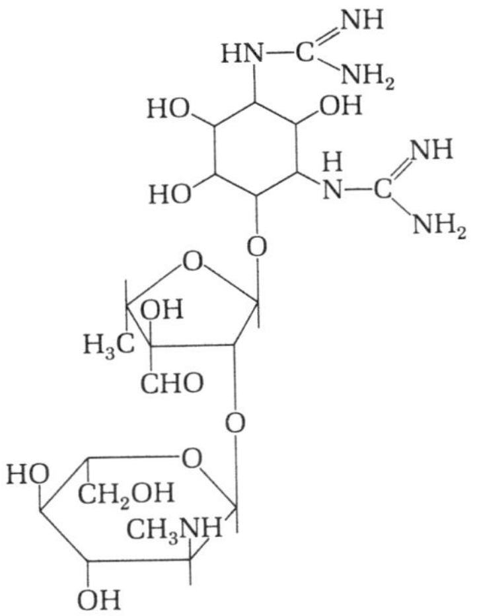

streptomycin

This molecule has the capacity to bind to the ribosomes which are involved in the protein synthesis process. As a result, the recognition process between the *m*RNA and the aminoacyl-*t*RNAs is disturbed. As a result, many of the code words are misread, and incorrect amino acids are inserted into the proteins synthesized. Most of these proteins fail to function normally, and the organism cannot survive.

Memory and learning

In Chapter 7 it was pointed out that ordered polymers such as proteins and nucleic acids could serve as information storehouses for cells. Nucleic acids, at least, do perform this function. This leaves us with an important question: How is the information gained in a learning process stored in brain cells? Although we have no definite answer to this question as yet, it is tempting to speculate that laying down the memory trace, or *engram,* involves the storage of information as newly synthesized nucleic acids or proteins (or both species might be involved). In support of this idea, it has recently been demonstrated that learning and memory are seriously inhibited in experimental animals by agents that inhibit protein synthesis. To be effective, these agents must be present at or shortly following the time of the learning process itself. Administration after a certain time has elapsed has no effect on memory. Perhaps, during this period the information has been stored in the form of a new protein molecule. It is possible that the molecular basis of forgetfulness involves the destruction of the molecules in which the information is coded.

Glossary

Adenosine triphosphate (ATP) the energy currency of the living cell; the form in which metabolic energy is stored and employed to drive endergonic processes.

Aminoacyl-AMP an intermediate in protein synthesis in which an amino acid is covalently linked to AMP. This is the first intermediate in the synthetic process.

Aminoacyl-*t*RNA the second intermediate in protein synthesis, in which an amino acid is covalently linked to a *specific* *t*RNA molecule.

Cellulose a polysaccharide composed of repeating glucose units. Cellulose is an important component of plant cell walls and is the most abundant organic compound on earth.

Cytochromes a series of proteins containing one or more heme groups which serve as electron carriers along the electron transport chain.

E_0' standard reduction potentials referred to a standard state of pH 7.

Electron transport chain an organized series of electron transfer agents, including nicotinamide nucleotides, flavin nucleotides, and cytochromes, along which are passed electrons derived from organic molecules and destined for terminal acceptance by molecular oxygen or other oxidizing agent. Sites for synthesis of ATP are located along the electron transport chain.

Flavin nucleotides electron transfer coenzymes, such as FAD, which are derived from riboflavin. The flavin nucleotides frequently are involved in oxidation of reduced nicotinamide nucleotides and reduction of oxidized cytochromes.

Genetic code the dictionary permitting translation of information coded in nucleic acids into the sequence of amino acids along a polypeptide chain. Specifically, the set of triplet codons specifying individual amino acids.

Inherited diseases those diseases that are passed from parent to progeny as a result of defects in the genetic information of the parent(s).

Metabolism the complete set of chemical reactions of a living system.
Nicotinamide (niacin) a vitamin required by man for the synthesis of the nicotinamide nucleotides.
Nicotinamide nucleotides complex molecules derived from nicotinamide which serve as electron transfer agents. They are frequently involved in the oxidation of organic molecules and the reduction of flavin nucleotides.
Polysaccharides polymers derived from sugars as monomers.
Respiration the transfer of electrons from (oxidation of) organic molecules used as fuels to molecular oxygen.
Riboflavin a vitamin required by man for the synthesis of flavin nucleotides.
Streptomycin a complex antibiotic that functions by interfering with the process of protein synthesis.

Study problems

1. Write a complete stoichiometric equation for the addition of one amino acid to a growing polypeptide chain. Given that the hydrolysis of a peptide bond is exergonic by about 1 kcal/mole, calculate the standard free-energy change for this reaction. What is the equilibrium constant for this reaction?
2. Suppose that a protein contains 101 amino acids in a linear chain and that the hydrolysis of each peptide bond is exergonic by 1 kcal/mole. Calculate the standard free-energy change and equilibrium constant for the complete hydrolysis of the protein to the individual amino acids.
3. Calculate the standard free-energy change for each of the following oxidation-reduction reactions at pH 7:

   a. $\underset{\text{pyruvate}}{CH_3CO{-}COO^-} + NADH \xrightleftharpoons{+H^+} \underset{\text{lactate}}{CH_3{-}CHOH{-}COO^-} + NAD^+$

   b. cyt $a(Fe^{2+})$ + cyt $c(Fe^{3+}) \rightleftharpoons$ cyt $a(Fe^{3+})$ + cyt $c(Fe^{2+})$

   c. 4cyt $a(Fe^{2+}) + O_2 \rightleftharpoons$ 4cyt $a(Fe^{3+}) + H_2O$
4. What are the codons for each of the following amino acids: Ile, Pro, Ser, Arg?
5. For which amino acids do each of the following triplets code: CAU, GUU, ACG, UGA? In each of these cases, what is the corresponding (complementary) triplet in the DNA from which the *m*RNA was derived?
6. What would be the consequences for protein structure of each of the following base changes in DNA?

   a. TAA ⟶ TGA   c. ATG ⟶ GTG
   b. GGT ⟶ GAT   d. ACG ⟶ ACT

Suggestions for further reading

I. M. Klotz *Energy Changes in Biochemical Reactions*, Academic, New York, 1967.
A. L. Lehninger *Bioenergetics*, Benjamin, Menlo Park, Calif., 1965.
A. Lwoff *Biological Order*, M.I.T. Press, Cambridge, Mass., 1965.
A. Szent-Gyorgyi *Bioenergetics*, Academic, New York, 1957.

22

Life processes: Some kinetic considerations

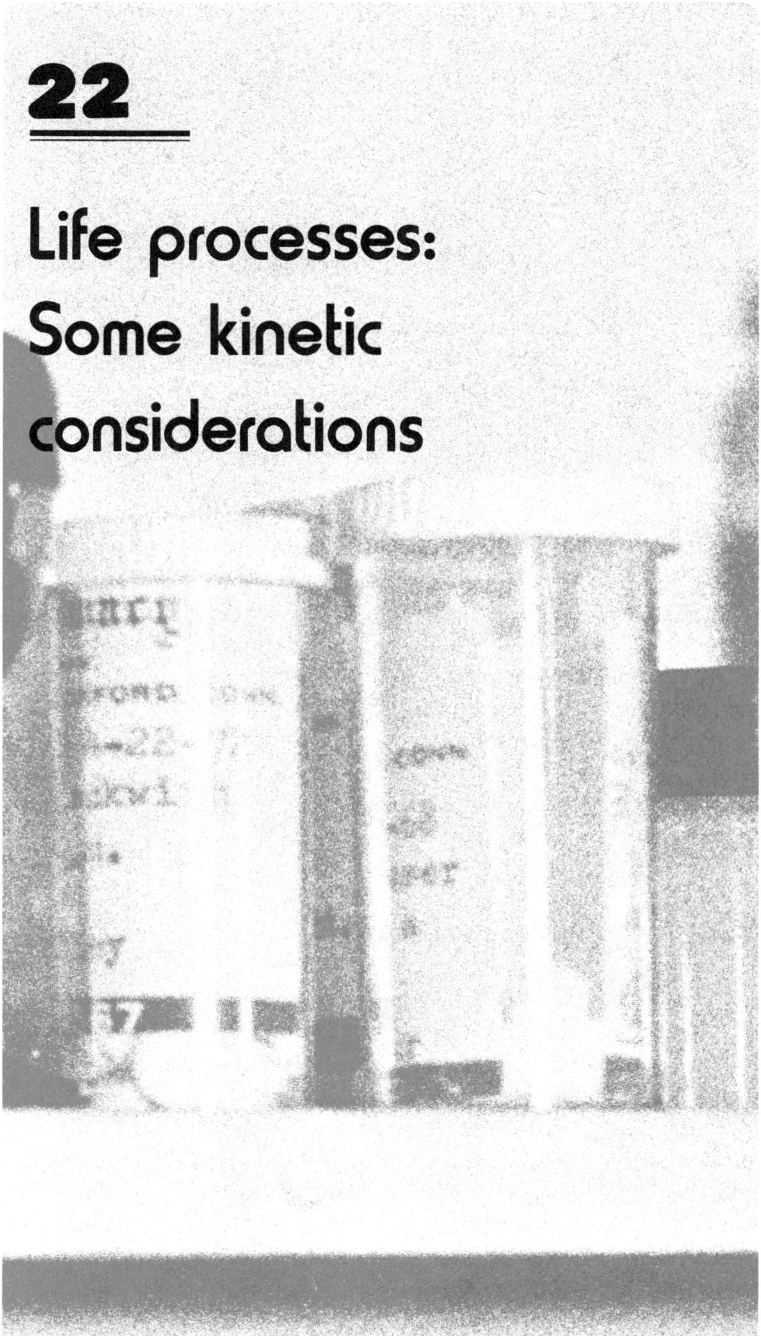

An early view of a life process:
. . . yeast . . . had a suctorial snout with which they devoured sugar, a stream of alcohol issued from the anus at the same time that carbon dioxide bubbled forth from enormous genital organs.

Annalen, **29,** 100 (1839)

In the last chapter we outlined how the free-energy changes in oxidation-reduction reactions are used to drive the formation of ATP, the energy currency of the cell. Utilization of this energy provides, in turn, the driving force required for the synthesis of such crucial molecules as proteins, nucleic acids, and carbohydrates. Moreover, ATP provides for other energy requirements as well. But the capacity to trap metabolic energy and to use it for life-supporting reactions is far from sufficient to satisfy all the needs for life sustenance. Basically, the set of reactions discussed thus far provides the *thermodynamic* basis of life. But is there a *kinetic* basis as well?

A little reflection will suffice to indicate that control of rates of chemical reactions must be crucial to the living organism. Not only are the unmodified rates of those reactions involved in energy generation and utilization much too slow to provide for the requirements of life, but the rates of such reactions must be subject to control to permit the organism to respond to change and stress. To meet the necessary requirements, organisms have developed the most exceptional catalysts known—the subclass of protein molecules called **enzymes.** Let us begin our discussion by examining the central characteristics of catalysis itself. Later, we shall turn to the special properties of enzymes.

CATALYSIS OF CHEMICAL REACTIONS

In our earlier discussion we have already encountered one example of catalysis: promotion of the synthesis of polyethylene by a transition metal complex (Chapter 7). This reaction illustrates the principal features of catalysis. Specifically, **catalysis** is the acceleration of a chemical reaction through the influence of a specific chemical species, the **catalyst,** without chemical alteration of the catalyst. Since the catalyst emerges from the process unchanged, one molecule of catalyst can participate in the chemical conversion of a great many substrate molecules. Note that catalysis of a chemical reaction influences the rate only; the equilibrium position for the reaction is unchanged. From this it follows that if a catalyst increases the rate of conversion of A to B, then it must increase the rate of conversion of B to A by an equal amount. Were this not so, the ratio of A to B at equilibrium would be changed. Thus, catalysis is a kinetic phenomenon, not a thermodynamic one. A free-energy diagram for a reaction with and without catalysis is provided in Figure 22-1.

Catalysis is exceedingly important, not only for the purposes of living systems but for commercial ones as well. The importance of catalysis is brought to mind by considerations developed in Chapter 9; specifically, many thermodynamically allowed reactions (i.e., reactions with large equilibrium constants) proceed extremely slowly. Therefore, to make use of such reactions, we must find ways to increase their rate. One way to accomplish this is to raise the temperature of the reaction mixture. While this is frequently a useful strategy, it is also one that fails in many instances. We need to remember that equilibrium constants are a function of temperature, and a reaction that is favorable at low temperatures may not be favorable at higher ones. Moreover, the use of high temperatures may cause one or more of the substrates or products to decompose or may result in the occurrence of unwanted side reactions. It follows, then, that for many reactions, discovery of suitable catalysts is of the greatest importance.

Examples of the use of catalytic procedures industrially are legion. One of the most important is provided by the oil industry. Crude oil as it is pumped from wells must undergo marked chemical modifications and fractionations before it can serve many human needs. Initial modifications of crude oil are effected by **distillation,** separation into fractions based on boiling-point differences. Low-boiling fractions may be useful for gasoline, higher-boiling ones for lubricants, and so forth. But each fraction must then be chemically altered through rearrangement and reorganization of the carbon chains prior to use. This is effected through the use of catalysts of many kinds. For example, the fraction of crude petroleum that boils in the range 20–200°C contains mostly hydrocarbons with four to ten carbon atoms. This would be a very inferior gasoline, however. Much better gasoline is produced by **reforming** this complex mixture of compounds under the influence of appropriate catalysts. This results in numerous isomerization reactions without change in molecular weight, leading to a product with a much higher octane number and, hence, better "antiknock" properties. Catalysts are used in other ways in gasoline manufacture as well. For example, the bulk of crude petroleum boils at temperatures higher than 200°C and consists of hydrocarbons having more than ten carbon atoms. Such compounds are not suitable for use as gasoline. However, this fraction of petroleum may be **cracked,** reduced to hydrocarbons having lower molecular weights and boiling points, again by the use of catalysts. Altogether, about 12 million barrels of crude oil a day are processed employing catalytic procedures. The economic impact of this industry is enormous. Another example of the use of catalysts is in the design and construction of antismog devices being considered for automobile exhaust systems. These devices employ various catalysts in efforts to convert hydrocarbons, carbon

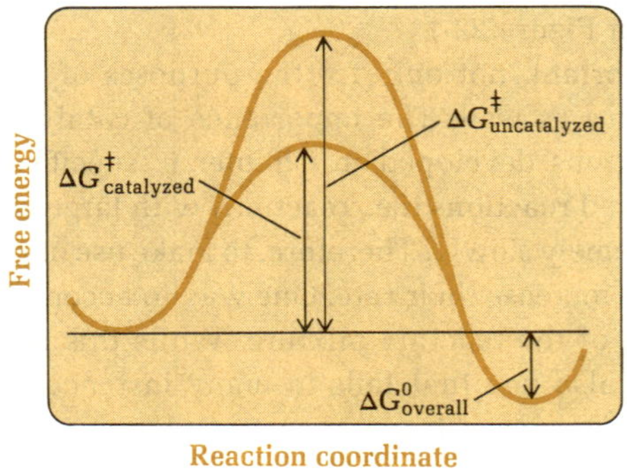

Figure 22-1 Free-energy profiles for an elementary process in the presence and absence of catalysis. Note that catalysis leaves the overall free-energy change for the reaction unchanged but markedly reduces the free energy of activation.

monoxide, and nitrogen oxides into nontoxic products prior to their release into the atmosphere, where they are involved in photochemical smog generation (see Chapter 18).

There are many kinds of catalysts and catalysis; happily there are some principles that unify all of these. On one hand, we have **heterogeneous catalysis.** This involves the use of a solid catalyst but reactants that are liquids or gases. In contrast, there is **homogeneous catalysis,** in which catalyst and substrates form a single phase. Finally, we may distinguish **enzymatic catalysis,** to which we shall turn somewhat later.

In each of these cases catalysis occurs via a similar pathway in which one or more reactants form a complex with the catalyst. For example, in a simple unimolecular isomerization, we may formulate the reaction process in the following way:

$$\text{A} + \text{Cat} \rightleftharpoons \text{A}\cdot\text{Cat} \quad (22\text{-}1)$$

$$\text{A}\cdot\text{Cat} \rightleftharpoons \text{B}\cdot\text{Cat} \quad (22\text{-}2)$$

$$\text{B}\cdot\text{Cat} \rightleftharpoons \text{B} + \text{Cat} \quad (22\text{-}3)$$

Note that reactant A must diffuse to the catalyst and adsorb onto it. It then undergoes chemical transformation to product B while bound to the catalyst. Finally, the product must desorb from the catalyst and diffuse away from it. In terms of this formulation, we can recognize the requirements for a successful catalytic process. First, the reactant(s) must have the capacity to complex with the catalyst. Second, the reactant when complexed with the catalyst must be more reactive than the reactant alone; that is, A · Cat must be more reactive than A. Finally, the product of the reaction must not complex too strongly with the catalyst. If it does, the catalyst will eventually be tied up as B · Cat, and the reaction will come to a halt.

Perhaps the central question for the understanding of catalysis is the nature of the substrate when adsorbed onto the catalyst. Let us consider this matter in relation to a couple of important catalytic processes. One of the first examples of catalysis to be discovered was the catalytic hydrogenation of carbon compounds containing double bonds. In the 1890s it was found that a number of metals, including nickel, palladium, and others, greatly accelerated these reactions:

$$H_2 + CH_2{=}CH_2 \xrightarrow{Ni} CH_3{-}CH_3 \qquad (22\text{-}4)$$

It has subsequently been shown that hydrogen forms complexes with the surfaces of these metals. Why does this complexation result in an enhanced capacity of hydrogen to add to carbon-carbon double bonds? A related example is provided by the **Haber process,** the catalytic reduction of nitrogen to ammonia in the presence of iron:

$$3H_2 + N_2 \xrightarrow{Fe} 2NH_3 \qquad (22\text{-}5)$$

Again we can ask why adsorption of hydrogen onto the surface of iron metal increases its capacity to add to the triple bond of molecular nitrogen.

There are actually two kinds of adsorption processes. The simpler is termed **physiosorption** and consists of the physical complexation of catalyst and reactant without marked chemical alteration of the reactant. The more dramatic type is termed **chemisorption** and is characterized by the formation of a **surface chemical compound** in which the structure and properties of the reactant will generally have changed greatly. Turning to the hydrogenations of Equations (22-4) and (22-5), we can ask which of these two possibilities applies in these cases. The answer to this question comes from an ingenious experiment. What happens if we expose the surface of, for example, nickel to a mixture of H_2 and D_2? If physiosorption occurs, we shall not expect a chemical reaction between the two. But if chemisorption occurs, a chemical reaction involving the surface compounds may be possible. When this experiment is carried out, the formation of H—D, an isotopically mixed variety of molecular hydrogen, is observed. This can be accounted for in terms of chemisorption involving the dissociation of the H_2 and D_2 followed by their recombination as shown in Figure 22-2. The crucial point is that the free energy of adsorption of hydrogen onto the metal surface is sufficiently great to cause the rupture of the H—H bond and the formation of atomic hydrogen bonded to the metallic surface. This form of hydrogen is a much stronger reducing agent than is molecular hydrogen itself. Catalysis results.

In Chapters 8 and 9 we discussed two crucial parameters of chemical reactions: to what *extent* they occur and at what *rate*. There remains one additional matter of great importance: How *specific* (selective) is the reac-

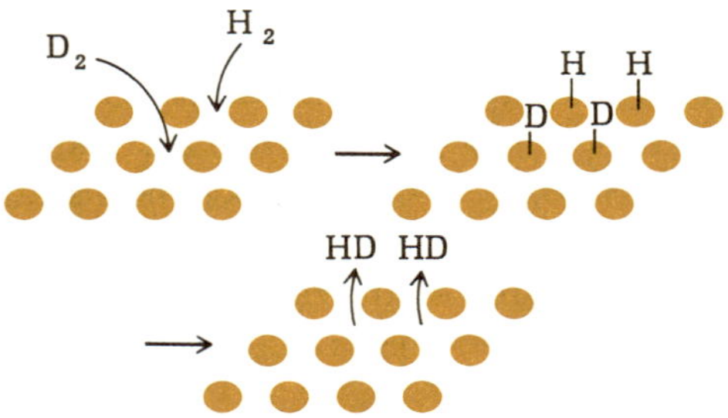

Figure 22-2 The process of chemisorption on the metallic surface. Individual atoms of the metal are shown as ovals. The interaction between H_2 and D_2 is viewed as involving the cleavage of the hydrogen bond and formation of hydrogen-metal bonds. Reformation of hydrogen–hydrogen bonds in desorption may result, as shown, in the formation of HD. It is the formation of this product that provides the evidence for chemisorption rather than physiosorption. [From V. Hansel and R. L. Burwell, Jr., *Sci. Amer.* **225,** 46 (December, 1971). Copyright © 1971 by Scientific American, Inc. All rights reserved.]

tion? That is, does it occur to form a single product, two products, or many? For most purposes, the most desirable reactions are the most specific ones, since they yield the maximum amount of the desired product with minimum problems of purification. Therefore, the utility of a catalyst is in part determined by its specificity. Does it promote just that reaction of interest?

As we shall see in the next section, enzymes are the most specific catalysts known. But chemists have succeeded in making some very specific catalysts of their own. For example, **Zeigler-Natta catalysts** of the type employed in the synthesis of polyethylene (Chapter 7) are also used in the synthesis of other polymers, including polypropylene. Individual catalysts have been designed to catalyze specifically the formation of isotactic, syndiotactic, and atactic polypropylene. Thus, by choosing the appropriate catalyst, we can direct the synthesis of the type of polymer desired.

With these considerations in mind, let us turn to the enzymes themselves—the most remarkable of catalysts.

SOME PROPERTIES OF ENZYMES

Essentially all chemical reactions that occur in living systems occur naturally at rates that are very much too slow to sustain life. In fact, these rates generally need to be promoted millions of times to gain this goal. It is, therefore, clear that living systems must contain catalysts to increase the rates of these reactions—the **enzymes.** The ability of living systems to promote chemical reactions has been known for a very long time, although it is difficult to pinpoint exactly when this discovery was made. Perhaps it is fair to go back as far as 1783, when Spallanzani proved that the gastric juices of falcons could liquefy meat. This scientist proved the point by simply placing the meat samples inside tiny wire cages to which a string was attached and feeding them to falcons. At later times, he was able to withdraw the cages and observe that the meat had disappeared. As early as 1810, Planche succeeded in isolating an enzyme, although in a very impure form, from plant roots. In the course of the remainder of the nineteenth century, a substantial number of additional enzymes were isolated, and studies of their properties were initiated. However, quantitative studies of the action of enzymes had to wait until the 1920–1930 era, in which good methods for the study of the rates of chemical reactions became available. It was in this period that the exceptional catalytic ability of enzymes was first clearly recognized. For example, in 1923, Willstätter and Polinger were able to demonstrate that a preparation of a peroxidase could activate 1000 times its own weight of H_2O_2 per second at 20°C! This has since proved to be an underestimate of the true catalytic activity, since the preparation used by these workers was not pure. It was in this same era that the chemical nature of enzymes was first established. In 1926 J. B. Sumner succeeded in obtaining the enzyme **urease** in crystalline form and demonstrated that it was a protein. This classic contribution to biochemistry, as is so often the case with revolutionary discoveries, was received with a response that ranged from criticism to derision. But in the following years M. Kunitz and J. H. Northrup succeeded in obtaining several additional enzymes in crystalline form, and all proved to be proteins. The protein nature of enzymes is now so well established that it is included as part of the definition of the word enzyme.

Enzymes are exceptional catalysts in four respects. First, they are exceedingly efficient. The catalytic efficiency of enzymes may be appreciated by realizing that, under optimal conditions, most enzymatic reactions proceed 10^8 to 10^{11} times more rapidly than the corresponding nonenzymatic reactions. Looked at in another way, most enzymes will metabolize 1000 to 10,000 substrate molecules per molecule of enzyme per minute. A really exceptional case is provided by **catalase,** an enzyme that promotes the decomposition of hydrogen peroxide:

$$H_2O_2 \longrightarrow H_2O + \tfrac{1}{2}O_2 \qquad (22\text{-}6)$$

One molecule of catalase is capable of decomposing more than *1 million* molecules of hydrogen peroxide per minute at room temperature. Looked at in still another way, many reactions that occur only slowly at elevated temperatures in the presence of high concentrations of acids or bases in the absence of enzymes occur rapidly at room temperature at pH 7 in their presence. The hydrolysis of the peptide bonds of the proteins themselves provides a good example. In the absence of enzymes, this reaction is ordinarily carried out in sealed tubes containing 6*M* HCl at a temperature of 120°C for about 48 hr. Using enzymes, much the same results can be accomplished at 25°C, pH 7, in a few minutes.

Table 22-1 Some data relevant to the specificity of fumarase

Active substances

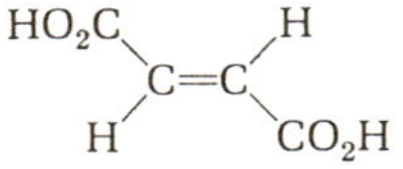

fumaric acid

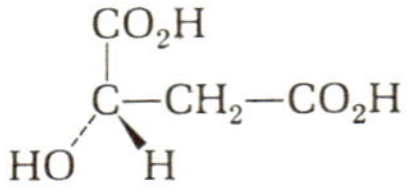

L-malic acid

Inactive structural analogs

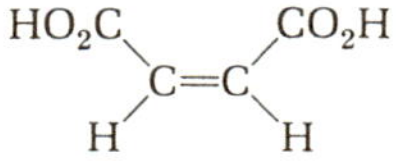

maleic acid

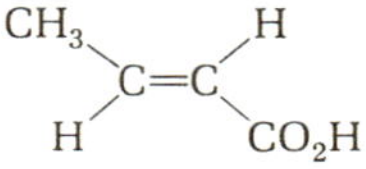

crotonic acid

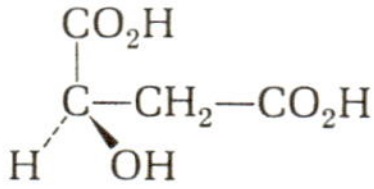

D-malic acid

$HO_2C—C{\equiv}C—CO_2H$

acetylenedicarboxylic acid

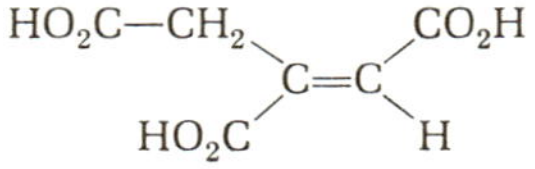

trans-aconitic acid

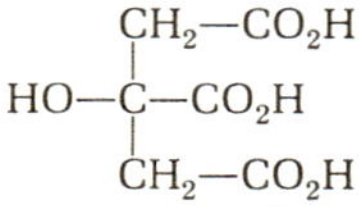

citric acid

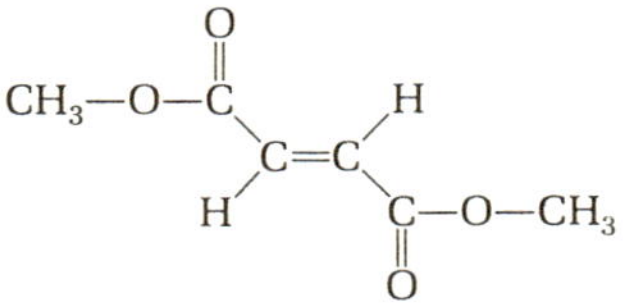

fumaric dimethyl ester

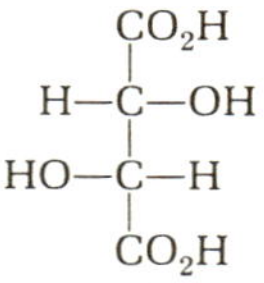

D-tartrate

Second, enzymes are very specific catalysts. That is, each enzyme generally catalyzes only one type of chemical reaction and uses only a very selected range of substrates for that reaction. A good example is provided by urease, Sumner's enzyme, which catalyzes the following reaction:

$$NH_2—\overset{\overset{\displaystyle O}{\|}}{C}—NH_2 + H_2O \longrightarrow 2NH_3 + CO_2 \qquad (22\text{-}7)$$

This enzyme will catalyze only one other reaction, the decomposition of hydroxyurea, in which one of the atoms of hydrogen is replaced by the hydroxyl function. Another good example, and one which is more instructive, is provided by the enzyme **fumarase,** which catalyzes the addition of water to fumaric acid to yield specifically L-malic acid:

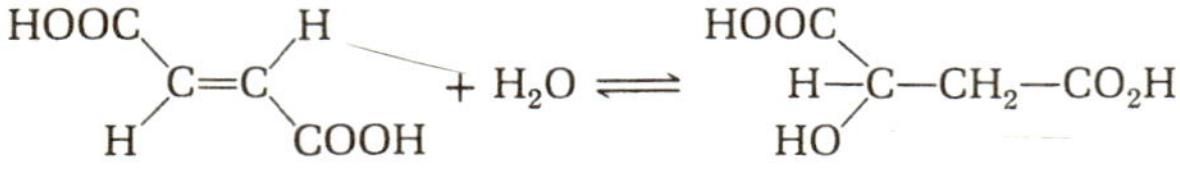

L-malic acid (22-8)

Now, with this reaction in mind, let us modify the structure of fumaric and malic acids in simple ways and see what happens to the enzymatic reaction velocity. A number of such structurally related substances are collected in Table 22-1. None of these substances is a substrate for fumarase; that is, fumarase will not catalyze the addition of water to any of them at a detectable rate. Of particular note are the cases in which the *trans* double bond of fumaric acid is converted to the *cis* configuration, maleic acid; to a triple bond, acetylenedicarboxylic acid; and in which the stereochemistry of L-malate is inverted, D-malate. Specificity of this type clearly demands a close interaction between substrate and enzyme. In fact, it is known that substantially all enzymatic reactions do occur via a preliminary step that involves the binding of the substrate to the enzyme. This interaction and the ones that follow in the catalytic process are very specific and must indicate that the geometries of the substrate and enzyme molecules bear a complementary relationship to each other, just as a key bears a complementary structural relationship to a lock.

As a group, enzymes catalyze a remarkable spectrum of chemical reactions even though they are markedly specific individually. Thus, the scope of reaction types that are enzyme-catalyzed includes most of those types known to chemistry, including hydrations, isomerizations, rearrangements, eliminations, substitutions, polymerizations, epimerizations, racemizations, and oxidation–reduction reactions.

Finally, the catalytic activity of enzymes is subject to control by interactions with small molecules which are frequently involved as substrates or products in reaction sequences in which that enzyme participates. A little reflection will indicate that the first three properties listed above are essentials for the maintenance of metabolism, which requires marked but specific catalysis of a variety of reactions. This last point may not be so obvious as a prerequisite for an orderly metabolism, but it is as crucial as the others.

The specificity and efficiency of enzymatic catalysis requires that formation of the complex between enzyme and substrate be uncommonly specific. Prior to examining directly the chemistry involved in enzymatic catalysis, you may wish to review the nature of the forces that must be involved in the binding of enzyme and substrate (Chapter 7). These same forces underlie the interaction of a great many biologically active molecules with their receptor sites in living systems. We turn attention to these matters at the conclusion of the chapter.

CARBOXYPEPTIDASE

Carboxypeptidase is an enzyme involved in the digestive process in mammals. It is secreted into the small intestine as a component of the pancreatic juice, and its function is to help catalyze the process of hydrolysis of proteins. The structure of this protein and its complex with glycyl tyrosine

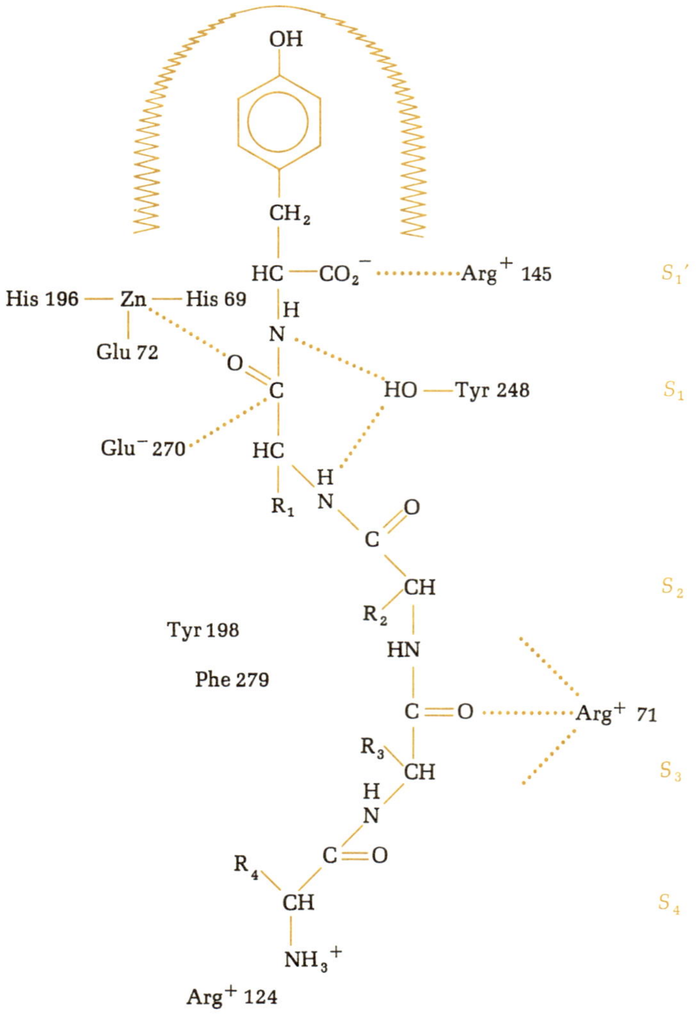

Figure 22-3 A schematic view of the interactions between carboxypeptidase and a typical substrate. Note the number of interactions that result in bringing the peptide bond into the vicinity of those groups involved in the catalytic process: Zn, Tyr^{248}, and Glu^{270}. (Courtesy of Prof. W. N. Lipscomb.)

has been discussed in some detail beginning on p. 230 (note especially Figures 7-17 and 7-18. Remember that carboxypeptidase does contain one atom of zinc per molecule of protein.

Carboxypeptidase catalyzes the hydrolysis of proteins in a special way: It promotes the hydrolysis of the peptide bond that is adjacent to the terminal *free* carboxyl group. Thus, this enzyme attacks a specific end of the polypeptide chain and works its way down the chain one amino acid residue at a time. The hydrolyses are most rapid when the terminal residue has a bulky hydrophobic side chain.

X-Ray diffraction studies of carboxypeptidase (pp. 230–235) and its complexes with low molecular weight substances permit us to understand the environment of a substrate bound to the carboxypeptidase surface. This environment is shown in a schematic way in Figure 22-3. Note the many interactions between enzyme and substrate: (1) Several amino acids are involved in formation of a pocket into which a bulky hydrophobic side chain can fit; (2) arginine at position 145 in the polypeptide chain, Arg^{145}, makes an ionic bond with the terminal carboxylate ion; (3) the zinc atom interacts with the carbonyl oxygen of the peptide bond to be hydrolyzed; (4) Tyr^{248}, Arg^{71}, and perhaps other residues as well form hydrogen bonds to the substrate at several positions. As a result of all these interactions, the positioning of the substrate on the enzyme surface is quite exact, bringing the bond to be hydrolyzed into close proximity with those groups that aid in the process. Let us now focus our attention on these groups and their role in the catalytic process.

Viewed in the simplest possible way, the reaction to be catalyzed can be represented as

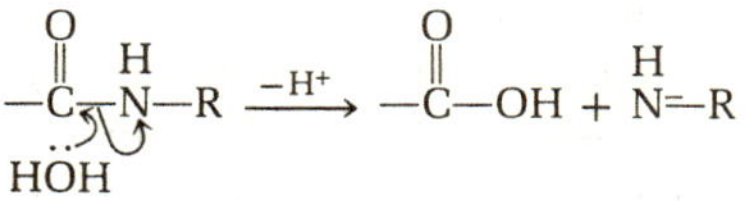

Thus, a molecule of water attacks the peptide bond, and the amino moiety is displaced as an amide anion. How might we help this reaction along? There are actually several possibilities and, as we shall see, carboxypeptidase makes use of all of them. Basically, we can either make the peptide bond more reactive or increase the reactivity of the water molecule. Turning to the latter point first, let us note that removal of a proton from water converts it into a hydroxide ion, a much more reactive species than water itself. Generally, reactivity is related to basicity: The more basic a given species, the more reactive it is. Thus, hydroxide ion is more reactive than water, amide ions, RNH^-, are more reactive than amines, and so on. Consequently, we expect the rate of attack of hydroxide ion on the peptide bond to be several orders of magnitude faster than that for attack of water.

Hydroxide ion, however is not a particularly attractive reagent for enzyme-catalyzed reactions, since these reactions occur in the body near pH 7, conditions in which the concentration of this reactant is vanishingly small. There is an intermediate possibility, however. Suppose we imagine a proton being removed from water as the water molecule attacks the peptide bond so that the water molecule becomes an *incipient* hydroxide ion:

H—O—H	<	$\overset{\delta-}{\text{B}\cdots\text{H}\cdots\text{O—H}}$	<	$^-$OH
water		an incipient hydroxide ion		hydroxide ion

We would expect that the reactivity of water would be increased in this way, but that the reactivity level would not reach that of hydroxide ion. These expectations have been confirmed experimentally a number of times. In carboxypeptidase, the function of B in the structure above, that is, the group that removes a proton from the attacking water molecule, is probably served by Glu^{270}, which exists as a carboxylate anion—COO^-.

The chelation of the carbonyl hydroxyl of the substrate by the zinc atom will result in removal of some charge from the carbonyl carbon atom. This will have the effect of increasing the susceptibility of that carbon atom to attack of reagents that have unshared electron pairs, such as water and hydroxide ion:

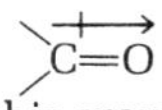

ordinarily, this group has a dipole moment with carbon at the positive end

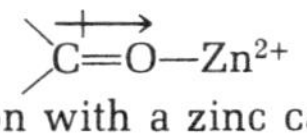

chelation with a zinc cation increases the positive character of the carbon atom

Thus, this single metal atom provides an important part of the overall catalytic process.

Finally, let us note that expulsion of the amine part of the peptide as an anion is not a favorable process, since nitrogen does not tolerate a negative charge very well. We might disperse that negative charge by adding a proton to the nitrogen atom either prior to breaking the bond or during the actual bond-breaking process in a manner analogous to that for converting water into an incipient hydroxide ion. Carboxypeptidase uses the latter alternative, with Tyr^{248} acting as the species that donates the proton to the departing amine function. Taken together, we can visualize the bond-breaking process as:

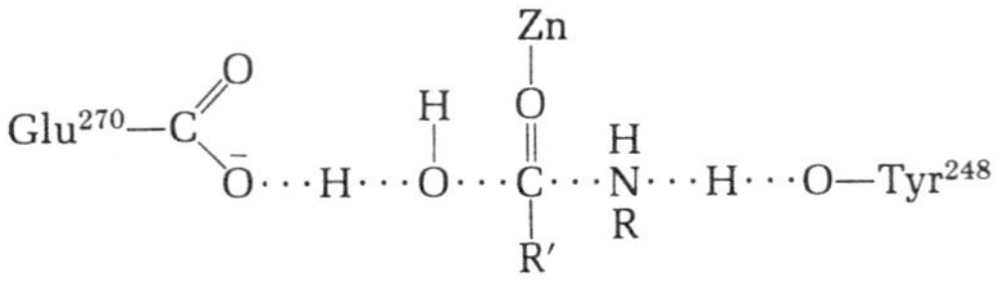

where the dotted lines indicate covalent bonds in the process of formation or rupture. The whole architecture of the carboxypeptidase molecule is devoted to achieving the proper three-dimensional arrangement between those groups of the enzyme directly involved in catalysis and the bond in the substrate that is to be hydrolyzed. By bringing several important groups together in the proper way in space, carboxypeptidase, as well as other enzymes, is able to achieve its striking catalytic efficiency.

The kind of arguments that we have developed to explain enzymatic catalysis can be generalized to account for other important biological functions as well. Let us begin by citing some examples of the ways in which drugs exert their effects.

SOME COMMENTS CONCERNING HOW DRUGS WORK

In the preceding pages we have outlined how the complementary shapes of substrate and enzyme permit them to interact in a very specific way to provide for enzymatic catalysis. What would happen if we designed a molecule that looked almost like the normal substrate for an enzyme but was changed so that it could not react? Provided that the substrate analog is not too different from the normal substrate structurally, we might expect it to bind to the enzyme in nearly the same way as does the normal substrate. This should have the effect of *inhibiting* the enzyme since, while occupied by the analog, it would be unable to bind and catalyze some reaction of the substrate. Schematically:

$$\text{Enz} + \text{Sub} \rightleftharpoons \text{Enz}\cdot\text{Sub} \longrightarrow \text{Enz} + \text{products}$$

+
Sub′
⇅
Enz · Sub′ (enzymatically inactive)

(Enz = enzyme)
(Sub = normal substrate)
(Sub′ = substrate analog)

It is in just this way—by inhibiting some crucial enzyme—that many drugs exert their effects. Others may function in a related way by binding to other types of biological receptors. As a specific case, let us briefly consider the chemistry of the **sulfa drugs**, once (until penicillin) the most potent weapons against many types of infections. The sulfa drugs were, in fact, the initial major group of useful antibacterial drugs and have certainly been responsible for saving a great many lives.

One of the typical sulfa drugs is sulfanilamide:

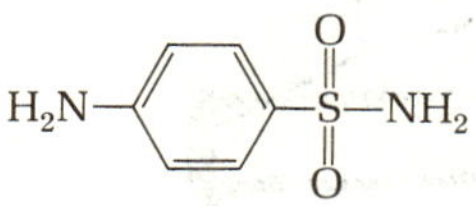

This substance is structurally related to *p-aminobenzoic acid,* which forms part of the structure of the biochemically important molecule **folic acid:**

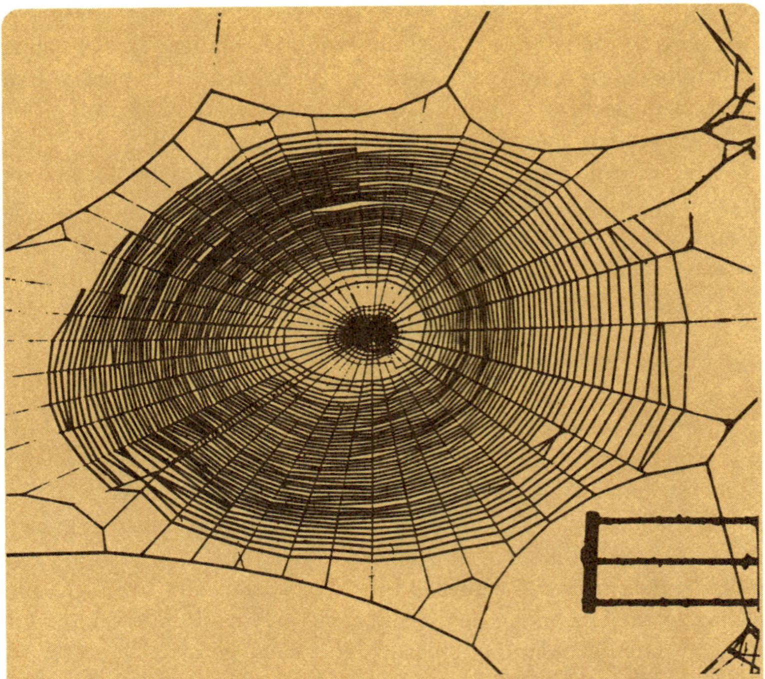

Amphetamines are powerful stimulant drugs that affect the behavior of both man and simpler organisms. At the top is a photograph of the normal web spun by the cross spider; at the bottom is a web spun by the same spider under the influence of an amphetamine drug. (Courtesy of Peter M. Witt, North Carolina Department of Mental Health.)

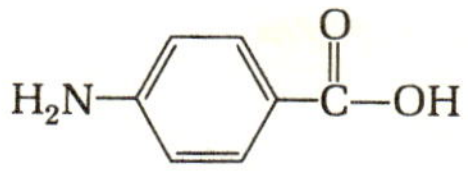

p-aminobenzoic acid

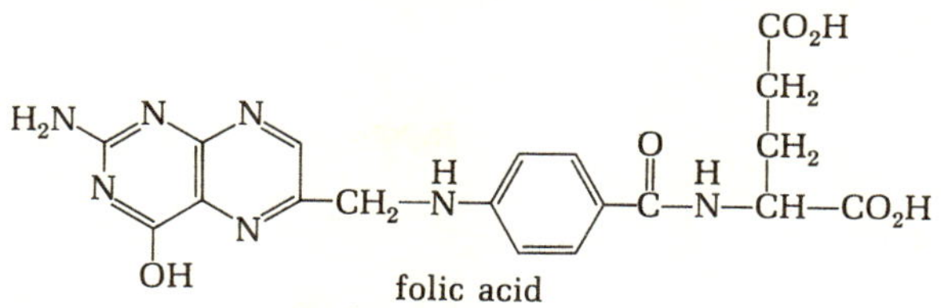

folic acid

Folic acid (in a slightly altered form) is responsible for indispensable functions in the chemistry of bacteria and higher forms alike. During the synthesis of folic acid, p-aminobenzoic acid is directly involved as a precursor. Sulfanilamide acts as an inhibitor of the enzyme that incorporates p-aminobenzoic acid into the folic acid molecule. As a result, bacteria cannot produce enough folic acid to provide for their requirements and die. But mammals do not have the capacity to synthesize this molecule in the first place (it must be supplied in the diet), and are hence not directly affected by the drug. Consequently, sulfanilamide, and other sulfa drugs, are selectively toxic to bacterial cells.

A rather closely related example involves drugs that are important antitumor agents: aminopterin and methotrexate (amethopterin):

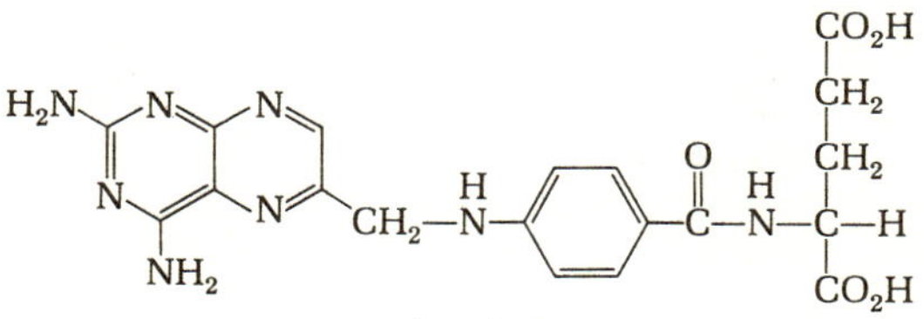

aminopterin

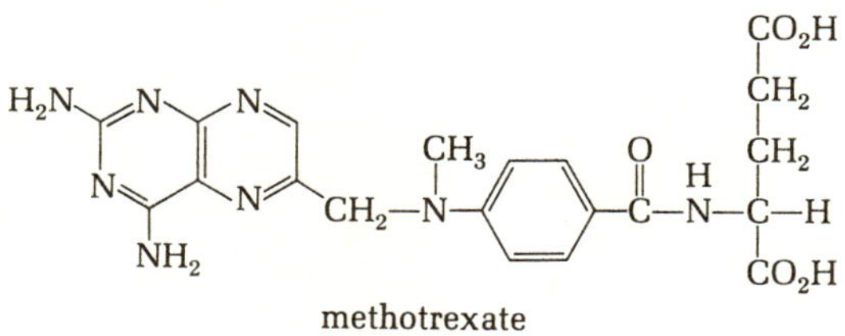

methotrexate

Comparing these structures with that for folic acid establishes the close structural relationships. One of the most important of the biochemical functions of folic acid is its participation in the biosynthesis of nucleic acids. Both aminopterin and methotrexate associate strongly with an enzyme necessary to convert folic acid to a biochemically active form called tetrahydrofolic acid. As a result, the rate of synthesis of this substance is greatly reduced, and nucleic acid synthesis is essentially shut off. This affects both normal cells and cancer cells but does the most damage to the latter ones. One distinguishing feature between normal and cancer cells is the rapid

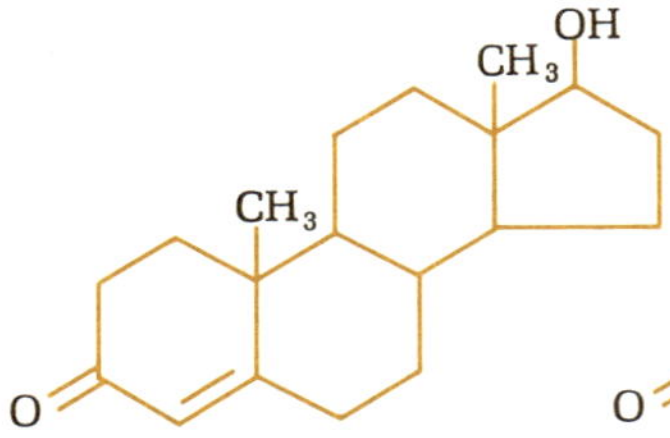

Testosterone—a normal sex hormone occurring in both males and females.

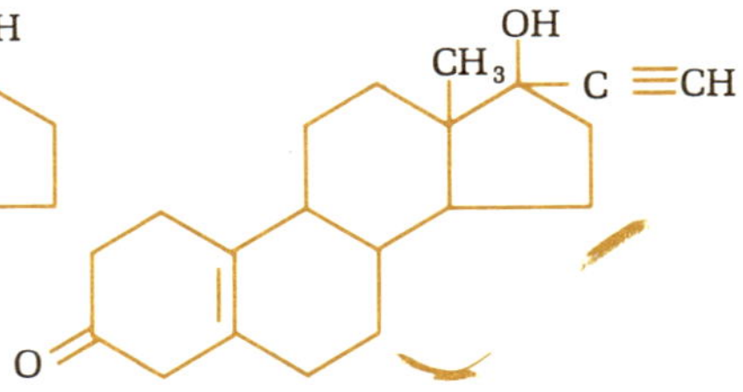

Norethynodrel—a common component of "the pill" (specifically Enovid), which acts to prevent fertility in the female.

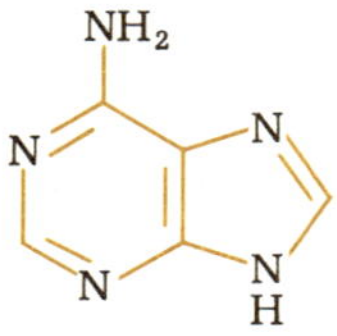

Adenine—a normal constituent of nucleic acids.

6-Mercaptopurine—a useful antagonist of nucleic acid synthesis and an important antitumor drug.

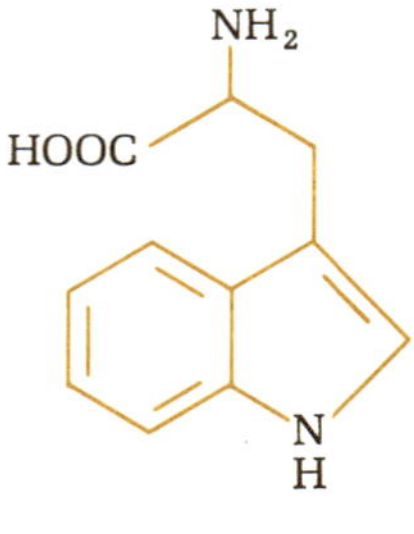

Tryptophan—an amino acid and common constituent of proteins.

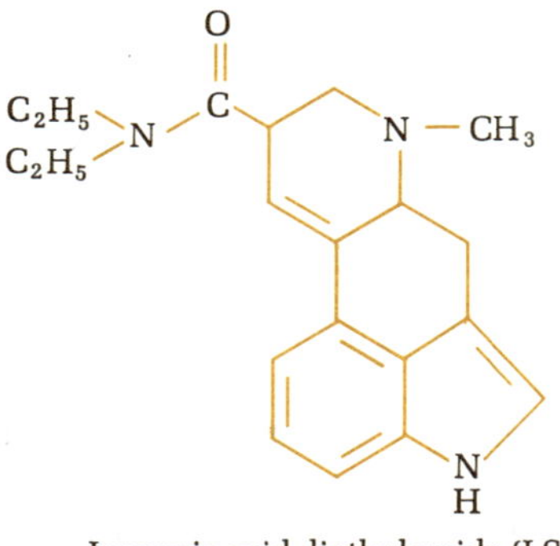

Lysergic acid diethylamide (LSD)—a powerful hallucinogenic drug.

Figure 22-4 A collection of comparisons between the structures of normal metabolities and certain drugs.

rate with which the latter multiply: Each time a cell divides, it must duplicate its supply of DNA. Hence, the more rapidly dividing cells are more sensitive to the effects on nucleic acid synthesis than are the more slowly dividing ones. Consequently, methotrexate and aminopterin selectively kill cancer cells.

It is, of course, true that not all drugs are closely related structurally to normal metabolites. But a great many are, and we can understand in general terms how these drugs work in terms of the background just developed. In Figure 22-4, some additional structural relationships between normal metabolites and drugs that antagonize their action are provided.

SENSORY RECEPTION

There are in nature a very large number of responses of living systems to stimuli in the form of specific molecules. In man, for example, the senses of **olfaction** (smell) and **gustation** (taste) depend on the interaction of certain molecules with specific receptor sites in the sense organs. In their basic aspects, these interactions are probably not very different from those of substrates with their enzymes. In fact, we see much the same type of specificity in both sets of reactions. In lower organisms, many aspects of behavior—feeding response, alarm reactions, sexual attraction, and the like—are also elicited by specific molecules. All of these reactions are markedly specific. To understand the point as clearly as possible, let us consider some examples.

The chemical basis of taste

Perhaps the earliest classification of tastes is that of Linnaeus proposed in 1754. He defined 11 kinds of taste: sweet, sour, sharp, salty, bitter, fatty, insipid, astringent, viscous, aqueous, and nauseous. In 1865 this list was trimmed quite a bit when Wundt suggested that just six categories would suffice to classify tastes: salt, sour, sweet, bitter, alkaline, and metallic. In later years the last two of these have been dropped in favor of the first four. Thus, while the number of recognized chemical elements in nature has increased from just four (fire, earth, air, and water) to more than 100, the basic number of recognized tastes necessary to describe nature has decreased to just four. In 1922 Skramlik at Freiburg University chose four compounds as models of the four tastes: quinine (bitter), sodium chloride (salty), tartaric acid (sour), and glucose (sweet).

The precise nature of the taste receptor is not known (but see below). The cells in which the receptors are located in the taste cells that make up the taste buds are shown diagrammatically in Figure 22-5. In an unknown

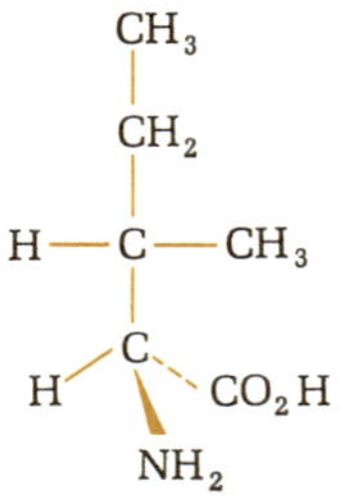

D-Isoleucine

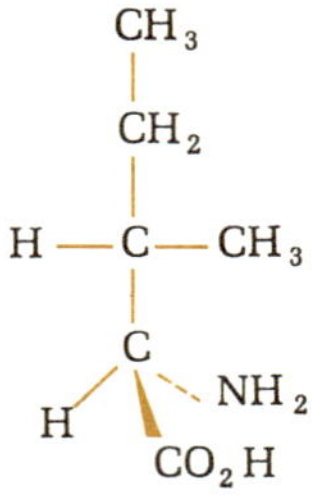

L-Isoleucine

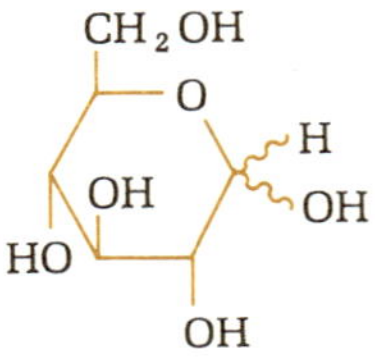

D-Glucose
(sweet)

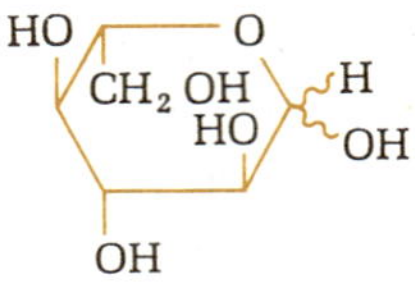

L-Glucose
(salty)

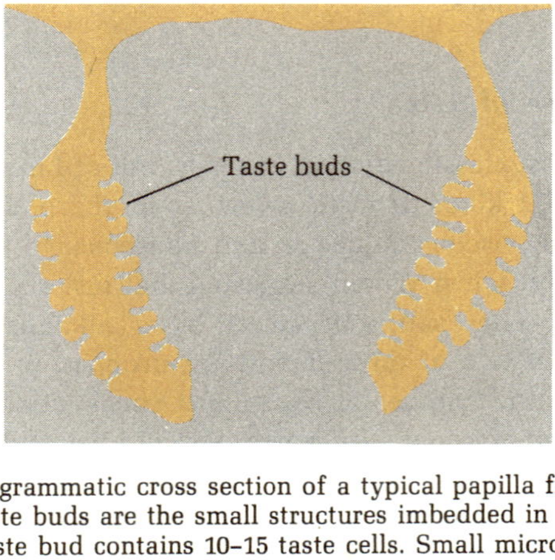

Figure 22-5 Diagrammatic cross section of a typical papilla from a human tongue. The taste buds are the small structures imbedded in the walls of the trench; each taste bud contains 10–15 taste cells. Small microvilli project from these cells and are presumed to be the actual site at which compounds trigger the taste sensation.

Table 22-2 The gustatory properties of salts[a]

Salts	Taste
Group I NaCl, KCl, NH_4Cl, LiCl RbCl, NaBr, NH_4Br, LiBr, NaI, Na_2CO_3, KNO_3, Na_2SO_4	Salty
Group II CsCl, RbBr, CsBr, KI, RbI, CsI, $MgSO_4$	Bitter
Group III Salts of beryllium and lead	Sweet
Group IV KBr, NH_4Br	Salty-bitter

[a]Many of these salts are poisonous.

way, the arrival of a compound at these taste cells triggers a sensory reaction that underlies the phenomenon of taste.

Correlation of taste with molecular structure is exceptionally complicated. For example, not all salts taste salty. Salts are classified in terms of their taste in Table 22-2. Although most of the common salts are "salty," note that several are classified bitter, sweet, and salty-bitter.

Among the various tastes, the most difficult to correlate with structure and those that are most sensitive to structural variations are sweet and bitter. Let us consider several such examples, beginning with the amino acids (see Table 7-3). The taste of amino acids depends critically on their stereochemistry: D-isoleucine is sweet, but L-isoleucine is bitter. Similarly, both D-valine and D-leucine are sweet, although L isomers are classified as flat to bitter.

Sugars also have tastes that are very sensitive to the nature of the precise molecular architecture. For example, D-glucose is sweet, but the L isomer is slightly salty. Moreover, α-D-mannose is sweet, but β-D-mannose is bitter. Clearly, the smallest of structural changes can lead to marked differences in taste. Hence, it follows that the nature of the interaction between the compounds and the taste receptors must be quite precise. Several additional examples reflecting behavior of this type are collected in Figure 22-6.

The taste sensation is almost certainly initiated by the reversible absorption of the stimulating ion or molecule to a protein in the receptor cell membrane. The magnitude of the taste response increases with increasing concentration of the stimulating species, but the increase is not linear since it is possible to saturate the receptor sites with stimulator molecules.

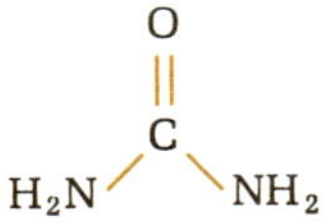

Urea
(bitter)

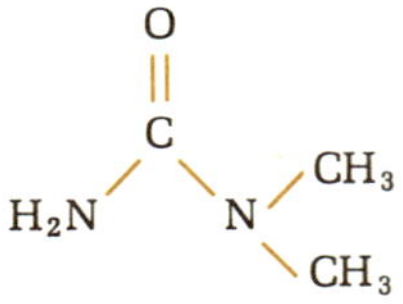

Dimethylurea
(sweet)

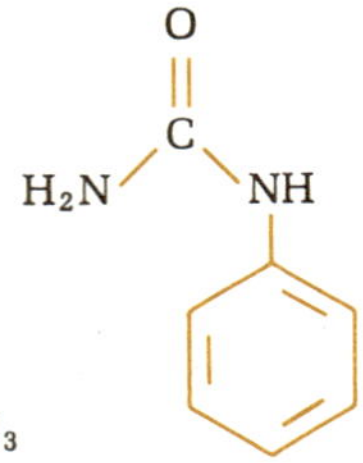

Phenylurea
(bitter)

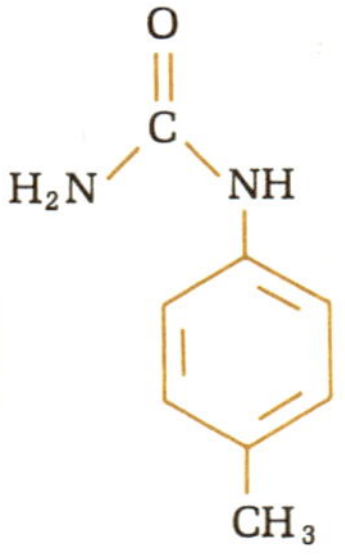

Tolylurea
(sweet)

Urea and derivatives

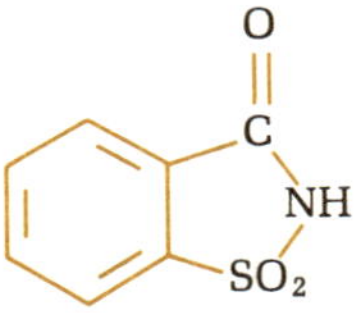

Saccharine
(306 times sweeter
than sucrose)

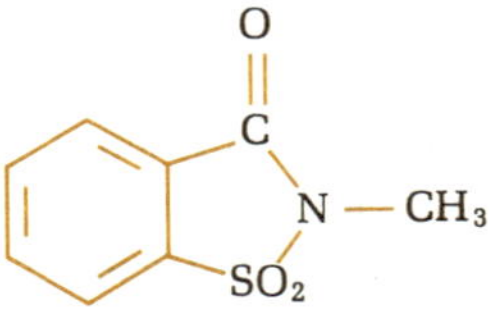

N-Methylsaccharin
(tasteless)

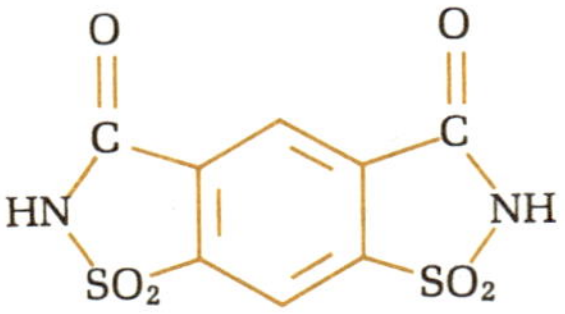

(very bitter)

Saccharine and derivatives

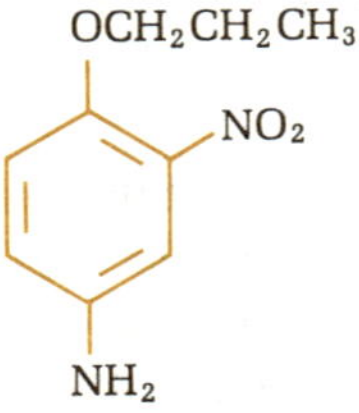

(tasteless)

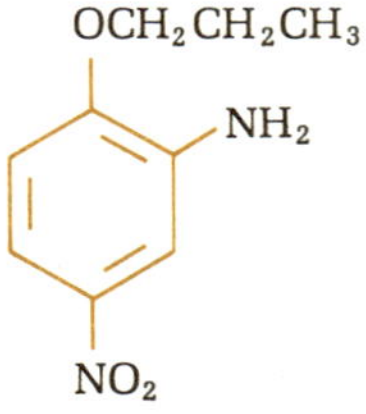

(4000 times sweeter
than sucrose)

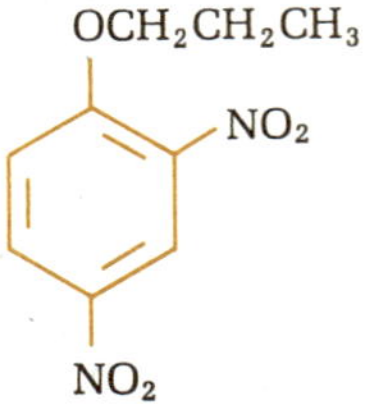

(bitter)

Some nitro-amino-aryl ethers

Figure 22-6 Examples of the relationship between molecular structure and taste.

Chemoreceptor proteins involved in evoking the taste sensation have been isolated and partially purified in several cases. Proteins that are sweet-sensitive and bitter-sensitive have been isolated from bovine taste buds, for example. The sweet-sensitive protein is now known to have a molecular weight near 150,000, and other of its physical and chemical properties are under study. On the basis of its behavior in solution, it seems to undergo a shape change when complexed with a sweet-tasting substance, perhaps altering the permeability barrier of the biological membrane with which it is associated and thus initiating a nerve impulse.

Olfaction: The sense of smell

Like taste, the sense of smell must reflect how molecules of various shapes interact with one or more chemoreceptors in biological systems. The nature of these chemoreceptors is not known, although it has been theorized that they may include biological membranes whose permeability properties are changed through interaction with smell-producing substances.

Smells have been classified into seven categories: ethereal, camphoraceous, musky, floral, minty, pungent, and putrid. For the first five of these classes, standard compounds have been chosen to represent the odors. These molecules have quite distinct shapes, as illustrated in Figure 22-7. It is these shapes that must, in large part, determine the nature of the interaction with the chemoreceptor sites.

Smell involves the detection of compounds in the gas phase. The ability to sense such compounds is not confined to man and higher animals, but is rather general in nature. Some of the lower organisms make important use of their sense of "smell." The biological responses including feeding response, sexual attraction, alarm and fright, and the like. Let us consider just one such case in terms of the chemistry involved.

One of the most striking examples of the influence of stereochemistry on biological activity comes from studies on the sex attractant of the female silkworm moth, *Bombyx mori*. These females emit a 16-carbon alcohol, called *bombykol*, which is detected by males who then are able to fly to the female. This alcohol contains two double bonds; consequently, there are four *cis–trans* isomers: *trans–trans, trans–cis, cis–trans* (which is normal bombykol), and *cis–cis*. The relative biological activity of these isomers is indicated in Table 22-3. Notice that change in stereochemistry about either of the double bonds in normal bombykol results in a diminution of biological activity by *at least* nine orders of magnitude! Whatever the structure of the chemoreceptor site in the male silkworm moth may be, it is quite clear that it is remarkably specific for normal bombykol. Man, perhaps unfairly, now uses sex attractants to lure unsuspecting males of specific, undesirable species to their doom. By using potent insecticides in traps and

Table 22-3 The relationship between stereochemistry and biological activity for the female sex attractant of the silkworm moth (*Bombyx mori*)

Isomer	Effective concentration (10^{-6} g/ml) in petroleum ether (50% of males show response)
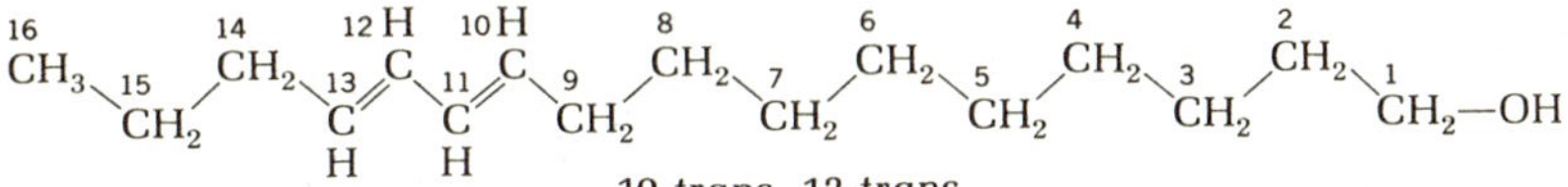 10-*trans*, 12-*trans*	10
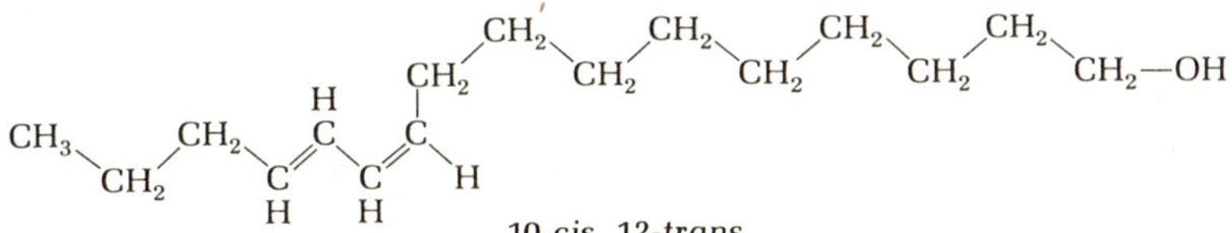 10-*cis*, 12-*trans*	0.001
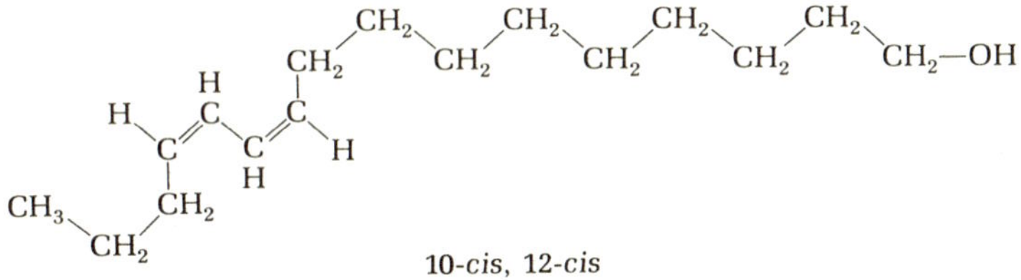 10-*cis*, 12-*cis*	1
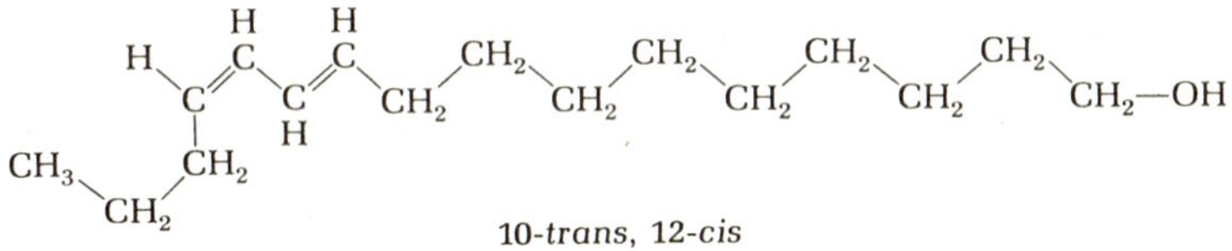 10-*trans*, 12-*cis*	0.000000000001

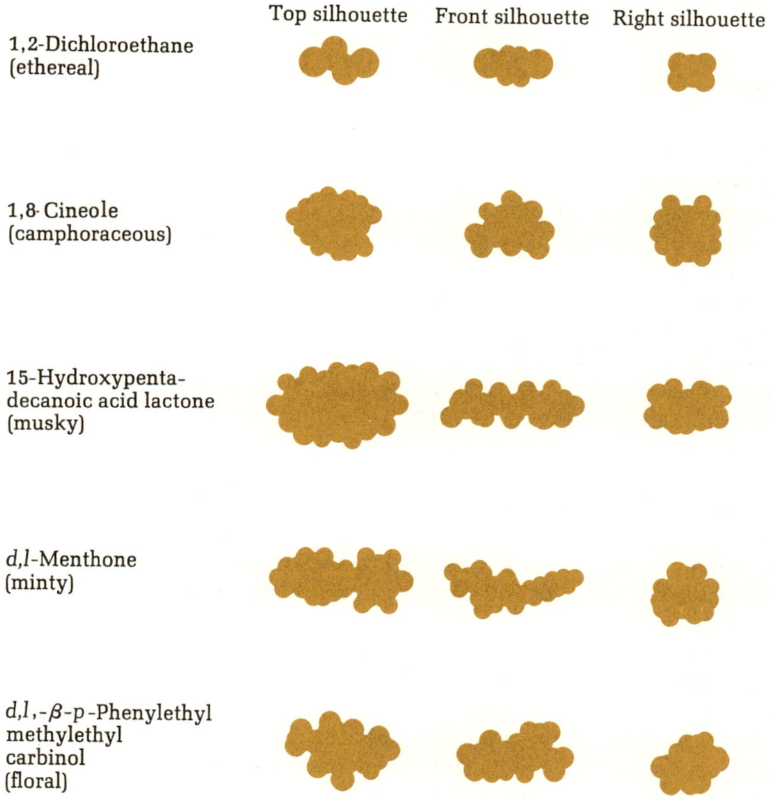

Figure 22-7 Molecular model silhouettes of the standard compounds chosen to represent five of the seven odor classes.

baiting the traps with the specific attractant of a particular species, the irradication of many insect pests can be achieved without harming other (perhaps beneficial) insects. Furthermore, general contamination of the biosphere by the insecticide is unnecessary.

CONCLUDING REMARKS

The above example should more than suffice to indicate the general biological importance associated with shapes of small molecules. The precise three-dimensional architecture underlies participation in enzymatic reactions, utility as a drug or chemotherapeutic agent, and ability to elicit a response at a sensory receptor. Clearly, the shape is important because this, in turn, determines the nature of the noncovalent interactions that the compound can make with biological receptors, generally.

Glossary

Bombykol a 16-carbon alcohol which is the sex attractant of the female silkworm moth.
Carboxypeptidase an important enzyme in the digestive process which functions by cleaving one amino acid at a time from the carboxyl terminus of polypeptide chains.
Catalase a remarkably efficient enzyme that catalyzes the decomposition of hydrogen peroxide.
Catalysis acceleration of a chemical reaction under the influence of an agent, the catalyst, which emerges unchanged from the reaction.
Catalyst an agent that accelerates the rate of a chemical reaction but undergoes no net chemical change itself in the process.
Chemisorption adsorption of a reagent to a catalyst surface with formation of a surface chemical compound, which involves a fundamental change in the structure and properties of that reagent.
Cracking in the process of refining crude petroleum, the cleaving of large hydrocarbon molecules into smaller ones, either by heating or by the use of catalysts.
Distillation fractionation of a mixture of liquids based on differences in their boiling points.
Enzymatic catalysis catalysis that is the result of the action of an enzyme.
Enzyme a protein having catalytic powers.
Folic acid an important agent for many biochemical reactions whose synthesis and/or function can be antagonized by sulfa drugs and certain antitumor drugs.
Fumarase an enzyme that catalyzes the addition of water to fumaric acid to yield L-malic acid specifically.
Gustation the sense of taste.
Haber process catalytic reduction of nitrogen to ammonia under the influence of metallic iron.
Heterogeneous catalysis use of a solid catalyst to promote reactions in a liquid or gaseous phase.
Homogeneous catalysis catalytic processes occurring in a single phase.
Olfaction the sense of smell.
Physiosorption the physical adsorption of a reagent onto the surface of a catalyst; does not involve a profound change in the nature of the adsorbed species.
Reforming in the refining of crude petroleum, a process in which hydrocarbons are isomerized to other hydrocarbons without a change in molecular weight.
Sulfa drugs a class of antibacterial drugs structurally related to *p*-aminobenzoic acid, which function by inhibiting the synthesis of folic acid.

Surface chemical compound a chemical compound formed as a result of adsorption of a reagent onto some surface, usually of a catalyst.
Urease an enzyme that catalyzes the hydrolysis of urea into ammonia and carbon dioxide.
Zeigler-Natta catalysts catalysts based on transition metals which are widely employed in the synthesis of addition polymers of olefins.

Study problems

1. Outline briefly all the advantages and disadvantages that you can think of for the use of catalysts as opposed to high temperatures to promote chemical reactions.
2. Living systems contain a great many enzymes, each specific for a certain chemical reaction. Explain why nature has chosen to do things in this way rather than simply making a single enzyme capable of catalyzing all the necessary reactions.
3. Many enzymes in the human body are potentially destructive in that they have the capacity to decompose important molecules or structures. For example, the enzymes, such as carboxypeptidase, that catalyze the hydrolysis of proteins could destroy proteins needed to maintain life, including other enzymes. How many ways can you think of that such destruction can be avoided? That is, how can such enzymes function in digestion without destroying the body's own proteins?
4. Suppose that we formulate the pathway for an enzyme-catalyzed pathway as follows:

$$\mathrm{A + E \overset{K}{\rightleftharpoons} A \cdot E}$$

$$\mathrm{A \cdot E \xrightarrow{k} P + E}$$

A is the reactant, P is the product, and E is the enzyme. The formation of A · E is considered to be rapid, and the breakdown of this complex is the rate-determining step.

a. Draw a free-energy diagram consistent with this pathway supposing that K is greater than one and that the overall reaction is exergonic.
b. Derive a rate law showing how the initial reaction velocity depends on the concentration of reactant and enzyme.
c. Supposing that the product P has a measurable affinity for the enzyme, that is, that formation of E · P from E and P occurs with a large equilibrium constant, how will the reaction velocity change as P accumulates?

5. Suppose that H_2, D_2, and T_2 are allowed to equilibrate with a metal surface at which chemisorption of hydrogen occurs. List all of the products that you would expect to find at equilibrium.
6. From the data in Table 8-2, calculate the enthalpy change for synthesis of ammonia by the Haber process. From the data in Table 8-4, calculate the equilibrium constant for this reaction. From these two pieces of data, calculate the entropy change for this reaction.

7. In addition to catalyzing the hydrolysis of proteins, carboxypeptidase catalyzes the hydrolysis of many small molecules. Several possibilities are listed below. In the case of each of these molecules, evaluate its suitability as a carboxypeptidase substrate, clearly justifying your line of reasoning.

a. 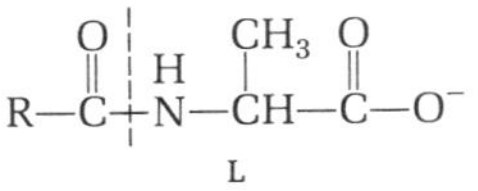

b. 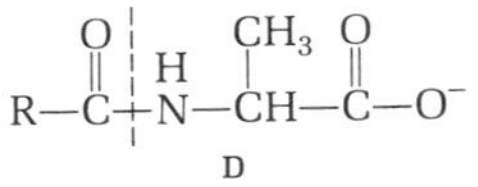

c. 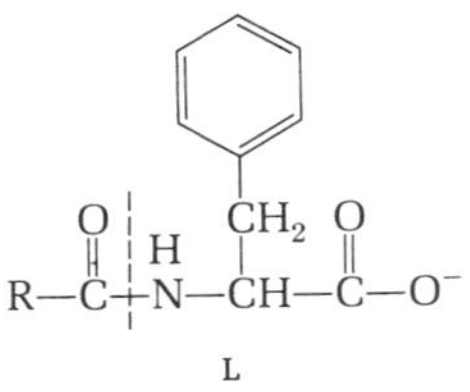

d. 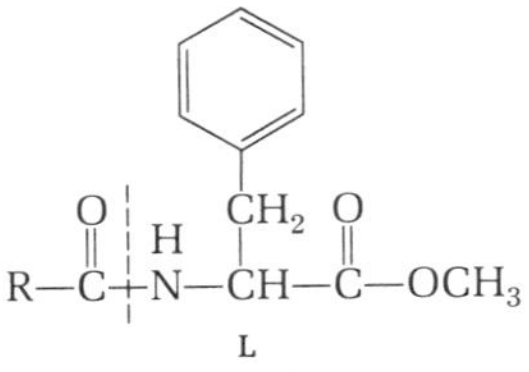

e. 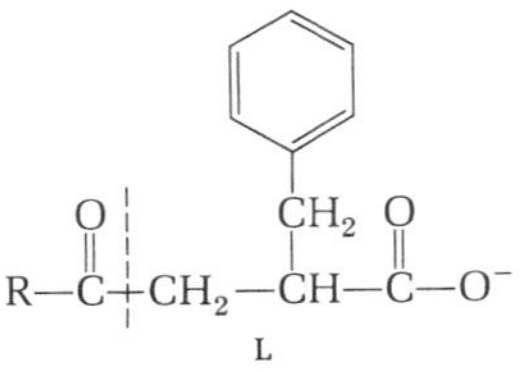

Suggestions for further reading

J. E. Amoore *Molecular Basis of Odor,* Thomas, Springfield, Ill., 1970.

M. Beroza and E. F. Knipling "Gypsy Moth Control with the Sex Attractant Pheromone," *Science,* **177,** 19 (1972).

V. H. Cheldelin and R. W. Newburgh *The Chemistry of Some Life Processes,* Van Nostrand Reinhold, New York, 1964.

R. E. Dickerson and I. Geis *The Chemistry and Function of Proteins,* Harper & Row, New York, 1970.

V. Haensel and R. L. Burwell, Jr. "Catalysis," *Sci. Am.,* **225,** 46 (December, 1971).

D. M. Locke *The Agents of Life,* Crown, New York, 1969.

K. E. Maxwell *Chemicals and Life,* Dickenson, Encino, Calif., 1970.

H. S. Rhinesmith and L. A. Cioffi *Macromolecules of Living Systems,* Van Nostrand Reinhold, New York, 1968.

E. Sondheimer and J. B. Simeone *Chemical Ecology,* Academic, New York, 1970.

C. M. Williams "Third Generation Pesticides," *Sci. Am.,* **217,** 13 (July, 1967).

E. O. Wilson *Insect Societies,* Harvard University Press, Cambridge, Mass., 1971.

Epilogue: A look back and a look forward

And the end of all our exploring will be to arrive where we started; and to know the place for the first time.

T. S. Eliot,
Little Gidding

All books must come to a close, and this one, too, has reached its end. In the very first chapter we briefly examined what chemists do and what is covered by the field of chemistry. When you first read that chapter we could do little more than allude to the many detailed topics that together constitute the body of chemistry. Now that we have concluded this introduction to chemistry, a rereading of that chapter might well be worthwhile. For example, a brief paragraph on page 3 suggests that the development of new drugs is an important task for chemists. In Chapter 22, some hints were provided concerning the development of useful drugs. However, we cannot claim to have given detailed description of how drugs act or how the chemist is guided in his search for them, nor have we described the actual chemistry involved in their synthesis. On page 6 we referred to the intriguing problem of the origin of the elements in the cosmos, and to the processes that may have occurred shortly after the creation of the universe. This topic has been treated in much more detail in Chapter 16. Even so, our discussion has been necessarily brief and has omitted much of the detailed picture presented by contemporary specialists.

Indeed, our entire study has touched upon only the high points that we consider most important for students beginning their study of chemistry. Each of the many subjects we have described constitutes a field in which individual research chemists may expend years, perhaps a lifetime of effort. It is perhaps significant that one of the authors of this book has devoted nearly 25 years to the study of boron and hydrogen compounds. This field represents the major research interest of only about 100 research chemists distributed throughout the world; it has received only brief mention in this book. The second author has spent the last 15 years studying the kinetics for diverse simple reactions related to certain reactions catalyzed by enzymes; these topics have also been dealt with only briefly in this book. In another instance, one entire group of research workers may devote several years to the determination of the sequence of amino acids in a single enzyme, whereas another group may be entirely engaged in x-ray studies intended to provide detailed information about the same molecule. Both groups inevitably show a lively interest in, and enter into prolonged and spirited speculation about, the relationship between the structure of the enzyme and the mechanism of its reaction. Again, one group of workers may be interested in the basic chemistry of nitrogen oxides in dilute solution in the earth's atmosphere, while a quite separate group may devote all its energies to the problem of minimizing these compounds in an automobile's exhaust. In writing this text, however, we have been almost totally concerned with describing the high points of chemistry, as a geographer might describe a mountain range. Now, individual mountain climbers appreciate the geographer's description but require a much more detailed knowledge of every nook, cranny, and crevice if those climbers wish to attain the peak.

Where will you go from here? Some of you may become chemists and perhaps devote your lives toward providing the essential chemical tools for progress in the control of disease. You may turn your future energies to more detailed study of topics we have only been able to introduce. Others may enter the fields of biology or medicine. Your primary future studies may be in the fields of organic chemistry and biochemistry, since it is now clear that much further progress in these areas will depend upon a better understanding of the chemistry involved in life processes. Those of you with interests in other physical sciences such as physics or geology may well concentrate instead on the fields of inorganic, analytical, or physical chemistry. A better understanding of the new fields of electronics or of how the world developed and continually changes rests heavily upon such knowledge. Of course, many of you will eventually enter fields not largely based on science. As educated citizens you will nevertheless be called upon to participate in repeated decisions involving the utilization of science.

Whatever your goal, the authors wish you well. We hope that this book has helped you on your way to your chosen goals and has, at least, convinced you that there is something worth knowing about chemistry.

Appendix

Table I Universal constants

N_0	Avogadro's number	6.02252×10^{23} molecules/mole
h	Planck's constant	6.6256×10^{-27} erg sec/molecule, or g cm^2/sec molecule
$\mathfrak{F}$	Faraday constant	96,487.0 A sec (or C)/mole 23.06 kcal/eq V
$k = R/N_0$	Boltzmann constant	1.38054×10^{-16} erg/°K molecule, or g cm^2/sec^2 °K molecule
R	Gas constant	1.9872 cal/°K mole, 0.082053 atm/°K mole, 8.3143×10^7 ergs/°K mole, or g cm^2/sec^2 °K mole
c	Speed of light (vacuum)	2.997925×10^{10} cm/sec
e	Electron charge	4.80298×10^{-10} esu, or (g cm^3/sec^2)$^{1/2}$ 1.69210×10^{-19} C
m_e	Electron rest mass	9.1091×10^{-28} g
e	Base of natural	2.71828
π	logarithms	3.14159
ln_e x		2.3036 $\log_{10}$ x

Table II Units and their symbols

ampere	A
angstrom	Å
atmosphere	atm
atomic mass unit	amu
calorie	cal
centimeter	cm
coulomb	C
cubic centimeter	cm^3
cubic foot	ft^3
cubic meter	m^3
day	day
degree Celsius (centigrade)	°C
degree Fahrenheit	°F
degree Kelvin	°K
dyne	dyne
electron volt	eV
electrostatic unit	esu
erg	erg
einstein	einstein
faraday	$\mathfrak{F}$
foot	ft
gallon	gallon
gram	g
gram-atom	g-atom
hour	hr
inch	in.
kilocalorie	kcal
kiloelectron volt	keV
kilogram	kg
kilowatt	kW
liter	liter
megawatt	MW
meter	m
mile	mile
miles per hour	mph
milligram	mg
milliliter	ml
millimeter	mm
millimole	mmole
million electron volts	MeV
minute	min
mole	mole
molal (concentration)	*m*
molar (concentration)	*M*
nanometer	nm
normal (concentration)	*N*
pH	pH
pK	pK
pound	lb
parts per hundred million	pphm
parts per million	ppm
second	sec
square centimeter	cm^2
square foot	ft^2
square meter	m^3
ton	ton
volt	V
watt	W
year	year

Table III Logarithms

N	0	1	2	3	4	5	6	7	8	9
10	0000	0043	0086	0128	0170	0212	0253	0294	0334	0374
11	0414	0453	0492	0531	0569	0607	0645	0682	0719	0755
12	0792	0828	0864	0899	0934	0969	1004	1038	1072	1106
13	1139	1173	1206	1239	1271	1303	1335	1367	1399	1430
14	1461	1492	1523	1553	1584	1614	1644	1673	1703	1732
15	1761	1790	1818	1847	1875	1903	1931	1959	1987	2014
16	2041	2068	2095	2122	2148	2175	2201	2227	2253	2279
17	2304	2330	2355	2380	2405	2430	2455	2480	2504	2529
18	2553	2577	2601	2625	2648	2672	2695	2718	2742	2765
19	2788	2810	2833	2856	2878	2900	2923	2945	2967	2989
20	3010	3032	3054	3075	3096	3118	3139	3160	3181	3201
21	3222	3243	3263	3284	3304	3324	3345	3365	3385	3404
22	3424	3444	3464	3483	3502	3522	3541	3560	3579	3598
23	3617	3636	3655	3674	3692	3711	3729	3747	3766	3784
24	3802	3820	3838	3856	3874	3892	3909	3927	3945	3962
25	3979	3997	4014	4031	4048	4065	4082	4099	4116	4133
26	4150	4166	4183	4200	4216	4232	4249	4265	4281	4298
27	4314	4330	4346	4362	4378	4393	4409	4425	4440	4456
28	4472	4487	4502	4518	4533	4548	4564	4579	4594	4609
29	4624	4639	4654	4669	4683	4698	4713	4728	4742	4757
30	4771	4786	4800	4814	4829	4843	4857	4871	4886	4900
31	4914	4928	4942	4955	4969	4983	4997	5011	5024	5038
32	5051	5065	5079	5092	5105	5119	5132	5145	5159	5172
33	5185	5198	5211	5224	5237	5250	5263	5276	5289	5302
34	5315	5328	5340	5353	5366	5378	5391	5403	5416	5428
35	5441	5453	5465	5478	5490	5502	5514	5527	5539	5551
36	5563	5575	5587	5599	5611	5623	5635	5647	5648	5670
37	5682	5694	5705	5717	5729	5740	5752	5763	5775	5786
38	5798	5809	5821	5832	5843	5855	5866	5877	5888	5899
39	5911	5922	5933	5944	5955	5966	5977	5988	5999	6010
40	6021	6031	6042	6053	6064	6075	6085	6096	6107	6117
41	6128	6138	6149	6160	6170	6180	6191	6201	6212	6222
42	6232	6243	6253	6263	6274	6384	6294	6304	6314	6325
43	6335	6345	6355	6365	6375	6385	6395	6405	6415	6425
44	6435	6444	6454	6464	6474	6484	6493	6503	6513	6522
45	6532	6542	6551	6561	6571	6580	6590	6599	6609	6618
46	6628	6637	6646	6656	6665	6675	6684	6693	6702	6712
47	6721	6730	6739	6749	6758	6767	6776	6785	6794	6803
48	6812	6821	6830	6839	6848	6857	6866	6875	6884	6893
49	6902	6911	6920	6928	6937	6946	6955	6964	6972	6981
50	6990	6998	7007	7016	7024	7033	7042	7050	7059	7067
51	7076	7084	7093	7101	7110	7118	7126	7135	7143	7152
52	7160	7168	7177	7185	7193	7202	7210	7218	7226	7235
53	7243	7251	7259	7267	7275	7284	7292	7300	7308	7316
54	7324	7332	7340	7348	7356	7364	7372	7380	7388	7396

Table III (continued)

N	0	1	2	3	4	5	6	7	8	9
55	7404	7412	7419	7427	7435	7443	7451	7459	7466	7474
56	7482	7490	7497	7505	7513	7520	7528	7536	7543	7551
57	7559	7566	7574	7582	7589	7597	7604	7612	7619	7627
58	7634	7642	7649	7657	7664	7672	7679	7686	7694	7701
59	7709	7716	7723	7731	7738	7745	7752	7760	7767	7774
60	7782	7789	7796	7803	7810	7818	7825	7832	7839	7846
61	7853	7860	7868	7675	7882	7889	7896	7903	7910	7917
62	7924	7931	7938	7945	7952	7959	7966	7973	7980	7987
63	7993	8000	8007	8014	8021	8028	8035	8041	8048	8055
64	8062	8069	8075	8082	8089	8096	8102	8109	8116	8122
65	8129	8136	8142	8149	8156	8162	8169	8176	8182	8189
66	8195	8202	8209	8215	8222	8228	8235	8241	8248	8254
67	8261	8267	8274	8280	8287	8293	8299	8306	8312	8319
68	8325	8331	8338	8344	8351	8357	8363	8370	8376	8382
69	8388	8395	8401	8407	8414	8420	8426	8432	8439	8445
70	8451	8457	8463	8470	8476	8482	8488	8494	8500	8506
71	8513	8519	8525	8531	8537	8543	8549	8555	8561	8567
72	8573	8579	8585	8591	8597	8603	8609	8615	8621	8627
73	8633	8639	8645	8651	8657	8663	8669	8675	8681	8686
74	8692	8698	8704	8710	8716	8722	8727	8733	8739	8745
75	8751	8756	8762	8768	8774	8779	8785	8791	8797	8802
76	8808	8814	8820	8825	8831	8837	8842	8848	8854	8859
77	8865	8871	8876	8882	8887	8893	8899	8904	8910	8915
78	8921	8927	8932	8938	8943	8949	8954	8960	8965	8971
79	8976	8982	8987	8993	8998	9004	9009	9015	9020	9025
80	9031	9036	9042	9047	9053	9058	9063	9069	9074	9079
81	9085	9090	9096	9101	9106	9112	9117	9122	9128	9133
82	9138	9143	9149	9154	9159	9165	9170	9175	9180	9186
83	9191	9196	9201	9206	9212	9217	9222	9227	9232	9238
84	9243	9248	9253	9258	9263	9269	9274	9279	9284	9289
85	9294	9299	9304	9309	9315	9320	9325	9330	9335	9340
86	9345	9350	9355	9360	9365	9370	9375	9380	9385	9390
87	9395	9400	9405	9410	9415	9420	9425	9430	9435	9440
88	9445	9450	9455	9460	9465	9469	9474	9479	9484	9489
89	9494	9499	9504	9509	9513	9518	9523	9528	9533	9538
90	9542	9547	9552	9557	9562	9566	9571	9576	9581	9586
91	9590	9595	9600	9605	9609	9614	9619	9624	9628	9633
92	9638	9643	9647	9652	9657	9661	9666	9671	9675	9680
93	9685	9689	9694	9699	9703	9708	9713	9717	9722	9727
94	9731	9736	9741	9745	9750	9754	9759	9763	9768	9773
95	9777	9782	9786	9791	9795	9800	9805	9809	9814	9818
96	9823	9827	9832	9836	9841	9845	9850	9854	9859	9863
97	9868	9872	9877	9881	9886	9890	9894	9899	9903	9908
98	9912	9917	9921	9926	9930	9934	9939	9943	9948	9952
99	9956	9961	9965	9969	9974	9978	9983	9987	9991	9996

Index

B

C

D

E

F

G

H

I

J

K

L

M

N

Q

R

W

X

Y

Z

Periodic table of the elements

Period	IA	IIA	IIIB	IVB	VB	VIB	VIIB	VIIIB	
1	1 H 1.008								
2	3 Li 6.939	4 Be 9.012							
3	11 Na 22.990	12 Mg 24.31	Transition elements						
4	19 K 39.102	20 Ca 40.08	21 Sc 44.956	22 Ti 47.90	23 V 50.94	24 Cr 51.996	25 Mn 54.94	26 Fe 55.85	27 Co 58.93
5	37 Rb 85.47	38 Sr 87.62	39 Y 88.91	40 Zr 91.22	41 Nb 92.91	42 Mo 95.94	43 Tc 99	44 Ru 101.07	45 Rh 102.91
6	55 Cs 132.91	56 Ba 137.34	57–71 La–Lu Rare earths	72 Hf 178.49	73 Ta 180.95	74 W 183.85	75 Re 186.2	76 Os 190.2	77 Ir 192.2
7	87 Fr 223	88 Ra 226.05	89–103 Ac–Lw Actinides						

Rare earths	57 La 138.91	58 Ce 140.12	59 Pr 140.91	60 Nd 144.24	61 Pm 145	62 Sm 150.35	63 Eu 151.96
Actinides	89 Ac 227	90 Th 232.04	91 Pa 231	92 U 238.03	93 Np 237	94 Pu 242	95 Am 243